MESOSCALE METEOROLOGY AND FORECASTING

Mesoscale Meteorology
and Forecasting

EDITED BY

PETER S. RAY

A M S

American Meteorological Society
Boston
1986

ISBN 0-933876-66-1

American Meteorological Society
45 Beacon Street, Boston, MA 02108

CONTENTS

EXTERNALLY FORCED CIRCULATIONS

MESOSCALE MODELING

Chapter 31. The Use of Computers for the Display of Meteorological Information

PREFACE

This volume is an extension of the effort to provide notes for the Intensive Course on Mesoscale Meteorology and Forecasting, given in the summer of 1984 under the cosponsorship of the American Meteorological Society (AMS), the Cooperative Institute for Mesoscale Meteorological Studies (CIMMS), and the National Oceanic and Atmospheric Administration (NOAA). The purpose of the course, and book, is to supplement a practicing meteorologist's experience with new material and ideas on mesoscale processes and to acquaint the student meteorologist with the broad range of topics embraced by the classification *mesoscale meteorology*.

Each phase of organizing and editing this book presented new challenges. The most enjoyable was the opportunity to work with thoroughly competent and professional colleagues.

In addition to thanking the authors, the editor acknowledges especially the significant contribution from more than 60 anonymous reviewers. Every chapter was improved by the uniformly careful reviews. Many authors also expressed special admiration for typists who showed unusual skill in converting their nearly illegible scribbling to a well-presented text. Kim Banner, Charles Clark, Chris Derrenbacker, Joan Kimple, Joseph Klemp, Sandra McPherson, Sharon Ray, Dewey Rudd, and Hope Hamilton each made special contributions to the completion of this project. Institutionally, Florida State University's Department of Meteorology, and Supercomputer Computations Research Institute, CIMMS, NSSL and ERL, and NCAR contributed. Finally, special recognition is due to the Environmental Research Laboratories editors in Boulder: Lindsay Murdock, who skillfully edited all chapters and worked with me over many months to present the best text possible, and Christine Sweet, whose assistance in review, especially of mathematical notation, was invaluable.

CHAPTER 1

Overview and Definition of Mesoscale Meteorology

Kerry A. Emanuel

1.1. Introduction

For a substantial fraction of the modern history of meteorology, practitioners of the science have been fond of classifying the highly various and complex phenomena of atmospheric flow according to the physical scales of apparently coherent structures that appear generally or intermittently within the flow. The conscientiousness and vigor applied to scale classification have advanced to the point where the layman, listening to a contemporary discussion of atmospheric motions and "scale interaction," might conclude that the atmosphere is somehow quantized and the scales are discrete. In reality, of course, the spectrum of atmospheric motions is smooth and continuous between the limits imposed by the mean free path of molecules on the short end and the circumference of the Earth on the large.

Historically, definitions of scales have arisen from essentially three sources: observations of atmospheric phenomena, sizes of observational networks, and theoretical inferences. Some connection can usually be made between definitions based on the first and the third of these, but definitions based entirely on utility may bear no relation to the others. Mesoscale meteorology may be said to involve both the study of more or less ubiquitous energy transfer processes and the special events that occur locally and intermittently as a result of topographic forcing and mesoscale instabilities.

1.2. Empirical Definitions of Scale

1.2.1. Visual Observations

The first motions of the scales of atmospheric phenomena undoubtedly arose from visual observations of processes operative on spatial or temporal scales not too different from the physical size and lifespan of human beings. For example, we immediately notice that cumulus clouds are roughly as broad as they are high, and by comparing their dimensions with those of objects of known sizes, such as mountains, we deduce that cumulus clouds have dimensions on the order of a kilometer. One also observes that such clouds can form and decay in abut 20 minutes. Similarly, residents of middle and high latitudes notice that weather changes on a time scale of roughly one or several days, and that seasons progress on an annual cycle. The latter two observations are strictly temporal and frame-relative, though, and further information is needed to make inferences regarding related spatial dimensions.

Although humans must have known about the space and time scales of cumulus-cloud-sized phenomena and the time scales of larger storms for many thousands of years, knowledge of the spatial scales of such phenomena as cyclones and planetary waves is surprisingly recent. In principle, what is now called synoptic analysis might have been performed in rudimentary fashion as long ago as there existed civilizations covering a large area and possessing some form of clock to record the time of observations. It is a mystery why the Romans, for example, did not attempt to collect observations of wind direction and perhaps speed that were simultaneously recorded at various locations in their empire, in view of the importance of storms to naval and other shipping operations.

In any event, serious attempts to compile observations on the scale of nations and continents did not begin until 400 or 500 years ago. The first of these compilations seems to have concerned more or less steady circulations, notably the trade winds. (Early interest in the trade winds was no doubt motivated by commerce, as their name implies.) The character of the surface trades was deduced principally from ship observations collected over long periods of time; some notion of the circulation aloft was deduced from cloud observations. The comparative success of these observations led naturally to attempts to define the middle- and high-latitude flow on the basis of time-average observations; in hindsight these efforts held up progress in understanding the general circulation for perhaps 200 years.

Although there were a few notable attempts to define the spatial structure of temporally variable large-scale systems as early as the 17th century, the first real success did not occur until the turn of this century when the Norwegians capitalized on the natural spacing of European cities to set up an observation array and thereby deduce characteristics of cyclones and fronts at the surface. For reasons that are not entirely clear, but must be related to these early synoptic analyses, phenomena on the scale of cyclones came to be referred to as "synoptic-scale" processes. ("Synoptic," derived from the Greek *synoptikos*, literally refers to a general view of the whole and in no way implies a particular scale; neither, as is sometimes thought, does it imply simultaneity.)

It is immediately apparent from weather maps that the "lows" and "highs" have a characteristic spacing and that isobars, even for closely spaced data, are more often than not quite smooth. Closer inspection also reveals the presence of sharp pressure troughs accompanied by discontinuities in wind and temperature gradient; time-series synoptic analysis of observations at a single point confirms the nearly discontinuous character of some fronts. The especially perceptive will notice that a time sequence of hemisphere weather maps at higher levels in the atmosphere frequently seems to represent a superposition of mobile storm-scale waves upon a much larger, slowly varying train of waves of planetary scale. In addition, the latter are the only obvious feature of weather maps of the higher wintertime stratosphere.

In view of these features of weather maps, it is not at all surprising that those who work with such maps every day come to view the atmosphere almost as a quantum superposition of various discrete scales of motion: planetary waves, synoptic-scale storms, and fronts. All of these are fairly well resolved by observations spaced on the scales of cities and collected at hourly or daily periods. In addition, we know there are much smaller scale phenomena that we can see with the eye or with instruments such as radar and satellite-borne cameras.

1.2.2. Spectral Analysis

Those not content with what the eye perceives can attempt to quantify the degree to which the atmosphere is really composed of more or less discrete scales. One method of doing this is spectral analysis. As Fourier discovered, any smooth distribution of variables can be represented by a series of sine and/or cosine waves; for example,

$$a_1 \sin \frac{\pi x}{L} + a_2 \sin \frac{2\pi x}{L} + a_3 \sin \frac{3\pi x}{L} + \cdots = \sum_n a_n \sin \frac{n\pi x}{L}, \qquad (1.1)$$

where the a_n's are constant coefficients, and L represents the largest resolved scale (say half the circumference of the Earth). One can determine the sizes of the various coefficients a_n by fitting the series to observations. The size of each a_n in some sense tells us how important the scale L/n is in the observations. Figure 1.1 shows, in essence, the size of the coefficients a_n, where the observations are of the square of the wind speed and the plot is a function of frequency rather than wavelength. The information in this figure partially supports the "quantum view": although the spectrum shows some energy at all scales of motion, there are strong peaks at frequencies ranging from a few days (the synoptic scale) to a few weeks (the planetary

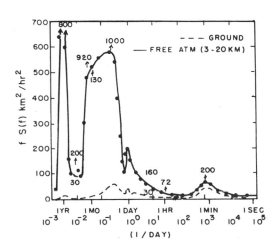

Figure 1.1. Average kinetic energy of west-east wind component in the free atmosphere (solid line) and near the ground (dashed line). Numbers indicate maximum values of kinetic energy. (After Vinnichenko, 1970.)

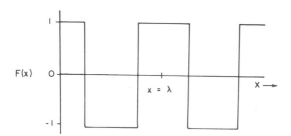

Figure 1.2. A periodic function $F(x)$ with wavelength λ.

scale). There are also peaks at 1 year and 1 day and a smaller peak at a few minutes, though the latter may be an artifact of the analysis. Nevertheless, the spectrum is continuous and the peaks are not entirely isolated from one another; note particularly that there is no gap between the planetary and synoptic scales.

One must be very cautious in interpreting spectra, however. A time spectrum like the one shown in Fig. 1.1 may look very different from a space spectrum for the same phenomenon; the relation between space and time scales depends partially on how fast structures move by the observation point, and not all phenomena move at the same speed. Even without these problems, spectra are sometimes misleading. Fig. 1.2 shows a hypothetical distribution of a variable in space which may, for example, vaguely represent the record of liquid water measured by an aircraft flying through a field of cumulus clouds. The subjective human eye perceives this and forces the question, "Why is the wavelength λ and why are the edges so sharp?" The spectral decomposition of the field in Fig. 1.2 is shown in Fig. 1.3. Although there is a prominent spectral peak at $L = \lambda$, there are other peaks at smaller wavelengths as well. Not much can be said about the character of the field

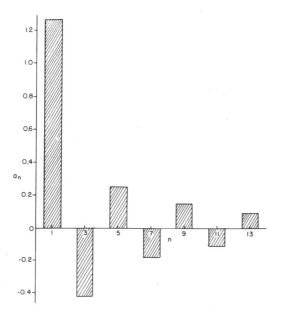

Figure 1.3. Coefficients a_n of the spectral representation of the function $F(x)$ shown in Fig. 1.2. The spectral representation has the form $\sum_{n=1}^{\infty} a_n \cos \frac{2n\pi}{\lambda}$, and the coefficients a_n are given by

$$ a_n = \begin{cases} 0 & \text{if } n \text{ is even}, \\ \frac{4}{n\pi}(-1)^{\frac{n-1}{2}} & \text{if } n \text{ is odd}. \end{cases} $$

except that λ is a prominent wavelength. One advantage of the human eye is that it is acutely aware of the obvious.

1.3. Scale Definitions Based on Utility

Forty years after synoptic analyses of the Norwegian school had provided a comparatively complete description of the surface cyclone and attendant fronts, radar was first applied to meteorology. Radar effectively extended the degree to which the human eye could observe precipitating clouds simultaneously from a single observation point. In the early days of radar meteorology, the discoveries involved more the substance of particular phenomena than their form; the naked eye had already discerned the basic form of events such as thunderstorms, squall lines, and the eyes of hurricanes. Nevertheless, there were hints of a set of phenomena partially encompassed by a typical radar display that had not previously been seen or imagined. Perhaps most prominent among these were hurricane spiral rainbands, though there were additional indications of forms not quite encompassed by the radar but too small to resolve by conventional observations on the scale of the spacing between cities. The term *mesoscale* was coined by the radar meteorologist Ligda (1951) to describe phenomena of this intermediate scale. This scale definition differs from earlier ones in being somewhat speculative, since it was based on what had not been observed, rather than on observed features.

In the last two decades or so, experimental arrays of observation stations have been set up in an attempt to resolve the intermediate scales. Most of these experiments, such as the National Severe Storms Project, GATE, and SESAME were designed to investigate various aspects of moist convection, particularly its interaction with larger scale flows. In attempting to resolve so-called scale interactions it seemed natural to devise a series of imbedded arrays, each with a different average spacing between observations. GATE, for example, used arrays on three scales and SESAME used "storm-scale" and "regional-scale" networks. Definitions of such scales are based more on practical considerations than on a clear perception of the existence of special spatial scales in the atmosphere; since it is not practical to cover large areas with dense arrays, the best one can do is resolve part of the area with a dense network and the rest with a coarser array. Even so, it was not long before utility scale definitions such as these were confused with somewhat more physical definitions, and soon Ligda's mesoscale had been quantized into a plethora of subdivisions, each capable of interacting with all the others. Today it seems that the degree of quantization of the mesoscale is limited only by the human imagination and the length of the Greek alphabet.

A serious question confronting the mesoscale meteorologist is whether there are really inherent scales in the atmosphere that one might reasonably use to define mesoscale. Put a little differently, do there exist ordered processes in the atmosphere that generate kinetic energy on scales within Ligda's mesoscale domain (does a natural mesoscale exist), or does the "mesoscale" really consist only of a smooth, continuous, and uninteresting spectrum of disordered motions, which serve to transfer energy between energy-producing processes on the small scale and on the cyclone scale?

1.4. The Physical Significance of Scales of Motion

1.4.1. Definition of System

A system must in some sense be distinct in space and time, and the transfer of physically relevant quantities across the boundaries of the system must be understood. Clearly it is not difficult to call the entire atmosphere a system, having as it does more or less distinct boundaries (across which, however, there are fluxes of radiation, heat, momentum, etc.)

Circulation systems smaller than the entire general circulation must be more carefully defined. A cyclone is not at all isolated from the surrounding atmosphere, yet we perceive it in some sense as a separate entity. However, it is usually profitable to define such a system as a perturbation on a known much-larger-scale circulation, with the clear understanding that the latter evolves much more slowly than the perturbation upon it. A cumulus cloud, for example, can be thus partially isolated from the surrounding atmosphere, which presumably changes much more slowly than does the cloud itself and its immediate environment.

1.4.2. Scales Associated With Atmospheric Systems

Certain atmospheric scales of motion are more or less obvious; for example, the peak at 1 day in Fig. 1.1 is clearly related to diurnal solar forcing. Other scales such as the time and length scales of cyclones are not so obviously related to other natural time or length scales, and further inquiry into the dynamics of such systems is necessary to understand the scales in which those systems occur.

The first and most obvious constraint on the spatial scales of processes in a fluid system is the geometry of the boundaries of the system. The circumference of the Earth provides an upper bound on the horizontal scale of motions in the atmosphere; at the same time it is not meaningful to talk about fluid motions comparable with or smaller than the mean free path of molecules.

The vertical scale of a circulation system will also be limited on its upper end by scales related to the effective depth of the atmosphere. One choice is the density scale height, which is merely the depth over which the mean density of the atmosphere decreases by a factor e; in the Earth's atmosphere this is about 10 km. Density plays an important role in constraining the vertical depth of circulation systems in the troposphere since the tropopause, although it does fluctuate, is basically determined by density-dependent radiational characteristics of the atmosphere: above the tropopause, ozone heating maintains a statically stable lapse rate whereas below, surface heating would result in an unstable lapse rate (except in polar regions and middle latitudes in winter) were it not for the continuing action of convection. The temperature structure of the tropical troposphere is approximately determined by an equilibrium between the stabilizing influence of moist convection and the destabilizing tendency of radiative processes. It would not be surprising to find a peak in the spectrum of tropical cumulus cloud depths corresponding to the height of the tropopause (though there are usually other peaks as well).

Special time and space scales may exist, corresponding to normal modes of oscillation of a fluid system; there may also be characteristic instabilities of the system that take on certain length and time scales. Other scales may be externally forced, but the response of the system to the external influences will depend in part on the free oscillation characteristics of the system.

Two kinds of "special scales" appear in some very simple global-scale flows. One kind is based upon the small-amplitude instabilities that may appear in the flow; the other is based on special wave frequencies that determine the character of stable small-amplitude oscillations. One might expect these scales to play an important part in the character of the flow.

Large Richardson Number

As a classic example, consider the special scales of oscillation and instability in the temporal and zonal mean circulation of the atmosphere in one hemisphere. As an extreme simplification, take the flow to have constant vertical shear and stratification. This may severely distort or eliminate certain oscillations and instabilities in the system; moreover, in this case the state of "the system" is largely determined by the fluxes associated with the perturbations upon it.

Unless the shear is very strong compared with the stratification, essentially one type of instability may occur in the system, which, if viscosity is ignored, can be locally described in terms of

$$\overline{U}_z, N, f, \beta \ , \tag{1.2}$$

where $\overline{U}_z$ is the vertical shear of the zonal wind, N is the buoyancy frequency (whose square is proportional to the stratification), f is the Coriolis parameter, and β is its meridional gradient. A detailed analysis of this system reveals an instability that can be associated with cyclones and anticyclones; this instability has an intrinsic vertical scale

$$h = \frac{f^2 \overline{U}_z}{N^2 \beta} \tag{1.3}$$

and an intrinsic horizontal scale

$$L = \frac{f \overline{U}_z}{N \beta} \ , \tag{1.4}$$

which for typical conditions in the Earth's atmosphere have values of roughly 18 km and 1800 km, respectively. Although the values of these parameters vary with space and time, peaks in the spectrum of atmospheric phenomena can be expected at roughly this scale. For an observer at a fixed point on the Earth, the systems will move by with a period

$$T \sim \frac{2\pi L}{U_o} \simeq \frac{2\pi L}{\overline{U}_z h} = \frac{2\pi N}{f \overline{U}_z} \ , \tag{1.5}$$

which is about 3 days. Notice that this corresponds nicely to a peak in the spectrum in Fig. 1.1, but does not explain the high number of events with periods ranging up to weeks.

Small Richardson Number

At the other end of the spectrum are various small-scale instabilities, which can occur when the shear becomes locally strong or the stratification locally weak, so that the Richardson number ($\equiv N^2/\overline{U}_z^2$) is less than $1/4$. In the Earth's atmosphere, the background rotation (f) is generally not influential on the time scales of these small-scale instabilities, so the important parameters are N, $\overline{U}_z$, and the depth (D) of the region over which the Richardson number is small or negative. Since D is the only really influential length scale, it determines the magnitude of all spatial dimensions of the disturbances; the time scale is N^{-1} if N is positive or $(-N)^{-1}$ if N is negative. The latter case corresponds to convection, which occurs more or less continuously in the planetary boundary layer over warm oceans, over continents in sunny weather conditions, and through the depth of the troposphere in the form of cumulus clouds in most of the tropics and further poleward in spring and summer. Convection and small-scale turbulence probably account for the peak at several minutes observed on some frequency spectra (e.g., Fig. 1.1).

Intermediate Richardson Number

Further analysis of the stability of simple zonal flows also reveals the possibility of disturbances of intermediate scale which may occur when the Richardson number is less than 1; these tend to occur in bands parallel to the thermal wind and are essentially a form of convection driven by a combination of buoyancy and Coriolis accelerations. This instability has been called symmetric instability although it is here called slantwise convection. Stability analysis shows that length scales associated with this form of convection are of order $\overline{U}_z D/f$, where D is the depth of the layer of instability, and the time scale is roughly

$$T \sim f^{-1} \tag{1.6}$$

Unlike the disturbances discussed previously, slantwise convection does not appear to be common in the atmosphere but seems to occur very intermittently in association with saturated regions of baroclinic cyclones. There is no spectral peak in Fig. 1.1 at the time scale represented by (1.6).

Table 1.1 summarizes scales according to known instabilities of large-scale zonal flow, when "system" is defined as the zonal and time average flow of the atmosphere. System might also be defined as a known state of large-scale atmospheric flow containing developed waves; examination of its stability could be extended to distinctly smaller and faster disturbances. In either case a clear separation between what we call the large-scale and what we define as smaller scale disturbances is vital for the concept of a system to have some meaning. It is instructive to illustrate a counter-example where

Table 1.1. Scale classification based on instabilities of planetary zonal flow

Type	Vertical scale	Horizontal scale	Time scale	Rossby number	Frequency of occurrence in Earth's atmosphere
Baroclinic	$f^2 \overline{U}_z / N^2 \beta$	$f \overline{U}_z / N\beta$	$2\pi N / f\overline{U}_z$	$1/\sqrt{\text{Ri}}$	Nearly ubiquitous poleward of 30° latitude
Slantwise convection	D	$\overline{U}_z D / f$	f^{-1}	2π	Highly intermittent
Cumulus convection	D	D	N^{-1}	$2\pi N / f$	Ubiquitous over tropical oceans, intermittent elsewhere
Boundary-layer turbulence	h_b	h_b	h_b / U^*	$2\sigma U^* / h_b$	Ubiquitous in boundary layer

Key: D = Depth of unstable layer
h_b = Depth of planetary boundary layer
f = Coriolis parameter
$\overline{U}_z$ = Scale for vertical shear of zonal wind
β = Meridional gradient of f
N = Scale for buoyancy frequency
Ri = Richardson number $\equiv N^2 / \overline{U}_z^2$
U^* = Friction velocity

clear separation between scales is not possible. Consider a box filled with air and maintain the bottom of it at a temperature larger than that maintained along its top. If the temperature difference is large enough, the convection in the box will be very turbulent, and it can be shown that away from the walls only the very smallest eddies will be directly affected by molecular viscosity. Then, aside from the length scale of the box itself, there are no special scales in the system except those of the eddies that are so small that viscosity becomes important. Subdivision beyond box-scale eddies and small-scale eddies for which molecular diffusion is important would be entirely arbitrary. Between the two extremes there exists a whole spectrum of eddies of intermediate scale whose effect is to generate kinetic energy from buoyancy and absorb it from larger scales of motion and then shunt it to smaller and smaller scales until it is ultimately destroyed (turned into heat) by viscosity. Or, according to L.F. Richardson,

> Big whirls have little whirls which feed on their velocity
> and little whirls have lesser whirls and so on to viscosity.

The point is that, aside from the box-scale eddies and the dissipation-scale eddies, there are no identifiable scales of special significance.

1.4.3. Classification of Systems by Physical Characteristics

Amount of Instability

The difference between the convection in a box and the mean zonal flow with waves discussed earlier is that, in the latter, the system is weakly unstable and supports a limited spectrum of waves, whereas the former is highly unstable and a much more complete spectrum of eddies exists. An attempt can be made to classify systems according to the degree of disorder present in them. A completely stable flow varies in space and time in a manner directly determined by the initial and boundary conditions, and represents a very high degree of order. A system that is barely unstable will generally contain a very well defined train of disturbances with a specific wavelength that is partially or wholly determined by the character of the mean flow. (Such a flow will contain a very sharp spectral peak.) The flow is still very highly ordered and predictable, and it is quite meaningful to refer to a special scale. As the degree of instability increases, disturbances of different wavelengths will appear together with the main disturbances, and a spectrum will show many peaks. Even so, the collection of disturbances may be regular and an observer at a fixed point might still see a perfectly periodic, if complex, progression of events. Eventually, as instability increases, the flow will cease to be periodic and thus easily predictable, and a spectrum will look much more continuous, although peaks corresponding to the main scales of instability may continue to exist. Finally, the "turbulent box" is achieved in which the motions are extremely chaotic and no spectral peaks occur other than those determined by the geometry of the box.

The atmosphere outside the boundary layer probably represents an intermediate state in terms of disorder. Instabilities are prominent at certain scales, but the spectrum appears to be smooth and continuous in between the peaks. What is not well understood is how kinetic energy is transferred from the scales at which it is generated to the "inter-peak" scales and then to its ultimate demise.

Response to Forcing

Another way to identify scales of special physical significance is to examine the free stable modes of oscillation of atmospheric systems. One way of doing this is to examine the response of a system to some specified forcing. Figure 1.4 shows the vertical wavenumber of a disturbance caused by a constant zonal flow of speed $\overline{U}$ passing over sinusoidal corrugations of wavelength $2\pi/k$. The flow has constant buoyancy frequency N and occurs on a β plane. The solid lines show the oscillatory part of the vertical structure of the disturbances; the dashed lines denote exponential growth or decay. Four regimes are distinguishable depending on the wavelength of the corrugations relative to certain intrinsic scales. For wavelengths less than $2\pi\overline{U}/N$ (Doppler-shifted frequencies greater than N) the disturbance decays exponentially with height and is therefore trapped near the ground. For wavelengths between $2\pi\overline{U}/N$ and $2\pi\overline{U}/f$ (frequencies between N and f) vertical propagation can occur and the waves can propagate to large altitudes (the exponential part of the

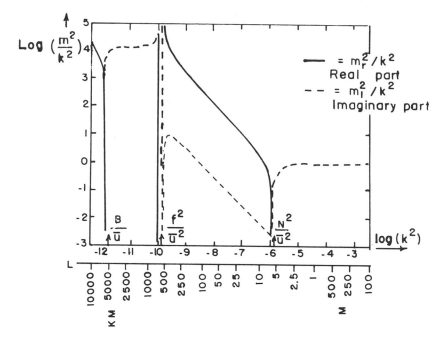

Figure 1.4. Natural log of the square of the real part (solid line) and imaginary part (dashed line) of the vertical wavenumber m normalized by the horizontal wavenumber k of the topography; these wavenumbers pertain to the linear response of constant flow $\overline{U}$ over topography, and they are plotted as a function of k. (After Wipperman, 1981.)

vertical structure is actually an increase with height related to the decay of mean density with altitude). For wavelengths between $2\pi\overline{U}/f$ and $2\pi\sqrt{\overline{U}/\beta}$ (frequencies between f and $\sqrt{\overline{U}\beta}$) the disturbances are once again trapped; for wavelengths greater than $2\pi\sqrt{\overline{U}/\beta}$ vertical propagation is allowed.

The two kinds of disturbances that can propagate vertically in this flow are inertia-gravity waves (frequencies between N and f) and Rossby waves (frequencies greater than $\sqrt{\overline{U}\beta}$). Thus in a constant west wind blowing over an irregular mountain range, planetary-scale Rossby waves and small-scale gravity waves in the stratosphere would be expected, with little energy in other modes (unless a local instability excites other wavelengths).

Lagrangian Time Scales

A convenient and physically meaningful way of categorizing the scale of disturbances in a system is by comparing the Lagrangian time scale (the time over which particles accelerate) with some intrinsic time scale in the whole system, for example the "pendulum day" $(2\pi/f)$. In the case of wave-like oscillations, the Lagrangian time scale is simply the time it takes for a particle to move through one wavelength. For nonwavy instabilities, such as convection, it may be defined as the time it takes for a particle to move through the whole disturbance. The ratio of the pendulum day to the Lagrangian time scale is a kind of Rossby number, which may in turn be used to characterize

**Table 1.2. Lagrangian time scales and Rossby numbers
of selected phenomena**

Phenomenon	T	Ro
Zonal mean circulation	$2\pi a/\overline{U}$	$\overline{U}/af$
Planetary (stationary) Rossby waves	$2\pi/\sqrt{\overline{U}\beta}$	$\sqrt{\overline{U}\beta}/f$
Cyclones and anticyclones	$2\pi\sqrt{\mathrm{Ri}}/f$	$1/\sqrt{\mathrm{Ri}}$
Classical fronts	$2\pi\sqrt{\mathrm{Ri}}/f$	$1/\sqrt{\mathrm{Ri}}$
Sea and land breezes	1 day	$1/2\sin\psi$
Slantwise convection	$1/f$	2π
Tropical cyclones	$2\pi R/V_T$	V_T/Rf
Inertia-gravity waves	$2\pi f^{-1}$ to $2\pi N^{-1}$	1 to N/f
Thunderstorms and cumulus clouds	N_w^{-1}	$2\pi N_w/f$
Kelvin-Helmholtz waves	$2\pi N^{-1}$	N/f
PBL turbulence	$2\pi h_b/U^*$	U^*/fh_b
Tornadoes	$2\pi R/V_T$	V_T/fR

Key:
a ≡ Radius of Earth
$\overline{U}$ ≡ Scale for mean zonal velocity
f ≡ Coriolis parameter
β ≡ Meridional gradient of f
U^* ≡ Friction velocity scale
h_b ≡ Scale for PBL (planetary boundary layer) depth
N ≡ Scale for mean buoyancy frequency
N_w ≡ Scale for moist buoyancy frequency
Ri ≡ $N^2/\overline{U}_z^2$
R ≡ Radius of maximum wind scale
V_T ≡ Maximum tangential wind scale
ψ ≡ Latitude

the degree of geostrophy of the flow. Table 1.2 shows typical Lagrangian time scales for a variety of disturbances together with Rossby numbers based on these time scales. Figure 1.5, which partially follows the work of Orlanski (1975), classifies scales on the basis of Lagrangian time scales of stable and unstable disturbances.

Two aspects of Fig. 1.5 are unconventional: (1) Fronts of the classical Norwegian-school type appear at the same Lagrangian scale as cyclones. This is because, although very strong spatial gradients of quantities exist at fronts, the parcels themselves undergo only slow changes except perhaps in regions of turbulence right at the front. The geostrophic momentum equations upon which the semi-geostrophic system is founded are based on the smallness of the Lagrangian Rossby number as defined here. (2) Tropical cyclones occur at the short end of the mesoscale. This is because the Lagrangian time scale

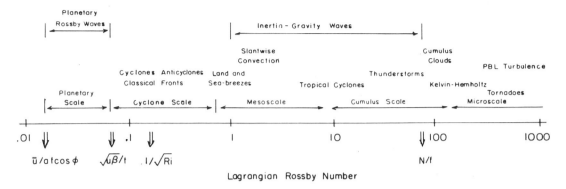

Figure 1.5. Forced and free atmospheric phenomena categorized on the basis of their characteristic Lagrangian Rossby number. Suggested scale definitions are given. See Table 1.2 for definitions of symbols.

of parcels circling the core can be quite short compared with a pendulum day. Surely the eye of hurricane is a "mesoscale" phenomenon, though the circulation usually extends out to what is called synoptic scale. But even a modest cumulus cloud drives a small circulation outside the cloud, which extends to the Rossby deformation radius ND/f. These points merely serve to illustrate the inherent ambiguities of scale classification, a naive procedure at best.

The scheme shown in Fig. 1.5 illustrates a classification of coherent phenomena based on some physically relevant property of the phenomena, rather than on some arbitrary length or time scale. Although it is useful in this respect, it may not be so useful for other purposes, such as devising observational networks to measure particular phenomena. For that purpose, it is obviously important to know the horizontal length scales and Eulerian (with respect to an observer on the ground) time scales. But in general, references to the scale of particular phenomena as if the scale had some physical significance are appropriate only if the scaling parameter itself is physically relevant.

1.5. Scale Interaction

"Scale interaction" is at once one of the most popular and one of the most ambiguous phrases in all the meteorological literature. In planetary-scale meteorology it generally refers to the interactions between the time and zonal average zonal flow and a fairly limited set of waves that are quantized by the circumference of the Earth. In turbulence theory it means the multitudinous interactions among a continuous spectrum of eddies of all sizes. Many mesoscale meteorologists think of scale interaction, probably mistakenly, in the quantum sense, that is, as a limited set of interactions among discrete scales. For example, "meso-γ-scale–meso-β-scale interactions" are often referred to as if the atmosphere were really so quantized. Evidence continues to mount (e.g., see Ch. 11) that on average the mesoscale is much more like a continuous spectrum of scales, which shunt energy between larger and smaller scales, though there are also local or intermittent events that actually generate kinetic energy on the mesoscale.

In a very simple system consisting of a slowly varying mean flow with a very weak disturbance upon it, the main scale interaction can be considered to be the influence of the mean flow upon the disturbance. As the perturbation becomes more substantial it exerts an increasing influence on the mean flow, and other scales of motion develop because of secondary instabilities. The scale interactions become more and more numerous, and the general degree of disorder becomes greater. In the most highly nonlinear systems such as fully turbulent flow the scales interacting in a complex way are extremely numerous, and an explicit description of their interaction becomes problematic; one then resorts to stochastic treatments of the interactions. Perhaps the most difficult flows to deal with are ones for which the degree of disorder is too great to allow explicit treatment of the dynamics but too small to allow a simple stochastic treatment.

A serious problem confronting the mesoscale meteorologist is the question of the degree of disorder of mesoscale motions. Can the mesoscale be treated as a manageable finite number of forced and free waves interacting with each other and the mean flow, or must it be considered a continuous spectrum of eddies whose interactions must be handled statistically? According to Lilly (1983) and others, on average the mesoscale is a bland region of disordered waves and eddies serving to shunt energy between the cumulus and cyclone scale, but locally and/or very intermittently a mesoscale "event" occurs that converts potential to kinetic energy on the mesoscale. If this is a qualitatively correct view, then the job of the mesoscale meteorologist is to ascertain the physical nature of the intermittent organized mesoscale events and to determine the way in which energy, heat, and momentum are transferred by both the intermittent ordered events and the more continuous disordered motions.

1.6. Organized Mesoscale Phenomena

Perhaps the most fundamental instinct of the scientist is to discern order in nature. The success of the science of meteorology lies in its determination of a degree of order in the apparent chaos of weather. The drive to see the order of natural processes is so strong that even very weak coherent structures in a generally random pattern are picked out by the human eye, to the degree that the importance of those coherent structures that do exist is often overstated. In distinguishing mesoscale organized phenomena from phenomena of other scales it is convenient to distinguish free motions from forced motions of a system. The latter result from the direct influence of boundaries on the system; the former result from natural instabilities of flows that owe their existence to forcing on a much larger scale.

1.6.1. Examples of Mesoscale Instabilities

Mesoscale atmospheric convection is probably the most widely discussed mesoscale phenomenon. Very little is known, however, about the factors that really determine the scale of convective systems. It has been known for a long time, for example, that vertical shear of the mean horizontal wind tends to organize free dry convection into long shear-parallel rows whose spacing is on the order of the depth of the convective layer. This is not a mesoscale spacing as mesoscale is defined here. This distinctly larger scale organization seems to show up when the convection is driven principally by latent heat release. In such a case, as shown by Bjerknes (1938), the most unstable mode of convection is one cloud surrounded by an infinite clear region. This is not what is usually observed. Analysis shows that a cumulus cloud elicits essentially two kinds of response: (1) a transient response consisting of inertia-gravity waves (frequency between N and f, by definition "mesoscale") which, in the absence of mean shear, rapidly propagate away from the cloud, both vertically and horizontally, and (2) a steady response which, in the absence of turbulence in the free atmosphere, consists of a synoptic-scale disturbance of dimensions NH/f, where H is the depth of the convection. This latter scale shows up clearly in the case of the mesoscale convective complex (MCC) which exhibits a synoptic-scale anticyclone in the high troposphere, but the response at this scale to weaker forms of moist convection is not so well observed or understood. Our present meager theoretical understanding suggests that, except for the gravity waves, the principal response to moist convection in an otherwise calm atmosphere should be at the cumulus and the synoptic scales, with no intervening mesoscale of special significance. This leaves unexplained such phenomena as mesoscale cellar convection, whose breadth may be as much as 50 times its depth.

As has already been pointed out, strongly baroclinic flows that are stable to upright moist convection may nevertheless support slantwise moist convection (e.g., Emanuel, 1983), which has horizontal dimensions $\overline{U}_z H/f$ and a time scale of f^{-1} and is thus the only known instability that is unambiguously mesoscale. There is in addition some evidence that cumulus convection in baroclinic atmospheres may excite circulations whose scale is $\overline{U}_z H/f$, but this has not been clearly demonstrated.

Tropical cyclones constitute another phenomenon with distinctly mesoscale characteristics, even though their areal extent may be quite a bit larger than what is generally thought of as mesoscale. They arise from what must be considered an instability based on air-sea interaction, though the precise mechanism has not been fully described. The same mechanism perhaps operates within very intense oceanic cyclones whose origins are otherwise baroclinic.

1.6.2. Forced Mesoscale Circulations

It has already been mentioned that flow over irregular topography will generally excite a spectrum of waves only some of which are capable of propagating appreciable vertical distances in the absence of vertical shear. In particular, inertia-gravity waves (Doppler-shifted frequencies between N and f) constitute the mesoscale response and are responsible for phenomena such as chinooks and severe downslope windstorms. The classical theory of lee waves and mountain waves is directed toward those waves at the small end of the gravity wave spectrum that are virtually uninfluenced by the Earth's rotation. Recent investigations have, on the other hand, revealed a wealth of interesting effects that result from Coriolis accelerations, and that influence the larger mesoscale waves. When flow over a mountain range is sufficiently weak or the mountains are sufficiently high, for example, the upstream flow becomes stagnant for a distance Nh/f from the base of the mountains, where h is the height of the mountains. Although this is dynamically a synoptic-scale distance, its typical scale in the Earth's atmosphere is only about 100 km and would therefore be considered a mesoscale phenomenon by observers. This blocking may, however, lead to distinctly mesoscale events such as coastal fronts.

Diurnally driven circulations such as sea breezes straddle the upper end of the mesoscale in terms of Lagrangian time scales. The diurnal period is less than or greater than a pendulum day, depending on whether one is located equatorward or poleward of 30° latitude; in the absence of dissipation the character of the sea breeze may be profoundly different in the two locations (Rotunno, 1983).

A striking aspect of both free and forced organized mesoscale circulations is their local character. The mesoscale instabilities that have been identified tend to be highly intermittent events, while the forced circulations occur at particular locations at certain times. This perhaps accounts for the absence of spectral peaks in the mesoscale range in long-period records of atmospheric motions. The provincial and ephemeral nature of organized mesoscale circulations presents special challenges for both the research meteorologist and the forecaster.

1.7. Conclusion

The natural philosopher presented with the complete spectrum of atmospheric variability notices very quickly a preference for certain time and length scales. The unaided eye perceives the physical scale of individual clouds as well as the time scale of diurnal and seasonal phenomena and the interval between storms in middle latitudes. Synoptic analyses based on diverse observations extended the capability of the unaided eye and quickly revealed the scale of such phenomena as cyclones and planetary waves. Terms such as "synoptic scale," "planetary scale," and "cumulus scale" crept into the meteorologist's vocabulary in recognition of the scales of special significance. The term "mesoscale" was first coined to describe scales that could not be observed with contemporary observing systems.

Assigning physical significance to the classical definitions of special scales depends in part on the degree of linearity of atmospheric processes. Stable and weakly nonlinear systems exhibit special scales at which processes are externally forced and at which instabilities preferentially develop; very strongly nonlinear systems may have a continuous and monotonic spectrum of motions ranging from the principal externally forced scales to the scales at which molecular dissipation becomes important. Scale categorization becomes difficult if not meaningless in the latter case.

In the limited context of very simple known stationary states of the atmosphere consisting of simple zonal flows with no meridional variation, it is possible to identify special scales of oscillation and instability with some classical scale definitions based on observation. The planetary scale is identified with Rossby waves large enough to propagate vertically in a constant mean zonal wind, and the synoptic scale and cumulus scale may be related to baroclinic and convective instabilities. The mesoscale may be said to involve processes with Lagrangian time scales between the period of a pure buoyancy oscillation and a pendulum day and includes inertia-gravity waves and slantwise convection.

The processes that generate kinetic energy on the mesoscale—topographic forcing and mesoscale instabilities—appear to be local and intermittent so that on average the mesoscale is characterized either by very little energy at all (a spectral gap) or by a continuous and monotonic spectrum representing energy cascading up from smaller scales or down from synoptic scales. The cascading processes operative on the mesoscale are poorly understood, in spite of some gains in our knowledge of topographically forced mesoscale circulations and certain mesoscale instabilities. It is a challenge for the observer, the theoretician, and the forecaster to gain a deeper understanding of the processes that generate energy on and transfer energy through the mesoscales.

REFERENCES

Bjerknes, J., 1938: Saturated ascent of air through a dry-adiabatically descending environment. *Quart. J. Roy. Meteor. Soc.*, **64**, 325–330.

Emanuel, K. A., 1983: The Lagrangian parcel dynamics of moist symmetric instability. *J. Atmos. Sci.*, **40**, 2368–2376.

Ligda, M. G. H., 1951: Radar storm observations. In *Compendium of Meteorology*, American Meteorological Society, Boston, 1265–1282.

Lilly, D. K., 1983: Stratified turbulence and mesoscale variability of the atmosphere. *J. Atmos. Sci.*, **40**, 749–761.

Orlanski, I., 1975: A rational subdivision of scales for atmospheric processes. *Bull. Amer. Meteor. Soc.*, **56**, 527–530.

Rotunno, R., 1983: On the linear theory of the land and sea breeze. *J. Atmos. Sci.*, **40**, 1999–2009.

Vinnichenko, N. K., 1970: The kinetic energy spectrum in the free atmosphere—one second to five years. *Tellus*, **22**, 158–166.

Wipperman, F., 1981: The applicability of several approximations in mesoscale modeling—a linear approach. *Contrib. Atmos. Phys.*, **54**, 298–308.

CHAPTER 2

Mesoscale Classifications: Their History and Their Application to Forecasting

T. T. Fujita

2.1. The Synoptic Scale

So-called synoptic meteorology is the study of meteorological data collected and transmitted by a network of weather stations. The *Glossary of Meteorology* (Huschke, 1959) defines "synoptic" as referring to "meteorological data obtained simultaneously over a wide area for the purpose of presenting a comprehensive and nearly instantaneous picture of the state of the atmosphere." Thus, to be synoptic, weather observations must be made simultaneously; the areal extent or horizontal scale of the station network is not specified.

Because of difficulties in collecting simultaneous data from extremely large areas, the coverage of early synoptic charts was limited to the practicable area of data collection, such as Europe, the United States, and Japan. The scale of the disturbances depicted on synoptic charts of such areas was identified as "synoptic-scale," implying that the term synoptic denotes a specific scale. Thus, the use of synoptic to indicate scale is now also in common practice. (See a related discussion in Ch. 1.) Disturbances that are too small to be depicted on synoptic charts are called subsynoptic disturbances.

An early weather chart (Fig. 2.1) includes a synoptic-scale cyclone over England, analyzed with vector winds, isobars, and isotherms. Analysts and forecasters followed the movement and the development of cyclonic swirls as seen in an early cyclone model (Fig. 2.2) by Fitz-Roy (1863). His model showed the existence of cold-air currents from the north, which are counteracted by warm-air currents from the south.

During the 1910s, the concept of warm and cold air masses was widely used by forecasters. The continuity of air mass boundaries was analyzed with dashed, dotted, and chained lines (Fig. 2.3). Modern air mass analyses originated with the Norwegian school in the 1920s.

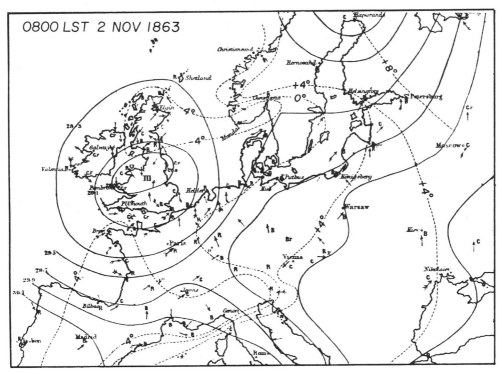

Figure 2.1. Meteorological chart of Europe at 0800 LST, 2 November 1863. (From *Journal of the Scottish Meteorological Society*, October 1868.) Isobars in inches and air temperature in degrees Celsius are drawn with solid and dashed lines, respectively. Symbols of station weather are C, cloudy; B, blue sky and few clouds; R, rain; r, rain during the past 24 hours.

The well-known model of extratropical cyclones by Bjerknes and Solberg was established in 1922. Since then, operational synoptic charts in Europe have been characterized by key features: warm and cold fronts, and precipitation areas closely related to these fronts (Fig. 2.4). Frontal analysis in the United States began in the 1930s, leading to the discovery of the prefrontal squall lines that often induce severe local storms. Recently, Kochin (1983) re-

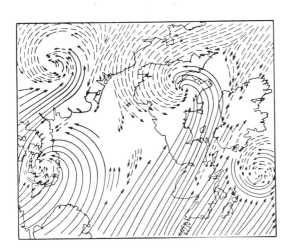

Figure 2.2. Fitz-Roy model of extratropical cyclones introduced by Petterssen (1956). Streamlines of polar air from the north are drawn with full lines, and those of tropical air from the south with dashed lines.

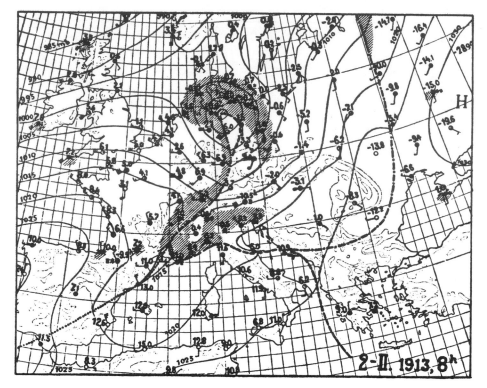

Figure 2.3. A synoptic map of Europe on 2 February 1913, before the advent of the cyclone model with warm and cold fronts. Airmass boundaries are shown, and precipitation areas are shaded. (From Chromow, 1942.)

analyzed a sequence of synoptic charts of the Blizzard of 1888, demonstrating the effectiveness of the frontal analysis in depicting the record-breaking New England snowstorm.

2.2. The Mesoscale

Isolated strange values of pressure, winds, etc., are often found on plotted synoptic charts, and are suspected to be errors. Some data are obviously in error; other data may represent true signatures of subsynoptic disturbances

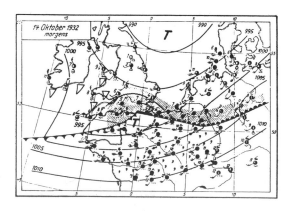

Figure 2.4. A wave cyclone analyzed by Van Mieghem, introduced by Chromow (1942); warm and cold fronts are shown. Isobars are in millibars; plotted temperatures are in degrees Celsius.

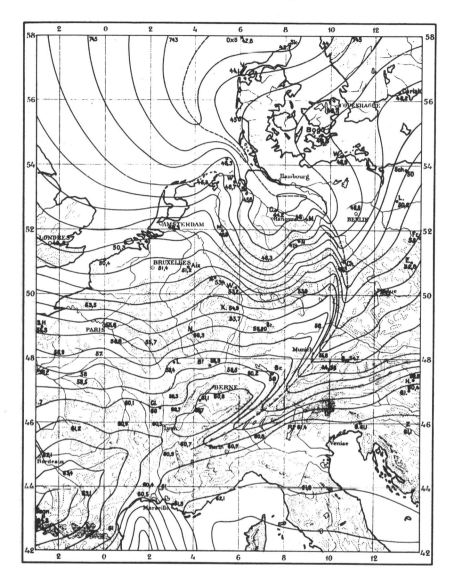

Figure 2.5. Detailed analysis of a squall line situation in Europe at 2100 LST, 27 August 1890 by Durand-Gréville (1892).

having spatial and temporal scales too small to be analyzed on synoptic charts.

Although the first analysis of the subsynoptic disturbance has not been identified, Durand-Gréville (1892) made a detailed analysis of the pressure field of a squall line in Europe extending from near Hamburg, Germany, to near Lyons, France (Fig. 2.5). He drew isobars for every millimeter of mercury, making use of 84 pressure values. The precise steps taken in the analysis are not clear. It is likely that the isochrones of the pressure surges were analyzed first, followed by the reconstruction of the pressure field that would result in the abrupt increases in pressure at successive stations.

Table 2.1. Early subsynoptic networks for studying thunderstorm circulation

Years	Location	Country	No. of stations	Spacing
1939–41	Lindenberg	Germany	19	3–20 km
1940	Maebashi	Japan	20	8–13 km
1946	Florida	United States	50	2 km
1947	Ohio	United States	58	3 km

Suckstorff (1938) described the feature of thunderstorm-induced highs by converting the station data (time section) into horizontal data (space section) in the direction of the system movement. After that, the analysis of subsynoptic disturbances was identified as a "micro-scale" study. A surface "micro-study" of a squall line (Williams, 1948), and "microanalytical" studies of a thunder nose (Fujita, 1950) and a cold front (Fujita, 1951) are examples.

As air traffic increased in the 1930s, thunderstorm and squall-line-related accidents and difficulties occurred in various parts of the world, necessitating operation of subsynoptic networks for a better understanding of severe local storms that escape detection by the synoptic network. Examples of these early subsynoptic networks are given in Table 2.1. Analytical results of the data from these networks were reported by Koschmieder (1955), Fujiwara (1943), and Byers and Braham (1949). Figure 2.6 presents examples of thunderstorm outflows depicted by these networks.

Weather radar was used extensively by the Thunderstorm Project (Byers and Braham, 1949) for identifying and mapping precipitation areas. For the first time in history, surface data were analyzed in relation to PPI echoes (nearly horizontal cross sections) and RHI echoes (nearly vertical cross sections) obtained simultaneously. The most important accomplishments by the Thunderstorm Project were to identify or confirm these characteristics of storms:

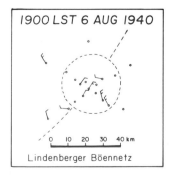
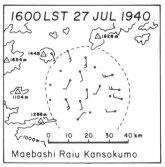

Figure 2.6. Thunderstorm outflows depicted by thunderstorm-related projects in (left) Germany, (center) Japan, and (right) the United States. For network details, see Table 2.1.

- The spatial scale of individual thunderstorm cells is 10 to 30 km.
- The life of an individual cell through three stages (cumulus, mature, dissipating) is approximately 30 min.
- Storms contain a strong downdraft in post-mature stages.
- Strong divergence in the surface wind is as high as 0.08 s^{-1}.

These characteristics, with dimensions of 10 to 100 km, cannot be depicted on an ordinary synoptic chart; they would merely appear as "noise."

On the basis of his research on radar storm echoes collected at the Massachusetts Institute of Technology (MIT) Weather Radar Research Project, Ligda (1951, p. 1281) made the foresighted statement,

> It is anticipated that radar will provide useful information concerning the structure and behavior of that portion of the atmosphere which is not covered by either micro- or synoptic-meteorological studies. We have already observed with radar that precipitation formulations which are undoubtedly of significance occur on a scale too gross to be observed from a single station, yet too small to appear even on sectional synoptic charts. Phenomena of this size might well be designated as mesometeorological.

The U.S. Signal Corps at Fort Monmouth, N.J., which was sponsoring the MIT project, recognized the importance of the mesometeorological scale for the operation of the U.S. Army. Two years later, Swingle and Rosenberg (1953) of the Signal Corps performed a mesometeorological analysis of a cold front.

An apparent increase in tornado frequencies in the 1950s was attributed to the U.S. Weather Bureau's effort to improve the tornado reporting system, public awareness, etc. In an attempt to provide better forecasting of tornadoes, Tepper related the pressure-jump mechanism (Tepper, 1950a) to the origin of tornadoes (Tepper, 1950b). He hypothesized that a tornado is likely to form at the intersection of two pressure-jump lines and travel with the intersection.

Early in the 1950s, a multiple-state network of microbarograph stations was established in the Midwest. Barograph traces from the network were collected and analyzed by the Severe Storms Project of the U.S. Weather Bureau. By 1954, however, it became evident that the correlation between tornadoes and pressure-jump lines was not as high as originally expected. As a result, the main emphasis of the project was shifted into the study of mesometeorological (i.e., mesoscale) systems, leading to a mesoscale analytical study of squall lines by Fujita (1955).

In order to describe mesoscale disturbances related to severe-storm activities for analysts and forecasters, Fujita *et al.* (1956) published a discussion of mesoanalysis in color. Their primary purpose was to use color, as it was used by the Norwegian School in the 1920s, to describe the rather complicated structure of mesoscale systems (mesosystems). At that time, the mesoscale was either being overlooked or intentionally ignored and smoothed in much of analysis, because mesoscale atmospheric motions are, in effect, "noise" superimposed upon synoptic-scale disturbances.

**Table 2.2. High-pressure, low-pressure, and frontal systems
on or near the ground in synoptic scale, mesoscale, and submesoscale**

System	Synoptic scale	Mesoscale	Submesoscale
High-pressure system	Anticyclone (accompanied by fine weather)	Mesohigh (accompanied by bad weather)	Pressure nose (accompanied by high winds)
Low-pressure system	Subtropical cyclone Tropical cyclone Hurricane Typhoon	Mesocyclone Tornado cyclone Mesolow Midget typhoon	Tornado vortex signature Tornado Suction vortex Dust devil
Frontal system	Warm front Cold front Stationary front Occluded front Dry front	Pressure-jump line Gust front Radar thin line Arc-cloud line Sea-breeze front	Downburst front Precipitation roll Microburst ring vortex

2.3. Characteristics of Convective Mesoscale and Submesoscale Phenomena

Mesoscale disturbances induced by convective activities (convective mesoscale systems) are roughly divided into high-pressure (mesohigh), low-pressure (mesolow), and frontal (mesofront) systems. These disturbances have been known for a long time, probably since Durand-Gréville's analysis of the 27 August 1890 squall line. His chart (Fig. 2.5) indicates a low-pressure system to the south of Hamburg, Germany, a high-pressure system over Switzerland, and another over southern France. A summary of high-pressure, low-pressure, and frontal systems with their dimensions of synoptic scale, mesoscale, and submesoscale is shown in Table 2.2.

A synoptic-scale high-pressure system is an anticyclone in which air subsides slowly, resulting in mostly fair weather. By contrast, most mesohighs are characterized by rain and bad weather, because they are caused by the rain-cooled air that descends to the ground, forming a dome of high pressure. Mesoscale high-pressure systems were studied by Suckstorff (1938) who used time variations of meteorological parameters measured during thunderstorms in Germany. In the United States, Brunk (1949) analyzed the pressure pulsation of 11 April 1944, pointing out that it was characterized by a sudden pressure increase followed by a rapid and significant pressure drop. Williams (1953) called the high-pressure region the elevation-type wave, and the low-pressure region, the depression-type wave. Isochrone charts that he analyzed revealed that the latter followed the former rather regularly.

Identifying the high-pressure dome accompanied by thunderstorms as a mesohigh, Fujita (1955; 1959) and Fujita *et al.* (1956) studied the formation of the depression that forms in the wake of a mesohigh. These studies led to the conclusion by Fujita (1963) that an initial high-pressure dome induces a

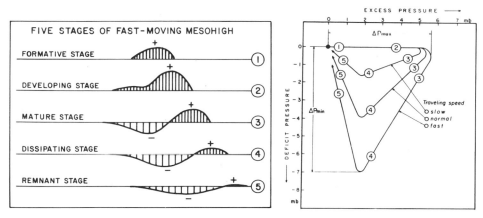

Figure 2.7. Five stages of a mesohigh that turns into a mesolow called a wake depression. The total lifetime is 6–12 h, so that a mesohigh formed during the night could remain as a mesolow the following morning. (From Fujita, 1963.)

mesolow within 1–2 h. Through the five stages shown in Fig. 2.7, what is initially a mesohigh turns into a mesolow.

The total time of the excess-deficit pressure cycle is 6–12 h, long enough for use of the cycle in a short-range forecast. The excess-deficit pressure (Fig. 2.7) is highly dependent upon the traveling speeds of mesohighs. If the mesohighs are stationary, practically no wake depression forms. However, a fast-moving mesohigh is characterized by a significant high- and low-pressure couplet. A model of mesohighs in relation to the position of a synoptic front is presented in Fig. 2.8.

The submesoscale high, called the pressure nose by Byers and Braham (1949) is very small and short-lived. Adjacent meso-gamma stations of the Thunderstorm Project rarely depicted the same pressure nose simultaneously. Recently, Fujita and Caracena (1977) hypothesized that the pressure nose is responsible for inducing downburst winds that endanger aircraft operations at low altitudes.

From time to time, tornadoes form inside mesolows. Brooks (1949) called the mesolow with a tornado in it the "tornado cyclone." During the 1950s, weather radar provided an effective means of detecting and displaying a rotation in a cloud, which appears as a hook echo. Radar displays of the hook echo were first photographed by Stout and Huff (1953). Since then hooks have been used in nowcasting and warning for tornadoes, sometimes even before a tornado forms. As reported by Garrett and Rockney (1962), the area represented by the hook is 2–5 km across, several times larger than the damage width of the tornado located inside the hook. Not all hook echo areas produce tornadoes, however. Schaefer (1960) and others reported no sign of a tornado inside some hook echoes. Nonetheless, the hook echo is an important predictor of tornadoes.

Use of Doppler radar in meteorological observations was first reported by Smith and Holmes (1961). Brown *et al.* (1971) observed with Doppler radar a submesoscale tight vortex inside a developing hook echo. When dual-

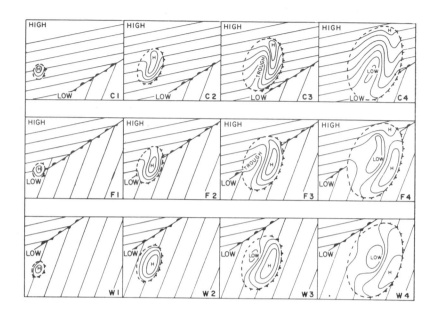

Figure 2.8. Model of a squall-line mesosystem: (upper) in the cold sector, (middle) on the front, and (lower) in the warm sector. Because the air temperature inside the pressure dome is colder than that behind the cold front, a synoptic-scale front does not intrude into the region of the mesoscale disturbance. (From Fujita, 1963.)

Doppler measurement became available for velocity computations, the three-dimensional structure of mesocyclones became evident. Ray *et al.* (1975) observed submesocyclone circulations, 3–5 km in diameter, embedded inside the parent mesocyclone. Brandes (1978) pointed out that a tight circulation, 2–4 km in diameter, existed at low levels, above a tornado on the ground. Burgess *et al.* (1976) developed the multimoment Doppler display for identifying the TVS (Tornado Vortex Signature) defined by Burgess *et al.* (1975). The TVS is 1 order of magnitude smaller than the overall dimension of the parent mesocyclone, 10–100 km in diameter.

The subtornado-scale vortex was called a suction spot by Fujita (1970), and revised to suction vortex in 1971. Single and multiple suction vortices were documented frequently during the 1970s. The diameter of suction vortices is 5–10 m. A large tornado could induce one to several suction vortices simultaneously. For case studies, see Fujita (1974; 1976; 1981) and Agee *et al.* (1977); for numerical simulation, see Rotunno (1981).

Mesoscale frontal (mesofront) systems have long been observed as dramatic events of pressure, temperature, and wind changes. Depending upon the emphasis, they are called pressure-jump lines, gust fronts, radar thin lines, arc-cloud lines, etc. Nevertheless, these terms all refer to the same aspect of the front of a thunderstorm outflow (Fig. 2.9).

The pressure-jump line is located along the leading edge of an advancing mesohigh. Tepper (1950a) attempted to explain the dramatic jump in pressure as a "hydraulic" jump. Later, Tepper (1955) theorized the impulsive

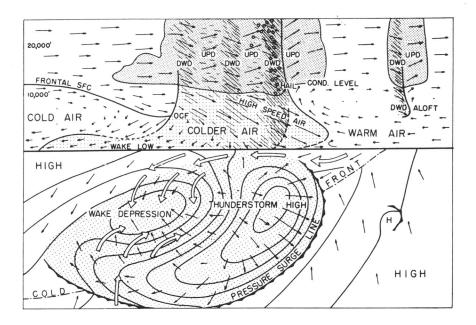

Figure 2.9. A vertical cross section and a plan view of a mesofront system (squall line). Arrows indicate the airflow relative to the ground. (From Fujita, 1955.)

addition of momentum as an indicator of pressure-jump lines that extend 10–300 km.

In many cases, winds behind the front of a mesohigh are strong and gusty. The term gust front signifies the band of gusty winds following the passage of a pressure-jump or pressure-surge line. Byers and Braham (1949) defined and analyzed the first-gust lines. By using tower data, Goff (1976) obtained vertical cross sections of gust fronts in Oklahoma. Using the NIMROD (Northern Illinois Meteorological Research On Downbursts) data, Wakimoto (1982) produced Doppler cross sections of gust fronts. These studies indicated the horizontal length of gust fronts to be 10–300 km, and the life of high winds 1–6 h.

Boundaries of thunderstorm outflow have been observed on radar PPI scopes as thin lines of echo moving out of thunderstorm areas. Ligda and Bigler (1958) called the thin line a cloudless cold front, and Harper (1960) called it a thin line of angel echo. No matter what the line may be called, it represents the leading edge of a mesoscale outflow, along which the detectable backscattering is caused by the strong gradient of refractive index. The horizontal extent of a thin line is 10–100 km, depending upon the size of the mesohigh and the radar coverage at low altitude.

Arc-cloud lines form frequently along the leading edge of an expanding mesohigh. Purdom (1973) found an excellent correspondence between mesohigh boundaries and arc-cloud lines. Making use of GOES (Geostationary Operational Environmental Satellite) imagery, Purdom (1976) emphasized the arc-cloud line as an effective forecasting tool. He noted specifically that

Figure 2.10. Two pictures (5 s apart) showing the curling motion of the dust cloud behind the front of a microburst outflow. (From Fujita, 1984; photos by Brian Waranauskas near Denver, Colo., on 15 July 1982.

new convection often begins at the intersections of two arc-cloud lines 10–500 km long.

Submesoscale fronts, which are much shorter than mesofronts both in lifetime and characteristic length, are not well known to forecasters because they are seldom reported. After the wind-shear study by Fujita and Caracena (1977), Caracena hypothesized that the submesoscale front is characterized by a vortex ring similar to that created by a cigarette smoker. This is illustrated in Fig. 2.10.

On the basis of this hypothesis and a sequence of photographs of an earlier microburst in Colorado, Fujita (1984) produced a graphic model of the Andrews Air Force Base microburst on 1 August 1983, which had induced a 67 m s^{-1} (130 kt) peak wind (Fig. 2.11).

2.4. Classifications of Meteorological Scales

Synoptic meteorology, including the study of cyclones and anticyclones, dominated the field of analytical meteorology from the 19th century to the 1930s. From the 1940s through the 1960s, the subsynoptic scale, regarded as the noise of the synoptic scale, nearly embraced the mesoscale. Since then, significant progress in mesometeoreology has been achieved by the use of geostationary satellite, Doppler radar, instrumented-aircraft, and advanced-mesonet data along with increased mathematical modeling capabilities.

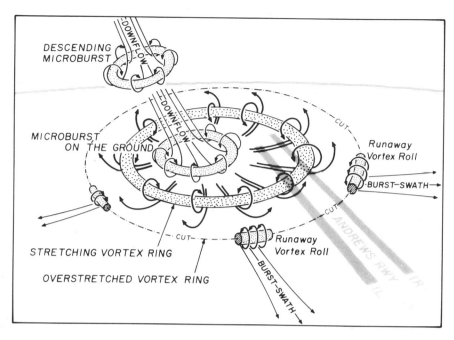

Figure 2.11. Stages of the Andrews Air Force Base microburst on 1 August 1983. An anemometer on the base recorded a scaled-off 130 kt peak wind. So far, it is the strongest microburst wind ever recorded. (From Fujita, 1984.)

In spite of the dramatic advancement of the field, the various definitions of the mesoscale have been rather arbitrary (Fig. 2.12). This growing research field has been apt to expand the scale limits on both sides of the original dimensions of 10–100 miles (16–160 km), depending on the scope and purpose of individual research projects.

Ligda's (1951) classification emphasized the importance of the mesoscale when weather radars began depicting subsynoptic features that could not possibly be analyzed on synoptic charts. Ligda also recognized the existence of microscale disturbances that are noises of mesoscale disturbances (Fig. 2.12).

Fujita (1963) subclassified the mesoscale into alpha (α), beta (β), and gamma (γ) scales (not shown in Fig. 2.12) on the basis of the average spacing of observation stations. Table 2.3 lists the stations of early mesoscale networks (mesonets) operated before 1962; Fig. 2.13 shows their distribution. The Thunderstorm Project network in Florida with 50 stations and 2 km separation was the densest of all. At the other extreme, the 210 microbarograph stations of the National Severe Storms Project (NSSP) were separated by ~60 km.

During the past 10 years, Doppler radars have been collecting data for mapping submesoscale disturbances with features as small as defined by a 150 m range gate spacing. Multiple Doppler analyses now permit researchers to construct "microsynoptic charts" of the airflow around the tornado environment.

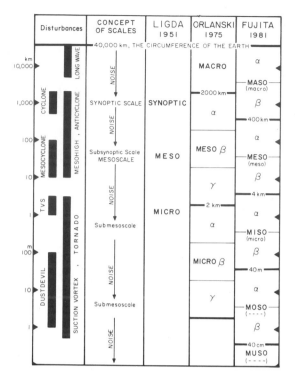

Figure 2.12. The horizontal scales of atmospheric disturbances in relation to the scales proposed by Ligda (1951), Orlanski (1975), and Fujita (1981). Note that a cyclone is the parent storm of a mesocyclone, which in turn is the parent storm of a tornado, an inducer of suction vortex. Scales of these parent cyclonic storms are separated by approximately 2 orders of magnitude.

Orlanski (1975) classified the mesoscale into α, β, and γ scales according to horizontal dimensions of meteorological systems (Fig. 2.12). These scales have been widely adopted, for reference purposes at least. Orlanski's mesoscale extends through 3 orders of magnitude, leaving the range of the macroscale between 2,000 and 40,000 km (1.3 in log scale). In spite of the

Table 2.3. Mesonets operated before 1962*

Year	Name of network	Classification	No.	Spacing (km)	Objectives
1962	NSSP	Alpha	210	60	Squall lines, pressure jumps
1960	Fort Huachuca	Beta	28	20	Operation of the Army
1961	NSSP	Beta	36	20	Structure of severe storms
1961	Dugway Proving Ground	Beta	15	15	Desert area research
1941	Lindenberger Böennetz	Beta	25	15	Squall lines, wind gusts
1960	New Jersey	Beta	23	10	Mesoscale pressure systems
1941	Muskingum basin	Beta	131	10	Rainfall, runoff
1940	Japanese thunder-storm	Beta	25	9	Structure of thunderstorms
1961	Flagstaff	Beta	43	8	Cumulonimbus convection
1947	Thunderstorm Proj., Ohio	Gamma	58	3	Thunderstorm convection
1961	Fort Huachuca	Gamma	17	3	Influence of orography
1946	Thunderstorm Proj., Florida	Gamma	50	2	Thunderstorm convection

*After Fujita (1963).

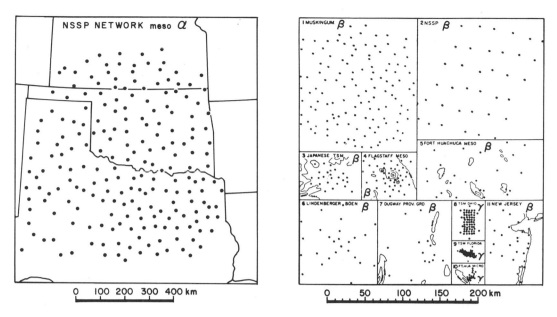

Figure 2.13. Alpha, beta, and gamma mesonets operated prior to 1962. Individual networks are listed in Table 2.3. (From Fujita, 1963.)

narrow macroscale range, there is an advantage in extending the mesoscale through 3 orders of magnitude. The center-scale grid (β) can be nested inside the coarser (α) grid, and a denser (γ) grid can be nested inside the center-scale grid. On the other hand, the expanded mesoscale classifies synoptic cyclones and anticyclones as mesocyclones and mesohighs. This classification contradicts the original concept of mesoscale disturbances as noises of synoptic-scale disturbances.

A more recent classification of meteorological scales (Fujita, 1981) follows two trends:

- The center of the mesoscale is still between 10 and 100 km.
- The dimensions of successive subdivisions decrease by a factor of 10, or 1 order of magnitude.

2.5. Classification Scales in Forecasting

Weather forecasters must predict disturbances that occur near the ground. In general, the smaller the scale of such disturbances, the shorter the duration (Orlanski, 1975). Table 2.4 shows both scale and duration of disturbances ranging from extratropical cyclone to dust devil. In Fig. 2.14, the values in Table 2.4 are placed within the Fujita scales (Fig. 2.12); each disturbance is represented by a rectangle marking the upper and lower limits of horizontal scales and durations.

Aviation forecasters are required to predict midair disturbances that may never descend to the ground. In Fig. 2.15 the scales and durations of middle- and high-level disturbances presented in Table 2.5 are placed within the Fujita scales. In accordance with the *Glossary of Meteorology* (Huschke,

Table 2.4. Weather systems on or near the ground

Disturbance	Scale	Duration	Max. wind
Extratropical cyclone	500–2000 km	3–15 days	55 m s^{-1}
Cold front	500–2000 km	3–7 days	25 m s^{-1}
Anticyclone	500–2000 km	3–15 days	10 m s^{-1}
Warm front	300–1000 km	1–3 days	15 m s^{-1}
Hurricane	300–2000 km	1–7 days	90 m s^{-1}
Tropical cyclone	300–1500 km	3–15 days	33 m s^{-1}
Tropical depression	300–1000 km	5–10 days	17 m s^{-1}
Dry front	200–1000 km	1–3 days	20 m s^{-1}
Midget typhoon	50–300 km	2–5 days	50 m s^{-1}
Mesohigh	10–500 km	3–12 h	25 m s^{-1}
Gust front	10–300 km	0.5–6 h	35 m s^{-1}
Mesocyclone	10–100 km	0.5–6 h	60 m s^{-1}
Downslope wind	10–100 km	2–12 h	55 m s^{-1}
Macroburst	4–20 km	10–60 min	40 m s^{-1}
Microburst	1–4 km	2–15 min	70 m s^{-1}
Tornado	30–3000 m	0.5–90 min	100 m s^{-1}
Suction vortex	5–50 m	5–60 s	140 m s^{-1}
Dust devil	1–100 m	0.2–15 min	40 m s^{-1}

1959), the angular wavenumber of long waves is taken to be in the range 1–5, and that of short waves, 5–10.

Figure 2.16 combines the information in Figs. 2.14 and 2.15. The nonlinear regression line shows that the average duration of 4–400 km disturbances is between 30 min and 1 day. The duration of 40 m–4 km disturbances is only 1–30 min. Disturbances >400 km last 1 day or longer and therefore can be depicted and followed on synoptic charts that are analyzed every 3–6 h.

It has been known also that the horizontal scale of a parent disturbance is approximately 2 orders of magnitude larger than that of a disturbance spawned by the parent disturbance. For instance, in the sequence extratropical cyclone, hook-echo cyclone, tornado, and suction vortex, each successive phenomenon is nearly 2 orders of magnitude smaller than the one that

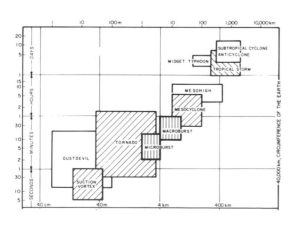

Figure 2.14. Horizontal scale vs. duration of near-ground disturbances, classified according to Fujita (1981).

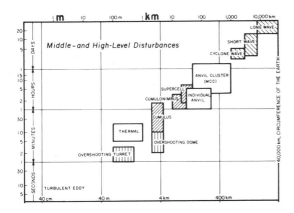

Figure 2.15. Horizontal scale vs. duration of middle- and high-level disturbances, classified according to Fujita (1981).

Table 2.5. Scale and duration of middle- and high-level disturbances

Disturbance	Horizontal scale	Duration	
Long wave	8000–40000 km	15+	days
Short wave	3000–8000 km	3–15	days
Cyclone wave	1000–3000 km	2–5	days
Jet stream	1000–8000 km	5–15	days
Low-level jet	300–1000 km	1–3	days
Jet streak	200–1000 km	2–5	days
Anvil cluster (MCC)	50–1000 km	3–36	h
Individual anvil	30–200 km	1–5	h
Supercell storm	20–50 km	2–6	h
Cumulonimbus	10–30 km	1–3	h
Cumulus	2–5 km	10–100	min
Overshooting dome	2–5 km	2–10	min
Tornado vortex signature	1–5 km	20–90	min
Overshooting turret	100–500 m	1–3	min
Thermal	100–1000 m	5–20	min
In-cloud turbulent eddy	10–100 m	Variable	

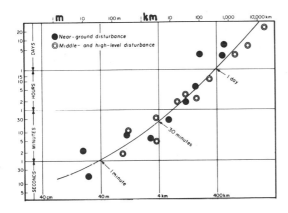

Figure 2.16. Horizontal scale vs. duration of all disturbances shown in Figs. 2.14 and 2.15. Circles represent the center points of rectangles in those figures.

spawned it. However, in classifications that separate scales by 3 orders of magnitude, two or more disturbances in a sequence are likely to be included in one scale, resulting, for example, in an extratropical cyclone and a mesocyclone together in one scale, and a tornado and a suction vortex together in another scale.

REFERENCES

Agee, E. M., J. T. Snow, F. S. Nickerson, P. R. Clare, C. R. Church, and C. A. Schaal, 1977: An observational study of the West Lafayette, Indiana, tornado of March 20, 1976. *Mon. Wea. Rev.*, **105**, 893–907.

Bjerknes, J., and H. Solberg, 1922: Life cycle of cyclones and the polar front theory of atmospheric circulation. *Geofys. Publ. Norske Vidensk.-Akad. Oslo,* **3** (1): 1–18.

Brandes, E. A., 1978: Mesocyclone evolution and tornadogenesis: Some observations. *Mon. Wea. Rev.*, **16**, 995–1011.

Brooks, E. M., 1949: The tornado cyclone. *Weatherwise*, **2**, 32–33.

Brown, R. A., W. C. Bumgarner, K. C. Crawford, and D. Sirmans, 1971: Preliminary Doppler velocity measurements in a developing radar hook echo. *Bull. Amer. Meteor. Soc.*, **52**, 1186–1188.

Brunk, I. W., 1949: Pressure pulsations of April 11, 1944. *J. Meteor.*, **6**, 181–188.

Burgess, D. W., L. R. Lemon, and R. A. Brown, 1975: Evolution of a tornado signature and parent circulation as revealed by single Doppler radar. Preprints, 16th Radar Meteorology Conference, Houston, Tex., American Meteorological Society, Boston, 99–106.

Burgess, D. W., L. H. Hennington, R. J. Doviak, and P. S. Ray, 1976: Multimoment Doppler display for severe storm identification. *J. Appl. Meteor.*, **15**, 1302–1306.

Byers, H. R., and R. R. Braham, Jr., 1949: The Thunderstorm. U.S. Department of Commerce, Weather Bureau, Washington D.C., 287 pp.

Chromow, S. P., 1942: Einführung in die synoptische Wetteranalyse. Springer-Verlag, Berlin, 532 pp.

Durand-Gréville, E., 1892: Les grains et les orages. *Ann. Centr. Meteor. France*, **1**, 249.

Fitz-Roy, R., 1863: *Weather Book: A Manual of Practical Meteorology.* London.

Fujita, T. T., 1950: Microanalytical study of thunder nose. *Geophys. Mag.* (Japan), **22**, 78–88.

Fujita, T. T., 1951: Microanalytical study of cold front. *Geophys. Mag.* (Japan), **22**, 237–277.

Fujita, T. T., 1955: Results of detailed synoptic studies of squall lines. *Tellus*, **4**, 405–436.

Fujita, T. T., 1959: Study of mesosystems associated with stationary radar echoes. *J. Meteor.*, **16**, 454–466.

Fujita, T. T., 1963: Analytical mesometeorology: A Review. In *Severe Local Storms*, Meteor. Monogr. 5(27), American Meteorological Society, Boston, 77–125.

Fujita, T. T., 1970: Lubbock tornadoes: A study of suction spots. *Weatherwise*, **23**, 160–173.

Fujita, T. T., 1974: Jumbo tornado outbreak of 3 April 1974. *Weatherwise*, **27**, 116–126.

Fujita, T. T., 1976: Close-up view of 20 March 1976 tornadoes: Sinking cloud tops to suction vortices. *Weatherwise*, **29**, 116–131.

Fujita, T. T., 1981: Tornadoes and downbursts in the context of generalized planetary scales. *J. Atmos. Sci.*, **38**, 1512–1534.

Fujita, T. T., 1984: Andrews AFB microburst. SMRP Res. Paper 205, University of Chicago, 38 pp.

Fujita, T. T., and F. Caracena, 1977: An analysis of three weather-related aircraft accidents. *Bull. Amer. Meteor. Soc.*, **58**, 1164–1181.

Fujita, T. T., H. Newstein, and M. Tepper, 1956: Mesoanalysis—an important scale in the analysis of weather data. U.S. Weather Bureau Research Paper 39., Supt. of Documents, U.S. Government Printing Office, Washington, D.C.

Fujiwara, S., 1943: Report of thunderstorm observation project. Japan Meteorological Agency, Tokyo, 248 pp.

Garrett, R. A., and V. D. Rockney, 1962: Tornado in Northern Kansas, May 19, 1960. *Mon. Wea. Rev.*, **90**, 231–240.

Goff, R. C., 1976: Vertical structure of thunderstorm outflows. *Mon. Wea. Rev.*, **104**, 1429–1440.

Harper, W. G., 1960: An unusual indicator of convection. *Mar. Obs.*, **30**, 36–40.

Huschke, R. E. (Ed.), 1959: *Glossary of Meteorology*, American Meteorological Society, Boston, 638 pp.

Kochin, P. J., 1983: An analysis of the Blizzard of '88. *Bull. Amer. Meteor. Soc.*, **64**, 1258–1272.

Koschmieder, H., 1955: *Ergebnisse der Deutschen Böenmessungen, 1939/41.* Friedr. Vieweg & Sohn, Braunschweig, 148 pp.

Ligda, M. G. H., 1951: Radar storm observation. In *Compendium of Meteorology*, American Meteorological Society, Boston, 1265–1282.

Ligda, M. G. H., and S. G. Bigler, 1958: Radar echoes from a cloudless cold front. *J. Meteor.*, **15**, 494–501.

Orlanski, I., 1975: A rational subdivision of scales for atmospheric processes. *Bull. Amer. Meteor. Soc.*, **56**, 527–530.

Petterssen, S., 1956: *Weather Analysis and Forecasting*. McGraw-Hill, New York, 428 pp.

Purdom, J. F. W., 1973: Meso-high and satellite imagery. *Mon. Wea. Rev.*, **101**, 180–181.

Purdom, J. F. W., 1976: Some uses of high-resolution GOES imagery in the mesoscale forecasting of convection and its behavior. *Mon. Wea. Rev.*, **104**, 1474–1483.

Ray, P. S., R. J. Doviak, G. B. Walker, D. Sirmans, J. Carter, and B. Bumgarner, 1975: Dual-Doppler observation of tornadic storms. *J. Appl. Meteor.*, **14**, 1521–1530.

Rotunno, R., 1981: A numerical simulation of multiple vortices. In *Intense Atmospheric Vortices*, L. Bengtsson and J. Lighthill (Eds.), Springer-Verlag, Berlin, 213–226.

Schaefer, V. J., 1960: Hailstorms and hailstones of the Great Plains. *Nubila*, **3**, 18–29.

Smith, R. L., and D. W. Holmes, 1961: Use of Doppler radar in meteorological observations. *Mon. Wea. Rev.*, **89**, 1–7.

Stout, G. E., and F. A. Huff, 1953: Radar records Illinois tornadogenesis. *Bull. Amer. Meteor. Soc.*, **34**, 281–284.

Suckstorff, G. A., 1938: Kaltlufterzeugung durch Niederschlag. *Z. Meteor.*, **55**, 287–292.

Swingle, D. M., and L. Rosenberg, 1953: Mesometeorological analysis of cold front passage using radar weather data. Proceedings, 4th Weather Radar Conference, Boston, Mass., American Meteorological Society, Boston, XI-5, 1–3.

Tepper, M., 1950a: Proposed mechanism of squall line: The pressure jump line. *J. Meteor.*, **7**, 21–29.

Tepper, M., 1950b: On the origin of tornadoes. *Bull. Amer. Meteor. Soc.*, **31**, 311–314.

Tepper, M., 1955: On the generation of pressure-jump lines by the impulsive addition of momentum to simple current systems. *J. Meteor.*, **12**, 287–297.

Wakimoto, R. M., 1982: The life cycle of thunderstorm gust fronts as viewed with Doppler radar and rawinsonde data. *Mon. Wea. Rev.*, **110**, 1060–1082.

Williams, D. T., 1948: A surface microstudy of squall-line thunderstorms. *Mon. Wea. Rev.*, **76**, 239–246.

Williams, D. T., 1953: Pressure wave observations in the central midwest, 1952. *Mon. Wea. Rev.*, **81**, 278–298.

CHAPTER 3

Forecast Problems: The Meteorological and Operational Factors

Daniel L. Smith, Fred L. Zuckerberg,
Joseph T. Schaefer, Glenn E. Rasch

3.1. Introduction

Weather forecasting is the practice of operational meteorology. Thus, forecasting problems can be separated into those that are meteorological, and those that are operational. Examples of the former are quickly learned by all students and become familiar to all. Mesoscale meteorological problems are dealt with extensively in other chapters of this volume. For example, our knowledge of the mesoscale and of interactions among mesoscale and larger (and smaller) systems is not complete; we have also not yet made the strides necessary to transfer knowledge gained by researchers into the material of routine forecasts. Problems of the operational sort, however, are reserved for those who, in addition to coping with the intricacies of the atmosphere, must formulate and communicate results of their efforts to users in an effective and timely manner. Problems concern availability, timeliness, and quality of observational data; time constraints on forecast preparation; the nature and reliability of communication systems available for forecast dissemination; and the makeup and requirements of the user community.

Forecast problems can be separated for analysis and discussion, but in practice they are inseparable and forecasters must deal with both kinds. Understanding the vagaries of the weather—its current and future states—and translating that understanding into a meaningful statement for the public (or other users) are the essence of weather forecasting. Questions posed more than 200 years ago by the German philosopher Immanuel Kant seem appropriate to this interpretation of the Forecast Problem: Kant asked "What can I know? And, on the basis of that knowledge, what ought I do?" Mesoscale meteorological systems will be described in detail in subsequent chapters, giving at least partial answers to the question, "What can I know?" This chapter, in order to provide a perspective often overlooked by meteorologists intent on understanding weather systems, considers answers to the second question, "On the basis of that knowledge, what ought I do?"

Perhaps the easiest way to gain an appreciation for the interrelationships among meteorological and operational problems is to examine case studies of several weather events. The four presented here illustrate specific aspects of the overall forecast problem. The events took place in different parts of the country and show that geography is a very real part of forecast problems. It should be obvious, however, that with perhaps small but significant differences, the events might occur in other locations (and introduce other kinds of forecast problems). The examples deal with weather events only as seen from the perspective of the public at large, rather than that of the aviation, agricultural, marine, or other communities. Special groups are not without forecast problems. However, it is left to the reader to apply the general comments that follow to specific areas of interest.

3.2. Case Studies

3.2.1. Case A. Heavy Rain and Flash Flooding

Late on a summer afternoon, a large and severe thunderstorm developed over southern Nevada. The storm was easily detected by satellite imagery as it formed at the point of intersection of two well-defined lines of convection—one oriented east-west, the other north-south. The storm was no surprise; earlier in the day, National Weather Service (NWS) offices with responsibility for the area had recognized the potential for heavy thunderstorms and worded their forecasts accordingly. The National Severe Storms Forecast Center in Kansas City had warned of the potential for severe activity in the area.

Once the storm developed, actions by NWS forecasters followed textbook procedures. Watches, warnings, and statements were issued in a timely manner; local public officials, law enforcement agencies, and the National Park Service were notified of threatening conditions; and volunteer weather observers were tapped for additional information. Forecasters were on top of the developing weather situation.

Then hurricane force winds from the thunderstorm caused a massive power failure. Even though supplied with emergency backup power, the local NWS office effectively lost telephone service, and with it all incoming reports and outgoing special information such as coordination messages. Community officials lost not only telephone, but also NOAA Weather Wire services. Power was lost for NOAA Weather Radio transmissions. Nevertheless, the word had gone out. Apparently.

What was unknown at the time was that the intense thunderstorm produced 6.5 inches of rain over an isolated mountainous area in only an hour, rain that, within a few hours, caused record-breaking flash floods in several valleys. Damage from flooding and the high winds was put at several tens of millions of dollars. Although miraculously there were no deaths in the storm, residents of the hardest hit areas reported that the storm struck "without warning."

3.2.2. Case B. Severe Local Storms

The weather pattern over the Central Plains contained significant synoptic features on a late spring day, but after careful analysis, the weather did not appear to be ominous. A slight chance of evening thunderstorms, possibly severe, was forecast in advance of an approaching cold front, but rapid movement of frontal and pressure systems, and strong low-level cold advection behind the front, appeared to preclude thunderstorms after dark. The forecast was strongly supported by numerical guidance, including very low point precipitation probabilities. When the evening upper-air data became available, they confirmed earlier forecasts of system development and motion. Network radars reported no echoes over the forecast area at 9:30 p.m.; an hour later thunderstorms with tops to 48,000 ft were reported. The storms continued to develop rapidly to 60,000 ft and moved quickly eastward. By 1:00 a.m., baseball-sized hail and winds to 90 mph were being reported, along with damage to mobile homes, trees, and power lines. A tornado struck around dawn. Many areas received 1 to 2 inches of rain in a short time.

3.2.3. Case C. Sea Breeze Showers

Synoptic analysis revealed little to distinguish the weather expected for the Fourth of July in a large coastal metropolitan area from that of a typical summer day. A stationary pressure system had established a steady prevailing flow that promised the usual sunny weekend with just widely scattered afternoon showers. The forecast for the local area read accordingly, including a 20% probability of rain for any location. Interest in the holiday weather was probably higher than usual, as many people eyed local beaches with plans for outdoor activities.

As expected, showers did develop during the afternoon—in fact, a few hours earlier than forecast. They were not widely scattered, however. The prevailing westerly flow retarded the reliable sea breeze along the beaches of this east coast area with the result that showers— and a few thunderstorms— were contained along a narrow coastal strip. Persistence of showers in the pronounced convergence zone created by opposing sea breezes and prevailing westerly flow produced sizable amounts of rain in some areas. Few areas along the coast escaped at least some rainfall. Lured to the beaches by the promise of the morning's fine weather and by the low-probability forecast, thousands of people returned home wet and more than a little unhappy. As if to add insult to injury, just a few miles inland not a trace of rain fell.

3.2.4. Case D. Heavy Snow

Forecasting snowstorms in the Northeast is always difficult. It is often not difficult to determine with confidence that snow will fall, but the questions are when, where, and how much? Details are important since the high population density ensures that any miss in the forecast will be noticed by many people. Fortunately, from a forecast standpoint, the storms occur often enough to provide material from which to build reasonably good local forecast models. Numerical and statistical models do relatively well in the Northeast because observational data and prior tracking are available for weather systems approaching from the west.

The forecast for a particular March storm seemed to be well in hand. Located initially over Indiana, a deepening surface low was forecast to move almost due east over New Jersey and to a position 200 mi offshore in 24 h. The likelihood of snow was obvious, and 6 inches were forecast over northern Massachusetts in keeping with well-proven rules that place heaviest precipitation 2.5°lat. to the left of the storm track. Numerical and statistical guidance reinforced this local forecast.

The computer's forecast of the 24 h position of the surface low was excellent. So was the local forecast rule, which placed the snow to the left of the storm track. Unfortunately, the 6 inches of snow blanketed most of New Jersey and New York City, not Massachusetts.

Why the error in the snow forecast? Unfortunately, at about the time the local forecast was made (late afternoon), the storm began a pronounced turn to the right, jumping to a position 200 miles south of the model's 12 h forecast position. From there it moved to a position very close to where it was forecast to be 12 h later. Missing the forecast track of the storm meant missing the location of heavy snow.

3.3. Discussion of Case Studies

Case A occurred in and around Las Vegas, Nev., in 1981. Even though forecasters were "on top" of a developing severe thunderstorm situation, those who most needed accurate forecasts and warnings failed to get them, for these reasons:

- Failure of communication and forecast dissemination systems at a critical time.

- Failure of observational systems to record and transmit to forecasters critical rainfall information.

- Failure of a community preparedness program, which might otherwise have ensured broader dissemination of critical information.

- Failure of forecasters to recognize the degree of heavy rain threat.

For Case A, these failures constituted the "forecast problem." Only the last one would generally be considered a meteorological problem; it relates to the forecasters' understanding of the weather system. The first three failures were due to a variety of operational considerations, many of which were beyond the control of the forecaster. Still, they are of major concern to the forecaster because there is little advantage in producing even a flawless forecast if it cannot be communicated in a timely manner to those who need it. In this case a variety of operational problems seriously degraded the effectiveness of the forecast.

Case B occurred over Kansas and Missouri in June 1982. There is little doubt that the main forecast problem that night was failure to recognize meteorological conditions that led to explosive development of a severe mesoscale weather system—a meteorological problem. A retrospective analysis of the storm indicates it had characteristics common to a phenomenon known as a derecho (Johns and Hirt, 1983). Furthermore, application of models developed for forecasting nocturnal thunderstorms yields clues to why this storm formed when and where it did.

There were operational problems as well. As the system developed, forecasters responded with updated forecasts and warnings, but these were not widely received. For a number of reasons (see, for example, Hoxit *et al.*,1978) significant weather systems—especially heavy rains—tend to occur after midnight. With the exception of warnings with alarms (local sirens or NOAA Weather Radios with alarm features), the public is generally out of touch with warnings that are issued late at night. This places a premium on communicating the best possible information prior to the time most people retire for the night—generally, in time for use as part of late night TV news and weather programs.

Not all forecast problems involve hazardous weather. If a measure of significance is the degree to which weather influences the routine or planned activities of individuals, then Case C is no less significant. The biggest problem was again meteorological. A better understanding of interactions between synoptic-scale (prevailing flow) and mesoscale (sea breeze) systems might have resulted in a forecast more specifically tailored for the area. The dynamics of sea/land breeze circulations are well known, as are interactions with larger scale systems. Sophisticated models that include a variety of local terrain effects are available. Even though such models may never pinpoint locations of individual showers, they can still be very useful for refining decisions concerning favored shower locations, areal coverage, intensity, timing, and movement. Precipitation forecasts, primarily probability forecasts, could be improved accordingly.

What of the operational aspects of the sea breeze forecast problem? Although Pielke (1977) suggested how better meteorology might lead to better forecasts (nowcasts), existing communication systems impose serious limitations on the forecaster's ability to tell the public what is known. Thus, unlike the first case, in which communication failures were the problem, this case revealed routine communication limitations.

Case D demonstrates again that even though the event may be mesoscale, the forecast problem is often larger. In this case the forecaster's dilemma was to determine the track of the synoptic-scale low, and then to pinpoint location of the expected snowfall. The geographic area made the forecast problem all the more complex. The Appalachian terrain combined with strong temperature and moisture gradients over the coastal states drastically affected progression of the system. The low essentially dissipated west of the mountains, and a new low formed in the lee. The result was a discontinuity in the storm track.

Existing operational models can account for effects of terrain with only limited success because of spatial resolution of the models. Forecasters must keep this in mind when dealing with model solutions for mountainous areas. Even if the model does well with the synoptic-scale system, it may be difficult to infer a sufficiently accurate storm track from the guidance because the solution is generally available only at 12 h intervals. Both problems were important in Case D. Solutions may be found in careful local analyses— hourly, if possible— with attention to features that might be expected in the particular event but are likely to be missed by numerical models. The fore-caster makes the largest contribution by improving on numerical guidance. The discontinuous storm track should not have been a major surprise. The problem was mainly to determine whether it would occur, and if so, to what effect. A careful on-going analysis quite possibly would have revealed first clues to what was happening, allowing a revision to the forecast in time to alert the public to a significant shift in the area of heaviest snow.

In this case, lack of local analyses was an operational problem. Provision of such analyses is often difficult for reasons that exist in most forecast offices or centers. Among these are limitations on availability of sufficient data, distractions that limit a forecaster's ability to focus on the problem at hand, and the tendency to place too much reliance on numerical guidance.

3.4. Analysis of Forecast Problems

The process of forming a weather forecast can be approached very simply by asking three questions: (1) What is going on? (2) Why is it happening? (3) How is it likely to change? The questions are answered through obser-vation, diagnosis, and forecasting, respectively. Table 3.1 summarizes the meteorological and operational issues that must be successfully addressed to provide answers to each of these questions. Other chapters deal with the meteorological problems and their solutions; this discussion emphasizes operational aspects.

Table 3.1. The process of producing an effective weather forecast

Activity	Forecasting Problems	
	Meteorological	Operational
Observation: What is going on?	Identifying scale of importance	Data: Availability, Quality, Representa-tiveness, Timeliness
Diagnosis: Why is it happening?	Adequacy of analysis techniques	Effective data assimilation
	Adequate diagnostic models	Forecaster focus
		Appropriate local use models
Forecasting: How is it going to change?	Adequate conceptual and prediction models	Adequate local analyses to comple-ment numerical guidance
	Understanding of scale interactions	Appropriate forecaster focus

3.4.1. Observation

The lack of observations sufficiently close together in time and space to define mesoscale systems is a critical meteorological problem. Surface observations are diminishing in number in many areas, and serious questions still exist about the reliability and accuracy of automated observations (both surface and upper air). At stake is our ability to observe events, the first step in understanding them. Even where routine observational networks may appear to be sufficiently dense (in time and space) for resolving mesoscale systems, data from such networks can be misleading. Gradients of convective rainfall, for instance, can be extremely tight, leading to false conclusions regarding rainfall maxima in areas from which even a number of reliable observations are available. Radar and satellite data can augment *in situ* observations, but at the present time, remotely sensed observations are usually available only as visual displays that are subjectively analyzed by the forecaster. Furthermore, not all weather forecast or warning offices have access to such data even in this form. Availability, accuracy, and timeliness of observations are all needs of the operational forecaster.

3.4.2. Diagnosis

Research efforts over the last few decades have resulted in significant improvements in diagnostic models for many mesoscale weather systems. New observational tools (Doppler radar, geostationary satellite imagery), automated analysis techniques, and extensive field studies (e.g., SESAME) continue to pay dividends. Obviously, there is still much to be learned, but we now know more than ever about the meteorology of significant weather systems on almost all scales. The extent to which this increase is knowledge has been translated into improvements in weather forecasts is, however, a subject of some debate. In some cases—the forecasting of tropical storm motion, for example—a definite plateau in skill seems to be reflected by objective measures of forecast accuracy. On the other hand, where the latest tools and techniques have been applied to operational forecasting of severe thunderstorms, results have been dramatic (see Sec. 3.5).

Forecasters will continue to rely heavily on researchers to develop new and improved diagnostic models and to provide the necessary solutions to meteorological problems. Meanwhile, a number of very real operational problems must also be considered to affect the diagnostic process. A major problem is a forecaster's inability to assimilate large quantities of data. Even though sufficient observations on the mesoscale are often lacking, there is no lack of other information at the modern forecast office. Observations, computer analyses, automated forecasts, radar, satellite, and telecommunications of every sort flow in to be sorted and assimilated by forecasters.

Data seldom are presented initially in the most usable form. The new forecaster, and many older ones, may become overwhelmed in rapidly developing weather situations. Unless special efforts are made to develop and maintain individual analytical skills, forecasters may simply not understand developing weather. Forecasters may come to place too much reliance on centralized (i.e., automated) diagnoses, which are not capable of resolving significant mesoscale systems (or intended to be, for that matter). Augulis (1978) was referring to a young forecaster "with little analytic and diagnostic experience" when he noted that "it is only natural that he accept the centrally prepared product more often than not ... ."

Even among the most experienced forecasters, because of the variety of ongoing tasks in most forecast offices, it is sometimes difficult to achieve and maintain focus on a single significant problem. By the time forecasters may realize where the problem lies (that is, by the time they identify the most significant weather system) and shift the focus of their efforts accordingly, they are already behind events. This is a situation hardly unique to weather forecasters. Making decisions, often quickly, by sorting through a mass of uncertain information, not all of which is even relevant to the problem at hand, is a way of life for military leaders, policy makers, and doctors, and to some degree is required of everyone. A growing science based on decision theory seems likely to help forecasters arrive at more objective decisions.

3.4.3. Forecasting

Diagnostic models have increased understanding of mesoscale (and other) systems, but the same models are seldom useful as forecast tools. Forecast models do exist, especially for synoptic and larger scales, but their utility for mesoscale systems is still largely unproved. Often their real-time use is hampered by lack of data or adequate computer resources. Small-scale models are needed "close to the action" at field forecast offices, if they are to be used for short-range forecasting. Even given sufficient data, running such models at central locations seems impractical at this time because of difficulties involved in focusing on problem areas and disseminating model results in a timely manner. Nevertheless, it seems safe to assume that use of operational mesoscale forecast models will increase. Lack of computer resources does not seem to be a major obstacle; microprocessors have already begun to proliferate in forecast offices. Candidates for operational models are waiting in the wings. Obtaining observational data and transferring technology from research into forecast offices appear to be much bigger tasks.

Finally, just as forecasters have difficulty focusing on mesoscale systems in the analysis process, the problem carries over into the making of forecasts. Anticipation of and response to a significant mesoscale weather system must be viewed in the broader context of overall responsibilities levied on forecasters. They may become over-reliant on automated guidance with the result that early clues to significant development on smaller scales are overlooked. Murphy (Golden *et al.*, 1978) summarized this aspect of the operational problem:

> The current system, as it has evolved, makes it difficult if not impossible for the forecaster to assimilate and weigh the available guidance information properly (particularly the guidance forecasts) and discourages him from using his training and experience to depart from this guidance.

3.4.4. Dissemination

There are other formidable hurdles on the way to improved forecasts. Significant strides may be expected in overcoming problems associated with the meteorology of developing weather systems, but translating what we know about the weather into useful information for the public, and getting that information to them, will remain problems to be overcome. In some instances, our understanding of what is going on, and what is likely to happen, exceeds our ability to communicate an effective message to users of the forecast because of limitations in dissemination systems. Consider the probability-of-precipitation (PoP) forecasts and the sea breeze problem of the case studies. PoP forecasts are an effective means of communicating the likelihood that any given point in a forecast area will receive precipitation, which is presumably what users want to know. The concept of forecasts for relatively small areas (zones) is a compromise by the NWS to provide locally specific forecast information without overtaxing communication systems. Even if meteorological skill allowed forecasters to determine where individual showers would develop, it is difficult to envision how they could communicate such information to all interested parties.

3.5. Progress in Solving Meteorological and Operational Forecast Problems

Efforts to resolve both meteorological and operational forecast problems have received increased emphasis in recent years. A number of factors may account for this, but the primary ones are most likely related to an enhanced realization of the economic impact of weather forecasts, both good and bad. More people are more exposed than ever to the vagaries of the weather. It has also become generally clear that improved forecasts should be possible; at least one may be entitled to such an inference from dramatic achievements in other sciences.

Progress has certainly been made. Developments are notable in four areas: new technology, technology transfer, automation of forecast office operations, and forecaster training. In addition, there has been an increased understanding of and appreciation for the weather and its effects, increased automation of mass communication services (exemplified by cable TV dissemination systems, some of which include an educational function), and increased levels of community response to forecast and warning services.

3.5.1. Development of New Technology

New observation systems such as Doppler radar, automation of conventional radar, enhanced and processed satellite imagery, remote atmospheric soundings (from the bottom up and the top down), and automated surface observations all are either operational or undergoing field evaluation with expected implementation within the decade. The use of even more powerful, and inexpensive computers to process, communicate, and display data from these systems is taken for granted. McIDAS (Man-computer Interactive Data Analysis System) and CSIS (Centralized Storm Information System) are just two examples—from university and government communities, respectively—of how the very latest technology can be successfully integrated into routine forecast operations. Other government and private groups are developing and applying similar integrated systems. (See Ch. 31).

The new technology is unquestionably improving our understanding of weather systems. It is also facilitating analysis procedures, as in the case of CSIS at the National Severe Storms Forecast Center. Understanding of tropical weather systems has been improved by the tools and techniques that have become available in recent years to researchers and forecasters at NOAA facilities in Miami.

3.5.2. *Increased Technology Transfer*

For the most part, operational forecasters have had only limited direct access to the latest technological developments. There have been obvious signs in recent years that this is changing. The Program for Regional Observing and Forecasting Services (PROFS), discussed in Ch. 31, is a major initiative designed especially for the purpose of evaluating latest technology (tools and techniques) in an operational setting, to help design the "forecast office of the future." New hardware may be slowly implemented in field offices because of significant costs involved in full evaluation and acquisition; nevertheless, field use will no doubt eventually occur. Meanwhile, it is essential that efforts be continued to capitalize on the results of new technology as soon as possible. Such "technology transfer" requires the best communication possible between researchers, who do have access to new tools, and operational forecasters who do not. The "technology" that is transferred can include ideas, operational procedures, models, or actual hardware.

Significant changes have been made in recent years in operational procedures for analysis and forecasting of severe thunderstorms by NWS forecasters as a result of studies by research scientists. New knowledge of thunderstorm structure has been communicated not only to forecasters, but through them to the public—to storm spotters who contribute significantly to the success of the NWS severe weather warning program. Lemon (1980) developed criteria for using radar to assess storm severity, which are based on latest observational and research knowledge of storm structure and development. These criteria have largely replaced older ones, which were based on indirect relationships, in many cases, between severity and appearance of storms on radar. This has led to improved warnings with fewer false alarms and higher likelihood of severe storm detection where new procedures have been applied.

Improvement in warning operations has nowhere been as dramatic as at the National Weather Service Forecast Office (WSFO) in Oklahoma City. Results were achieved by capitalizing on advantages the WSFO has over other offices by being virtually collocated with NOAA's National Severe Storms Laboratory (NSSL), in Norman, Okla.; by having real-time access to NSSL's Doppler radar output; and by having major computer enhancements to its own network radar. It is also significant that few other offices face exposure to so many severe storms. The result has been a well-developed severe weather preparedness program involving close interaction among NWS and State and local groups, along with well-trained storm spotters.

Through 1982, statistical analysis of warning performance from the WSFO showed no significant differences from performances of other offices. In late 1982, a program was initiated to determine how effectively the unique capabilities available to the WSFO could be applied to improving warnings. A dedicated Technology Transfer Specialist was assigned to the office. Intensive forecaster training was initiated on both tools and techniques, and existing computer software for the WSFO's automated radar system was refined. Provision was also made for limited real-time access to output from the NSSL Doppler radar. Results, measured in terms of improved warning operation and services, were outstanding. Warning accuracy continues to be maintained at a remarkably high level into 1985. Table 3.2 compares recent severe thunderstorm warning verification statistics with those from a representative earlier year. Notice a major increase in the probability of storm detection with implementation of new technology, and a continuing, dramatic drop in false alarms.

Improvements in the severe weather warning program in Oklahoma are attributable to several factors: improved understanding of severe storm structure (knowing what to look for); new technology for observing and diagnosing severe storms (improving chances of finding them); automated communication systems (getting the message out), and significantly enhanced follow-up procedures for determining what actually happened. Forecasters

Table 3.2. Severe thunderstorm warning verification for WSFO Oklahoma City

Year	POD	FAR	CSI	Counties per warning
1978	.47	.40	.36	2.95
1983	.84	.52	.44	1.30
1984	.86	.40	.54	—
1985	.82	.29	.61	—

Note: Data are for March, April, and May for each year.

Definitions:

Probability of Detection (POD) is the ratio of correctly forecast severe events to known severe events.

False Alarm Ratio (FAR) is the ratio of severe weather predictions that fail to verify, to total severe weather predictions.

Critical Success Index (CSI) is the ratio of successful predictions to the number of severe events occurring or forecast to occur. A value of unity represents a perfect set of forecasts.

should be, and generally are, involved in post-storm evaluations to determine how forecasts verified. This is a significant part of the forecaster's education.

Technology transfer involving operational forecasters and researchers at universities, national laboratories, and elsewhere, focuses on severe storms, excessive precipitation, marine forecast problems, agricultural programs, and other meteorological problems. Within NOAA many of these efforts stem from an Interagency Technical Exchange Conference held in 1982 (NOAA,1982). Grice and Maddox (1983) described one example of the result of close interagency cooperation on a particular forecast problem. Grice, a forecaster at WSFO San Antonio, Tex., developed and applied NOAA researcher Maddox's flash flood climatology to local problems, resulting in improved understanding and better forecasts of South Texas heavy rain events. Many similar examples of cooperative efforts involving forecasters could be cited. Other efforts in recent years have included extended visits by researchers to forecast offices, allowing them to view firsthand the application of meteorology to operational forecast problems. At the same time, forecasters have received training through seminars and informal discussions. Operational forecasters have also made extended visits to research laboratories, demonstrating the potential of such collaboration for transferring technology.

3.5.3. Automation of Field Operations

The most obvious example of the integration of new technology for the purpose of automating service operations is provided by AFOS (the National Weather Service's Automation of Field Operations and Services) implemented in the mid-1970s. AFOS was designed primarily as a modern communication tool for disseminating graphic and alphanumeric data to field offices and for facilitating the preparation and transmission of forecasts and warnings. The computer-based system makes it possible for the first time to access current meteorological data stored in the AFOS data base and generate a variety of new analyses not previously available to the forecaster. Forecasters are also able to present old data in new ways— animated displays of surface and upper-air charts, for example. As experience with AFOS-type systems grows, the result should be greater efficiency in the analytic and forecast preparation stages of office operations, resulting in more time available to forecasters for actually applying meteorological skills.

Weather forecasting has benefited from diminishing costs of computers through the provision of microcomputers to field offices for a variety of purposes. Systems are being used operationally for data acquisition (saving time and forecaster effort and making additional observations quickly available), for message composition and dissemination, and for data analysis. Through a cooperative effort involving NOAA, NASA, and the University of Florida, NWS forecasters in Tampa, Fla., use a microcomputer to receive processed, color-enhanced satellite imagery and run small-scale models that allow refinement of critical winter temperature forecasts for areas almost as small as individual citrus groves. Citrus industry officials as well as individual growers may receive on personal computers the detailed forecasts and satellite, radar, and other data they need for making weather-critical decisions.

3.5.4. Forecaster Training

A review of forecaster training needs was provided by Doswell *et al.* (1981). Since that time, there have been a number of training program developments within NWS. Efforts are under way to organize and manage a general forecaster training program, supply training for meteorological technicians, and provide specialized training for newly hired meteorologists. Similar efforts are under way in other forecast services, and all reflect the demand for improved training as a means of improving forecast operations.

An East Coast Workshop on Weather Forecasting and Forecast Dissemination was held at Raleigh, N.C., in 1979. A significant aspect of the workshop was that it was developed by and for operational forecasters, including representatives from NWS forecast offices and the private sector. Similar workshops have been held annually at NWS Southern Region WSFOs to deal with forecast problems related to the occurrence of heavy rains and flash flooding. On-station workshops for forecasters provide excellent refresher training, as well as means for implementation of new ideas and techniques. Offices conduct their own local workshops for staff members, often with outside visitors from universities and NOAA research laboratories. These forecaster training efforts have generally resulted in noticeable improvements in the accuracy, timeliness, and informational content of forecasts and warnings.

3.5.5. Dissemination

Dissemination and user understanding of forecast information have been identified as problem areas, so it is appropriate also to consider efforts made to alleviate any such problems. Table 3.3 (NOAA, 1975) demonstrates the effectiveness of a well-planned disaster preparedness program in a community, including public education concerning severe storms and warning services. Statistics in the table show the effect of solving the "forecast problem." A killer tornado struck Jonesboro, Ark., in 1968, and another struck 5 years later. In the interim there had been major public education efforts, and improved community preparedness programs were implemented. NWS's ability to communicate forecast information rapidly and effectively was also improved. The result was that deaths and injuries were drastically reduced, even though the destruction caused by the second tornado was much greater.

3.6 Conclusions

Severe storms do not develop in isolation from their larger scale surroundings, and should not be studied or viewed that way. Mesoscale systems interact strongly with systems at larger and smaller scales; thus the forecaster is cautioned against over-emphasis of events at a single scale. And not all weather occurs as a result of mesoscale systems. Forecast problems range from understanding the science to communicating what the forecaster knows. All elements must work for the system to function properly.

Table 3.3. Comparison of the readiness of Jonesboro, Ark., for tornadoes in 1968 and 1973, and the relative effects of the storms

Readiness

	1968	1973
NOAA Weather Wire Service	None	Statewide
Radio and TV dissemination of watches and warnings	None	Yes
Spotter network	None organized	Organized and in action
Educational programs	None	Extensive
Emergency operations center	None	Yes
Disaster drill procedures	None	Yes

Effects

	1968	1973	% Change
Population	25,000	29,000	+16
Deaths	34	2	−94
Injuries	458	246	−46
Hospitalizations	82	21	−74
Damage	$8 million	$50 million	+625

REFERENCES

Augulis, R. P., 1978: Observations. *Nat. Wea. Dig.*, 3, 32–33.

Doswell, C. A., L. R. Lemon, R. A. Maddox, 1981: Forecaster training—a review and analysis. *Bull. Amer. Meteor. Soc.*, **62**, 983–988.

Golden, J. H., C. F. Chappell, C. G. Little, A. H. Murphy, E. B. Burton, and E. W. Pearl, 1978: What should the NWS be doing to improve short-range weather forecasting? A panel discussion with audience participation. *Bull. Amer. Meteor. Soc.*, 59, 1334–1342.

Grice, G. K., and R. A. Maddox, 1983: Synoptic characteristics of heavy rainfall events in South Texas. *Nat. Wea. Dig.*, 8, 8–16.

Hoxit, L. R., R. A. Maddox, and C. F. Chappell, 1978: On the nocturnal maximum of flash floods in the central and eastern U.S. Preprints, Conference on Weather Forecasting and Analysis and Aviation Meteorology, Silver Spring, Md., American Meteorological Society, Boston, 52–57.

Johns, R. H., and W. D. Hirt, 1983: The derecho—A severe weather producing convective system. Preprints, 13th Conference on Severe Local Storms, Tulsa, Okla., American Meteorological Society, Boston, 178–181.

Lemon, L. R., 1980: Severe thunderstorm radar identification techniques and warning criteria. NOAA Tech Memo. NWS NSSFC–3 (NTIS#PB81–234809), 60 pp.

NOAA, 1975: A federal plan for natural disaster warning and preparedness: First supplement (FY 1976–1980). Dept. of Commerce, Washington, D.C., 101 pp.

NOAA, 1982: Interagency Technical Exchange Conference, First Report. NOAA National Weather Service, Silver Spring, Md., 33 pp.

Pielke, R. A., 1977: An overview of recent work in weather forecasting and suggestions for future work. *Bull. Amer. Meteor. Soc.*, 57, 423–430.

CHAPTER 4

Atmospheric Sounding Systems

J. H. Golden, R. Serafin
V. Lally, J. Facundo

4.1. Introduction

The first temperature soundings were made in 1749 in Glasgow, when a thermometer was raised on a kite. It was not until the invention of the balloon by Montgolfier in 1783 that measurements to a substantial height became possible. Gay-Lussac ascended to 7 km as early as 1804 and made accurate temperature and pressure measurements. Free balloons without observers came into use in the 1890s and reached altitudes up to 18 km. These balloons carried meteorgraphs, which recorded pressure and temperature, but retrieval of the sounding was delayed, often for days. By the end of World War I, balloons and kites were replaced by airplane soundings in which the pilot climbed to about 4 km with a meteorograph and then quickly landed to provide current data.

In 1927, the first balloon-borne "radiosonde" was flown, telemetering pressure, temperature, and humidity data back to the ground. By 1940, radiosondes had completely replaced the aircraft meteorographs for daily soundings. Pilot-balloons tracked by theodolite provided wind data above the surface. World War II introduced the radar and direction-finding receivers, which provided the first accurate measure of winds to stratospheric altitudes.

After World War II, a proliferation of radiosonde systems took place, and many countries developed their own systems as a matter of pride. The tendency now appears to be toward the use of a few standard systems that are reliable and reasonably priced. It is important that the several radiosonde systems in use throughout the world be compared periodically to permit global analyses without artificial discontinuities due to temperature or pressure bias. The World Meteorological Organization sponsors periodic intercomparisons of radiosonde systems, and two such intercomparisons were conducted in the United Kingdom and the United States, in 1984 and 1985.

4.2. Parameters and Accuracies in Upper-Air Data

4.2.1. Pressure Measurement

It is critical that radiosondes measure pressure accurately. Accuracy is required over a range from the surface to 35 km (1050 mb to 5 mb). The instrument used in the United States for 30 years consists of a temperature-compensated aneroid capsule that moves a lever arm across a commutator plate. The aneroid design and the lever arm provide five times the deflection per millibar at 50 mb as at 1000 mb. This design has proved to be reliable and accurate. The VIZ Corporation in the United States is the principal manufacturer of this "baroswitch." A preflight adjustment must be made to correct for offset from the original calibration, so that the baroswitch element is accurate to ±1 mb near the surface, ±2 mb in the 500 mb region, and ±1 mb at 10 mb. The baroswitch was convenient in the early days of electronics since it provided switching of the sensors and reference elements. Now that compact electronic circuitry is available, the bulky baroswitch is an impediment to the development of a nonhazardous radiosonde and reduction of the overall weight and size.

The capacitive transducer with an aneroid capsule has been much improved in the last few years and will be used on all radiosondes within a few years.

4.2.2. Temperature Measurement

Most European radiosondes have used a bimetallic element to measure temperature. The lag of such elements and the need for radiation correction have been serious deficiencies. The use of bimetallic elements was a matter of convenience, since the motion of the element was easily converted into a changing capacitance so that the temperature and pressure transducers were compatible. The Väisälä Company recently developed a small temperature sensor that uses a ceramic chip coated with metal electronics. In this type of sensor the capacitance between the electrodes varies with temperature. This "Thermocap" has made obsolete the use of bimetal elements as capacitive transducers for temperature.

The American radiosondes have used resistive transducers for temperature and humidity sensing since the development of the first American radiosonde by Diamond and Hinman in the late 1930s. The sensor now used in American sondes is a rod thermistor about 0.7 mm in diameter and 1 to 2 cm long. It is coated with a lead carbonate pigment to reduce solar heating. The rod has radiation errors of 1–2°C above 25 km because of its thickness and its high absorption in the infrared. The lag of the rod thermistor varies from about 4 s at sea-level to 20 s at 30 km for a ventilation rate of 4 m s^{-1}. This lag could be made negligible with the use of a small aluminized thermistor bead (which would also eliminate radiation errors), but the added cost has not been considered necessary to achieve acceptable accuracies (0.5°C to 20 km). A simple correction that should be made for all radiosonde temperature measurements is

$$T = T_m + \frac{\delta T}{\delta z}\frac{\delta z}{\delta t}\lambda , \qquad (4.1)$$

where T is the true temperature, T_m is the measured temperature, $\delta T/\delta z$ is the lapse rate, $\delta z/\delta t$ is the ascent rate of the balloon, and λ is the lag constant of the temperature sensor. For a 5 m s^{-1} ascent rate with a lapse rate of 0.006°C per meter and a lag constant of 10 s, the correction is 0.3°C.

4.2.3. Humidity Measurement

American sondes use a carbon humidity sensor—a thin coating of a fibrous material on a glass or plastic substrate. At high humidity the element and the carbon granules are less densely packed and there is a corresponding increase in resistance. The response of the film is rapid, but the substrate response is slow. Since the measured humidity is with respect to the temperature of the film, and the film temperature is locked to the substrate, errors are made in assuming that air temperature and substrate temperature are the same.

The issue is whether the humidity sensor measures the moisture content of the atmosphere with sufficient accuracy. Under controlled environments such as humidity chambers, the accuracy can be tested against a standard (traceable to the National Bureau of Standards). The accuracy of the sensor is typically 5–7% in relative humidity over most temperatures. Väisälä developed a faster response element, the HUMICAP, which has a capacitive output, but it is susceptible to collecting moisture on the sensor if it is not properly shielded. The 1984 intercomparison tests indicated that the HUMICAP often reads 90–95% relative humidity in clouds.

The current United States radiosonde humidity sensor has a systematic bias of 2–4% around saturation for temperature above freezing. This means that under certain conditions (for instance, when humidity measurements are used to predict cloud formation) the sensor may be measuring 96–98% instead of 100%. Forecasters and modelers should keep this in mind when using the current humidity data in their analyses. In 1985, the humidity equations were rederived to account for this bias.

4.2.4. Wind Measurement

In the early days of meteorological measurements, winds were measured by tracking pilot balloons (pibals), using an optical theodolite and an assumed ascent rate of the balloon.

The first successful radio direction-finding system was the SCR–658 developed by the U.S. Signal Corps during World War II. It operated at 400 MHz and used two operators to steer a large antenna array to determine the direction of the radiosonde transmitter. It was quickly followed by a superior system operating at 1680 MHz using an automatic tracking system. For more than 30 years this system, the AN/GMD-1, and its civilian counterpart, the WBRT, have been the basic sounding systems used by the National Weather Service in the United States.

The height of the radiosonde is determined by conversion of pressure readings to equivalent altitudes. This altitude of a pressure surface is computed using the hydrostatic equation (Sec. 4.3.1), and the error is typically 20 m at 10 km, increasing to 100 m at 30 km. The error is much more sensitive to temperature error than to pressure error. However, when it is necessary to determine the time when the balloon reached a specific altitude, the error in pressure measurement directly affects the altitude error. The pressure-altitude error is typically 50 m at 10 km increasing to 500 m at 30 km. The WBRT radiosonde uses computed altitude and elevation angle to determine horizontal distance to the sonde. At an elevation angle of 6°, the error in horizontal distance becomes 5 km for a 500 m error in altitude. Because of this, many U.S. stations use a transponder attachment to the WBRT system, which measures slant range to improve accuracy at low elevation angles.

Generally speaking, the wind speed and measurement directions in synoptic-scale geostrophic flow pattern are representative of the ambient atmosphere. However, sharp gradients tend to be smoothed. For example, the averaging techniques that are employed above 14 km for deriving wind information tend to underestimate the magnitude of the jet stream, sometimes by as much as 20%. These techniques may also dampen measurements of wind shears around the jets.

4.3. Important Derived Quantities

4.3.1. Pressure Surface Height

Existing radiosonde systems measure pressure to 1 or 2 mb and temperature to 0.5°C. In order to produce synoptic maps of the height of constant-pressure surfaces (the most basic tool of the meteorologist), the heights must be measured to an accuracy of a few tens of meters. For example, the height of the 200 mb surface varies by about ±500 m over the world from its typical value of 12 km. Errors in the height of this surface above a radiosonde station should average 20 m or less for accurate analysis of gradients. The error in altitude for a 2 mb pressure error is equivalent to 60 m. It is therefore not feasible to determine the height of a pressure surface to the necessary accuracy by direct measurements. However, another solution is available and that is through use of the hydrostatic equation:

$$dp = -\rho g dz , \qquad (4.2)$$

where ρ, p, g, and z are the density, pressure, gravitational acceleration, and height, respectively. This can be rewritten in the form

$$dz = \frac{RT^*}{g} d(\ln p) , \qquad (4.3)$$

where R is the gas constant and T^* is the virtual temperature.

To simplify the analysis of height errors a relationship can be defined:

$$D = z - z_p , \qquad (4.4)$$

where D is the altitude correction, and z_p is the computed altitude.

Equation (4.3) can be rewritten:

$$dz_p = -\frac{RT_p}{g} d(\ln p) , \qquad (4.5)$$

where T_p is defined as the measured temperature.

Combining (4.3) and (4.5) gives

$$dz = \frac{T^*}{T_p} dz_p .$$

(4.6)

Differentiating (4.4) and substituting (4.6) gives

$$\frac{dD}{dz} = \frac{T^* - T_p}{T^*} .$$

(4.7)

The error in height D is a function only of the height interval and the ratio of temperature error $(T^* - T_p)$ to T^*. For random errors in temperature, the error in altitude is not significant. Serious errors in pressure measurement affect the computed error only in the offset introduced to the temperature measurement. In an isothermal region, pressure errors will not produce errors in the computed height of pressure surfaces.

A systematic error in virtual temperature of 0.5°C from surface to 12 km will produce an error of about 24 m in the height of the 200 mb surface $(\Delta z = 12,000; T^* = 250)$:

$$D = \frac{0.5 \times 12,000}{250} .$$

(4.8)

Recent comparisons between the United States radiosonde and the precision radar at the NASA/Wallops Island test facility show that the heights calculated as a function of pressure and temperature have an rms difference on the order of 10 to 100 m, up to a height of 20 km (see Fig. 4.1). At 30 km, the rms error may be as high as 1 km (Norcross and Brooks, 1983, Fig. 5 and Appendix A). It should be emphasized that these errors introduce errors in the assignment of winds and temperatures to an altitude. Because of the power of the hydrostatic computation, these errors do not introduce large errors in the determination of the height of a pressure surface.

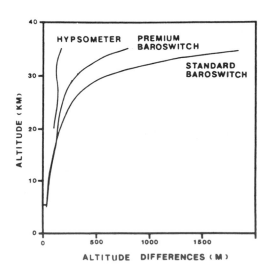

Figure 4.1. Differences (rms) between heights from precision radar and from a U.S. radiosonde using three commercial pressure devices. The standard baroswitch is in use in U.S. radiosondes. (After Norcross and Brooks, 1983.)

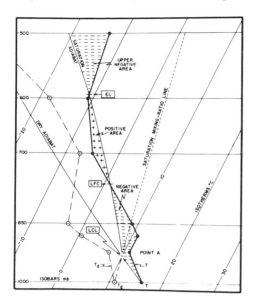

Figure 4.2. The positive and negative energy areas resulting from lifting the low-level parcel. (After USAF, 1961.) LCL is lifted condensation level; LFC is level of free convection; EL is equilibrium level (level where ascending parcel becomes colder than the environment). The larger the positive area, the better the chance for strong and deep convection conducive to severe weather. The negative area at the bottom must be overcome by forced lifting.

4.3.2. Stability Indices

In addition to the classical parameters found in the upper-air coded messages, i.e., stability indices, maximum winds, tropopause level, etc., a number of quantities have been derived from the coded data. Many of these parameters are readily derived from the plotted sounding on any of a host of sounding charts. Figure 4.2 indicates how the plotted sounders may be used to locate such parameters as LCL, LFC, and EL, as well as the positive and negative energy area useful in forecasting convection. Other quantities that may also be derived include θ_e, θ_w, θ_{max}, T^* and T_e.

A number of stability indices have been developed over the years, some, such as the K-Index, Total-Totals Index, and Showalter Index, enjoying wider use than others. The single station analysis also provides forecast information concerning such things as maximum temperature, extreme temperature, surface wind gusts, fog formation, precipitable water, and even an estimate of hail size. The interested reader is referred to Miller (1972), Galway (1956), and Fujita *et al.* (1970) for details on derived quantities from soundings.

4.4. Interpretation and Applications

4.4.1. Functional Precision

Functional precision is a measure of the reproducibility of a measurement and is defined as the root-mean-square (rms) of the measuring systems. Hoehne (1980) calculated the functional precision of recent NWS upper-air measurements by comparing simultaneous outputs from two identical radiosondes attached to the same balloon and tracked by two radiotheodolites. The data were taken from 50 weekly balloon flights in the spring of 1978. Hoehne noted that "the precision determined is the precision of the measurement, not that of any part of the instrumentation involved." Hoehne's functional precision results are summarized in Table 4.1 in three categories of

Table 4.1. Functional precision of measurements from identical radiosondes*

Quantity	Measured value	Bias
At same time of flight		
Pressure	±1.9 mb	0.0
Temperature	±0.67°C†	−0.14°C
Dewpoint depression	±3.67°C†	0.35°C
Height	±92.9 m†	−7.6 m
At same height		
Pressure	±0.7 mb	−0.1 mb
Temperature	±0.84°C†	−0.19°C
Dewpoint depression	±3.42°C†	0.38°C
Wind vector	±6.0 kt (±3.1 mps)	0.0
Wind speed (approximate)	±6.0 kt (±3.1 mps)	
Wind direction (approximate)	$\pm \cos^{-1} \dfrac{s + 6.05}{(s^2 + 12.5s + 53.4)^{\frac{1}{2}}}$	(s = wind speed)
At same pressure		
Height	±23.7 m†	−4.0 m
Temperature	±0.61°C†	−0.13°C
Dewpoint depression	±3.26°C†	0.35°C

*From Hoehne (1980).

†Precision taken from standard deviation because of bias introduced by heat and humidity of battery in upper sonde.

comparison. It is clear from Hoehne's (1980) study and others that concerted effort is needed to develop a faster, more accurate humidity sensor.

In another experiment, using data from a 300-m tower as reference, Kaimal *et al.* (1980) intercompared four balloon-borne radiosonde packages (GMD–1/VIZ, CORA, TDFS, and Airsonde). Excluding Airsonde data, for which rms differences were always higher, they found rms differences of < 1°C in dry-bulb temperature, 1.0–3.5°C in dewpoint temprature, 4% in relative humidity, and a few meters per second in wind data (supplied by only one sensor). These results are similar to an intercomparison between instruments when the average value of measurements at each level was used as a reference and the study extended to 3 km.

4.4.2. Representativeness

The issues of representativeness and reproducibility are the most crucial ones facing the research or operational meteorologist trying to use sounding

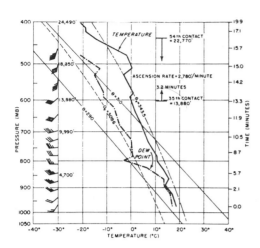

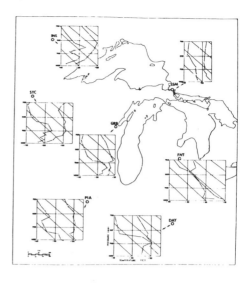

Figure 4.3a. A rare updraft sounding, taken at Columbia, Mo., at 0237 CST, 14 October 1954. Wind speeds to the nearest 5 kt (half barbs) and directions to the nearest 10° for the standard levels are given on the left. Time after release is given on the right. (From McComb and Beebe, 1956.)

Figure 4.3b. "Lake effect" soundings (air temperature and dewpoint) for 1200 GMT, 27 January 1971, at International Falls (INL), Sault Ste. Marie (SSM), St. Cloud (STC), Green Bay (GR), Flint (FNT), Peoria (PIA), and Dayton (DAY). Diagonal solid lines are selected dry adiabats. Temperatures are in °C. (From Baker, 1970.)

data. Where and when was the sounding taken, and how were the data processed? What local terrain, land/water, or prior mesoscale disturbances may have influenced the sounding characteristics? Representativeness is defined, after Nappo (1983), as "the extent to which a set of measurements in a space-time domain reflects the actual conditions in the same or different space-time domain taken on a scale appropriate for a specific application."

The NWS soundings in Fig. 4.3 illustrate both "typical" and "unusual" environments for important mesoscale weather phenomena. Any one sounding must be carefully examined as to its location, surrounding geomorphology, season of year, the likely synoptic and mesoscale environments, and sounding ascent "contamination" by entering clouds or precipitation. The latter effect is illustrated by the rare updraft sounding in Fig. 4.3a. All these factors plus radiation influence the end product, and therefore, the applicability of a given sounding for short-term nowcasting or use in initializing numerical prediction models. The winter "lake-effect" soundings in Fig. 4.3b show the importance of the Great Lakes in modifying continental polar air masses and thereby contributing to localized heavy snowfalls downstream.

A frequent approach in the past to solve the sounding representativeness problem for relatively rare mesoscale phenomena has been to construct composite or mean soundings. The operational meteorologist, however, is well advised to exercise caution in using such "mean" soundings. Mean soundings have been applied mostly as an aid to short-term and mesoscale forecasts of

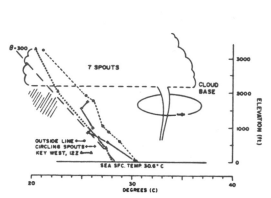

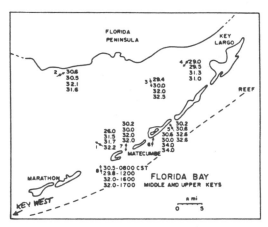

Figure 4.4a.   Temperatures obtained from a spirally ascending aircraft beneath (solid) and just outside (dashed) a cloud-line 15 n mi from Key West, which was producing a series of seven waterspouts during the time (~2100 GMT) of the sounding.   Temperatures obtained with rawinsonde at Key West (dotted) are for the same day at 1200 GMT. (From Golden, 1974.)

Figure 4.4b.   Sea surface temperatures (°C) showing the low-level differential heating mechanism along the Florida Keys.

tornado- or hail-producing thunderstorms in the Great Plains (e.g., Newton, 1980). A simplistic view used to be that the special organization and scale of the mesoscale "trigger" determined the subsequent type and scale of thunderstorm development, i.e., supercell, multicell, or squall line. Recent 3-D model results (Weisman and Klemp, 1982) suggest that the physical effects, especially variation of the shear vector with height (i.e., the wind hodograph) may be more important.

The 30 years of work on compositing upper-air data for mean soundings and mean height fields or flow patterns have had mixed results. These composites are useful in allowing the severe storm or winter storm forecaster to prepare "convective" outlooks or 0–12 h forecasts; however, it must also be remembered that there are important regional differences even in mean storm-sounding characteristics. Moreover, individual "precedence" or "proximity" soundings for hail, and/or tornado-producing thunderstorms exhibit wide variability, even within the same geographical area. The essence of this problem can be stated thus: A representative sounding that correctly describes the precedence or proximity near-environment of a tornadic or hail-producing thunderstorm may depend on the direction and distance of the sounding from the storm/tornado, time of year, geography, terrain, and especially type of mesoscale storm-initiating mechanism. The problem of representativeness is also illustrated in Fig. 4.4a. Representativeness may refer to time (in the case of evolution) or to location (space). Even though the aircraft sounding has less vertical resolution and accuracy than the NWS rawinsonde at Key West, it gives a more representative picture of the waterspout's near environment. Note especially the presence of a significant super-adiabatic temperature lapse rate in the subcloud layer of the aircraft sounding. This superadiabatic subcloud layer is well documented and is related to the strong, low-level differential heating mechanism that operates along the extensive envelope of very warm shallow water and the chain of islands (Keys) that extends WSW from the southern tip of the Florida peninsula (Fig. 4.4b). Superadiabatic layers more than a few tens of millibars thick are rare and generally do not persist long in the troposphere because of the absolute instability they imply (Hess, 1959); however, for waterspout and dust-devil environments some traditional sounding-smoothing procedures might eliminate the crucial, "representative" physical characteristic of the soundings for convective vortex generation.

4.4.3. In-Storm Observations

Another important consideration for the mesoscale forecaster is the assessment of in-storm or in-cloud soundings. Measurements of the interior thermodynamics and kinematics of severe thunderstorms were very difficult to obtain reliably up to 1970, because most balloon-borne sondes that are in a thunderstorm exit at some higher elevation, and supporting data to verify the rate of ascent (and the layers through which the sonde ascended) were lacking (a rare example is shown in Fig. 4.3a).

Davies-Jones and Henderson (1974) examined systematically a large number of rawinsonde ascents made in spring 1966–1973 by the National Severe Storms Laboratory in Norman, Oklahoma. Most of the surface mesoscale network soundings have been processed and recorded on magnetic tape (Barnes et al., 1971). In more than 4000 soundings, Davies-Jones and Henderson (1974) found 34 updraft soundings, which they defined as those wherein the balloons rose 5 m s^{-1} faster than the ascent rate in still air over vertical distances greater than 1 km. Figure 4.5 summarizes the mean updraft versus the mean environment profiles for all 34 Oklahoma storms. Study of these means and comparison with individual updraft soundings led to the confirmation of some previous observations and hypotheses, i.e., the existence of undiluted updraft cores (nearly pseudoadiabatic lapse rates up to middle levels), the warm core nature of the storms at middle and upper levels, and the weak vector wind shear in updrafts. Finally, the results substantiated Marwitz's (1973) observations that air constituting the updraft below and immediately above cloud base is generally cold relative to the environment and must therefore be rising under the influence of perturbation pressure gradient forces.

More recent aircraft measurements and multiple Doppler observations since 1975 have revealed new features and insights with respect to storm air flow circulation and inflow-outflow orientation (Ray et al., 1977; Ray et al., 1981). Collaborating with a NASA-sponsored mobile team of scientists during 1976, Sinclair (1983) surprisingly found that at middle levels (16,000 ft m.s.l.), upstream aircraft data indicated little or no environmental air inflow to the backside of the storm cells or line squalls investigated. This does not agree with other observational and theoretical severe storm models (e.g., Browning, 1964; Klemp et al., 1981). If Sinclair's (1983) measurements are

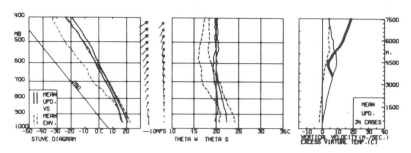

Figure 4.5. Profiles of mean updraft versus mean environment for 34 Oklahoma storms. (After Davies-Jones and Henderson, 1974.) (Left) Stuve diagram of mean updraft and environment soundings. The right solid line is the mean updraft temperature profile, the left solid line is the mean updraft dewpoint profile, and the dashed lines are similar curves for the mean environment. The 280°C dry adiabat is indicated. Arrows denote mean horizontal wind speed (proportional to length) and direction; height of observations is given by position of tip. Updraft winds are to the right of the environment ones. (Center) Mean profiles of θ_w and θ_s for updraft (both solid) and environment (both dashed); θ_w curves are to the left of corresponding θ_s curves. Pseudoadiabats are vertical lines. (Right) Profiles of mean updraft vertical velocity (solid), mean excess virtual temperature (dashed), and model vertical velocity (circles).

borne out by subsequent investigations, then the downdraft air below cloud base, which may have θ_e's similar to the upstream middle-level air, must in some cases have a different origin.

The proper application of sounding data to winter storms again depends on representativeness and can be affected by icing and large mesoscale variability in cloud type (see, for example, an excellent survey of results obtained by Houze and Hobbs, 1982). In a review of major East Coast snowstorms, Kocin and Uccellini (1984) presented an interesting summary of upper-level geopotential and wind pattern characteristics during the typical 60 h cyclogenetic period of 18 cases since 1960. Likewise, Danielsen (1968), Danielsen and Mohnen (1977), and Shapiro (1975, 1976, 1978) presented detailed isentropic cross-sectional and other mesoscale analyses that showed a strong association of tropopause folding and injection of stratospheric air with nearly concurrent intense cyclogenesis. Uccellini *et al.* (1984) and Bosart (1981) presented other interesting winter storm case studies.

4.4.4. Autocorrelation and Horizontal Variability Studies

Barnes and Lilly (1975) and Barnes (1979) used covariance analysis of sounding data to determine the statistical, mesoscale structure of meteorological fields, as a function of horizontal distance in the free atmosphere. The data used in the study were obtained in Oklahoma, during 2-month springtime periods in 1966, 1967, and 1968. Average station spacing of the ten sites common to 1966 and 1967 was 85 km (1966 had one additional station); however, the 1968 network of 10 stations covered a much smaller area, and average station spacing was only 39 km. The data were grouped into three categories according to weather conditions at the sounding times, ranging from nonstormy to stormy within the network. Barnes and Lilly

(1975) computed structure functions for parameters p, T, and q (mixing ratio), and wind components (u,v) at three levels: 1500, 3000, and 5700 m (or roughly at 850, 700, and 500 mb pressure levels). They summarized some of the results as percentages of total variance contained in mesoscales below 200 and 100 km (for sounding data at 1500 m). There was some question concerning the accuracy of humidity measurements (not corrected for radiation or thermal lag effects—see Betts *et al.*, 1974) and the possible influence of large vertical gradients in and near wave-disturbed inversions; nevertheless, a rather large 37% of total q variance associated with separations less than 200 km in nonstormy conditions is found. Results for higher altitudes show a similar or larger fraction of q-variance contained in meso-β scales for category-3 (stormy) conditions. In addition, during stormy conditions the structure functions for both wind components are nearly equal throughout the range 20–200 km, confirming that the flow is strongly divergent. These preliminary results indicate that sufficient variance exists within the mesoscale range of 100–200 km to require sounding station spacings of ~100 km to detect important severe thunderstorm "triggers" and interactions with the storm environments. More recently, Kelleher and Johnson (1983) made time-space correlation estimates of squall lines and severe storms, using NSSL surface mesonetwork data collected in 1973–1979, and found similar patterns and results.

Another recent structure function analysis of upper-air data was carried out by Fuelberg and Meyer (1982) to identify those synoptic and mesoscale wavelengths that were dominant during the Red River Valley tornado outbreak (10–11 April 1979). Data from 23 NWS rawinsonde sites and 16 special sites were combined to attain meso-α-scale resolution. Fuelberg and Meyer (1982) found that the synoptic-scale features dominated the fields of height, temperature, and mixing ratio during this observing period, and no major time variability was observed in the field structure functions. However, the structure for mixing ratio showed considerably more activity at the shorter wavelengths, which agrees with earlier results of Barnes and Lilly (1975). Mesoscale disturbances having wavelengths of 1000–1600 km are major components of the SESAME-I wind field. These wind features, especially those at 700 mb, show considerable time variability. Fuelberg and Meyer (1982) made the interesting comment that, for this tornado outbreak case, "NWS data alone provide a reasonably good representation of the important wavelengths during this particular period." If these preliminary results are extended and confirmed for other thunderstorm regions, there are also clear implications for future operational sonde-accuracy and resolution requirements, in addition to the trade-offs in network spacing.

4.5. Future Mesoscale Sounding Systems

4.5.1. In Situ Sensing Systems

The principle of windfinding with the use of navigational aids (navaids) is simple. A balloon or parachute, equipped with a navaid receiver, retransmits the navaid signals to a base station. The navaid signals are transmitted from a number of fixed stations through the sonde to the base station. The difference in time of arrival of the signals is used to determine the range difference between pairs of stations. Since the path from sonde to base station is identical for each transmitter, the measurement of range differences eliminates the common path from sonde to base station. The base station can, therefore, be in motion without introducing error to the wind computation. The technique is ideally suited for measuring winds below aircraft with a dropsonde or measuring winds from a moving ship with a balloon sonde.

The Omega global navigation system consists of eight VLF transmitters located around the globe. Windfinding is achievable with the use of three stations, but accuracy is improved by use of all stations that provide a readable signal. Care must be exercised to eliminate from computation any station whose signal over the longer great-circle path may be as strong as the short-path signal.

The eight worldwide Omega stations transmit sequentially over a 10 s period. Phase measurement errors due to noise and varying path lengths are 1°–5° for an individual sample. This corresponds to range errors of 60–300 m. Filtering techniques are used to provide wind accuracies of 1 to 2 m s^{-1} for 2 min averages under favorable conditions. Fine-scale wind structure cannot be obtained with Omega windfinding systems. However, accuracy does not deteriorate as the sonde moves away from the observer.

LORAN-C is a navigation system used to provide accurate ship location along the coastlines. A number of LORAN-C chains cover the Pacific, Atlantic, and Gulf Coasts, and the Aleutians. Only a few LORAN-C navaid systems have been built, but they can provide excellent wind accuracy if located in areas of good coverage. Complete coverage is not available across the central portions of the United States or in many other regions of the world.

In 1983 a new LORAN-C aircraft navigator was introduced which utilizes the signals from all LORAN-C chains within range of the receiver. Since the signals in the separate chains are coherent, it is possible to operate in a "cross-chain" mode for accurate wind computation across the continent even in areas where coverage is not feasible using single-chain systems. A new sounding system that uses the cross-chain mode is now planned by the National Center for Atmospheric Research (NCAR) as the standard system to obtain research quality data in the mesoscale programs of the next decade. The Department of Transportation has planned for an additional LORAN chain to eliminate the "mid-continent gap." This new chain will ensure superior wind-computation capability even with single-chain systems.

Wide-bodied jet aircraft now contain inertial navigation systems and flight data acquisition units, which sense and compute the meteorological parameters and present them in a digital cockpit display. The data are thus available for collection and retransmission.

The Geosynchronous Operational Environmental Satellites (GOES) operated by the United States, as well as the European and Japanese geostationary meteorological satellites, are equipped with data collection systems for retransmission of signals from fixed and moving platforms. In the usual mode of operation, signals from many sources (such as river-level gauges) are time-multiplexed to maximize use of the limited number of communication channels. Fortunately, a number of international channels have been assigned that are common to all U.S., European, and Japanese meteorological satellites. These channels can be time-multiplexed to permit several hundred aircraft to transmit, within their time-slots, data on winds, altitude, temperature, and pressure. The Aircraft-to-Satellite Data Relay (ASDAR) system consists of an 80 W transmitter at 402 MHz, a flush-mounted antenna, and a microprocessor unit for coding data extracted from the aircraft flight data acquisition unit.

The ASDAR unit was used extensively during the Global Weather Experiment in 1979. The World Meteorological Organization now has undertaken a program to improve the system capabilities and to standardize a system for use by all airlines that wish to cooperate and whose routes travel over data-sparse areas. The ASDAR system provides flight level winds accurate within 1–2 m s^{-1} and temperatures within $1°C$. As the use of ASDAR expands, it will become an important new element in the global weather-observing system. Since ASDAR is operating throughout a flight, it can be used to obtain vertical soundings both on takeoff and landing.

Aircraft dropwindsondes operate in the same fashion as a radiosonde, telemetering data back to the aircraft on pressure, temperature, and humidity. A parachute is used rather than a balloon, and the sonde must be designed to take high shock loadings on release. The dropsondes developed to date have been designed ruggedly; as a result, because of their weight and density, they are not safe to launch over populated land areas.

Until the development of the navigation-aid sonde, a number of expensive and abortive attempts were made to develop a dropsonde that could measure winds. With the navigation-air sonde, the problem disappears. At this time, there is only one dropsonde being used by the U.S. Air Force and NOAA. It was developed by NCAR for research programs associated with GARP—the Atlantic Tropical Experiment in 1974 and the Global Weather Experiment in 1979. It is used on larger aircraft, weights 1.8 kg, and operates with a complex data processor mounted in the aircraft. It operates only with Omega navigation signals.

Development is now under way at NCAR on a new dropwindsonde that can be safely launched over populated areas (weight, 0.35 kg), can be operated from light aircraft, and can operate with either Omega or LORAN-C (LORAN-C accuracy is a factor of 10 better than Omega accuracy wherever a LORAN-C network is within range).

4.5.2. Remote-Sensing Systems

The meteorological community throughout the world has been comfortable with and has made great strides with conventional balloon-borne radiosondes. Indeed, these systems have been, and remain, among the most fundamental of the observations upon which meteorologists base their analyses and their forecasts. Why then are remote-sensing sounding techniques of interest?

The reasons are many; some are purely pragmatic and some are scientifically substantive. The existing rawinsonde system of the United States has a limited lifetime. Costs of operation are high, and maintenance is difficult. More important, however, there are serious doubts that the existing system or any balloon-borne system can provide measurements on the smaller time and space scales necessary for accurate and reliable mesoscale modeling and prediction. How do remote-sensing systems help to mitigate this problem?

First, ground-based remote sensors operate continuously so that judicious time-to-space conversion may be used to extend the spatial coverage of fixed ground-based systems. Second, satellite systems are complementary and provide total areal coverage (at lower resolution and accuracy) to fill the gaps between the ground-based systems. Third, remote-sensing systems can be built to operate automatically without resident staff. Routine maintenance is all that is required.

Though attractive, remote sensors are not in greater use now because the technologies are not fully developed and tested. An important ancillary condition is that the meteorological community is familiar with balloon sounding data and as yet does not know how to make effective use of all of the remote-sensing capability that is available. Meteorologists are not ready to relinquish the proven rawinsonde soundings.

Remote-sensing techniques fall into two basic classes, active and passive. Active devices such as radars transmit some form of energy and sense a small fraction of the energy that is reflected from targets removed spatially from the transmitter. Monostatic systems are systems that transmit and receive from a single location; bistatic systems transmit and receive at separate locations. The energy radiated can take a number of forms: microwave, radio, infrared, and optical electromagnetic waves, or acoustic waves.

Passive sensors radiate no energy of their own, but instead "listen" to energy naturally radiated or reflected from the environment and interpret from these received signals the characteristics of the sources at a distance. The human eye and ear are two excellent examples of passive remote sensors. Radio direction finders represent another example. Radio telescopes such as the Very Large Array in New Mexico are another.

Wind Profilers

Perhaps the most thoroughly developed remote-sensing technique for sounding purposes is the wind profiler. These active devices are Doppler radars operating at VHF and UHF frequencies (30 MHz through 1000 MHz); they were developed from the resources of knowledge and experience in the radar meteorology and aeronomy communities. Ionospheric scientists developed UHF and VHF radars for studies of the structure of the stratosphere and the mesosphere. It became apparent, soon after these radars were put into use, that the systems were able to obtain sufficient backscattered energy in the troposphere to measure Doppler velocities accurately. Since 1980, there has been great interest in this technique for application to meteorology. NOAA's Wave Propagation Laboratory (WPL) (Strauch *et al.*, 1984)

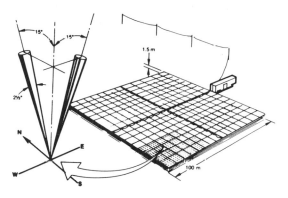

Figure 4.6. Design of the Platteville VHF wind-profiling radar. (After Hogg *et al.*, 1983b.)

and Aeronomy Laboratory (Balsley and Gage, 1982) led the way in advancing this technology for operational use. A schematic illustration of a 50 MHz wind profiler near Platteville, Colo., is shown in Fig. 4.6.

Wind profilers, like any Doppler radar, measure the Doppler frequency shift caused by the motion of the scatterers along the direction of the radar beam. The backscattered energy comes from the presence of turbulent eddies in the atmosphere at scales of the order of one-half the radar wavelength.

Essentially, the wind profiler works by calculating the wind vectors at various levels in the atmosphere from two or more beams separated by 15° from zenith in the north/south and east/west directions (see Fig. 4.6). The wind vectors can be derived in essentially real time, since each transmitted pulse samples all levels of the atmosphere, essentially simultaneously—a big advantage over current systems. Sampling can be averaged from as little as a few minutes to upwards of an hour or more. Table 4.2 compares some of the radars in development today.

Thermodynamic Profiling

Ground-based remote sensors for accomplishing thermodynamic profiling are passive radiometric devices operating in the microwave absorption bands for oxygen and water vapor.

Hogg *et al.* (1983a,b) described the fundamental principles of operation of remote profiling systems for temperature and water vapor. A two-channel radiometer operating at 22 GHz and 30 GHz is used for water vapor retrieval and liquid water. These two channels provide for measurements of the total integrated precipitable water vapor and liquid water in the vertical. A four-channel radiometer operating in the oxygen band (50–60 GHz) is used to estimate profiles of temperature for varying heights.

Statistical inversion techniques are used to convert the received radiation into identifiable profiles of temperature and humidity. A climatic data set of radiosonde ascents for time of year is used as a basis for the statistical analysis (James, 1983). A corresponding data set of profiler measurements correlate this set with the radiosonde data set to form a cross-correlation matrix. This matrix can then be used to compare observed remotely obtained profiles with those of the radiosonde ascent.

Table 4.2. Comparison of wind profiler radars under evaluation by NOAA's Environmental Research Laboratories

Radar parameter	VHF radar (Colorado network)	VHF radar (Platteville)	UHF radar
Frequency	49.80 MHz ($\lambda_0 = 6.002$ m)	49.92 MHz ($\lambda - 0 = 6.0054$ m	915 MHz ($\lambda_0 = 33$ cm)
Pulse width	3, 9 μs	16 μs	Variable, 0.6–4.8 μs
Pulse period	238, 672 μs	2400 μs	Variable, 20–500 μs
Antenna size	50 × 50 m (2 arrays)	100 ×100 m (3 arrays)	10 × 10 m (3 feeds)
Beam positions	15° off zenith to north; 15° off zenith to east	Zenith; 15° off zenith to north; 15° off zenith to east	Zenith; 15° off zenith to north; 15° off zenith to east
Peak power	30 kW (each beam)	12 kW (each beam)	6 kW
Average power	400 W (each beam)	80 W (each beam)	1500 W max.
Min. receiving range	1.7 km AGL	2.4 km AGL	$\simeq$300 m
Receiver range spacing	300, 900 m	1500 m	Variable, $\geq$100 m
Number of range locations	24, 18	13 (2.4–20.4 km AGL)	32
	Mode 1 (high spatial resolution)	Mode 1 (horizontal winds)	Mode 1 (horizontal winds)
Max. horizontal wind speed	±58.15 m s^{-1}	±75.46 m s^{-1}	$\geq$70 m s^{-1}
Horizontal wind resolution	±0.12 m s^{-1}	±0.29 m s^{-1}	< ±0.25 m s^{-1}
Dwell time	90 s	130 s	Variable, 10–90 s
	Mode 2 (high-altitude coverage)	Mode 2 (vertical winds)	Mode 2 (vertical winds)
Max. wind speed	±69.78 m s^{-1}	±2.44 m s^{-1}	$\geq$ ±10 m s^{-1}
Wind resolution	±0.14 m s^{-1}	±0.01 m s^{-1}	$\geq$ ±0.04 m s^{-1}
Dwell time	130 s	90 s	Variable, 10–90 s

A limitation to this form of analysis derives from use of climatic (or averaged) data. The results, when compared with actual ascents, indicate a "smoothing" of the temperature profile. NOAA's Wave Propagation Laboratory (WPL) has recently employed a variety of methods to improve the correlation with radiosonde soundings. Information from the wind profiler and other radars provide heights of the base of inversions and the tropopause height, which can all be incorporated into the temperature profile to sharpen inversions and stable layers. Further improvements can be made with the inclusion of satellite data, which are most accurate at higher altitudes.

The nearly continuous temperature data provide the forecaster with such things as the buildup and disintegration of the surface inversion, the possibility of computing stability indices in real time, and changes in the freezing and tropopause levels.

Compared with a radiosonde sounding, the humidity profiles have the smaller amount of resolution. However, the ability of the profiler to measure the total precipitable liquid water and vapor has numerous applications in forecasting and modeling. The timely analysis and dissemination of these data can provide information to pilots concerning aircraft icing (Hogg *et al.*, 1983a). Total liquid water is also necessary in many numerical weather prediction models, which may be improved with accurate measurements at periods less than 12 h (such data are not yet routinely available).

Acoustic Sounding

Incoherent and Doppler acoustic sounding techniques are now well developed, and systems are available commercially. Acoustic sounders are nothing more than acoustic "radars" or sodars that are designed for atmospheric use. They are useful for measuring boundary layer structure and depth (including inversion heights) and for measuring winds in the boundary layer. Acoustic sounders often are not sufficiently sensitive to make measurements throughout the depth of deep convective boundary layers or in nonturbulent conditions. These systems are, however, reasonably inexpensive and reliable. There has been some work in using acoustic sounder or profiler radar information to determine the heights of stable layers and inversions. Feeding just this information (i.e., inversion/stable-layer heights) back into the radiometrically derived temperature and moisture profiles gives improved representation of the inversion layers.

Lidars

Lidar, or optical radar, is another method for remote sensing of the atmosphere. Incoherent lidar, using scattering from aerosols and molecules, is useful for determining boundary layer depth and structure. Differential absorption lidar (DIAL) can be used to measure profiles of water vapor, temperature, and gaseous constituents of the atmosphere. Doppler lidar, operating in the CO_2 infrared band, can be used in a manner similar to that in which radars and wind profilers are used to measure atmospheric air motion (see, for example, Bilbro *et al.*, 1984). NOAA's Wave Propagation Laboratory and other agencies of the U.S. Government have conducted research on an orbiting satellite-based Doppler lidar system for making global measurements of winds (WINDSAT).

4.5.3. Hybrid Sounding Systems

There exist now an exciting array of new technological opportunities for new and better sounding systems. Satellite systems will provide contiguous coverage in all regions of the globe. Rawinsondes, using navigational aids, can be released from aircraft and ships of opportunity to provide high resolution soundings of temperature, moisture, and winds. Commercial aircraft are able to carry *in situ* sensors for measurements of winds, temperature, and moisture at flight altitudes, and, moreover, could be used to obtain many thousands of soundings daily, worldwide, during takeoffs and landings. Wind profilers will provide continuous, high-resolution measurements of winds and turbulence at all altitudes. All these systems can operate automatically or semi-automatically and provide digital information nearly in real time. And global communication systems exist to make these four-dimensional, high-quality data sets available to research and operational meteorologists alike, in all corners of the Earth.

It appears inevitable, therefore, that the sounding systems of the future will not rely on a single system or technique. Rather, the meteorological community can expect to have available to it a continuously updated collection of measurements made by a wide variety of sensors and from many different platforms. These measurements will be complementary to one another and will certainly lead to major new advances in basic understanding and in forecasting.

Future operational networks for mesoscale applications will need to address many of the issues outlined. Critical to the design of any future network is consideration of how all the kinds of systems, both present and planned, will mesh together into one compatible and cohesive data base of information. This is among the major challenges facing the mesoscale meteorological community in the coming decade.

REFERENCES

Baker, D. G., 1970: A study of high pressure ridges to the east of the Appalachian Mountains. Ph.D. Dissertation, Massachusetts Institute of Technology, Cambridge, 127 pp.

Balsley, B. B., and K. S. Gage, 1982: On the use of radars for operational wind profiling. *Bull. Amer. Meteor. Soc.*, **63**, 1009–1018.

Barnes, S. L., 1979: SESAME 1979 Field-Processed Rawinsonde Data from Supplementary Sites: April 10–June 8. Project SESAME, U.S. Department of Commerce, NOAA/ERL, Boulder, Colo., 236 pp.

Barnes, S. L., and D. K. Lilly, 1975: Covariance analysis of severe storm environments. Preprints, 9th Conference on Severe Local Storms, Norman, Okla., American Meteorological Society, 301–306.

Barnes, S. L., J. H. Henderson, and R. J. Ketchum, 1971: Rawinsonde observation and processing techniques at the National Severe Storms Laboratory. NOAA Tech. Memo. ERL NSSL-

53, Environmental Research Laboratories (NTIS#COM–71–00707), 246 pp.

Betts, A. K., F. J. Dugan, and R. W. Grover, 1974: Residual errors of the VIZ radiosonde hygristor as deduced from observations of subcloud layer structure. *Bull. Amer. Meteor. Soc.*, **55**, 1123–1125.

Bilbro, J., G. Fichtl, D. Fitzjarrald, M. Kraus, and R. Lee, 1984: Airborne Doppler lidar wind field measurements. *Bull. Amer. Meteor. Soc.*, **65**, 348–359.

Bosart, L. F., 1981: The Presidents' Day snowstorm of 18–19 February 1979: A subsynoptic-scale event. *Mon. Wea. Rev.*, **109**, 1542–1566.

Browning, K. A., 1964: Airflow and precipitation trajectories within severe local storms which travel to the right of the winds. *J. Atmos. Sci.*, **21**, 634–649.

Danielsen, E. F., 1968: Stratospheric-tropospheric exchange based on radioactivity, ozone, and potential vorticity. *J. Atmos. Sci.*, **25**, 502–518.

Danielsen, E. F., and V. A. Mohnen, 1977: Project Dustorm Report: Ozone transport, in-situ measurements, and meteorological analyses of tropopause folding. *J. Geophys. Res.*, **82**, 5867–5877.

Davies-Jones, R. P., and J. H. Henderson, 1974: Updraft properties deduced from rawinsoundings. NOAA Tech. Memo. ERL NSSL–72, Environmental Research Laboratories (NTIS#COM–75–10583/AS), 117 pp.

Fuelberg, H. E., and P. J. Meyer, 1982: A structure function analysis of the AVE-SESAME-1 period. Preprints, 12th Conference on Severe Local Storms, San Antonio, Tex., American Meteorological Society, Boston, 188–191.

Fujita, T. T., D. L. Bradbury, C. F. van Thullenar, 1970: Palm Sunday tornadoes of April 11, 1965. *Mon. Wea. Rev.*, **98**, 29–60.

Galway, J. G., 1956: The lifted index as a predictor of latent instability. *Bull. Amer. Meteor. Soc.*, **37**, 528–529.

Golden, J. H., 1974: Life cycle of Florida Keys' waterspouts. NOAA Tech. Memo. ERL–NSSL–70, Environmental Research Laboratories (NTIS#COM–74–11477/AS), 150 pp.

Hess, S. L., 1959: *Introduction to Theoretical Meteorology*. Holt, New York, 362 pp.

Hoehne, W. E., 1980: Precision of National Weather Service upper air measurements. NOAA Tech. Memo. NWS T&ED–16, National Weather Service (NTIS#PB81–108136), 23 pp.

Hogg, D. C., F. O. Guiraud, J. B. Snider, M. T. Decker, and E. R. Westwater, 1983a: A steerable dual-channel microwave radiometer for measurement of water vapor and liquid in the troposphere. *J. Clim. Appl. Meteor.*, **22**, 789–806.

Hogg, D. C., M. T. Decker, F. O. Guiraud, K. B. Earnshaw, D. A. Merritt, K. P. Moran, W. B. Sweezy, R. G. Strauch, E. R. Westwater, and C. G. Little, 1983b: An automatic profiler of the temperature, wind, and humidity in the troposphere. *J. Clim. Appl. Meteor.*, **22**, 807–831.

Houze, R. A., and P. V. Hobbs, 1982: Organization and structure of precipitating cloud systems. *Adv. Geophys.*, **24**, 225–315.

James, P. K., 1983: The WPL Profiler: A new source of mesoscale observations.

Meteor. Mag., **112**, 229–236.

Kaimal, J. C., H. W. Baynton, and J. E. Gaynor, 1980: The Boulder low-level intercomparison experiment. WMO Instruments and Observing Methods Rep. 3, 189 pp. (Preprint available as BAO Rep. 2, NOAA/ERL Wave Propagation Laboratory, Boulder, Colo.).

Kelleher, K. E., and K. W. Johnson, 1983: Time-space correlation estimates of squall-lines and severe storms. Preprints, 13th Conference on Severe Local Storms, Tulsa, Okla., American Meteorological Society, Boston, 33–36.

Klemp, J. B., R. B. Wilhelmson, and P. S. Ray, 1981: Observed and numerically simulated structure of a mature supercell thunderstorm. *J. Atmos. Sci.*, **38**, 1558–1580.

Kocin, P. J., and L. W. Uccellini, 1984: A review of major East Coast snowstorms. Preprints, 10th Conference on Weather Forecasting and Analysis, Tampa, Fla., American Meteorological Society, Boston, 189–198.

Marwitz, J. D., 1973: Trajectories within the weak echo regions of hailstorms. *J. Appl. Meteor.*, **12**, 1174–1182.

McComb, H. C., and R. G. Beebe, 1956: A thunderstorm sounding. *Mon. Wea. Rev.*, **84**, 107.

Miller, R. C., 1972: Notes on analysis and severe-storm forecasting procedures of the Air Force Global Weather Central. Air Weather Service Tech. Rep. 200 (Rev.), Air Weather Service, Scott Air Force Base, Ill., 190 pp.

Nappo, C. J., 1983: Methods of estimating meteorological representativeness. Preprints, 5th Symposium on Meteorological Observations and Instrumentation, Toronto, Canada, American Meteorological Society, Boston, 246–252.

Newton, C. W., 1980: Overview on convective storms systems. Proceedings, CIMMS Symposium, (Y.K. Sasaki, N. Monji, and S. Bloom, Eds.), University of Oklahoma, Norman, 3–107.

Norcross, G. A., and R. L. Brooks, 1983: Balloon-Borne Pressure Sensor Performance Evaluation Utilizing Tracking Radars, NASA (N83-34276), 64 pp.

Ray, P. S., K. K. Wagner, K. W. Johnson, J. J. Stephens, W. C. Bumgarner, and E. A. Mueller, 1977: Triple Doppler observations of a convective storm. *J. Appl. Meteor.*, **17**, 1201–1212.

Ray, P. S., B. C. Johnson, K. W. Johnson, J. S. Bradberry, J. J. Stephens, K. K. Wagner, R. B. Wilhelmson, and J. B. Klemp, 1981: The morphology of several tornadic storms on 20 May 1977. *J. Atmos. Sci.*, **38**, 1643–1663.

Shapiro, M. A., 1975: Simulation of upper-level frontogenesis with a 20-level isentropic coordinate primitive equation model. *Mon. Wea. Rev.*, **103**, 591–604.

Shapiro, M. A., 1976: The role of turbulent heat flux in the generation of potential vorticity in the vicinity of upper-level jet stream systems. *Mon. Wea. Rev.*, **104**, 892–906.

Shapiro, M.A., 1978: Further evidence of the mesoscale and turbulent structure of upper-level jet stream-frontal zone systems. *Mon. Wea. Rev.*, **106**, 1100–1111.

Sinclair, P. C., 1983: Severe storm air motion measurements by SSSP research aircraft. Preprints, 13th Conference on Se-

vere Local Storms, Tulsa, Okla., American Meteorological Society, Boston, 366–371.

Strauch, R. G., D. A. Merritt, K. P. Moran, K. B. Earnshaw, and J. D. Van de Kamp, 1984: The Colorado wind-profiling network. *J. Atmos. Oceanic Tech.*, **1**, 37–49.

Uccellini, L. N., P. J. Kocin, R. A. Petersen, C. H. Wash, K. F. Brill, 1984: The Presidents' Day cyclone of 18–19 February 1979: Synoptic overview and analysis of the subtropical jet streak influencing the precyclogenetic period. *Mon. Wea. Rev.*, **112**(1), 31–55.

USAF, 1961: Use of the Skew T, Log P Diagram in Analysis and Forecasting, Vol. I, Radiosonde Analysis. Air Weather Service Manual, USAF, 105–124.

Weisman, M. L., and J. B. Klemp, 1982: The dependence of numerically simulated convective storms on vertical windshear and buoyancy. *Mon. Wea. Rev.*, **110**, 504–520.

CHAPTER 5

Systems for Measurements at the Surface

Dennis W. Thomson

5.1. What Are Measurements and Observations?

Traditionally, meteorological measurements made at the Earth's surface have been called "surface observations." But few surface observations are observations of the surface; most are observations of atmospheric phenomena made by an observer on the surface. Strictly speaking, an observation is a noting. Thus, an observer can make a visual noting, for example, of sky color or cloud cover. Prior to the development and use of instruments such as barometers, thermometers, anemometers, and raingauges, much of the meteorological data base and many of the "forecasts" were the result of experience derived from such observations; *e.g.*,

> Red skies in the morning,
> Sailors take warning.
> Red skies at night,
> Sailors' delight.

Meteorologists today still heavily rely upon observations. However, for forecasting purposes most prefer the observational perspective provided by the imagery transmitted from geosynchronous or other satellites. The meteorologist's visual perspective has also been enormously expanded through the use of a variety of electromagnetic and acoustic remote sensing systems. The purpose of many sophisticated infrared-imaging, radar, lidar, and sodar systems is, fundamentally, to expand the spectral and dynamic response of the human eye.

In this discussion of atmospheric measurements made at the surface, it will be necessary to refer to "surface observations" that are neither observations nor measurements of the surface, and to "observing systems" that are incapable of observing anything. For example, RAMOS (Remote Automated Meteorological Observing System) is a widely used and reliable system for providing meteorological measurements. But to suggest by title that the sensing of temperature and other variables by RAMOS can be compared with

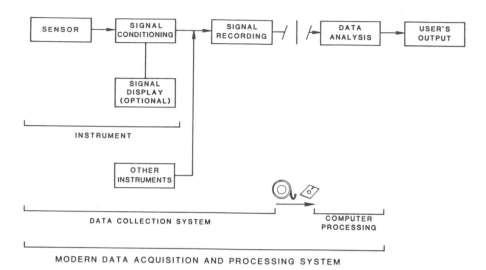

Figure 5.1. Typical components of meteorological measurement systems.

human observation is like equating the scribbling of a child to the beauty and intricacy of a master painting.

To make a measurement, an instrument is required. The instrument might be a meter stick or a radar, a clock or a theodolite. What is important is that the instrument can be used to establish the dimensions or quantity of something. In many cases it is the temporal or spatial rate of change (tendency or gradient, respectively) of that something which is of interest. For effective use of measurements, accepted standards are also essential, not only for reporting the measurements (and observations) but also for evaluating the performance and application of various instruments.

It is important to differentiate semantically between a sensor, an instrument, and a data collection or acquisition system (Fig. 5.1). The sensor is the part of an instrument that converts the state of the relevant meteorological variable into a signal (visual, mechanical displacement, electrical, etc.). Examples of sensors are the thermistor probe for temperature on a radiosonde, the aneroid chamber for pressure on a microbarograph, or the open X-band cavity in a refractometer that is used for sensing the radio refractive index. In a very simple instrument, such as a liquid-in-glass thermometer, sensor and instrument may be one. Typically now the sensor is only a small, critical part of a larger electronic instrument. A data collection or acquisition system normally consists of several instruments plus the means for signal manipulation and recording. A modern measurement system may also include user-tailored numerical models in order to optimize the processing of acquired signals.

It is sometimes useful to differentiate between *in situ* and remote measurements. If a sensor is submerged within the environment to be measured, (e.g., a thermometer in air), the measurements are *in situ* or direct. If, however, the measurements are obtained by processing signals from electromagnetic or acoustic radiation that has passed through the atmospheric region of interest and been scattered, absorbed, refracted, or Doppler-shifted, the measurements are remote or indirect. Sometimes the definition of remote depends on the user's point of view. Is a sonic anemometer a direct or a remote probe? Since its sampling path is small ($\simeq$40 cm) compared with the dimensions of the turbulent eddies ($\ell_{max} \simeq 4$ to 6 m) at the height at which the anemometer is usually mounted, most micrometeorologists would think of it as an *in situ* probe even though it operates on the basis of wind-velocity-dependent acoustic pulse propagation times. On the other hand, a path-averaging (300 m to 15 km path length) laser anemometer mounted at 2 to 10 m height (or between hilltops) would probably be considered a remote probe in the same sense as a radar, sodar, or lidar.

5.2. Changing Requirements for Surface Measurements

Since the inception of government-supported national weather services the principal purposes for performing meteorological measurements have been at least twofold: recording and analyzing notable weather conditions relevant to public interest, health, and safety; and establishing local climatological data bases. The latter has been of particular importance to agriculture. Meteorological records regarding variables such as first and last frosts, or rainfall, are essential for deciding, for example, the economic feasibility of fruit production, or of irrigation, in a given area.

For many years real-time weather data emphasized, and was essentially limited to, analysis of synoptic-scale features. The capabilities of many national and international weather data communication systems were ill-suited for regular transmission of measurements more frequent than hourly.

For mesoscale analysis and forecasting, it is essential not only that measurements of increased spatial and temporal resolution be available but also that new types of measurements be acquired in support of other, new, measurement and analysis activities. Table 5.1 contrasts by application traditional and mesoscale network data.

Table 5.1. Applications of network measurements

Application	Synoptic-scale network	Mesoscale network
Current weather	Hourly data Special observations	6–10 min data Special observations
Climatology	Daily (etc.) summaries	Daily (etc.) summaries, including special variables
Special variables	—	Fluxes: Radiative, Thermal, Moisture
Meso-β- and γ-scale features	Phenomenological or event reports only	Quantitative record of changing variables
Remote sensing baseline	—	Real-time surface reference for various tropospheric sounding systems

5.3. Variables of Interest

Surface observations or surface data traditionally have consisted of hourly notation of an altimeter setting (or pressure), temperature, dewpoint temperature (or wet bulb depression), wind speed and direction, precipitation type and amount, current weather including cloud height and coverage, and visibility. At many locations the process of sensing the meteorological variable, conditioning the signal, and recording or transmitting the data has, with the exception of cloud height and coverage data, been completely automated through the use of systems such as RAMOS. For example, in the weather observatory at Pennsylvania State University, a personal computer connected to a RAMOS system has been used for about 5 years to process and archive the station climate data automatically. Examples of newer automated measurement systems include ASOS (Automated Surface Observing System) for the National Weather Service (NWS), AWOS (Automated Weather Observation System) for the Federal Aviation Administration (FAA), and JAWOS (Joint Automated Weather Observing System), a system proposed to meet the combined needs of the Departments of Commerce, Defense, and Transportation. PAM II (Portable Automated Mesonet II) is a data acquisition system designed at the National Center for Atmospheric Research (NCAR) specifically for mesoscale network measurements (Brock and Saum, 1983).

Important additional capabilities planned for some of the new systems include automated thunderstorm detection, identification of weather type, and evaluation of cloud coverage. But the operational performance of weather identifiers and laser-based cloud amount sensors will continue to be evaluated through research even after the first field models are installed. Consider, for example, the possible errors in inferred cloud amount that a laser-based instrument without a sky-scanning capability might make. In some areas cumulus clouds are often organized into "streets" aligned nearly along the direction of the mean wind. A vertically pointing cloudiness sensor could indicate extended periods of either cloudy or clear air, depending upon where the street was positioned with respect to the stationary sensor for the measurement-averaging period. Thus, local climatological factors such as land-water boundaries, orography, and large-scale industrial developments may require small-scale arrays of instruments such as ceilometers, visibility meters, and radiometers for regionally representative measurements to be obtained.

Neither sensors of radiation nor sensors of other fluxes are included in the RAMOS, ASOS, AWOS, or JAWOS systems. Although there are plans for more than 1000 such measuring systems to be deployed nationally, the stated purpose is, basically, to improve the automated synoptic-scale measurement network. PAM II can easily accept a variety of standard, solar, and net radiation and micrometeorological momentum, thermal, and evaporative energy flux sensors. Solar radiation and net radiation are fundamentally important to the diurnal evolution of the planetary boundary layer. Surface-layer thermal and evaporative energy fluxes are important not only to boundary-layer development but also to the initiation and maintenance of convection. The new NWS and FAA systems will need such sensors. Otherwise, more than 1000 high-quality, automated measurement systems will be unable to transmit information essential to the analysis and understanding of mesoscale convective processes (SCPPD, 1985).

With respect to potential applications for flux measurements, it is worth noting that a large, research program is under way to develop methods for parameterizing the air-surface flux of acidic gases and particles in terms of meteorological fluxes (Hicks *et al.*, 1985). Acid fluxes are recognized as having an important effect on the environment; dry deposition may account for about half the total atmosphere-ground flux of acidic material, which is popularly known as acid rain.

5.4. Spatial and Temporal Scales

The total number of NWS, FAA, and DOD stations for surface weather measurements in the United States is $\sim$1300. These stations are spread over a land area of approximately 9.4 million km^2. Thus, at the present time, hourly measurements would appear to be available from stations with an "average" spatial separation of $\sim$85 km. In practice the total number of stations from which data can be used readily on an hourly basis for analysis of synoptic scale features is at most $\sim$400. For this network the "average" separation is approximately 155 km.

Historically, the existing surface measurement network has been at least adequate for synoptic-scale analysis and forecasting purposes. But meso-β scales of interest extend from 25 to 250 km. Furthermore, many of the mesoscale features of interest are convective systems, which can evolve significantly in a time period of less than an hour. Thus, use of nothing more than the existing network for mesoscale studies will result in serious spatial and temporal undersampling errors. For purposes of discussion undersampling is defined here as less than two measurements of a feature of interest within its characteristic length scale and time scale.

The costs, economic and logistical, of obtaining surface and upper-air measurements with satisfactory spatial and temporal resolution for application on the subsynoptic scale (say, the scale of convective precipitation) can be staggering. The problem is not simply one of the number of stations (2209 would be required for 50 km resolution in a 5.3 million km^2 domain) but also one of compatibility between the spatial resolution and sampling frequency. At 10 m s^{-1} a parcel will traverse 50 km in about 1.4 h. Thus, in order to observe changes in the parcel between grid points, hourly soundings might be prudent. At this rate in such a network 53,016 soundings per day would be required (about 55,000 radiosondes per year are currently launched in the whole contiguous United States).

A rainband in a mesoscale convective system (MCS) might be only 5 km wide. To resolve the distribution of precipitation in a 300 x 300 km MCS would require 3721 recording raingauges. To obtain adequate temporal resolution of the rain rate (say 5 min), a manually serviced gauge would require a 24 h chart and thus daily attention. Assuming 0.5 man-hour for transportation and service per gauge per day, more than 230 people would be

Table 5.2. Scales in various mesoscale domains, assuming a 50 × 50 uniform square grid array

Scale	Horizontal extent of domain (km)	Domain area (km²)	Vertical extent of domain (m)	Mesh size	Sample parcel speed (min) (m s⁻¹)	Sample parcel speed (max) (m s⁻¹)	Parcel transit time at min speed (h)	Parcel transit time at max speed (h)
γ {	2.5	6.25	50	50 m	1	10	0.1	1
β {	25	625	50	500 m	5	10	1.0	2
α {	250	62.5×10^3	H[1]*	5 km	10	50	2.0	10
	2500	62.5×10^6	H[1]*	50 km	10	50	20.0	100

*The depth of the atmospheric boundary layer.

involved in simply servicing the raingauges in a mesoscale network. Clearly automation is essential not only nationally for measurements of subsynoptic-scale features but even regionally if the need is for measurements of γ-scale features or events.

Anthes (1979) compiled a useful scale analysis relevant to regional-scale, air-quality modeling; Table 5.2 is in part an extension of his analysis.

The lower end of the meso-γ scale corresponds essentially to the scale plumes and convective clouds. Although the spatial and temporal scales of potential interest are within the resolution of acoustic sounding systems or high resolution FM-CW (frequency modulated-continuous wave) radars, the grid resolution is significantly finer than the spatial resolution of a typical clear-air radar ($\simeq$100 to 900 m) and also exceeds the radial resolution ($c\tau/2$, τ = pulse duration, c = speed of light) of modern pulsed Doppler radars. Surface or satellite-based remote sensing systems might be used for measurements of precipitation, wind, and state parameters in a meso-γ domain. But to make *in situ* measurements of corresponding resolution would require 2500 stations. In short, measurements, on a scale that could be resolved in a numerical model for the meso-γ scale are unlikely ever to be compiled, at least with traditional *in situ* methods. On the other hand, using a system such as Doppler sodar to probe the overlying boundary layer, one can obtain continuous time-height records with 50 m or better vertical resolution at those grid points where a system such as PAM II is installed for surface measurements, for the cost of about 100 radiosondes. Many scientists who are working with the remote sensing systems regard the following as a principal challenge for the community of mesoscale numerical modelers:

Design numerical diagnostic and prognostic models for optimal use of measurements from instrument systems that have temporal resolution of the order of the model time steps, but that are only sparsely distributed through the domain of interest.

On the meso-γ scale the mesh sizes Δs are such that contemporary measurement systems, such as the various radars, can define sub-grid-scale structure. One should not, however, confuse parcel transit time with the required sampling interval. To achieve approximately the same temporal as spatial resolution, the sampling interval should be only a fraction of the minimum parcel transit time. Six to 12 minutes is short enough that transient events such as gust fronts will be detected but long enough that means and standard deviations over the interval can be evaluated from data gathered at 1 min intervals. (Comparison of individual values with the mean and standard deviation is one useful method for rejecting spurious measurements.)

Although the spatial separation between surface stations for a meso-β-scale network might be 5 km, not an unreasonable sounding distance, 2500 stations would still be required to cover the entire area of interest. By increasing Δs to 10 km, as is suggested for an MCS (SCPPD, 1985), that number could be reduced to 625, still a factor of 3 larger than the compromise suggested for the STORM-Central meso-β array (Table 6.2 in SCPPD). If the phenomenon of interest (e.g., a foehn or regional-scale pollution) is linked to the underlying surface, the numbers might be improved by using the equivalent of a telescoping grid, which could exponentially decrease the station density in proportion to distance from the location of primary interest. However, Bartels and Skradski (1984) showed that a telescoping grid is not an option if the research or operational objective is analysis and prediction of widely geographically distributed MCSs.

The logistics and cost of deploying such a surface network are prohibitive. Although 625 stations with 20 channels of information recording each minute produce 7.5×10^5 data points each hour, this corresponds to less than a minute's worth of data for many modern remote sensing systems.

For the maximum length of the meso-α scale some of the numbers are misleading. Note first that a Δs of 50 km corresponds to the grid scale typically used in mesoscale models. Smaller terrain features are thus completely lost in such a model. For example, features such as terrain must be smoothed. In the eastern United States, the Appalachian Mountain-Valley complex is modeled as a large-scale smooth hill (see Fig. 5.2) instead of the complicated hollows and hills that are associated with phenomena such as damming and orographic convection.

A simplistic time-scale analysis for the meso-α scale, based on parcel transport, will be erroneous since the indicated 20 h minimum exceeds by a factor of 20 to 40 the time scale for important convective processes. It is convection and the evolution of convective systems that must be interpreted. Hence, for the meso-α scale, the time scale for both modeling and measurements must be determined on the basis of vertical rather than horizontal transport (Liu *et al.*, 1974).

In order to avoid aliasing (the generation of fictitious low frequencies) when a continuous signal is sampled, it is necessary to have at least two samples in the highest frequency present in the signal. A prudent systems engineer would prefer three. For application to MCSs, if significant evolution occurs in a 45 min period, at least three samples of the relevant variables

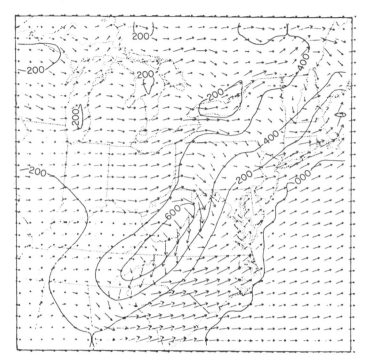

Figure 5.2. Contours of terrain elevation that have been smoothed and plotted at 200 m intervals, and superimposed wind field computed with a mesoscale model. (From Warner *et al.*, 1978.)

at the required spatial density should be acquired. In short, the maximum appropriate sampling interval should not exceed 15 min. Thus, even though it was possible to decrease the spatial density of observations required for meso-α analyses, since the evolution of convective systems occurs on the meso-β (or smaller) scale, the temporal resolution of measurements should be of the same order as the time steps currently used in mesoscale models (a few minutes). Thus, from a pragmatic, signal-processing point of view the use of radiosondes or dropsondes at 3 h or 90 min intervals is, at best, a limited-return investment and, at worst, could result in misleading measurements due to undersampling. Only instrumented aircraft or remote sensing systems can provide observations of the required temporal resolution.

In summary, the logistics and economics of *in situ* mesoscale networks are such that it will probably be possible to provide measurements "all the time" at only a few of the gridpoints.

5.5. Fluctuations and Trends of Individual Variables

It is useful to think that the changes of a meteorological variable constitute a signal. Many objective methods can be used to characterize or parameterize a signal, for example, evaluation of its spectrum. But sophisticated statistical analyses are often not practical to implement on a real-time basis. In effect, the event has passed before the necessary data base is complete. Even simple parameterizations of means, perturbations about the means, and tendencies for variables of mesoscale interest must be carefully performed.

The average or expected value of a random quantity x with probability distribution function $y(x)$ is

$$\overline{X} = \int_{-\infty}^{\infty} x[y(x)]dx \quad .$$

To specify the character of the departures from the mean, one normally estimates the variance:

$$\mathrm{Var}\underline{X} = \int_{-\infty}^{\infty} (x - \overline{X})^2[y(x)]dx \quad .$$

Both of these simple, widely used parameterizations are well defined only for a random signal x. Mesoscale signals are neither random nor determinate. Pressure changes over 15 min to 24 h periods are not randomly distributed about the mean; temperatures recorded over the course of a day at 2 m are not normally randomly distributed about the mean, nor is the depth of the atmospheric boundary layer; and so forth. Furthermore, because mesoscale features are embedded in larger synoptic-scale systems, data from arrays of instruments are also unlikely to be randomly distributed. The variance of an ensemble of pressure or temperature values may principally reflect, for example, the mean synoptic-scale gradients rather than contributions from mesoscale structure.

For a one-dimensional signal (e.g., voltage as a function of time) the standard instrumentation practice for estimating fluctuations is to use either an analog or digital low-pass filter to approximate the running mean. Signal fluctuations, then, correspond to the instantaneous differences between the voltage and the mean. Selection of the time constant for such a filter requires knowledge of the statistical (spectral) character of the signal. To define mesoscale perturbations, the analogous two-dimensional (2-D) estimate can be accomplished by using procedures such as linear regression or bivariate splines applied to the available synoptic-scale data to establish the mean distribution of the parameter of interest. Mesoscale perturbations would then be defined as the difference between the measured and the synoptically filtered data values. Maddox (1980), improving upon earlier work by Barnes (1964) and Doswell (1977), presented a useful technique for low-pass filtering mesoscale data sets. A simpler scheme, which can be used only for real-time analysis of data from a mesoscale network, is described by Wilson and Carpenter (1983).

Even the definition of trends or tendencies is not generally easy. The period over which differences should be calculated is phenomenon dependent. Consider, for example, the pressure trace that includes a barogram vee (Fig. 5.3). The trend, which could be calculated for any specified interval using linear regression or a low-pass filter, is very well defined from 0200 to 1400 on Tuesday. But use of that trend to predict a pressure at $\simeq$1600 would have resulted in about a 10 mb underestimate of the pressure actually

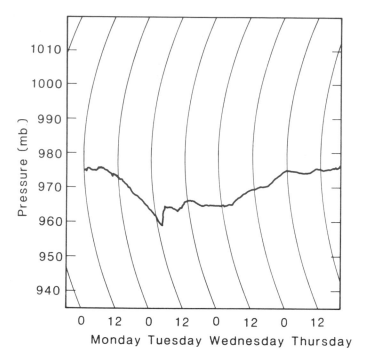

Figure 5.3. Barograph trace including a barogram vee. Record is from about 1300 LDT, on 7 May, through about 0700, 11 May 1984, at State College, Pa.

recorded at 1600. Such records as well as those obtained from windvanes at night, acoustic sounders, FM-CW and profiling radars, aircraft, and satellite camera images are continual reminders of the existence of discontinuities, both temporal and spatial, in atmospheric structure. Cubic and bivariate spline techniques are among those widely used to separate signals from noise or means from turbulence when records with characteristics of a turbulence time series are being objectively processed.

5.6. Specifications and Limitations of Surface-Based Sensors and Systems

Among the most used and misused terms in meteorology today are resolution, sensitivity, accuracy, and precision. The frequent misuse of these terms is not surprising, for even advertisements from instrument manufacturers and distributors use the terms in erroneous and nonsensical statements.

The resolution of an instrument or system is the smallest unambiguous change in a measured variable that will result in a detectable change at its indicator or output. If a 0.1°C change in temperature is required to produce readable change on a laboratory mercury-in-glass thermometer, its resolution is 0.1°C; if the digital display of an electronic thermometer is in increments of 1°F, then that is its resolution.

Sensitivity is the measured unit of change of output for a unit change of input. For a linear device, Output = Constant + Sensitivity × Input. For a nonlinear device the sensitivity will be a function of the input (or a function of another external variable, such as the temperature coefficient of a barometer or pyrheliometer). For either a linear or nonlinear device the calibration curve specifies the input-output relationship.

Accuracy is neither resolution nor sensitivity. Accuracy is the exactness with which an instrument or system will measure a variable with respect to an internationally accepted standard. Take, for example, a laboratory standard thermometer that is accurate to within 0.02°C absolute, "Traceable to the National Bureau of Standards." The accuracy of another thermometer referenced to it cannot be determined to any better than that value, even if the second thermometer has a resolution better than 0.0001°C.

The precision of an instrument is not to be confused with its accuracy. Precision is simply defined as an instrument's ability to produce the same result for a given, repeated measurement, without respect to an absolute standard. Thus, a digital voltmeter could precisely (repeatedly) determine a dc voltage to be 1.360 ± 0.001 V. However, if calibration against a voltage standard detected a 0.010 V offset or bias, a correction would have to be applied for the instrument to be accurate to within a millivolt.

If the only measurements of concern were those obtained with a single system such as PAM II or ASOS, the accuracy of the measurements might be of secondary interest. As long as the measurements were not compared with or used in conjunction with data from other sources, the uniformity of resolution and sensitivity of the individual units in the network and, finally, the operational precision of the units once in place, would be of primary concern. It is also important to recognize that both the natural inhomogeneity of the atmosphere and the underlying surface and the temporal fluctuations of the sensed variables resulting from turbulence can easily contribute as much as 2 orders of magnitude to the uncertainty of a given measurement. Temperatures in the turbulent boundary layer typically fluctuate (rms) about 0.3°–1.3°C. A good rule of thumb for the standard deviation (σ) of the velocity fluctuations is 30% of mean, but the magnitude of σ is strongly dependent upon the stability and character of the surface roughness, both at the site and upwind of it.

Various national weather and environmental monitoring services and, further, the World Meteorological Organization, have established detailed specifications for the installation of meteorological sensors (see, e.g., WMO, 1971). Consequently, most meteorological sensors are at least initially properly installed. But ordinarily, a sensor's environment is not controlled. Measurements from a sensor operated for many years in the same location may include short-term uncertainties resulting from natural biometeorological processes, such as feedback by evapotranspiration to surface temperatures, or longer term variations resulting from changes in land use (such as the change at O'Hare International Airport from rural to industrial since the airport was opened). Finally, regardless of the standards book, the quality of regular maintenance (and thus the quality of measurements) is highly dependent upon station personnel.

A measurement is not always representative of the surrounding environment. For example, on many summer evenings in central Pennsylvania the temperature at 2 m height next to the author's home, and at the same height above the surface but 75 m distant in the hollow below the house, will differ by about 10°F as a consequence of drainage flow. Both temperatures are right but neither is likely to be representative of the surrounding α, β, or γ mesoscales. Through complex terrain experiments, much experience has been gained in the past few years regarding the difficulties of siting instruments in fine-mesh networks. However, much remains to be learned.

As more fine-scale networks are established it should not be surprising if a number of individual stations appear to be climatologically anomalous. In the context of the network they may appear to be anomalous, but in reality they will not be. They will simply be reflecting the natural variability of the microclimate, a variability that biologists, phenologists, ecologists, and at least a few meteorologists have known about for many decades.

5.7 Requirements for Surface Measurements

For synoptic analysis purposes, surface measurements and observations have been used principally to define the state of the atmosphere and the position at the surface of the weather systems. The recorded measurements used by forecasters to prepare surface weather maps have included pressure, wind speed and direction, temperature, and humidity.

The preparation of more detailed and frequent mesoscale surface weather depictions, whether by hand or machine, will require that comparable "state" measurements be made and transmitted to the meteorologist. But the spatial and temporal nature of mesoscale weather phenomena is such that they not only depend upon day-to-day changes of the general state of the synoptic-scale environment but also critically depend upon the ongoing processes at the Earth's surface. Consequently, in order to forecast the evolution of a mesoscale system it will be essential to include measurements of those variables that are related to diurnal forcing and the effects of changes in terrain and land/water boundaries, and substantive changes in land use. Diurnal forcing is most strongly dependent upon the absorption and emission of solar and infrared radiation at the Earth's surface and the partitioning of the absorbed radiation between the evaporative, and thermal atmospheric and soil energy fluxes. It is convenient to think of the various energy fluxes as process variables for they largely control the diurnal evolution of the atmospheric boundary layer and associated convective processes.

Table 5.3 summarizes a wide variety of surface measurements. It is not a list of instruments and measurements or a tabulation of the technical specifications that are normally found in a system requirements document. Many currently available instrumentation systems can provide reliable, high-quality data. For the mesoscale forecaster, measurement adequacy will be limited not by the performance of the electronic measurement system but rather by the uncertainty about whether the available data are representative of the region of interest.

Note in Table 5.3 that none of the current operational or planned automated meteorological measurement systems includes sensors for flux variables other than solar radiation. Operational deployment of flux sensors

Table 5.3. Summary of surface-based measurements of meteorological variables for mesoscale application

Variable	Method operational	Method automated	Research status	Comments
Atmospheric State:				
Temperature	Yes	Yes	—	Site-limited representativeness
Humidity	Yes	Yes	—	Adequate sensors a continuing problem
Pressure	Yes	Yes	—	Reduction to sea leavel probably not optimum normalization
Wind speed	Yes	Yes	—	Siting critical
Wind direction	Yes	Yes	—	Siting critical
Visibility	Yes	Yes	—	Often site dependent
Present Weather:				
Cloud coverage	Almost	Almost	In process	Satellite coverage probably best
Cloud height	Yes	Yes		
Liquid precipitation	Yes	Yes	—	
Frozen precipitation	Almost	Almost		
Thunderstorm	Yes	Yes		Lightning detection networks useful
Process Related:				
Pressure tendency	Yes	Yes		
Radiation:				
Solar	Yes	Yes	—	Not widely deployed
IR	No	No	Available	
Net	No	Easily	Widely used	Ought to be deployed
Latent heat flux	No	Easily	Used	Improved hygrometric sensor needed
Sensible heat flux	No	Easily	Widely used	Ought to be deployed
Soil heat flux	No	Easily	Used	Probably not essential
Other:				
Max/Min temp	Yes	Yes	—	Site dependent
Wind gust and peak	Yes	Yes	—	Site dependent
Wind variance	Yes	Yes	—	Of aviation and air pollution interest
Precipitation accumulation	Yes	Yes	—	
Snow depth	—	—	Sensor needed	

could be costly, but numerical experiments certainly ought to be conducted to establish whether the lack of such measurements has the potential of fundamentally limiting the quality of selected mesoscale forecasts.

REFERENCES

Anthes, R. A., 1979: Meteorological aspects of regional-scale air quality modeling. *Advances in Environmental Science and Engineering*, Vol. I. J. R. Pfafflin and E. N. Ziegler (Eds.), Gordon and Breach, New York, 3–49.

Barnes, S. L., 1964: A technique for maximizing details in numerical weather map analysis. *J. Appl. Meteor.*, **3**, 396–409.

Bartels, D., and J. Skradski, 1984: Climatology of mesoscale convective systems and severe weather for the central United States. Appendix A to STORM-Central Phase, Preliminary Program Design, National Center for Atmospheric Research, Boulder, Colo.

Brock, F. V., and G. H. Saum (1983): Portable Automated Mesonet II. Proceedings, 5th Symposium on Meteorological Observations and Instrumentation, Toronto, Canada. American Meteorological Society, Boston, 314–320.

Doswell, C. A., III, 1977: Obtaining meteorologically significant divergence fields through the filtering property of objective analysis. *Mon. Wea. Rev.*, **105**, 885–892.

Hicks, B. B., D. D. Baldocchi, R. P. Hosker, Jr., B. A. Hutchison, D. R. Matt, R. T. McMillen, and L. C. Satterfield, 1985: On the use of monitored air concentrations to infer dry deposition. NOAA Tech. Memo. ERL–ARL–141, Environmental Research Laboratories (NTIS#PB86158409/AS), 84 pp.

Liu, M-K, J. H. Seinfeld, and P. M. Roth, 1974: Assessment of the validity of airshed models. Proceedings, 5th Meeting, Expert Panel on Air Pollution Modeling, No. 35, Adnish AEK, Roskilde, Denmark.

Maddox, R. A., 1980: An objective technique for separating macroscale and mesoscale features in meteorological data. *Mon. Wea. Rev.*, **108**, 1108–1121.

SCPPD (STORM-Central Phase, Preliminary Program Design), 1984: National Center for Atmospheric Research, Boulder, Colo., 148 pp.

Warner, T. T., R. A. Anthes, and A. L. McNab, 1978: Numerical simulations with a three-dimensional mesoscale model. *Mon. Wea. Rev.*., **106**, 1079–1099.

Wilson, F. W., Jr., and M. J. Carpenter, 1983: Portable automated mesonet: Real-time display capability. Proceedigs, 5th Symposium on Meteorological Observations and Instrumentation, Toronto, Canada, American Meteorological Society, Boston, 321–325.

WMO, 1971: *Guide to Meteorological Instrument and Observing Practices*. WMO-No. 8TP.3, Geneva.

CHAPTER 6

Principles of Radar

Donald Burgess
Peter S. Ray

6.1. Introduction

Radar was developed during World War II to identify and track warships and aircraft. In this application rain sometimes obscured the target and means were sought to mitigate rain effects and distinguish the signal from weather targets. However, in meteorology, characterizing and tracking storms is the primary use of radar. Many functions of operational meteorology depend on conventional C-band (5 cm wavelength) or S-band (10 cm) incoherent radars. The letters associated with the frequencies have their origin in wartime security needs—there is no logical designation. Their location in the frequency spectrum is marked in Fig. 6.1. Meteorologists tend to refer to the microwave portion of the electromagnetic spectrum in terms of wavelength. Since the 1970s and the implementation of high speed digital data processing to indicate the air motion toward or away from the radar, Doppler (or coherent) radar has played an increasingly large role in meteorology. It is certain to become the standard as new applications with increasingly sophisticated systems are developed. This chapter outlines some of the important principles and applications of radar. More background and detail can be found in texts by Skolnik (1970), Battan (1973), Doviak *et al.* (1986), and Doviak and Zrnic (1984).

6.2. Operating Principles and Characteristics

6.2.1. Design and Operation Basics

Electromagnetic waves can be described by their amplitude, phase, and polarization. Weather radar is based on the fact that electromagnetic waves interact with hydrometeors as they propagate through the atmosphere. When they encounter hydrometeors, a small fraction of their energy is scattered by the particles and a detectable amount of power is backscattered to a receiver. There are many radar designs; here only main elements are examined.

A basic radar consists of a transmitter to produce power at a known frequency; an antenna to focus the transmitted waves to a beam about 1°–2° wide and to receive the fraction of the power backscattered from the

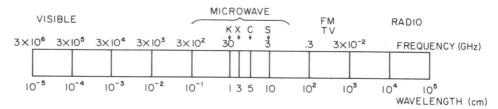

Figure 6.1. Frequency spectrum from visible to radio frequencies and corresponding scale in wavelength. The microwave region, important to meteorological radar, is highlighted. Most meteorological radars operate between 3 and 10 cm wavelength.

targets; a receiver to detect, amplify, and convert the reflected microwave signal into a low frequency signal; and some type of an indicator on which to display the detected signal. The principal difference between Doppler and conventional (incoherent) radar is that conventional radar does not provide phase information; it measures only received power from the amplitude of the received signal.

Most meteorological radars are pulse radars, the only type considered here. Pulsed radars are governed by a very accurate clock. This clock is used to synchronize the system, generating a train of pulses at the desired PRF (pulse repetition frequency). A typical PRF is about 1000 Hz, although this parameter is widely varied.

A conventional or incoherent radar uses an oscillator tube called a magnetron to generate high power pulses (~100 kW) at a prescribed frequency. The phase of each pulse from magnetrons is random. To measure the difference between the returned signal's phase and that of the transmitted signal, most pulsed Doppler radars use a power klystron amplifier tube to amplify (2 MW peak power) the signal to be transmitted.

The advantage of the klystron over the magnetron is that, because it serves only as an amplifier, it maintains phase coherence (or the same phase) for many pulses. A disadvantage is its relatively high cost, weight, and size. However, magnetrons can be used in Doppler radars by measuring the phase of each pulse and removing the effects of phase changes in the receiver. As a result, only first-trip echoes (see Sec. 6.2.5) give Doppler information unless phase of more than one magnetron pulse is used for processing received signals. Magnetrons do not produce as high output pulse power as klystrons.

The klystron amplifier is modulated by the pulse modulator to produce a pulse of the desired duration τ, typically 1 μs, corresponding to a packet of energy effectively 150 m long. (The velocity of propagation is 300 m μs^{-1}, but the 150 m effective length results because the path is from the radar to a target and back.) This is called the pulse volume depth. The high-power pulse goes through the T/R (transmit, receive) switch (or duplexer). The purpose of this device is to protect the receiver from damage by the high-power transmitted pulse. It connects the antenna to the transmitter when it is "on" and to the receiver when the transmitter is "off."

To detect returns at various ranges from the radar, the returning signals are gated or sampled periodically, usually about every microsecond, to obtain information about every 150 m in range. This sampling can go on until it is time to transmit the next pulse. If the PRF is 1000, then it is possible to acquire 1000 1 μs samples in between pulses. A sample point in time (which corresponds to a distance from the radar) is frequently called a range gate. The numbers used in the example below correspond to sampling in range a total distance of 150 km.

If f_d represents the frequency of the Doppler shift of the returned signal, the received signal has the frequency $f_t + f_d$ where f_t is the transmitted frequency. When the received signal is mixed with a sample of the transmitted frequency, the result is two frequencies, the sum and the difference. The difference frequency signal is used in the receiver.

When $f_d > 1/\tau$, the Doppler signal may easily be described from the information in a single pulse. For example, if the frequency were sufficiently high, you could simply count the number of zero crossings and $f_d \approx$ (number of crossings)$/(2\tau)$. But for an S-band radar with a pulse width of 1 μs, this implies velocities greater than 10^4 m s^{-1} $\sim$10 km s^{-1} $\sim$30,000 miles h^{-1}. In the case of microwave radar, the Doppler frequency shifts for meteorological targets are much lower than $1/\tau$ and the frequency change is detected only by examining the return of many pulses. The successive samples at each range gate form a discrete time series from which the velocity spectrum may be computed by using discrete Fourier transform methods.

6.2.2. Scattering

When a particle intercepts a radio wave, some of the energy is absorbed and some is scattered. The amount scattered in the backward direction depends upon the shape and dielectrical property of the scatterer (refractive index), and the ratio of the wavelength to the size of the scatterer. We are interested in the energy that is redirected back toward the sources of the incident radiation. Scattering mechanisms are not completely understood but include specular reflection from the front surface, waves that appear to be defracted around the sphere, and waves that, if not attenuated, undergo internal reflections before emerging in the backward direction toward the receiver. Since no meteorological targets scatter isotropically, it is convenient to define the backscattering cross section σ as the equivalent area required for an isotropic scatterer to return to a receiver the amount of power actually received. Regardless of the mechanisms, the complete solution for the backscattering cross section for spheres is expressed by the theory of Mie (1908) which takes the form of a series expansion:

$$\sigma = \frac{\pi D^2}{4\alpha^2} \left| \sum_{n=1}^{\infty} (-1)^n (2n+1)(a_n - b_n) \right|^2 , \tag{6.1}$$

where D is the drop diameter, $\alpha = \pi D/\lambda$ and is called the electrical size; λ is the incident radiation wavelength; and a_n and b_n are coefficients of the scattered field involving spherical Bessel and Hankel functions. These are functions of the scattering angle, electrical size, and complex refractive

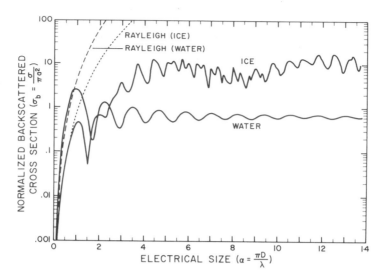

Figure 6.2. Backscatter cross sections for ice and water at 0°C. The Rayleigh approximation is indicated for both cross sections. The electrical size has been determined with a constant wavelength of 10 cm.

index. Among others, Stratton (1930) and Kerr (1951) further studied and extended scattering theory. The complex index of refraction is written

$$m = n - ik , \tag{6.2}$$

where n is the ordinary refractive index and k the absorption coefficient. The complex index of refraction of water varies with temperature and wavelength.

The normalized (with respect to drop cross-section area) backscatter cross section,

$$\sigma_b = \frac{4\sigma}{\pi D^2} , \tag{6.3}$$

has been calculated for a variety of refractive indices and sizes by numerous investigators. An example is shown in Fig. 6.2 for ice and water. Note that σ_b increases monotonically up to electrical sizes of about unity. Similar calculations have been made for more complex scatterers such as water-coated ice, nonspherical shapes, and composites of different materials. In each case the interpretation is somewhat more complex than the cases treated here. Reviews of scattering from complex meteorological targets can be found in Battan (1973) and Atlas (1964). For large enough electrical size the backscattering cross section approaches the geometric cross section, which is the limiting value of reflection off a flat plate. For water, which is fairly conductive at these wavelengths ($\sim$10 cm), this occurs at an electrical size near 20; for ice the electrical size must be greater than 1000. Because ice is more nearly a dielectric, there are many weakly attenuated internally reflected waves that complicate the cross-section appearance.

The region where the drop diameter is small compared with the wavelength is frequently called the Rayleigh or dipole region and lends itself to a simplified evaluation of σ. For $\alpha << 1$, terms greater than α^5 in the expansion (6.1) are ignored and only the b coefficient term in the summation contributes. By substitution into (6.3) the backscattered cross section can be written

$$\sigma = \frac{\lambda^2}{\pi}\alpha^6 \left|\frac{m^2-1}{m^2+2}\right|^2 \tag{6.4}$$

or

$$\sigma = \frac{\pi^5}{\lambda^4}\left|\frac{m^2-1}{m^2+2}\right|^2 D^6 . \tag{6.5}$$

Thus, the backscattered energy for small particles is proportional to the 6th power of the particle's diameter. This is how one would expect a dipole to radiate in a changing electric field.

6.2.3. Radar Range Equation

This section outlines the fundamental considerations necessary to calculate the returned power P_r from knowledge of the transmitted power P_t, the radar system, the range, and the characteristics of the target. A target of area A_t on the surface of a sphere of radius r intercepts an amount of power equal to $A_t P_t/4\pi r^2$. If the antenna focuses the energy, increasing the energy density one would get from an isotropic radiator by an amount G, and if the target radiates isotropically, the power received at the antenna will be

$$P_r = \frac{P_t G^2 \lambda^2 A_t \ell}{(4\pi)^3 r^4} , \tag{6.6}$$

where the effective receiving area of an antenna is $G\lambda^2/4\pi$ (Silver, 1951). All losses are contained in the term ℓ, including those along the two-way propagation path (including the radome). For meteorological targets, the average power received is the sum of the contributions of the individual particles, or

$$\overline{P}_r = \frac{P_t G^2 \lambda^2}{(4\pi)^3 r^4} \sum_{i=1}^n \sigma_i . \tag{6.7}$$

This is the received power averaged over many independent realizations of the relative position of the scatterers. Even if the radar does not yield phase information, the location of the individual scatterers determines the composite phase of the backscattered signal. The σ_i can be derived from computations similar to those in Sec. 6.2.2. Others, particularly Probert-Jones (1962), extended and corrected this relationship for beam shape, etc. The resulting equation is

$$\overline{P}_r = \frac{P_t G^2 \lambda^2 \theta\phi h\ell}{512(2\ln 2)\pi^2 r^2}\frac{1}{\Delta v}\sum_{\text{vol}}\sigma_i , \tag{6.8}$$

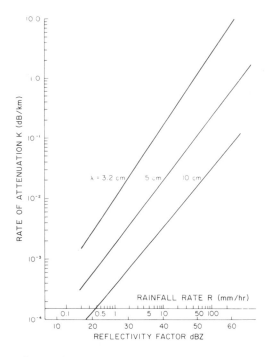

Figure 6.3. Rate of attenuation (one-way) for propagation through rain showers versus rainfall rate and reflectivity factor. The Laws and Parsons drop size distribution is assumed. (From Doviak and Zrnic, 1984.)

where θ and ϕ are the horizontal and vertical beamwidths in radians, and h is the pulse length. Beamwidths are defined as the angular distance where the transmitted power density has fallen to one-half its peak value.

In using (6.5) for σ_i in (6.8) it is common to replace $1/\Delta v \sum_{\text{vol}} D_i^6$ by the reflectivity factor Z. This is to be distinguished from the reflectivity, which is defined by $\sum_{\text{vol}} \sigma_i$ and which differs in the Rayleigh region by the factor

$$\frac{\pi^5}{\lambda^4} \left| \frac{m^2 - 1}{m^2 + 2} \right|^2 . \tag{6.9}$$

If the scattering is not well described by Rayleigh scattering, then the effective reflectivity factor Z_e is used in place of Z and represents the reflectivity factor under the assumption of Rayleigh scattering. Accepted units of reflectivity are millimeters to the 6th power per cubic meter.

Typical values of Z in storms range from 10^2 to 10^6 $(\text{mm})^6$ m^{-3}. It is convenient to express those numbers, and hence reflectivity, in decibels (ten times the logarithm to the base 10). Thus a reflectivity of 10^5 $(\text{mm})^6$ m^{-3} becomes 50 dBZ.

As the transmitted or backscattered radiation passes through the air, cloud, rain, snow, or hail, some of the energy is attenuated, i.e., absorbed or scattered in other directions. These losses are usually expressed in decibels per kilometer. The effect is most pronounced for wavelengths shorter than 10 cm. Battan (1973) contains a review of the theoretical and empirical estimates of attenuation. Figure 6.3 illustrates typical values of one-way attenuation for three wavelengths. The attenuation at 3.2 cm wavelength is about 6 times greater than at 5 cm and 40 times greater than at 10 cm wavelength. Attenuation estimates for mixed phase and irregularly shaped scatterers are more complicated to derive. However, the net result is that attenuation increases dramatically as wavelengths decrease from 10 cm. Attenuation, even for C-band radars, can approach 1 dB km^{-1} and can be >1 dB km^{-1} for X-band radar systems. It is not uncommon in storms containing regions of high reflectivity extending over several tens of kilometers that returned signals are reduced by more than 10 dB for C- or X-band radars.

6.2.4. Doppler Radar Principles

In Sec. 6.2.1 the basic operation of a pulsed Doppler radar was described. The result of phase detection of the backscattered signal, when the transmitted signal is used as a reference, is the bipolar video, or the I and Q (or inphase and quadrature phase) components. The radial wind speed (but not direction) is contained in either the I or Q component. If the Doppler shift is positive (indicating motion toward the radar), Q will lead I by $\pi/2$; if negative, Q will lag I by $\pi/2$. A succession of I and Q pairs from the same range form a time-series sample of the Doppler shifted signal. Using standard techniques of Fourier analysis it is possible to analyze this series as a power density spectrum (Blackman and Tukey, 1958) or "Doppler spectrum" $S(V)$, of velocities toward or away from the radar. In general, the longer the sampling time, the more samples, and the more resolution available in the Doppler spectrum.

The integral or area under the spectrum is the reflectivity and the spectrum's zeroth moment. The mean velocity $\overline{V}$ can be found from the spectrum's first moment:

$$\overline{V} = f_1(V) = \frac{\int\limits_{-\infty}^{\infty} V S(V) dV}{\int\limits_{-\infty}^{\infty} S(V) dV} , \tag{6.10a}$$

The general expression for higher order moments ($n \geq 2$) of the Doppler spectrum is

$$f_n(V) = \frac{\int\limits_{-\infty}^{\infty} (V - \overline{V})^n S(V) dV}{\int\limits_{-\infty}^{\infty} S(V) dV} , \tag{6.10b}$$

where n is the moment of the spectrum. The spectrum variance or second moment corresponds to $n = 2$.

The spectrum width can be related to (1) spread of the scatterers' terminal fall velocities (especially with the antenna at vertical incidence), (2) turbulence in the sampling volume (3) shear of the wind along or across the beam, and (4) antenna rotation rate. The use of the spectrum width or higher spectral moments in this summary is not stressed other than for its use in turbulence detection.

In the early 1970s Doppler radar signal processing was revolutionized by the use of autocovariance or pulse-pair processing algorithms that allow rapid and computationally efficient calculations of $\overline{V}$. The advantages are that with relatively modest hardware investment it is possible to rapidly obtain the mean velocity (and variance, with an additional investment) for large numbers of gates (e.g., 1000) along the radar beam. As the antenna turns, the radial wind component throughout the storm or target area is mapped. More recent advances in electronics have made the Fourier transform method competitive with the pulse-pair method. In the future the unique advantages of each will determine which should be applied in a given situation.

6.2.5. Velocity-Range-Wavelength Relationships

The pulses of transmitted energy are separated in space by a distance R, which is related to the Pulse Repetition Frequency (PRF) by

$$R = \frac{c}{\text{PRF}}, \tag{6.11}$$

where c is the radio wave propagation speed (3×10^8 m s^{-1}). When a pulse travels to and scatters from a cloud volume, a portion of the signal is returned to the radar. The maximum time available is time between pulses, or $1/\text{PRF}$. The pulse must travel out to and back from the maximum observable range in this time. This distance, $R_{\max}$, is only half the distance that light can travel in the time $1/\text{PRF}$. That is,

$$R_{\max} = \frac{c}{2\,\text{PRF}}. \tag{6.12}$$

For a PRF of 1000 s^{-1}, $R_{\max}$ corresponds to an unambiguous range of 150 km. However, an echo at a range of $R_{\max} + nc/(2\,\text{PRF})$ will be received by the radar at the same time, where n is the integer number of a previous pulse. All echoes at ranges greater than $R_{\max}$ therefore appear to be located at ranges less than $R_{\max}$ since the radar has no way of distinguishing between the return from the most recent pulse and more distant returns from a prior pulse. In a radar display, these distant echoes may be distinguished by their elongated appearance since their angular width is unchanged when displayed at their apparent range. This is illustrated in Fig. 6.4. The radar assumes it is receiving a return from the most recent pulse and indicates the echo position accordingly. It is possible to distinguish range echoes from those within "the first trip" by changing the PRF. Range echoes will appear to move as the PRF changes whereas the echoes within the first trip will remain stationary.

A fundamental sampling theorem states that to measure a frequency f_d it is necessary to sample at a frequency of at least $2f_d$. The sampling rate is the PRF; thus

$$2f_d = \text{PRF}. \tag{6.13}$$

Since the Doppler theorem states that the observed velocity is related to the frequency shift f by

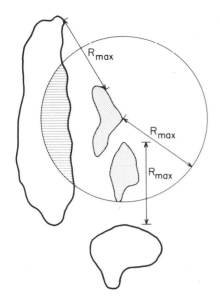

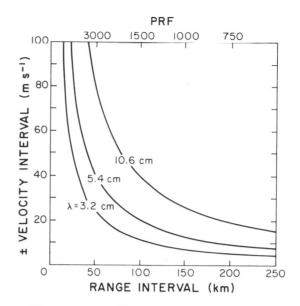

Figure 6.4. Schematic illustration of range-ambiguous echo. The heavy line defines the extent of the storm. The PRF-determined maximum range is labeled R_{max}. The echo area denoted by hatching is in its correct position. However, the radar processes the return so that echoes are displayed in the hatched and stippled areas. The correct position of the stippled area is radially outward a distance R_{max}.

Figure 6.5. The relationship between unambiguous velocity, range, wavelength, and PRF for three meteorologically important wavelengths.

$$V = -\frac{f\lambda}{2}, \tag{6.14}$$

it follows that the maximum unambiguous Doppler velocity, which defines the Nyquist interval, is given by

$$V_{max} = \pm\frac{(\text{PRF})\lambda}{4}. \tag{6.15}$$

Thus, an S-band (10 cm) Doppler radar operating at a PRF of 1000 s^{-1} has a maximum unambiguous velocity or Nyquist interval of ±25 m s^{-1}. An X-band radar (3 cm) operating at the same PRF would have an unambiguous velocity of ±8 m s^{-1}. All velocities exceeding these bounds are aliased and would be indistinguishable from the ones in the fundamental Nyquist interval if conventional Doppler processing is used.

Increasing the PRF to extend V_{max} decreases R_{max}. This is shown by combining (6.12) and (6.15) so that the relation

$$V_{max}R_{max} = \frac{\pm\lambda c}{8} \tag{6.16}$$

is obtained. Equation (6.16) is plotted in Fig. 6.5. It is clear that longer wavelength radars have the advantage that for a given PRF they enjoy larger unambiguous velocities. Another important advantage of longer wavelength radar is the mitigation of attenuation effects. It should be mentioned that there are a variety of techniques based on signal design or polarization switching which may also mitigate these effects in certain circumstances. Disadvantages of longer wavelength are less ability to detect very small particles, generally worse ground clutter problems, the requirement of larger antenna, and higher cost of the radar system.

6.3. Uses of Reflectivity Information

6.3.1. Thunderstorm Structure

The three-dimensional shapes of radar echoes relate to the configuration of rising and descending air currents and their interactions with the ambient air flow. The configuration of echoes changes with time as the drafts change intensity and become laterally aligned, and the ascending cyclonic and descending anticyclonic circulations become organized. This organization encourages the regenerative precipitation process within the storm; rates of rise and fall of air control the conversion of cloud moisture to rain and affect suspension and growth of hail aloft. These processes are intrinsic to the stability and duration of the radar echo. Radar reflectivity can be used to infer updraft strength and associated phenomena (tornado, wind, hail, turbulence, etc.) by observing storm structure. The reflectivity structure of ordinary (nonsevere) thunderstorms and severe thunderstorms can be differentiated (see Lemon, 1980; Grebe, 1982).

The Ordinary Thunderstorm

Structure and evolution of nonsevere thunderstorms were studied extensively in the Thunderstorm Project during the late 1940s. In that study, the fundamental kinematic and thermodynamic building blocks of thunderstorms were described as cells, relatively small regions of rather strong vertical air motion (Byers and Braham, 1949). It is also common to describe the precipitation produced by the updraft as a cell—a maximum of reflectivity factor surrounded by closed intensity contours on a radar display. Most thunderstorms consist of several cells in different stages of development and decay. The cell is generally first seen on radar between 3 and 6 km altitude. The mature stage is noted by the tallest tops and highest reflectivities. Finally, the dissipating stage begins as the downdraft, formed during the mature stage, spreads through the precipitation area. The end of the storm's life is the fallout of the precipitation previously aloft. The whole lifetime is about 30 minutes.

Depending on environmental conditions, rainfall can be moderate to heavy. In addition to rain, graupel and small hail may form aloft and reach the ground. Turbulence associated with the drafts rarely reaches moderate levels. The downdraft transports cold air to the surface where it forms a diverging pool of cool air. The leading edge of this pool is marked by a microscale cold front called a gust front. New cell formation frequently occurs somewhere on the flanks of the old cell in response to low-level convergence created by the gust front. Because of this regeneration, the total radar lifetime of the storm (complex of several cells—a nonsevere multicell storm) may be several hours. Nonsevere multicell storms account for the largest percentage of thunderstorms.

The Severe Thunderstorm

Since the Thunderstorm Project, and particularly since the late 1950s, research has focused on the severe thunderstorm. The key to the severe thunderstorm's structure appears to be updraft strength and its interaction with the environment.

The multicell severe storm consists of an organized, periodic sequence of severe cells, in which new cells develop preferentially on the right storm flank, and 2 to 4 cells in different stages of development are easily identified at any one time. The new cell does not lose its identity as it moves into the storm but becomes the mature storm as the older cells decay on the left storm flank. Since new cells are developing regularly, there is consistently a region of echo in middle levels on the right flank (the echo overhang), beneath which the echo is very weak or absent. Thus, below the overhang is the weak echo region. Because each new cell develops to the right of its predecessor, the storm as a whole moves to the right and more slowly than individual cells or the mean environmental winds. Details of this motion are well explained by Newton and Fankhauser (1964). The multicell severe storm can last several hours and produce much severe weather, but the severe phenomena are intermittent and usually not over a large area.

The supercell thunderstorm, although the least frequent, is the most severe and is responsible for a disproportionate amount of damage. Giant hail, strong surface winds, and long-lived tornadoes are not uncommon with these storms. In contrast to the severe storms previously discussed, this storm develops an intense updraft ($\sim$25–50 m s^{-1}) and downdrafts, which coexist for relatively long periods of time. Because of this stable configuration, storm lifetimes commonly range from 1 h to occasionally more than 6 h. They often move considerably to the right and more slowly than the mean environmental wind, but propagation is more continuous than discrete.

Also related to the stable organized character of the storm drafts are strikingly similar characteristics found from storm to storm. In plan view, the mature supercell exhibits a single cellular structure of roughly elliptical form. The weak echo region (WER) is persistent for long periods of time, and high-resolution scanning usually reveals a bounded weak echo region (BWER) near updraft center. A prominent feature seen in plan view at low levels is the pendant or hook echo. This feature, indicative of updraft rotation, is generally found in the right rear storm flank, extending at right angles to storm motion.

The thunderstorm that eventually reaches supercell proportions typically begins as a nonsevere multicell storm. Initially each cell's low-level echo, middle-level echo, and top are vertically stacked. With time, a middle-level echo overhang (WER) develops on the storm's right flank as the draft intensifies and the periodic production of new cells ceases. The transformation into a supercell is completed as the hook or pendant echo develops and the storm top shifts over the WER or BWER in a position rather far out on the right storm flank.

The distinction between multicell and supercell storms has been presented as sharp. In reality, recent research suggests that a spectrum of storm types exists. More detailed examinations of radar data depict some multi-cellular structure even for the classic single, large convective cell that appears to propagate continuously for long periods of time. In fact, Foote and Frank (1983) suggested that many storms belong in a new intermediate class called Westplains storm (weak evolution), which fits between multicell (strong evolution) and classic supercell (quasi-steady). In this categorization, the weak-evolution storm is produced by a single large updraft that undergoes gradual changes but remains connected. Separate reflectivity cores do not develop at low levels but may be inferred at middle and high levels. Individual reflectivity maxima are only weakly defined in an overall supercell pattern, but individual vertical velocity centers may be readily apparent in three-dimensional Doppler velocity analysis.

The squall line is defined to be any line or narrow band of active thunderstorms. It may be composed of nonsevere or severe multicells, supercells, or any combination of these. However, when a nearby continuous band of storms develops, the result is a significantly different structure from that described for isolated storms. Updrafts form a nearly continuous curtain along the advancing edge of the echo, rather than long the trailing edge, as in the isolated supercell or multicell storm. Downdrafts form in the precipitation echo to the rear of the leading updrafts. That portion of the line with the most intense gust front and updrafts is identified by a strong low-level reflectivity gradient, echo overhang, and shift of echo top from over the storm core to along the leading edge of the line. Often, the surface gust front surges out away from the leading edge of the radar echo and, if close enough to radar, can be observed as a thin line echo (Fig. 6.6). The radar return for the thin line echo comes from refractive index gradients of the air and discrete nonprecipitation wind-borne particles, such as insects.

The shape of the squall line echo band can sometimes be used as an indication of potential severe weather. A prominent bulge in the echo, documented by Nolan (1959), has long been known to be associated with squall line tornadoes and high winds. More recently, Fujita (1981) pointed out that bow-shaped echoes are related to strong diverging winds at low levels. Some bow echoes evolve into a comma shape that features a rotating head. Such systems probably reveal the existence of a mesoscale low pressure area. Many times the bow or comma echo is part of a larger echo band and appears as a bulge in the echo.

It is not uncommon for convection to begin as isolated storms, a few of which become severe multicell or supercell storms, and evolve into a squall line as more echoes form and their outflows combine into a gust front of

Figure 6.6. Squall line with thin line echo out ahead of the line. (Taken from NSSL WSR–57 radar, 30 May 1976.)

considerable horizontal extent. A nonsevere multicell storm evolves into a severe supercell storm, which collapses and becomes part of a squall line. The collapse phase occurs when the main storm updraft region dramatically weakens and greatly enhanced downdrafts are noted. It is observed on radar as a sharp decline in storm top, maximum reflectivity aloft, and vertical extent of the WER and BWER. The collapse stage is a prime time for severe windstorms and production of strong tornadoes. The storm may split, one part moving to the left of the mean winds and the other to the right of the mean winds. The right mover is usually a classic supercell. The left mover is a mirror image of the right mover, i.e., the updraft, WER, and gust front are all on the left storm flank. Well-defined splitting storms are rare but are usually associated with severe weather.

6.3.2. Motion and Storm Winds

The future location of existing radar echoes is of great interest for improved short-range forecasts (nowcasts) and warnings of hazardous weather. The advent of digital radar output and access to computer processing has made it possible to improve on the longtime practice of echo tracing of PPI radar displays. Computer algorithms can track echoes automatically. Once obtained, the centroids or cell boundaries are tracked and future positions are forecast by extrapolation.

Incoherent radars map reflectivity and, if their resolution volume is sufficiently small and reflectivity estimates are accurate, they can track prominent reflectivity structures (smaller than the cells discussed previously) to map winds within storms (Crane, 1979). One attractive technique still under development uses correlation analysis between gridded reflectivity data at two different times. Data from the initial array at the first time are compared with all similar sized arrays at the second time over some search area defined by the time lag and a maximum acceptable travel velocity. The patterns within the initial array and all second arrays are correlated, and the pair with the highest correlation coefficient defines the motion vector. The procedure is repeated for other initial arrays so that many vectors are determined over the entire storm area.

6.3.3. Estimating Rainfall and Determining Hydrometeor Phase

It has long been hoped that radar could be used to measure area-averaged rainfall in such a manner that it would augment or even replace standard rain-gauge networks. Several measurement techniques that use various properties of radar signals have been suggested. First, empirical studies indicate a relationship between reflectivity factor Z and rainfall rate R; most rainfall estimation from radar is based on this concept (Brandes and Wilson, 1986). Second, because the degree of rain attenuation is a function of wavelength, rainfall may be determined by measuring the attenuation using a short wavelength, and comparing it with the radar return from a longer wavelength radar that suffers little attenuation (Doviak and Zrnic', 1984). Finally, it has been proposed that measurements of Z at two different polarizations of the transmitted electric field (horizontal and vertical) give further information on raindrop sizes and rainfall rate (Seliga and Bringi, 1976).

Radar Reflectivity Factor and Raingauge Methods

As shown earlier, the reflectivity factor Z is defined as the 6th power of drop diameter summed over all drops in a unit volume. Thus the reflectivity factor can be given in terms of the drop size distribution:

$$Z = \frac{1}{\Delta(\text{vol})} \sum_{i=1}^{N} D_i^6 = \int N(D) D^6 \, dD \,, \qquad (6.17)$$

where $N(D)$ is the number of drops at diameter D. One can see that Z is dependent on the drop size distribution. Many studies of drop size distribution for stratiform and convective precipitation have been made. They generally find an exponential size distribution that can be written in the form proposed by Marshall and Palmer (1948):

$$N(D) = N_0 e^{-\Lambda D} \qquad (\text{cm}^{-4})$$

$$\Lambda = 41 R^{-0.21} \qquad (\text{cm}^{-1}) \qquad (6.18)$$

$$N_0 = 0.08 \qquad (\text{cm}^{-4}) \,.$$

The median volume diameter D_0 is defined as the diameter of a raindrop such that half the water volume is contained in drops larger than D_0. D_0 may be used in the drop size distribution expression since it is an easily identifiable physical attribute of the cloud. Then the exponential expression is

$$N(D) = N_0 e^{-3.67D/D_0} . \tag{6.19}$$

Integration of (6.17) using (6.18) and (6.19) gives

$$Z = \int_0^\infty D^6 N_0 \exp\left[\frac{-3.67D}{D_0}\right] dD \simeq \left(\frac{D_0}{3.67}\right)^7 N_0 6! . \tag{6.20}$$

Rainfall rate R is usually measured as depth of water per unit time. Therefore, rainfall rates are of the form

$$R = \frac{\pi p}{6} \int_0^\infty D^3 N(D) V_t(D) dD , \tag{6.21}$$

where $V_t(D)$ is the terminal velocity for each diameter. Using (6.19) and (6.21) and the appropriate form of V_t (see Doviak and Zrnic', 1984), we get

$$R = \frac{\pi N_0 D_0^4}{(3.67)^4}\left[9.65 - \frac{10.3}{(1 + 163 D_0)^4}\right](3.6 \times 10^6)\text{mm h}^{-1} . \tag{6.22}$$

Unfortunately, the expressions for Z and R, (6.20) and (6.21), are two-parameter expressions; N_0 and D_0 are unknowns. The system cannot be solved from the measurement of Z alone. Hence, the Z-R relationship is not unique. There has been much empirical study to seek appropriate values for N_0 and D_0. Combining (6.20) and (6.22) gives a result of the form

$$Z = AR^b . \tag{6.23}$$

In this form, A or b must be estimated independent of Z or assumed constant. The most common solution follows the original work of Marshall and Palmer (1948) for stratiform precipitation:

$$Z = 200 R^{1.6} . \tag{6.24}$$

The radar-derived solution for R suffers from the errors in the estimation of N_0 or D_0. Additional errors result when vertical air motions exceed raindrop terminal velocity (V_t), particularly in thunderstorms. Thus, even accurate estimation of water in the cloud may not lead to a good estimate of what actually reaches the ground. Also, the above discussion has treated all precipitation particles as liquid even though mixed precipitation phases are common (especially for thunderstorms). A few large hailstones may dominate the Z measurement but contribute nothing to R.

Since radar does have the advantage of sampling large areas easily, meteorologists have suggested calibrating the radar estimate with gauges to obtain the most useful results (Wilson, 1970). Brandes (1975) suggested a technique whereby gauges can be used to adjust radar-estimated rainfall. This procedure minimizes the effects of choosing an inappropriate Z-R relationship, partially accounts for unsampled vertical velocity, and reduces the need for perfect radar calibration. After radar estimates of rain are obtained, rain-gauge to radar (G/R) ratios are calculated at each gauge location, using radar data from within a fixed radius about the gauge. The G/R data are used to determine a field of adjustment factors, which are applied to the radar data. This first-guess field is iterated with new G/R ratios until the final radar estimates agree with gauge estimates at all gauge locations. Brandes (1975) shows that radar measurements corrected by gauges have significantly improved accuracy where density is as low as one gauge per 1600 km^2.

Attenuation Methods

There is a relationship between drop size spectra and attenuation rate. The attenuation is proportional to the absorption and scatter cross sections of each drop within the sampling volume. Radar meteorologists have observed a consistent relationship between microwave attenuation and rainfall rate. Thus, accurate measures of attenuation might be used to obtain reasonable estimates of rainfall rate R without prior knowledge of drop size distribution $N(D)$.

Because of strong attenuation at short wavelengths (i.e., 1 cm), longer wavelengths (3–10 cm) have been used. Rainfall rate estimates have been poor with single-wavelength methods, because of insufficiently accurate radar measurements and an insufficient knowledge of the scattering volume population. Alternatively, a two-wavelength method has been proposed in which one wavelength should involve Rayleigh scattering ($\lambda > 16D$) and the other should be outside the region described by Rayleigh scattering, where the full Mie scattering equation must be used. From the differences in measured reflectivity factors, one can deduce drop size parameters and rainfall rate (Wexler and Atlas, 1963).

It also has been proposed that large values of differential reflectivity from two wavelengths could be used to detect the presence of hail (Atlas and Ludlam, 1961). There has been uncertainty associated with dual wavelength "hail signals" over the past two decades because of measurement problems and the accuracies necessary for reliable results.

Dual Polarization Method

This technique uses echo intensity information at two orthogonally polarized waves. Each transmitted pulse has alternate horizontal and vertical polarization. The basis for the dual polarization scheme is the observation that drops are not spherical but have an oblate spheroid shape with a horizontal major axis. Thus for particles satisfying Rayleigh scattering, we expect larger echo power for horizontally polarized waves. Remembering that Z is proportional to D^6, we have

Table 6.1. Relationship of reflectivity factor and differential reflectivity to hydrometeor type

Particle type	Reflectivity factor (dBZ)	Differential (Z_{DR} in dB)
Hail	High (60)	Low (0.5)
Rain	Moderate (45)	High (2)
Drizzle	Low (20)	Low (0.5)
Snow	Low (30)	Moderate (1)

$$Z_{DR} = 10 \log \left(\frac{Z_H}{Z_V} \right) \approx \left(\frac{b}{a} \right)^6 \text{(dB)} , \qquad (6.25)$$

where Z_{DR} is differential reflectivity, subscripts H and V are the horizontal and vertical polarization, respectively, and a and b are the drop diameter in the vertical and horizontal polarization directions, respectively. We can see that Z_{DR} is uniquely proportional to raindrop size. Thus, we can directly determine Λ or D_0 [see (6.18)]. The N_0 can be determined from measurement of Z_H once we have a value for D_0. This means that R can be estimated from reflectivity measurements alone, i.e., without an unknown second parameter as in (6.20) or (6.22).

Perhaps the most important use of Z_{DR} is in determining hydrometeor phase. It appears encouraging that estimates of Z and Z_{DR} can be used together to separate different types of precipitation particles. This hydrometeor phase discrimination can be used to identify regions of hail, where usual rainfall estimation techniques should not be used. In general, the relationships in Table 6.1 hold.

Values of Z_{DR} larger than 3.5 dB appear to be associated with melting ice crystals near the freezing level aloft and may serve as a useful indicator of that important temperature boundary.

6.4. Uses of Doppler Velocity Information

6.4.1. Signatures

A Doppler radar that is scanning in azimuth at a constant elevation angle maps an inverted conical surface with the radar at its apex. Assuming that the wind field is nearly uniform over the radar, a great deal of insight may be gained about the structure of the wind field. When the antenna is pointed in the direction of wind flow, the maximum velocity will be observed; when the radar is pointing orthogonal to the wind direction, the Doppler velocity will be zero. If and only if the wind has a maximum with height, a pair of closed contours of opposite signs will be observed. The orientation gives the direction of the jet; height is determined from the range and the radar's elevation angle.

Similarly, a simple model of a wind field often applies to other cases of interest. The flow within a severe storm, particularly one associated with strong updrafts and areas of rotation, can be approximated by a rotational and divergent flow. This has been modeled by a Rankine (1901) combined vortex where the velocity is described as

$$v = c_1 r \qquad 0 < r < R_c$$
$$v = c_2/r \qquad R_c < r \, , \qquad (6.26)$$

where c_1, c_2, and R_c are parameters that distinguish different flow fields. A rotational flow field with a solidly rotating core of radius R_c implies that the tangential component v_t increased linearly to its maximum value at R_c. Beyond R_c the rotational component decreases as the reciprocal of the distance. A radar scanning such a flow field would see two maxima of opposite signs oriented in the azimuthal direction, as portrayed in Fig. 6.7a. (In this figure and throughout this discussion, Doppler radar is considered to be well below the bottom of the page.)

The pattern seen in Fig. 6.7a is associated with mesocyclones and is often the precursor circulation to the formation of a tornado. A typical core radius is 3 km, and peak tangential winds are 25 m s^{-1}. If the vortex is imbedded in a uniform wind field the contour pattern will remain unchanged but the value of the contour will change. The Rankine profile may also be used to model axisymmetric divergent winds. Again, the flow toward (away from) the sink (source) region increases linearly to R_c and decreases in inverse proportion to distance beyond R_c. Figure 6.7b illustrates the case for such a divergent flow. Note the similarity to the pattern for the rotational case (Fig. 6.7a); the outstanding feature in the pattern is shifted 90°. This type of pattern is often found at the top of storms where divergence predominates. These and other flows often occur in combination.

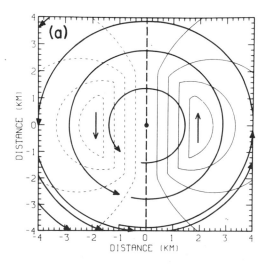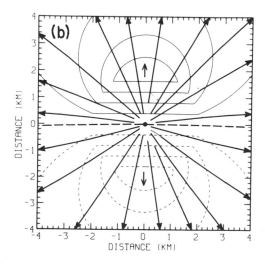

Figure 6.7. (a) Vortex flow (heavy streamlines) and corresponding single Doppler velocity signature (thin lines); radar is south of flow field. (From Wood and Brown, 1983.) (b) Divergent flow (heavy streamlines) and corresponding single Doppler velocity signature (thin lines); radar is south of flow field. (From Wood and Brown, 1983.)

6.4.2. Tornado Recognition

Although one Doppler radar measures only the radial component of wind, its spatial distribution can identify regions of small-scale circulation (see Fig. 6.7a for idealized circulation signature). Unfortunately, the recognition of thunderstorm vortices is not feasible if the resolution volume size is significantly larger than the vortex radius. When the beamwidth exceeds the core radius by a factor of 3, less than one-half of the true maximum velocity is detected and failure to recognize the vortex may result. The recognition problem is further complicated by discrete azimuthal sampling which, at random, might miss taking a sample near the peak of the velocity distribution. Thus, average size tornadoes cannot reasonably be detected much beyond 100 km range by a radar with a 0.8° beamwidth. On the other hand, mesocyclones, the larger circulation parent to the tornado, can easily be detected out to 230 km with the same beamwidth. The signature associated with tornadoes is called the tornado vortex signature (TVS), not tornado signature, because it is seen above cloud base and often before the tornado funnel below cloud base forms.

Mesocyclone Signature

Not all mesocyclones produce tornadoes. During the Joint Doppler Operational Project (JDOP), when considerable attention was paid to verification, it was found that roughly 50% of all mesocyclones produced verified tornadoes. All storms with mesocyclones were likely to be severe; more than 90% produced some kind of surface damage (hail, wind, or tornado). Data from 45 well-observed signatures indicate few well-defined boundaries between mesocyclones that produced no tornadoes, weak or moderate tornadoes, or violent tornadoes. There seems to be a tendency for tornadic mesocyclones to have higher rotational velocity and stronger shear. All storms with violent tornadoes had high shear that resulted from strong rotational velocity and small core diameter.

With single Doppler, the mesocyclone inner core (solid rotation) is easily measured but the outer region of potential flow is difficult to define from the radar measurement. The larger surrounding region of cyclonic swirl is resolved in multi-Doppler wind fields (see Fig. 6.12). It is possible for more than one core to exist simultaneously within the same region of cyclonic swirl. The multiple cores do not, however, occur at random. Series of cores occur in certain mesocyclones in a predictable way and act as the mesocyclone propagation mechanism. The mesocyclone forms after the storm echo is mature and near its peak height. The first mesocyclone core has a relatively long organizing and mature stage. The second core organizes as the first begins dissipating. Second and succeeding cores have extremely short organizing stages; they form quickly over a large depth and have relatively short mature stages as evolution proceeds rapidly. One storm possessed six successive cores and persisted for nearly 5 h. An example of a radar observation of a tornadic storm with mesocyclone is shown in Fig. 6.8.

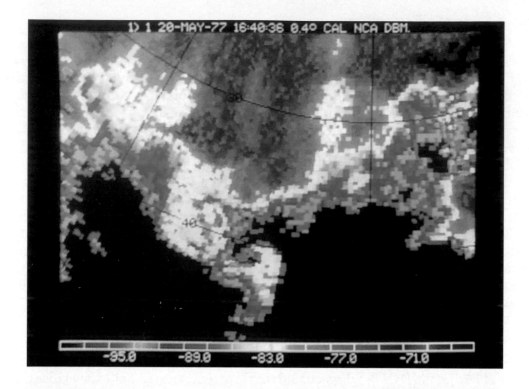

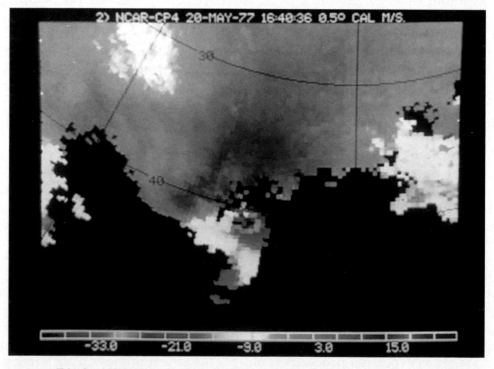

Figure 6.8. Display of Doppler radar data from a tornadic storm that occurred near Fort Cobb, Okla., 20 May 1977: (Top) reflectivity; (bottom) Doppler velocity. Elevation angle is 0.0° and radar is located at the figure top. Data are from NCAR's 5 cm wavelength CP–4 radar. The display was generated from the ATD/FOF RDSS display.

A conceptual model of mesocyclone evolution in horizontal section reveals the existence of a meso-cold front (gust front) that wraps cyclonically about the mesocyclone core. The core evolution closely resembles synoptic cyclone development: during the first-core mature stage, the gust front accelerates around the right flank; occlusion occurs and as the mesocyclone warm sector separates from the first core its dissipation begins; strong convergence localizes at the point of occlusion, and a second vortex core organizes. The second vortex core organizes so rapidly because of the vorticity-rich environment within which it forms. The ultimate result of such continued evolution is a tornado family with a recurrence interval of approximately 40 min.

Tornado Vortex Signature (TVS)

The TVS was discovered in Doppler observations of the Union City, Okla., tornado (Brown *et al.*, 1978). It was detected as an anomalous region of high shear within the mesocyclone. The signature appears when two velocity profiles are overlaid, one being small (tornado) and one being large (mesocyclone). Note that the TVS would stand out as a higher shear region only for strong and wide tornadoes at relatively short range. If azimuthal sampling were 1° or less, the high shear signature should be contained in adjacently sampled radials. Criteria for TVS recognition, based on these guidelines, were developed during JDOP. TVSs detected with these criteria (more than a dozen) were always associated with tornadoes or funnel clouds.

It is important to emphasize that not all tornadoes, even at close range, will have a TVS. For example, it is rare for tornadoes along gust fronts (usually small and short-lived) to have TVSs. Also, significant tornadoes at moderate-to-long ranges will not have a detectable TVS. The longest range for a TVS observation was 180 km for an unusually large tornado in western Oklahoma (16 May 1977).

Figure 6.9a is an example of the wind field within a tornadic storm. The contoured single Doppler observations from Ray *et al.* (1981) reveal a large couplet (mesocyclone signature) with an embedded high shear location (TVS). This is derived from the data shown in Fig. 6.8. The model parameters used by Wood and Brown (1983) to simulate this field are shown schematically in Fig. 6.9b; the results of the simulation are in Fig. 6.9c. Note that considerable convergence at this low elevation angle is superimposed with the rotation. The similarity between the observed field and the simulation is striking, showing that the major flow components have been captured.

Mesocyclone signatures and TVSs, as well as strong wind and hail signatures (Secs. 6.4.4 and 6.4.5), were tested during JDOP. The Doppler signatures were verified to work well in identifying severe-storm phenomena and to have superiority over conventional radar. Forecasts from the Oklahoma City Weather Service Forecast Office (OKC WSFO) based on conventional radar data, and JDOP advisories based on Doppler radar data were compared in terms of the critical success index (CSI). Results are shown in Table 6.2. Analysis reveals that Doppler's advantages are increased leadtime for torna-

**Table 6.2. Analysis of forecasts based on conventional and
on Doppler radar data (JDOP Staff, 1979)**

Conventional (OKS WSFO warning)	Doppler (JDOP advisory)
1978 Severe Storms	
POD=.47	POD=.70
FAR=.40	FAR=.16
CSI=.36	CSI=.62
LT=13.6 min	LT=15.4 min
1977 + 1978 Tornadoes	
POD=.64	POD=.69
FAR=.63	FAR=.25
CSI=.30	CSI=.56
LT=2.2 min	LT=21.4 min

Probability of Detection (POD) = $X/(X+Z)$
False Alarm Ratio (FAR) = $Y/(X+Y)$
Critical Success Index (CSI) = $X/(X+Y+Z)$
   where
   X = No. of forecast severe events that occur.
   Y = No. of forecast severe events that do not occur.
   Z = No. of forecast nonsevere events that occur severe.
LT = Lead time between advisory/warning and event.

does, reduced false alarm rates for tornadoes and severe thunderstorms, and improved probability of detection for severe thunderstorms.

6.4.3. Wind Profiling and Prestorm Measurements

When a radar antenna is tilted above horizontal, increasing range implies increasing height, and a profile of wind with height can be obtained. The observed radial velocity patterns for different combinations of wind speed and direction profiles with height are shown in Fig. 6.10. Since the radar beam is tilted above its horizontal position, increasing range from the radar implies greater height as well as distance from the radar. This is how a height profile of winds is obtained. The winds along the zero radial velocity contour are perpendicular to the radar beam axis. It is possible to employ a simple model of the wind field to fix the observed radial component.

An example of this technique applied to real data is shown in Fig. 6.11. There is an S-shaped zero velocity contour in Fig. 6.11a suggesting winds that veered with height. Near the edge of the display a slight backing is indicated, suggesting cold air advection. Wood and Brown (1983) took the profile given in Fig. 6.11b and, by fitting it with a sixth-order polynomial, arrived at an expression for wind speed and direction for up to 6.7 km height. The maximum wind speed was 41 m s^{-1}. The result is the strikingly similar synthetic Doppler wind field shown in Fig. 6.11c.

Since the measurement accuracy is good, divergence estimates can also be obtained employing the velocity-azimuth display (VAD) technique (see Battan, 1973). Vertical velocities applicable to a vertical column over the radar may be computed from upward integration of the mass continuity equation. This rather complete set of wind measurements is available from Doppler radar in any widespread precipitation system (considerable echo

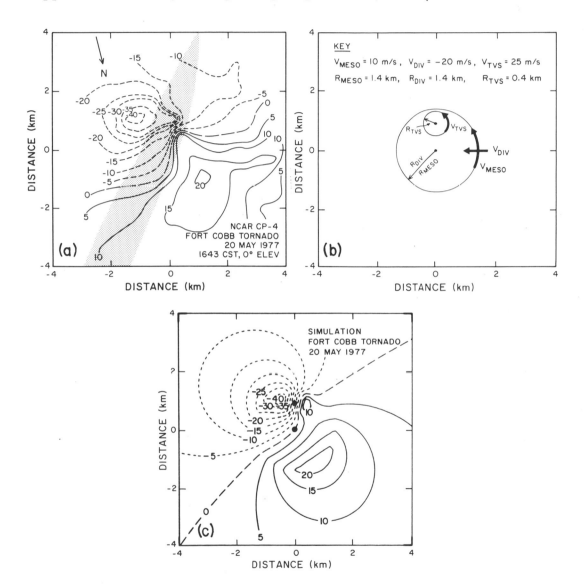

Figure 6.9. (a) Single Doppler signature for Fort Cobb tornado and parent mesocyclone on 20 May 1977. Radar is located below the bottom of the figure. The mesocyclone center is at the center of the grid. Doppler velocities (m s^{-1}) are positive (negative) for flow away from (toward) the radar. (From Ray *et al.*, 1981.) (b) Parameters used to simulate the single Doppler velocity data in Fig. 6.9a. (c) Simulation of parent circulation in which the TVS is embedded as in Fig. 6.9a. Solid (short dashed) contours represent positive (negative)values of single Doppler velocities (m s^{-1}). Long dashed contour is zero Doppler velocity. Dots mark TVS and mesocyclone center. (From Wood and Brown, 1983.)

WIND SPEED PROFILE

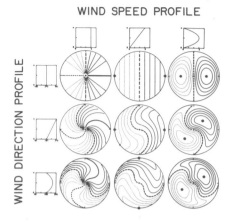

Figure 6.10. Doppler velocity patterns (constant elevation PPI scans) for various vertical profiles of wind speed and direction. The radar is located in the center of the display. Solid lines represent flow away from the radar (positive) and dashed lines flow toward the radar (negative). (From Wood and Brown, 1983; see also Fig. 28.2, this volume.)

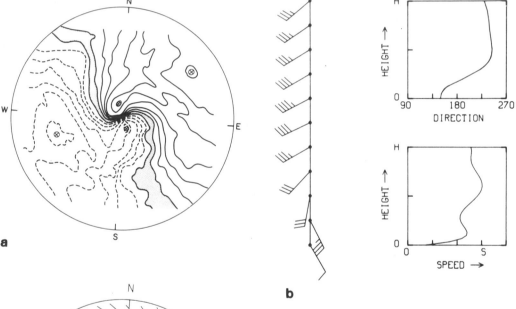

Figure 6.11. (a) Elevated-Doppler velocity pattern measured at 0905 PST on 7 February 1978 near Sacramento, Calif. Velocity extremes are measured at circled x's. (After Wilson et al., 1980.) (b) Modeled wind speed and direction profiles based on Doppler derived winds. (From Wilson et al., 1980.) (c) Simulated Doppler velocity pattern based on Fig. 6.11a. Heavy dashed line is zero velocity contour. Solid contours are velocities away from radar; dashed contours are velocities toward radar. Velocity extremes are measured at circled x's (From Wood and Brown, 1983.)

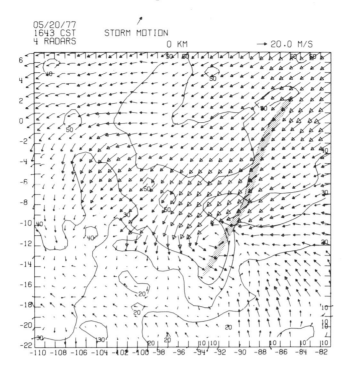

Figure 6.12. Surface wind field during tornado near Fort Cobb, Okla., on 20 May 1977. Damage path is indicated by stippling. Data from four Doppler radars were used in this synthesis. Reflectivity is contoured in increments of 10 dBZ. All winds are storm relative.

must be present in all quadrants of the radar). This technique appears particularly suited for winds from the precipitation systems associated with extratropical and tropical cyclones (Wilson *et al.*, 1980) but may not work well in convective storm areas where echoes may be widely scattered (Battan, 1973).

Sensitive Doppler radars with moderate power also have some wind-measuring capability in the optically clear air. Whenever turbulence mixes air in which there are gradients of temperature and water vapor, fluctuation of the refractive index results. The fluctuations cause small but faint echoes to be returned. Also, particulate matter in the atmosphere (e.g., concentrations of insects) contributes to what are called clear-air echoes. These echoes are strongest in the daytime planetary boundary layer where considerable turbulent mixing is taking place. This fact has special significance to forecasting because detailed observations are available prior to the formation of thunderstorms, whose locations may be predicted. Areas of convergence show up in single Doppler PPI presentations and through observation of local deepening of the boundary layer. VAD analysis may also be performed on clear-air echoes, as in the widespread precipitation cases discussed above. A time series of measurements of divergence and derived vertical velocity is particularly useful in ascertaining the probability of deep convection.

The newest application of clear-air Doppler measurement involves dividing the PPI display into many volumes for which wind field parameters are estimated. Providing that the wind field is linear, a multivariate regression analysis may be used to retrieve wind parameters from single Doppler data. The advantage of this technique is that horizontal winds and divergence may be obtained on small scales over a broad region, whereas VAD analysis gives profiles applicable over a broad area centered at the radar. The analysis technique, called volume velocity processing (VVP), produces wind fields that can be used to detect air mass boundaries and wind flow near locations where convection might form (Koscielny et al., 1982).

6.4.4. Strong Low-Level Winds

Single Doppler radar does not detect all hazardous, low-level (near-ground) winds. Techniques available to profile winds with broad-scale flows in a relatively quiescent environment have already been discussed. Of concern here are the strong gusty winds observed near convective activity. When the direction of the wind is parallel to the radar antenna beam, its true magnitude will be sensed. Winds blowing at some angle to the beam will have only their radial (along-beam) component measured. In the extreme case, flow will be perpendicular to the beam axis where no component is seen. Fortunately, the sensed component decreases as a cosine function, and a large fraction of the flow magnitude is detected for viewing angles up to 45°. During JDOP, slightly less than half of the damaging wind storms could be confirmed as severe (radial component >25 m s^{-1}) but the scale size of the outflow signature (20 km) was large enough that some evidence of the gust front and outflow was visible on Doppler nearly all the time.

Small-scale wind maxima associated with narrow downdrafts (called downbursts and microbursts by Fujita, 1981) are even harder to detect than the large outflows just discussed. According to Wilson et al. (1984) microbursts have small horizontal extent (1–6 km) and reach their maximum intensity in less than 10 minutes. Also, the strength of the outflow is frequently asymmetric so relative maxima are highly dependent on viewing angle. Maximum horizontal winds in microbursts may be confined to heights of 200 m, well below the radar horizon of even a nearby radar.

6.4.5. Hail

A single Doppler radar can measure the divergence signature at the updraft summit (see Fig. 6.7b for idealized divergence signature). Assuming that the divergence pattern is axisymmetric, the magnitude of divergence may be equated to updraft speed and potential for producing hail.

Qualitative estimates of divergence near the storm summit were used in part to estimate storm severity during JDOP. JDOP participants concluded that an increase in divergence magnitude was likely to be the first indicator that a storm was becoming severe and was a good estimator of hail potential. Witt and Nelson (1984) expanded these early findings into a more formal relationship between divergence magnitude and hailstone size. Their data indicate a surprisingly linear relationship between velocity change across the outflow and reported hail diameter. The study further finds that hailswath width is related to the size of the high radial shear area.

6.4.6. Turbulence (Aircraft Safety)

The advent of pulse-pair processors during the 1970s made it possible to routinely calculate the first and second moments of the Doppler spectrum at large numbers of sampling points in real time. The second-moment (variance) estimate can be used to help locate turbulent areas in and around storm systems. The parameter usually chosen for display is the square root of the variance—the spectrum width (σ_v)—which is a measure of velocity dispersion about the mean within the sampling volume.

As mentioned in Sec. 6.2.4, several spectral-broadening mechanisms may contribute to the total velocity spectrum width. That is,

$$\sigma_v^2 = \sigma_s^2 + \sigma_t^2 + \sigma_a^2 + \sigma_d^2 , \qquad (6.27)$$

where σ_s^2 is due to shear, σ_t^2 to turbulence, σ_a^2 to antenna rotation, and σ_d^2 to differential drop fall speeds. Shear is defined as the velocity change on a scale larger than the resolution volume, and turbulence is the energetic eddy motion on scale sizes smaller than the radar resolution volume. For elevation angles less than 20° and rotation rates less than 8° s^{-1}, calculations indicate that σ_d^2 and σ_a^2 are small and can be ignored. Thus we can assume that broad spectral widths are caused by shear and turbulence. Examination of σ_s^2 shows it to be a minor contribution unless shear exceeds 10^{-3} s^{-1}. This condition is most likely along low-level outflow boundaries, in the interface between updrafts and downdrafts and in association with mesocyclones. Since shear can be measured from Doppler data, turbulence can be calculated. It is found that turbulence, assumed isotropic and homogeneous, is related to eddy dissipation rate. Assuming an eddy dissipation rate of 1 m^2 s^{-3} (a value found in severe thunderstorms), the spectrum width due to turbulence alone is approximately 6 m s^{-1}.

During a series of experiments in the 1970s, aircraft equipped to make wind and turbulence measurements were flown through thunderstorms being scanned by Doppler radar. The results indicate a relationship between turbulence measured by gust velocity and Doppler spectrum width. Spectrum widths in excess of 4 m s^{-1} are associated with moderate and greater turbulence. The correlation coefficient (0.4), however, is not as large as might be expected because of a large number of false alarms (i.e., $\sigma_v > 4$ m s^{-1} and no aircraft-reported turbulence).

One result of the turbulence detection work has been the ability to identify the most likely turbulence region of a storm. For many storms, large spectrum width (turbulence) occurs in the gradient between updraft and downdraft. According to Zrnic' and Lee (1982), turbulence is produced by horizontal shear of the vertical wind.

6.5. Multiple Doppler Radar Analysis Techniques

With three independent (orthogonal) directions of air motion to be determined, at least three independent relationships (or equations) must be used to determine the wind at any point. Additional information results in an over-determined problem, which allows the most nearly correct answer to be estimated through the method of least squares (Draper and Smith, 1966). In the application discussed here, the radar does not actually measure the wind motion at a point; it measures the average speed of the hydrometeors (liquid water or ice) toward or away from the radar in each sampling volume. The hydrometeor falls with respect to the air around it and the radar measures a vertical component W which is the sum of the vertical air motion w and the particle terminal fallspeed V_t:

$$W(x, y, z) = w + V_t \ . \tag{6.28}$$

The variable V_t requires yet another relation (for a total of at least 4) to form a unique solution. Normally this is provided by an empirical formula relating reflectivity and terminal fallspeed.

A radar located at (x_i, y_i, z_i) observes a point at the position defined by coordinates (x, y, z). The point P is at a distance R_i from the radar given by

$$R_i = [(x - x_i)^2 + (y - y_i)^2 + (z - z_i)^2]^{1/2} \ . \tag{6.29}$$

The azimuth angle, usually measured in a clockwise direction of the antenna from north, is given by θ, and the elevation angle, defined as the antenna position above the horizontal, is given by ϕ. Usually the air motions u, v, and w are associated with the $\hat{x}$, $\hat{y}$, and $\hat{z}$ directions, respectively.

Many of the basic principles in deriving wind fields from the combined use of separated Doppler radars are given by Armijo (1969). Most methods incorporate a terminal velocity and reflectivity relationship and the anelastic form of the mass continuity equation

$$\frac{\partial u}{\partial x} + \frac{\partial v}{\partial x} + \frac{\partial w}{\partial x} = \kappa w \ , \tag{6.30}$$

where κ is a constant that approximates the logarithmic vertical gradient of air density $\partial[\ln(\rho)]/\partial z$. The measured particle motions, V_{ri}, along the radar's beam axis is used with the above relationships to determine the orthogonal wind components:

$$u \cos \phi_i \sin \theta_i + v \cos \phi_i \cos \theta_i + W \sin \phi_i = V_{ri} \ . \tag{6.31}$$

Through standard trigonometric relationships, (6.31) can alternatively be written

$$u(x - x_i) + v(y - y_i) + W(z - z_i) = R_i V_{ri} \ . \tag{6.32}$$

6.5.1. Methodology for Two Doppler Radars

As derived by Armijo (1969), the solution for the combined equation set (6.28, 6.30, and 6.31) results in a linear, inhomogeneous, hyperbolic partial differential equation. The natural coordinate for the solution is a cylindrical coordinate system. These techniques have been documented by Miller and Strauch (1974), Ray et al. (1975), and Doviak et al. (1976; 1986).

Brandes (1977) and Ray *et al.* (1980) pointed out that the horizontal wind components can be determined in terms of the Doppler radar measured radial wind component V_{ri} and the vertical wind w:

$$u = Aw + B$$
$$v = Cw + D , \qquad (6.33)$$

where A and C are functions of distances, and B and D also include the radial wind components V_{ri}. Usually B and D are greater than A and C. In practice this is solved by using an initial estimate of w for (6.33) and then using the derived horizontal wind components in (6.30) to compute a new value for w. This process is repeated with updated wind component values until the adjustments are less than measurement error. Air density decreasing with height suppresses error when the integration proceeds downward from the storm top. This error reduction must be balanced with the possibility of greater uncertainty in the upper boundary condition than in the lower boundary condition. Generally, deep convection is reasonably steady state, and it is advantageous to integrate downward. Analysis of these effects can be found in Ray *et al.* (1980). The spatial distribution of the errors generated by the uncertainty in the estimates of V_{ri} is discussed in Ray *et al.* (1979). Since there is no measurement of the wind component perpendicular to the baseline joining the two radar locations, the uncertainty in that direction becomes unbounded. In general, the more nearly orthogonal the direction in which the measurements are obtained, the smaller the error. This yields a minimum error at points where the angle defined by the intersection of the radar beams is near 90°.

When more than two radars are available, an analogous equation set in the same form as (6.33) can be derived using standard least-squares methodology (Draper and Smith, 1966). The resulting equation set and the distribution of error are discussed in Ray and Sangren (1983) for combinations of two to nine radars. Of course, for only two radars, the equations reduce to the same form given by Brandes (1977) and Ray *et al.* (1980).

Following this initial analysis, wind components are prescribed at all grid points that were sampled by at least two radars. However, the vertical velocities may depart substantially from reasonable values at the boundary away from where the integration was started. Following Ray *et al.* (1980), the integrated horizontal divergence from the surface to Z_T can be specified as a constant C, usually zero or a few meters per second:

$$C = \int_0^{Z_T} \left(\frac{\partial u}{\partial x} + \frac{\partial v}{\partial y} \right) dz = - \int_0^{Z_T} \left(\frac{\partial w}{\partial z} dz \right) . \qquad (6.34)$$

The result is that the horizontal wind components are adjusted so that when they are integrated, both new boundary conditions are satisfied. The adjustment of each horizontal wind component is governed by the magnitude of its error in comparison with others in the same vertical column.

The adjustments are

$$u = u^0 + \frac{2}{2\alpha^2}\frac{\partial\lambda}{\partial x}$$

$$v = v^0 + \frac{1}{2\beta^2}\frac{\partial\lambda}{\partial x} \; , \tag{6.35}$$

where u^0 and v^0 are the initial grid point estimates, u and v are the adjusted estimates, and α^2 and β^2 are proportional to the reciprocal of the error in u^0 and v^0. The error due to geometrical relationship of the grid point to the radars is given in Ray and Sangren (1983). The λ are solutions found through the calculus of variation using (6.34) as a constraint that must be satisfied. It varies only in the horizontal direction. The vertical velocity is found by integrating (6.30) using the estimates found in (6.35).

The degree of adjustment of wind components at each grid point is dependent on the presumed relative error (α^2, β^2) and on λ, which contains information on the residual error in the integrated horizontal divergence (6.30). However, a special case exists when α^2 and β^2 are constant in the vertical direction. In this case the adjustment of the horizontal wind components $(u - u^0$ and $v - v^0)$ is constant and the adjustment of the horizontal divergence is

$$D - D^0 = \frac{w' - w^0}{Z_T} \; , \tag{6.36}$$

where w' is the specified lower (or upper) boundary condition and w^0 the vertical velocity that was computed. In this case the adjustment of the divergence is constant for all heights. The adjustment at any height ℓ can be found to be

$$W_\ell - W_\ell^0 = -\int_0^{Z_\ell}(D - D^0)dz$$

$$W_\ell - W_\ell^0 = -\frac{Z_\ell}{Z_T}(W' - W_T^0) \; . \tag{6.37}$$

This means the adjustment is linear with height.

6.5.2. Methodology for Three Doppler Radars

The introduction of a third radar closes the system of equations (6.28) and (6.31) without the need to use the continuity relation (6.30). However, because the errors in the vertical component are proportional to $(1/z)^2$, many analysts use only the horizontal wind component as part of the solution and obtain w from integrating the continuity relation. Note that in either approach the terminal velocity V_t does not enter, removing that aspect of uncertainty.

Another approach was first described by Ray et al. (1978) and expanded in Ray et al. (1980). In this approach the solution for three or more Doppler radars is found directly, but in such a way that the u, v, w wind components at a grid point are derived as the first term in a Taylor expansion of the wind field in the near vicinity of the grid point. The wind field is further required to satisfy constraints, especially that of mass continuity (6.30). Constraints are satisfied only in the region between grid points. After the field has been determined this way, the whole field is adjusted so that the mass continuity relationship holds in its discretized form. These relationships are conveniently expressed by using techniques of variational calculus.

The methodology outlined for three Doppler radars was used with data such as those displayed in Fig. 6.9 and with data from three other radars to construct the three-dimensional wind field during tornado occurrence. The winds at the surface and the tornado damage path are displayed in Fig. 6.12. The circulation in which the tornado is imbedded is evident as the "hook" in the reflectivity pattern. It is the hook pattern alone that is seen in a conventional radar.

REFERENCES

Armijo, L., 1969: A theory for the determination of wind and precipitation velocities with Doppler radars. *J. Atmos. Sci.*, 26, 566–569.

Atlas, D., 1964: Advances in radar meteorology. In *Advances in Geophysics*, 10, Academic Press, New York, 318–478.

Atlas, D., and F. H. Ludlam, 1961: Multiwavelength radar reflectivity of hailstorms. *Quart. J. Roy. Meteor. Soc.*, 37, 523–534.

Battan, L. J., 1973: *Radar Observations of the Atmosphere.* University of Chicago Press, Chicago, Ill.

Blackman, R. B., and J. W. Tukey, 1958: *The Measurement of Power Spectra.* Dover, New York.

Brandes, E. A., 1975: Optimizing rainfall estimates with the aid of radar. *J. Appl. Meteor.*, 14, 1339–1345.

Brandes, E. A., 1977: Flow in severe thunderstorms observed by dual-Doppler radar. *Mon. Wea. Rev.*, 105, 113–120.

Brandes, E. A., and J. W. Wilson, 1986: Measuring storm rainfall by radar and rain gage. In *Thunderstorms: A Social, Scientific, and Technological Documentary.* Vol. 3: *Instruments and Techniques for Thunderstorm Observation and Analysis* (E. Kessler, Ed.), Second ed., University of Oklahoma Press, Norman, Okla. (in press).

Brown, R. A., L. R. Lemon and D. W.

Burgess, 1978: Tornado detection by pulsed Doppler radar. *Mon. Wea. Rev.*, 106, 29–38.

Byers, H. R., and R. R. Braham, Jr., 1949: *The Thunderstorm.* U. S. Government Printing Office, Washington, D.C., 287 pp.

Crane, R. K., 1979: Automatic cell detection and tracking. *IEEE Trans. Geosci. Electron*, GE–17, 250–262.

Doviak, R. J., and D. Zrnic', 1984: *Doppler Radar and Weather Observations.* Academic Press, Orlando, Fla., 480 pp.

Doviak, R. J., P. S. Ray, R. G. Strauch, and L. J. Miller, 1976: Error estimation in wind fields derived from dual-Doppler radar measurements. *J. Appl. Meteor.*, 15, 868–878.

Doviak, R. J., D. Sirmans, and D. Zrnic', 1986: Weather radar. In *Thunderstorms: A Social, Scientific, and Technological Documentary.* Vol. 3: *Instruments and Techniques for Thunderstorm Observation and Analysis* (E. Kessler, Ed.), Second ed., University of Oklahoma Press, Norman, Okla. (in press).

Draper, W. R., and H. Smith, 1966: *Applied Regression Analysis.* Wiley and Sons, 407 pp.

Foote, G. B. and H. W. Frank, 1983: Case study of a hailstorm in Colorado. Part III: Airflow from triple Doppler measurements. *J. Atmos. Sci.*, 40, 686–707.

Fujita, T. T., 1981: Tornadoes and downbursts in the context of generalized planetary scales. *J. Atmos. Sci.*, **38**, 1511–1534.

Grebe, R., 1982: An outline of severe local storms with the morphology of associated radar echoes. NOAA Tech Memo. NWS TC-1, National Weather Service (NTIS#PB83-114454), 80 pp.

JDOP Staff, 1979: Final report on the Joint Doppler Operational Project (JDOP), 1976-1978. NOAA Tech. Memo. ERL NSSL-86, Environmental Research Laboratories (NTIS#PB80-107188/AS), 84 pp.

Kerr, D. E., 1951: *Propagation of Short Radio Waves*. McGraw-Hill, New York.

Koscielny, A. J., R. J. Doviak and R. Rabin, 1982: Statistical considerations in the estimation of wind fields from single Doppler radar and application to prestorm boundary layer observations. *J. Appl. Meteor.*, **21**, 197–210.

Lemon, L. R., 1980: Severe thunderstorm radar identification techniques and warning criteria. NOAA Tech. Memo. NWS NSSFC-3, National Weather Service (NTIS#PB81-234809), 60 pp.

Marshall, J. S., and W. M. Palmer, 1948: The distribution of raindrops with size. *J. Meteor.*, **5**, 165–166.

Mie, G., 1908: Beiträge zur Optik trüber Medien, speziell kolloidaler Metallösungen [Contribution to the optics of suspended media, specifically colloidal metal suspensions]. *Ann. Phys.*, **25**, 377–445.

Miller, L. J., and R. G. Strauch, 1974: A dual Doppler radar method for the determination of wind velocities within precipitating weather systems. *Remote Sens. Environ.*, **3**, 219–235.

Newton, C. W., and J. C. Fankhauser, 1964: On the movement of convective storms, with emphasis on size discrimination in relation to water budget requirements. *J. Appl. Meteor.*, **3**, 651–668.

Nolan, R. H., 1959: A radar pattern associated with tornadoes. *Bull. Amer. Meteor. Soc.*, **40**, 277–279.

Probert-Jones, J. R., 1962: The radar equation in meteorology. *Quart. J. Roy. Meteor. Soc.*, **88**, 485–495.

Rankine, W. J. M., 1901: *A Manual of Applied Mechanics*. 16th ed., Charles Griff and Company, London, 574–578.

Ray, P. S., and K. L. Sangren, 1983: Multiple-Doppler radar network design. *J. Climate Appl. Meteor.*, **22**, 1444–1454.

Ray, P. S., R. J. Doviak, G. B. Walker, D. Sirmans, J. Carter, and B. Bumgarner, 1975: Dual-Doppler observations of a tornadic storm. *J. Appl. Meteor.*, **14**, 1521–1530.

Ray, P. S., K. K. Wagner, K. W. Johnson, J. J. Stephens, W. C. Bumgarner, and E. A. Mueller, 1978: Triple-Doppler observations of a convective storm. *J. Appl. Meteor.*, **17**, 1201–1212.

Ray, P. S., J. J. Stephens, and K. W. Johnson, 1979: Multiple Doppler radar network design. *J. Appl. Meteor.*, **15**, 706–710.

Ray, P. S., C. L. Ziegler, W. Bumgarner, and R. J. Serafin, 1980: Single- and multiple-Doppler radar observations of tornadic storms. *Mon. Wea. Rev.*, **108**, 1607–1625.

Ray, P. S., B. C. Johnson, K. W. Johnson, J. S. Bradberry, J. J. Stephens, K. K. Wagner, R. B. Wilhelmson, and J. B. Klemp, 1981: The morphology of several tornadic storms on 20 May 1977. *J. Atmos. Sci.*, **38**, 1643–1663.

Seliga, T. A., and V. N. Bringi, 1976: Potential use of radar differential reflectivity measurements at orthogonal polarizations for measuring precipitation. *J. Appl. Meteor.*, **15**, 69–76.

Silver, S., 1951: *Microwave Antenna Theory and Design*. McGraw-Hill, New York.

Skolnik, M. I., 1970: *Radar Handbook*. McGraw-Hill, New York.

Stratton, J. A., 1930: The effects of rain and fog upon the propagation of very short radio waves. *Proc. Inst. Electr. Engineers*, **18**, 1064–1075.

Wexler, R., and D. Atlas, 1963: Radar reflectivity and attenuation of rain. *J. Appl. Meteor.*, **2**, 276–280.

Wilson, J., 1970: Integration of radar and raingage data for improved rainfall measurement. *J. Appl. Meteor.*, **9**, 489–497.

Wilson, J., R. Carbone, H. Baynton, and R. Serafin, 1980: Operational application of meteorological Doppler radar. *Bull. Amer. Meteor. Soc.*, **61**, 1154–1168.

Wilson, J., R. D. Roberts, C. Kessinger, and J. McCarthy, 1984: Microburst wind structure and evaluation of Doppler radar for airport wind shear detection. *J. Climate Appl. Meteor.*, **23**, 898–915.

Witt, A., and S. P. Nelson, 1984: The relationship between upper-level divergent outflow magnitude as measured by Doppler radar and hailstorm intensity. Preprints, 22nd Conference on Radar Meteorology, Zurich, Switzerland, American Meteorology Society, Boston, 108–111.

Wood, V. T., and R. A. Brown, 1983: Single Doppler velocity signatures: an atlas of patterns in clear air widespread precipitation and convective storms. NOAA Tech. Memo. ERL NSSL–95, Environmental Research Laboratories (NTIS-#PB84–155779), 71 pp.

Zrnic, D. S., and J. T. Lee, 1982: Pulsed Doppler radar detects weather hazards to aviation. *J. Aircraft*, **19**, 183–190.

CHAPTER 7

The Use of Satellite Data for Mesoscale Analyses and Forecasting Applications

Roderick A. Scofield
James F. W. Purdom

7.1. Introduction

A new era in meteorological observation and forecasting was ushered in with the launching and orbital operations of the TIROS–1 (Television and Infrared Observation Satellite) weather satellite on 1 April 1960. TIROS successfully demonstrated the ability to provide images of the Earth's cloud cover over remote areas of the planet on a vast scale in a timely manner from an Earth-orbiting spacecraft. Today the TIROS–N/NOAA (A–D) polar orbital satellites circle the Earth twice a day at an altitude of approximately 850 km. The instrument payload consists of the following:

- AVHRR (Advanced Very High Resolution Radiometer)—5-channel imaging radiometer producing visible (VIS) and infrared (IR) pictures

- TOVS (TIROS Operational Vertical Sounder) made up of three instruments:
  - HIRS/2 (High-resolution Infrared Radiation Sounder)
  - MSU (Microwave Sounding Unit)
  - SSU (Stratospheric Sounding Unit)

- SEM (Space Environment Monitor)

- DCS (Data Collection System)

The VIS channel (0.55–0.90 μm) and IR channel (10.5–11.5 μm) have resolutions at the subpoint of 1 km.

The SMS (Synchronous Meteorological Satellite), evolved from the ATS–1 (Advanced Technology Satellite) and launched in 1974, provided NOAA a powerful tool to observe the weather on a continuous basis from geostationary orbit. The orbit is at an altitude of 22,000 miles; picture frequency is normally on a half-hourly basis. The SMS prototypes were replaced by the GOES (Geostationary Operational Environmental Satellite) in 1975. The VIS imagery (0.55 μm) and IR imagery (10–12.5 μm) have resolutions of 1 km and 8 km, respectively. Further improvements were introduced with GOES–4 in 1980, providing dual-capability visible and infrared imaging and atmospheric temperature sounding.

Additional operational weather satellites include the polar orbital Defense Meteorological Satellite Program (DMSP), the European geostationary Meteosat system, Japan's Geostationary Meteorological Satellite (GMS), and the geostationary Indian National Satellite (INSAT) system.

7.2. Remote Soundings

Sounding instruments now carried on the NOAA satellites are the beneficiaries of many years of development, mostly on the Nimbus satellites during 1969–1976. From 1973 to 1978 the NOAA satellites carried the first operational sounder, which was succeeded by the current TOVS system with the launch of TIROS–N. Altogether, 18 instruments preceded the TOVS, two of them were microwave radiometers and the others infrared devices. Technological advances had a large role in instrument designs, but new ideas and requirements were also influential.

One polar satellite can provide up to 10,000 soundings per day over the whole globe; two polar satellites provide global observations four times per day. The TOVS provides dense, consistent measurements that are coherent both in time and space. TOVS observations on a global scale can be automatically processed and delivered as input to numerical analysis within operational time constraints. For these reasons, TOVS data are now the prime satellite input to synoptic-scale numerical weather analysis.

Indirect soundings are based on multispectral measurements of radiation from the Earth's atmosphere. The radiation emitted by the Earth's atmosphere at a spectral frequency results from contributions along the line of view from the satellite instrument. Radiation from lower levels is partly absorbed at higher levels and re-emitted there in proportion to the absorption; if the two levels are at different temperatures, then the radiation will be depleted or enhanced depending upon the sense of the temperature differences. This gives rise to a characteristic quantity known as spectral radiance, the radiant energy flowing through a unit area in a given direction per unit of time per unit of spectral frequency per unit of solid angle.

Spectral radiance is thus the sum of the contributions along the viewing path, given by the expression

$$R = R_0 t_0 + B(t/x)dx , \qquad (7.1)$$

in which B is the radiance of a black emitting surface (Planck radiance) at the atmospheric temperature, t is the optical transmittance of the atmosphere, and x is a variable such as distance or pressure; the term $R_0 t_0$ is the contribution by the Earth's surface if the transmittance at a given frequency is not zero there. Equation (7.1) is a statement that the radiance observed from a satellite is a weighted mean of the blackbody radiances B in which the weight is given by t/x, known hereafter as the weighting function. The use of the blackbody function B relies upon a characteristic of the atmosphere known as local thermodynamic equilibrium, or LTE, in which certain statistical properties of molecules are maintained. Equation (7.1) expresses only the effect of atmospheric gases in cloudless areas.

Retrieval processes can be assigned to two classes: probabilistic, in which the temperature at each level in the atmosphere is inferred from statistically based relations between soundings and observations (Smith and Woolf, 1976); and inversion, where (7.1) is mathematically inverted to produce B as a function of x (Wark and Fleming, 1966; Rodgers, 1976). Each technique has its weak and strong points. For example, the probabilistic method is usually more stable, matching radiosonde measurements with satellite radiances so that inferred soundings have the same statistical properties as the basic sample. But in the process there is no assurance that (7.1) holds true. Inversion methods overcome this objection, but are subject to errors from assumptions of the shape of the profile and from systematic errors in our knowledge of transmittances.

Soundings are generally obtained in cloudless atmospheres. However, microwave measurements are useful in the retrieval process in cloudy atmospheres.

An excellent, comprehensive description of methods for extracting quantitative data from satellites (remote soundings, winds, and temperatures) and for applying the data to numerical analysis and prediction is contained in Winston (1979).

7.3. Applicability of Satellite Data

7.3.1. VIS and IR Imagery

With GOES data, a "reporting station" exists every 1 km when VIS data are used, and every 8 km with IR data. The clouds and cloud patterns in a satellite image may be thought of as a visualization of mesoscale meteorological processes. When that imagery is viewed in animation, the movement, orientation, and development of important mesoscale features can be observed, adding a new dimension to mesoscale reasoning. Furthermore, animation provides observations of convective behavior at temporal and spatial resolutions compatible with the scale of the mechanisms responsible for triggering deep and intense convective storms.

7.3.2. Sounding Data

The use of satellite sounding data in mesoscale applications is currently the subject of intense investigation. Considerable effort has been expended comparing individual satellite soundings with those from rawinsondes. For mesoscale applications, that might not be a meaningful approach. The two data sets are inherently different, each with its strengths and weaknesses. From rawinsondes we have 90 minutes of point observations (as the balloon rises) with high vertical resolution and poor spatial resolution. Each 90 minute point observation set is taken by a different sensor, and sets of observations are taken at 12 h intervals (over the United States). Doswell and Lemon (1979) looked in detail at certain atmospheric parameters in severe storm development. They encountered a major sampling problem: the lack of resolution in the operational sounding data base (radiosonde spacing is approximately 400 km at 12 h intervals) relative to the size of the phenomena being predicted. From GOES VIS and IR spin scan radiometer (VISSR) atmospheric sounder (VAS) data we have nearly instantaneous observations through a column in the atmosphere, moderate vertical resolution, high spatial resolution (in the absence of clouds), and one uniformly calibrated sensor making all of the measurements. With GOES–VAS, sounding data may be taken over an area the size of the United States at hourly intervals.

For many mesoscale applications, the most important quantity is the gradient of atmospheric parameters and their changes in time. This is one area where the new observational data from VAS, in combination with data from other sources, should help lead to a better understanding of mesoscale atmospheric processes.

7.4. Differential Heating

An interesting observation concerning local trigger mechanisms for convective development is the strong influence exerted by differential heating. The land-sea breeze is a well-understood differential heating phenomenon whose effects are routinely observed in satellite imagery. Another is the inhibiting effect that early morning cloud cover has on afternoon thunderstorm development. Arc-cloud lines, the driving force for deep convection over the southeast United States in summertime, represent another trigger mechanism due to differential heating.

It is important to realize that these phenomena are mechanisms which establish local convergence zones that are the result of differential heating. Other properties of the atmosphere help determine the effectiveness of those local mechanisms in their ability to generate new deep convection. They include instability, large-scale dynamics, and the trajectory of the low-level air with respect to the convergence zone (amount of time an air parcel will experience vertical motion in the local forcing region).

7.4.1. Sea, Lake, and River Breezes

Although convective cloud development due to terrain influences may often be very complicated when viewed with high-resolution GOES VIS imagery, many of the cloud patterns are easier to understand. This is because the 1 km resolution of the imagery is close to the cumulus cloud scale, and the frequent interval between pictures allows convective development to be observed from its earliest stages through maturity.

The land-sea breeze, a consequence of differential heating between land and adjacent water, has been one of the most widely studied terrain phenomena (Haurwitz, 1947; Estoque, 1962; Pielke, 1973, 1974, and Ch. 22 in this volume). various factors influence the development of the land-sea breeze. Among them are (1) the shape of the coast line, (2) the direction and strength of the gradient winds, (3) friction, (4) the Coriolis effect, (5) the stability of the air mass, and (6) the land and water temperature difference. Items 4–6 are not discussed here although infrared data are applicable to item 6 and VAS sounding data have a potential relative to item 5.

Different curvatures in a coast line cause areas of convergence or divergence along the land-sea breeze front, thus leading to a local strengthening or weakening of cumulus activity along that front. As pointed out by Pielke (1974), "Local maxima in vertical motion form in regions where the curvature of the coast line accentuates the horizontal convergence created by the differential heating between land and water." Additionally, a small peninsula is generally an area of earlier strong convective development along the land-sea breeze front because the breezes formed along opposing shores merge near the peninsula's center.

Convective development due to land and water interfaces also occurs regularly around lakes (Lyons, 1966). The factors influencing convective development are the same as those for the land-sea breeze. However, the low-level wind field becomes increasingly important, the smaller the lake involved.

Although the effects of oceans, lakes, and even cities (Changnon, 1976) on convective development have received considerable attention, relatively little work has been done concerning the effects of rivers and adjacent moist or swampy areas on convective development. GOES imagery has shown that rivers and their adjacent areas at certain times exert a considerable influence on convective development. The influence has been most noticeable when the low-level winds are either nearly calm (generally less than 5 m s^{-1}) or blowing parallel to the river.

7.4.2. Early Morning Cloud Cover

Weak Synoptic Forcing

Purdom and Gurka (1974) discussed the effects of early morning cloud cover on afternoon thunderstorm development under conditions of weak synoptic-scale forcing; the situation is similar to that of the land-sea breeze, with the first shower clouds forming in the clear region near the boundary of the early morning cloud cover—a sort of cloud-breeze front. Additionally, they found that the slower heating rate in the early cloudy areas helped keep those regions free from convection for most of the day. Figure 7.1 is a good example of this phenomenon.

Although the effect of early cloud cover may most easily be thought of as a simple differential heating mechanism, for mesoscale applications our reasoning must extend beyond that point. Unlike the land-sea breeze regime, the cloud field is constantly changing in character: this can affect the development of instability since varying amounts of insolation will lead to differences in heating and mixing over the land area.

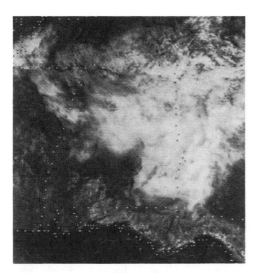 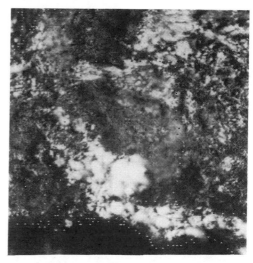

Figure 7.1. GOES-East 1 km VIS images, 27 May 1977 at 1530 and 1930 GMT. Convective development over Alabama is due to early cloud cover. These images show the effect early cloud cover can have on afternoon thunderstorm development. Note that the early clear region over southwest Alabama becomes filled with strong convection during the day, whereas the early cloud region over the remainder of the state evolves into mostly clear skies. The strongest activity later in the day develops in the "notch" of the clear region in south-central Alabama, as one might expect from merging cloud breeze fronts.

Strong Synoptic Forcing

Early morning cloud cover can also have an important role in helping the stage for intense convection under conditions of strong synoptic-scale forcing. Purdom and Weaver (1982) showed the importance of mesoscale boundary interactions in focusing tornado activity in the Red River Valley area on 10 April 1979. Figure 7.2 shows the location of one of the mesoscale frontal boundaries that helped focus that activity. The most probable cause of that meso-front was differential heating due to the cloudy (stratus) region to the east versus the clear area to the west. The mechanism that led to the development of the meso-front, and subsequent focusing of tornadic activity, also played an important role in the development of instability in the warm sector. As the mesoscale frontal boundary moved eastward, mostly clear skies developed in the warm sector between the boundary and the stratiform overcast to the east. During that period Abilene (ABI) became clear while Stephenville (SEP) maintained its cloud cover. An analysis of surface static energy (Darkow, 1968) pointed to strong potential instability at both ABI and SEP (Purdom and Weaver, 1982). However, time series of horizontal cross sections from mesoscale rawinsonde data showed a marked decrease in the amount of negative buoyant energy at ABI and only a slight decrease at SEP; these changes are related to changes in cloud cover. Low-level air, similar in character to that near ABI, fed the tornadic storm system and thus allowed intense thunderstorms to develop.

Figure 7.2a. GOES-East 1 km VIS image, 10 April 1979 at 2126 GMT.

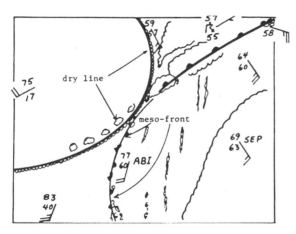

Figure 7.2b. Analysis of significant features and cloud patterns in Fig. 7.2a.

Using a numerical boundary layer model, McNider and colleagues (McNider *et al.*, 1984) examined a case in which early cloud cover played an important role in the generation of a squall line when synoptic-scale forcing was present. They found shading due to cloud cover to be fundamentally important in the formation of a local baroclinic zone and the development of low-level convergence.

The effect of early morning cloud cover can be complicated. It can act to set up a baroclinic zone (or reinforce an existing one) through differential heating. At the same time, it can affect the local destabilization of an airmass: (a) If skies clear too quickly, moisture may be mixed to great depths, making the region unsuitable for supporting strong moist convection; (b) if the area remains cloudy, thunderstorms moving into the region might dissipate or weaken considerably because of the negatively buoyant low-level air; (c) if the area clears an hour or two before thunderstorms move into it, sufficient heating and mixing at low level may have occurred, priming the local air mass to support explosive convection. Use of VAS sounding channel data should aid in the assessment of the convective potential of such situations for mesoscale forecasting.

7.5. Thunderstorm Outflow

7.5.1. Arc-Cloud Lines

During the decade following the Thunderstorm Project, detailed analyses of mesoscale systems associated with thunderstorms showed a sharp pressure rise accompanying the systems' arrival (Fujita, 1955; Fujita and Brown, 1958). Other phenomena noted with the passage of such systems were a wind shift that was often accompanied by an increase in wind speed, a temperature decrease, and in many instances rainfall. Although it was generally recognized that evaporation of precipitation was an important factor in the production of the higher pressure and sharp pressure rise associated with such systems, Fujita (1959) first quantitatively investigated that production. He showed that evaporation of raindrops in the downdraft was responsible for the development of a cold dome of air beneath the thunderstorm and the resultant mesoscale high-pressure system. At the leading edge of this "mesohigh" was the thunderstorm gust front. The subsidence of the cold dome is due to the cold dome's greater hydrostatic pressure, which causes it to sink and spread outward in an attempt to reach equilibrium with the surrounding environment.

It was Purdom (1973) who first pointed out that in satellite imagery, the leading edge of the mesohigh appears as an arc-shaped line of convective clouds moving out from a dissipating thunderstorm area. The arc-shaped cloud line is normally composed of cumulus, cumulus congestus, or cumulonimbus clouds. These arc-cloud lines play a part in controlling the development and evolution of deep convection through convective-scale interactions. Often the arc-cloud lines in the satellite image are not detected by radar.

Figure 7.3 shows the characteristic appearance of arc-cloud lines in satellite imagery. Figure 7.4 shows a mesoscale analysis/nephanalysis of those arc-cloud lines, for which 15-min-interval satellite images were used. In this case, interaction between the arc cloud and frontal boundary in Oklahoma produced a large severe thunderstorm.

7.5.2. Convective-Scale Interaction

NESDIS and NASA, recognizing the importance of geostationary satellite data in understanding convective development, have operated the GOES system in a special rapid interval (3–5 min) imaging mode on selected days during convective seasons since 1975. Those unique data sets have allowed observation of convective behavior at temporal and spatial resolutions compatible with the scale of the physical mechanisms responsible for triggering deep and intense convective storms. Analyses of movies made from these data show that convective-scale interaction is of primary importance in determining the development and evolution of deep convection (Bohan, 1981). Such interaction occurs when thunderstorm-produced arc-cloud lines merge or intersect with other convective lines, areas, and boundaries. Pertinent conclusions from studies using rapid interval imagery can be summarized as follows:

- Thunderstorm-produced outflow boundaries (arc clouds) are of primary importance in the formation and maintenance of strong convection.

- Outflow boundaries may maintain their identity as arc clouds for several hours after the convective array that produced them has dissipated.

- Deep convective development along an arc-cloud line is favored where two arcs intersect or where an arc moves into a convectively unstable region.

Figure 7.3. GOES-East 1 km VIS image, 26 May 1975 at 2000 GMT.

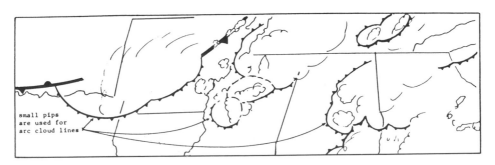

Figure 7.4. Analysis of arc-cloud lines in Fig. 7.3.

• In a weakly forced atmosphere, arc boundary interactions determine the location of the majority of new thunderstorms that occur toward the end of the day.

The convective-scale interaction process is fundamental in the evolution and maintenance of deep convective activity, and until recently, it was observed only by satellite. Thunderstorm evolution that appears as a random process when it is observed by conventional radar is often found to be well ordered when viewed in time lapse satellite imagery.

7.6. Mesoscale Convective Systems

7.6.1. Squall Line Development

Generally, organized convergence lines that trigger strong convection (such as fronts, drylines, or pre-frontal convective lines) are detectable in satellite imagery prior to deep convective development along them (Purdom, 1976). An example of a frontal system that develops into a severe squall line is shown in Fig. 7.5. Early squall lines of this type are routinely detected in GOES imagery prior to deep convective development on them and their detection by radar. In this case, a Pacific front that extended from eastern South Dakota and Nebraska into central Kansas developed into a line of severe thunderstorms as it moved into Minnesota and Iowa later in the

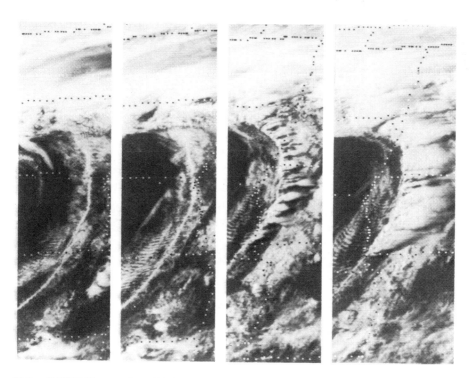

Figure 7.5. GOES-East 1 km VIS images, 14 June 1976 at 1900 GMT, 2000 GMT, and 2300 GMT; a typical example of squall line development.

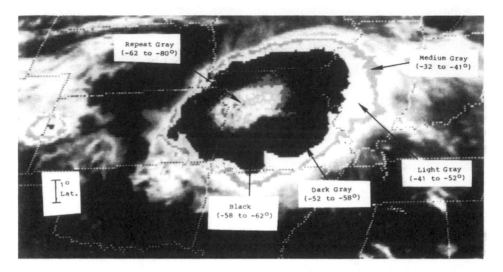

Figure 7.6. GOES-East enhanced IR image of a typical mesoscale convective complex (MCC) over the central United States on 13 August 1982 at 0730 GMT.

day. Near the time of the latest image shown in Fig. 7.5, large hail and funnel clouds were reported in South Dakota, a tornado injured six people in Minnesota, and a weak tornado was reported in Iowa.

7.6.2. Mesoscale Convective Complexes

Although squall lines generally form under conditions of moderate to strong synoptic-scale forcing, a different type of highly organized mesoscale convective system, which forms under conditions of weak synoptic-scale forcing, was documented by Maddox (1980). These spring and summertime convective weather systems, which occur most often over the central United States during the late evening and nighttime hours, have been given the name Mesoscale Convective Complex (MCC). MCCs appear to be convectively driven; although they appear to have many characteristics similar to those of tropical convective systems, their dynamics are not well understood. MCCs have been observed to interact with and modify the larger scale environment in which they are embedded. By influencing the larger scale environment they affect downstream weather long after their demise.

As with arc-cloud lines, MCCs were not recognized prior to observations afforded by GOES. Figure 7.6 shows the characteristic appearance of an MCC in satellite imagery. It should be noted that MCCs produce much of the beneficial rainfall that occurs over the central plains of the United States. However, they are also prolific producers of severe weather.

7.7. Using VIS Imagery to Isolate Severe Convective Storms

7.7.1. Squall Line Boundary Interaction

Organized convergence lines that trigger strong convection (drylines, fronts, etc.) are detectable in GOES imagery before deep convection forms.

Under proper dynamic forcing, when the thunderstorms that form along such lines interact with other boundaries, severe storms develop (Miller, 1972). For example, on 6 May 1975 strong tornado activity that began in northeast Nebraska developed southeastward into the Omaha area along a squall line and warm frontal intersection. Figure 7.7 shows how these features as identified in satellite imagery could be tracked during the outbreak. The developing squall line appeared as an organized line of convective clouds along a surface windshift line separating moist and dry air. The warm frontal boundary appeared as another organized line of convection separating warm moist air embedded in southerly flow from slightly cooler and drier air in southeasterly to easterly flow. Other clues helpful in the precise location of

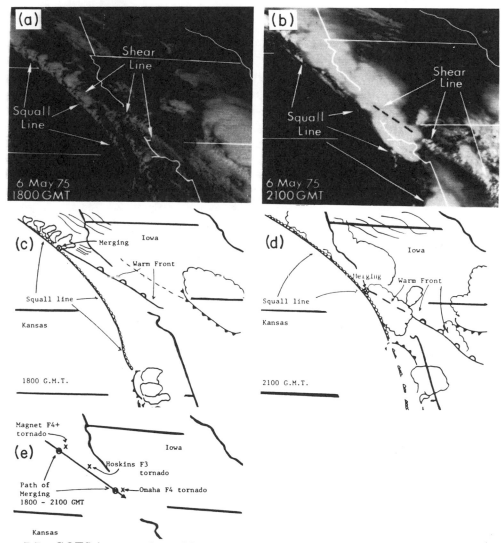

Figure 7.7. GOES imagery for 6 May 1975, the day of the Omaha tornado, at (a) 1800 GMT and (b) 2100 GMT. (c,d) Significant features in (a), and (b), derived by combining information from surface observations and cloud patterns in GOES imagery. (e) Tornado locations relative to the path of merging between the squall line and the warm frontal boundary.

the warm-frontal boundary may be found by inspecting the change in cloud type across it (cumulus versus stratus), as well as through cloud-tracked winds.

7.7.2. Prior Convective Activity in Severe Storm Development

Arc clouds and their convective-scale interactions are a natural part of the convective cloud genesis and evolution process. They occur anywhere and, because of the strong vorticity and convergence that they produce, are often associated with severe storm development. An extreme example was the storm associated with the most destructive tornado in the history of Wyoming (Parker and Hickey, 1980). Figure 7.8 is a satellite view and analysis of the mesoscale convective environment prior to the Chicago tornadoes of 13 June 1976. Tornadic activity began shortly after the west-to-east arc-cloud line south of Lake Michigan moved north and interacted with the storm over the Chicago area. Numerical cloud models (Klemp and Rotunno, 1983) support satellite imagery indications of the importance of arc-cloud lines (outflow) in producing tornadoes.

Arc-cloud lines are not the only satellite indicator of previous convection. Figure 7.9 is a good example of how satellite imagery can be used to identify a

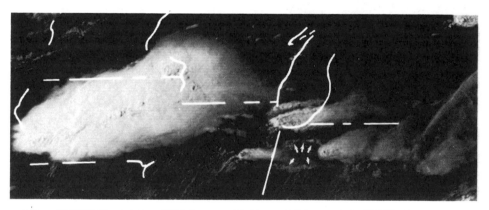

Figure 7.8a. GOES-East 1 km VIS image, 13 June 1976 at 2245 GMT, showing mesoscale convective development prior to the Chicago tornadoes.

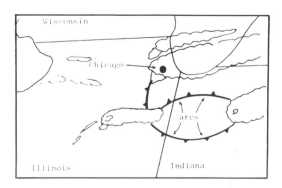

Figure 7.8b. Analysis of significant features and cloud patterns in Fig. 7.8a.

transition zone between an unstable air mass and a more stable mesoscale air mass generated by earlier thunderstorm activity. Note the cumulus cloudiness (unstable) at A and the stable wave-cloud-dominated air at B. Both are low-level cloud features. In this particular case the wave-dominated air had been stabilized by early morning thunderstorms in the area. The low-level air mass boundary that existed between these waves and cumulus streets was instrumental in the development of tornadic storms upon interaction with a cold frontal zone.

The importance of prior convection in setting the stage for tornadic storm activity was vividly demonstrated by a storm system that moved through the southeastern United States on 28 March 1984. Thunderstorm activity that had moved through the area during the early morning hours produced a well-defined outflow boundary that extended west from South Carolina across Georgia and into Alabama. This boundary moved slowly northeast during the day, and by midafternoon (1930 GMT) was evident in satellite imagery as a line of organized cumulus congestus clouds extending across South Carolina into North Carolina (Fig. 7.10a). The large storm in northeast Georgia lies at the junction of the convergence boundary and a mesolow. That storm developed into a large supercell, which tracked east-northeast along the boundary, producing most of the intense killer tornado activity (Fig. 7.10b; see also Fig. 7.11).

Note that, for both 27 March and 28 March, strong synoptic-scale forcing set the stage for severe storm development. However, the major meteorological difference between 28 March (a long-track tornadic supercell producing killer tornadoes) and 27 March (numerous reports of severe activity but no major tornadoes) was the existence of a well-organized low-level convergence

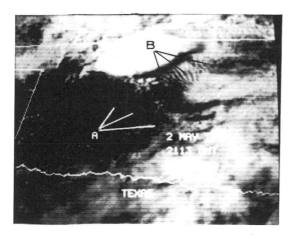

Figure 7.9a. GOES-East 1 km VIS image, 2 May 1979 at 2030 GMT, showing transition zone between (A) an unstable and (B) a stable air mass.

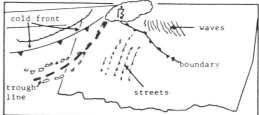

Figure 7.9b. Analysis of significant features and cloud patterns in Fig. 7.9a.

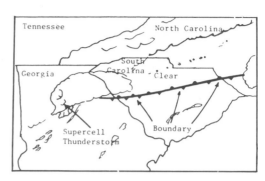

Figure 7.10a. GOES-East 1 km VIS image, 28 March 1984 at 2030 GMT. Arrows indicate the outflow boundary.

Figure 7.10b. Analysis of significant features and cloud patterns in Fig. 7.10a.

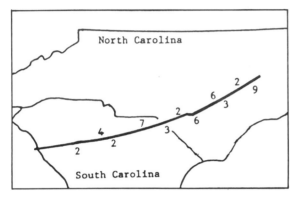

Figure 7.11. Path of supercell that produced significant tornadoes, and the numbers of deaths associated with those tornadoes for 28 March 1984.

boundary and mesolow to help support the supercell's growth and development.

7.7.3. Vertical Wind Shear

It is well known that vertical wind shear has an important role in determining the character of storms that evolve in a mesoscale environment (Newton, 1963). Furthermore, recent numerical cloud-modeling studies have shown the importance of vertical wind shear in the formation of rotation in growing thunderstorms (Klemp and Wilhelmson, 1978; Bleckman, 1981). However, a severe storm environment is one in which both the dynamic and thermodynamic characteristics of the atmosphere are changing on mesoscale space and time domains. Determination of vertical wind shear in regions of growing cumulus clouds using rapid scan satellite imagery is a feasible method of studying mesoscale variations in that parameter over large areas.

After thunderstorms have developed, the imagery may be animated relative to the thunderstorm, and the flows at different levels with respect to the storm may be inspected. Purdom *et al.*(1982) used GOES-East images of a storm (11 April 1979, 0015 GMT) to study this phenomenon; Fig. 7.12a shows the satellite image, and Fig. 12b shows the flow relative to the storm, derived from 30-min-interval data at low, middle, and high levels. It is interesting to note how closely the flow compares with that proposed by Browning (1964) for severe storms that travel to the right of the wind, and

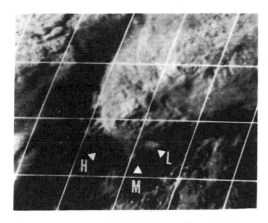

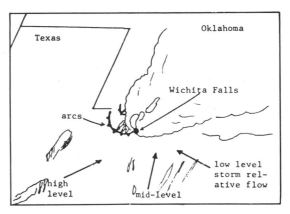

Figure 7.12a. GOES-East 1 km VIS image, 11 April 1979 at 0015 GMT. H, M, and L indicate regions of high-, middle- and low-level flow relative to thunderstorm.

Figure 7.12b. Analysis of cloud-relative flow and significant features in Fig. 7.12a.

storm-relative proximity soundings (Maddox, 1976). Being able to diagnose storm-relative flow has important implications for defining mesoscale regions that are favorable for the production of rotating storms and severe weather.

7.7.4. Overshooting Tops

Overshooting thunderstorm tops as detected in satellite VIS imagery, and their relation to severe weather, continue to be investigated. Early work in this area was done by Pearl (1974) and Fujita *et al.*(1976); photographs of overshooting tops taken from airplanes were correlated with overshooting tops detected in satellite imagery and severe weather. Pearl's study used ATS data, and Fujita's study used the higher resolution data available from SMS (GOES). Hasler (1981) found similar results, correlating severe events with overshooting tops; he used stereo measurements of cloud top characteristics obtained by combining GOES-East and GOES-West data.

Overshooting tops are generally associated with updraft regions that penetrate well above the equilibrium level (EL) as defined by the top portion of the spreading thunderstorm anvil. Although overshooting tops are often associated with intense convection, there is one serious limitation in using them as storm intensity indicators: they are detectable for only a few hours in the early morning or late afternoon when the sun is low enough on the horizon for the tops to cast shadows on the underlying anvil.

7.8. Using IR Imagery

7.8.1. Combining IR Imagery and Radar Data

For almost 40 years, radar reflectivity has been the standard for remotely assessing a thunderstorm's intensity; use of satellite IR data is relatively new. Both satellite and radar data have important roles. Satellite data provide information on cloud top mean vertical growth rates, cloud top temperature, and anvil expansion rates to help assess storm intensity; conventional radar data provide information about reflectivity and volumetric echo properties and their changes in time.

The utility of satellite and radar data in thunderstorm analysis has been the subject of limited investigation. Negri and Adler (1981) found that the location of satellite-defined thunderstorms coincided with radar echo locations and that radar reflectivity correlated with satellite-based estimates of intensity. Purdom et al.(1982) and Green and Parker (1983) found similar results concerning satellite-defined thunderstorm tops and those defined by radar. Reynolds and Smith (1979) demonstrated the concept of a composite radar and satellite display to study severe storm development and convective rainfall. Green and Parker (1983) used an improved product to develop a composite satellite IR/Doppler radar image for the purpose of evaluating radar echo changes in time as they relate to infrared cloud top behavior.

Those studies and one by Fujita (1982) produced these findings:

- When storms are forming in a line, with one storm downwind from another, the downwind storm's cloud top temperature may appear warmer than it should because of wake cirrus masking. Furthermore, to be able to follow the evolution of these tops and to observe this masking requires satellite imagery at 3 min intervals.

- The coldest satellite cloud top temperature in an anvil may not necessarily overlie the low-level core of highest radar reflectivity.

- When satellite IR data are used to measure cloud top growth characteristics, cirrus masking may make results less accurate for certain storms. Radar Constant Altitude Plan Position Indicator displays help to ensure that the correct infrared anvil area is being matched with the correct echo.

- The horseshoe cold ridge, or "V notch" observed in satellite IR imagery is an indicator of intense local convection. Storms whose anvils exhibit that characteristic should be closely monitored for severe weather.

7.8.2. Clues in IR Imagery for Severe Storm Identification

Since the earliest days of geostationary meteorological satellites, various cloud top properties have been measured in attempts to diagnose storm severity. Using ATS imagery, Sikdar et al.(1970) and others related thunderstorm anvil growth rates to the occurrence of severe weather. Those results showed promise; however, further developments awaited the arrival of GOES and its more quantifiable infrared data. Using GOES IR rapid scan data, Adler and Fenn (1977, 1979) correlated the areal expansion rate of isotherms colder than the tropopause with severe storm occurrence. They found that both the divergent growth of areas within various IR contours and the decrease of temperatures in the main convective growth were twice as fast for thunderstorms producing severe weather (4.9 m s^{-1}) as for ordinary area thunderstorms.

Figure 7.13 represents the rapid growth of a severe storm at Miles City, Mont., that produced winds of >80 km and 2 inch hail. Pryor (1978) used rapid scan GOES IR data to relate severe storms with cloud top temperatures colder than the environmental tropopause temperature. He found that only those thunderstorms with IR cloud top temperatures colder than the tropopause produced severe weather. In a similar study, Reynolds (1979) found a positive relationship between the occurrence of hail and IR tops colder than the tropopause. Weaver and Purdom (1983) noticed a displacement between the position of the overshooting tops on the VIS imagery versus the location of the cold top on IR. In nearly every case, the cold top was found to be slightly upwind of the overshooting top.

Fujita (1978) characterized anvil tops of severe storms on the basis of the pattern of equivalent blackbody temperature as shown in enhanced GOES IR imagery. This signature appears as a cold "V" shape in the anvil top, with marked downstream warming from the center of the V. Figure 7.14 shows a detailed analysis of GOES infrared data for the storm that produced the Grand Island, Nebr., tornadoes; note the cold V (horseshoe ridge), the intense overshooting cold tops, and the downstream warm region (wake). This type of anvil signature was also documented by Reynolds (1979) for severe hailstorms. Using operational GOES data, McCann (1981) showed a good correspondence between the enhanced V-notch signature and severe weather. Downbursts on the larger scale (macroburst of 2.5 mi or larger outflow size) are associated with intense convection. Intense downdrafts sometimes occur in conjunction with tornadic storms (Forbes and Wakimoto,

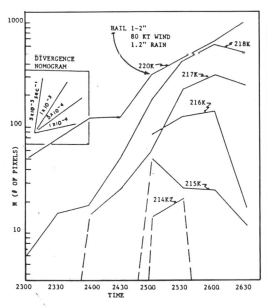

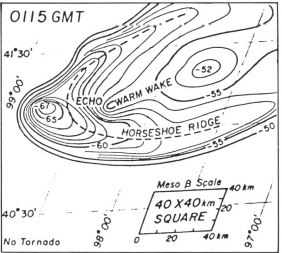

Figure 7.13. Temporal variation of the number of pixels enclosed by the given isotherm for temperatures colder than the environmental tropopause, for 18 July 1978. (From Adler and Fenn, 1977.)

Figure 7.14. Isotherms of IR temperature of the Grand Island tornado cloud at 0115 GMT, drawn at 1°C intervals. Horseshoe ridge (cold) is shown with dashed arc. Warm wake is on the downwind side of radar echoes. (From Fujita, 1981.)

Table 7.1. VAS Instrument Characteristics*

Spectral channel	Central wavelength (μm)	Absorbing constituent	Weighting Functions		Surface or cloud emission effect
			Peak level (mb)	Representative thickness (mb)	
1	14.7	CO_2	40	150–10	Usually none
2	14.5	CO_2	70	200–30	None below 500 mb
3	14.2	CO_2	300	500–10	None below 800 mb
4	14.0	CO_2	450	800–300	Weak
5	13.3	CO_2	950	Sfc–500	Moderate
6	4.5	CO_2	850	Sfc–500	Moderate
7	12.7	H_2O	Surface	Sfc–700	Strong
8	11.2	Window	Surface	—	Strong
9	7.2	H_2O	600	800–400	Weak at surface
10	6.7	H_2O	450	700–250	None at surface
11	4.4	CO_2	500	800–100	Weak
12	3.9	Window	Surface	—	Strong

*From Zehr and Green (1984).

1983). Although the V-notch signature is a good positive indicator of severe storm activity, it is not always present with tornadic storms.

7.9. Sounding Information From GOES–VAS

Sounding information from GOES–VAS can detect small significant temporal variations in atmospheric temperature and moisture (Smith *et al.*, 1981). The sounder on GOES–VAS has seven CO_2 channels, three water vapor channels, and two window channels; instrument characteristics that relate to vertical resolution are shown in Table 7.1.

GOES–VAS can be operated in one of three modes: (1) a normal VISSR mode as with GOES, (2) a dwell sounding mode, which uses all or most of the sounder channels, and (3) a multispectral imaging mode in which VISSR data plus one or two additional VAS channels are sampled. The three possible modes of operation are not independent; when the satellite is operated in one mode, data from the other modes are not available. Data from the VAS channels may be viewed individually or combined to produce mesoscale soundings. Both methods are applicable to mesoscale convection.

7.9.1. Mesoscale Moisture Information

The moisture structure of the atmosphere, both in the horizontal and vertical, is an important factor in storm development (Doswell and Lemon, 1979): significant variance exists within the mesoscale range below 100 to 200 km. Using sounding information from polar-orbiting satellites, Hillger and Vonder Haar (1979, 1981) were able to extract moisture information at

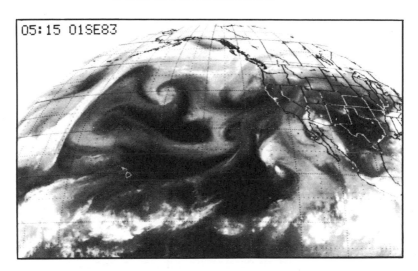

Figure 7.15. 6.7 μm channel image, for 1 September 1983 at 0515 GMT, showing moist, cool areas (light tones) and warmer, drier areas (dark tones).

a much finer resolution than was available from conventional data sources. Although GOES–VAS is still in the evaluation phase (the system will become operational in mid-1986) similar results concerning mesoscale moisture distribution have been demonstrated (Smith *et al.*, 1981; Petersen and Mostek, 1982).

GOES–VAS has three water vapor channels (Table 7.1). The energy received in each of those channels is a function of the amount of water vapor, its distribution and temperature within a portion of a column in the atmosphere, and surface (except 6.7 μm) and cloud effects. Except where there is significant low-level water vapor, the 12.7 μm band and the 11.2 μm band produce similar information. Therefore, it is valid to find the difference in energy between those two bands. That difference, termed the "split window," represents low-level water vapor. According to Chesters *et al.*(1982),

> The VAS split window differentiates those areas in which water vapor extends over a deep layer and is more able to support convective cells from those areas in which the water vapor is confined to a shallow layer and is therefore less able to support convection.

This value is very useful over land during the afternoon when there will be a large difference between signals from the split window channels. However, in the evening when the land temperature has cooled (or over the ocean) the signal differential is again small and a meaningful product is difficult to derive.

The 6.7 μm channel image (Fig. 7.15) shows regions of middle-level moisture and clouds. The moister and cooler areas and the warmer and drier areas are readily detected. These features are related to both synoptic and mesoscale advection and vertical motion. When viewed in time lapse, they exhibit excellent spatial and temporal continuity. Strong baroclinic regions such as jet streams and vorticity maxima can often be easily identified in cloud-free regions by the sharp moisture gradient detected in the 6.7 μm image (Anderson et al., 1982).

Anthony and Wade (1983) demonstrated the capability of GOES–VAS data in severe weather analysis. VAS can provide useful temporal and spatial continuity in monitoring mesoscale stability trends. In addition, VAS water vapor imagery can be used to monitor the progression or evolution of jet streams, jet maxima, and dry areas aloft (dry slots). If the dry area surges downwind of moist, unstable lower levels and the middle- and upper-level flow is at least slightly diffluent, then severe thunderstorms will probably develop.

7.9.2. Retrievals From VAS

Initial research with GOES–VAS data concentrated on deriving vertical profiles of temperature and moisture based on observed radiances. Details of the retrieval algorithms were given by Smith (1983). Since that initial research, significant results have been achieved in deriving objective mesoscale convective forecast parameters from VAS retrieval information (Smith et al., 1984). VAS-derived products that are being provided in real time to the National Severe Storms Forecast Center in Kansas City include (1) upper and lower wind analyses derived from VAS thermal field gradient winds and cloud drift winds, (2) analyses of lifted index, (3) 850 mb and 500 mb temperature fields, (4) total precipitable water, (5) a thermodynamic stability index similar to a total-totals index, and (6) a statistical probability estimate of severe weather potential. Products 1, 2, 3, and 6 are presented at 80 km resolution; products 4 and 5 are presented at the full 7 km GOES–VAS resolution as image products.

Although VAS-derived products have demonstrated their usefulness for diagnosing the convective storm environment, questions still must be answered concerning the utility of individual VAS soundings for such analyses. VAS-derived soundings contain information on the atmosphere's thermal and moisture characteristics at a spatial and temporal resolution never before available. Now the question is, "How do we best use this information?"

Clouds have historically been viewed as a source of contamination to satellite sounding data. However, work is now under way to add information for the cloudy regions. In one approach, information on the atmosphere's thermal and moisture structure is being fine tuned according to the cloud type and structure revealed in GOES VIS and IR imagery. In a similar approach (Smith et al., 1984), cloud drift winds from the "cloud-contaminated regions" are being blended with thermal information from VAS to improve regional-scale analyses.

7.10. Precipitation Estimation Techniques

Precipitation information is a primary requirement of hydrologists and agriculturalists around the world. Also of utmost importance is the need to estimate areas of heavy precipitation prior to issuance of flash flood or winter storm warnings and special weather statements. Hydrologists, meteorologists, and river forecasters use precipitation estimates as an aid in their evaluation or prediction of flood potential.

7.10.1. VIS and IR Techniques

The GOES spacecraft, which provides half-hourly coverage of that part of the Earth viewed, is the primary source of precipitation estimates. For those areas where geostationary satellite data are not available, however, polar-orbiter data are used. Other data sources used in estimating precipitation include NMC (National Meteorological Center) analyses and forecasts, conventional atmospheric soundings, radar data, satellite-derived soundings, and rainfall climatologies. Satellite-derived precipitation estimation techniques ranging in uses from the mesoscale to the climate scale are covered by Barrett and Martin (1981).

There are two basic types of VIS and IR precipitation estimation techniques: cloud history and indexing. VIS and IR methodologies infer precipitation amounts for specific cloud features. Cloud history works best where geostationary satellite data are available. The frequent views from geostationary satellites allow the life cycle of a cloud to be followed and precipitation estimates to be computed for each stage of the cloud's development. In contrast, cloud indexing is the principal method used when only polar-orbiter satellite data are available; only two pictures a day can be obtained from a single polar-orbiting satellite. Cloud indexing involves characterizing a cloud by an index number according to its appearance in imagery and then using a look-up table or regression equation to estimate the precipitation from the cloud. Both the cloud history and cloud indexing methods have procedures for modifying the estimates for different climates and environments.

Estimates of rainfall are a function of the following:

- Dominant organization of the synoptic weather.

- Proportion of sky that is cloud covered.

- Intensity of rain, which is influenced by the types of cloud present, cloud top temperature/cloud growth, overshooting tops, mergers, etc.

The intended use of the precipitation analysis determines the size of the area over which the estimates are computed: small areas for heavy rain, large areas for climate.

7.10.2. Passive Microwave Techniques

The primary advantage of microwave measurements is their ability to probe through clouds, rain being the major source of attenuation for the "window" frequencies below 50 GHz. Furthermore, over low-emissivity sea surfaces the brightness temperature measurements in clearer areas are highlighted against the more emissive, warmer measurements in precipitating regions. The large contrast (>50 K) in brightness temperature between the rain and its surroundings has stimulated much interest in applying microwave radiometry over oceans for determining rainfall. Satellite and aircraft instruments have been used to obtain rain data for tropical storms (Allison *et al.*, 1974; Adler and Rodgers, 1977; Jones *et al.*, 1981). The global distribution of rainfall over the oceans was also attempted from satellite measurements (Rao *et al.*, 1976).

Single-Frequency Technique

In most situations it is sufficient to relate rain intensity directly to increase in brightness temperature over oceans. Effects such as those of sea surface winds on emissivity or the contributions due to cloud and water vapor absorption are of second order and can usually be neglected.

The more variable and higher emissivity of land presents problems because of the similarity in brightness temperature of rain and land. Also, decreases in surface emissivity due to wet ground and snow cover can produce false precipitation signatures if not accounted for (Rodgers and Siddalingaiah, 1983). Only in the case of heavy thunderstorms can the rain measurements be much lower (>50 K) than the surrounding brightness temperatures. The enhancement for the intense storms is principally due to the scattering of upwelling radiation by large ice particles at the tops of rain layers. This scattering effect increases with increasing frequency, and was first noted during observation of convective storms over the United States, using scanning multi-channel microwave radiometers (SMMRs)(Spencer *et al.*, 1983). Brightness temperatures as low as 163 K were obtained at the center of an intense storm for the highest frequency (37 GHz) channel.

In the case of stratiform rain the contribution due to scattering is small. To identify rain one must use the smaller difference between the thermal emission from precipitation drops and that from the land background. Dual frequency and/or polarization techniques are being developed to enhance the precipitation signature over land by minimizing the effect of surface emissivity on the microwave measurements. These approaches use statistical correlations between emissivities at different frequencies or between the two polarizations.

Frequency Screening

For many surfaces the emissivity has either very little frequency dependence or increases with frequency. This is true for wet and dry soil, vegetation, open water, melting snow, lake ice, and new sea ice. In contrast, brightness temperatures of rain cells decrease with increasing frequency; i.e., temperatures are due to scattering and thermal emission. It is therefore possible to isolate the precipitation signature by measuring the difference between the brightness temperatures at two frequencies—$T_b(v_1) - T_b(v_2)$, where $v_2 > v_1$—and displaying only the positive values. However, surfaces such as dry snow have a frequency response similar to that of rain, and would not be screened out. This problem can be alleviated by viewing sequential measurements and associating the rapid changes with precipitation. Also,

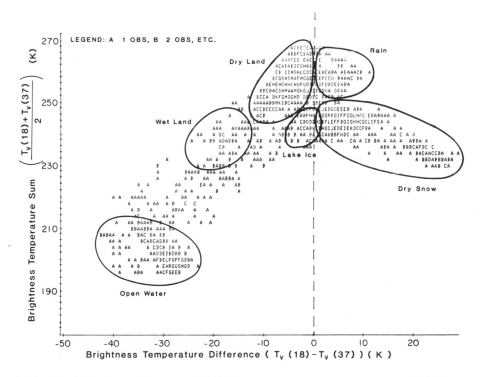

Figure 7.16. Typical cluster diagram of brightness temperature sum versus brightness temperature difference; 18 and 37 MHz measurements were used.

the brightness temperature sum at the two frequencies— $T_b(v_1) + T_b(v_2)$— can be used to make the final discrimination (i.e., stratiform rain has higher brightness temperatures than dry snow).

Separating snow cover from rainfall requires the use of the brightness temperature sum as well as the difference. For example, the cluster diagram of Fig. 7.16 is based on the January measurements from the Gulf of Mexico to Hudson Bay. The clusters corresponding to dry snow and to rain both lie in the positive quadrant. Fortunately, they are separated vertically by brightness temperature. The negative quadrant contains clusters formed by dry soil, wet land along the Mississippi Delta, and open water in the Great Lakes and Gulf of Mexico. The scattered points between the wet land and open water correspond to fields of view containing mixtures of different surfaces, e.g., land and water. The small cluster labeled lake ice represents data from over Hudson Bay.

Polarization Algorithm

Frequency screening uses the correlation between emissivities at different frequencies to enhance the precipitation signature. Alternatively, one can employ polarization information to minimize the effect of emissivity on measurements. The vertical and horizontal emissivity components are bound by the curves representing a specular and diffuse surface. To minimize the brightness temperature variations due to surface variations, it is sufficient to combine the vertical T_v and horizontal T_h polarization measurements according to the linear transformation $(T_v - BT_h)/1 - B)$. The parameter B physically represents the linear slope of vertically polarized to horizontally polarized emissivity and can be estimated from plots of T_v against T_h for land and water surfaces (Weinman and Guetter, 1977). It is found that B has a value near 0.52 at 37 GHz. By using the radiative transfer equation, it can be shown that the transformed measurements approximately represent the brightness temperature of a unity emissivity surface (Grody, 1984). As such, the weighted measurements show a large contrast between the atmospheric emission (and scattering) due to precipitation and the higher background emission.

7.10.3. Heavy Precipitation and Flash Flood Forecasting

VIS and IR data are being used operationally for analyzing and forecasting heavy precipitation events. Satellite-derived precipitation estimates and 3 h precipitation trends for convective systems (Scofield and Oliver, 1977; Scofield, 1984; and Spayd and Scofield, 1984a), extratropical cyclones (Scofield and Spayd, 1984b), and tropical cyclones (Spayd and Scofield, 1984b) are computed on the NESDIS Interactive Flash Flood Analyzer (IFFA) and transmitted to NWS Forecast offices, NWS offices, and River Forecast centers. The National Hurricane Center uses digitized infrared satellite data for forecasting tropical cyclone rainfall at landfall (Jarvinen and Griffith, 1981).

The operational NESDIS Convective Storm Technique gives half-hourly or hourly rainfall estimates for convective systems by using GOES IR and high-resolution VIS images. The technique is designed for deep convective systems that occur in tropical air masses with high tropopauses, and it uses IR images (Fig. 7.6) digitally enhanced according to the Mb Curve designed to help estimate convective storm intensity. Estimates of convective rainfall are computed by comparing changes in cloud character between two consecutive images. The technique is divided into three main steps (Fig. 7.17).

The Convective Storm Technique has been modified so that the temperature of the convection computed from a sounding is compared with the observed cloud-top temperature. This computed temperature (called the Equilibrium Level) is the best measure of the expected anvil temperature and should be used for examining the anvil growth rates. Cloud top temperatures equal to or colder than the computer temperature would indicate heavier rainfall rates than warmer ones.

Accurate verification of operationally produced rainfall estimates for deep convective rainfall events is very difficult because of the extreme variations in the temporal and spatial distribution of convective precipitation. Studies of thunderstorm events over dense raingauge networks have shown that variations of up to 2 inches of rainfall over 1 n mi range in a half-hour period do occur.

Nevertheless, a verification study of the operationally produced rainfall estimates from May through July 1984 was completed by comparing the total maximum rainfall estimates for a storm (usually a period of 1 to 6 h)

CONVECTIVE STORM TECHNIQUE

STEP 1

RAINFALL IS COMPUTED ONLY FOR THE <u>ACTIVE</u> PORTION OF THE THUNDERSTORM SYSTEM:

THE FOLLOWING ARE CLUES FOR HELPING TO MAKE THIS DECISION.

. IR TEMPERATURE GRADIENT IS TIGHTEST AROUND STATION END OF ANVIL FOR A THUNDERSTORM SYSTEM WITH VERTICAL WIND SHEAR (IR).
. STATION IS LOCATED NEAR THE CENTER OF THE ANVIL WITH A TIGHT, UNIFORM IR TEMPERATURE GRADIENT AROUND ENTIRE ANVIL FOR A THUNDERSTORM SYSTEM WITH NO VERTICAL WIND SHEAR (IR).
. AN OVERSHOOTING TOP IS OVER THE STATION (VIS AND IR).
. ANVIL IS BRIGHTER AND/OR MORE TEXTURED (VIS).
. FROM COMPARING LAST TWO PICTURES: STATION IS UNDER HALF OF ANVIL BOUNDED BY EDGE WHICH MOVES LEAST (IR).
. STATION IS NEAR 300-MB UPWIND END OF ANVIL (IR, SKIP THIS CLUE IF NO UPPER AIR DATA AVAILABLE).
. STATION IS NEAR THE AREA OF LOW-LEVEL INFLOW (VIS).
. STATION IS LOCATED UNDER A RADAR ECHO.

STEP 2

HALF-HOURLY RAINFALL ESTIMATES IN INCHES ARE COMPUTED FROM THE FOLLOWING FACTORS:

FACTOR 1

CLOUD-TOP TEMPERATURE AND CLOUD GROWTH FACTOR [IR].
DETERMINE AMOUNT THAT THE COLDEST CLOUD TOPS INCREASED WITHIN HALF-HOUR.

	>2/3° LAT	>1/3° ≤2/3° LAT	≤1/3° LAT OR SAME	AREAL DECREASE OF SHADE OR WARMING FROM WHITE TO RPT GRAY OR WITHIN THE RPT GRAY	COLDEST TOPS 1 OR MORE SHADES WARMER
MED GRAY (-32 TO 41°)	0.25	0.15	0.10	0.05	↑
LT GRAY (-41 TO -52°)	0.50	0.30	0.15	0.10	
DK GRAY (-52 TO -58°)	0.75	0.40	0.20	0.15	
BLACK (-58 TO -62°)	1.00	0.60	0.30	0.20	↓
RPT GRAY*(-62 TO -80°)	1-2.00	0.60-1.00	0.30-0.60	0.30	
WHITE (BELOW 80°)	2.00	1.00	0.60	0.40	0.10

*COLDER REPEAT GRAY SHADES SHOULD BE GIVEN HIGHER RAINFALL ESTIMATES.

OR

DIVERGENCE ALOFT FACTOR* [IR AND 200-MB ANALYSIS].

MED GRAY	LT GRAY	DK GRAY	BLACK	RPT GRAY	WHITE
0.15	0.30	0.40	0.60	0.60-1.00	1.00

*IR IMAGERY SHOWS EDGES OF THUNDERSTORM ANVIL ALONG THE UPWIND END FORMING A LARGE ANGLE OF BETWEEN 50-90 DEGREES POINTING INTO THE WIND; 200-MB ANALYSIS OFTEN SHOWS THESE STORMS JUST DOWNWIND FROM WHERE THE POLAR JET AND SUBTROPICAL JET SEPARATE.

GO TO FACTOR 2

Figure 7.17a. Step 1 and Step 2 (Factor 1) of technique to estimate convective rainfall. Temperatures in Factor 1 are in centigrade. Data to be used in the analyses are indicated in the parentheses.

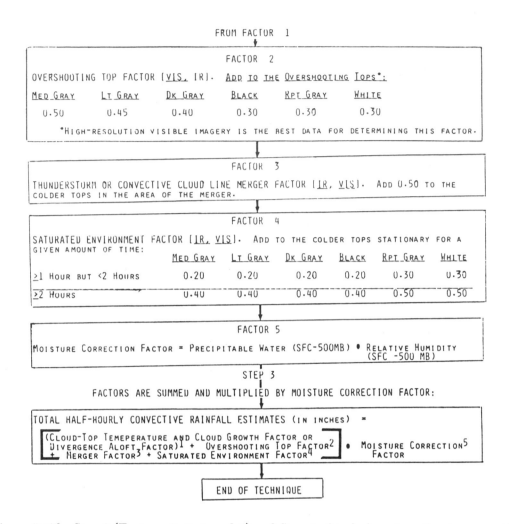

Figure 7.17b. Step 2 (Factors, 2, 3, 4, and 5) and Step 3 of technique to estimate convective rainfall. Data to be used in the analyses are indicated in the parentheses.

with the 24-hourly cooperative rainfall reports available the next day. Events that included less than two or three rainfall reports per county, and events in which additional rain fell outside the time rainfall estimates were produced, were eliminated; 268 events remained.

The verification results showed that the average error of the rainfall estimates for a storm total precipitation event is about 30%. The absolute and average errors both increase as the magnitude of the event increases. For relatively small events (3.9 inches or less) there is a tendency to overestimate the event, and as the magnitude of the event increases there is a distinct tendency to underestimate. If a rainfall estimate is 4 inches or more, there is an excellent probability that at least 4 inches will be observed. Improved heavy precipitation estimates are possible when passive microwave data are combined with VIS and IR data.

7.11. Cloud-Tracked Winds

Geostationary satellites provide the frequent coverage and the animated display of cloud motion that are needed to derive cloud-tracked winds. Quantitative estimates of wind derived from cloud motion cr cloud drift have become an important satellite product. These estimates not only supplement other forms of wind measurement, but also provide the only wind information over many areas. In most cases, winds are derived operationally at least twice a day from each satellite.

Researchers also make use of satellite winds for any number of specialized purposes. Usually they derive their own winds to satisfy a particular need. For instance, they may wish to derive winds more frequently than twice a day, in greater density, or over much shorter picture intervals than the usual 30 min (down to 3 min intervals) in order to define mesoscale wind fields and variations. Mesoscale wind fields have great potential for studying the precursor severe storm environment, the tropical cyclone environment, and other mesoscale phenomena.

Conceptually, deriving winds from cloud motion is simple. It involves tracking a cloud over two or more images and calculating the Earth distance traversed in a given time and the directional angle. Accurate knowledge of the Earth locations at the first and last picture are critical to the calculation. But the accuracy required is really related more to relative registration of images than absolute location. Motion must be measured relative to a fixed Earth background. Of course, accurate absolute location can become important for certain meteorological purposes; for instance, the precision of Earth location is more important in mesoscale analysis than in global scale analysis.

Several strategies exist for determining height. Most seek height of the cloud top since that is what the satellite observes from above; there are exceptions. Height determination methods include approximation, climatology, shadows, stereoscopy, cloud-top temperature, multiple channels, and multiple CO_2 channels.

Comparisons with rawinsondes have been the most popular approach, probably because balloon measurements are the most readily available and because they have long represented the standard for conventionally measured profiles of wind, temperature, and humidity. Mean absolute vector differences between cloud and balloon winds are generally estimated (Serebreny *et al.*, 1970; Hubert and Whitney, 1971; Suchman and Martin, 1976; Whitney, 1983) to range anywhere from 2 to 5 m s^{-1} at low levels and from 3 to 10 m s^{-1} or sometimes more at high levels. These ranges probably relate to variations in the severity of the editing and selection procedures, the approach of the measurement technique, and variations in climatic character of the wind depending on the geographical area and season dominating each study.

For a most detailed and comprehensive discussion of cloud-tracked winds, readers are referred to Hubert (1979).

Satellite gradient winds are calculated using geopotential height fields analyzed from satellite soundings such as those measured by the VAS. The soundings afford the advantage of providing vertical distributions of wind as rawinsondes do rather than just single-level vectors as cloud and water-vapor tracking do. Moreover, the technique works best in a clear atmosphere where satellite soundings are most reliable. Gradient wind data can complement cloud motion data whenever motions derived from tracking moisture variations fail.

7.12. Future Data Resources and Satellite Systems

The geostationary satellite has the unique ability to observe the atmosphere (sounders) and its cloud cover (visible and infrared) frequently and from the synoptic scale down to cloud scale. This ability to provide frequent, uniformly calibrated data sets over a broad range of meteorological scales places the geostationary satellite at the very heart of the understanding of mesoscale weather development.

As GOES–VAS data become available with greater regularity, significant breakthroughs in predicting mesoscale phenomena will result. Short-range forecasting algorithms that use GOES–VAS data will be developed. In the near future, satellite data (including passive microwave data) will be combined in real time with radar data and surface and upper-air observations. If short-range forecasting is to improve, a primary objective for the future is to have the ability to integrate, in an operational environment, satellite information with other data and technologies (Doppler radar, vertical profiles, etc.). New geostationary operational satellites (GOES-Next, successor to VAS) will have (1) an independent imaging and sounding system rather than the current system, which is limited to either imaging or sounding at any given time; (2) increased location accuracy for analysis and forecasting of small-scale meteorological phenomena; (3) improved monitoring of moisture, and improved vertical resolution and accuracy in soundings; (4) expanded data collection system to include search-and-rescue capabilities. The first flight of GOES-Next is expected to take place in 1990.

REFERENCES

Adler, R. F., and D. D. Fenn, 1977: Satellite-based thunderstorm intensity parameters. Preprints, 10th Conference on Severe Local Storms, Omaha, American Meteorological Society, Boston, 8–15.

Adler, R. F., and E. B. Rodgers, 1977: Satellite-observed latent heat release in a tropical cyclone. *Mon. Wea. Rev.*, 105, 956–963.

Allison, L. J., E. B. Rodgers, T. T. Wilheit, and R. W. Fett, 1974: Tropical cyclone rainfall as measured by the Nimbus 5 electrically scanning microwave radiometer. *Bull. Amer. Meteor. Soc.*, 55, 1074–1089.

Anderson, R. K., J. J. Gurka, and S. J. Steinmetz, 1982: Application of VAS multispectral imagery to aviation forecasting. Preprints, 9th Conference on

Weather Forecasting and Analysis, Seattle, American Meteorological Society, Boston, 227–234.

Anthony, R. W., and G. S. Wade, 1983: VAS operational assessment findings for spring 1982/1983. Preprints, 13th Conference on Severe Local Storms, Tulsa, American Meteorological Society, Boston, 323–328.

Barrett, E. C., and D. W. Martin, 1981: *The Use of Satellite Data in Rainfall Monitoring.* Academic Press, New York, 340 pp.

Bleckman, J. B., 1981: Vortex generation in a numerical thunderstorm mode. *Mon. Wea. Rev.*, 109, 1061–1071.

Bohan, W. A., 1981: The importance of thunderstorm outflow boundaries in the development of deep convection. WAB 456, 16 mm color sound movie, The Walter A. Bohan Co., 2026 Oakton Street, Park Ridge, Ill., 29 minutes.

Browning K. A., 1964: Airflow and precipitation trajectories within severe local storms which travel to the right of the winds. *J. Atmos. Sci.*, 21, 634–639.

Changnon, S. A., 1976: Effects of urban areas and echo merging on radar echo behavior. *J. Appl. Meteor.*, 15, 561–570.

Chesters, D., L. W. Uccellini, and A. Mostek, 1982: VISSR Atmospheric Sounder (VAS) simulation experiment for a severe storm environment. *Mon. Wea. Rev.*, 110, 198–216.

Darkow, G. L., 1968: The total energy environment of severe storms. *J. Appl. Meteor.*, 7, 199–205.

Doswell, C. A. III, and L. R. Lemon, 1979: An operational evaluation of certain kinematic and thermodynamic parameters associated with severe thunderstorm environments. Preprints, 11th Conference on Severe Local Storms, Kansas City, American Meteorological Society, Boston, 397–402.

Estoque, M. A., 1962: The sea breeze as a function of the prevailing synoptic situation. *J. Atmos. Sci.*, 19, 244–250.

Forbes, G. S., and R. M. Wakimoto, 1983: A concentrated outbreak of tornadoes, downbursts, and microbursts and implications regarding vortex classification. *Mon. Wea. Rev.*, 111, 220–235.

Fujita, T. T., 1955: Results of detailed synoptic studies of squall lines. *Tellus*, 7, 405–436.

Fujita, T. T., 1959: Precipitation and cold air production in mesoscale thunderstorm systems. *J. Meteor.*, 16, 454–466.

Fujita, T. T., 1978: Manual of downburst identification for Project NIMROD. SMRP Research Paper 156, University of Chicago, 103 pp.

Fujita, T. T., 1981: Mesoscale aspects of convective storms. Proceedings, IAMAP Symposium: Nowcasting: Mesoscale Observations and Short-Range Prediction, Hamburg, European Space Agency, 3–10.

Fujita, T. T., 1982: Infrared, stereo-height, cloud-motion, and radar-echo analysis of SESAME-day thunderstorms. Preprints, 12th Conference on Severe Local Storms, San Antonio, American Meteorological Society, Boston, 213–216.

Fujita, T. T., and H. A. Brown, 1958: A study of mesosystems and their radar echoes. *Bull. Amer. Meteor. Soc.*, 39, 538–554.

Fujita, T. T., G. S. Forbes, and T. A. Umenhofer, 1976: Close-up of 20 March 1975 tornadoes: Sinking cloud tops to suction vortices. *Weatherwise*, 29, 115–132.

Green, R. N., and H. A. Parker, 1983: Application of satellite and radar data to severe thunderstorm analysis. Preprints, 13th Conference on Severe Local Storms, Tulsa, American Meteorological Society, Boston, 190–193.

Grody, N. C., 1984: Precipitation monitoring over land from satellites by microwave radiometry. Proceedings, International Geoscience and Remote Sensing Symposium (IGARSS '84), Strasbourg, France, 27–30 August, European Space Agency SP–215, 417–423.

Hasler, A. F., 1981: Stereographic observations from geosynchronous satellites: An important new tool for the atmospheric sciences. *Bull. Amer. Meteor. Soc.*, 62, 194–212.

Haurwitz, B., 1947: Comments on the sea breeze circulation. *J. Meteor.*, 4, 1–8.

Hillger, D. W., and T. H. Vonder Haar, 1979: Analysis of satellite infrared soundings of the mesoscale using statistical structure and correlation functions. *J. Atmos. Sci.*, 36, 287–305.

Hillger, D. W., and T. H. Vonder Haar, 1981: Retrieval and use of high-resolution moisture and stability fields from Nimbus 6 HIRS radiances in preconvective situations. *Mon. Wea. Rev.,* **108**, 1010–1028.

Hubert, L. F., 1979: Wind derivation from geostationary satellites. Quantitative Meteorological Data From Satellites, J. S. Winston (Ed.), WMO Technical Note 166, 33–59.

Hubert, L. F., and L. F. Whitney, Jr., 1971: Wind estimation from geostationary satellite pictures. *Mon. Wea. Rev.,* **99**, 665–672.

Jarvinen, B. R. and C. G. Griffith, 1984: Forecasting rainfall in tropical cyclones using digitized infrared satellite data. Unpublished NOAA/ERL NHC Technical Paper, Miami, 12 pp.

Jones, W., P. G. Black, V. E. Delnore, and C. T. Swift, 1981: Airborne microwave remote-sensing measurements of Hurricane Allen. *Science,* **214**, 274–280.

Klemp, J. B., and R. Rotunno, 1983: A study of the tornadic region within a supercell thunderstorm. *J. Atmos. Sci.,* **40**, 359–377.

Klemp, J. B., and R. R. Wilhelmson, 1978: The simulation of three-dimensional convective storm dynamics. *J. Atmos. Sci.,* **35**, 1070–1096.

Lyons, W. A., 1966: Some effects of Lake Michigan upon squall lines and summertime convection. SMRP Research Paper, 57, Dept. Geophys. Sci., University of Chicago, 22 pp.

Maddox, R. A., 1976: An evaluation of tornado proximity wind and stability data. *Mon. Wea. Rev.,* **104**, 133–142.

Maddox, R. A., 1980: Mesoscale convective complexes. *Bull. Amer. Meteor. Soc.,* **61**, 1374–1387.

McCann, D. W., 1981: The enhanced-V, a satellite observable severe storm signature. NOAA Tech. Memo. NWS NSSFC-4, National Severe Storms Forecast Center, Kansas City, Mo. (NTIS-#PB81-230328), 31 pp.

McNider, R. T., G. Jedlovec, and G. Wilson, 1984: Data analysis and model evaluation of the initiation of convection on 24 April 1982. Preprints, 10th Conference on Weather Forecasting and Analysis, Clearwater, American Meteorological Society, Boston, 543–549.

Miller, R. C., 1972: Notes on analysis and

severe-storm forecasting procedures of the Air Force Global Weather Central. AWS Technical Report 200 (rev.), Air Weather Service (MAC), U.S. Air Force (NTIS#AD-744042), 190 pp.

Negri, A., and R. Adler, 1981: Relation of satellite-based thunderstorm intensity to radar-estimated rainfall. *J. Appl. Meteor.,* **20**, 66–78.

Newton, C. W., 1963: Dynamics of severe convective storms. In *Severe Local Storms,* Meteor. Monogr. 5(27), American Meteorological Society, Boston, 33–58.

Parker, W. T., and R. D. Hickey, 1980: The Cheyenne tornado of 16 July 1979. *Natl. Wea. Dig.,* **5**, 45–62.

Pearl, E. W., 1974: Characteristics of anvil-top associated with the Poplar Bluff tornado of May 7, 1973. SMRP Research Paper 119, University of Chicago, 12 pp.

Petersen, R. A., and A. Mostek, 1982: The use of VAS moisture channels in delineating regions of potential convective instability. Preprints, 12th Conference on Severe Local Storms, San Antonio, American Meteorological Society, Boston, 168–171.

Pielke, R., 1973: A three-dimensional numerical model of the sea breezes over south Florida. NOAA Tech. Memo. ERL WMPO-2, Environmental Research Laboratories, Boulder, Colo. (NTIS#COM-73-11307/8), 136 pp.

Pielke, R., 1974: A three-dimensional numerical model of the sea breezes over south Florida. *Mon. Wea. Rev.,* **102**, 115–139.

Pryor, S. P., 1978: Measurement of thunderstorm cloud-top parameters using high-frequency satellite imagery. M. S. Thesis, Dept. of Atmospheric Science, Colorado State University, Fort Collins, 101 pp.

Purdom, J. F. W., 1973: Meso-highs and satellite imagery. *Mon. Wea. Rev.,* **101**, 180–181.

Purdom, J. F. W., 1976: Some uses of high resolution GOES imagery in the mesoscale forecasting of convection and its behavior. *Mon. Wea. Rev.,* **104**, 1474–1483.

Purdom, J. F. W., and J. G. Gurka, 1974: The effect of early morning cloud cover on afternoon thunderstorm development. Preprints, 5th Conference on Weather Forecasting and Analysis, St. Louis, American Meteorological Society, Boston, 58–60.

Purdom, J. F. W., and J. F. Weaver, 1982: Nowcasting during the 10 April 1979 tornado outbreak: A satellite perspective. Preprints, 12th Conference on Severe Local Storms, San Antonio, American Meteorological Society, Boston, 467–470.

Purdom, J. F. W., R. N. Green, and H. A. Parker, 1982: Integration of satellite and radar data for short range forecasting and storm diagnostic studies. Preprints, 9th Conference on Weather Forecasting and Analysis, Seattle, American Meteorological Society, Boston, 51–55.

Rao, M. S. V., W. V. Abbott, III, and J. S. Theon, 1976: Satellite-derived global oceanic rainfall atlas (1973 and 1974). NASA SP–410, Washington, D.C. (NTIS#N77-19709).

Reynolds, D., 1979: Observations and detection of damaging hailstorms from geosynchronous satellite digital data. Preprints, 11th Conference on Severe Local Storms, Kansas City, American Meteorological Society, Boston, 181–188.

Reynolds, D. A., and E. Smith, 1979: Detailed analysis of composited digital radar and satellite data. *Bull. Amer. Meteor. Soc.*, 60, 1024–1037.

Rodgers, C. D., 1976: Retrieval of atmospheric temperature and composition from remote measurements of thermal radiation. *Rev. Geophys. Space Phys.*, 14, 609–624.

Rodgers, E., and H. Siddalingaiah, 1983: The utilization of Nimbus-7 SMMR measurements over land. *J. Clim. Appl. Meteor.*, 22, 1753–1763.

Scofield, R. A., 1984: The NESDIS operational convective precipitation estimation technique. Preprints, 10th Conference on Weather Forecasting and Analysis, Clearwater Beach, American Meteorological Society, Boston, 171–180.

Scofield, R. A., and V. J. Oliver, 1977: A scheme for estimating convective rainfall from satellite imagery. NOAA Tech. Memo. NESS–86, National Earth Satellite Service, Washington, D.C. (NTIS-#PB-270762/8G1), 47 pp.

Scofield, R. A., and L. E. Spayd, Jr., 1984: A technique that uses satellite, radar and conventional data for analyzing and short-range forecasting of precipitation from extratropical cyclones. NOAA Tech. Memo. NESDIS–8, National Environmental Satellite, Data, and Information Service, Washington, D.C. (NTIS#PB85-164994), 51 pp.

Serebreny, S. M., E. J. Wiegman, and R. E. Hadfield, 1970: Further comparison of cloud motion vectors with rawinsonde observations. Final Report, ESSA Contract No. E210–69(N), Stanford Research Institute, Menlo Park, Calif., 60 pp.

Sikdar, D. N., V. E. Suomi, and C. E. Anderson, 1970: Convective transport of mass and energy in severe storms over the United States—An estimate from geostationary altitude. *Tellus*, 22, 521–532.

Smith, W. L., 1983: The retrieval of atmospheric profiles from VAS geostationary radiance observations. *J. Atmos. Sci.*, 40, 2025–2035.

Smith, W. L., and H. M. Woolf, 1976: The use of eigenvectors of statistical covariance matrices for interpreting satellite sounding radiometer observations. *J. Atmos. Sci.*, 33, 1127–1140.

Smith, W. L., V. E. Suomi, W. P. Menzel, H. M. Woolf, L. A. Stromousky, H. E. Revercomb, C. M. Hayden, D. N. Erickson, and F. R. Mosher, 1981: First sounding results from VAS-D. *Bull. Amer. Meteor. Soc.*, 62, 232–236.

Smith, W. L., G. S. Wade, W. P. Menzel, V. E. Suomi, F. J. Fox, C. S. Velden, and J. S. LeMarshall, 1984: Nowcasting—advances with McIDAS III. Preprints, Nowcasting II Symposium, Norrköping, Sweden, European Space Agency, Paris, 433–438.

Spayd, L. E., Jr., and R. S. Scofield, 1984a: An experimental satellite-derived heavy convective rainfall short range forecasting technique. Preprints, 10th Conference on Weather Forecasting and Analysis, Clearwater Beach, American Meteorological Society, Boston, 400–408.

Spayd, L. E., Jr., and R. S. Scofield, 1984b: A tropical cyclone precipitation estimation technique using geostationary satellite data. NOAA Tech. Memo. NESDIS–5, National Environmental Satellite, Data, and Information Service, Washington, D.C. (NTIS-#PB84-226703), 36 pp.

Spencer, R. W., W. S. Olson, Wu Rongzhang, D. Martin, J. A. Weinman, and D. A. Santek, 1983: Heavy thunderstorms observed over land by the Nimbus-7 scanning multichannel microwave radiometer. *J. Clim. Appl. Meteor.*, **22**, 1041–1046.

Suchman, D., and D. W. Martin, 1976: Wind sets from SMS images: An assessment of quality for GATE. *J. Appl. Meteor.*, **15**, 1265–1278.

Wark, D. Q., and H. E. Fleming, 1966: Indirect measurement of atmospheric temperature profiles from satellites: I. Introduction. *Mon. Wea. Rev.*, **94**, 351–362.

Weaver, J. F., and J. F. W. Purdom, 1983: Some unusual aspects of thunderstorm cloud top behavior on May 11, 1982. Preprints, 13th Conference on Severe Local Storms, Tulsa, American Meteorological Society, Boston, 154–157.

Weinman, J. A., and P. J. Guetter, 1977: Determination of rainfall distributions from microwave radiation measured by the Nimbus-6 ESMR. *J. Appl. Meteor.*, **16**, 437–442.

Whitney, L. F., 1983: International comparison of satellite winds—an update. *Adv. Space Res.*, **2**, 1265–1277.

Winston, J. S. (Ed.), 1979: Quantitative Meteorological Data From Satellites. WMO Technical Note 166, Geneva, Switzerland, 102 pp.

Zehr, R. M., and R. N. Green, 1984: Mesoscale applications of VAS imagery. Preprints, Conference on Satellite Meteorology/Remote Sensing and Applications, Clearwater Beach, American Meteorological Society, Boston, 94–98.

CHAPTER 8

Operational Objective Analysis Techniques and Potential Applications for Mesoscale Meteorology

Ronald D. McPherson

8.1. Introduction

Objective analysis is a process by which meteorological observations distributed in space and time, and from different observing systems, are combined with other information—predictions from previous analyses, or perhaps climatology—to form a numerical representation of the state of the atmosphere. This representation takes the form of digital values of the pressure, temperature, wind, and moisture at a set of regularly spaced grid points covering the domain of interest, or as the coefficients of series of expansions representing the variables.

The descriptor "objective" appears in the first paper published on this topic, by Panofsky (1949). Panofsky was seeking to distinguish meteorological analyses done manually from those by computer. The former tend to be not reproducible for a given set of data, but vary depending upon the skill and judgment of the analyst. The latter are objective in the sense that the method follows a prescribed set of rules, or "model," and will produce the same analysis given the same set of data. Prescribing the set of rules, however, is not objective and varies from one technique to another. Nevertheless, the term "objective analysis" is widely accepted, and has become more or less synonymous with automated analysis methods using large computers.

Analyses produced by objective methods are used by meteorologists for diagnostic purposes and as initial conditions for numerical weather prediction models. The latter is the focus of this chapter. Indeed, objective analysis methodology evolved as a result of requirements imposed by numerical weather prediction. At the beginning of the numerical weather prediction era, initial conditions for the early numerical predictions were obtained from manually analyzed meteorological charts. Values of the analyzed parameters were laboriously interpolated to the intersections of the grid point lattice upon which the finite-difference model equations were to be solved, and then manually entered on punched cards. It became evident that automation of this process would be essential for operational numerical weather prediction to be successful.

The Joint Numerical Weather Prediction Unit (JNWPU), supported by the U.S. Air Force, Navy, and Weather Bureau, was established in July 1954 and first began making scheduled numerical predictions in April 1955. Subjective analyses for these predictions were provided by the National Weather Analysis Center. By July 1955, experiments were in progress using an automated procedure, or objective analysis. An operational version was implemented on 10 October 1955.

The papers and other documents describing the first few years of operational numerical weather prediction reveal much concern with incorporating certain principles of subjective meteorological analysis into the automated procedures. Among them are spatial coherence, temporal continuity, and adherence to dynamic constraints. Early investigators wished to generate analyses that vary smoothly in space, depict a reasonable evolution of the atmosphere in time, and exhibit approximate hydrostatic and geostrophic equilibrium between the geopotential, temperature, and wind. Consideration of these principles has continued in subsequent years as objective analysis methodology has evolved.

That evolution has been driven principally by two interrelated developments: rapid progress in prediction modeling and advances in observing technology. The first numerical predictions made by the JNWPU were severely limited in domain and in vertical resolution, and the upper-air data base consisted of data from radiosonde stations distributed over Northern Hemisphere continents. Only a few reports from the Ocean Station Vessels were available for oceanic regions. Operational prediction models are now global, with much enhanced resolution in both the vertical and horizontal dimensions, and with quite sophisticated modeling of physical processes such as radiation and precipitation. The data base is still founded upon the radiosonde network, but has been augmented by other observing systems such as space-based remote temperature sensors and instruments aboard modern commercial aircraft.

Both of these developments have been associated with the analysis and prediction of large-scale atmospheric motions—those with characteristic length scales greater than 2000 km and with periods of several days. The extension of numerical analysis and prediction techniques to smaller scales—e.g., the meso-α scale (200 km to 2000 km)—has not been seriously considered, in part because of inadequate understanding of the physical and dynamical processes associated with the phenomena, and in part (and closely related) because a suitable observational system does not exist. In recent years, however, comprehension of the physical and dynamical processes has been greatly enhanced through the development and use of sophisticated prediction models. An excellent review of the current status of mesoscale numerical modeling efforts has been provided by Anthes (1983). Several mesoscale models, e.g., those of Anthes and Warner (1978), Perkey (1976), and Kaplan *et al.* (1982), have demonstrated an ability to simulate many aspects of meso-α-scale phenomena. Reliable prediction of those phenomena from real, daily-varying initial conditions has not yet been achieved, at least in part because of an inadequate observing system. The smaller scales and shorter life cycles of meso-α events require much closer spacing of observational locations and much higher frequency of observations in time. Only at the surface does a network exist that even approximately satisfies the requirements, and surface data alone are not sufficient.

Advances in observing technology have begun to indicate the possibility of alleviating this deficiency over the next decade. An operational meso-α-scale observing network over the United States now appears possible. These systems, among others, are expected to contribute:

- Profiler. The Profiler has two components: a clear-air Doppler radar to measure wind profiles from near the surface to the tropopause, and a passive upward-looking microwave radiometer for profiles of temperature and moisture. Profiles once each hour, or more frequently, may be obtained with both components.

- VAS (VISSR Atmospheric Sounder). Passive radiometers on geostationary space platforms can provide temperature and humidity soundings with good horizontal and temporal resolution, but poor vertical resolution.

- Aircraft-borne instruments. Inertial navigation systems on wide-bodied commercial aircraft, together with modern communications, have made possible the automatic acquisition of wind and temperature information from a large number of aircraft platforms. These data are available at cruise altitude with a frequency of eight reports each hour, or about every 100 km at typical aircraft speeds. Higher frequency reporting is possible on ascent and descent, thus approximating soundings.

- Doppler radar. NOAA will soon deploy new surveillance radars to replace the WSR–57 sets now in use throughout the United States. The new radars can also be used to monitor winds in the vicinity of the radar site.

Each new observing system has advantages and disadvantages. Each has its own measurement capabilities and error characteristics. No single system is sufficient. Instead, a national meso-α observing system will no doubt include several of the new systems, as well as familiar balloon-borne measurements. The task of mesoscale objective analysis will be to transform information from these disparate sources into mathematically regular and meteorologically complete sets of numbers that represent the state of the atmosphere—a challenging task indeed.

8.2. Basic Concepts in Objective Analysis

To illustrate the principal concepts in objective analysis it suffices to consider only the geopotential at first, and later the wind.

8.2.1. Analysis Methods

There are two basic methods of analyzing meteorological observations (geopotential, temperature, wind, etc.). The first is often called the "grid point" method, where the analyzed value Z^a at a discrete point g is given by a linear combination of observations that are nearby in space and time:

$$Z_g^a = a_1 Z_1^o + a_2 Z_2^o + \cdots + a_n Z_n^o , \qquad (8.1)$$

where Z_1^o, Z_2^o, etc., are observed values at the several stations around point g, and the a_i's are coefficients that determine the influence of each observation on the analyzed value. The analysis then consists of determining the coefficients a_i in the linear combination for each point of the analysis grid. An array of analyzed values of Z results at the discrete points of the analysis grid.

A second basic method of objective analysis represents an analyzed field as a series expansion:

$$Z_g = a_1 f_1 + a_2 f_2 + \cdots + a_M f_M , \qquad (8.2)$$

where Z_g represents the departure of the field from its mean value. The f_i are a set of orthogonal functions, for example, a cosine series:

$$f_j = \cos \frac{2\pi j x}{L} , \; j = 1 , \; M . \qquad (8.3)$$

Once again, the analysis consists of determining the coefficients a_i, but in this case by "fitting" the mathematical representation to the observed data Z^o (expressed also as a departure from the mean value) by, for example, a least-squares technique. The quantity

$$S = \sum_{i=1}^{N} \left[Z^o(i) - \sum_{j=1}^{M} a_j f_{ij} \right]^2 \qquad (8.4)$$

is minimized by differentiating with respect to each of the a_j and equating the result to zero. This yields a set of M linear equations in the M coefficients a_j that can be solved by standard methods. Once the a_j are determined, the analyzed value Z_g [e.g., (8.2)] may be obtained at any point where the f_{ij} are defined, simply by evaluating (8.2). Thus, if the functions of f_{ij} are continuous, the representation of the meteorological variable is also continuous. This method of objective analysis is often called "surface fitting," because the series description of the variable in (8.2) can be thought of as a surface, albeit a very complicated one, that is "fitted" to the observations in some way.

Associated with the two basic methods of analysis are two forms of representation: the discrete form, in which the analysis is a set of numbers at discrete points in space and time; and the spectral form, in which the analysis is represented by a series expansion such as (8.2). The spectral form has not been widely used for the limited-area applications in mesoscale meteorology.

The first objective analysis method in meteorology was of the surface-fitting type, devised by Panofsky (1949). Somewhat later, an extension of this method was developed by Gilchrist and Cressman (1954) and was used as the first operational objective analysis by the JNWPU. The method fitted a conic section locally—i.e., to a small set of observations in the vicinity of each analysis point. Around the boundaries between data-rich and data-sparse areas, the local fits were not well behaved, and produced unreasonable analyses. The method was replaced by a grid point method adapted by Cressman (1959) after earlier work by Bergthorssen and Döös (1955). The surface-fitting method reappeared in the British Meteorological Office analysis (Dixon *et al.*, 1972), and in the global analysis devised by Flattery (1971). Flattery used a series expansion on the set of eigenfunctions of Laplace's tidal equation, called Hough functions. These are defined globally, and so are not susceptible to the earlier difficulties of the local surface-fitting methods.

8.2.2. Resolution

The smallest resolvable scale represents the maximum resolution in an objective analysis. For the surface-fitting method, this scale is defined by the truncation M in (8.4). In the grid point method, the maximum resolution depends on the spacing of the discrete points of the analysis grid. To illustrate this, the solid line in Fig. 8.1a presents a trace of 500 mb height around latitude 40°N, represented as a continuous curve. Suppose only a very coarse analysis grid can be afforded, one with points every 60°, and the "analysis" consists of simply connecting the "observation" points by line segments, as shown by the dashed line in Fig. 8.1a. Clearly, only the largest features are suggested, and these are grossly distorted. Decreasing the spacing between analysis points to 40°, as in Fig. 8.1b, and then to 30° (Fig. 8.1c) and finally to 20° (Fig. 8.1d) obviously improves the representation. The example could be extended by further reductions in spacing. The limiting value is twice the distance between adjacent points of the analysis grid; thus the minimum resolvable scale on the analysis grid has a wavelength of two increments. Similarly, the minimum resolvable scale in the observations is twice the average spacing between the observing locations.

The "effective" resolution of an objective analysis—that is, the smallest scale resolvable with some degree of accuracy—is generally less than the maximum resolution. The definition of accuracy in this context is rather imprecise, but it depends in part on the intended purpose of the analysis. For example, it might be useful for diagnostic purposes to include the minimum resolvable scale (the two-grid-increment wave) in the analysis. For initial conditions for a numerical prediction model, however, the two-grid-increment wave must be removed because of the large truncation error in the finite-difference approximations to derivatives on such small scales. As a general rule, the spectrum at the small end of the scale is represented with less accuracy, and below some scale the representation becomes too inaccurate to use for a given purpose.

When the purpose is to provide initial conditions for a prediction model, the effective resolution is governed by the accuracy required to approximate derivatives in the model. It has been shown (Gerrity *et al.*, 1972) that calculation of the advective terms in the equations of motion can yield accurate estimates of nonlinear interactions for waves sampled 10 times or more per wavelength. Less frequent sampling leads to a rapid decrease in accuracy.

On a chart display of an objective analysis, resolution limitations are frequently described by synoptic meteorologists as missing "details." Cyclonic systems may not be as intense as the data indicate, gradients may be somewhat underestimated, wind speed extremes in jet streams may be too slow, or isobars may be smoother than warranted by the data.

It is worth noting that a greater number of computations will be required if the resolution of the grid is increased. Thus, high-resolution objective analysis methods may place severe demands on computer resources, especially in an operational environment where schedule deadlines are ever present. The practical limit on analysis resolution must also take computational economics into account.

8.2.3. Sampling

In Fig. 8.1a, the spacing between points is 60° longitude, and the analysis grid starts at the Greenwich meridian. Suppose, however, that the grid is shifted 40°, but retains the same spacing; the resulting "analysis" is the

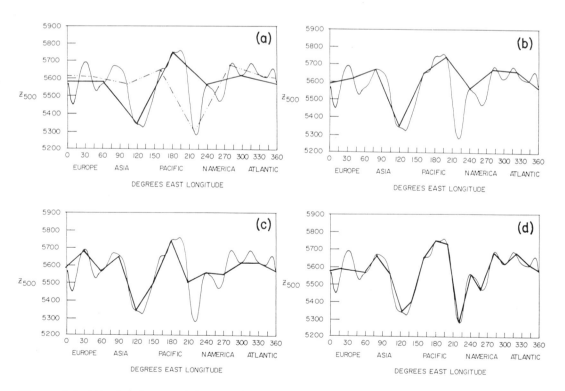

Figure 8.1. Simulated 500 mb height (thin solid line) around latitude 40°N, and grid point "analysis" (heavy line) with various scales of resolution. (a) Analysis grid points separated by 60° starting at 0° or 40°E (dash-dot line); (b) grid points separated by 40°; (c) grid points separated by 30°; (d) grid points separated by 20°.

dash-dot line in Fig. 8.1a. The large difference in the representations on a 0°-based analysis grid and one based on 40°E arises from "sampling" the 500 mb height trace at different points. Clearly, such sensitivity of the analysis to the location of the analysis grid is not a desirable characteristic. Sampling problems are most vexing when there is a great mismatch between the resolution of the measurements and that of the analysis, as in Fig. 8.1a; the 500 mb height trace exhibits more variability, and shorter scales, than the 60° analysis grid can represent. If, for example, the 20° analysis grid in Fig. 8.1d is shifted 10°, the differences are much smaller, except with respect to the shorter-scale features.

Practical sampling problems are easily illustrated. For example, a radiosonde ascent just inside a cumulonimbus and one in the cloud-free environment around it will produce very different soundings. For the principal use of objective analysis discussed in this chapter—the generation of initial conditions for prediction models—practical guidance in reducing the impact of sampling problems may be obtained from numerical considerations. If 10 samples per wavelength is taken as a reasonable criterion, then sampling difficulties may be eased by ensuring that both the observation system and analysis grid are matched to provide 10 measurements per characteristic length of the smallest meteorological phenomenon to be included in the analysis. Observations affected by smaller scales must be treated as noise, even if the observations are accurate; if the impact of smaller scales is very large, as in the example of the radiosonde ascending inside a cumulonimbus, the observation must be discarded as "unrepresentative" of the phenomena being analyzed.

Sampling is a much more complicated problem than has been indicated here, but a detailed discussion is beyond the scope of this chapter. A more general treatment of sampling theory may be found in Panofsky and Brier (1963).

8.2.4. *Dynamic Balance*

As noted in Sec. 8.1, adherence to dynamic constraints is a principle of subjective analysis that is also desirable in objective methods. For example, since the atmosphere is observed to be in approximate hydrostatic and geostrophic balance, at least for large scales, it follows that analyses of the atmosphere ought to portray the same behavior. This principle is incorporated explicitly in most objective analysis systems by imposing mathematical constraints in the analysis equations.

The relationship between the mass field and the motion field is incorporated in objective analysis methods typically by modifying the analysis equation, e.g., (8.1), to make use of gradient information. Thus, grid point values of geopotential height are determined with

$$Z_g^a = \sum_{i=1}^{N} a_i Z_i^o + \sum_{j=1}^{M} b_j u_j^o + \sum_{k=1}^{P} c_k v_k^o \,, \tag{8.5}$$

where the observed wind, u_j^o and v_k^o, is related to the height gradient ΔZ through the geostrophic relationship

$$u_j^o = -\frac{g}{f}\frac{\Delta Z^o}{\Delta y} \quad , \quad v_k^o = \frac{g}{f}\frac{\Delta Z^o}{\Delta x} \; . \tag{8.6}$$

This device appears in the work of Gilchrist and Cressman (1954), Berg-thorsson and Döös (1955), and Cressman (1959) and, as noted in Sec. 8.2.7, in the optimum interpolation method as well. Its importance is threefold: (1) It increases the amount of data available to the analysis by allowing wind observations to be used in the mass analysis and vice versa; (2) it allows some decrease in the minimum definable wavelength; and (3) it results in analyzed mass and motion fields that are approximately balanced.

Especially when the analyses are to be used as initial conditions for primitive equation prediction models, dynamic balance is extremely important. The model atmosphere attempts to restore any imbalance by a process called geostrophic adjustment. The theory of this process states that the adjustment occurs as a function of scale and latitude. For short horizontal scales, deep vertical scales, and relatively low latitudes the mass field adjusts to the wind field; for large horizontal scales, shallow vertical scales, and high latitudes, the reverse occurs. The mechanism by which this adjustment occurs is the generation of gravity waves, which move rapidly away from the region of the initial balance and dissipate over time. Thus, the first few hours of the prediction are contaminated by "noise" which may be of sufficient amplitude to be detrimental to the prediction. More importantly, the imbalance and subsequent adjustment process may result in the loss of important information from data that were included in an unbalanced way.

Typically, dynamic constraints incorporated directly into objective analysis methods are linear in form; more complicated constraints that account for nonlinearities, e.g., departures from geostrophy, impose severe mathematical or computational difficulties that make their use unattractive or unfeasible. More general constraints are frequently used in a post-analysis adjustment called initialization. It is possible to engineer a combination of analysis and initialization adjustments (Williamson and Daley, 1983).

The result of including a linear dynamic constraint between the mass and the motion fields in the objective analysis is a representation in which the analyzed wind approximately follows the contours of the geopotential analysis. Depending upon the details of the constraint, this may result in wind speeds that are faster than observed in regions of cyclonic curvature. The analyzed winds are mostly rotational, and typically very weak in divergence. Experience has shown that for larger scale models and forecast periods beyond 12–24 h, the initial divergence pattern is not very important; the prediction model tends to generate its own divergence pattern no matter what the initial value. One manifestation of this is that prediction models tend to underestimate precipitation rates and accumulations in the early hours of the prediction, because the divergence and vertical motion patterns require some time to develop fully.

8.2.5. Use and Effect of a "First Guess"

Objective analysis commonly starts with an initial estimate of the atmospheric structure and makes corrections based on available data. The analysis equation (8.1) then becomes

$$Z_g^a = Z_g^g + \sum_{i=1}^{N} a_i (Z^o - Z^g)_i , \tag{8.7}$$

where

$$Z_g^a = \text{analyzed value of } Z \text{ at grid point } g$$

$$Z_g^g = \text{estimated, or "first guess" of } Z \text{ at point } g$$

$$(Z^o - Z^g)_i = \text{difference between the observation } Z^o \text{ and the first guess } Z^g \text{ interpolated to the } i^{\text{th}} \text{ observation location.}$$

All modern operational systems use as a first guess a short-period prediction based on an analysis a few hours earlier. This practice allows the effect of past data to influence the analysis, thus contributing to temporal continuity. It also contributes to spatial coherence of the analysis. Most objective analysis systems assume at least tacitly that the first guess is quite accurate and thus in need of only small corrections. This implies that the prediction model supplying the first guess must be quite good, and that there are no areas completely devoid of data for long periods.

8.2.6. Treatment of Imperfect Data

All observations are imperfect to some degree. Errors are of two kinds: "measurement" error, which arises from imperfections in the measuring instrument, and the "unrepresentativeness" or sampling error referred to earlier, in which a measurement may be accurate in itself but reflects phenomena of a scale too small to be adequately resolved by either the data network or the analysis grid. The magnitude of observational errors ranges from very small to too large to be accepted by the analysis. Minimizing the influence of these errors is a primary task of the analysis. Redundancy in the data base is essential for this task.

For errors that are relatively small in magnitude, the analysis itself acts to reduce the effect of observational errors, if the errors are random. Figure 8.2 illustrates this. In the example depicted, the "true" height of the 500 mb surface to be analyzed is a constant 5640 m. The six observations represent the truth plus an error of 10 m, which is random in the sense that any report is equally likely to be too high or too low. If the "analysis" method is a simple arithmetic average, the effect of independent error is removed. In more realistic examples the effect is reduced only by the averaging inherent in the analysis. It can be shown (see, for example, Bergman and Bonner, 1976) that the error in the analysis is inversely proportional to the square root of the number of reports affecting the analysis at each point, provided the observational errors are random.

Small errors that are systematic (e.g., if one temperature observation is too cold, its neighbors are likely also to be too cold) cannot be removed or reduced in this way, as shown in the lower part of Fig. 8.2. Here, all six observations are depicted as being 10 m too low. Application of the simple averaging analysis method yields an analyzed value that is also 10 m too low. Systematic errors in an observing system should be minimized to the extent possible prior to the analysis. Observing systems that are prone to systematic errors are those that make a large number of measurements over an area, or a sequence of observations, with the same instrument. The satellite-based remote sounder is an example of the former. The radiosonde is an example of the latter. Radiosonde temperature measurements are, in general, not subject to systematic errors, other than high-level contamination due to solar radiation effects, but errors accrue in geopotential "measurements" calculated from radiosonde temperature profiles; the numerical integration of the hydrostatic equation to obtain geopotential values is calculated as a sequence in the vertical, each value depending on the one previously computed at the next lower level. As illustrated in Sec. 8.2.7, if the statistical characteristics of these systematic errors are known, their effects can be accommodated in some analysis methods.

Large errors must be identified prior to the analysis and corrected if possible but discarded if uncorrectable. This process, usually referred to as quality control, depends heavily on redundancy—being able to compare an observation with others nearby—and on dynamic consistency. At the U.S. National Meteorological Center and most other operational numerical weather prediction (NWP) centers, there are three separate steps:
(1) Incoming reports are checked for various mechanical errors—improper coding, incorrect location, transmission errors. Meteorological content of some observations is also examined in this step. Each radiosonde profile is checked for hydrostatic consistency by comparing the reported geopotential with values recalculated from the reported temperature profile. Data identified as questionable are ultimately examined by a meteorologist prior to the analysis. Only radiosonde data are treated in this step, because of the redundancy inherent in each radiosonde profile.
(2) All reports are subjected to comparison with the latest available prediction. This "gross error check" is generally done with much caution, for it is

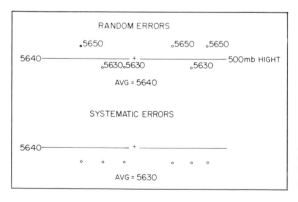

Figure 8.2. Illustration of the effect of random errors (upper) and systematic errors (lower) in objective analysis.

precisely in the areas where the prediction (first guess) is most in need of correction by data that large, apparently suspicious differences may occur. Consequently, this step is used primarily to delete reports that are not climatologically possible and to identify potentially suspicious reports for further examination.

(3) The internal consistency check, sometimes called the buddy check, compares each report with near neighbors. Generally it proceeds by computing an interpolated value of the observed variable at the location of the datum being checked, from observations nearby but excluding the datum being checked. If the difference is larger than some threshold acceptance value, the datum is deleted. It is best if the internal consistency check uses all the available information and makes use of dynamic consistency principles as well. Even so, automated quality control still is a very difficult process. For example, it is very clear that observations with systematic errors will tend to elude the defenses. If the errors are not large enough to be caught in the gross error check, they will tend to support each other in the internal consistency check. No completely adequate solution for the problem has yet been formulated.

8.2.7. Optimum Interpolation

Optimum interpolation (OI) currently enjoys great favor in the numerical weather prediction community. It incorporates in a systematic framework most of the basic concepts discussed here, and appears likely to remain in widespread use for some time. The method was apparently first suggested by A. Eliassen (1954). Gandin (1963) developed the method over a number of years, and his name is often associated with it in the literature. Eddy (1964) was also one of the early developers of this method. By the middle 1970s, the method was being introduced by a number of research and operational data assimilation systems, stimulated by anticipation of the Global Weather Experiment in 1979. Numerous articles concerning the method and its application have appeared in the literature; see, for example, Schlatter (1975); Rutherford (1972); Bergman (1979); Lorenc(1981).

Univariate Application

Consider, as an example, the analysis of geopotential Z_g^a and assume that two observations of geopotential Z_1^o, Z_2^o are available. The analysis equation is (8.7), rewritten here as

$$Z_g^a = Z_g^g + a_1(Z^o - Z^g)_1 + a_2(Z^o - Z^g)_2 \ . \tag{8.8}$$

Assume that the true value Z^T can be subtracted from both sides of the equation. The "truth" is not known, of course, but something may be known about its statistics, and that will prove useful in the analysis method. Equation (8.8) becomes

$$Z_g^T - Z_g^a = (Z_g^T - Z_g^g) - [a_1(Z^o - Z^g)_1 + a_2(Z^o - Z^g)_2] \ . \tag{8.9}$$

The left side of the equality is the analysis error z^a, and the first term on the right is the error in the guess z^g. Optimum interpolation determines the weights a_1 and a_2 by minimizing the mean-square analysis error:

$$\frac{\partial}{\partial a_i}\overline{(z^a)}^2 = 0 = \frac{\overline{\partial}}{\partial a_i}\left[z^g - \overline{\left(a_1 Z_1 + a_2 Z_2\right)}\right]^2 , \qquad (8.10)$$

where the overbar indicates an average over many realizations, and the notation

$$Z_i = (Z^o - Z^g)_i \qquad (8.11)$$

has been introduced for convenience. Performing the differentiation leads to a system of two equations,

$$\begin{aligned}(Z_1 Z_1)a_1 + (Z_1 Z_2)a_2 &= Z_1 z^g \\ (Z_2 Z_1)a_1 + (Z_2 Z_2)a_2 &= Z_2 z^g ,\end{aligned} \qquad (8.12)$$

in the unknown weights a_1, a_2. The elements appearing in (8.12) are statistical quantities to be discussed in more detail below. It is convenient at this point to note that

$$Z = Z^o - Z^g = (Z^T + \varepsilon) - Z^g , \qquad (8.13)$$

where ε is defined as the observational error in Z^o. Rearranging leads to

$$Z = Z^T - Z^g + \varepsilon = z^g + \varepsilon , \qquad (8.14)$$

where z^g is the error in the guess. The superscript g will henceforth be understood unless explicitly denoted otherwise. Equations (8.12) then become

$$\begin{aligned}\left(\overline{z_1 z_1} + \overline{\varepsilon_1 \varepsilon_1}\right) a_1 + \left(\overline{z_1 z_2} + \overline{\varepsilon_1 \varepsilon_2}\right) a_2 &= \overline{z_1 z} \\ \left(\overline{z_2 z_1} + \overline{\varepsilon_2 \varepsilon_1}\right) a_1 + \left(\overline{z_2 z_2} + \overline{\varepsilon_2 \varepsilon_2}\right) a_2 &= \overline{z_2 z} ,\end{aligned} \qquad (8.15)$$

and terms of the form $\overline{\varepsilon z}$ have been neglected. The elements are defined as

$\overline{z_1 z_1} = \sigma_i^2(z) =$ variance of the guess error at the ith location

$\overline{\varepsilon_1 \varepsilon_1} = \varepsilon^2 =$ observational error variance associated with the ith observation

$\overline{z_i z_j} =$ covariance of the guess error at location i with guess error at location j

$\overline{\varepsilon_i \varepsilon_j} =$ covariance of the observation error of the ith observation with that of the jth observation

$\overline{z_i z} =$ covariance of the guess error at the location of the ith observation with the guess error at the analysis point.

If these statistical quantities are known, (8.15) can be solved for the weights (a_1, a_2), and the analysis will be optimum in the sense that the mean square analysis error will be minimized.

In practice, the statistical quantities are not perfectly known, and thus the analysis is not strictly optimum. Nevertheless, it has proved to be a useful method. Some of its most useful characteristics can be noted by examination of (8.15). Note, for example, that the presence of the observational error variance terms means that different error characteristics of the two reports are accounted for by different values of ε_1 and ε_2. Also, the presence of the observational error covariance terms $\overline{\varepsilon_i \varepsilon_j}$ provides the mechanism for treating data with known systematic errors, such as satellite temperature soundings. Terms such as $\overline{z_i z_j}$ account for data distribution; the analysis recognizes that two observations close together are not necessarily independent pieces of information.

Modeling the Forecast Error Covariance

Although it is possible to collect statistics on the various elements in (8.15), it is more convenient to model them by a mathematical expression. A commonly used expression for the covariance of the geopotential guess error at two points is

$$\overline{z_i z_j} = \sigma_i \sigma_j \exp\!\left(-k S_{ij}^2\right) , \qquad (8.16)$$

where

$$
\begin{aligned}
\sigma_i &= \text{standard deviation of guess error at point } i \\
\sigma_j &= \text{standard deviation of guess error at point } j \\
S_{ij} &= \text{distance between points } i \text{ and } j \\
k &= \text{constant.}
\end{aligned}
$$

The guess error covariance model is the essence of the method. Its scale characteristics are governed by the shape of the covariance function (8.15); the smaller the value of k, the broader the covariance function and the larger the minimum scale representable. This relationship is of clear importance to mesoscale applications. A cogent discussion of the scale response characteristics of this method may be found in Hollingsworth (1984).

Multivariate Application

The example used thus far illustrates only univariate analysis, in which observations of geopotential are used to analyze geopotential. The procedure is also applicable to the multivariate case. If, for example, observation number two in (8.8) is a wind report, then

$$Z_g^a = Z_g^g + a_1 Z_1 + a_2 (U^o - U^g)_2 . \qquad (8.17)$$

This would modify (8.15) to be

$$\left(\overline{z_1 z_1} + \overline{\varepsilon_1 \varepsilon_1}\right) a_1 + \left(\overline{z_1 u_2}\right) a_2 = \overline{z_1 z}$$

$$\left(\overline{u_2 z_1}\right) a_1 + \left(\overline{u_2 u_2} + \overline{\varepsilon_2 \varepsilon_2}\right) a_2 = \overline{u_2 z} , \qquad (8.18)$$

where now

$\overline{z_1 u_2}$ = covariance of the guess error in geopotential at point 1 with the guess error in the u-component at point 2

$\overline{u_2 z}$ = covariance of the guess error u-component at location 2 with the geopotential guess error at the analysis point

$\overline{u_2 u_2}$ = variance of the wind guess error at point 2.

It is necessary only to model these terms, and (8.18) can be solved just as (8.15) can be. Common practice assumes that the guess errors in height and wind are related geostrophically. Evidence shows this is approximately true in middle and high latitudes for large scales. With this assumption, it is possible to show that

$$\overline{u_i z_j} = -\frac{g}{f}\frac{\partial}{\partial y}(\overline{z_i z_j}) \ . \tag{8.19}$$

With the additional assumption of hydrostatic equilibrium between guess errors in temperature and geopotential, a complete set of guess error covariances between temperature, geopotential, and wind can be derived from the specification of $\overline{z_i z_j}$ in (8.16). Thus observations of any of the basic variables can influence the analysis of the others, subject to sound dynamical constraints.

The preceding discussion has used as an example a situation in which two observations are available. This can be generalized to any number of reports. If N observations are to be used, then (8.15) or (8.18) becomes a system of N equations (rather than two) in the N unknown weights $a_1, a_2, \ldots, a_N$. A system like this must be formulated and solved at each point of the three-dimensional analysis grid. The demand on computational resources thus becomes very large as either N or the number of analysis points (resolution) increases.

Summary of OI Chararcteristics

The optimum interpolation analysis method provides a systematic framework for blending observations of differing error characteristics with other information such as recent predictions. More accurate data receive more weight in the analysis. Data distribution is accounted for, by assigning less weight to observations clustered together. Observations characterized by systematic errrors receive less weight in univariate analyses than do those with random errors.

The method is in principle spatially coherent, given a reasonable data distribution, and incorporates temporal continuity through the use of a short-period prediction from the preceding analysis as a first guess. Dynamic constraints are incorporated into the modeled guess error covariance function. The scale response of the analysis is governed by the shape of that function.

8.3. Requirements for Mesoscale Objective Analysis

8.3.1. Data Base

The first requirement is that of an adequate database. In the discussion of sampling consideration (Sec. 8.2.3), it is suggested that 10 samples per characteristic dimension are sufficient to resolve a meteorological feature of interest. Consider, for example, the Mesoscale Convective Complex (MCC), widely discussed in the literature (see, e.g., Maddox, 1980). Infrared images from satellites suggest that the characteristic length scale of a mature MCC is highly variable, but for this argument a value of 1000 km is reasonable. The vertical extent of a vigorous MCC is approximately 20 km, and its typical lifetime might be half a day. Many MCCs can be tracked for longer periods, but exhibit variations that are at least in part related to the diurnal cycle.

To satisfy the criterion of 10 samples per characteristic dimension would require an observing system with approximately 100 km horizontal spacing, at least 2 km vertical resolution, and the capability of producing about 10 profiles per 12 h, or roughly a sounding every hour. Such a network could accurately resolve MCCs and other features with the characteristic dimensions noted. It could also resolve smaller-scale features, but with less accuracy. The smallest resolvable scale would be the lower limit of the meso-α range (200 km), but the accuracy there would be too poor for numerical weather prediction.

8.3.2. Resolution

To match the characteristics of the observing system, the objective analysis grid will require 100 km spacing in the horizontal, 2 km vertical spacing, and the capability of ingesting data every hour. The suggested horizontal spacing is based on the assumption that the meteorological quantities are continuous and vary smoothly across the domain. It is well known, however, that vigorous mesoscale convective activity is often associated with boundaries between air masses having different characteristics. Gradients of the temperature, pressure, wind, and moisture may be very large across those boundaries, although quite shallow in the vertical. If it proves important to objectively analyze these near-discontinuities, 100 km resolution of the analysis grid will prove insufficient.

Likewise, vertical gradients under conditions of strong stratification are thought to be important for the prediction of onset and severity of convection. This is especially true of moisture. The suggested vertical resolution of the analysis grid may not be sufficient. At the least, it may be necessary to have greater resolution near the surface and the tropopause, with possibly lower resolution elsewhere.

Analyses of meso-α-scale data at hourly intervals would undoubtedly be useful for operational diagnostic activities such as "nowcasting" (Browning, 1982). There is increasing interest in the concept of maintaining a frequency-updated analysis of wind and temperature for aviation purposes. A 1 h update interval would be sufficient. There are, however, potentially serious numerical problems associated with such frequent updating. A discussion of these rightly belongs with consideration of four-dimensional data assimilation, and is thus beyond the scope of this chapter.

8.3.3. First Guess

Because the proposed 100 km horizontal spacing is still rather coarse, it will be necessary to interpolate between data points. The best way to accomplish this is to use the atmosphere's governing equations to assist, that is, to use a sophisticated prediction model integrated from the previous analysis. It seems likely that such models will be more sensitive and vigorous than are global NWP models, because of higher resolution and more complicated physical parameterizations. Errors of phase or amplitude that are minor in global models may be amplified in a mesoscale model. Relatively small errors of position in the first guess could thus require large corrections in the analysis. As noted previously, most analysis methods work best when the required corrections are small.

8.3.4. Dynamic Balance

The evolving methodology for large-scale objective analysis has had at least one relatively constant feature: the use of simple linear dynamic constraints. In the OI analysis method (Sec. 8.2.7) it is assumed that the errors in the first-guess mass and motion fields are related geostrophically. This is not strictly valid even for large-scale flows, especially in curved flow, and is likely to be even less valid for meso-α-scale objective analysis.

For example, if a prediction begins with inital conditions that include a vigorous MCC, any errors in a short-period prediction in the vicinity of the MCC are more likely to be caused by errors in the model's physical parameterizations than by the model's treatment of large-scale dynamics. Therefore, errors in the mass and motion predictions will probably not be related by a simple linear constraint. This circumstance will probably place more emphasis on reflecting the data than on adhering to balanced mass-motion laws.

8.3.5. Moisture

In large-scale numerical weather prediction, it has been sufficient to treat the moisture analysis somewhat casually, especially for predictions beyond a day or so. This is because the moisture parameter in a model responds quickly to dynamic forcing and modeled sources and sinks. After a day or two of integration, dynamically forced vertical motion centers and convergent flow usually combine to organize the moisture into the patterns desired by the model, even if the initial humidity field was everywhere constant.

Experience with meso-α-scale models indicates that simulations of convection and associated precipitation in short-period predictions are extremely sensitive to the details of the initial moisture field. For meso-α phenomena, it is therefore important to produce an accurate and detailed analysis of the moisture field.

8.3.6. *Treatment of Terrain and Surface Conditions*

Many important mesoscale phenomena are initiated and modulated by the effects of orography and surface conditions. For example, sea breezes and mountain-induced convection are triggered by variations in terrain. There is observational evidence from satellite imagery that the edges of areas covered by dense fog in the early morning are likely regions for convection in the afternoon. Differential heating in foggy and adjacent clear areas leads to weak thermal boundaries' being established. There is also evidence that edges of areas where rain has left wet soil are preferred regions of afternoon convection.

Consequently, it will be important to analyze carefully with respect to terrain, including as much detail as is resolvable by the analysis grid. It will also be necessary to provide analyses of soil moisture, surface temperature, and surface humidity.

8.4. An Operational Objective Analysis System Suitable for Mesoscale Applications

A meso-α-scale observing system does not now exist, and therefore a matching objective analysis system has not been developed. Even so, some improvements in objective analysis procedures are possible with the current data base. The National Meteorological Center (NMC) has developed a new Regional Analysis and Forecasting System (RAFS) to provide more accurate guidance predictions of heavy precipitation events. The RAFS prediction model is a primitive-equation grid-point model with three fully interactive nested grids (Phillips, 1979; Hoke and Phillips, 1981). The outer grid covers the Northern Hemisphere and has a resolution of 360 km at 60°N; the innermost grid covers most of North America and adjacent oceans with a 90 km mesh. Figure 8.3 depicts the vertical structure of the RAFS model, the NMC Limited-area Fine Mesh (LFM) model, and the Global Spectral Model.

The analysis component of the RAFS is a step in the evolution of objective analysis methodology that will eventually lead to a full meso-α-scale objective analysis with most of the characteristics described in Sec. 8.3.

With recognition that a mesoscale observing system is not available, the goals of the RAFS are quite modest:

- Increase resolution, especially in the vertical.

- Use more of the available data, especially the significant-level radiosonde data.

- Relax the linear mass/motion constraint.

- Improve the treatment of terrain.

The first recognizes that considerable vertical resolution is currently available in radiosonde data, but is not used operationally. Analysis of mandatory-level data is sufficient for coarse vertical resolution of most current operational prediction models, but not for advanced models. The second seeks to match the resolution of the analysis to that of the data, at least in the vertical. In the absence of good theoretical guidance, the third aims at shifting the emphasis from achieving quasi-geostrophic balance to carefully reflecting the wind observations. Finally, no longer analyzing on specified (the mandatory levels) isobaric surfaces, even when they are below ground, is the fourth.

Table 8.1 summarizes the characteristics of the RAFS analysis. Because the prediction model is hemispheric, the analysis grid also covers the hemisphere. The horizontal resolution is not very different from that of the LFM analysis, since the spacing of the data base is unchanged. The vertical resolution is improved, however, from 10 mandatory isobaric levels to the 16 levels of the prediction model, in the terrain-following σ coordinate. Complete radiosonde profiles are preprocessed to yield geopotential and wind at 25 mb intervals; thus the significant-level data are used directly to enhance the vertical resolution.

The objective analysis has two parts. The first is based on multivariate optimum interpolation. It uses as a first guess a 6 h forecast from the NMC Global Spectral Model. Using up to 30 pieces of information at each analysis point, the analysis corrects the geopotential and wind, subject to a weak geostrophic constraint on the correction fields. The analysis is done at all points of the 1.5° latitude by 2° longitude analysis grid (except at high latitudes) and in all 16 prediction model layers. Moisture (specific humidity) is updated in the lowest 10 layers. The result is a representation of wind and

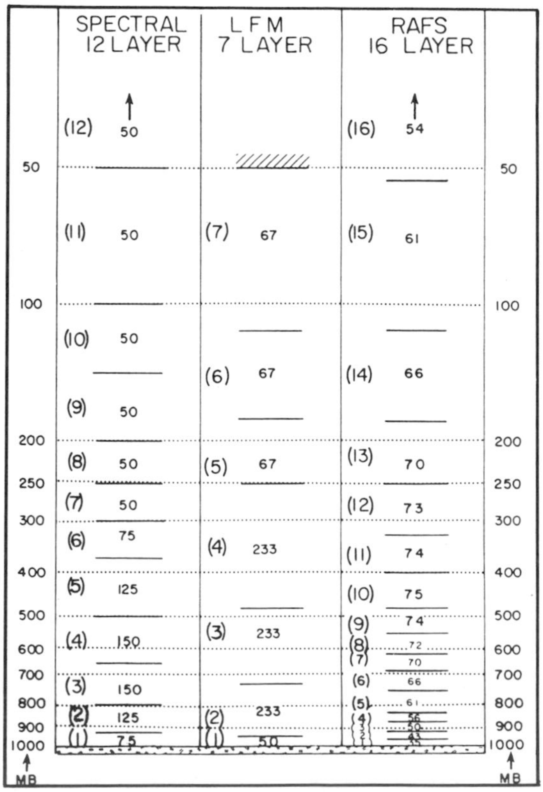

Figure 8.3. Vertical structure of the three NMC prediction models.

geopotential that is in approximate large-scale balance and reflects the data to a suitable degree.

The second step in the analysis procedure uses the result of the first step as a first guess. Beginning with this broadly balanced representation, refinements are added with a successive-corrections analysis on isentropic surfaces. This step is done univariately so as to improve the fidelity with which the wind and temperature observations are reflected. Vertical resolution is 30 isentropic surfaces from just above the boundary layer up to the tropopause. The isentropic update is done only on a subset of the analysis grid over North America, since it requires the high vertical resolution of the radiosonde network. Meshing of the two analysis steps is necessary around all boundaries, both the lateral and the upper and lower. Moisture is also refined during this step, on all 30 isentropic surfaces.

The analysis that results exhibits an underlying large-scale balance, but reflects ageostrophic motions when they are apparent in the data. The vertical structure of the thermal and moisture fields is represented with fidelity. Gradients are depicted more accurately, yet the fields themselves are spatially coherent and synoptically reasonable.

Figures 8.4 and 8.5 illustrate the characteristics of the regional analysis system. The former shows the 500 mb height and vorticity for 1200 GMT, 18 February 1979, the famous Presidents' Day snowstorm case. Of interest

Table 8.1. Characteristics of RAFS analysis

Characteristic	Optimum Interpolation	Isentropic
Domain	180 × 60 1.5°lat. × 2° long. Northern Hemisphere	Subset over contiguous U. S.
Vertical resolution	16 layers of the prediction model	30 isentropic levels
Analysis variables*	Z, U, V, Q (12 layers)	D_p, U, V, Q
First guess	12-level global 6 h forecast	OI analysis
Interpolation	3-D multivariate optimum interpolation	2-D successive corrections
Mass/Motion balance	Geostrophic	Uncoupled
Data selection	30 closest reports, including all significant-level data	All data within influence radius
Data weighting	Recognizes forecasts and observational errors	Distance dependent

*Z = geopotential; U = eastward wind component; V = northward component; Q = specific humidity; D_p = difference in pressure between adjacent isentropic surfaces.

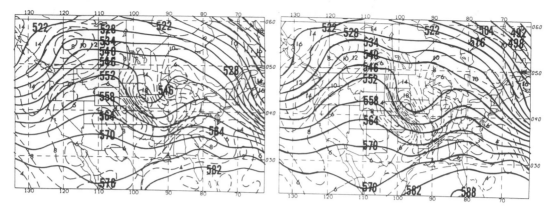

Figure 8.4. 500 mb height (solid lines, km) and vorticity (dashed lines, $10^{-5}s^{-1}$) for 1200 GMT, 18 February 1979. (Left) Regional OI analysis; (Right) isentropic update.

is the short-wavelength disturbance over the upper midwestern states, which 24 h later triggered explosive cyclogenesis along the middle Atlantic coast. Figure 8.4 displays the analysis resulting from the OI first step and the result of the isentropic update of the analysis. The enhancement of gradients and extrema is especially apparent in the vorticity patterns.

Figure 8.5 shows a sequence of cross sections along a line from southern California to Nova Scotia, for a case in April 1981. The first guess (GES)

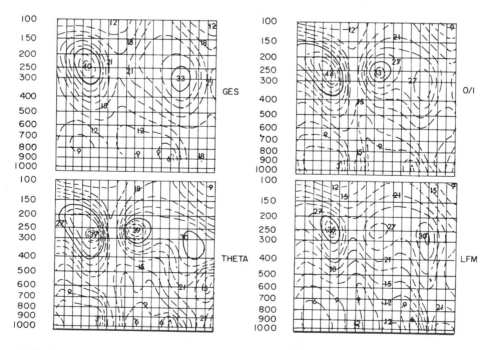

Figure 8.5. Cross sections of wind speed (m s^{-1}) along a line from San Diego, Calif., to near Halifax, Nova Scotia, for 1200 GMT, 22 April 1981. (GES) First guess; (O/I) regional OI; (THETA) isentropic update; (LFM) operational analysis.

depicts the broad features of the jet structure. Details are added by the OI, especially on the east side of the trough lines, where double maxima are analyzed. The isentropic update (THETA) enhances this further, bringing the maximum immediately east of the trough line above 45 m s^{-1}. A nearby wind observation at St. Cloud, Minn., at 250 mb, showed 48 m s^{-1}. For comparison, the same cross section from the LFM analysis has a much smoother and less active analysis.

Although the analysis method described here represents a step forward, much improvement is needed. The system does not treat surface processes at all, and of course it cannot treat hourly profiles from the proposed meso-α observing network.

REFERENCES

Anthes, R., 1983: Regional models of the atmosphere in middle latitudes. *Mon. Wea. Rev.*, **111**, 1306–1335.

Anthes, R., and T. Warner, 1978: Development of hydrodynamic models suitable for air pollution and other mesometeorological studies. *Mon. Wea. Rev.*, **106**, 1045–1078.

Bergman, K., 1979: Multivariate analysis of temperatures and winds using optimum interpolation. *Mon. Wea. Rev.*, **107**, 1423–1444.

Bergman, K., and W. Bonner, 1976: Analysis error as a function of observation density for satellite temperature sounds with spatially correlated errors. *Mon. Wea. Rev.*, **104**, 1308–1316.

Bergthorsson, P., and B. Döös, 1955: Numerical weather map analysis. *Tellus*, **7**, 329–340.

Browning, K. (Ed.), 1982: *Nowcasting*. Academic Press, New York, 256 pp.

Cressman, G., 1959: An operational objective analysis system. *Mon. Wea. Rev.*, **87**, 367–374.

Dixon, R., E. Spackman, I. Jones, and A. Francis, 1972: The global analysis of meteorological data using orthogonal polynomial base functions. *J. Atmos. Sci.*, **29**, 609–622.

Eddy, A., 1964: The objective analysis of horizontal wind divergence fields. *Quart. J. Roy. Meteor. Soc.*, **90**, 424–440.

Eliassen, A., 1954: Provisional report on calculation of spatial covariance and autocorrelation of the pressure field. Norske Videnskaps-Akademi i Oslo. Institutt vor Vaer-og Klimaforskning, Rep. 5, 11 pp. Reprinted in *Dynamic Meteorology: Data Assimilation Methods*. L. Bengstsson, M. Ghil, and E. Kallen (Eds.), Applied Mathematical Sciences **36**, Springer-Verlag, New York, 1981, 319–330.

Flattery, T., 1971: Spectral models for global analysis and forecasting. Proceedings, 6th Technical Exchange Conference, U.S. Naval Academy, Sept. 1970, Air Weather Service Technical Report 242, 42–54.

Gandin, L., 1963: Objective analysis of meteorological fields. Gidrometeorolicheskoe Istadel'stvo, Leningrad. Translated from Russian, Israel Program for Scientific Translation, Jerusalem (1965), 242 pp.

Gerrity, J., R. McPherson, and P. Polger, 1972: On the efficient reduction of truncation error in numerical weather prediction models. *Mon. Wea. Rev.*, **100**, 637–643.

Gilchrist, B., and G. Cressman, 1954: An experiment in objective analysis. *Tellus*, **6**, 309–318.

Hoke, J., and N. Phillips, 1981: Recent improvements to the Nested Grid Model of the National Meteorological Center. Preprints, 5th Conference on Numerical Weather Prediction, Monterey, Calif., American Meteorological Society, Boston, 188–190.

Hollingsworth, A., 1984: Spectral response of an optimal interpolation analysis. In Lecture Note No. 2.2, European Centre for Medium Range Weather Forecasts, Reading, U.K., 37–84.

Kaplan, M., J. Zach, V. Wong, and J. Tuccillo, 1982: Initial results from a mesoscale atmospheric simulation system and comparisons with the AVE-SESAME I data set. *Mon. Wea. Rev.*, **110**, 1564–1590.

Lorenc, A. C., 1981: A global three-dimensional multivariate statistical interpolation scheme. *Mon. Wea. Rev.*, **109**, 701–721.

Maddox, R., 1980: Mesoscale convective complexes. *Bull. Amer. Meteor. Soc.*, **61**, 1374–1387.

Panofsky, H., 1949: Objective weather map analysis. *J. Meteor.*, **6**, 386–392.

Panofsky, H., and G. Brier, 1963: *Some Applications of Statistics to Meteorology.* The Pennsylvania State University, University Park, Pa., 224 pp.

Perkey, D., 1976: A description and preliminary results from a fine-mesh model for forecasting quantitative precipitation. *Mon. Wea. Rev.*, **104**, 1513–1526.

Phillips, N., 1979: The Nested Grid Model. NOAA Tech. Rep. NWS–22, National Weather Service, Silver Spring, Md. (NTIS#PB-299046/3GA), 80 pp.

Rutherford, I., 1972: Data assimilation by statistical interpolation of forecast error fields. *J. Atmos. Sci.*, **29**, 809–815.

Schlatter, T., 1975: Some experiments with a multivariate statistical objective analysis scheme. *Mon. Wea. Rev.*, **103**, 617–627.

Williamson, D., and R. Daley, 1983: A unified analysis-initialization technique. *Mon. Wea. Rev.*, **111**, 1517–1536.

CHAPTER 9

Fronts and Jet Streaks: A Theoretical Perspective

Howard B. Bluestein

9.1. Introduction

It is an extremely curious observation that in our atmosphere there are narrow zones of intense temperature gradient and strong wind. The purpose of this chapter is to explain from a theoretical perspective the existence of bands of concentrated gradients in temperature and momentum, and the reason distributions of temperature and momentum are not more uniform.

A front is often defined as an "elongated" zone of "strong" temperature gradient and relatively large static stability and cyclonic vorticity. (It may also be defined in terms of density or moisture, or derived variables like potential temperature or equivalent potential temperature.) "Strong" means at least an order of magnitude larger than the typical synoptic-scale value of $10°$ (1000 km) [1]. Godson (1951) calls a gradient this intense a hypergradient. A zone whose length is roughly half an order of magnitude or more larger than its width is "elongated."

A jet is an intense, narrow, quasi-horizontal current of wind that is associated with strong vertical shear (Reiter, 1961, p.1). "Intense" usually means at least 30 m s^{-1} for the upper portions of the troposphere (Berggren et al., 1958) and 15 m s^{-1} for the lower portions of the troposphere. A "narrow" current is one whose width is approximately one-half to one order of magnitude less than its length. "Strong" vertical shear is at least 5–10 m s^{-1} km^{-1}, i.e., at least one-half to one order of magnitude larger than synoptic-scale shear. An isotach maximum embedded within a jet is called a jet streak (Palmén and Newton, 1969, p.199).

Fronts and jets are hybrid phenomena because both are characterized by length scales that are functions of direction and differ by as much as an order of magnitude. Thus, for example, it would at first glance be expected that a proper explanation of the formation and behavior of fronts and jets whose length scales are 1000 km and width scales are 100 km requires use of both synoptic-scale, quasi-geostrophic dynamical principles and mesoscale dynamical principles. Fronts whose length scales are 100 km and whose

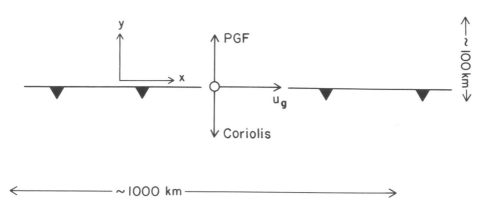

Figure 9.1. The horizontal scales along and across a front. The balance of forces across the front is depicted.

width scales are 10 km are usually dependent on topography, and require reference to mesoscale and in some cases even smaller scale dynamics for proper explanations. Nonhydrostatic effects, however, are not considered here.

If the length of a front or jet is on the order of 1000 km, then the Rossby number is usually reasonably small for flow along the front or jet so that geostrophic balance is approximately maintained across but not along the front or jet (Hoskins, 1971) (Fig. 9.1). Therefore, long, narrow zones of strong temperature gradient are associated with jets and jet streaks through the thermal wind relation, and hence a dynamical explanation of fronts also explains jets (Palmén, 1951). (This, of course, may not be altogether true for smaller scale fronts and jets.) Therefore fronts and jets are discussed together here.

The importance of fronts and jets in the atmosphere is immense because meteorological conditions vary widely across them. Hence, the ability to forecast the weather often involves detailed knowledge of the motion of fronts and their structure on the mesoscale. Furthermore, the location and structure of jets are especially important to the aviation industry; clear-air turbulence in the strongly sheared region of a jet and upper-level front (Shapiro, 1974; Kennedy and Shapiro, 1980) is a hazard to aircraft. Upper-level fronts are sometimes responsible for the transport of ozone and other materials from the stratosphere into the troposphere (Danielsen, 1968; Shapiro, 1980). Also, the occurrence of severe thunderstorms is often related to the position of fronts and jets (Fawbush et al., 1951).

9.2. Frontogenesis

9.2.1. The Front as a Discontinuity in Temperature

Suppose that a front is described as a boundary between two different air masses as in polar-front theory (Bjerknes, 1919). If density characterizes an air mass, then the density is discontinuous across the frontal boundary. In order that the acceleration due to the pressure-gradient force be finite across

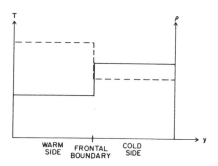

Figure 9.2. The front as a discontinuity in density (solid line) and temperature (dashed line).

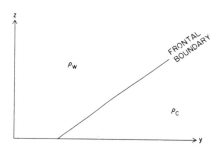

Figure 9.3. A sloping frontal boundary in an incompressible atmosphere. Subscripts w and c indicate warm and cold sides of front.

the front, the pressure must be continuous across the front. According to the ideal gas law, then, temperature in addition to density must be discontinuous (Fig. 9.2).

Since

$$dp = \frac{\partial p}{\partial y} \, dy + \frac{\partial p}{\partial z} \, dz \tag{9.1}$$

along the frontal boundary (Fig. 9.3), it follows that in a hydrostatic atmosphere (where $\partial p / \partial z = -\rho g$)

$$\frac{dz}{dy} = \frac{(\partial p / \partial y)_c - (\partial p / \partial y)_w}{g(\rho_c - \rho_w)} \, , \tag{9.2}$$

where the subscripts w and c refer to the warm and cold sides of the front. Then if $dz/dy \neq 0$ there must be a discontinuity in the cross-front pressure gradient across the frontal boundary. In particular, the cross-front pressure gradient is larger in value (but not greater in magnitude necessarily) on the cold side. If the along-the-front pressure gradient is non-zero, the isobars are "kinked" or bent at the front (Fig. 9.4).

Using (9.2) and the along-the-front component of the geostrophic-wind relation in height coordinates

$$u_g = \frac{-1}{f\rho} \frac{\partial p}{\partial y} \, , \tag{9.3}$$

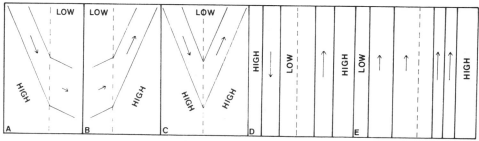

Figure 9.4. Examples of pressure or height fields (solid lines) associated with frontal boundaries (dashed lines). (From Petterssen, 1956, p. 199.)

where f is the Coriolis parameter, gives

$$\frac{dz}{dy} = \frac{f_w \rho_w u_{gw} - f_c \rho_c u_{gc}}{g(\rho_c - \rho_w)} \ . \tag{9.4}$$

Since the discontinuity in pressure gradient across the front is more significant than the discontinuity in density and variation in Coriolis parameter across the front,

$$u_{gw} - u_{gc} > 0 \ , \tag{9.5}$$

and there must be cyclonic geostrophic shear vorticity across the front (Fig. 9.4), (Petterssen, 1956, pp. 199–200). Unfortunately, the geostrophic vorticity is infinite at the front, an obviously unrealistic consequence of this model (except perhaps for the situation in which a strong elongated tornado is embedded in the front!).

The model does, however, reasonably specify the slope of a front dz/dy, according to Margules' formula, which expresses the thermal wind relation:

$$\frac{dz}{dy} \approx \frac{f\overline{T}}{g} \frac{(u_{gw} - u_{gc})}{T_w - T_c} \ , \tag{9.6}$$

where $\overline{T}$ is the mean temperature across the front. One cannot naively infer that strong fronts (large $T_w - T_c$) have smaller slopes than weak fronts, because strong fronts (it will soon be seen) are likely to have larger values of vorticity (i.e., large values of $u_{gw} - u_{gc}$ across the front). The front must slope toward the cold air or else there would be static instability (Fig. 9.5). For given values of temperature contrast and vorticity, the slope of a front is shallower at low latitudes than it is at middle latitudes.

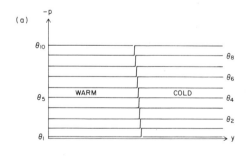

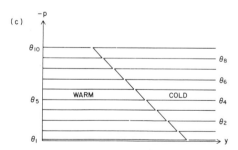

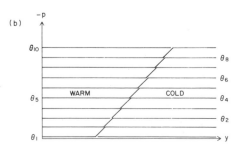

Figure 9.5. The vertical distribution of potential temperature (solid lines) across a frontal boundary that (a) is vertically oriented; (b) slopes toward the "cold" air; (c) slopes toward the "warm" air.

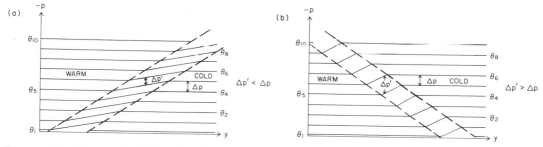

Figure 9.6. The vertical distribution of potential temperature (solid lines) across a frontal zone (dashed lines) that (a) slopes toward the "cold" air; (b) slopes toward the "warm" air.

9.2.2. The Front as a Discontinuity in Temperature Gradient

Fronts are usually modeled as a discontinuity in temperature gradient (Reed, 1955; Palmén, 1948) (Fig. 9.6). The temperature, however, is continuous across the front. Thus, a front is regarded as a finite zone and not as the infinitesimal boundary usually depicted on weather maps. The zone of strong horizontal temperature gradient may tilt with height toward the cold or the warm air mass. The rare latter case (Fig. 9.6b) has been observed in the United States by Kreitzberg and Brown (1970) and Danielsen (1974), and in the United Kingdom by Browning and Monk (1982). However, although the zone of temperature gradient can slope toward the warm air, it is characterized by low static stability rather than high static stability, and therefore should probably not be called a front. A zone of temperature gradient that slopes toward the cold air, on the other hand, is characterized by high static stability, and may be called a front (Fig. 9.6a). The configuration in Fig. 9.6b, because of the low static stability, would be very susceptible to cumulus convection if adequate moisture were available.

The strength of a front may be defined as the magnitude of the horizontal temperature gradient. It is convenient to define the strength of a front as the magnitude of the horizontal potential temperature gradient on a constant-pressure surface since potential temperature is conserved for adiabatic processes.

9.2.3. Kinematics and Thermodynamics of Two-Dimensional Frontogenesis

The formation of a front is called frontogenesis, and the decay of a front is called frontolysis. Petterssen (1936) described these processes quantitatively with the frontogenetical function defined as

$$F = \frac{D}{Dt}|\nabla_p\theta| \, , \tag{9.7}$$

where

$$\frac{D}{Dt} = \frac{\partial}{\partial t} + \boldsymbol{v}\cdot\nabla_p + \omega\frac{\partial}{\partial p}$$

and the subscript p means "on a pressure surface." The frontogenetical function quantifies the amount of change in potential temperature gradient following air-parcel motion. Although positive values of F do not necessarily indicate that a front will form, and negative values of F do not indicate that an existing front will dissipate, a description of the atmosphere in terms of F has proved to be a useful concept for understanding frontal processes. (Strictly speaking, one should consider F in a reference frame moving with the front. Although a front may neither intensify nor weaken, an air parcel experiences frontogenesis as it enters the frontal zone and frontolysis as it leaves it. In practice it is difficult to work in a reference frame moving with the front because a front's motion vector usually varies as a function of location along the front.)

Consider, for example, a frontal zone aligned along the x-axis and whose isentropes (potential temperature isotherms) are parallel to the front and whose wind field has no variations along it. (In the real atmosphere there is usually also a $\partial u/\partial x$ contribution. To simplify the problem and to focus on the significant physical mechanisms responsible for frontogenesis this contribution is not considered in this section. It is considered later, however.) Suppose also that the temperature on a constant-pressure surface decreases to the north, i.e., with increasing y (Fig 9.7). The thermodynamics equation is

$$\frac{D\theta}{Dt} = \left(\frac{p_0}{p}\right)^\kappa \frac{1}{C_p} \frac{dQ}{dt} \, , \tag{9.8}$$

where $\kappa = R/C_p$, and dQ/dt includes latent heat exchange, radiation, and turbulent heat fluxes. From (9.7) it is found that

$$F = \frac{D}{Dt}\left(-\frac{\partial\theta}{\partial y}\right)_p = \left(\frac{\partial v}{\partial y}\right)_p \left(\frac{\partial\theta}{\partial y}\right)_p + \left(\frac{\partial\omega}{\partial y}\right)_p \frac{\partial\theta}{\partial p}$$
$$- \frac{1}{C_p}\left(\frac{p_0}{p}\right)^\kappa \frac{\partial}{\partial y}\left(\frac{dQ}{dt}\right)_p \, . \tag{9.9}$$

The first term $(\partial v/\partial y)(\partial\theta/\partial y)$ (on a constant pressure surface) represents the kinematic effect of confluence (diffluence) on the quasi-horizontal temperature gradient. Negative $\partial v/\partial y$, confluence, acts to increase $|\nabla_p\theta| = -\partial\theta/\partial y$ (Fig. 9.7a), whereas positive $\partial v/\partial y$, diffluence, acts to decrease $|\nabla_p\theta|$ (Fig. 9.7b). Confluence and diffluence contribute to both convergence and horizontal (nondivergent) deformation. The first term represents the thermodynamic effect of a horizontal gradient in temperature advection. Thus, cold advection on the cold side and warm advection on the warm side increase the temperature gradient.

The second term, $(\partial\omega/\partial y)(\partial\theta/\partial p)$ represents kinematically the tilting of the vertical potential temperature gradient $(\partial\theta/\partial p)$ onto the horizontal (Fig. 9.8). The second term also represents thermodynamically a horizontal gradient of adiabatic temperature change due to a horizontal gradient in vertical motion. In a statically stable atmosphere, rising motion and its associated adiabatic cooling on the cold side, and sinking motion and its associated adiabatic warming on the warm side, increase the temperature gradient.

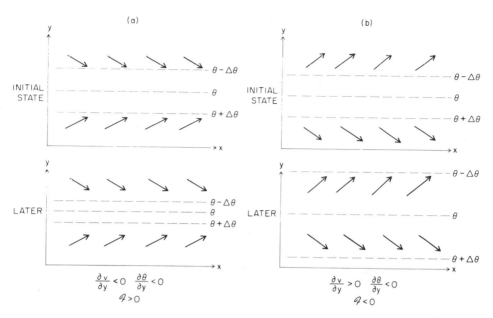

Figure 9.7. The effects of (a) confluence (frontogenesis) and (b) diffluence (frontolysis) acting on a quasi-horizontal potential temperature (dashed lines) gradient.

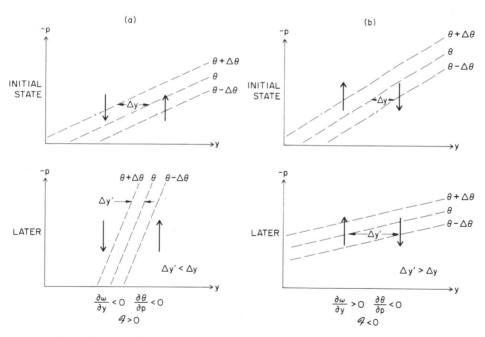

Figure 9.8. The effects of tilting on a vertical potential temperature (dashed lines) gradient: (a) frontogenesis; (b) frontolysis.

The third term, $(-1/C_p)\,(p_0/p)^\kappa\,(\partial/\partial y)(dQ/dt)_p$ represents a horizontal variation in diabatic heating. For example, heating of the warm, clear side of a front by the sun during the day, without heating on the cold, cloudy side, is

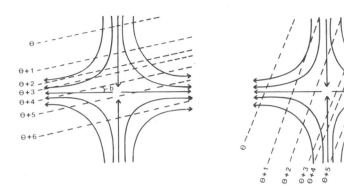

Figure 9.9. The relationship between the relative orientation of the axis of dilatation and the potential temperature isotherms (dashed lines) as measured by angle b, and frontogenesis. (Left) Frontogenesis $(0° < b < 45°)$ $F > 0$. (Right) Frontolysis $(45° < b < 90°)$ $F < 0$. (After Petterssen, 1956, p. 204.)

a frontogenetical process. At night the effect of longwave cooling is dominant, so that cooling on the clear, warm side, with restricted cooling on the cold, cloudy side, represents a frontolytical process. Bluestein (1982) suggested that differential heating over regions of snow cover and bare ground could act frontogenetically. The efficiency of a horizontal variation in diabatic heating in contributing to F increases with height by means of the factor $(p_0/p)^\kappa$.

The kinematic effects of the horizontal component of the wind field can be examined in more detail if the effects of tilting and diabatic heating are neglected, and if the restriction that the wind not vary along the front is removed. Bergeron (1928) and Petterssen (1936) showed that if the x axis is oriented along the axis of dilatation, then the effect of confluence (diffluence) and downstream changes in wind speed on the frontogenetical function is

$$F = \frac{|\nabla_p \theta|}{2} (D \cos 2b - \delta) , \qquad (9.10)$$

where D is the resultant deformation, δ is the horizontal divergence, and b is the angle measured from the axis of dilatation to the potential temperature isotherms passing through the point under consideration (see Appendix). Thus, the effect of horizontal deformation alone is to promote frontogenesis when the axis of dilatation lies within 45° of the isotherms, and to promote frontolysis when the axis of dilatation lies between 45° and 90° of the isotherms (Fig. 9.9). Convergence acts frontogenetically, and divergence acts frontolytically.

9.2.4. Kinematics and Thermodynamics of Three-Dimensional Frontogenesis

With the advent of a network of upper-air observations, the concept of a frontal zone was broadened to include vertical gradients of potential temperature (Miller, 1948). The complete three-dimensional frontogenetical function is therefore defined as

$$F = \frac{D}{Dt}|\nabla\theta|$$

$$= \frac{1}{|\nabla\theta|}\left\{\frac{\partial\theta}{\partial x}\left[\underbrace{\frac{1}{C_p}\left(\frac{p_o}{p}\right)^\kappa\frac{\partial}{\partial x}\left(\frac{dQ}{dt}\right)}_{1} - \underbrace{\frac{\partial u}{\partial x}\frac{\partial\theta}{\partial x}}_{2} - \underbrace{\frac{\partial v}{\partial x}\frac{\partial\theta}{\partial y}}_{3} - \underbrace{\frac{\partial w}{\partial x}\frac{\partial\theta}{\partial z}}_{4}\right]\right.$$

$$+ \frac{\partial\theta}{\partial y}\left[\underbrace{\frac{1}{C_p}\left(\frac{p_o}{p}\right)^k\frac{\partial}{\partial y}\left(\frac{dQ}{dt}\right)}_{5} - \underbrace{\frac{\partial u}{\partial y}\frac{\partial\theta}{\partial x}}_{6} - \underbrace{\frac{\partial v}{\partial y}\frac{\partial\theta}{\partial y}}_{7} - \underbrace{\frac{\partial w}{\partial y}\frac{\partial\theta}{\partial z}}_{8}\right]$$

$$\left.+ \frac{\partial\theta}{\partial z}\left[\underbrace{\frac{p_o{}^\kappa}{C_p}\frac{\partial}{\partial z}\left(p^{-\kappa}\frac{dQ}{dt}\right)}_{9} - \underbrace{\frac{\partial u}{\partial z}\frac{\partial\theta}{\partial x}}_{10} - \underbrace{\frac{\partial v}{\partial z}\frac{\partial\theta}{\partial y}}_{11} - \underbrace{\frac{\partial w}{\partial z}\frac{\partial\theta}{\partial z}}_{12}\right]\right\} \qquad (9.11)$$

in which the thermodynamics equation was used to eliminate $D\theta/Dt$, and height is used as the vertical coordinate so that the vertical component of the gradient of θ can be neatly combined with $\nabla_z\theta$. (The reader should verify that (9.9) and (9.10) are special cases of (9.11).) Terms 1, 5, and 9 are the diabatic terms; 2, 3, 6, and 7 are the horizontal-deformation terms; 10 and 11 are the vertical-deformation terms; 4 and 8 are the tilting terms; and 12 is the vertical divergence term. The physical interpretation of each term is just a generalization of the physical interpretation given to each term in (9.9). Reed and Sanders (1953), Newton (1954), and Sanders (1955) were the first to use observations to evaluate some of the terms in the frontogenetical function.

9.2.5. *Dynamics of Surface Frontogenesis*

Quasi-Geostrophic Frontogenesis

Consider (9.9) (the x-axis is oriented along the isotherms as earlier) applied at a level surface so that the kinematic lower boundary condition is $\omega \approx 0$. Furthermore, suppose that diabatic heating is neglected. The two-dimensional frontogenetical function is

$$F = \frac{D}{Dt}\left(-\frac{\partial\theta}{\partial y}\right) = \left(\frac{\partial v}{\partial y}\right)\left(\frac{\partial\theta}{\partial y}\right) . \qquad (9.12)$$

If $v = v_g$ and the confluence represented by $-\partial v_g/\partial y$ is held fixed through time t, then

$$\left(-\frac{\partial\theta}{\partial y}\right)_t = \left(-\frac{\partial\theta}{\partial y}\right)_{t=0}\exp\left(-\frac{\partial v_g}{\partial y}t\right) . \qquad (9.13)$$

That is, the initial horizontal gradient will increase exponentially with an e-folding time of $(-\partial v_g/\partial y)^{-1}$, which is on the order of 10^5 s, or one day. In other words, it would take about 2.5 days for the temperature gradient to increase by one order of magnitude. With diffusion, this could be even longer.

The dynamical effects of confluence acting on a temperature gradient were first discussed by Namias and Clapp (1949). Suppose the atmosphere that initially is in thermal wind balance does not adjust to the increase in horizontal temperature gradient induced by geostrophic deformation. Then there is no ageostrophic wind component (and no ω aloft). In addition, if the β-effect, friction, and diabatic heating are neglected, then the equation of motion is

$$\frac{D_g \boldsymbol{v}_g}{Dt} = 0 \tag{9.14}$$

and the thermodynamics equation is

$$\frac{D_g \theta}{Dt} = 0 , \tag{9.15}$$

where

$$\frac{D_g}{Dt} = \frac{\partial}{\partial t} + \boldsymbol{v}_g \cdot \boldsymbol{\nabla} .$$

The thermal wind relation is

$$\frac{\partial \boldsymbol{v}_g}{\partial(-p)} = \frac{R}{f_o p}\left(\frac{p}{p_0}\right)^\kappa (\hat{k} \times \boldsymbol{\nabla}_p \theta) . \tag{9.16}$$

(Vertical derivatives are expressed with respect to $-p$ rather than p so that, e.g., $\partial v/\partial(-p)$ may be interpreted as shear in the upward direction.) It is easily shown from (9.14), (9.15), and (9.16) (Hoskins *et al.*, 1978) that

$$\frac{D_g}{Dt}\left[\frac{\partial \boldsymbol{v}_g}{\partial(-p)}\right] = -\frac{R}{f_o p}\left(\frac{p}{p_0}\right)^\kappa \frac{D_g}{Dt}(\hat{k} \times \boldsymbol{\nabla}_p \theta) . \tag{9.17}$$

The effect of geostrophic deformation acting to increase the horizontal temperature gradient is therefore to force the atmosphere from thermal wind balance. That is, the temperature gradient becomes too large for the vertical shear.

The effect of the geostrophic "disturbance" may conveniently be expressed through the $\boldsymbol{Q}$ vector (Hoskins *et al.*, 1978),

$$\boldsymbol{Q} = -\frac{R}{\sigma p}\left(\frac{p}{p_0}\right)^\kappa \begin{pmatrix} \frac{\partial \boldsymbol{v}_g}{\partial x}\cdot\boldsymbol{\nabla}_p\theta \\ \frac{\partial \boldsymbol{v}_g}{\partial y}\cdot\boldsymbol{\nabla}_p\theta \end{pmatrix} , \tag{9.18}$$

where the static stability parameter $\sigma = \sigma(p) = -[R\overline{T}(p)/p]\,[\partial \ln \overline{\theta}(p)/\partial p]$. If the coordinate system is rotated so that the isentropes lie along the x-axis, and the y-axis points in the direction of $-\boldsymbol{\nabla}_p\theta$, then

$$\boldsymbol{Q} = -\frac{R}{\sigma p}\left(\frac{p}{p_0}\right)^\kappa \begin{pmatrix} \frac{\partial v_g}{\partial x}\frac{\partial\theta}{\partial y} \\ \frac{\partial v_g}{\partial y}\frac{\partial\theta}{\partial y} \end{pmatrix} = \begin{pmatrix} Q_1 \\ Q_2 \end{pmatrix} . \tag{9.19}$$

One useful property of the Q vector is that it is Galilean invariant; that is, it is independent of the motion of the coordinate system.

The quasi-geostrophic equation of motion without friction and the β-effect is

$$\frac{D_g \boldsymbol{v}_g}{Dt} = -f_0 (\hat{k} \times \boldsymbol{v}_a) , \qquad (9.20)$$

where $\boldsymbol{v}_a$ is the ageostrophic wind component. Advection of momentum by the ageostrophic part of the wind and by ω are ignored. In addition, local changes in the ageostrophic part of the wind and advection of ageostrophic momentum by the geostrophic part of the wind are ignored.

Equation (9.20) may be combined with the quasi-geostrophic thermodynamics equation

$$\frac{D_g \theta}{Dt} + \omega \frac{\partial \bar{\theta}(p)}{\partial p} = 0 , \qquad (9.21)$$

where ω is the vertical velocity and for which advection of temperature by the ageostrophic part of the wind has been neglected, to obtain

$$\nabla_p \omega + \frac{f_0^2}{\sigma} \frac{\partial \boldsymbol{v}_a}{\partial (-p)} = -2Q . \qquad (9.22)$$

So, for example, for the case in which geostrophic confluence increases the horizontal temperature gradient, $\partial v_g / \partial y < 0$ and $\partial \theta / \partial y < 0$. If there is no shear along the front, then $\partial v_g / \partial x = 0$, and therefore $Q_1 = 0$ and $Q_2 < 0$ [see (9.19)]. Then (9.22) becomes

$$\frac{\partial \omega}{\partial y} + \frac{f_0^2}{\sigma} \frac{\partial v_a}{\partial (-p)} = -2 Q_2 > 0 . \qquad (9.23)$$

The ageostrophic adjustment to the temperature field is represented by $\nabla_p \omega$, and the adjustment to the wind field is represented by $(f_0^2 / \sigma) [\partial \boldsymbol{v}_a / \partial (-p)]$. The adjustments to the temperature and wind fields are equally important when

$$|\nabla_p \omega| \sim \left| \frac{f_0^2}{\sigma} \frac{\partial \boldsymbol{v}_a}{\partial (-p)} \right| . \qquad (9.24)$$

The magnitude of the left-hand side of (9.24)

$$|\nabla_p \omega| \sim \frac{\hat{\omega}}{L} , \qquad (9.25)$$

where $\hat{\omega}$ and L are the order of magnitude of vertical velocity and horizontal scale. From continuity and the hydrostatic equation, it is seen that

$$\frac{\hat{\omega}}{\rho g H} \sim \frac{V_a}{L} , \qquad (9.26)$$

where ρ, H, and V_a are the order of magnitude of density, vertical scale, and ageostrophic wind speed. Then from (9.25) and (9.26) we find that (9.24) holds when

$$\frac{\rho g H V_a}{L^2} \sim \frac{f_0^2 V_a}{\sigma(\rho g H)} \ . \tag{9.27}$$

Since

$$\sigma = -\frac{1}{\rho} \frac{\partial \ln \bar{\theta}(p)}{\partial p} = \frac{1}{\rho^2 g} \frac{\partial \ln \bar{\theta}(p)}{\partial z} = \frac{N^2}{(\rho g)^2} \ , \tag{9.28}$$

where N^2 is the square of the Brunt-Väisälä frequency, then

$$L \sim \frac{NH}{f_0} = L_R \ , \tag{9.29}$$

the Rossby radius of deformation (Rossby, 1938), which is the distance beyond which the ageostrophic adjustment process is negligible. For typical values of $N^2 \sim 10^{-4}$ s^{-2}, $f_0 \sim 10^{-4}$ s^{-1}, $H \sim 10^4$ m, and $L \sim 10^6$ m, $L_R \sim 10^6$ m = 1000 km. Therefore for synoptic-scale motions* $\partial w/\partial y > 0$ and $(f_0^2/\sigma)[\partial v_a/\partial(-p)] > 0$ in a statically stable atmosphere if $Q_2 < 0$. Taking into account continuity, it is seen that there is subsidence on the cool side and rising motion on the warm side. This acts to decrease the horizontal temperature gradient according to (9.21),

$$\frac{\partial}{\partial y}\left(\frac{D_g \theta}{Dt}\right) = \frac{\partial \bar{\theta}(p)}{\partial p} \frac{\partial \omega}{\partial y} \ , \tag{9.30}$$

or, in other words, to bring the atmosphere back toward thermal wind balance by adjusting the temperature field back toward what it was. Furthermore there is southerly ageostrophic wind aloft, and a northerly ageostrophic wind below. This acts to increase the geostrophic shear according to (9.20),

$$\frac{\partial}{\partial(-p)}\left(\frac{D_g u_g}{Dt}\right) = f_0 \frac{\partial v_a}{\partial(-p)} \ , \tag{9.31}$$

or, in other words, to bring the atmosphere back toward thermal wind balance in adjusting the wind field by increasing the vertical shear to accommodate the increased temperature gradient. Therefore ω and v_a constitute an ageostrophic, thermally direct circulation that obeys LeChatelier's principle and acts to restore the atmosphere to geostrophic and hydrostatic balance.

The adjustment process is depicted in Fig. 9.10 for the case in which the temperature gradient is changed impulsively. Consider a meridional temperature gradient such that it is cold to the north and warm to the south. If the temperature gradient is suddenly increased by deformation, then hydrostatically the thickness to the south increases while the thickness to the north decreases, so that the isobars become more sloped.

---

*   In fact, Hoskins and West's (1979) definition of a front includes a statement that the length scale along the front is on the order of the Rossby radius of deformation.

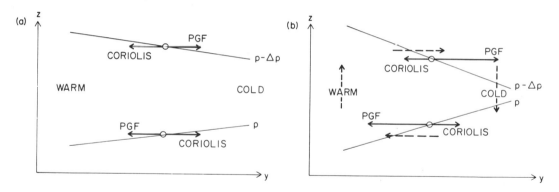

Figure 9.10. The development of an ageostrophic circulation during frontogenesis. Solid lines are isobars; arrows are forces. (a) Initial state of geostrophic balance; (b) imbalance of forces after horizontal potential temperature gradient has been increased. Dashed lines represent ageostrophic circulation.

Since the horizontal pressure gradient has been increased, there is now an imbalance of forces and a net force acting to the north aloft and to the south below. Thus air accelerates in these directions and, because of continuity, air rises and cools adiabatically on the warm side and sinks and warms adiabatically on the cold side. The horizontal temperature gradient is therefore decreased. When the air has been in motion long enough to be affected by the Earth's rotation, the northward-moving air aloft is deflected to the right and becomes westerly, while the southward-moving air below is deflected to the right and becomes easterly. Thus, the vertical shear is increased.

It should be pointed out that the Q vector points in the same direction as the ageostrophic wind below and toward the region of rising motion, and in the direction opposite to the ageostrophic wind above and away from the region of sinking motion (Fig. 9.11). From (9.18) we find that

$$Q = \frac{R}{\sigma p}\left(\frac{p}{p_0}\right)^{\kappa}\frac{D_g}{Dt}(\nabla_p\theta) \ . \tag{9.32}$$

So Q is proportional to the rate of change of temperature gradient following geostrophic motion. Furthermore, the frontogenetical function is related

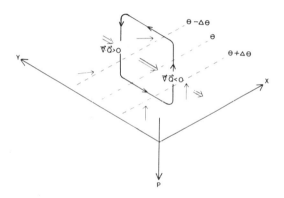

Figure 9.11. The relationship among Q, v_a, and ω for the frontogenetical configuration of v and θ shown in Fig. 9.7a. Bold arrows represent the Q field; arrows represent the geostrophic wind field; closed streamline represents the vertical circulation.

to Q as follows:

$$F = \frac{\sigma p}{R} \frac{1}{|\nabla_p \theta|} (\nabla_p \theta \cdot Q) \ . \qquad (9.33)$$

Therefore if Q and $\nabla_p \theta$ are in the same direction, the frontogenetical function is positive and there is a thermally direct circulation, and vice versa (Hoskins and Pedder, 1980).

Since the geostrophic forcing and the ageostrophic response oppose each other, it is necessary to know which, if either, is dominant. At the ground or at the tropopause where $\omega \approx 0$, we would expect the geostrophic forcing to dominate. In the midtroposphere, where ω can be significant, it would be expected that the forcing and response could be equally significant. These processes may be illustrated by considering the variation of the frontogenetical function's behavior with respect to height.

At the ground $\omega = 0$, so frontogenesis can proceed according to (9.12) and (9.13). Above the ground $\omega \neq 0$, and the tilting of the vertical temperature gradient onto the horizontal $[(\partial \omega / \partial y)(\partial \theta / \partial p) < 0]$ opposes frontogenesis. Therefore, the presence of the level lower boundary, in the absence of diabatic heating and diffusion, allows the frontogenetical process to occur most rapidly at the surface. The strength of the surface temperature gradient is ultimately limited by diffusion.

From the equation of continuity it is known that there is convergence at the surface under the rising branch of the circulation. Petterssen and Austin (1942) pointed out that this convergence acting on Earth's vorticity can produce cyclonic vorticity, and furthermore must be the only source for cyclonic vorticity at the surface. On a level surface there is no tilting. Vorticity advection cannot produce vorticity; it only moves vorticity from one place to another. Furthermore, friction acting on a uniform, level surface cannot produce vorticity; it in fact destroys vorticity if the eddy coefficient of diffusion is constant. Quasi-geostrophic frontogenesis therefore requires that a convergent, cyclonic windshift line form at the surface on the warm side of a frontal zone (Hoskins and West, 1979). In fact, Kirk (1966) has suggested that the fundamental process of frontogenesis is the concentration of vorticity.

Quasi-geostrophic models (both analytical and numerical) have been used by Stone (1966b), Williams and Plotkin (1968), Williams (1968, 1972), and Mudrick (1974) to study frontogenesis. Hoskins and Bretherton (1972) summarized the weaknesses of quasi-geostrophic frontogenesis as follows:

(a) The frontogenetical process at the ground is relatively slow, and intense gradients do not form in a finite time.
(b) The frontal zone does not tilt with height (Fig. 9.12).
(c) The field of relative vorticity contains regions of large anticyclonic as well as cyclonic vorticity.
(d) Regions of static instability may be produced (Fig. 9.12).

Williams (1967) referred to quasi-geostrophic fronts as pseudo-fronts.

Quasi-geostrophic principles can therefore account for some, but not all, of the features associated with surface fronts. Eliassen (1962, 1984) pointed

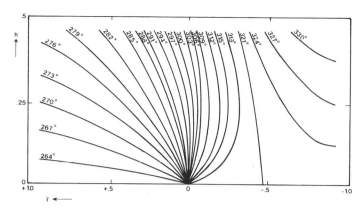

Figure 9.12. Vertical cross section through a quasi-geostrophic "pseudo-front." Solid lines are potential temperature isotherms; the abscissa and ordinate represent scaled horizontal and vertical coordinates. (From Stone, 1966b.)

out that many have argued that quasi-geostrophic theory cannot be applied to the study of fronts because the Rossby number, the ratio of the magnitude of the parcel acceleration to the magnitude of the Coriolis acceleration,

$$\mathrm{R}_0 = \frac{|D\boldsymbol{v}/Dt|}{|-f_0\left(\hat{k}\times\boldsymbol{v}\right)|} \sim \frac{U}{fL}\,, \tag{9.34}$$

where U and L are wind and length scales, is too large for motions across the front. However, he suggested that if local $(\partial\boldsymbol{v}/\partial t)$ and convective $(\boldsymbol{v}\cdot\nabla)\boldsymbol{v}$ accelerations are of comparable magnitude and opposite sign, then the conventional definition of the Rossby number is not appropriate, and since air parcels may reside in the frontal zone for a long time and parcel accelerations are small, the narrow width of the front is not necessarily relevant. The development of semi-geostrophic theory (Hoskins, 1975) led to an explanation of why quasi-geostrophic theory has been successfully used under conditions in which it was not clear whether it was valid.

The Geostrophic Momentum Approximation and Semi-Geostrophic Frontogenesis.

The total wind may be expressed as the sum of the geostrophic and ageostrophic components:

$$\boldsymbol{v} = \boldsymbol{v}_g + \boldsymbol{v}_a \tag{9.35}$$

The frictionless form of the equation of motion can be written as

$$\frac{D\boldsymbol{v}}{Dt} = -f\,\hat{k}\times\boldsymbol{v}_a \tag{9.36}$$

$$\boldsymbol{v}_a = \frac{1}{f}\,\hat{k}\times\frac{D\boldsymbol{v}}{Dt}. \tag{9.37}$$

Then from (9.35) and (9.37) it follows that

$$v = v_g + \frac{1}{f} \hat{k} \times \frac{Dv}{Dt}. \tag{9.38}$$

By substituting (9.35) and (9.37) into (9.38) the following is obtained:

$$v = v_g + \frac{1}{f} \hat{k} \times \left[\frac{D}{Dt} \left(v_g + \frac{1}{f} \hat{k} \times \frac{Dv}{Dt} \right) \right]. \tag{9.39}$$

Then

$$v - v_g = v_a = \frac{1}{f} \hat{k} \times \frac{Dv_g}{Dt} - \frac{1}{f^2} \frac{D^2 v}{Dt^2}. \tag{9.40}$$

This recursive process could be continued and an infinite series would result (Hoskins, 1975). For time scales longer than $1/f$, $|(1/f)(D^2 v/Dt^2)| \ll |Dv_g/Dt|$ and hence the $(1/f^2)(D^2 v/Dt^2)$ term may be neglected (Hoskins, 1982).

The equation of motion subject to the geostrophic momentum approximation is

$$v_a = \frac{1}{f} \hat{k} \times \frac{Dv_g}{Dt} \tag{9.41a}$$

or

$$\frac{Dv_g}{Dt} = -f \left(\hat{k} \times v_a \right), \tag{9.41b}$$

where

$$\frac{D}{Dt} = \frac{\partial}{\partial t} + (v_g + v_a) \cdot \nabla_p + \omega \frac{\partial}{\partial p}.$$

Equation (9.41) is similar to the quasi-geostrophic equation of motion (9.20), except that ageostrophic and vertical advection of geostrophic momentum are retained. In other words, the geostrophic momentum approximation, which was first introduced by Eliassen (1948), is essentially the substitution of the individual rate of change of geostrophic momentum for the individual rate of change of momentum.

The thermodynamics equation is

$$\frac{D_p \theta}{Dt} + \omega \frac{\partial \theta}{\partial p} = 0, \tag{9.42}$$

where

$$\frac{D_p}{Dt} = \frac{\partial}{\partial t} + (v_g + v_a) \cdot \nabla_p.$$

That is, in contrast to the quasi-geostrophic thermodynamics equation (9.21), advection of temperature by the ageostrophic part of the wind is retained.

For simplicity it will be assumed that the front is oriented along the x axis. Furthermore, suppose the front is straight, i.e., there are accelerations only along it, so that $Dv_g/Dt \approx 0$ and hence from (9.41) $u_a \approx 0$ (Shapiro, 1981); assume that there is no curvature vorticity so that $\partial v_g/\partial x = 0$, and also, neglect variations in f. After combining the x-components of (9.41) and (9.42) and using the equation of continuity and assuming that the thermal wind relation holds, it is found that (Sawyer, 1956; Eliassen, 1962; Shapiro, 1981)

$$\frac{R}{f_0 p}\left(\frac{p}{p_0}\right)^\kappa \frac{\partial}{\partial y}\left(-v_a \frac{\partial \theta}{\partial y} - \omega \frac{\partial \theta}{\partial p}\right) + \frac{\partial}{\partial p}\left[-v_a\left(f_0 - \frac{\partial u_g}{\partial y}\right) + \omega \frac{\partial u_g}{\partial p}\right]$$

$$= 2\left(\frac{\partial u_g}{\partial p}\frac{\partial v_g}{\partial y} - \frac{\partial v_g}{\partial p}\frac{\partial u_g}{\partial y}\right) - \frac{R}{C_p f_0 p}\frac{\partial}{\partial y}\left(\frac{dQ}{Dt}\right). \tag{9.43}$$

This diagnostic equation relates the two dependent variables v_a and ω to the geostrophic wind field, the horizontal temperature field, diabatic heating, static stability, and absolute vorticity. The ageostrophic circulation lies in the y-p plane only. Finding solutions to (9.43) is simplified if a vertical stream function ψ is defined such that

$$v_a = -\frac{\partial \psi}{\partial p} \tag{9.44a}$$

and

$$\omega = \frac{\partial \psi}{\partial y}. \tag{9.44b}$$

Then, from (9.44) and (9.16), (9.43) becomes

$$\frac{\partial^2 \psi}{\partial y^2}\left[-\frac{\partial \theta}{\partial p}\frac{R}{f_0 p}\left(\frac{p}{p_0}\right)^\kappa\right] + \frac{\partial^2 \psi}{\partial y \partial p}\left(2\frac{\partial u_g}{\partial p}\right) + \frac{\partial^2 \psi}{\partial p^2}\left(f_0 - \frac{\partial u_g}{\partial y}\right)$$

$$= 2\frac{R}{f_0 p}\left(\frac{p}{p_0}\right)^\kappa\left(\frac{\partial \theta}{\partial y}\frac{\partial v_g}{\partial y} + \frac{\partial \theta}{\partial x}\frac{\partial u_y}{\partial y}\right) - \frac{R}{C_p f_0 p}\frac{\partial}{\partial y}\left(\frac{dQ}{dt}\right). \tag{9.45}$$

This linear, second-order equation, known as the Sawyer-Eliassen equation, relates the dependent variable ψ to the other independent variables; it is elliptic and therefore always has unique solutions if

$$\left(\frac{\partial u_g}{\partial p}\right)^2 + \left[\frac{R}{f_0 p}\left(\frac{p}{p_0}\right)^\kappa\right]\frac{\partial \theta}{\partial p}\left(f_0 - \frac{\partial u_g}{\partial y}\right) < 0. \tag{9.46}$$

It has been shown that the following is a sufficient condition for dynamic instability (Solberg, 1933) and also a sufficient condition for (9.45) to be hyperbolic (Eliassen, 1959):

$$\left(f_0 - \frac{\partial u_g}{\partial y}\right)\frac{\partial \theta}{\partial(-p)} < 0; \tag{9.47}$$

that is, Ertel's potential vorticity is negative. In a statically stable atmosphere, the instability will occur only if absolute vorticity is negative. Equation (9.47) in isentropic coordinates is equivalent to (9.46), so that a necessary and sufficient condition for instability is that Ertel's potential vorticity

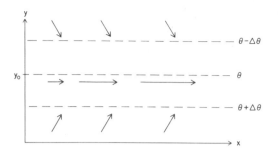

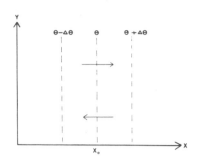

Figure 9.13. The frontogenetical action of geostrophic stretching deformation on the cross-front potential temperature (dashed lines) gradient $(\partial\theta/\partial y)$. Arrows indicate wind direction. $\partial u_g/\partial x > 0$ and $\partial\theta/\partial y < 0$ at $y = y_0$; $\partial u_g/\partial x - \partial v_g/\partial y > 0$.

Figure 9.14. The frontogenetical action of geostrophic shearing deformation on the along-front potential temperature (dashed lines) gradient $(\partial\theta/\partial x)$. Arrows indicate wind direction.

on an isentropic surface be negative (Hoskins, 1974). A necessary and sufficient condition for instability in height coordinates (Stone, 1966a; Hoskins, 1974) is that

$$\mathrm{Ri} = \frac{g\partial\ln\theta/\partial z}{(\partial u_g/\partial z)^2} < \frac{f_0}{f_0 - (\partial u_g/\partial y)}. \tag{9.48}$$

When the Richardson number falls to a value below the fraction of Earth's vorticity contained in the absolute vorticity, symmetric instability (also called inertial instability) can occur. "Symmetric" denotes that perturbations do not vary along the basic current (Stone, 1966a). For typical values of relative vorticity in the atmosphere, the sufficient condition for symmetric instability is that $\mathrm{Ri} < 1$. Symmetric instability is favored by high vertical shear (low Ri), large $\partial u_g/\partial y$ (anticyclonic shear), and low static stability. Hoskins (1974), Emanuel (1979), and Bennetts and Hoskins (1979) have hypothesized that mesoscale convective lines may be forced by the circulations triggered by symmetric instability along fronts.

The right-hand side of (9.45) represents forcing:

- Forcing due to changes in the across-the-front temperature gradient caused by geostrophic stretching deformation along the front (Fig. 9.13):

$$2\frac{R}{f_0 p}\left(\frac{p}{p_0}\right)^\kappa\left(\frac{\partial v_g}{\partial y}\frac{\partial\theta}{\partial y}\right).$$

- Forcing due to changes in the across-the-front temperature gradient as geostrophic shearing deformation tilts the along-the-front temperature gradient into the cross-front direction (Fig. 9.14):

$$2\frac{R}{f p_0}\left(\frac{p}{p_0}\right)^\kappa\left(\frac{\partial u_g}{\partial y}\frac{\partial\theta}{\partial x}\right).$$

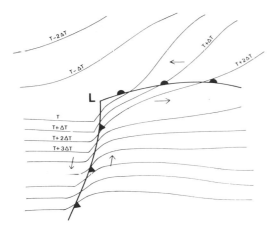

Figure 9.15. Frontolysis along the warm front and frontogenesis along the cold front due to shear. Solid lines are isotherms; arrows indicate wind direction. (After Gidel, 1978.)

- Forcing due to differential diabatic heating:

$$-\frac{R}{C_p f_0 p}\frac{\partial}{\partial y}\left(\frac{dQ}{dt}\right)\ .$$

Some differences in the structure of warm fronts and cold fronts have been explained in terms of the second kind of forcing (Eliassen, 1962; Gidel, 1978). Along warm fronts its effect is often frontolytical, and along cold fronts its effect is often frontogenetical (Fig. 9.15).

Eliassen (1962) showed that if (9.45) is transformed into a coordinate system in which the abscissa is

$$m = u_g - f_0 y, \tag{9.49}$$

then the resulting diagnostic equation for ψ is

$$\left[\frac{\partial}{\partial m}\left(p'\frac{\partial\psi}{\partial m}\right)\right]_p + \left(\frac{\partial^2\psi}{\partial p^2}\right)_m = \frac{2\frac{R}{f_0 p}\left(\frac{p}{p_0}\right)^\kappa\left(\frac{\partial\theta}{\partial y}\frac{\partial v_g}{\partial y} + \frac{\partial\theta}{\partial x}\frac{\partial u_g}{\partial y}\right)}{f_0 - \frac{\partial u_g}{\partial y}}\ , \tag{9.50}$$

where

$$p' = 4\left[\left(\frac{\partial u_g}{\partial p}\right)^2 + \frac{R}{f_0 p}\left(\frac{p}{p_0}\right)^\kappa\frac{\partial\theta}{\partial p}\left(f_0 - \frac{\partial u_g}{\partial y}\right)\right]\ .$$

An important result of this coordinate transformation is that the operator in (9.50) does not have mixed derivatives, and so the vertical circulation in (m,p)-space does not tilt with height for a point source or constant forcing (Fig. 9.16). However, in physical space (y,p) the circulation tilts along the m surface, whose slope

$$\frac{\partial m}{\partial p} = \frac{\partial u_g}{\partial p} = \frac{R}{f_0 p}\left(\frac{p}{p_0}\right)^\kappa\frac{\partial\theta}{\partial y} \tag{9.51}$$

is proportional to the horizontal temperature gradient. The quantity m has been called the absolute momentum of an air parcel because

$$\frac{Dm}{Dt} = \frac{Du_g}{Dt} - fv = -\frac{\partial \Phi}{\partial x} , \qquad (9.52)$$

and hence m is like momentum in a nonrotating reference frame. Since

$$-\frac{\partial m}{\partial y} = f_0 - \frac{\partial u_g}{\partial y} = f_0 + \zeta_g \qquad (9.53a)$$

and

$$\frac{\partial m}{\partial p} = \frac{\partial u_g}{\partial p} , \qquad (9.53b)$$

m has also been called the stream function for absolute vorticity in the y-p plane (Hoskins and Bretherton, 1972). Thus the circulation cell tilts along the vector line of the three-dimensional absolute vorticity.

The intensity of the v_a-ω vertical circulation depends upon the strength of the forcing. Positive forcing (from a point source or constant in space) is accompanied by clockwise circulation in the y-$(-p)$ plane (Fig. 9.16), and vice versa. If the absolute vorticity $(f_0 - \partial u_g/\partial y)$ is small compared with the static stability $\partial \theta / \partial(-p)$, then in (9.45)

$$\frac{\partial^2 \psi}{\partial p^2} \sim \frac{\partial v_a}{\partial p} \gg \frac{\partial^2 \psi}{\partial y^2} \sim \frac{\partial \omega}{\partial y}$$

and the circulation is composed mostly of the v_a branch. Similarly, if the absolute vorticity is large compared with the static stability $\partial \theta / \partial(-p)$, then the circulation is composed mostly of the ω branch. In other words, if the inertial stability is small compared with the static stability, horizontal motions dominate and vertical motions are inhibited. If the inertial stability is large compared with the static stability, vertical motions dominate and horizontal motions are inhibited. It is also noted that the effect of the geostrophic forcing function

$$2\frac{R}{f_0 p} \left(\frac{p}{p_0}\right)^\kappa \left(\frac{\partial \theta}{\partial y}\frac{\partial v_g}{\partial y} + \frac{\partial \theta}{\partial x}\frac{\partial u_g}{\partial y}\right)$$

in (9.50) is enhanced by low values of absolute vorticity, i.e., by weak inertial stability.

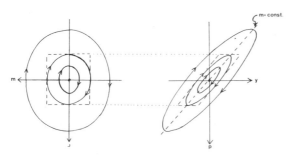

Figure 9.16. Streamlines around a positive point source in the m-p plane (left) and in the y-p plane (right). (After Eliassen, 1962.)

It is instructive to compare (9.45) with its quasi-geostrophic counterpart (Shapiro, 1982). If diabatic heating is neglected and $\partial\theta/\partial p$ is a function of p only, then the quasi-geostrophic form of (9.45) is

$$\frac{\partial^2\psi}{\partial y^2}\left[-\frac{\partial\bar{\theta}(p)}{\partial p}\frac{R}{f_0 p}\left(\frac{p}{p_0}\right)^\kappa\right]+\frac{\partial^2\psi}{\partial p^2}f_0$$

$$=2\frac{R}{f_0 p}\left(\frac{p}{p_0}\right)^\kappa\left(\frac{\partial\theta}{\partial y}\frac{\partial v_g}{\partial y}+\frac{\partial\theta}{\partial x}\frac{\partial u_g}{\partial y}\right). \tag{9.54}$$

The ellipticity (and stability) condition for (9.54) is that $\partial\bar{\theta}/\partial p < 0$, i.e., that the atmosphere be statically stable. The forcing function on the right-hand side of (9.54) is equivalent to that on the right-hand side of (9.45). However, since (9.54) does not contain a mixed derivative on the left-hand side, the circulation does not tilt with height. The terms responsible for the tilt in the geostrophic momentum approximation formulations (9.43) and (9.45) are the cross-front ageostrophic advection of temperature $(-v_a\,\partial\theta/\partial y)$ and the vertical advection of geostrophic momentum $(-\omega\,\partial u_g/\partial p)$. Furthermore, the coefficient of the $\partial^2\psi/\partial p^2$ term in (9.54) does not contain the effects of relative vorticity as it does in (9.45). Thus in the quasi-geostrophic system the inertial stability has a lower bound (expressed as f_0), and consequently, v_a cannot be much more important than ω unless the static stability is extremely high.

It is concluded that one effect of the ageostrophic circulation is to tilt the frontal zone (see bottom of Fig. 9.17). Without tilting, frontogenesis can be accompanied by a production of static instability on the warm side of the front (see top of Fig. 9.17). Furthermore, ageostrophic temperature advection hastens the frontogenetical process at the surface where $\omega\approx 0$. Another way of saying this is that convergence at the surface under the rising branch of the circulation acts frontogenetically, and augments the effects of geostrophic deformation (9.10). Thus synoptic-scale, geostrophic deformation is regarded as initiating frontogenesis, whereas the completion and maintenance of the front is due to the vertical circulation (Eliassen, 1959). However, in the special case of coastal frontogenesis (Bosart *et al.*, 1972; Bosart, 1975, 1981) the initiation of frontogenesis is due to ageostrophic deformation.

During quasi-geostrophic frontogenesis, regions of large anticyclonic as well as cyclonic vorticity are produced at the surface, since, according to the vorticity equation, divergence and convergence act only on f. However, divergence and convergence act on $\zeta_g + f$ at the surface in the geostrophic-momentum form of the vorticity equation [obtained from (9.41)]:

$$\frac{D}{Dt}(\zeta_g+f)=-\delta(\zeta_g+f)+\hat{k}\cdot\left(\frac{\partial\boldsymbol{v}_g}{\partial p}\times\boldsymbol{\nabla}\omega\right). \tag{9.55}$$

Consequently, the rate at which anticyclonic vorticity is produced in a region of divergence is less than the rate at which cyclonic vorticity is produced in

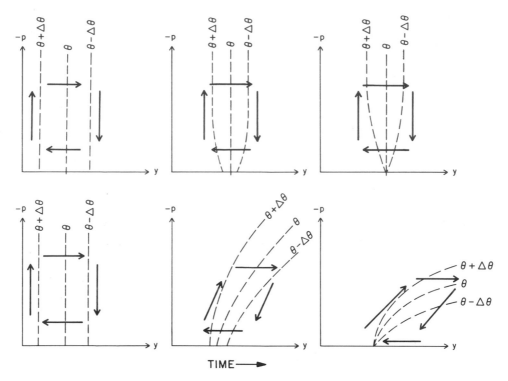

Figure 9.17.   Quasi-geostrophic frontogenesis (top) and frontogenesis governed by the geostrophic momentum approximation (bottom). Dashed lines are potential temperature isotherms; arrows represent ageostrophic vertical circulation. In top row, the vertical circulation does not affect distribution of surface isotherms; only geostrophic deformation affects distribution of isotherms. In bottom row, ageostrophic circulation does affect distribution of isotherms.

a region of convergence, and hence the strip of cyclonic vorticity is more prominent.

Hoskins (1975) has shown that if the dynamical equations containing the geostrophic momentum approximation [9.41 and 9.42] are transformed into a coordinate system in which air parcels move along with the geostrophic wind, i.e., geostrophic coordinates (Yudin, 1955), then the new equations are similar in form to the quasi-geostrophic equations in the new coordinate system. In other words, let

$$\frac{DX}{Dt} = u_g \quad \text{and} \quad \frac{DY}{Dt} = v_g \ , \tag{9.56}$$

where X and Y are the horizontal geostrophic coordinates. From (9.41)

$$\frac{Du_g}{Dt} = f(v - v_g) \quad \text{and} \quad \frac{Dv_g}{Dt} = -f(u - u_g) \ ,$$

where

$$u_g = u + \frac{1}{f}\frac{Dv_g}{Dt} = \frac{DX}{Dt} \tag{9.57a}$$

and

$$v_g = v - \frac{1}{f}\frac{Du_g}{Dt} = \frac{DY}{Dt} \; . \tag{9.57b}$$

Integrating (9.57) with respect to time gives

$$X = x + \frac{v_g}{f} \tag{9.58a}$$

and

$$Y = y - \frac{u_g}{f} \; . \tag{9.58b}$$

Note that Eliassen's m surface is proportional to the y coordinate in geostrophic space. The equations (9.41) and (9.42) expressed in geostrophic coordinates are called the semi-geostrophic equations. The semi-geostrophic equations are as follows (Hoskins and Draghici, 1977):

$$\frac{D_g v_g}{Dt^*} = -f_0(\hat{k} \times \boldsymbol{v}_a^*) \tag{9.59a}$$

$$\frac{D_g \theta}{Dt^*} = -\omega \frac{\partial \theta}{\partial P} \; , \tag{9.59b}$$

where

$$\frac{D_g}{Dt^*} = \frac{\partial}{\partial t^*} + u_g \frac{\partial}{\partial X} + v_g \frac{\partial}{\partial Y} \tag{9.59c}$$

$$\boldsymbol{v}_a^* = \boldsymbol{v}_a + \frac{\omega}{f_0{}^2}\boldsymbol{\nabla}_P\left(\frac{\partial \Phi^*}{\partial P}\right) \tag{9.59d}$$

$$\Phi^* = \Phi + \frac{1}{2}(u_g{}^2 + v_g{}^2) \tag{9.59e}$$

$$\boldsymbol{v}_g = f_0\,(\hat{k} \times \boldsymbol{\nabla}_P \Phi^*) \tag{9.59f}$$

$$P = p \tag{9.59g}$$

$$t^* = t \tag{9.59h}$$

$$\frac{\partial u_a^*}{\partial X} + \frac{\partial v_a^*}{\partial Y} + \frac{\partial \omega^*}{\partial P} = 0 \tag{9.59i}$$

$$\omega^* = \left(\frac{f_0}{f_0 + \varsigma_g}\right)\omega \; . \tag{9.59j}$$

Some mathematical consequences of the geostrophic coordinate transformation are discussed by Blumen (1981). The coordinates X, Y, Φ^*, and t^* are referred to as semi-geostrophic space; the coordinates x, y, Φ, and t are referred to as physical or real space. Note that the geopotential in semi-geostrophic space is like the Bernoulli function for geostrophic motion, less the $C_p T$ term. These equations show that despite the small length scale across a frontal zone, the frontogenetical process in a sense can be considered to be quasi-geostrophic along m surfaces.

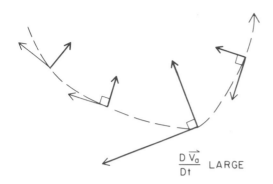

Figure 9.18. An ageostrophic wind field v_a (arrows) associated with a sharply curved parcel trajectory (dashed line with arrowhead). Centripetal acceleration experienced by air parcel at various points along the trajectory is depicted by bold arrows.

Geostrophic coordinates in effect stretch the length scale across the front so that, as Eliassen (1984) has pointed out, air parcels reside in the frontal zone for a relatively long time, and hence parcel accelerations are small compared with the acceleration from the Coriolis force. Hoskins and West (1979) showed that the surface frontogenetical function in semi-geostrophic space does not contain the divergence term [as in (9.10)]. They found that the effects of deformation and convergence are separate, and that the resultant deformation in semi-geostrophic space is f/ς times that in physical space, and the axis of dilatation is oriented in the same direction. In summary, the semi-geostrophic equations allow us to study a mesoscale process using synoptic-scale dynamics in a coordinate system in which air parcels move with the geostrophic wind.

Shapiro (1982), however, pointed out that the rapid formation of convective systems may cause the atmosphere to be thrown out of geostrophic balance in an impulsive way, so that the geostrophic momentum approximation (9.41) is not valid. Furthermore, if the parcel trajectories are sharply curved, the centripetal acceleration of the ageostrophic wind can be significant (Fig. 9.18) (Shapiro and Kennedy, 1981). Uccellini et al. (1984) showed that the geostrophic momentum approximation can also fail for straight flow if the parcel accelerations are so strong that the Rossby number is close to 1.

Friction and Latent Heat Release

Keyser and Anthes (1982) showed that convergence due to surface friction (i.e., Ekman-like pumping) drives a vertical velocity jet above the surface frontal trough. The frictional convergence furthermore acts frontogenetically at the surface (9.10). They also showed that behind the frontal zone the vertical shear of the ageostrophic wind, which is frictionally induced, acts to diminish the static stability in the boundary layer and enhance the static stability at the top of the boundary layer (Fig. 9.19). Differential temperature advection by the ageostrophic component of the wind ahead of the front acts to increase the static stability in the boundary layer and decrease the static stability at the top of the boundary layer (Fig. 9.19). It is thought that although friction is not of primary importance in frontogenesis, it does limit the strength of the front when the temperature gradient becomes large.

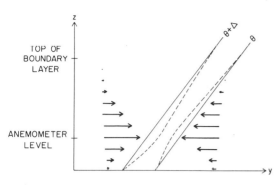

Figure 9.19. Effect of boundary layer on stability within a frontal zone. Arrows are frictionally induced ageostrophic wind component normal to frontal zone. Solid and dashed lines are initial and later potential temperature. Vertical gradient of potential temperature (static stability) changes because of $-(\partial\theta/\partial z)(\partial v/\partial z)$ $(\partial\theta/\partial y)$, vertical deformation (Miller, 1948).

When a frontal zone becomes very intense, the Richardson number becomes so low that small-scale motions become important and the resulting turbulent diffusion acts frontolytically (Hoskins and Bretherton, 1972). Williams (1974) showed that frontogenetic processes (e.g., deformation acting on the temperature field) can be balanced by diffusion so that a steady state is achieved.

Fultz (1952) and Faller (1956) were able to produce front-like structures in their laboratory models without the aid of latent heat release. Latent heat release is therefore not necessary for frontogenesis. However, Eliassen (1962) argued that latent heat release in the rising branch of the vertical circulation could strengthen the vertical circulation, and could therefore enhance the frontogenetical process (as more convergence is produced under the rising branch).

Williams *et al.* (1981), using a numerical model, and Thorpe (1984), using an analytical quasi-geostrophic model, showed that latent heat release could in fact act frontogenetically aloft. Evaporative cooling, which occurs when precipitation falls through dry air below on the cold side of fronts, could also act frontogenetically (Browning and Pardoe, 1973). Ross and Orlanski (1978) found numerical-model evidence that a moist front could excite gravity waves, which could trigger convection ahead of the front.

"External" forcing can sometimes be important. For example, the effects of differential surface diabatic heating of cold, continental air flowing over warm water may be responsible for the convergence (ageostrophic deformation) that results in the formation of the coastal front (Ballentine, 1980; Bosart, 1981).

Finally, it is possible that friction and diabatic heating could act cooperatively through the conditional instability of the second kind (CISK) mechanism along the band of cyclonic vorticity associated with an old frontal zone to produce a vertical circulation even after frontogenesis has ceased (Hoskins, 1972). Bluestein (1977) suggested that this mechanism may account for some tropical cloud bands.

9.2.6. *The Dynamics of Middle and Upper Tropospheric Frontogenesis*

Unlike surface frontogenesis, for which vertical motions do not play a direct role because of the kinematic lower boundary condition (at the surface if it is level), middle and upper tropospheric frontogenesis (also referred

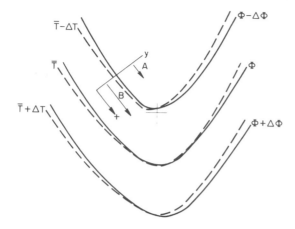

Figure 9.20.   Quasi-geostrophically induced thermally indirect circulation upstream from a trough in the baroclinic westerlies. Solid lines are height contours; dashed lines are mean isotherms; large + is vorticity maximum; V_T is the thermal wind vector (indicated by arrows).

At A: $-\vec{V}_T \cdot \vec{\nabla}_p(\zeta_g + f) < 0 \Rightarrow \omega > 0$

At B: $-\vec{V}_T \cdot \vec{\nabla}_p(\zeta_g + f) \ll 0 \Rightarrow \omega \gg 0$

to as "internal" frontogenesis [Buzzi *et al.*, 1981]) is influenced by vertical motions [see (9.9)]. The effect of adiabatic warming or cooling should be significant where vertical motion and static stability are relatively large. Since the magnitude of vertical motions is greatest in the middle troposphere and static stability is the highest at the tropopause, the combined effects of vertical motion and static stability should be largest in the upper troposphere. Vertical motions are suppressed at the tropopause because of large static stability in the stratosphere; that is, the tropopause functions as an upper boundary (albeit not rigid) to the middle and upper tropospheric front. Thus the tropopause in a sense plays a role in upper tropospheric frontogenesis that is similar to the role played by the ground in surface frontogenesis.

According to (9.9), if diabatic processes are ignored, the middle and upper tropospheric front could be formed either through the action of deformation or convergence acting on the temperature field and/or through a horizontal variation in vertical motion (if the horizontal variation in static stability is ignored). If the level of nondivergence is in the middle troposphere, it is expected that convergence does not play a role there. However, convergence aloft (above sinking motion) could be important. If the action of geostrophic deformation on a temperature field is regarded as forcing, a primary process, then the vertical-motion field is a response, a secondary process.

On the basis of observational studies, Reed and Sanders (1953) and Bosart (1970) suggested that a horizontal variation in subsidence that is part of a thermally indirect vertical circulation ($\partial \omega / \partial y < 0$ for $\partial \theta / \partial y < 0$) could initiate middle and upper tropospheric frontogenesis. Bosart (1970) hypothesized that the horizontal variation in subsidence was due to a quasi-geostrophic, horizontal variation in vorticity advection decreasing with height. Including the effects of temperature advection, Bosart's hypothesis can be generalized by considering the horizontal variation in advection of anticyclonic vorticity by the thermal wind (Trenberth, 1978) (Fig. 9.20). As long as the static stability and the forcing responsible for the horizontal variation in subsidence remain constant, frontogenesis will occur at a relatively slow rate according to (9.9):

$$\frac{D}{Dt}\left(-\frac{\partial \theta}{\partial y}\right) = \frac{\partial \omega}{\partial y}\frac{\partial \theta}{\partial p}. \tag{9.60}$$

However, the geostrophic-momentum form of the vorticity equation (9.55) includes the tilting term. The effect of tilting of the vertical shear onto the horizontal is to increase the cyclonic vorticity (Fig. 9.21) and therefore to increase the advection of anticyclonic vorticity by the thermal wind on the cyclonic side of the basic current. Consequently the horizontal variation in subsidence increases, tilting further increases the cyclonic vorticity, and hence a positive feedback mechanism is possible that can lead to much more rapid frontogenesis (Mudrick, 1974). The role of the tilting of vorticity in middle and upper tropospheric frontogenesis is therefore similar to the role of convergence (ageostrophic deformation) acting on the temperature field in surface frontogenesis: Both increase the rate of frontogenesis over what is possible quasi-geostrophically.

Shapiro (1970, 1981) was the first to argue that the geostrophic momentum approximation could be used to explain, consistent with observations, the vertical circulation in a middle or upper tropospheric front. Suppose the Sawyer-Eliassen equation (9.45) is used to diagnose qualitatively the horizontal variation in subsidence resulting from a given forcing (Shapiro, 1982). If the coordinate system is rotated so that the x axis is parallel to the isotherms at the origin of the coordinate system, it is found that only when $(\partial \theta/\partial y)(\partial v_g/\partial y) < 0$ is there the response of $\partial \omega/\partial y < 0$ needed for frontogenesis. A negative $(\partial \theta/\partial y)(\partial v_g/\partial y)$ represents a frontolytical geostrophic

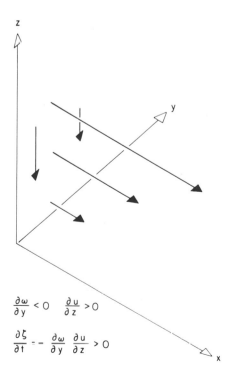

$\frac{\partial \omega}{\partial y} < 0 \quad \frac{\partial u}{\partial z} > 0$

$\frac{\partial \zeta}{\partial t} = -\frac{\partial \omega}{\partial y}\frac{\partial u}{\partial z} > 0$

Figure 9.21. The effects of tilting of vertical shear onto the horizontal by the thermally indirect circulation depicted in Figure 9.20. Horizontal arrows are geostrophic wind; vertical arrows are vertical velocity.

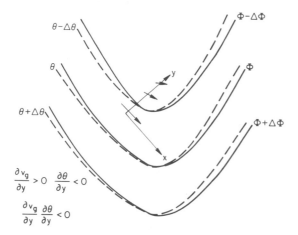

Figure 9.22. The frontolytical effect of the geostrophic wind field on the cyclonic side of the flow upstream from a trough in the baroclinic westerlies. Arrows are geostrophic wind; solid lines are height contours; dashed lines are potential temperature isotherms. This frontolytical geostrophic forcing results in a frontogenetical, thermally indirect circulation according to (9.45) and (9.54).

forcing due to deformation. Figure 9.22 shows that this can happen just upstream from a cyclonic vorticity maximum. [Here $\partial v_g / \partial x > 0$ and hence curvature may have to be taken into account, and (9.45) cannot be used. In this case $(\partial v_g / \partial x)(\partial \theta / \partial y) < 0$, which is frontolytical, since the across-the-front temperature gradient is being tilted onto the along-the-front direction, and the contribution to $\partial \omega / \partial y$ is therefore negative also. Curvature effects are discussed in detail in Newton and Trevisan (1984 a,b).]

Since the vertical shear increases when a midtropospheric front forms, there is the possibility that the Richardson number (9.48) could drop below the critical value (0.25) necessary for the development of turbulence resulting from Kelvin-Helmholtz instability (Shapiro, 1974). Since the midtropospheric front is found in a region of subsidence, any resulting turbulence is likely to be in clear air. Shapiro (1974) demonstrated that anticyclonically curved trajectories and anticyclonic curvature increasing with height (or cyclonic curvature decreasing with height) reduce the value of Ri.

Strong subsidence underneath the jet can lead to the downward advection of the tropopause (Reed, 1955). In the vicinity of a zonally oriented jet, $\partial^2 u_g / \partial p^2 < 0$. It follows from the thermal wind relation (9.16) that

$$\frac{\partial^2 u_g}{\partial p^2} = -\frac{1}{f_0}\frac{\partial \sigma}{\partial y} \ . \tag{9.61}$$

Therefore the static stability increases with latitude across a zonal jet. Thus the tropopause decreases with height across the jet from south to north in the Northern Hemisphere. Subsidence can result in tropopause folding (Fig. 9.23) if the tropopause is advected downward at an angle such that it is eventually found at two levels (Reed, 1955).

If the standard definition of the tropopause in terms of the vertical change in static stability is used, then there is a problem with the concept of the advection of the tropopause since stability is not conserved for adiabatic motions (Danielsen, 1968). However, potential vorticity is conserved, and it is therefore useful to define the tropopause in terms of the potential vorticity or the magnitude of its gradient. The following is a component of Ertel's

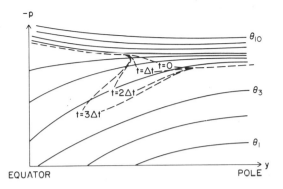

Figure 9.23. Tropopause folding. Solid lines are potential isotherms; dashed line is tropopause.

potential vorticity (P) defined in terms of static stability and absolute vorticity on an isentropic surface in a hydrostatic atmosphere (Reed and Sanders, 1953):

$$P = (\varsigma_\theta + f)\frac{\partial \theta}{\partial(-p)} . \tag{9.62}$$

The potential vorticity is large in the stratosphere because $\partial\theta/\partial(-p)$ is large (La Seur, 1974). In the troposphere, however, P is small because $\partial\theta/\partial(-p)$ is relatively small and ς_θ is small or even anticyclonic owing to the westerly vertical shear and the poleward sloping of the isentropic surfaces.

9.2.7. The Relationship Between Frontogenesis and Cyclogenesis

Polar-front theorists recognized that fronts can play an important role in cyclogenesis (Bjerknes, 1919). Since the front is a region of strong temperature gradient, it is a zone in which temperature advection can be quite strong. If there is localized warm advection over a level surface, then from quasi-geostrophic theory it is found that there must be rising motion, surface convergence, an increase in surface vorticity, and a drop in surface pressure. Thus, there is a dynamical reason why fronts should be preferred regions for cyclogenesis in the absence of inhibition from other forcing functions.

On the other hand, deformation that is associated with the checkerboard of surface highs and lows is responsible for the initiation of frontogenesis. Therefore there is an intimate relationship between surface cyclogenesis and frontogenesis: each may be responsible for the other.

Eliassen (1966) noted that synoptic-scale waves in the middle-latitude baroclinic westerlies are due to baroclinic instability, and do not depend upon the existence of fronts. Therefore he suggested that the production of the surface highs and lows, which is a result of the baroclinic instability process, leads to the formation of fronts and, further, to the formation of smaller-scale cyclones along the fronts. Thus, baroclinic instability is thought to be the primary process, and frontogenesis a secondary process.

Hoskins and West (1979) discussed the ways in which frontogenesis depends upon the baroclinically unstable mean state. They found that in the case of a broad upper-level westerly jet there is a "primary" cold front and a "secondary" warm front (Fig. 9.24, left). If the surface flow has cyclonic vorticity on the equatorward side, a "primary" warm front forms, with no

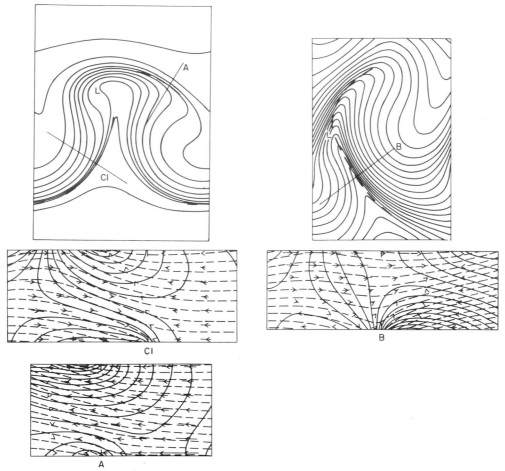

Figure 9.24. (Left) Surface map (top) showing isotherms (solid lines) every 2.5°, center of low-pressure area (*L*). Vertical sections (middle and bottom) show "primary" cold front (C1) and "secondary" warm front (A). Dashed lines are 4° potential temperature isotherms; solid lines are component of wind into and out of section every 5 m s^{-1}; arrows indicate wind field relative to zonally propagating baroclinic wave in plane of sections. (Right) Surface map and vertical section for "primary" warm front. (After Hoskins and Heckley, 1981.)

cold front (Fig. 9.24, right). The primary cold frontal zone is steep and is associated with a deep layer of cyclonic vorticity. The primary warm frontal zone slopes less and is associated with a shallower layer of cyclonic vorticity. The primary warm front is like an occlusion in that the temperature gradient reverses across it, and the warmest air present is along it. Hoskins and Heckley (1981) postulated that the dynamical reason why cold fronts and warm fronts differ structurally stems from the forward (eastward) tilt of the temperature wave in a growing baroclinic wave at low levels.

Keyser and Pecnick (1985a, b) numerically investigated the relationship between typical flow patterns associated with baroclinic waves during various stages in their life cycles and middle and upper tropospheric frontogenesis. They found that, in the case of cold advection at upper levels from a "digging" trough, a front forms near the tropopause and builds downward with time. In the case of warm advection downstream from a "lifting" trough, frontogenesis occurs throughout the troposphere.

9.3. Jet Streaks

9.3.1. The Formation of Jet Streaks

Jet streaks are found within the polar front jet and subtropical jet, i.e., in jets aloft near the tropopause in middle latitudes. The intensity of an isotach field for flow along a jet that is oriented in the x-direction is given by $\nabla^2 u$, which is negative in a jet streak. The jetogenetical function may be defined as

$$J = \frac{D}{Dt}(-\nabla^2 u) \, , \tag{9.63}$$

the rate of change of the intensity (three-dimensional Laplacian) of the jet streak following air-parcel motion. It is analogous to Petterssen's frontogenetical function. [Strictly speaking, (9.63) should be expressed in a coordinate system moving along with the jet streak; however, (9.63) should be useful for the analysis of jet streaks just as (9.11) has been useful for the analysis of fronts.] So

$$J = -\nabla^2 \left(\frac{Du}{Dt} \right) + \nabla^2 (\boldsymbol{v} \cdot \nabla u) - \boldsymbol{v} \cdot \boldsymbol{\nabla}(\nabla^2 u) \, . \tag{9.64}$$

By substituting the x-component of the equation of motion in (9.64) and assuming that ∇u is uniform in the neighborhood of the air parcel ($\nabla u = 0$ in the center of a jet streak), it is found that

$$J = f \nabla^2 v_a - 2\beta \frac{\partial v_a}{\partial y} \, , \tag{9.65}$$

where v_a is the ageostrophic component of the wind in the y-direction.

For typical values of wind speed in middle latitudes, the first term is dominant. The jetogenetical function is positive wherever there is a local maximum in the cross-jet component of the ageostrophic wind. This happens, for example, when there is a thermally direct circulation below and a thermally indirect circulation aloft (Fig. 9.25). It would therefore be expected, for example, that the superposition of a layer of vertical motions acting frontogenetically over a layer of horizontal deformation acting frontogenetically on the temperature field is a favorable configuration for jet-streak formation. This configuration may be realized when an upper tropospheric front and jet approach a surface front below (Shapiro, 1982).

9.3.2. Kinematics of Jet Streaks

The wind speed in a jet streak is faster than the speed with which the jet streak moves. Therefore the region downstream from a jet streak where air parcels decelerate is called the exit region; the region upstream from a jet streak where air parcels accelerate is called the entrance region. The

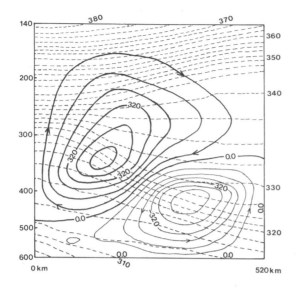

Figure 9.25. Thermally indirect circulation aloft over a thermally direct circulation below (but above 600 mb) inferred from the Sawyer-Eliassen equation in a jet-front system. Solid and closely dotted lines represent vertical stream function; widely dotted lines are potential temperature isotherms. (After Shapiro, 1981.)

further subdivision of a jet streak into left and right sectors is often used, and is valid for an observer facing downstream. The left sides of the entrance and exit regions are often called the left-rear and left-front quadrants, respectively; similarly, the right sides of the entrance and exit regions are often called the right-rear and right-front quadrants. To keep the geostrophic wind field quasi-nondivergent, the entrance region and exit region of a jet streak in the geostrophic wind field are confluent and diffluent, respectively (Fig. 9.26). Shapiro (1982) pointed out that jet streaks can form when a trough in the westerlies becomes in phase with a ridge at lower latitudes. This configuration is like the opposite of a high-over-low block (Rex, 1950).

In natural coordinates the inviscid equations of motion are as follows (Shapiro and Kennedy, 1981):

$$\frac{DV}{Dt} = -\frac{\partial \Phi}{\partial s} = f V_{an} \tag{9.66a}$$

$$\frac{V^2}{R_t} = -fV - \frac{\partial \Phi}{\partial n} = -f V_{as} , \tag{9.66b}$$

where V is the wind speed, n refers to the direction normal and to the left of the flow, s refers to the direction of the flow, and R_t is the radius of curvature of the parcel trajectory (a positive R_t represents cyclonic curvature). A

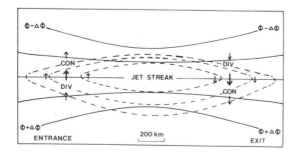

Figure 9.26. The ageostrophic motions (arrows) and associated convergence (CON) and divergence (DIV) patterns in the vicinity of a straight jet streak. (From Shapiro and Kennedy, 1981.)

straight jet streak is one in which the dynamical effects of curvature are negligible; i.e., R_t is large, so that the centripetal acceleration on the left-hand side of (9.66b) is unimportant in comparison with the Coriolis and pressure-gradient terms. In a straight jet streak the ageostrophic component of the wind is normal to the flow. Thus, another measure of the straightness of a jet streak is $V_{an}/|v_a|$ (Bluestein and Thomas, 1984), where a value of 1 represents perfect straightness and a value of 0 represents circular flow.

9.3.3. Vertical-Motion Field Near Jet Streaks

There are several ways to infer the vertical-motion field in the vicinity of a jet streak. Consider the frictionless form of the vorticity equation. If the twisting terms are small compared with the divergence term,

$$\frac{D(\varsigma + f)}{Dt} = -\delta(\varsigma + f). \qquad (9.67)$$

For example, an air parcel moving through the left side of the exit region of a straight jet streak experiences a decrease in relative vorticity, since it is moving away from the region of strong cyclonic shear on the left side of the jet streak (Riehl *et al.*, 1952, pp. 35–36). Since the change in relative vorticity overwhelms the change in Earth's vorticity experienced by the air parcel, the parcel must experience divergence (Fig. 9.27).

Similarly, an air parcel entering the jet streak on the left side experiences convergence. If the absolute vorticity is positive, then there is divergence in the entrance region on the right side, and convergence in the exit region on the right side. (If the jet streak is so intense that absolute vorticity is negative on the right side, the reverse is true; however, this represents an inertially unstable configuration.)

Another way to understand the divergence field is from the perspective of energy conservation (Namias and Clapp, 1949). An air parcel decelerates in the exit region, loses kinetic energy, and gains potential energy, which is consistent with its moving across contours toward higher heights. Convergence

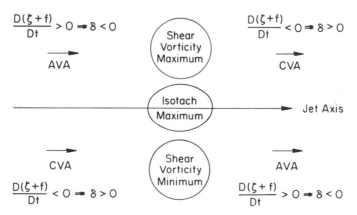

Figure 9.27. Relation between vorticity changes experienced by an air parcel moving through a jet streak, and divergence.

is found to the right and divergence is found to the left. A similar argument can be made for the entrance region. The transverse flow associated with jet streaks was first verified by Murray and Daniels (1953).

Suppose that $\omega = 0$ at the tropopause, at or near the level of the jet streak, and at the surface. Then from continuity considerations it is seen that there must be upward motion and convergence somewhere below the jet-streak level on the left side of the exit region and the right side of the entrance region. Similarly, there must be sinking motion and divergence somewhere below the jet-streak level on the right side of the exit region and the left side of the entrance region.

The divergence itself is due to part of the ageostrophic component of the wind; the geostrophic component of the wind and part of the ageostrophic component of the wind are nondivergent. Therefore more insight may be obtained by considering geostrophic and ageostrophic effects separately, and therefore the equations of motion are expressed as follows:

$$v_a = \frac{1}{f} \hat{k} \times \left\{ \frac{\partial v_g}{\partial t} + \frac{\partial v_a}{\partial t} + [(v_g + v_a) \cdot \nabla] (v_g + v_a) + \omega \frac{\partial v_g}{\partial p} + \omega \frac{\partial v_a}{\partial p} \right\} . \tag{9.68}$$

Suppose the portion of the ageostrophic wind that is represented by $(1/f)(\partial v_g/\partial t)$ is called the isallohypsic wind component, since

$$\frac{1}{f} \frac{\partial v_g}{\partial t} = \frac{1}{f^2} \hat{k} \times \nabla_p \left(\frac{\partial \Phi}{\partial t} \right) \tag{9.69}$$

accounts for the isallohypsic gradient, i.e., the changing cross-stream gradient of the height field. If height rather than pressure is used as the vertical coordinate, then

$$\frac{1}{f} \frac{\partial v_g}{\partial t} = \frac{1}{\rho f^2} \hat{k} \times \nabla_z \left(\frac{\partial p}{\partial t} \right) + \frac{1}{(\rho f)^2} (\hat{k} \times \nabla_z p) \nabla \cdot \rho v . \tag{9.70}$$

In this case, $[(1/\rho f^2) \hat{k}] \times [\nabla_z (\partial p/\partial t)]$ is called the isallobaric wind component (Brunt and Douglas, 1928), since it accounts for the isallobaric gradient, i.e., the changing cross-stream gradient of the pressure field. (Some define the isallobaric wind in terms of the changing cross-stream gradient of the pressure-gradient force.) The terms in (9.68) representing $(v \cdot \nabla)v$ are known as the inertial-advective terms (Uccellini and Johnson, 1979). The inertial-advective terms should be important in the upper troposphere where v and $\nabla|v|$ are large (Bjerknes, 1951). The isallobaric wind component should be important in the lower troposphere where wind speeds are relatively low and therefore air parcels have a longer period of time to adjust to the changing pressure gradient.

Quasi-Geostrophic Diagnosis

The quasi-geostrophic form of the momentum equation neglects ageostrophic and vertical advection, geostrophic advection of the

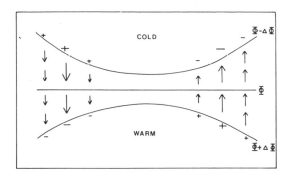

Figure 9.28. Height contours and Q vectors for a jet streak entrance and exit region. Signs of $\nabla_p \cdot Q$ and ω are indicated by $-$ and $+$. (From Hoskins *et al.*, 1978.)

ageostrophic component of the wind, and local changes in the ageostrophic wind, to obtain

$$v_a = \frac{1}{f} \hat{k} \times \left[\frac{\partial v_g}{\partial t} + (v_g \cdot \nabla) v_g \right] . \tag{9.71}$$

The vertical motion can be obtained by kinematically integrating (9.71), or from the quasi-geostrophic ω-equation in Cartesian coordinates (Hoskins *et al.*, 1978):

$$\left(\nabla_p^2 + \frac{f_0^2}{\sigma} \frac{\partial^2}{\partial p^2} \right) \omega = -2\nabla_p \cdot Q - \frac{R}{\sigma p} \left(\frac{p}{p_0} \right)^\kappa \frac{\partial}{\partial y} \left(\beta y \frac{\partial \theta}{\partial x} \right) , \tag{9.72}$$

where Q is defined in (9.19).

At the tropopause the horizontal temperature gradient reverses sign, so that $\nabla_p \theta = 0$, and hence $Q, \nabla_p \cdot Q$, and ω are zero. Below the tropopause, where the atmosphere is nearly equivalent barotropic, it is easy to see that if the β-effect is ignored (it is zero in the case of a zonally oriented jet streak), then $\nabla_p \cdot Q$ is negative on left side of the exit region and right side of the entrance region; similarly, $\nabla_p \cdot Q$ is positive on the left side of the entrance region and the right side of the exit region. Therefore, since a Laplacian operator acting on a variable vertical-tends to change its sign (Holton, 1979, p. 137), then according to (9.72) ω is negative (rising motion) on the left side of the exit region and the right side of the entrance region; furthermore, ω is positive (sinking motion) on the left side of the entrance region and the right side of the exit region (Fig. 9.28). Thus, the quasi-geostrophically deduced vertical-motion field is qualitatively identical to that deduced from the vorticity equation (9.67). Bluestein and Thomas (1984), however, have shown that in the case of an intense jet streak the quasi-geostrophic diagnosis is not even qualitatively accurate because of sharply curved parcel trajectories.

Semi-Geostrophic Diagnosis

The semi-geostrophic form of the momentum equation in real space is like the quasi-geostrophic form, except that ageostrophic and vertical advection of geostrophic momentum are retained. Therefore

$$v_a = \frac{1}{f} \hat{k} \times \left[\frac{\partial v_g}{\partial t} + (v_g + v_a) \cdot \nabla v_g + \omega \frac{\partial v_g}{\partial p} \right] . \tag{9.73}$$

The vertical-motion field can be determined by kinematically integrating (9.73), or the ω-field obtained from the semi-geostrophic ω-equation (Hoskins and Draghici, 1977) (the quasi-geostrophic ω-equation in geostrophic coordinates) can be transformed back into physical space. Using the former method, Bluestein and Thomas (1984) showed that unlike the quasi-geostrophic diagnosis, the semi-geostrophic diagnosis of an intense jet streak can be qualitatively correct. Quantitative errors may be attributed to the disregard of the advection of ageostrophic momentum, which can be important in sharply curved flow.

9.3.4. Propagation

The movement of a jet streak may be interpreted quasi-geostrophically, using the height-tendency equation (Holton, 1979, p. 131):

$$\left(\nabla_p^2 + \frac{f_0^2}{\sigma}\frac{\partial^2}{\partial p^2}\right)\frac{\partial \Phi}{\partial t} = -f_0 \mathbf{v}_g \cdot \nabla_p (\varsigma_g + f)$$

$$+ \frac{f_0^2 R}{\sigma}\frac{\partial}{\partial(-p)}\left[\frac{1}{p}\left(\frac{p}{p_0}\right)^{\kappa}(-\mathbf{v}_g \cdot \nabla_p \theta)\right] . \qquad (9.74)$$

If the upper troposphere is equivalent barotropic, then cyclonic vorticity advection on the left side of the exit region and right side of the entrance region is accompanied by height falls; similarly, anticyclonic vorticity advection on the right side of the exit region and the left side of the entrance region is accompanied by height rises. Qualitatively at least, it is easily seen that this height-tendency field is consistent with a downstream propagation of the jet streak (Fig. 9.29).

Petterssen's formula (Petterssen, 1956, p. 48) for the speed of a trough or ridge,

$$c = -\frac{\frac{\partial}{\partial x}\left(\frac{\partial \Phi}{\partial t}\right)}{\frac{\partial^2 \Phi}{\partial x^2}} , \qquad (9.75)$$

where the x-axis is oriented along the jet streak, can be used to test quantitatively the validity of the quasi-geostrophic propagation mechanism, since a jet streak is often defined along a trough axis or ridge axis by the location where a trough axis and ridge axis join; the trough lies to the left of the main

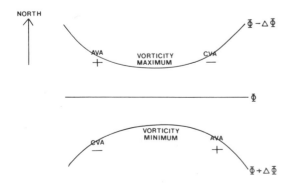

Figure 9.29. Propagation of a jet streak in the geostrophic wind field by vorticity advection. CVA and AVA denote cyclonic and anticyclonic vorticity advection, respectively; height tendencies are indicated by $+$ and $-$.

current when viewed looking downstream. The movement of the jet streak can also be determined using the isallohypsic field obtained from (9.75) and from the semi-geostrophic height-tendency equation (the quasi-geostrophic height-tendency equation in geostrophic coordinates) transformed back to physical space.

9.3.5. Coupling of Upper-Level Jet Streak to the Wind Field Below

Fawbush *et al.* (1951) observed that one of the synoptic conditions associated with tornado development is the intersection of the low-level projection of an upper-level jet with the axis of a low-level moisture ridge. The low-level moisture ridge is oriented along and is a result of the low-level jet (Beebe and Bates, 1955). Beebe and Bates hypothesized that the divergence field accompanying the superposition of intersecting jets could be associated with the upward motion necessary to release potential instability.

Uccellini and Johnson (1979) argued that the vertical transverse circulation in the exit region of an upper-level jet superposed above a low-level jet could increase the amount of potential instability through the following mechanisms:

- A decrease in ageostrophic temperature advection with height (warm advection below and cold advection aloft)
- Ageostrophic advection of moisture at low levels and ageostrophic advection of dry air aloft.

The following could also result in an increase in low-level potential instability:

- Cooling the air adiabatically at middle levels in the rising branch of the circulation.

Bluestein and Thomas (1984) further suggested that moisture convergence under the rising branch of the vertical circulation could be responsible for maintaining the supply of moisture necessary for growing cumulus towers. Another possible causal link between jet streaks and severe convection is through inertia-gravity waves generated when the mass and momentum fields cannot adjust rapidly enough at the entrance and exit regions (Van Tuyl and Young, 1982). Gravity waves might be able to lift air enough to release potential instability (Uccellini, 1975).

Uccellini and Johnson (1979) demonstrated that the vertical circulation associated with a jet streak could in fact be coupled to the flow at low levels, so that the intersection of the upper-level jet and low-level jet is not just a chance event. In particular, they suggested that a low-level jet could form underneath and at a significant angle to the exit region of an upper-level jet streak as a result of the isallobaric contribution to the ageostrophic wind. The isallobaric wind is associated with the pressure-fall/pressure-rise couplet at the surface in the plane of the vertical circulation. It is linked to the vertical motion coupled through the accompanying surface divergence.

Shapiro (1982) (Fig. 9.30) hypothesized that when an upper-level jet streak and front are uncoupled from a low-level jet and surface front, the thermally indirect circulation aloft associated with the upper-troposphere

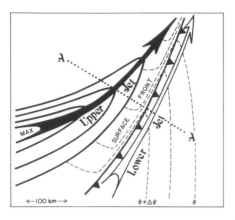
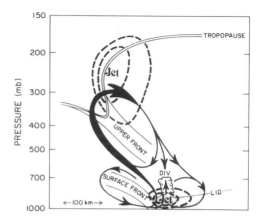

Figure 9.30a. Vertically uncoupled upper- and lower-tropospheric jet-front systems and their associated secondary circulation. (Left) Upper jet-front exit region displaced to the west of a surface front and low-level jet. Heavy solid lines and solid arrow show upper jet isotachs and axis; open arrow shows lower jet axis. Thin dashed lines indicate surface potential temperature. (Right) Cross section along the line AA', left. Heavy dashed lines are upper and lower jet isotachs; double thin line marks potential vorticity tropopause; moist layer is under the line labeled LID. Streamlines with heavy arrows show forced secondary circulation. (From Shapiro, 1982).

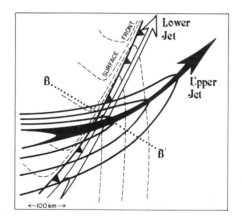
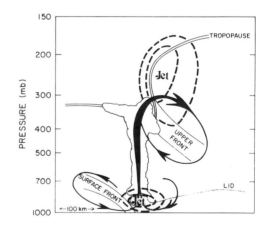

Figure 9.30b. Vertically coupled upper- and lower-tropospheric jet-front systems and their associated secondary circulations. (Left) Upper jet-front exit situated above the surface front and low-level jet. (Right) Cross section along the line BB', left. Isopleths and arrows are as in Fig. 9.30a. (From Shapiro, 1982.)

front and exit region of the jet streak and the thermally direct circulation associated with the surface front can act to produce subsidence and there-fore stabilization of the potentially unstable air ahead of the front. However, when the upper-level jet streak and front are coupled to a low-level jet streak and surface front, the thermally indirect circulation aloft and thermally di-rect circulation below can act to produce ascending motion and therefore destabilization of the potentially unstable air ahead of the front. Uccellini *et al.* (1984) have presented observational evidence that an upper-level jet streak and low-level jet streak can in fact play a role in the development of heavy precipitation.

Appendix: Derivation of the Two-Dimensional Frontogenetical Function for Horizontal Flow

$$F = \frac{D_p}{Dt} |\nabla_p \theta|$$

$$= \frac{\partial}{\partial t} |\nabla_p \theta| + \boldsymbol{v} \cdot \nabla_p |\nabla_p \theta|$$

$$= \frac{\partial}{\partial t} \left[\left(\frac{\partial\theta}{\partial x}\right)^2 + \left(\frac{\partial\theta}{\partial y}\right)^2 \right]^{1/2} + \boldsymbol{v} \cdot \nabla_p \left[\left(\frac{\partial\theta}{\partial x}\right)^2 + \left(\frac{\partial\theta}{\partial y}\right)^2 \right]^{1/2}$$

$$= \frac{1}{|\nabla_p \theta|} \left[\frac{\partial\theta}{\partial x} \frac{D_p}{Dt}\left(\frac{\partial\theta}{\partial x}\right) + \frac{\partial\theta}{\partial y} \frac{D_p}{Dt}\left(\frac{\partial\theta}{\partial y}\right) \right]^{1/2}, \tag{9.A1}$$

where

$$\frac{D_p}{Dt} = \frac{\partial}{\partial t} + u\frac{\partial}{\partial x} + v\frac{\partial}{\partial y}.$$

Now

$$\frac{\partial}{\partial x}\left(\frac{D_p\theta}{Dt}\right) = \frac{D_p}{Dt}\left(\frac{\partial\theta}{\partial x}\right) + \frac{\partial u}{\partial x}\frac{\partial\theta}{\partial x} + \frac{\partial v}{\partial x}\frac{\partial\theta}{\partial y} = 0$$

and

$$\frac{\partial}{\partial y}\left(\frac{D_p\theta}{Dt}\right) = \frac{D_p}{Dt}\left(\frac{\partial\theta}{\partial y}\right) + \frac{\partial u}{\partial y}\frac{\partial\theta}{\partial x} + \frac{\partial v}{\partial y}\frac{\partial\theta}{\partial y} = 0$$

if the atmosphere is adiabatic.

Substituting the latter two equations into the former gives

$$F = \frac{1}{|\nabla_p \theta|}\left[-\left(\frac{\partial\theta}{\partial x}\right)^2\frac{\partial u}{\partial x} - \frac{\partial\theta}{\partial y}\frac{\partial\theta}{\partial x}\frac{\partial v}{\partial x} - \frac{\partial\theta}{\partial x}\frac{\partial\theta}{\partial y}\frac{\partial u}{\partial y} - \left(\frac{\partial\theta}{\partial y}\right)^2\frac{\partial u}{\partial y} \right]. \tag{9.A2}$$

But

$$\frac{\partial u}{\partial x} = \frac{1}{2}(\delta + D_1) \qquad \frac{\partial v}{\partial x} = \frac{1}{2}(\varsigma - D_2)$$

$$\frac{\partial u}{\partial y} = -\frac{1}{2}(\varsigma + D_2) \qquad \frac{\partial v}{\partial y} = \frac{1}{2}(\delta - D_1)$$

where

$$\delta = \frac{\partial u}{\partial x} + \frac{\partial v}{\partial y}, \varsigma = \frac{\partial v}{\partial x} - \frac{\partial u}{\partial y}, D_1 = \frac{\partial u}{\partial x} - \frac{\partial v}{\partial y}, \text{ and } D_2 = \frac{\partial v}{\partial x} + \frac{\partial u}{\partial y}.$$

Then

$$F = -\frac{1}{2\,|\nabla_p\theta|}\left[\left(\frac{\partial\theta}{\partial x}\right)^2(\delta + D_1) + \left(\frac{\partial\theta}{\partial x}\right)\left(\frac{\partial\theta}{\partial y}\right)(\varsigma - D_2)\right.$$

$$\left. - \left(\frac{\partial\theta}{\partial x}\right)\left(\frac{\partial\theta}{\partial y}\right)(\varsigma + D_2) + \left(\frac{\partial\theta}{\partial y}\right)^2(\delta - D_1)\right]. \tag{9.A3}$$

Rotating the coordinate system so that $D_2 = 0$, i.e., so that the axis of dilatation or axis of contraction lies along the x-axis, gives

$$F = -\frac{1}{2}\,|\nabla_p\theta|\left\{\delta + \frac{D_1\left[\left(\frac{\partial\theta}{\partial x}\right)^2 - \left(\frac{\partial\theta}{\partial y}\right)^2\right]}{|\nabla_p\theta|^2}\right\}. \tag{9.A4}$$

Let α be the angle $\nabla_p\theta$ makes with the x-axis. Then

$$\cos\alpha = \frac{\partial\theta/\partial x}{|\nabla_p\theta|} \quad \text{and} \quad \sin\alpha = \frac{\partial\theta/\partial y}{|\nabla_p\theta|}.$$

The angle between the isotherms and the x-axis is b, where $\alpha = b + 90°$. It follows that

$$\begin{aligned}
F &= -\frac{1}{2}\,|\nabla_p\theta|\left[D_1(\sin^2 b - \cos^2 b) + \delta\right] \\
&= \frac{1}{2}\,|\nabla_p\theta|\,(D\cos 2b - \delta),
\end{aligned} \tag{9.A5}$$

where the subscript 1 has been dropped, so that D is the resultant deformation in the new coordinate system.

REFERENCES

Ballentine, R. J., 1980: A numerical investigation of New England coastal frontogenesis. *Mon. Wea. Rev.*, **108**, 1479–1497.

Beebe, R. G., and F. C. Bates, 1955: A mechanism for assisting in the release of convective instability. *Mon. Wea. Rev.*, **83**, 1–10.

Bennetts, D. A., and B. J. Hoskins, 1979: Conditional symmetric instability—a possible explanation for frontal rainbands. *Quart. J. Roy. Meteor. Soc.*, **105**, 945–962.

Bergeron, T., 1928: Über die dreidimensional verknüpfende Wetteranalyse. *Geofys. Publ.*, **5**, 1–111.

Berggren, R., W. J. Gibbs, and C. W. Newton, 1958: Observational characteristics of the jet stream. A survey of the literature. WMO, TN No. 19.

Bjerknes, J., 1919: On the structure of moving cyclones. *Geofys. Publ.*, **1**, 1–8.

Bjerknes, J., 1951: Extratropical cyclones. *Compendium of Meteorology*, American Meteorological Society, Boston, 577–598.

Bluestein, H. B., 1977: Synoptic-scale deformation and tropical cloud bands. *J. Atmos. Sci.*, **34**, 891–900.

Bluestein, H. B., 1982: A wintertime mesoscale cold front in the Southern Plains. *Bull. Amer. Meteor. Soc.*, **63**, 178–185.

Bluestein, H. B., and K. W. Thomas, 1984: Diagnosis of a jet streak in the vicinity of a severe weather outbreak in the Texas Panhandle. *Mon. Wea. Rev.*, **112**, 2501–2522.

Blumen, W., 1981: The geostrophic coordinate transformation. *J. Atmos. Sci.*, **38**, 1100–1105.

Bosart, L. F., 1970: Mid-tropospheric frontogenesis. *Quart. J. Roy. Meteor. Soc.*, **96**, 442–471.

Bosart, L. F., 1975: New England coastal frontogenesis. *Quart. J. Roy. Meteor. Soc.*, **101**, 957–978.

Bosart, L. F., 1981: The Presidents' Day snow storm of 18–19 February 1979: A subsynoptic-scale event. *Mon. Wea. Rev.*, **109** , 1542–1566.

Bosart, L. F., Vaudo, C. J., and J. H. Helsdon, Jr., 1972: Coastal frontogenesis. *J. Appl. Meteor.*, **11**, 1236-1258.

Browning, K. A., and G. A. Monk, 1982: A simple model for the synoptic analysis of cold fronts. *Quart. J. Roy. Meteor. Soc.*, **108**, 435–452.

Browning, K. A., and C. W. Pardoe, 1973: Structure of low-level jet streams ahead of mid-latitude cold fronts. *Quart. J. Roy. Meteor. Soc.*, **99**, 619–638.

Brunt, D., and C. K. M. Douglas, 1982: On the modification of the strophic balance. *Mem. Roy. Meteor. Soc.*, **3**, 22.

Buzzi, A., A. Trevisan, and G. Salustri, 1981: Internal frontogenesis: A two-dimensional model in isentropic, semi-geostrophic coordinates. *Mon. Wea. Rev.*, **109**, 1053–1060.

Danielsen, E. F., 1968: Stratospheric-tropospheric exchange based on radiactivity, ozone, and potential vorticity. *J. Atmos. Sci.*, **25**, 502–518.

Danielsen, E. F., 1974: The relationship between severe weather, major dust storms and rapid large-scale cyclogenesis (I). *Subsynoptic Extratropical Weather Systems: Observations, Analysis, Modeling and Prediction* (Vol. II), National Center for Atmospheric Research, Boulder, Colo., 215–225.

Eliassen, A., 1948: The quasi-static equations of motion. *Geofys. Publ.*, **17**(3), 1–44.

Eliassen, A., 1959: On the formation of fronts in the atmosphere. *The Atmosphere and the Sea in Motion* (B. Bolin, Ed.), Rockefeller Inst. Press, New York, 277–287.

Eliassen, A., 1962: On the vertical circulation in frontal zones. *Geofys. Publ.*, **24**, 147–160.

Eliassen, A., 1966: Motions of intermediate scale: Fronts and cyclones. *Advances in Earth Science* (P. M. Hurley, Ed.), M.I.T. Press, Cambridge, Mass., 111–138.

Eliassen, A., 1984: Geostrophy. *Quart. J. Roy. Meteor. Soc.*, **110**, 1–12.

Emanuel, K. A., 1979: Inertial instability and mesoscale convective systems. Part I: Linear theory of inertial instability in rotating viscous fluids. *J. Atmos. Sci.*, **36**, 2425–2449.

Faller, A. J., 1956: A demonstration of fronts and frontal waves in atmospheric models. *J. Meteor.*, **13**, 1–4.

Fawbush, E. J., R. C. Miller, and L. G. Starrett, 1951: An empirical method of forecasting tornado development. *Bull. Amer. Meteor. Soc.*, **32**, 1–9.

Fultz, D., 1952: On the possibility of experimental models of the polar front wave. *J. Meteor.*, **9**, 379–384.

Gidel, L. T., 1978: Simulation of the differences and similarities of warm and cold surface frontogenesis. *J. Geophy. Res.*, **83**, 915–928.

Godson, W. L., 1951: Synoptic properties of frontal surfaces. *Quart. J. Roy. Meteor. Soc.*, **77**, 633–653.

Holton, J. R., 1979: *An Introduction to Dynamic Meteorology*, Academic Press, New York, 391 pp.

Hoskins, B. J., 1971: Atmospheric frontogenesis models: some solutions. *Quart. J. Roy. Meteor. Soc.*, **97**, 139–153.

Hoskins, B. J., 1972: Discussions. *Quart. J. Roy. Meteor. Soc.*, **98**, 862.

Hoskins, B. J., 1974: The role of potential vorticity in symmetric stability and instability. *Quart. J. Roy. Meteor. Soc.*, **100**, 480–482.

Hoskins, B. J., 1975: The geostrophic momentum approximation and the semi-geostrophic equations. *J. Atmos. Sci.*, **32**, 233–242.

Hoskins, B. J.,, 1982: The mathematical theory of frontogenesis. *Ann. Rev. Fluid Mech.*, **14**, 131–151.

Hoskins, B. J., and F. P. Bretherton, 1972: Atmospheric frontogenesis models: Mathematical formulation and solution. *J. Atmos. Sci.*, **29**, 11–37.

Hoskins, B. J., and I. Draghici, 1977: The forcing of ageostrophic motion according to the semi-geostrophic equations and in an isentropic coordinate model. *J. Atmos. Sci.*, **34**, 1859–1867.

Hoskins, B. J., and W. A. Heckley, 1981: Cold and warm fronts in baroclinic waves. *Quart. J. Roy. Meteor. Soc.*, **107**, 79–90.

Hoskins, B. J., and M. A. Pedder, 1980: The diagnosis of middle latitude synoptic development. *Quart. J. Roy. Meteor. Soc.*, **106**, 707–719.

Hoskins, B. J., and N. V. West, 1979: Baroclinic waves and frontogenesis. Part II: Uniform potential vorticity jet flows—cold and warm fronts. *J. Atmos. Sci.*, **36**, 1663–1680.

Hoskins, B. J., I. Draghici, and H. C. Davies, 1978: A new look at the ω-equation. *Quart. J. Roy. Meteor. Soc.*, **104**, 31–38.

Kennedy, P. J., and M. A. Shapiro, 1980: Further encounters with clear air turbulence in research aircraft. *J. Atmos. Sci.*, **37**, 986–993.

Keyser, D., and R. A. Anthes, 1982: The influence of planetary boundary layer physics on frontal structure in the Hoskins-Bretherton horizontal shear model. *J. Atmos. Sci.*, **39**, 1783–1802.

Keyser, D., and M. J. Pecnick, 1985a: A two-dimensional primitive equation model of frontogenesis forced by confluence and horizontal shear. *J. Atmos. Sci.*, **42**, 1259–1282.

Keyser, D., and M. J. Pecnick, 1985b: Diagnosis of ageostrophic circulations in a two-dimensional primitive equation model of frontogenesis. *J. Atmos. Sci.*, **42**, 1283–1305.

Kirk, T. H., 1966: Some aspects of the theory of fronts and frontal analysis. *Quart. J. Roy. Meteor. Soc.*, **92**, 374–381.

Kreitzberg, C. W., and H. A. Brown, 1970: Mesoscale weather systems within an occlusion. *J. Appl. Meteor.*, **9**, 417–432.

LaSeur, N., 1974: Selected subsynoptic features of particular interest: An introduction. *Subsynoptic Extratropical Weather Systems: Observations, Analysis, Modeling, and Prediction*, Vol. I (notes from J. Cahir), National Center for Atmospheric Research, Boulder, Colo., 1–7.

Miller, J. E., 1948: On the concept of frontogenesis. *J. Meteor.*, **5**, 169–171.

Mudrick, S. E., 1974: A numerical study of frontogenesis. *J. Atmos. Sci.*, **31**, 869–892.

Murray, R., and S. M. Daniels, 1953: Transverse flow at entrance and exit to jet streams. *Quart. J. Roy. Meteor. Soc.*, **79**, 236–241.

Namias, J., and P. F. Clapp, 1949: Confluence theory of the high tropospheric jet stream. *J. Meteor.*, **6**, 330–336.

Newton, C. W., 1954: Frontogenesis and frontolysis as a three-dimensional process. *J. Meteor.*, **11**, 449–461.

Newton, C. W., and A. Trevisan, 1984a: Clinogenesis and frontogenesis in jet-stream waves. Part I: Analytical relations to wave structure. *J. Atmos. Sci.*, **41**, 2717–2734.

Newton, C. W., and A. Trevisan, 1984b: Clinogenesis and frontogenesis in jet-stream waves. Part II: Channel model numerical experiments. *J. Atmos. Sci.*, **41**, 2735–2755.

Palmén, E., 1948: On the distribution of temperature and wind in the upper westerlies. *J. Meteor.*, **5**, 20–27.

Palmén, E., 1951: The aerology of extratropical disturbances. *Compendium of Meteorology*, American Meteorological Society, Boston, 599–620.

Palmén, E., and C. W. Newton, 1969: *Atmospheric Circulation Systems*. Academic Press, New York, 603 pp.

Petterssen, S., 1936: A contribution to the theory of frontogenesis. *Geofys. Publ.*, **11**, 1–27.

Petterssen, S., 1956: *Weather Analysis and Forecasting*, Vol. I, Motion and Motion Systems, (second edition). McGraw-Hill, New York, 428 pp.

Petterssen, S., and J. M. Austin, 1942: Fronts and frontogenesis in relation to vorticity. Papers in Physical Oceanography and Meteorology, M.I.T. and Woods Hole Oceanographic Institution, Cambridge and Woods Hole, Mass., 37 pp.

Reed, R. J., 1955: A study of a characteristic type of upper-level frontogenesis. *J. Meteor.*, **12**, 226–237.

Reed, R. J., and F. Sanders, 1953: An investigation of the development of a mid-tropospheric frontal zone and its associated vorticity field. *J. Meteor.*, **10**, 338–349.

Reiter, E. R., 1961: *Jet-Stream Meteorology*. Univ. of Chicago Press, Chicago, 515 pp.

Rex, D. F., 1950: Blocking action in the middle troposphere and its effect upon regional climate. I. *Tellus*, **2**, 196–211.

Riehl, H., and J. Badner, J. E. Hovde, N. E. LaSeur, L. L. Means, W.C. Palmer, M. J. Schroeder, L. W. Snellman, and others, 1952: *Forecasting in Middle Latitudes*. Meteor. Monogr., **1**(5), American Meteorological Society, Boston, 80 pp.

Ross, B. B., and I. Orlanski, 1978: The circulation associated with a cold front. Part II: Moist case. *J. Atmos. Sci.*, **35**, 445–465.

Rossby, C.-G., 1938: On the mutual adjustment of pressure and velocity distributions in certain simple current systems, II. *J. Mar. Res.*, **1**, 239–263.

Sanders, F., 1955: An investigation of the structure and dynamics of an intense surface frontal zone. *J. Meteor.*, **12**, 542–552.

Sawyer, J. S., 1956: The vertical circulation at meteorological fronts and its relation to frontogenesis. *Proc. Roy. Soc. London.*, **A234**, 346–362.

Shapiro, M. A., 1970: On the applicability of the geostrophic approximation to upper-level frontal-scale motions. *J. Atmos. Soc.*, **27**, 408–420.

Shapiro, M. A., 1974: A multiple structured zone-jet stream system as revealed by meteorologically instrumented aircraft. *Mon. Wea. Rev.*, **102**, 244–253.

Shapiro, M. A., 1980: Turbulent mixing within tropopause folds as a mechanism for the exchange of chemical constituents between the stratosphere and troposphere. *J. Atmos. Sci.*, **37**, 994–1004.

Shapiro, M. A., 1981: Frontogenesis and geostrophically forced secondary circulations in the vicinity of jet stream-frontal zone systems. *J. Atmos. Sci.*, **38**, 954–973.

Shapiro, M. A., 1982: *Mesoscale Weather Systems of the Central United States*. CIRES, Univ. of Colo./NOAA, Boulder, Colo., 78 pp.

Shapiro, M. A., and P. J. Kennedy, 1981: Research aircraft measurements of jet stream geostrophic and ageostrophic winds. *J. Atmos. Sci.*, **38**, 2642–2652.

Solberg, H., 1933: Le Mouvement d'inertie de l'atmosphère stable et son role dans la théorie des cyclones. Meteor. Assoc. U.G.G.I. (memoir), DuPont Press, Lisbon, 66–82.

Stone, P. H., 1966a: On non-geostrophic

baroclinic stability. *J. Atmos. Sci.*, **23**, 390–400.

Stone, P. H., 1966b: Frontogenesis by horizontal wind deformation fields. *J. Atmos. Sci.*, **23**, 455–465.

Thorpe, A. J., 1984: Convective parameterization in a quasi-geostrophic diagnostic model of fronts. *J. Atmos. Sci.*, **41**, 691–694.

Trenberth, K. E., 1978: On the interpretation of the diagnostic quasi-geostrophic omega equation. *Mon. Wea. Rev.*, **106**, 131–137.

Uccellini, L. W., 1975: A case study of apparent gravity wave initiation of severe convective storms. *Mon. Wea. Rev.*, **103**, 497–513.

Uccellini, L. W., and D. R. Johnson, 1979: The coupling of upper and lower tropospheric jet streaks and implications for the development of severe convective storms. *Mon. Wea. Rev.*, **107**, 682–703.

Uccellini, L. W., P. J. Kocin, R. A. Petersen, C. H. Wash, and K. F. Brill, 1984: The Presidents' Day cyclone of 18–19 February 1979: Synoptic overview and analysis of the subtropical jet streak influencing the pre-cyclogenetic period. *Mon. Wea. Rev.*, **112**, 31–55.

Van Tuyl, A. H., and J. A. Young, 1982: Numerical simulation of nonlinear jet streak adjustment. *Mon. Wea. Rev.*, **115**, 2038–2054.

Williams, R. T., 1967: Atmospheric frontogenesis: A numerical experiment. *J. Atmos. Sci.*, **24**, 627–641.

Williams, R. T., 1968: A note on quasigeostrophic frontogenesis. *J. Atmos. Sci.*, **25**, 1157–1159.

Williams, R. T., 1972: Quasi-geostrophic versus non-geostrophic frontogenesis. *J. Atmos. Sci.*, **29**, 3–10.

Williams, R. T., 1974: Numerical simulation of steady-state fronts. *J. Atmos. Sci.*, **31**, 1286–1296.

Williams, R. T., and J. Plotkin, 1968: Quasi-geostrophic frontogenesis. *J. Atmos. Sci.*, **25**, 201–206.

Williams, R. T., L. C. Chou, and C. J. Cornelius, 1981: Effects of condensation and surface motion on the structure of steady-state fronts. *J. Atmos. Sci.*, **38**, 2365–2376.

Yudin, M. I., 1955: Invariant quantities in large-scale atmospheric processes. *Tr. Glav. Geogiz. Obser.*, No. 55, 3–12.

CHAPTER 10

Atmospheric Fronts:
An Observational Perspective

Daniel Keyser

10.1. Introduction

Atmospheric fronts may be defined as sloping zones of pronounced transition in the thermal and wind fields. Consequently, they are characterized by a combination of relatively large horizontal temperature gradients, static stability, absolute vorticity (horizontal wind shear), and vertical wind shear. When depicted on quasi-horizontal surfaces, fronts appear as long, narrow features in which the along-front scale is typically an order of magnitude greater than the cross-front scale (1000–2000 km compared with 100–200 km). Therefore, according to the scale definitions of Orlanski (1975), the along-front dimension is meso-α (200–2000 km), and the cross-front dimension is meso-β (20–200 km). The scale of the middle-latitude baroclinic waves in which fronts are embedded ranges from meso-α to macro-β or synoptic (2000–10,000 km). Fronts are shallow phenomena, with depths typically 1–2 km. Because of this geometrical configuration, horizontal variations in the cross-front direction greatly exceed those in the along-front direction.

The definition of fronts is here limited to features fitting the above description and generated by internal dynamical processes (such as deformation in horizontal and vertical planes), and in which the effect of the Earth's rotation (the Coriolis force) is significant. This restriction is intended to eliminate a number of lineal phenomena from explicit consideration, such as internal gravity waves, gravity or density currents, squall lines, and rainbands. Front-like features owing their existence to spatial variations in the radiative properties of the surface (e.g., land vs. sea or snow-covered vs. snow-free regions) are ruled out as well. According to this restricted definition, fronts form within a period of several days from smooth, synoptic-scale variations in the thermal and wind fields. They also can develop at all levels in the atmosphere, although their existence is favored by the presence of boundaries such as the Earth's surface and the tropopause. Consequently, fronts will be designated "surface" or "low-level" and "upper-tropospheric" or "upper-level," according to this distinction.

Surface fronts are important meteorologically because of their association with cloud and precipitation patterns, their correspondence with rapid local changes in the weather, and their frequent occurrence in middle-latitude circulation regimes. They also provide the background or environment for instabilities that may be responsible for generating smaller-scale weather phenomena. An example of current research interest is conditional symmetric instability for the initiation of rainbands (e.g., Emanuel, 1983).

Upper-tropospheric fronts are accompanied by jet streams and superimposed jet streaks, which are dynamically important in forcing vertical motions involving phenomena and processes occurring on scales ranging from those of middle-latitude cyclones to severe convection. Upper-level fronts are regions of intense small-scale (meso-β and smaller) mixing by a variety of phenomena such as gravity waves, Kelvin-Helmholtz billows, and patches of turbulent eddy motions, all of which are generically referred to as clear-air turbulence (CAT). Developing upper-level fronts are also regions of "organized" (meso-α scale) vertical motions, often in the form of subsidence on the order of 10 cm s^{-1}. Both modes of vertical transport promote the exchange of mass, including chemical trace constituents, between the stratosphere and troposphere. The location and intensity of upper-level fronts and jets are of considerable importance for the economical routing of aircraft in terms of fuel consumption and because of the potential safety hazards presented by CAT.

Additional material on observations of fronts can be found in Petterssen (1956, pp. 189–213), Palmén and Newton (1969, pp. 237–272), Frank and Barber (1977), and Shapiro (1983). Theoretical aspects are covered by Hoskins (1982) and Ch. 9 in this volume .

10.2. Historical Overview

In her monograph, *The Thermal Theory of Cyclones*, Kutzbach (1979) traced through the nineteenth and early part of the twentieth century the history of the theory that middle-latitude cyclones are driven primarily by the release of latent heat in ascending air. The so-called thermal theory was supplanted by the polar-front cyclone model of the Bergen School following World War I (Bjerknes, 1919; Bjerknes and Solberg, 1921, 1922). In the latter theory, available potential energy provided by the temperature contrast between polar and tropical air masses at the polar front replaced latent heat release as an initial energy source for cyclogenesis. Precipitation patterns were related to upward motions arising from the relative motion of air with respect to inclined frontal surfaces. Fronts were considered to be sloping, material surfaces characterized by zero-order discontinuities in temperature, density, and along-front wind (or cross-front pressure gradient). Cross sections of potential temperature for the wedge model of frontal structure are presented in Fig. 9.5 (Ch. 9, this volume).

10.2.1. The Wedge Model of Fronts

In view of the importance of fronts in explaining the distribution of vertical motions in the polar-front theory, their relationship is reviewed briefly.

The frontal slope is given by Margules' formula for dry, inviscid, hydrostatic flow (Palmén and Newton, 1969, pp. 168–173):

$$\frac{\partial h}{\partial y} = \left(\frac{f\overline{T}}{g}\right)\left(\frac{u_{gW} - u_{gC}}{T_W - T_C}\right) . \tag{10.1}$$

In (10.1), h is the altitude of the frontal surface above the ground, y increases toward colder air, $\overline{T}$ is the mean temperature of the cold and warm air masses, $u_g[= -(\rho f)^{-1}(\partial p/\partial y)]$ is the along-front component of the geostrophic wind, and the subscripts W and C refer to warm and cold air. This formulation is an expression of thermal wind balance for a "two-layer" atmosphere, and its best known implication is that cyclonic shear is required across the front $[u_{gC} < u_{gW}$ or $(\partial p/\partial y)_C > (\partial p/\partial y)_W]$ for the surface of discontinuity to have a statically stable configuration ($\partial h/\partial y > 0$). Various examples of the pressure and wind fields for fronts satisfying the conditions in (10.1) appear in Fig. 9.4 of Ch. 9, taken from Petterssen (1956, p. 199).

Given the expression for frontal slope, $\partial h/\partial y$, the vertical motion immediately above or below the interface h can be determined from

$$w = (v - c)\frac{\partial h}{\partial y} . \tag{10.2}$$

Equation (10.2) is a consequence of the kinematic boundary condition that the parcels on each side of h cannot cross it; i.e., the flow relative to the movement of the front must be parallel to h. In (10.2), v is the cross-front velocity component and c is the speed of the front. It is assumed that h does not vary in the along-front (x) direction and that the frontal surface does not change shape with time. The vertical velocity difference across h is formed by taking the difference of expressions for (10.2) applying at the cold and warm sides of h:

$$w_W - w_C = (v_W - v_C)\frac{\partial h}{\partial y} . \tag{10.3}$$

Bergeron (1937) classified cases in which $w_W > w_C(v_W > v_C)$ as anafronts and cases in which $w_W < w_C(v_W < v_C)$ as katafronts. These cases are characterized respectively by post- and pre-frontal cloud bands. In summary, provided that the distributions of temperature, horizontal wind, and front speed $(T_W, T_C; u_{gW}, u_{gC}; v_W, v_C; c)$ are known, (10.1) and (10.2) allow the diagnosis of the vertical velocity field on either side of the front.

It is commonly perceived that the existence of fronts was not recognized systematically until their identification by the Bergen School. This perception is reinforced by examining standard textbooks written prior to the publication of the polar-front theory (e.g., Henry *et al.*, 1916) in which analyses of isotherms, often displaying sharp contrasts, are shown without mention of frontal discontinuities or zones. The recent paper by Namias (1983) recounting the resistance of U.S. Weather Bureau management to accepting the ideas of the Bergen School during the 1920s and 1930s substantiates the apparent novelty of the frontal concept. Nevertheless, Kutzbach (1979) assembled considerable evidence that fronts had been known to researchers at least since the mid-nineteenth century. Examples include the explanation by Loomis* in 1841 on the formation of rain by cold air undercutting and lifting warm air; Fitzroy's storm model of 1863 displaying the confluence of polar and tropical air currents; analyses of squall lines by Durand-Gréville (1892) and Margules (1903); and analyses of front-like features referred to as "line-squalls" by Lempfert and Corless in 1910. The originality of the polar-front cyclone model lies not in the discovery of fronts, but in their incorporation into a conceptual model that (1) could be used for routine synoptic analysis, (2) provided a physical basis for diagnosing areas of rising motion and precipitation, and (3) contributed an observational basis for theoretical studies of cyclogenesis. Additional original aspects include the introduction of warm and occluded fronts (the latter credited to Bergeron) and the generalization of the model to describe the life cycle of middle-latitude cyclones.

The polar-front cyclone model and the related methods of frontal analysis became increasingly established and refined during the 1920s and 1930s. Examples are shown in Fig. 10.1 of elaborate schematics devised to illustrate the cloud and precipitation patterns associated with frontal surfaces. Figures 10.1a and 10.1b depict fast- and slow-moving cold fronts in which the precipitation band is pre- and post-frontal. The warm front (Fig. 10.1c) displays cloud and precipitation patterns that resemble those of the slow-moving cold front in the sense that an observer of the warm frontal passage would note a similar sequence of events occurring in reverse order compared with the slow-moving cold front. The cloud patterns for the fast- and slow-moving cold fronts are suggestive of Bergeron's (1937) description of katafronts and anafronts alluded to previously.

10.2.2. The Trend Toward a Zone Model of Fronts

The replacement of the wedge model of a front with the modern definition of a transition zone occurred with the introduction of routine upper-air observations in the 1930s. Studies such as those of Bjerknes and Palmén (1937), Palmén (1948), Palmén and Nagler (1948), and Palmén and Newton (1948) revealed the existence of sloping transition zones above the surface rather than abrupt discontinuities. Figure 10.2 is the well-known composite by Palmén and Newton (1948) of 12 cross sections obtained during December 1946, illustrating the structure of the polar jet at the tropopause and its associated tropospheric frontal zone. This analysis reflects the prevailing concept at that time of a deep polar front separating polar and tropical air. Consistent with the polar-front concept, the frontal zone is not analyzed above 400 mb, where the horizontal temperature contrast becomes diffuse, although a region of cyclonic wind shear extends into the stratosphere. The

---

* See Kutzbach (1979) for this and subsequent historical references. Brief historical discussions are given in Petterssen (1956, pp. 214–217) and Palmén and Newton (1969, pp. 119–122).

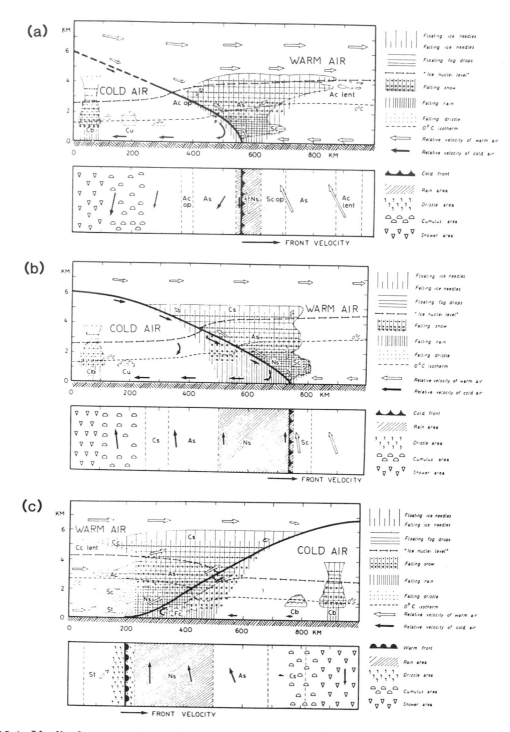

Figure 10.1. Idealized cross-section and plan views of cloud and precipitation distributions associated with (a) fast-moving cold fronts, (b) slow-moving cold fronts, and (c) warm fronts. Arrows represent front-relative air motions. (From Godske *et al.*, 1957, pp. 528–529.)

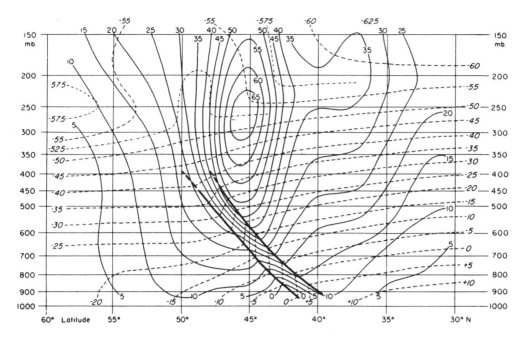

Figure 10.2. Cross section along 80°W of the temperature (°C, dashed) and the zonal component of the geostrophic wind (m s⁻¹, thin solid) averaged over 12 cases in December 1946. Heavy solid lines indicate the boundaries of the polar front. (From Palmén and Newton, 1948.)

analysis represents a break from the past in the sense that radiosonde data are used to portray a zone extending well above the surface.

The relevance of the wedge model was diminished also with the advent of quasi-geostrophic theory, applied diagnostically to infer the vertical motion patterns and rates of development associated with cyclogenesis (e.g., Sutcliffe, 1947), and theoretically to studies of baroclinic instability originating with Charney (1947) and Eady (1949). In this approach, atmospheric quantities vary continuously, so that fronts are defined as regions of pronounced gradients. Consequently, the zone model is ideally suited to this viewpoint. A noteworthy difference of baroclinic instability theory from the polar-front theory is that fronts develop in response to wave amplification rather than preceding it. In the polar-front theory, cyclogenesis was viewed as an instability of frontal surfaces of discontinuity. Prior to the formulation of baroclinic instability theory, considerable theoretical research had been performed concerning the stability of frontal surfaces to small perturbations. This research was reviewed by Eliassen (1966), who concluded that the instability of frontal surfaces may be responsible for the growth of short, shallow cyclone waves (e.g., the weak waves often observed on quasi-stationary fronts in noncyclogenetic situations), but that baroclinic instability is the dominant process responsible for intense middle-latitude cyclones. The wedge model also has served as a basis for theoretical studies of the influence of friction on surface frontal structure (Ball, 1960; Rao, 1971).

When the zone model replaced the wedge model of fronts, a revised diagnostic approach for determining vertical motions became necessary. In the continuous case, the vertical motion is contained within the ageostrophic circulation, which is two-dimensional and confined to cross-front vertical planes for the case of a straight frontal zone. A diagnostic equation developed by Sawyer (1956) and Eliassen (1962) [Eq. (9.45) in Ch. 9] relates the ageostrophic circulation to patterns of geostrophic deformation acting on cross- and along-front thermal contrasts and to frictional and diabatic processes. The underlying principle of the diagnostic approach is that the vertical shear of the along-front wind component remains in thermal wind balance with the cross-front temperature gradient, which is also assumed in the wedge model. The diagnostic approach is based on a set of physical approximations formalized in the semi-geostrophic theory (Hoskins, 1975), which is a generalization of the well-established quasi-geostrophic theory. In addition to the original papers by Sawyer and Eliassen, discussion of the diagnosis of ageostrophic circulations in frontal zones is given by Shapiro (1981), Hoskins (1982), and Ch. 9 of this volume.

Another avenue of research pertinent to the zone concept of a front is isolating kinematic processes leading to the intensification of gradients of conservative quantities such as potential temperature. For example, Petterssen (1956, pp. 189–213) determined that in the absence of horizontal divergence and vertical motions, horizontal deformation fields tend to concentrate potential temperature contrasts with respect to a point on a parcel trajectory, provided that the angle between the axis of dilatation of the deformation field and the isentropes is less than 45°. [See Eq. (9.10) and the related discussion in Ch. 9.] The influence of deformation in stretching and concentrating broad regions of fluid into long, narrow zones is graphically illustrated in the classic result of Welander (1955) shown in Fig. 10.3. This figure depicts the 36 h evolution of a passive tracer, initially in the form of a checkerboard pattern, in response to a flow field typical of those observed in the mid-troposphere within middle latitudes. The flow evolution was simulated numerically with a two-dimensional barotropic model. Impressively realistic frontal features also are observed in the laboratory experiments of Fultz (1952) and Faller (1956), in which baroclinic eddies resembling middle-latitude cyclones develop in differentially heated, rotating fluids.

A simplified form of Petterssen's kinematic frontogenesis equation (Miller, 1948) partitions the total temporal rate of change of the cross-front component of the potential temperature gradient into effects associated with diabatic heating and the components of the vector velocity. The equation is given by

$$\frac{d}{dt}\left(-\frac{\partial\theta}{\partial y}\right) = -\frac{\partial\dot\theta}{\partial y} + \frac{\partial u}{\partial y}\frac{\partial\theta}{\partial x} + \frac{\partial v}{\partial y}\frac{\partial\theta}{\partial y} + \frac{\partial w}{\partial y}\frac{\partial\theta}{\partial z}, \qquad (10.4)$$

where $\dot\theta$ is the diabatic heating rate. The positive y axis is directed toward colder air, and the minus sign on the left of (10.4) ensures consideration of a positive quantity. The y direction corresponds to the cross-front direc-

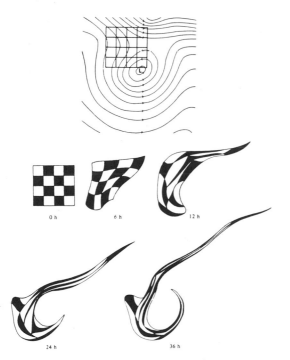

Figure 10.3. Graphic illustration of the deformation of a passive tracer in a numerical simulation of a 500 mb flow pattern with a barotropic model over a 36 h period. (From Welander, 1955.)

tion, which is not necessarily identical to that of the potential temperature gradient; along-front variations are possible. The first term on the right of (10.4) is the effect of differential diabatic heating in modifying $-\partial\theta/\partial y$. The second term, illustrated in Fig. 9.14 in Ch. 9, is the effect of horizontal shear in rotating along-front potential temperature gradients into the cross-front direction. The third term contains the effects of confluence and convergence (Figs. 9.7 and 9.13 in Ch. 9) in concentrating pre-existing cross-front gradients of potential temperature. The last term in (10.4) is usually referred to as the tilting term (Fig. 9.8 in Ch. 9), and quantifies the effect of a cross-front gradient of adiabatic warming or cooling in modifying $-\partial\theta/\partial y$. Equations analogous to (10.4) for static stability $(\partial\theta/\partial z)$, horizontal wind shear $(-\partial u/\partial y)$, and vertical wind shear $(\partial u/\partial z)$ can be derived as well.

A limitation to the kinematic approaches of Petterssen and Miller is that they can be applied only diagnostically to obtain patterns of instantaneous rates of change of frontal quantities along parcel trajectories. The kinematic formulations cannot reveal how the total frontogenetical tendencies are partitioned among local temporal changes and advective changes. Although the kinematic approaches are useful for physically describing frontogenetical processes, one still must isolate mutual interactions between the wind and thermal fields and diabatic and frictional processes. These issues require dynamical approaches capable of treating the time-dependent evolution of frontal systems. Dynamical investigations of frontogenesis in idealized flow patterns were initiated with the quasi-geostrophic equations by Stone (1966), Williams and Plotkin (1968), and Williams (1968), and generalized with the semi-geostrophic equations by Hoskins and Bretherton (1972). Parallel investigations with the primitive equations were begun by

Williams (1967). Dynamical investigations of fronts, from both diagnostic and theoretical perspectives, continue to form an active area of research in mesoscale meteorology.

10.3. Surface Frontal Structure

The shift of the conceptual model of fronts from an abrupt discontinuity in the thermal and wind fields to a transition zone having been established, the detailed structure of surface fronts documented in several observational case studies is considered. Theoretical studies have shown convincingly that surface fronts can be generated by geostrophic deformation patterns along with their induced ageostrophic circulations, in the absence of diabatic and frictional processes. Nevertheless, the detailed structure of observed low-level fronts will be shown to be affected by the turbulent mixing of heat and momentum in the planetary boundary layer (PBL) in the presence of a no-slip lower boundary, and additional diabatic processes such as sensible heating at the surface and latent heating and evaporative cooling.

10.3.1. Illustrative Case Studies

The Sanders Analyses

Using surface airways data and a spatially dense network of 6-hourly radiosonde and pilot balloon observations, Sanders (1955) subjectively analyzed an intense surface frontal zone situated over the south central United States. Figures 10.4 and 10.5 present a surface chart and cross section for

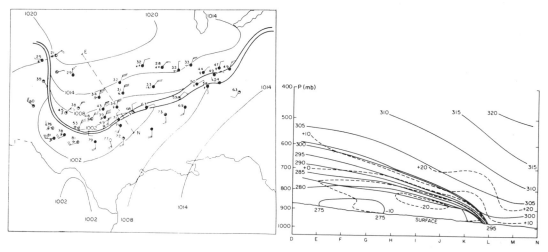

Figure 10.4. Surface analysis for 0330 GMT, 18 April 1953. Dashed line EN indicates position of vertical cross section in Fig. 10.5; heavy solid lines denote boundaries of frontal zone; light solid lines are isobars of sea-level pressure (contour interval, 6 mb). Plotted reports follow conventional station model. (From Sanders, 1955.)

Figure 10.5. Distribution of potential temperature (light solid lines, contour interval 5 K) and wind component (dashed lines, contour interval 10 m s^{-1}) normal to cross section EN in Fig. 10.4 for 0300 GMT, 18 April 1953. Heavy solid lines indicate boundaries of frontal zone. Distance between adjacent letters on horizontal axis is 100 km. (From Sanders, 1955.)

the frontal zone at about 0300 GMT, 18 April 1953. Note that the front is not associated with a well-defined cyclone or cyclogenesis. In terms of frontal properties such as horizontal potential temperature gradient (Fig. 10.5), cyclonic vorticity, and convergence (Fig. 10.6), the intensity of the front is greatest at or near the ground and diminishes rapidly with altitude. The front is barely detectable above the 600 mb level. The cross-frontal scale is least at the ground (a few kilometers according to Sanders [1955, p. 546], despite the broader scale indicated on Fig. 10.5) and increases with elevation. The front also is steepest at the ground, and its slope decreases with elevation. The air within the frontal zone is characterized by large static stability and vertical wind shear, whereas the cold air behind and beneath the frontal zone is well-mixed vertically in terms of potential temperature and along-front wind. Ahead of the front, static stabilities and vertical wind shears are moderate. The vertical motion pattern (Fig. 10.6) indicates a narrow plume of ascent ($w \sim 25$ cm s^{-1}) in the warm air above the intersection of the frontal zone with the surface. The pattern of ascent in the warm air behind the surface frontal position, associated with widespread cloudiness and generally light precipitation (Fig. 10.4), is suggestive of an anafront (Fig. 10.1b).

Applying a version of (10.4) in which diabatic effects and the along-front component of the horizontal potential temperature gradient are neglected, Sanders found the confluence term (Fig. 10.7a) to be strongly frontogenetical for parcels in the lowest parts of the zone and to decrease in magnitude with elevation. The tilting term (Fig. 10.7b) is strongly frontogenetical in the warm air ahead of the frontal zone, where the horizontal gradient of the vertical motion is large, and frontolytical within the zone, where the static stability is large. The combined effect of the horizontal and vertical motions (Fig. 10.7c) indicates the dominance of the tilting term in the pre-

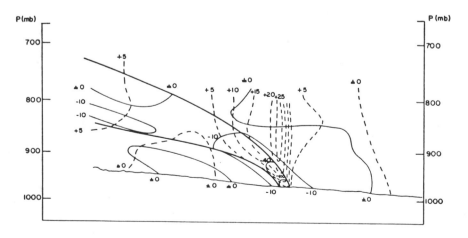

Figure 10.6. Distribution of horizontal divergence (light solid lines, selected contours in units of 10^{-5} s^{-1}) and vertical motion (dashed lines, contour interval 5 cm s^{-1}) for part of cross section EN at 0300 GMT, 18 April 1953. Heavy solid lines indicate frontal boundaries. (From Sanders, 1955.)

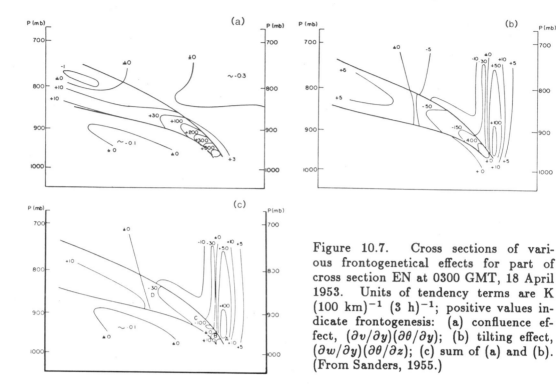

Figure 10.7. Cross sections of various frontogenetical effects for part of cross section EN at 0300 GMT, 18 April 1953. Units of tendency terms are K $(100\ \text{km})^{-1}\ (3\ \text{h})^{-1}$; positive values indicate frontogenesis: (a) confluence effect, $(\partial v/\partial y)(\partial\theta/\partial y)$; (b) tilting effect, $(\partial w/\partial y)(\partial\theta/\partial z)$; (c) sum of (a) and (b). (From Sanders, 1955.)

frontal warm air as well as within the zone, except near the surface where the confluence term is stronger.

Figure 10.7c also contains an approximate parcel trajectory relative to the motion of the front indicated by path ABCD. If the front is assumed to be approximately steady-state, the trajectory reveals that warm air is entrained into the frontal zone between A and B with the cross-front potential temperature gradient increasing through tilting and confluence. Between B and D, frontolysis due to tilting occurs, consistent with the decrease of the cross-front potential temperature gradient along the parcel path. This Lagrangian perspective is consistent with the configuration of the potential temperature field, exhibiting strongest cross-front gradients at the surface and rapid weakening with altitude. A significant conclusion is that the frontal zone is not a material boundary: parcels are ingested into its warm side. From an Eulerian perspective, frontal properties are generated in a localized region near the surface and are advected up and back toward the cold air to form the frontal zone. This Eulerian interpretation minimizes the strong frontogenetical tilting effect in the warm air ahead of the frontal zone. Palmén and Newton (1969, p. 261) suggested that neglected differential heating due to condensation in the region of ascent would tend to counteract the tilting effect, resulting in a much smaller total frontogenetical tendency in the warm air than is indicated in Fig. 10.7c.

The Ogura-Portis Analyses

The recent case study of Ogura and Portis (1982) documented the structure of a frontal zone associated with deep convection. They presented objec-

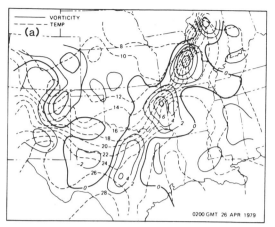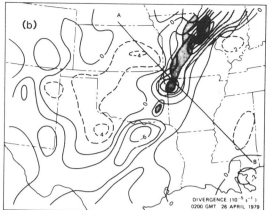

Figure 10.8. Objective analyses for 0200 GMT, 26 April 1979, of (a) surface temperature (contour interval 2°C, dashed), relative vorticity (contour interval 2×10^{-5} s^{-1}, solid), and (b) surface divergence (contour interval 2×10^{-5} s^{-1}). Radar echoes at 0235 GMT, 26 April 1979, are indicated in (b) by shading. Line AB indicates position of cross sections in Figs. 10.9 and 10.10. (From Ogura and Portis, 1982.)

tive analyses of surface airways data and 3-hourly radiosonde observations from the SESAME AVE-III (Severe Environmental Storms and Mesoscale Experiment —Atmospheric Variability Experiment) observing network. Radiosonde station spacing of 300 to 400 km limits the minimum resolvable horizontal scale of the objective analyses to widths broader than those determined subjectively by Sanders. Nevertheless, the analyses establish the meso-α-scale structure associated with the frontal zone and also possess resolution and structural detail similar to present-day research mesoscale models (e.g., Anthes *et al.*, 1982). At 0000 GMT, 26 April 1979, a surface frontal zone extended southwestward through Missouri, southeastern Oklahoma, and into Texas, with a line of convection corresponding to the surface frontal position. Figure 10.8 depicts the frontal zone through objective analyses of surface temperature, relative vorticity, and divergence for 0200 GMT, 26 April 1979. The frontal zone is best defined from northeastern Oklahoma northeastward, in the region covered by radar echoes at 0235 GMT, 26 April 1979 (Fig. 10.8b).

Figure 10.9 consists of cross sections along line AB shown in Fig. 10.8b. The potential temperature and along-front wind component (Fig. 10.9a) indicate the low-level frontal zone behind a low-level potential temperature maximum, consistent with the presence of the surface warm tongue (Fig. 10.8a). Low-level northerly and southerly jets are situated behind and ahead of the surface front. An inversion layer between 900 and 800 mb sloping back over the cold air is evident as in Sanders' analysis (Fig. 10.5). The vertical wind shear is distributed through the depth of the troposphere, and culminates in a jet core at the tropopause. Although the baroclinic structure is deep, the relative vorticity (Fig. 10.9b) is maximized at low levels and the tropopause, reflecting the introductory remarks (Sec. 10.1) concerning the prevalence of fronts at these "boundaries." The overall structure of the wind and potential temperature fields in Fig. 10.9a is suggestive of an idealized baroclinic wave (Hoskins and Heckley, 1981), which is consistent with the orientation of the surface front beneath the southwesterly flow ahead of an upper-level trough.

The kinematically derived vertical motion pattern (Fig. 10.9c) displays subsidence behind and well in advance of the front and a sloping region of ascent overlying the axis of maximum relative vorticity. The low-level feature within this region of ascent probably corresponds to the vertical velocity maximum Sanders detected in the PBL. The maximum at 400 mb directly above the surface frontal position is hypothesized by Ogura and Portis to be related to the squall line. The transverse flow relative to the frontal zone (Fig. 10.9d), summarized schematically in Fig. 10.10, indicates well-defined low-level flow in the warm air toward the front, part of which rises within the squall line and the remainder of which traverses the frontal zone at low levels. The midtropospheric portion of the transverse circulation behind the

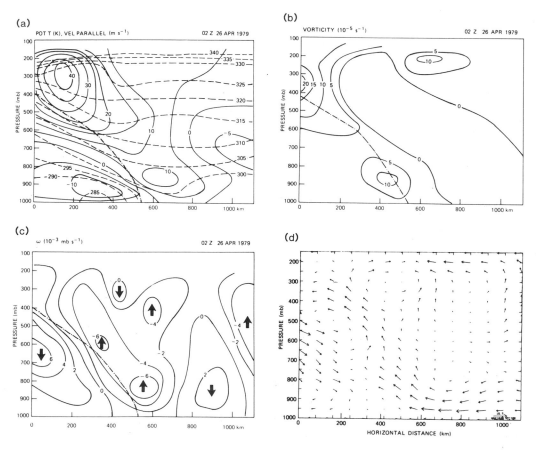

Figure 10.9. Cross sections for 0200 GMT, 26 April 1979, of (a) potential temperature (K, dashed) and velocity component normal to the cross section (m s^{-1}, solid); (b) relative vorticity (10^{-5} s^{-1}); (c) pressure-coordinate vertical velocity (10^{-3} mb s^{-1}); (d) transverse flow relative to the motion of the surface front. The dot-dashed line in (a)–(c) is the axis of maximum relative vorticity. (From Ogura and Portis, 1982.)

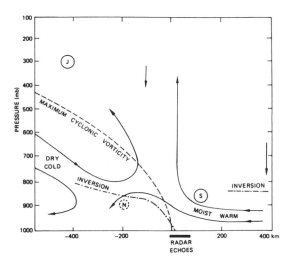

Figure 10.10. Schematic representation of the structure of the cold front. N, S, and J respectively denote the northerly and southerly low-level jets and the upper-level jet in Fig. 10.9a. The dashed line indicates the axis of maximum relative vorticity, and the dot-dashed line connects the position of the maximum surface potential temperature gradient to the statically stable layer behind the surface front. (From Ogura and Portis, 1982.)

convection is likely to be related to the presence of the upstream baroclinic wave and upper-level jet.

The Shapiro Analyses

A third set of analyses of a surface frontal zone over the Southern Plains (Shapiro, 1983) is shown in Figs. 10.11 and 10.12. As in Sanders' case, the analyses are subjective and are based on surface airways and radiosonde data. This case is presented to demonstrate the dramatic influence that differen-

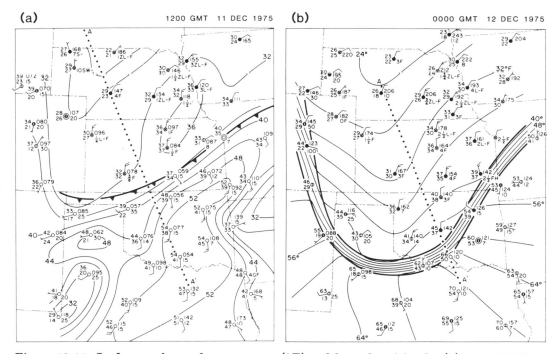

Figure 10.11. Surface analyses of temperature (°F) and frontal position for (a) 1200 GMT, 11 December 1975 and (b) 0000 GMT, 12 December 1975. Plotted reports follow conventional station model. Dotted lines AA′ indicate positions of cross sections in Fig. 10.12 at their respective times. (From Shapiro, 1983.)

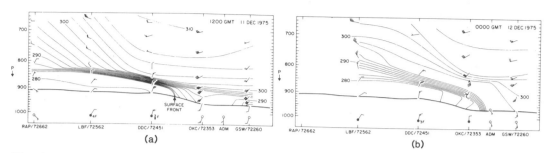

Figure 10.12. Cross sections of potential temperature (K) along paths indicated in Fig. 10.11 for (a) 1200 GMT, 11 December 1975 and (b) 0000 GMT, 12 December 1975. Upper-air wind directions and speeds (kt) are indicated using standard plotting convention; surface observations are entered at the lower portion of the figures. (From Shapiro, 1983.)

tial surface fluxes of sensible heat can exert on low-level frontal structure. Widespread stratiform cloudiness is situated behind the surface front over cold, moist low-level air, while the warm air is relatively clear (Fig. 10.11). The morning (1200 GMT, 11 December 1975) surface chart and cross section (Figs. 10.11a and 10.12a) indicate a relatively diffuse surface temperature pattern as radiative cooling in the clear air on the warm side of the frontal zone, which is identified by the shift in wind direction, has reduced the cross-frontal temperature contrast. The apparent extension of the frontal inversion into the warm air (Fig. 10.12a) may be due to a combination of nocturnal radiative cooling and vertical turbulent mixing in the southerly flow ahead of the front. Twelve hours later, a pronounced surface temperature gradient has formed (Figs. 10.11b and 10.12b), and the pre-frontal nocturnal inversion has been eradicated in response to sensible heating in the PBL. Despite the overcast skies, slightly superadiabatic lapse rates are indicated in the post-frontal PBL. The overall potential temperature structure in Fig. 10.12b is strikingly similar to that presented by Sanders (Fig. 10.5).

10.3.2. Physical Processes Affecting Surface Frontal Structure

Common structural aspects emerge from the three analyses of surface cold fronts, all of which are based on surface airways and radiosonde data. Aspects to be explained include

- the observation that low-level fronts are strongest at the surface and weaken rapidly with altitude,

- the narrow plume of rising warm air above the surface frontal position,

- the statically stable layer capping the post-frontal PBL,

- the well-mixed or slightly unstable potential temperature stratification in the post-frontal PBL, and

- the weakly stable stratification in the pre-frontal PBL.

The rapid weakening with altitude of frontal properties such as the cross-frontal potential temperature gradient and vorticity (cross-front wind shear) is explained by Sanders (1955) in terms of the vertical distributions of the frontolytical effect of the vertical velocity field and the frontogenetical effect of the horizontal wind field. Horizontal convergence, which contributes to frontogenesis in the potential temperature field and is required to increase vorticity, is maximized near the surface and diminishes with altitude. Continuity considerations require the vertical velocity to increase with altitude in the convergent region from a typically small surface value imposed by the presence of the Earth's surface as a rigid lower boundary (see Fig. 10.6). Consequently, within the frontal zone frontolytical tilting effects are maximized above the surface and the frontogenetical confluent and convergent effects, which are maximized at or near the surface, diminish with altitude. As stated previously, the net result is a rapid decrease of frontogenesis relative to a parcel with altitude within the frontal zone, which from an Eulerian perspective translates into the low-level generation of frontal properties and their subsequent transport up and back toward the cold air.

The above features of surface frontogenesis are confirmed by analytic semi-geostrophic models in which diabatic and frictional processes are neglected (e.g., Hoskins and Bretherton, 1972; Blumen, 1980). A limitation to these simplified theoretical models, however, is that they do not yield a narrow jet of rising motion above the surface front shown by Sanders and suggested by Ogura and Portis. A numerical study by Keyser and Anthes (1982), in which a no-slip lower boundary condition and a parameterization of PBL fluxes of heat and momentum are introduced to an idealized frontal formulation treated by Hoskins and Bretherton and Blumen, produces such a jet of vertical motion. In the Keyser and Anthes model, frictionally induced low-level convergence is the dominant contribution to this feature. Frictionally induced convergence also maintains the low-level vorticity field against the dissipative effects of friction itself (Petterssen and Austin, 1942) and enhances the cross-front potential temperature gradient.

The frictional convergence in the Keyser and Anthes model is a consequence of the depletion of along-front momentum (note low-level jets in Figs. 10.5 and 10.9a) by the downward turbulent flux of momentum toward the surface, where a no-slip boundary condition is imposed. The depletion of along-front momentum in the PBL by friction causes the along-front winds to become subgeostrophic and deviate toward the front, which coincides with a pressure trough. Consequently, ageostrophic inflow toward the front is generated and is maximized in the vertical within the PBL. Although observational verification of the structure of this ageostrophic PBL inflow is lacking, the model results are plausible in view of the pronounced cross-isobar flow often noted at the surface in the vicinity of fronts (e.g., Fig. 10.4).

The remarkable static stability of the post-frontal capping inversion identified in Sanders' and Shapiro's cross sections may be due in part to vertically differential cooling due to long-wave radiation. The tops of post-frontal stratiform clouds often correspond to the base of the capping inversion. This situation is comparable with an idealized situation treated theoretically by Staley (1965), in which the mixing ratio decreases rapidly upward from the base of an inversion. The maximum cooling (~ 2 K day^{-1}) is found to occur at the base of the inversion, which strengthens the inversion and destabilizes the region immediately below its base. Another plausible mechanism, identified in the modeling study of Keyser and Anthes (1982), is differential thermal advection within the vertically sheared ageostrophic frictional inflow directed toward the front in the post-frontal PBL. This process results in potentially cooler air within the PBL underrunning the warmer air overlying the PBL. The vertical extent of the capping inversion may be related to the strong vertical wind shear in this region (see Fig. 10.5). Using Doppler radar observations, Browning et al.(1970) inferred the presence of turbulent mixing due to shear (Kelvin-Helmholtz) instability in a warm frontal inversion, structurally similar to those shown in this section. The overall effect of the turbulent mixing should be to spread the isotachs and isentropes vertically, limiting the sharpness of the inversion and increasing its depth.

Vertically differential advection in the lower part of the PBL is suggested by Brundidge (1965) to contribute to the post- and pre-frontal static stability patterns within the PBL. Vertically sheared ageostrophic frictional inflow toward the frontal zone results in potentially colder air overrunning warmer air near the surface in the post-frontal PBL and warmer air overrunning cooler air in the pre-frontal PBL. The result is a slightly unstable or neutral potential temperature stratification in the cold air and a stable stratification in the warm air (see Figs. 10.5 and 10.12b). Another mechanism contributing to the post- and pre-frontal static stability patterns is upward and downward fluxes of sensible heat, generated as cold air is advected over warm ground behind the front and warm air is advected over cold ground ahead of the front (Sanders, 1955). Further discussion of the effects of sensible heating on the development and structure of surface fronts appears in the observational studies of Kousky (1967) and Koch (1984) and the numerical study of Pinkerton (1978).

This discussion of the physical processes that affect the structure of surface fronts has focused primarily on modifications imposed by the turbulent mixing of heat and momentum within the PBL. The effects of diabatic heating due to phase changes of water substance also require consideration. A primary effect of latent heat release is to counteract adiabatic cooling in regions of saturated ascent in the warm air ahead of fronts, resulting in reduced frontolysis due to tilting and increased cross-front thermal gradients within the frontal zone. This effect has been documented observationally (Rao, 1966) and numerically (Williams et al., 1981; Hsie et al., 1984) to contribute to midtropospheric frontogenesis. Including latent heating in these numerical simulations produces stronger upward motions above the PBL, which take on a banded configuration if the heating is sufficiently intense. The frictionally induced ageostrophic inflow in the PBL toward the front is enhanced by the latent heating, resulting in increased low-level convergence and cyclonic vorticity within the PBL in the frontal zone. The cooperative frontogenetical effect of frictional convergence and latent heating, hypothesized from theoretical considerations by Eliassen (1959), is also suggested from the results of Doppler radar studies of the structure of cold fronts (e.g., Browning and Pardoe, 1973), to be discussed subsequently.

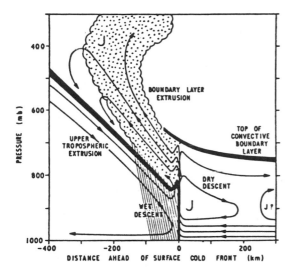

Figure 10.13. Schematic depiction of the transverse flow relative to the motion of a cold anafront. Thick lines represent the frontal zone and the top of the convective boundary layer, defined as a region in which wet-bulb potential temperature is well mixed in the vertical. Stippling indicates saturated ascent. (From Browning and Pardoe, 1973.)

Sanders' and Shapiro's subjective analyses of cold fronts can at best resolve features with horizontal scales of several tens of kilometers. In studies of cold anafronts crossing the British Isles, K. A. Browning and co-investigators (Browning and Harrold, 1970; Browning, 1971, 1974; Browning and Pardoe, 1973) used a combination of conventional and Doppler radar, as well as surface and radiosonde data, to establish a microscale description of cold frontal structure, in which air motions are resolved down to a scale of hundreds of meters. Their analyses of airflow relative to a moving front, represented schematically in Fig. 10.13, reveal an extremely narrow region of ascent in the form of forced line convection corresponding to the position of the surface front. The line convection is roughly 500 m wide, and the intensity of the updraft is on the order of 10 m s^{-1}. More gradual slantwise ascent is depicted in the midtroposphere above the frontal zone. The line convection, which is usually confined below 700 mb, is fed by strong inflow of warm moist PBL air within the lowest 100 mb adjacent to the surface. The inflow is hypothesized to be driven by frictional convergence and latent heat release in the line convection, the latter of which maintains the frontal pressure trough. A low-level jet is indicated in the warm air immediately ahead of the surface front. This jet, which may be several thousands of kilometers long, forms the boundary of an airstream referred to as the warm conveyor belt (Harrold, 1973), which transports substantial amounts of heat and moisture northward in the warm sectors of middle-latitude cyclones (see Sec. 10.5). Finally, evaporative cooling of rain falling into the post-frontal cold air (a process noted earlier by Oliver and Holzworth [1953]) is suggested to exert a role in the dynamics of the transverse circulation.

The phenomenon of line convection at cold fronts studied by Browning et al.(1973) was identified earlier with conventional radar by Wexler (1947) and Kessler and Wexler (1960). This phenomenon in particular and frontal rainbands in general have been studied extensively with Doppler radar (e.g., Carbone, 1982; Hobbs and Persson, 1982; Parsons and Hobbs, 1983). An interesting finding is the similarity of the leading edge of synoptic cold fronts to cold outflows forming thunderstorm gust fronts. Both may be described as the atmospheric analogy to gravity or density currents observed in laboratory models. The existence of this structure does not appear to require the presence of precipitation, which is common to all studies utilizing Doppler radar. Gravity current structure is suggested in the case studies of dry cold fronts in Australia by Clarke (1961), and in the thermal and velocity fields describing a dry surface cold front passing the Boulder Atmospheric Observatory tower (Shapiro, 1984; Shapiro et al., 1985). The Shapiro studies document a cross-frontal scale of 200 m and vertical velocities exceeding 5 m s^{-1}. It is intriguing that the microstructure of surface cold fronts contains the near discontinuities in the wind and thermal fields postulated in the original wedge models. An important distinction, however, is that on the microscale Coriolis forces are insignificant and the dynamics are controlled by solenoidal (pressure gradient), inertial, and frictional forces. In the wedge model, Coriolis and pressure gradient forces are assumed to nearly balance. Therefore, the wedge model may be criticized for applying synoptic-scale dynamics to microscale structure.

The zone concept of frontal structure and the results of the mesoscale studies presented in this section should not be considered to be invalidated or superseded by the recent results of Doppler radar studies. The mesoscale structure of surface cold fronts may be considered to be a spatially averaged representation of features occurring on a scale smaller than that resolvable from the data. It is possible that if surface frontal zones described in this section could be examined with the magnification afforded by microscale data sets, they would also reflect the structure indicated in recent studies. For example, the ascending plume on the order of 10 cm s^{-1} associated with a cross-frontal scale of 10 km in Sanders' case would increase to 10 m s^{-1} for a cross-frontal scale of 100 m, provided that the depth scales and cross-front differences in front-normal velocity remain similar. Although the zone representation of surface fronts is appropriate to mesoscale numerical prediction models and mesoscale data networks, a microscale representation may be more relevant to explaining and understanding the triggering and maintenance of convection. The narrow cloud lines associated with surface fronts evident on satellite imagery (e.g., Janes et al., 1976; Woods, 1983; Koch, 1984; Shapiro et al., 1985) appear to represent microscale rather than mesoscale updrafts. Finally, the extent to which surface fronts are characterized by microscale structure remains unknown. This structure may evolve from synoptic-scale or mesoscale variations, but the details of the transition process from a frontal zone to a feature resembling a gravity current remain undocumented observationally and poorly understood theoretically.

10.4. Upper-Level Frontal Structure

Through the systematic analysis of radiosonde data, Reed and Sanders (1953) identified regions of sharp thermal contrast and cyclonic wind shear in the mid-troposphere that neither extend to the surface nor are connected with surface frontal features. Further analysis of these phenomena (Reed,

1955) led to the concepts of upper-level frontogenesis and tropopause folding, a process in which a thin slice of stratospheric air is extruded into the mid-troposphere, occasionally as far down as 700 to 800 mb. In contrast to surface frontogenesis in which horizontal deformation is dominant, upper-level frontogenesis is controlled by tilting effects associated with the subsidence of stratospheric air. This description of upper-level frontogenesis was revolutionary in that it challenged the notion of a deep polar front separating polar and tropical air masses and it proposed that stratospheric air could be ingested deep into the troposphere. Reed's ideas, including their implications and the research they inspired, are reviewed in this section.

10.4.1. *Documentation of Tropopause Folding and Upper-Level Frontal Structure*

Figures 10.14–10.17 depict the remarkably intense development of an upper-level wave and surface cyclone over the eastern United States for the 48 h period between 0300 GMT, 13–15 December 1953, analyzed by Reed (1955). This case is fortuitous in its intensity and its location within an area of widespread radiosonde coverage. Upper-level frontogenesis and wave amplification begin as a short-wave feature in the 500 mb height and temperature fields enters the northwestern United States (Fig. 10.14). Dramatic amplification occurs, accompanied by continued frontogenesis, as the short wave propagates southeastward upstream from the axis of a diffluent, tilted trough, signaling continued upper-level development and low-level cyclogenesis (Fig. 10.15). By the end of the 48 h period (Fig. 10.16), the trough is confluent with the strongest winds on its east side, indicative of lifting and eventual weakening. An intense, narrow frontal zone stretches from the base of the 500 mb trough northeastward to a short-wave trough located over the

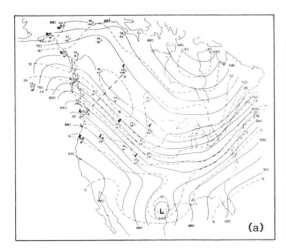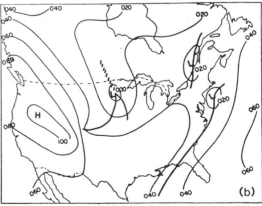

Figure 10.14. Analyses for the (a) 500 mb and (b) 1000 mb pressure surfaces at 0300 GMT, 13 December 1953. Thin solid lines are height contours (ft); thin dashed lines in (a) are isotherms (°C). Thick solid lines indicate frontal boundaries in (a) and positions in (b); dot-dashed line in (a) denotes the position of cross section appearing in Fig. 10.18. Selected observations are plotted in (a) according to the standard station model. (From Reed, 1955.)

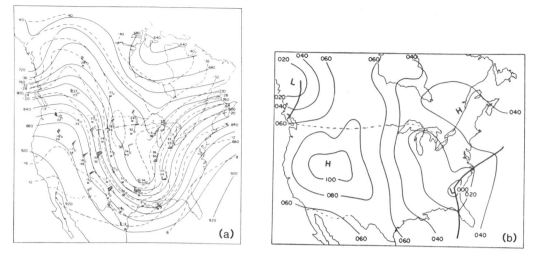

Figure 10.15. Analyses for the 500 and 1000 mb pressure surfaces (as in Fig. 10.14) for 0300 GMT, 14 December 1953. Dot-dashed line in (a) denotes the position of cross section appearing in Fig. 10.19. (From Reed, 1955.)

surface cyclone. The flow pattern at 0300 GMT, 15 December 1953, is also depicted on the 300 K isentropic surface (Fig. 10.17), which lies within the frontal zone. The isobaric topography of the 300 K surface, which depicts the thermal field (analogous to temperature on a pressure surface), and the isotach pattern reveal that the frontal zone occupies a much larger scale when depicted on an isentropic rather than a pressure surface. Recognition that upper-level fronts are more readily resolved with conventional radiosonde

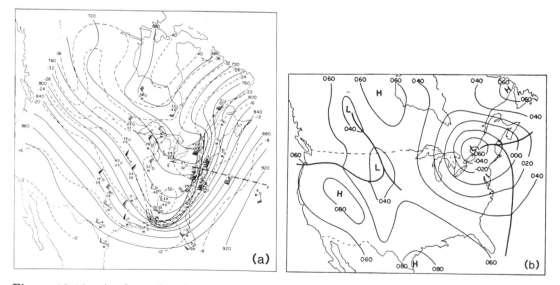

Figure 10.16. Analyses for the 500 and 1000 mb pressure surfaces (as in Fig. 10.14) for 0300 GMT, 15 December 1953. Dot-dashed line in (a) denotes the position of cross section appearing in Fig. 10.20. (From Reed, 1955.)

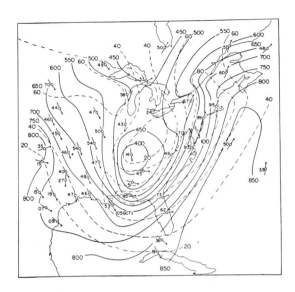

Figure 10.17. Analysis for the 300 K isentropic surface at 0300 GMT, 15 December 1953. Thin solid lines are isobars (mb), and thin dashed lines are isotachs (kt). Observed wind directions are represented by arrows, and observed wind speeds (kt) are entered to the left of the stations. (From Reed, 1955.)

data when viewed from an isentropic perspective motivated subsequent efforts in objective analysis and numerical modeling in isentropic coordinates.

Figures 10.18–10.20 show vertical cross sections of potential temperature and thermally derived geostrophic wind speeds normal to the cross sections, the locations of which are indicated on the corresponding 500 mb charts. The sequence of events indicates the initial steepening and descent of the tropopause accompanied by increases of mid- and upper-tropospheric baroclinicity and static stability, and stratospheric cyclonic

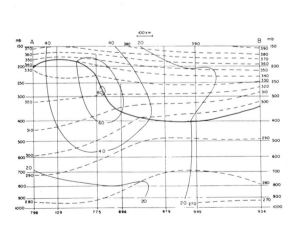

Figure 10.18. Cross section along line AB in Fig. 10.14a. Thin dashed lines are isentropes (K), and thin solid lines are isotachs (m s^{-1}) of the geostrophic wind normal to the cross section. Thick solid line depicts the tropopause. (From Reed, 1955.)

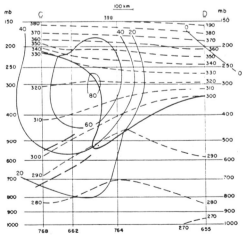

Figure 10.19. Cross section along line CD in Fig. 10.15a. Notation as in Fig. 10.18 except for heavy dashed lines, which represent the boundaries of a statically stable layer. (From Reed, 1955.)

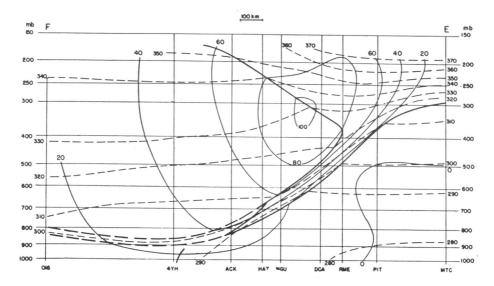

Figure 10.20. Cross section along line EF in Fig. 10.16a. Notation as in Figs. 10.18 and 10.19. (From Reed, 1955.)

wind shear (Fig. 10.19). By the end of the period (Fig. 10.20), an intense frontal zone exhibiting large static stability, baroclinicity, and cyclonic shear (on pressure surfaces) extends from the upper troposphere to nearly 800 mb. The cyclonic shear is weaker in the stratosphere, where the isotachs spread apart with height.

Reed used potential vorticity, defined as

$$P_\theta = -(\varsigma_\theta + f)\frac{\partial \theta}{\partial p} \,, \tag{10.5}$$

where ς_θ is the relative vorticity evaluated on an isentropic surface, to distinguish between stratospheric and tropospheric air. Potential vorticity is conserved along parcel trajectories in the absence of diabatic or frictional processes. Characteristic values of potential vorticity are about an order of magnitude larger in the stratosphere than in the troposphere, primarily because of the static stability difference (see Fig. 10.18). Because the tropospheric portion of the frontal zone depicted in Fig. 10.20 contained values of potential vorticity typical of the stratosphere, Reed concluded that the zone consisted of air of recent stratospheric origin. He further demonstrated that the large values of cross-front potential temperature gradient and cyclonic wind shear resulted from tilting due to a cross-stream gradient of subsidence, which is maximized on the warm-air side of the developing frontal zone.

The depiction of the frontal zone in Fig. 10.20 has been referred to as the waterspout model because of its suggestive geometrical pattern. The assumption of more-or-less-uniform potential vorticity within the stratosphere and the tropospheric frontal zone places a constraint on the wind and po-

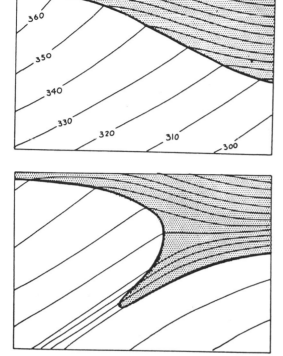
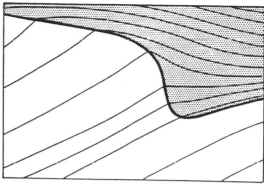

Figure 10.21.   Schematic depiction of upper-level frontogenesis and tropopause folding. Thin solid lines are isentropes; thick solid lines indicate the tropopause. Stratospheric air is indicated by stippling. (From Danielsen, 1964.)

tential temperature analyses. The large values of potential vorticity in the tropospheric part of the frontal zone are associated primarily with large static stability. Absolute vorticities are relatively moderate as isotachs coincide approximately with sloping isentropic surfaces. In the lower stratosphere, the static stability is similar to that in the tropospheric part of the front, but the isentropes have a much smaller slope. Consequently, the isotachs must spread apart and occupy a larger cross-frontal scale in order to maintain values of potential vorticity comparable with those in the tropospheric part of the frontal zone.

A number of objections were raised concerning the plausibility of the tropopause folding process proposed by Reed. The first was that the process would result in the folding of isentropes and the possible generation of superadiabatic lapse rates (see Staley [1960, p. 596] for an allusion to this problem). The process proposed by Danielsen (1964) (Fig. 10.21) responds to the objection by pointing out the necessity for potential temperature variation initially along the tropopause (evident in Fig. 10.18) for the occurrence of upper-level frontogenesis and tropopause folding. Further objections were raised because the tropopause folding hypothesis challenged the concept that a polar front separates polar from tropical air (see Fig. 10.2). The implication was that intense frontal zones could be generated within a polar air mass through differential adiabatic warming associated with subsidence. The notion of tropopause folding also challenged the assumption that the tropopause is a material surface separating stratospheric from tropospheric air. The folding process provides a mechanism for the transport of stratospheric air down to the mid- and lower troposphere, where it can eventually mix with tropospheric air (the folding process was not assumed to be reversible and exactly conservative with respect to potential vorticity). The tropopause folding hypothesis explained how radioactive fallout from the stratosphere resulting from atmospheric testing of nuclear weapons could be observed at the surface (Danielsen, 1964) and exposed the biological risks of such testing. It would take the evidence of aircraft measurements of radioactivity in tropopause folds (e.g., Danielsen, 1964, 1968) and the results of additional case studies (e.g., Reed and Danielsen, 1959; Staley, 1960; Bosart, 1970; Shapiro, 1970) to bring about the general acceptance of the tropopause folding concept in the meteorological community.

Figure 10.22 represents the transverse circulation associated with tropopause folding hypothesized by Danielsen (1968) and compatible with present-day considerations. The circulation pattern depicts thermodynamically indirect and direct cells centered on the cold and warm sides of the front, forming a confluence zone in the vertical plane between the cells. This confluence or "vertical deformation" zone is the region of upper-level frontogenesis and tropopause folding. Although the thermal structure connected with the schematic representation of the upper-level frontal zone in Fig. 10.22 is not shown, an essential aspect of upper-level frontogenesis is the orientation of the vertical deformation zone such that subsidence is maximized on the warm side of the developing frontal zone in the mid- and upper troposphere. This configuration of vertical motion not only results in a frontogenetical tilting term in (10.4), but also provides for the downward transport of frontal properties generated within the developing zone. Since the warm side of the midtropospheric portion of an upper-level front tends to be situated beneath the core of the associated jet, an explanation of upper-level frontogenesis and tropopause folding must account for midtropospheric subsidence maximized beneath the jet core. Several dynamical explanations have been proposed for the frontogenetical vertical deformation pattern. One explanation is based on the transverse ageostrophic circulations in the confluent entrance region of a straight jet streak (Fig. 10.23); another emphasizes vertical motions associated with along-stream ageostrophic circulations arising from variations in parcel trajectory curvature as dictated by the gradient wind (Fig. 10.24).

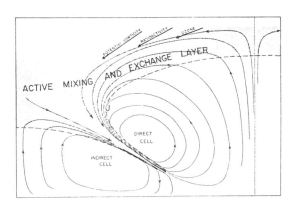

Figure 10.22. Schematic illustration of transverse circulations conducive to upper-level frontogenesis and tropopause folding. (From Danielsen, 1968.)

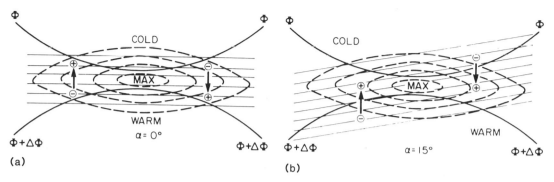

Figure 10.23. Idealized upper-level flow patterns leading to frontogenesis from (a) the confluence mechanism and (b) the cold advection mechanism. Thin solid lines are isotherms or isentropes; thick solid lines are geopotential height of a constant pressure surface. Thick dashed lines are isotachs. Arrows depict transverse ageostrophic wind components at the level of the jet; circled plus and minus signs indicate positions and signs of extrema in the midtropospheric pressure-coordinate vertical motion field. (From Shapiro, 1983.)

In the situation illustrated in Fig. 10.23a (postulated by Namias and Clapp [1949]), where the isotherms are parallel to the axis of the jet, theoretical models of upper-level frontogenesis (Hoskins, 1971, 1972) reproduce the initial stages of tropopause folding as in Fig. 10.19, but fail to generate a deep tropopause fold. The absence of significant tropopause folding and upper-level frontogenesis is related to the orientation of the transverse ageostrophic circulation, which is centered about the midtropospheric frontal zone. The result is subsidence and ascent maximized off to the cold and warm sides of the frontal zone, producing a frontolytical tilting effect. Shapiro (1981) proposed orienting the isotherms as in Fig.10.23b, so that cold advection occurs along the jet. On the basis of deductions from the Sawyer-Eliassen diagnostic equation for the transverse ageostrophic circulation (referred to in Sec. 10.2 and discussed in Ch. 9, this volume), Shapiro hypothesized that the thermodynamically direct cell centered about the jet axis in the absence of along-jet thermal advection (Fig. 10.23a) would shift to the warm air (as in Fig. 10.22), resulting in subsidence below the jet. Numerical simulations (Keyser and Pecnick,1985a, b) confirm Shapiro's hypothesis by producing descent of lower stratospheric air into the midtroposphere and better defined mid- and upper-tropospheric frontal structure than in Hoskins (1971, 1972). In the curvature hypothesis (Fig. 10.24), subsidence is maximized

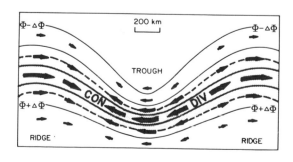

Figure 10.24. Schematic representation of the ageostrophic flow (heavy arrows) and associated patterns of convergence (CON) and divergence (DIV) associated with curvature variations for a uniform jet stream (between dashed lines) within a stationary synoptic-scale wave. Solid lines indicate geopotential height of a constant pressure surface; dashed lines are isotachs. (From Shapiro and Kennedy, 1981.)

beneath the jet stream near the inflection in the height contours upstream of
the axis of an upper-level trough. Newton and Trevisan (1984a, b) presented
theoretical arguments and numerical model evidence for tropopause folding and upper-level frontogenesis due to subsidence related to along-stream
variations in curvature. Although both explanations are applicable to observations of the folding process within the northwesterly flow upstream of
the trough of an amplifying baroclinic wave (Figs. 10.15 and 10.19), their
relative importance remains unresolved.

10.4.2. Evidence of Turbulent Processes in Upper-Level Frontal Zones

The direct probing of upper-level fronts with instrumented aircraft has resulted in modifications and extensions of prior concepts based on radiosonde
data. In the first of a series of studies of upper-level frontal structure using
combined aircraft and radiosonde observations, Shapiro (1974) reported that
earlier aircraft studies (e.g., Danielsen, 1964, 1968) were limited by the relative inaccuracies in extracting winds from Doppler navigation techniques.
The subsequent introduction of inertial navigation systems provided a degree
of accuracy in the winds sufficient for meteorological research. Once again,
the advent of a new technology afforded the possibility of re-examining conventional concepts concerning frontal structure. Major new results have been
the direct documentation of the meso-β-scale structure of upper-level fronts
and clearer inferences concerning the nature and effect of CAT.

Figures 10.25 and 10.26 depict an upper-level frontal zone and jet streak
subjectively analyzed by Shapiro (1978) using aircraft and radiosonde data.
A difference from Reed's analysis (Fig. 10.20) is the concentration of the
cyclonic shear in the stratosphere into a meso-β-scale zone continuous with
the tropospheric frontal layer, rather than over a meso-α-scale as shown
by Reed. The evidence for the scale of the cyclonic shear zone is in the
original wind speed traces from tropospheric (465 mb) and stratospheric
(355 mb) traverses of the frontal zone (not shown), which reveal a cross-
front scale of 100 km. On the basis of the temporal and spatial resolution
available from the European radiosonde network, Berggren (1952) had analyzed stratospheric cyclonic shear over a 100 km scale, but his approach was
dropped by Reed (1955) and Reed and Danielsen (1959) in view of the lack of
mesoscale horizontal resolution in the North American radiosonde network
and the assumption of a more-or-less-uniform distribution of stratospheric
potential vorticity in the frontal layer.

The observation of a meso-β-scale zone of cyclonic shear in the stratosphere implies a distribution of potential vorticity exhibiting anomalously
large values in the vicinity of the stratospheric cyclonic shear zone relative to background stratospheric values (Fig. 10.27a). The distribution of
ozone (Fig. 10.27b), which can be considered a conservative quantity over
the lifetime of an upper-level front, suggests that potential vorticity is generated locally. (If potential vorticity were conservative, its distribution would
match that of the ozone much more closely.) The proposed nonconserva-
tion of potential vorticity had been suggested by Staley (1960), but does
not contradict the tropopause folding hypothesis, which is confirmed by the

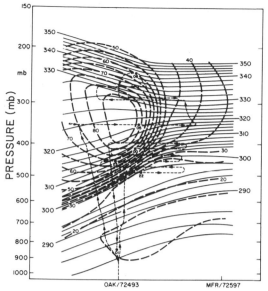

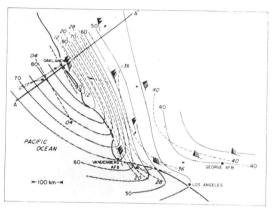

Figure 10.25. Composite cross-section analysis for 0000 GMT, 16 April 1976, along line AA′ in Fig. 10.26. Heavy dashed lines indicate wind speed (m s⁻¹); solid lines are potential temperature (K). Flight path is denoted by light dashed lines; solid circles are times (GMT). Horizontal distance between OAK and MFR is about 500 km. (From Shapiro, 1978.)

Figure 10.26. Analyses of wind speed at 287 mb (m s⁻¹, solid) and ozone concentration (pphm vol⁻¹, heavy dashed lines) derived from Sabreliner data taken between 0615 and 0700 GMT, 16 April 1976. Flight tracks are indicated by light dashed and dot-dashed lines. Line AA′ is the projection for the cross section in Fig. 10.25. (From Shapiro, 1978.)

downward-directed tongue in the ozone pattern. The next step is to ascertain the source of the potential vorticity in the cyclonic shear zone in the lower stratosphere.

The prognostic equation for potential vorticity P_θ, defined in (10.5), is

$$\frac{dP_\theta}{dt} = -(\varsigma_\theta + f)\frac{\partial \dot{\theta}}{\partial p} + \frac{\partial \theta}{\partial p}\left[\boldsymbol{k} \cdot \left(\nabla_\theta \dot{\theta} \times \frac{\partial \boldsymbol{V}}{\partial \theta}\right)\right]$$
$$- \frac{\partial \theta}{\partial p}\left[\boldsymbol{k} \cdot (\nabla_\theta \times \boldsymbol{F_r})\right], \qquad (10.6)$$

where $\boldsymbol{k}$ is the vertical unit vector, ∇_θ is the horizontal gradient operator on surfaces of constant potential temperature, and $\boldsymbol{F_r}$ is the friction term in the vector momentum equation. (Setting $\dot{\theta}$ and $|\boldsymbol{F_r}| = 0$ confirms the conservative nature of potential vorticity.) Detailed discussion of (10.6) can be found in Staley(1960) and Gidel and Shapiro (1979). Shapiro (1976) applied (10.6) at the level of maximum wind (LMW) in the stratospheric zone of cyclonic shear (roughly 355 mb in Figs. 10.25 and 10.27a), negating the middle term on the right of (10.6). Speculating from the basis of a scale analysis, Shapiro neglected the friction term and focused on the first term, which involves the vertical gradient of diabatic heating due to vertical mixing by CAT. The diabatic heating rate may be expressed as

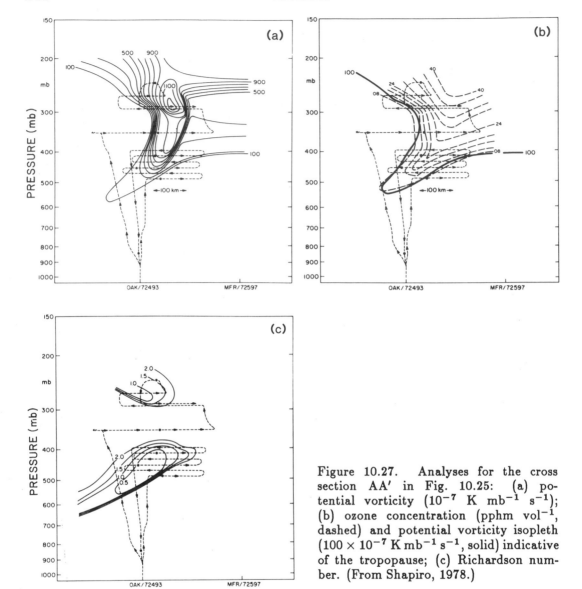

Figure 10.27. Analyses for the cross section AA′ in Fig. 10.25: (a) potential vorticity $(10^{-7}$ K mb^{-1} s$^{-1})$; (b) ozone concentration (pphm vol^{-1}, dashed) and potential vorticity isopleth $(100 \times 10^{-7}$ K mb^{-1} s^{-1}, solid) indicative of the tropopause; (c) Richardson number. (From Shapiro, 1978.)

$$\dot{\theta} = -\frac{1}{\rho}\frac{\partial}{\partial z}\left(\rho\overline{w'\theta'}\right) = g\frac{\partial}{\partial p}\left(\rho\overline{w'\theta'}\right) , \qquad (10.7)$$

where ρ is air density, primed quantities indicate eddy motions, and the overbar denotes a suitably defined average.

The vertical eddy flux of potential temperature, $\overline{w'\theta'}$, is expected to be large in regions of CAT, which is favored in regions of small Richardson number (Ri),

$$\mathrm{Ri} = \frac{\frac{g}{\theta}\frac{\partial \theta}{\partial z}}{\left|\frac{\partial \boldsymbol{V}}{\partial z}\right|^2}. \qquad (10.8)$$

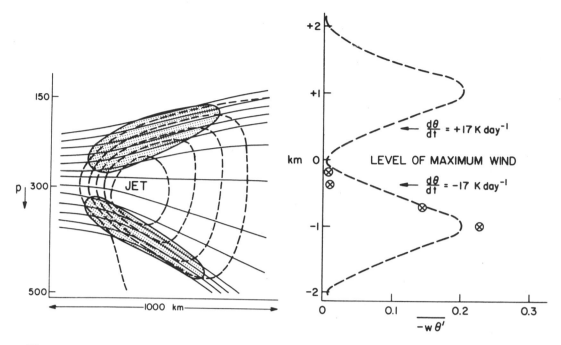

Figure 10.28. Schematic illustration of regions of clear-air turbulence (stippled) in the vicinity of an upper-level jet core and frontal zone. Solid and dashed lines respectively indicate potential temperature and wind speed. (From Shapiro, 1976.)

Figure 10.29. Vertical profile of the downward eddy flux of potential temperature $(-\overline{w'\theta'}$ in m s^{-1} K) expected on the cyclonic shear side of an upper-level jet. Circled x's indicate observed fluxes from Kennedy and Shapiro (1975). (From Shapiro, 1976.)

The distribution of calculated values of Ri is given in Fig. 10.27c for the frontal zone exhibited in Fig. 10.25. The Richardson number is minimized in the shear zones above and below the jet, implying maximum positive values of $-\overline{w'\theta'}$ in the regions stippled in Fig. 10.28. The hypothesized vertical distribution of $-\overline{w'\theta'}$ in Fig. 10.29 applies to the cyclonic shear side of the jet core and indicates stabilization and potential vorticity generation at the LMW where the curvature of the profile of $-\overline{w'\theta'}$ is maximized. This inference is based on substituting (10.7) into the first term on the right of (10.6). The effect of CAT is to produce warming (cooling) in the 1-km-thick layer above (below) the LMW through a convergence (divergence) of the vertical eddy flux of potential temperature $[\dot\theta \simeq -\partial(\overline{w'\theta'})/\partial z]$. The vertical distribution of diabatic heating due to CAT acts to inhibit the vertical spreading of the isentropes at the LMW, which would have to occur if potential vorticity were conserved during the frontogenetical scale contraction of the cyclonic shear zone to 100 km. Direct measurements of the eddy flux of potential temperature indicate that the mechanism involving CAT is sufficiently intense to account for an exponential "doubling" time of potential vorticity of about 10 h, which is comparable with the time required for air parcels to pass through an upper-level frontal layer (Shapiro, 1978).

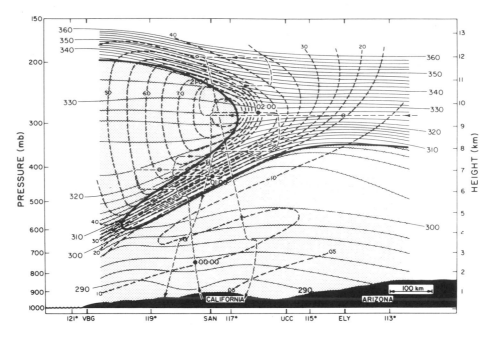

Figure 10.30. Cross section through a tropopause fold for 0000 GMT, 13 March 1978. Potential temperature (K) is indicated by thin solid lines; wind speed (m s^{-1}) by thick dashed lines; Sabreliner flight track by thin dashed lines; tropopause defined in terms of potential vorticity (100 × 10^{-7} K mb^{-1} s^{-1}) by thick solid line. The troposphere is stippled. (From Shapiro, 1980.)

The foregoing mechanism can be shown to contribute to decreases in static stability and potential vorticity within the frontal shear layers, which are regions of intense mixing (note the curvature of the $-\overline{w'\theta'}$ profile 1 km above and below the LMW in Fig. 10.29). Such a decrease in static stability was described by Browning (1971) to have occurred within an upper-tropospheric frontal layer over a time scale of 1 h as a result of an episode of Kelvin-Helmholtz billows, which were detected by radar. The overall effect of CAT is to limit the horizontal and vertical scale contractions of sloping upper-level frontal layers to their observed values of roughly 100 km and 1.5 km, respectively. An implication of this discussion is that the folded tropopause is not a material surface, but is a region of active mixing. Furthermore, parcels are not restricted from crossing the fold; generation of potential vorticity at the LMW provides a mechanism by which parcels may enter the stratosphere from the troposphere (Fig. 10.27a).

Further considerations of stratospheric-tropospheric exchange are pursued in Shapiro (1980). Figure 10.30 is a cross section through a tropopause fold, based on composited aircraft and radiosonde data, and exhibits patterns similar to those in Fig. 10.25. Figures 10.31 and 10.32 depict concentrations of ozone and condensation nuclei, which respectively serve as tracers for the stratosphere and troposphere. The ozone distribution provides evidence of stratospheric air in the fold as discussed in reference to Fig. 10.27b, but the presence and vertical decrease of condensation nuclei demonstrate mixing of

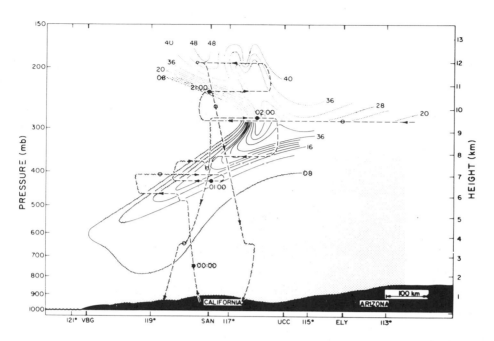

Figure 10.31. Analysis of ozone concentration (pphm vol^{-1}) for the cross section in Fig. 10.30. Tropopause from Fig. 10.30 is indicated by the boundary between the clear and stippled regions. (From Shapiro, 1980.)

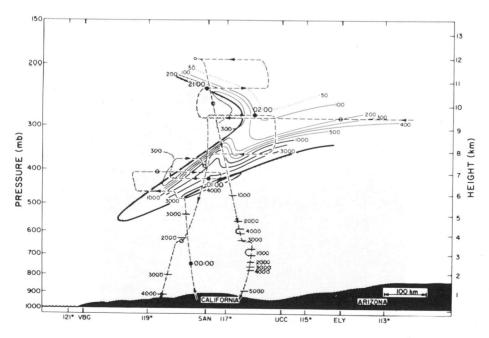

Figure 10.32. Analysis of concentrations of condensation nuclei (particles cm^{-3}, thin solid lines) for the cross section in Fig. 10.30. Thick solid line is the isopleth indicating an ozone concentration of 12 pphm vol^{-1} from Fig. 10.31; tropopause is indicated as in Fig. 10.31. (From Shapiro, 1980.)

tropospheric and stratospheric air within the fold. The significance of the observations presented in Figs. 10.31 and 10.32 is that chemical constituents produced at the Earth's surface can enter the stratosphere at tropopause folds. At the time Shapiro's results were published, there was substantial interest in identifying processes by which chlorofluoromethanes could enter the stratosphere and participate in the chemical consumption of ozone, which acts to filter the ultraviolet portion of the spectrum of incoming solar radiation. A broader perspective on the stratospheric-tropospheric exchange problem is provided in the review by Reiter (1975).

10.5. Relationship of Fronts to the Structure of Middle-Latitude Cyclones

An appealing attribute of the polar-front cyclone model is the close relationship between the cloud patterns and low-level frontal boundaries. With the shift in emphasis from frontal discontinuities to baroclinic zones, the routine availability of upper-air data, and the introduction of satellite imagery to synoptic meteorology, the polar-front concepts relating cloud patterns and frontal boundaries have undergone revision. Conceptual models have been introduced that relate cloud patterns within middle-latitude cyclones to three-dimensional airflow patterns. Such conceptual models are useful because they facilitate the interpretation and diagnosis of satellite imagery relative to conventional synoptic analyses of the wind, thermal, and moisture fields, and because they provide a basis for subjectively refining distributions of cloud and precipitation predicted by operational numerical models.

Examples of models relating cloud patterns to the airflow through middle-latitude cyclones have appeared in a number of studies, including Browning et al.(1973), Harrold (1973), and Browning (1974). These papers are noteworthy for their synthesis of data from a variety of sources to describe and interpret the structure and mechanism of precipitation systems in the context of the larger-scale flow patterns occurring within middle-latitude baroclinic disturbances. Many of the findings of these three cited works are incorporated into Carlson's (1980) conceptual model for the airflow in mature middle-latitude cyclones, which will be discussed subsequently (Fig. 10.33). Abundant schematics and interpretations relating cloud patterns in satellite imagery to flow features and patterns identifiable from the operational surface and upper-air analyses available from the National Meteorological Center appear in Weldon (1979).

Carlson's approach utilizes relative isentropic analysis (Green et al., 1966), in which streamlines are constructed relative to a translating feature identifying a middle-latitude baroclinic system (such as a surface low center or midtropospheric vorticity maximum). The relative streamlines are constructed on surfaces of dry- and wet-bulb potential temperature, respectively, for regions that are unsaturated and regions in which air is experiencing saturated, large-scale slantwise ascent. The system is assumed to be steady-state so that relative streamlines correspond to trajectories in the moving frame of reference. The steady-state assumption permits the vertical motion to be expressed as

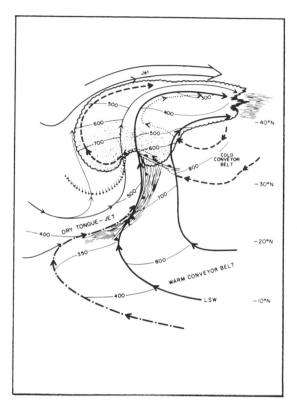

Figure 10.33. Airflow relative to middle-latitude cyclone. Heavy solid streamlines depict warm conveyor belt; dashed line represents cold conveyor belt (drawn dotted where it lies beneath the warm conveyor belt or dry airstream); dot-dashed line represents flow of air originating at middle levels in tropics. Thin solid streamlines pertain to dry air that originates at upper levels west of the trough. Thin solid lines denote the heights of the airstreams (mb) and are approximately normal to the direction of the respective air motion. (Isobars are omitted for the cold conveyor belt where it lies beneath the warm conveyor belt or beneath the jet stream flow.) Scalloping marks the region of dense clouds in upper and middle level layers; stippling indicates sustained precipitation; streaks denote thin cirrus. Small dots with tails mark the edge of the low-level stratus. The major upper-tropospheric jet streams are labeled Jet, and Dry Tongue-Jet. The limiting streamline for the warm conveyor belt is labeled LSW. (From Carlson, 1980.)

$$w = (\boldsymbol{V} - \boldsymbol{C}) \cdot \nabla_\theta h \,, \qquad (10.9)$$

where $\boldsymbol{C}$ is the phase speed of the system and h the height of an isentropic surface (note that θ_w should be substituted for θ in saturated regions). Consequently, the orientation of the relative streamlines with respect to the isentropic topography implies the sign and intensity of the vertical motion. Furthermore, the relative streamline pattern usually reveals several distinct airstreams of differing original altitude and geographical location. These airstreams tend to account for the shape and orientation of cloud features, such as the characteristic comma structure of middle-latitude baroclinic systems. Cloud edges usually coincide with airstream boundaries, which are often associated with low-level and upper-level fronts.

The conceptual model appearing in Fig. 10.33 displays three airstreams: (1) the warm conveyor belt, originating from easterly flow in the subtropical or tropical boundary layer; (2) a cold conveyor belt, originating from the easterly flow at low levels around a cold anticyclone to the north or northeast of the center of the surface cyclone; and (3) a dry airstream of subsided air, originating at upper levels west of the upper trough associated with the cyclone. The low-level boundary of the cold and warm conveyor belts corresponds to the warm front, whereas the boundary between the dry airstream and warm conveyor belt coincides with the cold front at lower levels and enhanced baroclinicity in the midtroposphere. This boundary may correspond to the deep fronts analyzed by Newton (1958) and Palmén

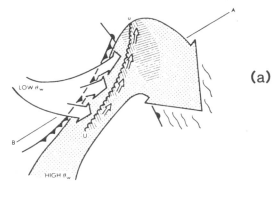

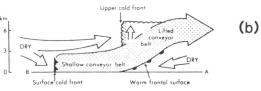

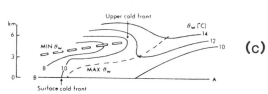

(a)

(b)

(c)

Figure 10.34. Schematic portrayal of a katafront (a) in plan view, and (b) and (c) in vertical section. The broad stippled arrow in (a) and (b) represents the conveyor belt (high θ_w) flow, which, relative to the moving system, travels ahead of the surface cold front before rising and turning to the right above a warm frontal zone. The arrows entering from the left of the diagram represent dry air (low θ_w), which overruns the conveyor belt after having descended from the cold side of the upper-tropospheric jet stream. At the leading edge of the overrunning dry airflow, air ascends convectively from the top of the conveyor belt to generate a major rainband along the upper cold front (cusped line Uu), which may extend from within the conventional warm sector to a position ahead of the surface warm front. (From Browning and Monk, 1982.)

and Newton (1969, pp. 335–338) on the east sides of well-developed upper troughs associated with mature middle-latitude cyclones. It is possible that such a deep frontal zone originates through tropopause folding at upper levels within the northwesterly flow upstream from the axis of a developing trough prior to cyclogenesis, and subsequently is transported around the base of the trough into southwesterly flow, where it becomes oriented above the cold front. (This process is compatible with the sequence of events depicted by Reed [1955]; see Figs. 10.14–10.16.) Upper-level frontogenesis also can be expected in the confluence zone forming the sharp northern edge of the warm conveyor belt. The presence of a jet in this confluent region is in accord with the hypothesis of Namias and Clapp (1949) concerning the formation and maintenance of upper-level jets.

The conceptual models of airflow in middle-latitude cyclones revealed by relative isentropic analysis may be considered an elaboration of the three-dimensional character of the polar-front cyclone model. Airstreams are analogous to air masses in the sense that they originate outside the cyclone in vastly different regions of the atmosphere in terms of latitude as well as altitude. Airstream boundaries correspond to frontal zones, analogous to air mass boundaries and frontal discontinuities. Vertical motions are determined by flow relative to isentropic topography (10.9) rather than frontal topography (10.2). The similarity between (10.9) and (10.2) is notable. In terms of the vertical motion field, the zone concept of fronts is to the modern cyclone models what the wedge concept is to the polar-front cyclone model.

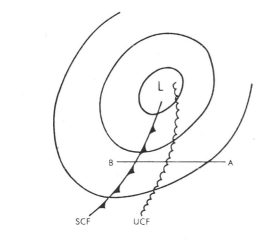

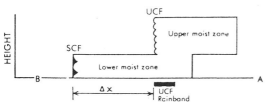

Figure 10.35. Schematic summary of the split-front model portrayed in Fig. 10.34. SCF and UCF denote the surface and upper cold fronts. The UCF is often better regarded as a transition zone in the moisture field. (From Browning and Monk, 1982.)

Browning and Monk (1982) used concepts from the relative isentropic approach in their interpretation of a common signature in the cloud and precipitation patterns associated with frontal systems affecting the British Isles (Figs. 10.34 and 10.35). Referred to as the split-front model, it seeks to explain the orientation of deep clouds and embedded rainbands well ahead of the position of the surface cold front (Fig. 10.34a). The raised wall of cloud indicated along the line Uu consists of air rising convectively from the top of the warm conveyor belt, which is lifted as it flows above the warm frontal zone (Fig. 10.34b). The rear edge of the deep cloud is referred to as an upper cold front, which separates the moist air rising out of the warm conveyor belt from a dry westerly airstream. Whether the boundary called the upper cold front by Browning and Monk strictly fits the definition of a front proposed in Sec. 10.1 is unclear, since they did not present wind analyses. At any rate, the upper cold front consists of a sharply defined boundary in the moisture field. Figs. 10.34c and 10.35 contain additional schematic depictions of this feature.

Browning and Monk suggested that their description of cold frontal structure conforms to the concept of a katafront (Bergeron, 1937) introduced in the context of the wedge model as described in Sec. 10.2.1 (Fig. 10.1a). Motivation for preserving this terminology stems in part from the extensive study by Sansom (1951), in which 50 cold fronts crossing the British Isles were classified and found to conform to Bergeron's anafront and katafront designations. Anafronts were shown to be sharply defined in terms of a drop in surface temperature and a wind shift, and characterized by extensive post-frontal precipitation. Katafronts were shown to be diffuse and best identified by a drop in relative humidity following pre-frontal precipitation. Sansom further determined that over the life cycle of a middle-latitude cyclone the cold front is initially an anafront along its length, and acquires katafront characteristics near the center of the developing cyclone, which spread along its entire length as the cyclone matures. The cases considered by Browning and Monk in formulating the split-front model involve mature cyclones, consistent with Sansom's observation. Although the terminology is rarely used today, anafronts and katafronts are often apparent in satellite imagery. Examples can be found in visible and infrared satellite images of cloud bands respectively ahead of and behind analyzed surface frontal positions.

10.6. Future Directions of Research

Changing observing technologies are an important necessary condition for the evolution of conceptual models of the structure of surface and upper-level fronts in relation to middle-latitude cyclones and baroclinic waves. Modifications to the polar-front cyclone model of the Bergen School have followed the introduction of upper-air observations and satellite imagery to synoptic meteorology. Significant revisions have involved generalizing the concept of frontal discontinuities to frontal zones, and interpreting cloud and precipitation patterns in terms of three-dimensional airflows. The theory of baroclinic instability formalizes the shift in emphasis from earlier ideas that middle-latitude cyclones develop on pre-existing low-level fronts to the idea that cyclones develop within regions of pre-existing baroclinicity spread over a deep layer. Furthermore, according to baroclinic instability theory, low-level fronts develop in response to the horizontal deformation patterns implicit in the structure of the amplifying disturbance.

Another modification to earlier ideas has resulted from the identification of upper-level fronts, which owe their existence to patterns of differential subsidence associated with the evolution of baroclinic waves. The synoptic concept of a polar front that separates polar and tropical air through the depth of the troposphere has become a statistical concept, evident in long-time averages of middle-latitude tropospheric flow patterns. The tropopause is no longer considered a material surface as the concept of stratospheric-tropospheric exchange has become established. Finally, the structure of surface and upper-level fronts and the processes and mechanisms contributing to their development have been identified in case studies and reproduced in theoretical and numerical models.

Despite the foregoing advances, our knowledge and understanding of fronts remains incomplete and fragmentary. A number of areas of research deserve consideration. Observational studies of surface fronts in the meteorological literature have focused on the structure of cold fronts. Case studies documenting the detailed structure of warm fronts are rare, although cross sections of warm fronts have appeared in studies of the structure of precipitation systems associated with middle-latitude cyclones (e.g., Browning et al., 1970, 1973). Modern case studies illustrating occluded fronts and the occlusion process are virtually nonexistent, although the cross sections shown by McClain and Danielsen (1955) conform to the structure envisioned by Bergeron. Furthermore, there is an absence of detailed case studies of the evolution and structure of surface fronts in the context of the life cycle of middle-latitude cyclones. It should be noted that the detailed analyses of cold fronts by Sanders (1955) and Shapiro (1983) shown in Sec. 10.3 involve nondeveloping situations. Satellite imagery of frontal cloud bands has revived interest in the anafront and katafront classifications originally proposed by Bergeron (1937). The general applicability of this distinction and Sansom's (1951) findings should be reexamined and reconciled in terms of zone concepts of frontal structure and vertical circulations.

There is a continuing need for investigations of the detailed structure of fronts and their relationship to smaller-scale phenomena and processes. Both surface and upper-level fronts have been shown to provide the environmental background for Kelvin-Helmholtz instability contributing to CAT. The search has begun for dynamical instabilities accounting for the banded and cellular organization of precipitation systems associated with moist fronts within middle-latitude cyclones. Another problem is documenting the microscale structure of surface fronts and determining the applicability of the gravity current model in describing their structure. The transition process from a meso-β-scale zone to a microscale feature needs to be identified observationally and explained theoretically. Finally, there is considerable interest in examining the interaction of surface fronts with topography. As an example, the process of cold-air damming (Richwien, 1980) along the eastern slopes of mountain barriers is crucial to the existence and evolution of backdoor cold fronts in the eastern United States (Carr, 1951) and the "Southerly Busters" (e.g., Baines, 1980) occurring along the east coast of Australia. The initiation of New England coastal frontogenesis (Bosart *et al.*, 1972; Bosart, 1975), which is favored along a line from Providence to Boston in advance of coastal cyclones, appears to involve cold-air damming as well.

Another area of research involves considering the evolution of upper-level and surface fronts from a unified perspective, rather than independently of each other. The case study approach requires focusing on specific phenomena or physical processes in order to be tractable. Consequently, the studies cited here (as well as the chapter's organization) tend to leave the impression that surface and upper-level processes are unrelated. There is a need for research emphasizing the relationships and interactions between upper-level and surface fronts in the context of the initiation and organization of mesoscale convective systems and the evolution of baroclinic waves. Hypothesized schematics reflecting this goal are given by Shapiro (1983), and this philosophy is evident in the study concerning the organization of severe convective storms by Uccellini and Johnson (1979). Practical issues related to such research include numerically simulating the interactions between upper-level jets and fronts with mesoscale convective systems and determining the extent to which numerical weather prediction models must resolve upper-level frontal-scale phenomena and processes in order to accurately predict the evolution of baroclinic waves and related cyclogenesis events.

Progress in frontal research has been hindered by limitations in the spatial coverage and temporal resolution of operational radiosonde networks.

The application of a network of remote-sensing devices such as VHF and UHF Doppler radars (see Larsen and Röttger [1982] for a review), which provide wind data with high vertical and temporal resolution, would permit the nearly continuous monitoring of frontal phenomena. The capability exists of remotely sensing patterns of vertically integrated ozone from satellites, which permits the detection of tropopause folding (Uccellini *et al.*, 1985) and the position, and perhaps the intensity, of jet streaks (Shapiro *et al.*, 1982). It is tempting to speculate that the above novel remote-sensing technologies, in combination with aircraft, radiosondes, and conventional radar, may lead to advances in our knowledge of fronts comparable with those discussed in this chapter.

A major objective of observational research should be the formulation of conceptual models of the structure and evolution of middle-latitude baroclinic waves and cyclones that are sufficiently general to incorporate surface and upper-level fronts and cloud and precipitation patterns. A related goal is explaining the essential features and processes of the conceptual models using a spectrum of dynamical approaches, ranging from simplified analytic models capable of isolating specific mechanisms to complex numerical models capable of faithfully reproducing observed atmospheric structures and processes (Hoskins, 1983). Such advances in the conceptual and theoretical descriptions of baroclinic waves and cyclones have the potential of eventually translating into improvements and refinements in numerical guidance and subjective forecasts of frontal weather.

REFERENCES

Anthes, R. A., Y.-H. Kuo, S. G. Benjamin, and Y.-F. Li, 1982: The evolution of the mesoscale environment of severe local storms: Preliminary modeling results. *Mon. Wea. Rev.*, 110, 1187–1213.

Baines, P. G., 1980: The dynamics of the Southerly Buster. *Aust. Meteor. Mag.*, 28, 175–200.

Ball, F. K., 1960: A theory of fronts in relation to surface stress. *Quart. J. Roy. Meteor. Soc.*, 86, 51–66.

Bergeron, T., 1937: On the physics of fronts. *Bull. Amer. Meteor. Soc.*, 18, 265–275.

Berggren, R., 1952: The distribution of temperature and wind connected with active tropical air in the higher troposphere and some remarks concerning clear air turbulence at high altitude. *Tellus*, 4, 43–53.

Bjerknes, J., 1919: On the structure of moving cyclones. *Geofys. Publ.*, 1(1), 1–8.

Bjerknes, J., and E. Palmén, 1937: Investigations of selected European cyclones by means of serial ascents. *Geofys. Publ.*,

12(2), 1–62.

Bjerknes, J., and H. Solberg, 1921: Meteorological conditions for the formation of rain. *Geofys. Publ.*, 2(3), 1–60.

Bjerknes, J., and H. Solberg, 1922: Life cycle of cyclones and the polar front theory of atmospheric circulation. *Geofys. Publ.*, 3(1), 1–18.

Blumen, W., 1980: A comparison between the Hoskins-Bretherton model of frontogenesis and the analysis of an intense surface frontal zone. *J. Atmos. Sci.*, 37, 64–77.

Bosart, L. F., 1970: Mid-tropospheric frontogenesis. *Quart. J. Roy. Meteor. Soc.*, 96, 442–471.

Bosart, L. F., 1975: New England coastal frontogenesis. *Quart. J. Roy. Meteor. Soc.*, 101, 957–978.

Bosart, L. F., C. J. Vaudo, and J. H. Helsdon, Jr., 1972: Coastal frontogenesis. *J. Appl. Meteor.*, 11, 1236–1258.

Browning, K. A., 1971: Radar measurements of air motion near fronts. *Weather*, **26**, 320–340.

Browning, K. A., 1974: Mesoscale structure of rain systems in the British Isles. *J. Meteor. Soc. Japan*, **52**, 314–327.

Browning, K. A., and T. W. Harrold, 1970: Air motion and precipitation growth at a cold front. *Quart. J. Roy. Meteor. Soc.*, **96**, 369–389.

Browning, K. A., and G. A. Monk, 1982: A simple model for the synoptic analysis of cold fronts. *Quart. J. Roy. Meteor. Soc.*, **108**, 435–452.

Browning, K. A., and C. W. Pardoe, 1973: Structure of low-level jet streams ahead of mid-latitude cold fronts. *Quart. J. Roy. Meteor. Soc.*, **99**, 619–638.

Browning, K. A., T. W. Harrold, and J. R. Starr, 1970: Richardson number limited shear zones in the free atmosphere. *Quart. J. Roy. Meteor. Soc.*, **96**, 40–49.

Browning, K. A., M. E. Hardman, T. W. Harrold, and C. W. Pardoe, 1973: The structure of rainbands within a mid-latitude depression. *Quart. J. Roy. Meteor. Soc.*, **99**, 215–231.

Brundidge, K. C., 1965: The wind and temperature structure of nocturnal cold fronts in the first 1420 feet. *Mon. Wea. Rev.*, **93**, 587–603.

Carbone, R. E., 1982: A severe frontal rainband. Part I: Stormwide hydrodynamic structure. *J. Atmos. Sci.*, **39**, 258–279.

Carlson, T. N., 1980: Airflow through mid-latitude cyclones and the comma cloud pattern. *Mon. Wea. Rev.*, **108**, 1498–1509.

Carr, J. A., 1951: The east coast "backdoor" front of May 16–20, 1951. *Mon. Wea. Rev.*, **79**, 100–105.

Charney, J. G., 1947: The dynamics of long waves in a baroclinic westerly current. *J. Meteor.*, **4**, 135–163.

Clarke, R. H., 1961: Mesostructure of dry cold fronts over featureless terrain. *J. Meteor.*, **18**, 715–735.

Danielsen, E. F., 1964: Project Springfield Report. Defense Atomic Support Agency, Washington, DC 20301, DASA 1517 (NTIS#AD–607980), 97 pp.

Danielsen, E. F., 1968: Stratospheric-tropospheric exchange based on radioactivity, ozone and potential vorticity. *J. Atmos. Sci.*, **25**, 502–518.

Eady, E. T., 1949: Long waves and cyclone waves. *Tellus*, **1**(3), 33–52.

Eliassen, A., 1959: On the formation of fronts in the atmosphere. *The Atmosphere and the Sea in Motion* (B. Bolin, Ed.), Rockefeller Institute Press, 277–287.

Eliassen, A., 1962: On the vertical circulation in frontal zones. *Geofys. Publ.*, **24**(4), 147–160.

Eliassen, A., 1966: Motions of intermediate scale: Fronts and cyclones. *Advances in Earth Science* (P. M. Hurley, Ed.), M.I.T. Press, Cambridge, Mass., 111–138.

Emanuel, K. A., 1983: On assessing local conditional symmetric instability from atmospheric soundings. *Mon. Wea. Rev.*, **111**, 2016–2033.

Faller, A. J., 1956: A demonstration of fronts and frontal waves in atmospheric models. *J. Meteor.*, **13**, 1–4.

Frank, A. E., and D. A. Barber, 1977: Fronts and frontogenesis as revealed by high time resolution data. NASA/Marshall Space Flight Center, AL 35812, NASA RP-1005 (NTIS#N77-30705), 136 pp.

Fultz, D., 1952: On the possibility of experimental models of the polar-front wave. *J. Meteor.*, **9**, 379–384.

Gidel, L. T., and M. A. Shapiro, 1979: The role of clear air turbulence in the production of potential vorticity in the vicinity of upper tropospheric jet stream-frontal systems. *J. Atmos. Sci.*, **36**, 2125–2138.

Godske, C. L., T. Bergeron, J. Bjerknes, and R. C. Bundgaard, 1957: *Dynamic Meteorology and Weather Forecasting*. American Meteorological Society, Boston, and Carnegie Institution of Washington, Washington, D.C., 800 pp.

Green, J. S. A., F. H. Ludlam, and J. F. R. McIlveen, 1966: Isentropic relative-flow analysis and the parcel theory. *Quart. J. Roy. Meteor. Soc.*, **92**, 210–219.

Harrold, T. W., 1973: Mechanisms influencing the distribution of precipitation within baroclinic disturbances. *Quart. J. Roy. Meteor. Soc.*, **99**, 232–251.

Henry, A. J., E. H. Bowie, H. J. Cox, and H. C. Frankenfield, 1916: *Weather Forecasting in the United States*. U.S. Department of Agriculture, Weather Bureau, W.B. No. 583, 370 pp.

Hobbs, P. V., and P. O. G. Persson, 1982:

The mesoscale and microscale structure and organization of clouds and precipitation in midlatitude cyclones. Part V: The substructure of narrow cold-frontal rainbands. *J. Atmos. Sci.*, **39**, 280–295.

Hoskins, B. J., 1971: Atmospheric frontogenesis models: Some solutions. *Quart. J. Roy. Meteor. Soc.*, **97**, 139–153.

Hoskins, B. J., 1972: Non-Boussinesq effects and further development in a model of upper tropospheric frontogenesis. *Quart. J. Roy. Meteor. Soc.*, **98**, 532–541.

Hoskins, B. J., 1975: The geostrophic momentum approximation and the semi-geostrophic equations. *J. Atmos. Sci.*, **32**, 233–242.

Hoskins, B. J., 1982: The mathematical theory of frontogenesis. *Ann. Rev. Fluid Mech.*, **14**, 131–151.

Hoskins, B. J., 1983: Dynamical processes in the atmosphere and the use of models. *Quart. J. Roy. Meteor. Soc.*, **109**, 1–21.

Hoskins, B. J., and F. P. Bretherton, 1972: Atmospheric frontogenesis models: Mathematical formulation and solution. *J. Atmos. Sci.*, **29**, 11–37.

Hoskins, B. J., and W. A. Heckley, 1981: Cold and warm fronts in baroclinic waves. *Quart. J. Roy. Meteor. Soc.*, **107**, 79–90.

Hsie, E.-Y., R. A. Anthes, and D. Keyser, 1984: Numerical simulation of frontogenesis in a moist atmosphere. *J. Atmos. Sci.*, **41**, 2581–2594.

Janes, S. A., H. W. Brandli, and J. W. Orndorff, 1976: "The blue line" depicted on satellite imagery. *Mon. Wea. Rev.*, **104**, 1178–1181.

Kennedy, P. J., and M. A. Shapiro, 1975: The energy budget in a clear air turbulence zone as observed by aircraft. *Mon. Wea. Rev.*, **103**, 650–654.

Kessler, E., and R. Wexler, 1960: Observations of a cold front, 1 October 1958. *Bull. Amer. Meteor. Soc.*, **41**, 253–257.

Keyser, D., and R. A. Anthes, 1982: The influence of planetary boundary layer physics on frontal structure in the Hoskins-Bretherton horizontal shear model. *J. Atmos. Sci.*, **39**, 1783–1802.

Keyser, D., and M. J. Pecnick, 1985a: A two-dimensional primitive equation model of frontogenesis forced by confluence and horizontal shear. *J. Atmos. Sci.*, **42**, 1259–1282.

Keyser, D., and M. J. Pecnick, 1985b: Diagnosis of ageostrophic circulations in a two-dimensional primitive equation model of frontogenesis. *J. Atmos. Sci.*, **42**, 1283–1305.

Koch, S. E., 1984: The role of an apparent mesoscale frontogenetic circulation in squall line initiation. *Mon. Wea. Rev.*, **112**, 2090–2111.

Kousky, V. E., 1967: A case study of frontogenetic and frontolytic processes. M. S. thesis, Department of Meteorology, The Pennsylvania State University, University Park, PA 16802, 32 pp.

Kutzbach, G., 1979: *The Thermal Theory of Cyclones: A History of Meteorological Thought in the Nineteenth Century.* American Meteorological Society, Boston, 255 pp.

Larsen, M. F., and J. Röttger, 1982: VHF and UHF Doppler radars as tools for synoptic research. *Bull. Amer. Meteor. Soc.*, **63**, 996–1008.

McClain, E. P., and E. F. Danielsen, 1955: Zonal distribution of baroclinicity for three Pacific storms. *J. Meteor.*, **12**, 314–323.

Miller, J. E., 1948: On the concept of frontogenesis. *J. Meteor.*, **5**, 169–171.

Namias, J., 1983: The history of polar front and air mass concepts in the United States—an eyewitness account. *Bull. Amer. Meteor. Soc.*, **64**, 734–755.

Namias, J., and P. F. Clapp, 1949: Confluence theory of the high tropospheric jet stream. *J. Meteor.*, **6**, 330–336.

Newton, C. W., 1958: Variations in frontal structure of upper level troughs. *Geophysica*, **6**, 357–375.

Newton, C. W., and A. Trevisan, 1984a: Clinogenesis and frontogenesis in jet-stream waves. Part I: Analytical relations to wave structure. *J. Atmos. Sci.*, **41**, 2717–2734.

Newton, C. W., and A.Trevisan, 1984b: Clinogenesis and frontogenesis in jet-stream waves. Part II: Channel model numerical experiments. *J. Atmos. Sci.*, **41**, 2735–2755.

Ogura, Y., and D. Portis, 1982: Structure of the cold front observed in SESAME-AVE III and its comparison with the Hoskins-Bretherton frontogenesis model. *J. Atmos. Sci.*, **39**, 2773–2792.

Oliver, V. J., and G. C. Holzworth, 1953:

Some effects of the evaporation of widespread precipitation on the production of fronts and on changes in frontal slopes and motions. *Mon. Wea. Rev.*, **81**, 141–151.

Orlanski, I., 1975: A rational subdivision of scales for atmospheric processes. *Bull. Amer. Meteor. Soc.*, **56**, 527–530.

Palmén, E., 1948: On the distribution of temperature and wind in the upper westerlies. *J. Meteor.*, **5**, 20–27.

Palmén, E., and K. M. Nagler, 1948: An analysis of the wind and temperature distribution in the free atmosphere over North America in a case of approximately westerly flow. *J. Meteor.*, **5**, 58–64.

Palmén, E., and C. W. Newton, 1948: A study of the mean wind and temperature distribution in the vicinity of the polar front in winter. *J. Meteor.*, **5**, 220–226.

Palmén, E., and C. W. Newton, 1969: *Atmospheric Circulation Systems: Their Structure and Physical Interpretation*. Int. Geophys. Ser., **Vol. 13**, Academic Press, New York, 603 pp.

Parsons, D. B., and P. V. Hobbs, 1983: The mesoscale and microscale structure and organization of clouds and precipitation in midlatitude cyclones. XI: Comparisons between observational and theoretical aspects of rainbands. *J. Atmos. Sci.*, **40**, 2377–2397.

Petterssen, S., 1956: *Weather Analysis and Forecasting*, Vol. 1, *Motion and Motion Systems*. 2nd ed., McGraw-Hill, New York, 428 pp.

Petterssen, S., and J. M. Austin, 1942: Fronts and frontogenesis in relation to vorticity. *Pap. Phys. Oceanogr. Meteor.*, **7**(2), 1–37.

Pinkerton, J. E., 1978: Numerical experiments on boundary layer effects on frontal structure. Ph.D. thesis, Department of Physics and Atmospheric Science, Drexel University, Philadelphia, PA 19104, 214 pp.

Rao, G. V., 1966: On the influences of fields of motion, baroclinity and latent heat source on frontogenesis. *J. Appl. Meteor.*, **5**, 377–387.

Rao, G. V., 1971: A numerical study of the frontal circulation in the atmospheric boundary layer. *J. Appl. Meteor.*, **10**, 26–35.

Reed, R. J., 1955: A study of a characteristic type of upper-level frontogenesis. *J. Meteor.*, **12**, 226–237.

Reed, R. J., and E. F. Danielsen, 1959: Fronts in the vicinity of the tropopause. *Arch. Meteor. Geophys. Bioklim.*, **A11**, 1–17.

Reed, R. J., and F. Sanders, 1953: An investigation of the development of a midtropospheric frontal zone and its associated vorticity field. *J. Meteor.*, **10**, 338–349.

Reiter, E. R., 1975: Stratospheric-tropospheric exchange processes. *Rev. Geophys. Space Phys.*, **13**, 459–474.

Richwien, B. A., 1980: The damming effect of the southern Appalachians. *Natl. Wea. Dig.*, **5**, 2–12.

Sanders, F., 1955: An investigation of the structure and dynamics of an intense surface frontal zone. *J. Meteor.*, **12**, 542–552.

Sansom, H. W., 1951: A study of cold fronts over the British Isles. *Quart. J. Roy. Meteor. Soc.*, **77**, 96–120.

Sawyer, J. S., 1956: The vertical circulation at meteorological fronts and its relation to frontogenesis. *Proc. Roy. Soc. London*, **A234**, 346–362.

Shapiro, M. A., 1970: On the applicability of the geostrophic approximation to upper-level frontal-scale motions. *J. Atmos. Sci.*, **27**, 408–420.

Shapiro, M. A., 1974: A multiple structured frontal zone-jet stream system as revealed by meteorologically instrumented aircraft. *Mon. Wea. Rev.*, **102**, 244–253.

Shapiro, M. A., 1976: The role of turbulent heat flux in the generation of potential vorticity in the vicinity of upper-level jet stream systems. *Mon. Wea. Rev.*, **104**, 892–906.

Shapiro, M. A., 1978: Further evidence of the mesoscale and turbulent structure of upper-level jet stream-frontal zone systems. *Mon. Wea. Rev.*, **106**, 1100–1111.

Shapiro, M. A., 1980: Turbulent mixing within tropopause folds as a mechanism for the exchange of chemical constituents between the stratosphere and troposphere. *J. Atmos. Sci.*, **37**, 994–1004.

Shapiro, M. A., 1981: Frontogenesis and geostrophically forced secondary circulations in the vicinity of jet stream-frontal zone systems. *J. Atmos. Sci.*, **38**, 954–973.

Shapiro, M. A., 1983: Mesoscale weather systems of the central United States. *The National STORM Program: Scientific and Technological Bases and Major Objectives* (R. A. Anthes, Ed.), University Corporation for Atmospheric Research, P. O. Box 3000, Boulder, CO 80307, 3.1–3.77.

Shapiro, M. A., 1984: Meteorological tower measurements of a surface cold front. *Mon. Wea. Rev.*, **112**, 1634–1639.

Shapiro, M. A., and P. J. Kennedy, 1981: Research aircraft measurements of jet stream geostrophic and ageostrophic winds. *J. Atmos. Sci.*, **38**, 2642–2652.

Shapiro, M. A., A. J. Krueger, and P. J. Kennedy, 1982: Nowcasting the position and intensity of jet streams using a satellite-borne total ozone mapping spectrometer. *Nowcasting* (K. A. Browning, Ed.), Academic Press, New York, 137–145.

Shapiro, M. A., T. Hampel, D. Rotzoll, and F. Mosher, 1985: The frontal hydraulic head: A micro-α scale (~ 1 km) triggering mechanism for mesoconvective weather systems. *Mon. Wea. Rev.*, **113**, 1166–1183.

Staley, D. O., 1960: Evaluation of potential-vorticity changes near the tropopause and the related vertical motions, vertical advection of vorticity, and transfer of radioactive debris from stratosphere to troposphere. *J. Meteor.*, **17**, 591–620.

Staley, D. O., 1965: Radiative cooling in the vicinity of inversions and the tropopause. *Quart. J. Roy. Meteor. Soc.*, **91**, 282–301.

Stone, P. H., 1966: Frontogenesis by horizontal wind deformation fields. *J. Atmos. Sci.*, **23**, 455–465.

Sutcliffe, R. C., 1947: A contribution to the problem of development. *Quart. J. Roy. Meteor. Soc.*, **73**, 370–383.

Uccellini, L. W., and D. R. Johnson, 1979: The coupling of upper and lower tropospheric jet streaks and implications for the development of severe convective storms. *Mon. Wea. Rev.*, **107**, 682–703.

Uccellini, L. W., D. Keyser, K. F. Brill, and C. H. Wash, 1985: The Presidents' Day cyclone of 18–19 February 1979: Influence of upstream trough amplification and associated tropopause folding on rapid cyclogenesis. *Mon. Wea. Rev.*, **113**, 962–988.

Welander, P., 1955: Studies on the general development of motion in a two-dimensional, ideal fluid. *Tellus*, **7**, 141–156.

Weldon, R. B., 1979: Cloud patterns and the upper air wind field. Air Weather Service, Scott AFB, IL 62225, AWS/TR-79/003, 101 pp.

Wexler, R., 1947: Radar detection of a frontal storm 18 June 1946. *J. Meteor.*, **4**, 38–44.

Williams, R. T., 1967: Atmospheric frontogenesis: A numerical experiment. *J. Atmos. Sci.*, **24**, 627–641.

Williams, R. T., 1968: A note on quasi-geostrophic frontogenesis. *J. Atmos. Sci.*, **25**, 1157–1159.

Williams, R. T., and J. Plotkin, 1968: Quasi-geostrophic frontogenesis. *J. Atmos. Sci.*, **25**, 201–206.

Williams, R. T., L. C. Chou, and C. J. Cornelius, 1981: Effects of condensation and surface motion on the structure of steady-state fronts. *J. Atmos. Sci.*, **38**, 2365–2376.

Woods, V. S., 1983: Rope cloud over land. *Mon. Wea. Rev.*, **111**, 602–607.

CHAPTER 11

Instabilities

Douglas K. Lilly

11.1. Introduction

The concepts of stability and instability are of broad importance in all of physical science, and are appreciated intuitively by most scientists. Studies in atmospheric dynamics often consider the stability or instability of a flow in equilibrium. The concepts can also be extended to evolving but statistically steady flows, following Lorenz (1963). Suppose two initial states of the global atmosphere that differ only slightly from each other could be identified. The atmosphere is regarded as unstable if the two states diverge in their evolution, so that the difference between them becomes as large as the difference between any two widely separated states of the flow evolution, for example, the states on the same date in different years. The atmosphere is believed to be always unstable in this sense, a belief based on the evident instability of virtually all numerical prediction or simulation models and on the rapid divergence of near-analog pairs, which occasionally occur.

Stability and predictability are very closely related. The assumption is that, by discerning the existence of periodic, stable flow, one could learn to predict its evolution, whereas an unstable flow would be relatively unpredictable, because the initial state is imperfectly known. Since the atmosphere is evidently unstable, it should also be regarded as ultimately unpredictable. Nevertheless, some aspects of a generally unstable flow field, or some regions in space and time, may be locally stable and predictable, e.g., the tides or local sea-breeze regimes.

Three kinds of instability can grow in a horizontally uniform flow with vertical gradients of wind and/or buoyancy: (1) pure buoyant instability; (2) the inertial-buoyancy type called symmetric instability; and (3) the shearing type known as Kelvin-Helmholtz instability waves. The first and third occur over scales of tens to thousands of meters and produce much of the small-scale turbulence observed in the troposphere. The second, in its moist form, occurs over scales of tens to hundreds of kilometers and may be the immediate cause of many of the rain and snow bands usually associated with warm and occluded fronts.

11.2. Buoyant Instability

To review the analysis of buoyant or convective instability, first assume an incompressible fluid, like water, in which density is conserved in parcel motion. Assume also that the fluid is in a stationary equilibrium state and density varies linearly with height; i.e., $\bar{\rho} = \rho_0 + (\partial\bar{\rho}/\partial z)z$. Consider a small horizontal tube of fluid of density ρ_0 at the level $z = 0$. At this and all other levels, the fluid is in hydrostatic equilibrium; i.e., $\partial\bar{p}/\partial z = -\bar{\rho}g$. If this tube is displaced upward to some level z, the environmental pressure is unaffected by this displacement at least to a first approximation. Thus the pressure gradient across the tube in its new environment is $\partial\bar{p}/\partial z = -g[\rho_0 + (z\partial\bar{\rho}/\partial z)]$. The tube must then accelerate according to the vertical equation of motion.

$$\rho_0 \frac{dw}{dt} = -\frac{\partial\bar{p}}{\partial z} - g\rho_0 = g\frac{\partial\bar{\rho}}{\partial z}z \ . \tag{11.1}$$

Since $w = dz/dt$, this can be written in the form of an ordinary second-order differential equation with constant coefficients; i.e.,

$$\frac{d^2 z}{dt^2} = \left(\frac{g}{\rho_0}\frac{\partial\bar{\rho}}{\partial z}\right) z \ . \tag{11.2}$$

If the mean density gradient $\partial\bar{\rho}/\partial z$ is negative, then the tube tends to return toward its original level, showing that the original equilibrium is stable. If the initial displacement is to a level $z = h$, the solution to (11.2) is

$$z = h \cos Nt \ , \tag{11.3}$$

where $N^2 = -(g/\rho_0)(\partial\bar{\rho}/\partial z)$. If $\partial\bar{\rho}/\partial z > 0$, the tube accelerates upward, indicating an unstable equilibrium. For this case the solution is

$$z = h \cosh \nu t + h e^{\nu t} \text{ for } \nu t >> 1 \ , \tag{11.4}$$

where $\nu^2 = (g/\rho_o)(\partial\bar{\rho}/\partial z) = -N^2$. Here the tube accelerates away from its initial position and never returns.

The above is a classic example of fluid instability, which must be modified in a number of ways to apply to a real fluid, especially the atmosphere. In a compressible but unsaturated atmosphere the appropriate conserved buoyancy variable is potential temperature θ, rather than density, so that $N = [(g/\theta_0)(\partial\bar{\theta}/\partial z)]^{1/2}$, where θ_0 is a reference value of potential temperature. N is often referred to as the Brunt-Väisällä frequency. When condensation occurs, the release of latent heat changes, and in many cases reverses the sign of N^2, producing conditional instability.

The real motion of an unstable buoyant tube, or two-dimensional thermal, is complicated by friction and diffusion. These effects slow down and smooth out the thermal and also cause it to expand and become diluted by entraining part of its environment (see, e.g., Turner, 1973, Ch. 7). In addition, the assumption that the environmental pressure is unaffected by the thermal's buoyancy is not quite correct locally, because a pressure gradient must develop in the near environment to allow it to make way for the moving thermal. Part of the buoyant energy is then used to accelerate the environment. Nevertheless, the original analysis, corrected for compressibility and moisture effects, makes accurate stability predictions.

11.3. Inertial and Symmetric Instabilities

Fluid motions are stabilized by rotation, much as they are by a stable buoyancy gradient, because the pressure field that develops in cyclostrophic or geostrophic equilibrium tends to return a displaced parcel to its original position, as does the hydrostatic pressure field in the example of buoyant instability, when $N^2 > 0$. Consider, for example, a mean zonal flow balanced geostrophically with a north-south pressure gradient $\partial \overline{p} / \partial y = -\overline{\rho} f \overline{u}$, and an east-west-oriented tube of fluid capable of moving laterally through this environment. The quantity $M = u - fy$ is defined as a local approximation to absolute angular momentum, and it is conserved in the tangent plane equations of motion, since

$$\frac{dM}{dt} = \frac{d}{dt}(u - fy) = \frac{du}{dt} - fv = 0 \,, \qquad (11.5)$$

where friction is neglected and f is assumed locally constant. Thus a tube of fluid displaced northward increases its zonal velocity by f times its displacement.

If the environmental geostrophic pressure gradient does not similarly increase northward, the tube will be supergeostrophic and tend to return southward, as shown by the northward component equation of motion

$$\frac{dv}{dt} = f(u_g - u) \,. \qquad (11.6)$$

Thus, an environmental flow with a mean northward shear, $\partial \overline{u} / \partial y$, which is algebraically smaller than f, is stable to horizontal displacements. This condition can be generalized to two dimensions and defines the requirement for inertial stability, in the Northern Hemisphere, as

$$f + \frac{\partial v}{\partial x} - \frac{\partial u}{\partial y} = \varsigma_{\text{abs}} > 0 \,, \qquad (11.7)$$

where ς_{abs} is the vertical component of absolute vorticity.

In the case of a hydrostatically and geostrophically balanced mean flow with vertical (and possibly horizontal) shear, the effects of buoyancy and rotation combine and lead to a new kind of buoyant-inertial instability. We assume that the mean flow and potential temperature field satisfy the thermal wind equation; i.e.,

$$f \frac{\partial \overline{u}}{\partial z} = -\frac{g}{\theta_0} \frac{\partial \overline{\theta}}{\partial y}. \qquad (11.8)$$

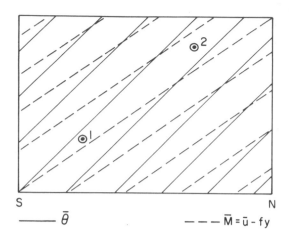

Figure 11.1. Schematic $y-z$ profile of potential temperature and absolute angular momentum under conditions of symmetric instability. Parcels moving along or parallel to the line between points 1 and 2 are unstable.

In Fig. 11.1, isopleths of hypothesized fields of mean potential temperature $\bar{\theta}$ and mean local eastward absolute angular momentum $\overline{M} = \bar{u} - fy$ are plotted. Both quantities are assumed to increase with height and decrease northward, which would be the normal situation in the Northern Hemisphere. If a fluid tube at position 1 is moved to 2, it will have a higher value of $\bar{\theta}$ and lower $\overline{M}$ than its environment. The tube is then both positively buoyant and sub-geostrophic in wind speed, so according to the previous analyses will accelerate upward and northward. This is true for any displacement trajectory lying within the acute angle formed by the intersections of the isolines of $\overline{M}$ and $\bar{\theta}$. Thus an instability can exist any time the $\bar{\theta}$ isolines slope upward more steeply than do those of $\overline{M}$, so that

$$\left.\frac{\partial z}{\partial y}\right|_{\bar{\theta}} > \left.\frac{\partial z}{\partial y}\right|_{\overline{M}}$$

allows instability, or

$$\left.\frac{\partial z}{\partial y}\right|_{\bar{\theta}} < \left.\frac{\partial z}{\partial y}\right|_{\overline{M}} \tag{11.9}$$

assures stability.

The second expression in (11.9) is evaluated by applying first the chain-law identities: $\partial z / \partial y\,|_{\bar{\theta}} = -\frac{\partial\bar{\theta}/\partial y}{\partial\bar{\theta}/\partial z}$, $\partial z/\partial y\,|_{\overline{M}} = -\frac{\partial\overline{M}/\partial y}{\partial\overline{M}/\partial z}$. Then with the aid of (11.8) the stability condition becomes

$$\frac{N^2}{(\partial\bar{u}/\partial z)^2}\left(1 - \frac{\partial\bar{u}/\partial y}{f}\right) > 1. \tag{11.10}$$

The first ratio is the Richardson number, so that upon neglecting lateral shear, the stability criterion is that the Richardson number be greater than unity. If the designations of the isopleths in Fig. 11.1 were reversed, so that the slope of the $\overline{M}$ lines was greater than that of the $\bar{\theta}$ lines, (11.10) would hold and no displacement direction would lead to instability.

Consideration of vertical and northward motion equations shows that for typical atmospheric stratifications, the displacement direction favored for fastest growth is nearly along the $\overline{\theta}$ isolines. The horizontal scale of Fig. 11.1 is much greater than the vertical scale, so that motions along an isentrope are nearly horizontal. The acceleration rate in the y-direction is that given by (11.6), which is of order $f^2 y$, since the conservation of M requires that a y-displacement produce a u-deviation of $-fy$. Thus the stability analysis predicts rather gently accelerated motions along tilted surfaces, with slopes similar to those of fronts and time scales of order f^{-1}, i.e., a few hours. This instability was initially investigated as a perturbation on a baroclinic circular vortex. Since, in that context, its motion fields are zonally symmetric, it was so designated in contrast to the azimuthally modulated classic baroclinic instability.

Inequality (11.10) normally holds in the atmosphere when N is the dry Brunt-Väisällä frequency. The potential importance of static stability became apparent upon recognition by Bennetts and Hoskins (1979) and Emanuel (1983a,b,c) that the conditionally destabilizing effects of condensation could be applied to convert the symmetric instability problem to that of conditional symmetric instability. As with conditional buoyant instability, the latent heat effects greatly complicate the analytic stability problem and it is helpful to use graphic aids, such as a thermodynamic diagram. Now, however, we collect the data for such a diagram along constant $\overline{M}$ surfaces. If a parcel moves upward along an $\overline{M}$ surface, becomes saturated, and then finds itself warmer than its environment, the conditions for symmetric instability have been met. Emanuel (1983b) showed how this technique could be used to deduce instability from synoptic data.

Figure 11.2 shows a west-east cross-sectional analysis of $\overline{M}$ and $\overline{\theta}_e$. $\overline{M} = \overline{v} + fx$ in this case, because the mean lateral temperature gradient is east-west, and x is chosen to be zero at the west edge of the cross section. Figure 11.3 shows a normal vertical sounding from Oklahoma City (OKC). The dashed lines are moist adiabats and are shallower than the actual sounding, indicating that the sounding is unconditionally stable at nearly all levels, though saturated near 750 mb. Figure 11.4 is a sounding constructed by plotting the temperature and dewpoint along the $\overline{M} = 50$ m s^{-1} surface,

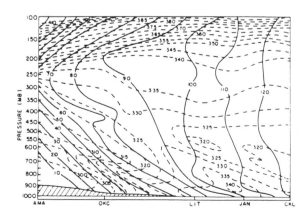

Figure 11.2. Cross section from Amarillo, Tex., to Centreville, Ala., at 0000 GMT, 3 December 1982. Solid lines denote M (m s^{-1}); dashed lines are θ_e (K). (From Emanuel, 1983b.)

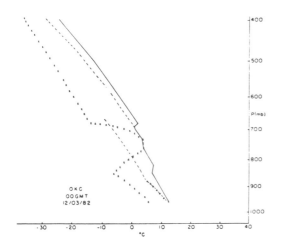

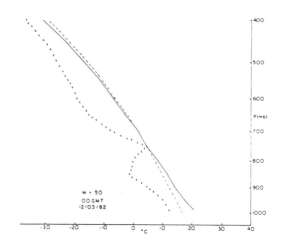

Figure 11.3. Oklahoma City sounding at 0000 GMT, 3 December 1982. Solid line shows temperature; crosses show dewpoint. Dashed lines are pseudo-moist adiabatic; dashed lines with circles are dry adiabats. (From Emanuel, 1983b.)

Figure 11.4. Sounding along $M = 50$ m s^{-1} surface constructed from the cross section in Fig. 11.2. (From Emanuel, 1983b.)

which crosses OKC at about the 750 mb level. Here the sounding is conditionally unstable or neutral at nearly all levels, indicating the likelihood of conditional symmetric instability. Satellite imagery showed cloud and precipitation bands in a generally stratiform cloud mass in the OKC area at this time. A more complete analysis of a precipitation band in New England has been carried out by Wolfsberg *et al.* (1986), and provides strong evidence for the importance of conditional symmetric instability.

11.4. Kelvin-Helmholtz Instability Waves

Laboratory experiments and observations of natural fluids, such as flow in a pipe or channel, supported by theoretical analysis, indicate that the lateral shear of parallel flows is a potent source of nonhydrostatic, nongeostrophic instability, leading to turbulence. The normal vertical shear of the horizontal wind is thus a source of flow instability, though it is generally not strong enough to overcome static stability. In places where the shear is unusually strong or the stable thermal stratification weak, the instability is realized and produces a disturbance form known as Kelvin-Helmholtz (K-H) waves. These disturbances have been observed visually as they affect cloud formations, through radar imagery, and in laboratory experiments arranged to produce them. As suggested by their designation as "waves," they commonly include several or many elements spaced periodically. In their simplest form they move approximately with the mean flow at the level of their greatest amplitude. Since they become unstable to smaller scale perturbations, their lifetime is ordinarily only a few minutes, after which they degenerate into less organized turbulence.

Apparently there is no adequate conceptual model of shearing turbulence based on parcel displacement analysis, nor is there any simple mechanistic model that predicts the correct stability criterion for K-H waves. Figure 11.5 illustrates the concept of instability produced at the shearing interface between two layers of uniformly moving fluid. The interface is regarded as a thin, initially horizontal layer of strong vorticity, i.e., vortex sheet. One may think of the vortex sheet as being composed of a large number of small discrete vortices, all rotating in the same direction. Their up-down motion components cancel each other out, but their left-right motion components reinforce each other and produce the net velocity difference from top to bottom of the sheet. If the sheet is perturbed sinusoidally, as shown, the vortex elements interact to produce velocity perturbations along the sheet. The vortex elements near points A and C produce leftward motion at the wave crest B, and similar elements produce rightward motion at the wave trough D. These motions advect the vortex sheet itself, and tend to converge and thicken it at points like A, and diverge and thin at points like C. The increased counterclockwise vorticity at A then tends to lift the crest and lower the trough, thus amplifying the wave, and the motion perturbations are further amplified.

If the sheet is infinitesimal in thickness, then all wavelengths are theoretically unstable, the smallest waves growing fastest. If the sheet has a finite thickness, however, as any real shear layer must, then it seems plausible that the fastest growing wave would be of a wavelength not shorter than the depth of the shear layer, since the motions induced by the mechanism described become weak if the displacement is less than the layer's thickness. Finally, if the lower layer has greater density or smaller potential temperature than the upper layer, the displacements are suppressed by buoyancy-restoring forces. These forces are strongest for large displacements and, therefore, large wavelengths. Again, an optimal wavelength related to the shear layer thickness is suggested. Detailed theoretical analyses by Miles (1961) and Howard (1961) show that instability cannot occur unless the Richardson number, as previously defined, is less than $\frac{1}{4}$.

A simple nonmechanistic energy argument due to Chandrasekhar (1961) leads to the same result. If two parcels of fluid of the same mass, initially separated by a distance δz, are interchanged in position, the potential en-

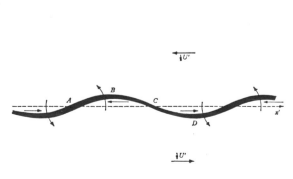

Figure 11.5. Growth of a sinusoidal disturbance of a vortex sheet with positive vorticity normal to the paper. The local strength of the sheet is represented by the thickness of the sheet. The arrows indicate the directions of self-induced movement in the sheet and show the accumulation of vorticity at points like A, and the general rotation about points like A, which together lead to exponential growth of the disturbance. (From Batchelor, 1967, pp. 515–516.)

ergy (per unit volume) is increased by an amount $g \; \delta\rho \; \delta z$, where $\delta\rho$ is the difference in density of the parcels, and the lower parcel is assumed to have the higher density. A linear density gradient is assumed, so that $\delta\rho = -(\partial\overline{\rho}/\partial z)\delta z$. The only source of that energy is the mean shear $\partial\overline{u}/\partial z$. The lower and upper parcels are assumed to have initial velocities U and $U + \delta U$, respectively, where $\delta U = (\partial\overline{u}/\partial z)\delta z$. The maximum reduction of kinetic energy occurs if the two parcels are assumed to have the same velocity $(U + \delta U)/2$ after exchange. The net change in kinetic energy, again per unit volume, is then

$$\frac{1}{2}\overline{\rho}\{U^2 + (U + \delta U)^2 - 2\left[(U + \delta U)/2\right]^2\} = \frac{1}{4}\overline{\rho}(\delta U)^2.$$

Energy is not released by the exchange, and therefore the flow must be stable, provided that

$$\overline{\rho}\frac{(\delta U)^2}{4} < g \; \delta\rho \; \delta z.$$

From the previous relationship of mean flow, this implies that

$$\frac{-(g/\overline{\rho})(\partial\overline{\rho}/\partial z)}{(\partial\overline{u}/\partial z)^2} > \frac{1}{4} \tag{11.11}$$

assures stability.

A simple physical realization of K-H waves was obtained by Thorpe (1971). A long, covered trough is filled part way with saltwater and the rest of the way with freshwater; care is taken to minimize mixing during the filling process. After a few hours a smooth, diffusive interface develops between the saltwater and freshwater, but they do not mix strongly because of net static stability of the interface. The trough is then tilted on a central fulcrum. The lower salty fluid, being denser, tends to flow downward and the upper fresh fluid upward, producing a strong shear layer in the middle. Figure 11.6 shows the development of instability waves, made visible by

Figure 11.6. Kelvin-Helmholtz instability of stratified shear flow. A horizontal rectangular tube is filled with water above colored brine. The fluids are allowed to diffuse for about an hour, after which the tube is quickly tilted 6°. The brine accelerates down the slope, and the water accelerates up the slope. Sinusoidal instability of the interface occurs after a few seconds and has here grown nonlinearly into overturning waves. (From Thorpe, 1971.)

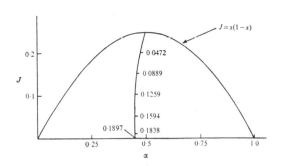

Figure 11.7. Stability boundary and curve of maximum growth rate for hyperbolic tangent profiles of velocity and density. Abscissa is wavenumber, made dimensionless by the half-width of the shear layer; ordinate is Richardson number at the origin. The region below the curve is unstable; maximum growth rates are shown. Growth rates are scaled by the maximum shear amplitude. (Hazel, 1972.)

dyeing the lower fluid. Initially looking like sinusoidal displacements, they quickly curl over and break, leading to turbulent mixing. Similar events have been observed in shear layers in the ocean and atmosphere. They occur much more often in the atmosphere than they can be visually observed, since cloud layers are usually not present at the critical location.

A rather large amount of theoretical research has been addressed to the problem of shearing instability and K-H waves. If the mean flow profile is known, it is possible to predict theoretically the wavelength of maximum instability and, by numerical simulation, the detailed evolution, up to the time full turbulent breakdown occurs. Figure 11.7 shows a typical stability curve, relating the wavelength of marginal instability to the minimum Richardson number. The most unstable wavelength is scaled by a nominal depth of the shear layer, and is about 6 times that depth.

In applying the theory to real atmospheric events, it is difficult to test or to utilize for prediction the critical Richardson number criterion, because that criterion need not be exceeded over a large region or period. The Richardson number is itself a notoriously unstable statistic, frequently ranging from 0 to $\pm\infty$. The prediction of the optimal wavelength is often satisfactory, but waves apparently similar to K-H waves are observed with considerably longer wavelengths than predicted.

This result has been rationalized in several ways. First it turns out that there is another set of unstable modes present in a fluid if the lower boundary is not too far away (compared with the wavelength). These modes enhance their instability by reflection of their amplitude off the boundary, which in the optimal case arrives in phase to reinforce the original disturbance. A second source of longer waves is subharmonic resonance. Waves with an original length of, say, L tend to interfere as they grow to large amplitude, so that they become transformed into a train of waves of 2L, 4L, and possibly greater lengths. Figure 11.8 shows a laboratory example of this process. It is also possible for two wave trains of different wavelength and period to interact as they pass through each other and produce new waves with wavenumbers and frequencies that are sums and differences of the original. This process, called resonant interaction, tends to mix and confuse the originally separate K-H and stable gravity waves, and is apparently responsible for the rather chaotic array of internal waves observed in the ocean and, probably, the upper atmosphere.

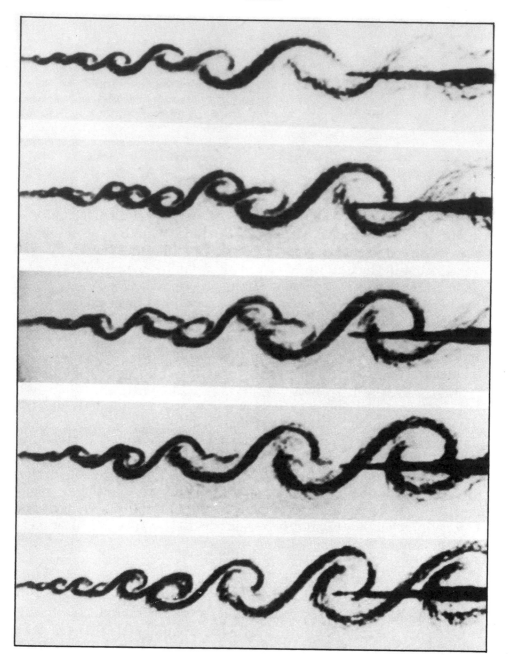

Figure 11.8. Subharmonic resonance. The small vortex rolls in the left part of the top frame are seen to combine into larger rolls in the later frames.

    The intensity of a K-H wave train and the turbulence produced from it is directly proportional to the magnitude of the velocity difference from which it derives its energy. Although K-H waves are ubiquitous in the normal nocturnal boundary layer, the total velocity differences of a few meters per second will not produce dangerous aircraft responses. The most intense waves are commonly found near the upper tropospheric jet stream or in

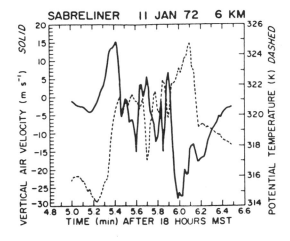

Figure 11.9. Profiles of vertical velocity (solid) and potential temperature (dashed) vs. time during a westward flight through an intense mountain wave over the Colorado Rockies. Three or four K-H instability waves are apparent between the major downdraft and updraft of the mountain wave. (From Lilly, 1978.)

association with strong mountain waves, both of which can produce locally strong shears on a scale suitable for generating instabilities. For example, during a flight through an extreme mountain wave near Boulder, Colo., Lilly (1978) observed a shearing instability zone at the 500 mb level produced by a total velocity difference of about 50 m s^{-1} over a depth of less than 1 km. The instability waves were several kilometers long, and the turbulence reached levels usually seen only in intense thunderstorms. Figure 11.9 shows the time record of vertical velocity and potential temperature during the aircraft's passage through this area.

11.5. The Effects of Flow Instabilities on Predictability

As indicated in Sec. 11.1, flow instabilities directly limit and eventually destroy predictability. This is apparently true even when the instability occurs on a scale much smaller than that of the prediction. The reason for this is the nonlinearity of the atmospheric equations of motion. The nature of the process is explained by Lorenz (1969) and Leith and Kraichnan (1972), using turbulence theory concepts. Their results indicate that even if the larger scales of motion (as defined by components of a Fourier analysis) were observed perfectly and the differential equations of their evolution perfectly understood and solved, the unknown motions on smaller scales would infect the larger scales with error and eventually destroy the prediction. The rate at which this infection proceeds depends on the statistical structure of the small-scale flow field, especially its energy spectrum.

The observed kinetic energy spectrum of the atmosphere at wavelengths between about 1000 and 4000 km approximately follows a -3 slope; that is, energy decreases according to the inverse cube of the wavenumber or increases according to the cube of the wavelength. If this rapid drop-off in energy continued to small scales of a few tens of kilometers and less, both the larger and smaller atmospheric scales would probably be predictable over a considerably longer time. In fact, the atmospheric kinetic energy spectrum for wavelengths less than 1000 km drops off at a rate more like $-\frac{5}{3}$ to -2, as discussed by Nastrom and Gage (1983) and Lilly and Petersen (1983).

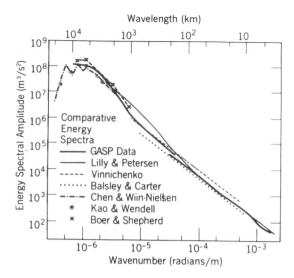

Figure 11.10. Several horizontal spectra of horizontal kinetic energy. (Adapted from Lilly and Petersen, 1983 by Nastrom and Gage, 1985.) The dashed and dotted curves have a $-\frac{5}{3}$ slope.

Figure 11.10, from Nastrom and Gage (1985) shows several estimates of the spectrum, for wavelengths greater than 2 km. This increased small-scale energy allows small-scale instability to contaminate the smaller synoptic scales rather quickly, probably in a day or two, and helps limit prediction accuracy. The reason for the relatively shallow spectral slope in the mesoscale domain is not fully understood, but theoretical explanations have been proposed by Gage (1979), Van Zandt (1982), and Lilly (1983).

REFERENCES

Batchelor, G. K., 1967: *An Introduction to Fluid Mechanics.* Cambridge University Press, London, New York, 615 pp.

Bennetts, D. A., and B. J. Hoskins, 1979: Conditional symmetric instability—A possible explanation for frontal rainbands. *Quart. J. Roy. Meteor. Soc.*, **105**, 945–962.

Chandresekhar, S., 1961: *Hydrodynamic and Hydromagnetic Stability.* Oxford University Press, London, 653 pp.

Emanuel, K. A., 1983a: The Lagrangian parcel dynamics of moist symmetric instability. *J. Atmos. Sci.*, **40**, 2368–2376.

Emanuel, K. A., 1983b: On assessing local conditional symmetric instability from atmospheric sounding. *Mon. Wea. Rev.*, **111**, 2016-2033.

Emanuel, K. A., 1983c: Conditional symmetric instability: A theory for rainbands within extratropical cyclones. In *Mesoscale Meteorology—Theories, Ob-* servations, and Models. D.K. Lilly and T. Gal-Chen (Eds.), D. Reidel, Dordrecht, Boston, Lancaster, 231–246.

Gage, K. S., 1979: Evidence for a $k^{-5/3}$ law inertial range in mesoscale two-dimensional turbulence. *J. Atmos. Sci.*, **36**, 1950–1954.

Hazel, P., 1972: Numerical studies of the stability of inviscid stratified shear flows. *J. Fluid Mech.*, **51**, 39–61.

Howard, L. N., 1961: Note on a paper of John W. Miles. *J. Fluid Mech.*, **10**, 509–512.

Leith, C. E., and R. H. Kraichnan, 1972: Predictability of turbulent flows. *J. Atmos. Sci.*, **29**, 1041–1058.

Lilly, D. K., 1978: A severe downslope wind and aircraft turbulence event induced by a mountain wave. *J. Atmos. Sci.*, **35**, 59–77.

Lilly, D. K., 1983: Stratified turbulence and the mesoscale variability of the atmosphere. *J. Atmos. Sci.*, **40**, 749–761.

Lilly, D. K., and E. L. Petersen, 1983: Aircraft measurements of atmospheric kinetic energy spectra. *Tellus*, **35A**, 379–382.

Lorenz, E. N., 1963: Deterministic nonperiodic flow. *J. Atmos. Sci.*, **20**, 130–141.

Lorenz, E. N., 1969: The predictability of a flow which possesses many scales of motion. *Tellus*, **21**, 289–307.

Miles, J. W., 1961: On the stability of heterogeneous shear flows. *J. Fluid Mech.*, **10**, 496–508.

Nastrom, G. D., and K. S. Gage, 1983: A first look at wavenumber spectra from GASP data. *Tellus*, **35A**, 383–388.

Nastrom, G. D., and K. S. Gage, 1985: Climatology of atmospheric wavenumber spectra observed by commercial aircraft. *J. Atmos. Sci.*, **42**, 950–960.

Thorpe, S. A., 1971: Experiments on the instability of stratified shear flows: Miscible fluids. *J. Fluid Mech.*, **46**, 299–319.

Turner, J. S., 1973: *Buoyancy Effects in Fluids.* Cambridge University Press, Cambridge, 368 pp.

Van Zandt, T. E., 1982: A universal spectrum of buoyancy waves in the atmosphere. *Geophys. Res. Lett.*, **9**, 575–578.

Wolfsberg, D. G., K. A. Emanuel, and R. E. Passarelli, 1986: Band formation in a New England winter storm. *Mon. Wea. Rev.*, **114**, 1552–1569.

CHAPTER 12

Gravity Waves

William H. Hooke

12.1. Introduction

Gravity waves are fundamentally important to the mesoscale dynamics of the Earth's atmosphere. Linearized gravity-wave theory provides an important conceptual framework for mesoanalysts and forecasters in interpreting observations and the results of numerical weather simulations and predictions. Gravity waves transport significant amounts of energy and momentum. They trigger convective storms and spawn clear-air turbulence. They interact strongly with one another, and, in the process, drive major spectral energy transfers in the atmosphere.

12.1.1. Characterization

The onset of convective instability in the atmosphere triggers a wide range of atmospheric responses, all too often of awesome destructive power, that command the attention and respect of anything and everyone in their path. The 1984 tornado outbreak in the Carolinas and the flash floods in Tulsa, Okla., are but two of many recent examples. Naturally enough, such events and the forecast problems they pose are a primary focus of meteorological forecasting.

Under most times and circumstances, however, the atmosphere is stably stratified—in part, precisely because of these atmospheric adjustments. Accordingly, the motions and circulations that can be sustained by the stable atmosphere are of considerable theoretical and practical interest. Fundamental to such circulations are the atmosphere's free oscillations, which occur on virtually all spatial and temporal scales.

On the synoptic and global scales, there are waves with periods in excess of a day that are strongly affected by the Earth's sphericity, rotation, and zonal circulation—the familiar Rossby and planetary waves of conventional meteorology. On time scales a submultiple of a solar or lunar day there are the atmospheric tides, excited either by solar heating, or by gravitational forcing of both solar and lunar origin. Although not so noticeable at temperate latitudes, where Rossby and planetary waves dominate the observations, tides are quite apparent in the hourly records kept in the tropics. On time and space scales less than several hours, and a thousand kilometers or so—the so-called mesoscale (see Ch. 1 for a definition couched in dynamic terms and Ch. 2 for a phenomenological definition) — the Earth can be considered locally plane, only relatively slowly rotating, and essentially incompressible. In this regime, the only preferred direction is the vertical, associated with the local gravitational force and the resulting stable stratification. Important wave motions exist on these scales that are referred to as gravity waves or buoyancy waves. (The text by Gossard and Hooke, 1975, provides a comprehensive review.)

Like all free oscillations, gravity waves can be excited in a number of ways, e.g., by atmospheric instabilities or by external forcing like that affecting airflow over irregular terrain. The waves so generated are of special interest in their own right. For example, gravity waves arising from atmospheric shear instability, or generated by penetrative convection of cumulus and cumulonimbus clouds, are responsible for much of the clear-air turbulence encountered by jet aircraft at cruising altitudes. Gravity waves generated by the airflow over mountains often produce downslope windstorms that go by a variety of names worldwide (chinook, foehn, bora, etc.). Accordingly, a discussion of the instabilities responsible for such wave generation and a description of mountain waves appear separately in Chs. 11 and 20.

12.1.2. Role of Gravity Waves in the Atmosphere

Gravity waves modulate every atmospheric variable—windspeed, temperature, density, humidity, pressure, chemical composition, particulate concentration, and airflow. In consequence, every sensor—whether *in situ*, ground-based, satellite-mounted, or remote—reveals wave motions in the records. Gravity waves are not isolated events whose occurrence is confined to those few cases in which atmospheric conditions favor observable cloud modulation by the waves; they are virtually ubiquitous.

With appreciation for the pervasive character of gravity waves has come an appreciation for their dynamical importance. The major dynamical role of gravity waves in the atmosphere includes (1) spectral energy transfer, (2) vertical and horizontal transport of momentum and energy from one region of the atmosphere to another, (3) generation of the clear-air turbulence that affects aircraft, and (4) triggering of instabilities that lead to the development of severe weather.

Spectral Energy Transfer

Spectral energy transfer is a subject well known from turbulence theory on the microscale. What is not so well appreciated is the role played by gravity waves in pre-conditioning atmospheric momentum and energy on synoptic scales—providing the initial breakdown in scales required to initiate turbulent spectral energy transfer. For the most part this gravity-wave energy transfer, like its turbulent counterpart, appears to be from larger scales to smaller scales. For numerical models this is a good thing, since precisely this assumption is made in all parameterizations of subgrid scale phenomena. There continue to be suggestions, however, (e.g., Chimonas and

Grant, 1984 a, b) that certain processes in shear instability may work to favor upscale energy transfer. Such findings, if confirmed, would have important consequences both for an understanding of atmospheric predictability on both the meso- and synoptic scales, and for the approach numerical models make to subgrid-scale parameterizations. (These would have to incorporate a kind of wild-card factor not now utilized to allow for random generation of resolvable-scale wave motions.)

Transport of Energy and Momentum

Gravity-wave transport of energy and momentum in the horizontal competes with Rossby wave transport and has received little attention to date, but transport of energy and momentum in the vertical is more evident. The so-called spectral gap in the mesoscale at tropospheric levels may reflect this ability of gravity waves. Waves of short period (less than the Brunt period defined in Sec. 12.2) are evanescent in the vertical; their energy remains trapped at the height of generation. Waves of long period, although free to propagate their energy both vertically and horizontally, propagate their energy very nearly horizontally for other reasons; the residence time of such wave energy in the troposphere is therefore also long. By contrast, gravity waves of intermediate period are free to propagate their energy vertically (apart from subtleties introduced by the background wind and temperature structure, as explained in Secs. 12.3.1–12.3.2). As a result, all else being equal, waves of intermediate period will be the exception, rather than the rule, in the observations.

The gravity-wave energy leaving the troposphere is small in relative terms, because of the troposphere's enormous mass. However, it represents a relatively large input to the stratosphere, and a strong energy input indeed to the rarefied mesosphere and upper atmosphere above. Those who focus on the troposphere are accustomed to think of the dynamical energy of the atmosphere as being a negligible fraction of its thermal energy, as generated by radiant heat inputs. By contrast, at thermospheric heights (above 100 km), up to 25% of the energy input is dynamical in origin, associated with wave energy transport from the lower atmosphere. (This energy transport can be even more dramatic in other astro-geophysical contexts. In the solar atmosphere, for example, waves generated in the solar photosphere, where T $\sim$ 5000 K, heat the solar corona above to temperatures in excess of 10^6K).

Generation of Clear-Air Turbulence

Gravity waves have practical importance in their own right. Gravity waves are the predominant cause of the clear-air turbulence (CAT) encountered by high-altitude aircraft flying in nominally stable air. Some of this CAT is generated locally, by shear instability, as explained in Ch. 11, but some is generated from below, by convective storms, which impulsively force oscillations of the stable environment as they develop.

Triggering of Instabilities

Finally, waves provide "triggers" for convective activity. They do this, for example, by modulating the atmospheric stability of elevated inversions on which they preferentially propagate. Convective instabilities are then able to first "punch through" weakened areas of such capping inversions, triggering convective development, that subsequently strongly modifies the ambient environment. Studies of such phenomena date back to Tepper (1954); Uccellini (1973) and Stobie *et al.*(1983) documented several cases of severe weather modulated by gravity waves.

12.1.3. *Additional Features*

Gravity waves are important to us for other than dynamical reasons: (1) They are by definition easy to forecast; (2) in some sense, they are the simplest, most fundamental motions that exist on the mesoscale; and (3) they are of enormous value as a conceptual aid in our thinking about mesoscale atmospheric dynamics.

Gravity waves motions represent the simplest of all forecast problems. Like Rossby waves or acoustic waves, they satisfy a dispersion equation. Once one has measured the temperature of the atmosphere, one can "forecast" that acoustic waves will propagate with the speed of sound. Rossby's development of the dispersion equation for the waves that bear his name was a major step forward in atmospheric forecasting on the synoptic scale. In a similar way, we can forecast the motion of gravity waves, working in part from direct measurements of their propagation, and in part from our knowledge of atmospheric temperature and wind structure. Even in complicated atmospheres, it is a straightforward procedure to compute the wave modes and to calculate the response of the atmosphere to the forcing.

Gravity waves are solutions to the linearized equation of motion on the mesoscale. They are thus the very simplest motions we can study on the mesoscale. They are the fundamental motions; they are building blocks from which a wide variety of circulations, some quite complicated, can be constructed through superposition.

Like their synoptic-scale counterparts, the linearized Rossby waves, gravity waves are so idealized a theoretical construct that their simple physics by itself can rarely be used to forecast in everyday practice. The complicating factors at work in the real atmosphere are overwhelming in number and impact. Thus, on the mesoscale as well as the synoptic, forecasters rely on numerical models that include these complications either explicitly or in parameterized form. However, once the numerical forecast is in hand, an understanding of gravity wave physics is of enormous value in evaluating and interpreting the model results. Looking at atmospheric dynamics on the mesoscale will help in developing a framework into which more specialized material can fit for purposes of forecasting.

12.2. Basic Gravity-Wave Physics

12.2.1. Convective Stability

Introductory meteorological texts usually consider stability in terms of the oscillation of a parcel of air that has been displaced vertically from its equilibrium position. Such a parcel is shown schematically in Fig. 12.1. At its initial equilibrium height z, it has the pressure, temperature, and density corresponding to its surroundings. Suppose now that the parcel is displaced vertically upward a small distance from its equilibrium position.

If the displacement were very rapid, it would occur before the parcel would have time to achieve pressure equilibrium with its surroundings. As pressure equilibrium was once again restored, this information would be communicated to the surrounding medium by the generation of a spectrum of sound waves, propagating at speeds the order of the thermal velocities of the molecules involved. Such a restoration is a familiar event following the sudden heating of an air column by lightning discharge, for example.

At the other extreme, it is possible to displace the imaginary parcel of air so slowly that not only is pressure equilibrium maintained, but also thermal conductivity maintains thermal equilibrium between the parcel and the surrounding medium (the equation of state in the form of the ideal gas law also requires a density equilibrium).

However, there is a broad intermediate range of parcel displacements, including most of the cases of practical interest, for which pressure equilibrium is maintained, but thermal equilibrium is not established except in a layer of negligible thickness around the periphery of the parcel. (Such displacements are termed adiabatic.)

Under conditions of unstable stratification, the rising parcel would have a higher temperature than its surroundings, despite the cooling resulting from work done in expanding. Pressure equilibrium, together with the ideal gas law, would imply that the parcel is less dense than the surrounding medium, so that it would accelerate from its initial position, setting into motion the chain of events normally associated with the generation of turbulence and/or convective storms. This case (and variations associated with the effects of latent heat release from water vapor condensation or with the entrainment of surrounding air) receives a great deal of attention because of implications for severe weather development.

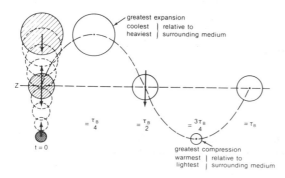

Figure 12.1. Parcel motion in stably stratified air.

By contrast, under the statically stable atmospheric conditions of interest here, the parcel has a lower temperature than its surroundings. The parcel is denser than the surrounding medium. Its rate of ascent does not accelerate, but slows, until finally the parcel begins to sink. When it returns to its initial height, it is not at rest, but moving (with whatever kinetic energy had been imparted to it initially). Thus it overshoots. It continues to remain in pressure equilibrium, but compresses adiabatically, and becomes increasingly warmer and less dense than its surroundings. Its rate of descent slows; the parcel finally stops, then begins to rise once more. This oscillation would continue indefinitely but for the effect of dissipative forces, which have been ignored in this simplified discussion.

Note that the oscillation occurs with a very clearly defined frequency N given by

$$N^2 = \frac{g}{\Theta_0} \frac{d\Theta_0}{dz} , \qquad (12.1)$$

where g is the acceleration of gravity and Θ_0 is the potential temperature of the undisturbed atmosphere. N is the so-called Brunt-Väisälä frequency, named after its discoverers. The more stable the atmosphere (the larger $d\Theta_0/dz$), the greater this frequency. For an isothermal atmosphere, the Brunt period

$$\tau_B \equiv \frac{2\pi}{N} \qquad (12.2)$$

is roughly 5 min. For the troposphere as a whole (in which temperature is decreasing with increasing height) $\tau_B \sim 8$ min.

Just as the atmosphere communicates pressure disequilibrium at any point through the emission of acoustic waves, it signals the density disequilibrium through the generation of gravity waves. Unlike pressure disequilibria, however, which are communicated through the medium with the speed of sound, density disequilibria are typically communicated at a much smaller speed (because the only means of communicating this information lies in the relatively slow sinking and rising of the air parcels involved).

12.2.2. Energy and Phase Propagation

Figure 12.2 shows schematically the propagation, in a shearless and otherwise isothermal atmosphere, of phase and energy by a gravity wave "packet" (a disturbed region of the atmosphere of limited dimensions, the kind that might be generated by an impulsive atmospheric source located somewhere below and to the left of the diagram proper). At $t = t_1$, the packet is located in the lower left-hand corner of the diagram. Within the packet itself, the wave-associated motions exhibit the amplitude and phase pattern shown. Note that the effect of the wave generation is to produce shears. At some later time $t_2 > t_1$, the wave packet (region of disturbance) has moved upward and to the right, away from the source region, in the direction of the arrow labeled "wave energy propagation." (At $t_3 > t_2$, the

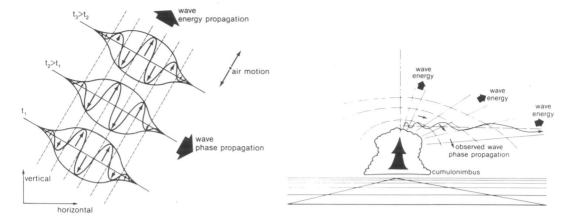

Figure 12.2. Gravity-wave energy and phase propagation.

Figure 12.3. Gravity-wave generation by penetrative convection.

wave energy and packet propagation have proceeded farther still.) Within the packet itself, however, the phase pattern of the velocity disturbance has moved downward and to the right, in the direction of the arrow labeled "wave phase propagation."

This nature of gravity-wave phase and energy or packet propagation is important in wave generation by convective systems (Fig. 12.3). A thermal, or cumulonimbus, acts as a point source of gravity waves when it encounters stable overlying air (in the thermal case, a boundary layer inversion; in the cumulonimbus case, the stratosphere itself). Short-period gravity waves are observed at locations nearly directly above the point source; longer period waves are observed at greater distances. Waves generated in this circumstance are dispersive, and the packet continues to change in appearance as it propagates away from the source region, the larger waves propagating rapidly and shorter waves propagating slowly.

As a result, the wave frequency is directly tied to the angle of wave propagation relative to the vertical. Recall that a parcel of air moving vertically experiences a vertical restoring force proportional to the acceleration of gravity, and thus oscillates with the Brunt-Väisälä frequency. A parcel of air moving the same distance, but off-vertically, experiences a lesser restoring force in its plane of motion, proportional to the cosine of the angle its motion makes with the vertical; hence its frequency of oscillation varies as the square root of this cosine.

12.2.3. Wave Generation

Gravity waves are generated in one of two ways: by shear instability, or by external forcing. In the case of wave generation by shear instability, energy of shear flow is converted into wave energy contained in (initially small) fluctuations in the wave field. Although statically stable, shear-flow wind fields may have sufficient kinetic energy differentials across the shear zone itself, so that air parcels, once displaced vertically, can extract some of this available kinetic energy and accelerate away from their equilibrium position. Dividing stable and unstable shear-flow regimes is the so-called Richardson number criterion, which states that

$$\mathrm{Ri} \equiv \frac{\frac{g}{\Theta}\frac{d\Theta}{dz}}{\left(\frac{dU}{dz}\right)^2} < \frac{1}{4} \tag{12.3}$$

is a prerequisite for unstable flow and turbulence. This criterion can be derived through formal mathematical arguments (Miles, 1961; Howard, 1961) or through parcel arguments analogous to those advanced for static instability (Hines, 1971).

External forcing can be either thermal or dynamical, or some combination of the two. For example, airflow over mountainous terrain forces vertical motion of the parcel involved. Airflow over differentially heated areas (land-sea boundaries, for example) induces similar circulation patterns. Similarly, airflow around or over cumulus clouds or thermal plumes launches gravity waves in the surrounding stably stratified air. Wave-wave interaction is another important wave source in some regions of the spectrum.

12.2.4. Unified Wave Theory

It is important to understand that acoustic waves, gravity waves, and Rossby waves represent idealized motions—limiting forms of the actual waves observed in the Earth's atmosphere, which must necessarily experience simultaneously the effects of atmospheric compressibility, gravity and density stratification, and the Earth's rotation, although in varying degrees.

It is possible by starting with the Navier-Stokes equations in their full generality, to develop a generalized dispersion equation that encompasses all these wave motions, plus several more not considered here (the atmospheric tides, for example, as well as thermal and viscous waves). It is then possible to see the precise conditions that must be satisfied in order to recover any of the above limiting forms. It also becomes evident in such an analysis that each of the wave types contains certains aspects of the others. For example, in the Rossby wave dispersion equation, the prominent anisotropy is that imposed by the Earth's rotation. However, when Rossby waves propagate vertically, they behave much like their gravity wave counterparts discussed here.

12.3. Complicating Factors

12.3.1. Temperature Structure

In the real atmosphere, a number of factors complicate the basic gravity-wave picture. Since most of them have thus far defied comprehensive analysis, and since observations reveal that simple theory does not account for the motions observed in the real atmosphere, the complicating factors are subjects of active research. One of these is the effect on wave propagation of atmospheric temperature structure, whether in the vertical or the horizontal. This structure affects the Brunt-Väisälä frequency, and hence the index of refraction of the medium. The result is wave refraction, as well as partial or total wave energy trapping. Temperature structure on all scales is important here, ranging from the gross temperature structure of the troposphere-stratosphere-mesosphere-thermosphere to ground-based and capping boundary-layer inversions.

Temperature structure greatly complicates the theoretical analyses. Although the equations governing gravity-wave motion in an isothermal atmosphere have constant coefficients, the equations describing wave motion in the real atmosphere do not. The equations then no longer allow simple plane wave solutions.

The mathematical difficulties thus encountered are surmounted by one of several techniques common to other branches of wave physics:

- Ray tracing methods. These are valid when the temperature variation is small over a single wavelength; they make use of the fact that the simple dispersion equation holds locally by replacing $N^2(z,t)$ (denoting time-varying vertical temperature structure).

- Multi-layer models. The temperature is held constant within each layer, and matching of interfacial boundary conditions determines the wave solution within each layer.

- Analytical solutions, obtained for special profiles, which may approximate realistic atmospheric temperature profiles.

- A variety of specialized numerical techniques.

All these techniques serve to illustrate wave refraction, wave energy trapping, and partial ducting; to varying degrees they serve to identify the trapped modes and in many cases lead to good comparison with observation. This last point is an important one. All else being equal, one can anticipate that trapped modes will dominate the observations, since freely propagating modes decay in energy as $1/r^2$, where r is radial distance from the source.

Inversions are primary sites of wave energy trapping and ducting. The Brunt-Väisälä frequency is relatively high in such inversion layers. Any wave components having frequencies less than the maximum value of N in such regions but greater than the values of N that prevail both above and below will be internal within the duct but evanescent outside, and so trapped there. Figure 12.4 illustrates this, for the WKB (ray tracing) approximation.

It should also be noted that atmospheric temperature structure affects gravity wave propagation far more strongly than it does acoustic waves. For acoustic waves, variations in the speed of sound through the lower atmosphere amount to 10% or less; corresponding variations in the refractive index for gravity waves can amount to an order of magnitude or more.

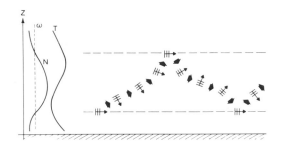

Figure 12.4. Gravity-wave trapping by inverted atmospheric temperature structure. The small arrows indicate phase propagation direction; the cross marks indicate the associated phase fronts and location of the gravity-wave packet. Broad arrows indicate direction of packet motion.

12.3.2. Wind Structure

Atmospheric wind structure introduces analogous complications, producing similar wave refraction and trapping, and its effects can be handled in exactly the same way as the effects of temperature structure, with one important exception—wave encounters with critical levels, treated in Sec. 12.3.3. In the WKB approximation, wind structure enters the equations through changes in the so-called wave intrinsic frequency, i.e., the frequency measured by an observer moving with the local background wind speed.

One important feature of atmospheric wind structure is the vertical profile of horizontal wind, particularly the horizontal structure in convergence zones. Such zones affect acoustic wave propagation only slightly, because these waves have phase speeds high in comparison with the speed of winds they encounter. For gravity waves, however, wave phase speeds can often be comparable with the wind fields in which they propagate, so that the effects of horizontal variation in wind speeds can be dramatic. For example, a wave propagating in the vicinity of a thunderstorm gust front may find its propagation entirely halted by the advancing gust front. Since the wave energy and momentum fluxes remain finite behind such a convergence zone, the result is a rapid convergence of wave energy and momentum, large-amplitude vertical motions in the gust-front region itself, and spectral transfer of the wave energy and momentum into the mean flow. Such processes are especially marked in colliding gust fronts, and underlie the extraordinary potential of such collision sites for severe weather development (e.g., Purdom and Marcus, 1982).

12.3.3. Critical Levels

So-called critical levels mark heights at which the component of the background wind speed in the direction of wave propagation matches the horizontal wave phase speed. At wave critical levels, the vertical wavelength becomes vanishingly small, and the amplitudes of wave-associated fluctuations in horizontal velocity and vertical shear tend to infinity. Full-wave calculations show that wave energy penetration of slowly varying (large Richardson number) shear layers is slight.

Confirmatory WKB calculations of the approach of wave packets to the critical layer show why this is so; they reveal that such wave packets approach the critical level only asymptotically (Fig. 12.5). The slow approach to the critical level, coupled with the increasing wave amplitudes, implies that the

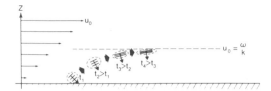

Figure 12.5. Gravity-wave approach to a critical layer (see legend, Fig. 12.4).

usual linearized models of wave motion must break down there. Critical-level encounters are consequently difficult to treat theoretically.

The difficulties are compounded when the shears are large (small Richardson number), as they often are. Full-wave analyses of such cases reveal wave overreflections; i.e., the wave reflected from the layer has a larger amplitude than the incident wave itself. In effect, this conclusion (and the concepts that stem from it) suggests that the shear layer is in fact unstable with respect to the generation and development of gravity waves. The two subjects—wave overreflection and wave generation by shear instability—are thus intimately related. This matter is treated in more detail in Ch. 11.

12.3.4. Atmospheric Moisture

Moisture in the atmosphere introduces further complications. The simplest modification is a trivial one—it is a change in the Brunt-Väisälä frequency introduced because water vapor and dry air have different molecular weights. The more significant modification is the role played by moisture in triggering convective instability during updraft phases of the wave motion in situations of marginal stability. Again, this subject is treated in more detail in Ch. 11; some of the original work can be found in papers by Lalas and Einaudi (1973) and Einaudi and Lalas (1974).

In wave trains representing superposition of several spectral components, it is possible to achieve conditions under which the dispersive effects operating in the wave packet are precisely canceled by the nonlinear interaction, so that the resultant wave form can propagate without change in shape for long distances. Such waves are known as solitary waves, a term that stems from first observations in the middle 1800s of water waves (in canals) that consisted of a single elevation, without corresponding regions of depression. Despite their rather restrictive definition, waves of this class have been invoked by several authors, e.g., Christie *et al.*(1977), to explain boundary layer phenomena associated with low-altitude wind shear that represents a hazard to jet aircraft takeoffs and landings.

12.3.5. Energy and Momentum Exchanges

It became apparent early, from studies of the upper atmosphere (where wave amplitudes are large and their dynamics have greater impact), that wave energy and momentum transfer are important atmospheric processes. The earliest theoretical attempts to examine such transports were based on calculating energy and momentum fluxes as products of the relevant first-order qualities. The resultant energy specifications were incomplete, and it has taken several decades to develop consistent second-order treatments of this problem (e.g., Bretherton, 1969; Whitham, 1974; Dunkerton, 1980).

Nonlinear effects become still more complex and difficult when viewed in terms of wave-wave interactions. At any given time, the unstable atmosphere experiences a broad spectrum of atmospheric motions, all of which interact to a greater or lesser degree. Most of our understanding of such phenomena comes from the work of oceanographers. They have studied these phenomena with respect to the "fully developed" seas that follow the onset of strong surface winds, and the excitation of internal waves in the thermocline (the oceanographic analogy to the atmosphere's boundary-layer capping inversion) by wind waves.

It happens that gravity waves are able to interact through so-called "resonant triads" of waves. The more interesting and novel of these exchanges are those that are upscale. For example, Chimonas and Grant (1984 a, b) showed that it is possible for pairs of Kelvin-Helmholtz waves of small scale, generated by shear instability in a thin shear layer, to excite internal gravity waves of large dimension. Although the K-H waves themselves are confined to the narrow region surrounding the shear zone proper, the resulting gravity wave is free to propagate energy and momentum away from the shear zone.

Of considerable practical interest here is the forecasting of clear-air turbulence through the use of the Richardson number tendency equation, which equates the tendency of the larger-scale flow to reduce the Richardson number with the energy available for the production of turbulence (Roach 1970; Keller, 1984).

12.4. Relation of Gravity Waves to Mesoscale Weather

12.4.1. Apparent Initiation of Severe Convection Storms

Uccellini (1973) reported an analysis of surface weather reports, radar data, surface pressure perturbations, and surface convergence associated with an outbreak of severe convective storms across the north-central United States on 18 May 1971. He found that the intensity of the storms observed on that day pulsated in periods of 2–4 h, and that gravity waves, with an average period of 3 h and horizontal wave phase speeds of some 50 m s^{-1}, were a precursor to the convective storms and apparently acted to trigger these storms. The convective activity would reintensify, or new storm cells would develop, following passage of the gravity wave troughs; maximum rainfall intensity would coincide with ridge passage.

More detailed study showed that the gravity wave occurred over a much larger area than that subject to the convective activity. The storms broke out only where the ambient moisture distribution could support their development.

More recently, Stobie *et al.* (1983) carried out case studies of similar gravity wave events. Figure 12.6 shows the wave-initiated storm systems as revealed by enhanced infrared GOES satellite pictures. Figure 12.7 shows the wave motion revealed in surface pressure records throughout the Midwest. Figure 12.8 shows the track of the wave-storm motions. Stobie *et al.* calcu-

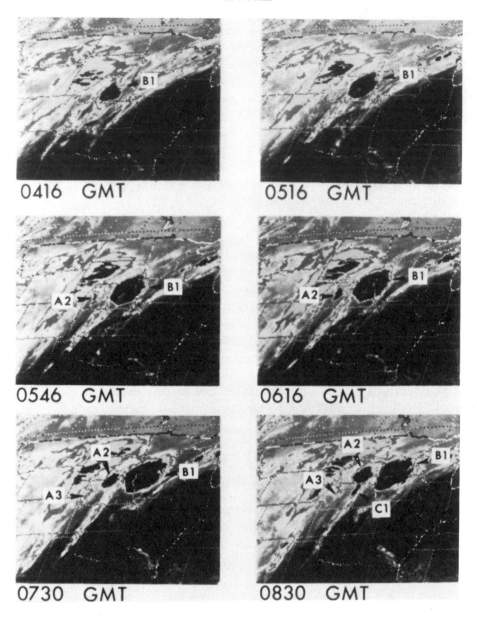

Figure 12.6a. Enhanced infrared satellite images of a wave-initiated storm system, 0416 to 0830 GMT, 9 May 1979. (From Stobie *et al.*, 1983.)

lated the wave motion expected theoretically from knowledge of the upper-air temperature and wind fields, finding good agreement with observation. Space does not permit a detailed discussion of the work here; however, the figures demonstrate that wave-modulated storm structures have a familiar appearance on satellite records, and that such structures are probably relatively common.

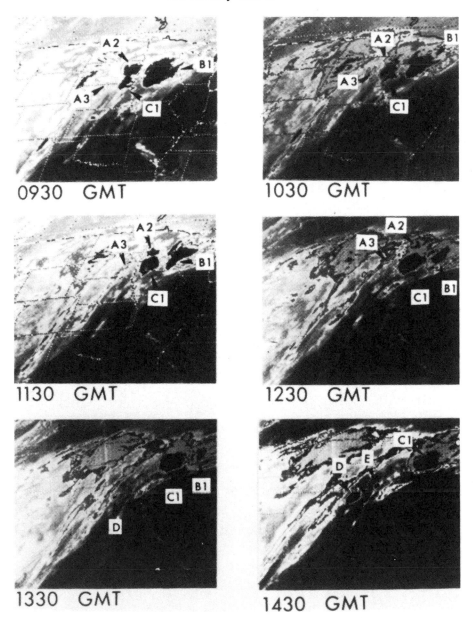

Figure 12.6b. Enhanced infrared satellite images of a wave-initiated storm system, 0930 GMT to 1430 GMT, 9 May 1979. (From Stobie *et al.*, 1983.)

12.4.2. Generation by Thunderstorm Outflow

Erickson and Whitney (1973) published a very interesting satellite photograph (Fig. 12.9) showing gravity waves extending over Texas, Louisiana, and Arkansas, generated by a thunderstorm outflow originating within Oklahoma. Ley and Peltier (1978a) published an interpretation couched in terms of wave generation by frontogenesis, which led to some spirited dialog (Erickson and Whitney, 1978; Ley and Peltier, 1978b).

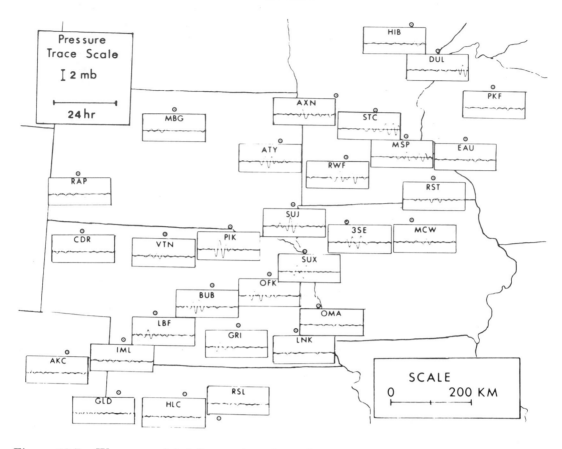

Figure 12.7. Wave-associated fluctuations in surface pressure for the 9 May 1979 event pictured in Fig. 12.6. (From Stobie *et al.*, 1983.)

Of interest here are several things. First, the squall line triggering along the front associated with the outflow itself is quite similar to the kind of event discussed by Purdom and colleagues over the years (e.g., Purdom and Marcus, 1982). What is unusual here is the train of wave motions (with wavelength the order of 10 km) following the frontal boundary. It should be

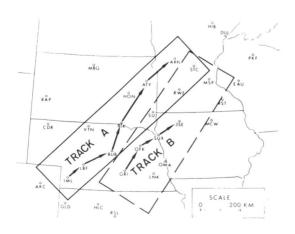

Figure 12.8. Observed track of the wave-storm motions, 9 May 1979. (From Stobie *et al.*, 1983.)

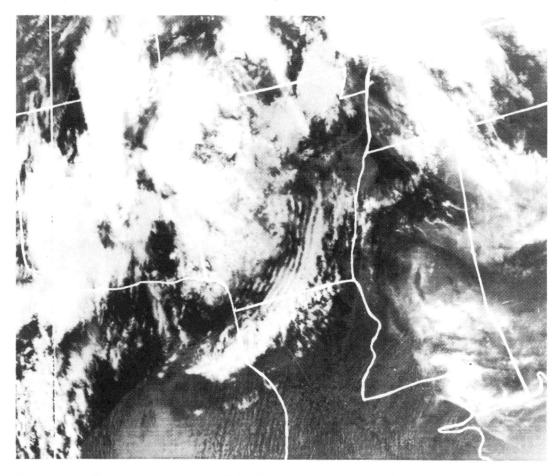

Figure 12.9. Gravity waves generated by thunderstorm outflow. (From Erickson and Whitney, 1973.)

noted that the cloud structure associated with the wave train following is weak compared with the first squall line. This may result in part from the lesser amplitude of subsequent cycles of the wave train. However, it more likely results because the first squall line "wrings out" what moisture there is in the atmosphere, leaving less moisture available for the wave cycles that follow. It could therefore be argued that wave trains such as that revealed here typically exist behind Purdom's "arc lines," but that the moisture supply is inadequate to reveal such structure in most cases.

REFERENCES

Bretherton, F. P., 1969: Momentum transport by gravity waves. *Quart. J. Roy. Meteor. Soc.*, 95, 213–243.

Chimonas, G., and J. R. Grant, 1984a: Shear excitation of gravity waves. 1. Modes of a two scale atmosphere. *J. Atmos. Sci.*, 41, 2269–2277.

Chimonas, G., and J. R. Grant, 1984b: Shear excitation of gravity waves. 2. Upscale scattering from the Kelvin-Helmholtz waves. *J. Atmos. Sci.*, 41, 2278–2288.

Christie, D. R., K. J. Muirhead, and A. L. Hales, 1977: On solitary waves in the atmosphere. *J. Atmos. Sci.*, **35**, 805–825.

Dunkerton, T., 1980: A Lagrangian mean theory of wave, mean-flow interaction with applications to nonacceleration and its breakdown. *Rev. Geophys. Space Res.*, **18**, 387–400.

Einaudi, F., and D. P. Lalas, 1974: Some new properties of Kelvin-Helmholtz waves in an atmosphere with and without condensation effects. *J. Atmos. Sci.*, **31**, 1995–2007.

Erickson, C. O., and L. F. Whitney, Jr., 1973: Gravity waves following severe thunderstorms. *Mon. Wea. Rev.*, **101**, 708–711.

Erickson, C. O., and L. F. Whitney, Jr., 1978: Comments on "Wave generation and frontal collapse." *J. Atmos. Sci.*, **35**, 2379.

Gossard, E. E., and W. H. Hooke, 1975: *Waves in the Atmosphere.* Elsevier, Amsterdam, New York, 456 pp.

Hines, C. O., 1971: Generalizations of the Richardson criterion for the onset of atmospheric turbulence. *Quart. J. Roy. Meteor. Soc.*, **97**, 429–439.

Howard, L. N., 1961: Note on a paper of John Miles. *J. Fluid Mech.*, **10**, 509–512.

Keller, J. L., 1984: Performance of a quantitative jet stream turbulence forecasting technique: the specific CAT risk (SCATR) index. Proc., AIAA 22nd Aerospace Sciences Meeting, AIAA Paper 84–0271, 7 pp.

Lalas, D. P., and F. Einaudi, 1973: On the stability of a moist atmosphere in the presence of a background wind. *J. Atmos. Sci.*, **30**, 795–800.

Ley, B. E., and W. R. Peltier, 1978a: Wave generation and frontal collapse. *J. Atmos. Sci.*, **35**, 3–17.

Ley, B. E., and W. R. Peltier, 1978b: Reply. *J. Atmos. Sci.*, **35**, 2380.

Miles, J. W., 1961: On the stability of heterogeneous shear flows. *J. Fluid Mech.*, **10**, 496–508.

Purdom, J. F. W., and K. Marcus, 1982: Thunderstorm trigger mechanisms over the southeast United States. Preprints, 12th Conference on Severe Local Storms, San Antonio, Tex., American Meteorological Society, Boston, 487–488.

Roach, W. T., 1970: On the influence of synoptic development on the production of high level turbulence. *Quart. J. Roy. Meteor. Soc.*, **96**, 413–429.

Stobie, J. G., F. Einaudi, and L. W. Uccellini, 1983: A case study of gravity waves - convective storms interaction: 9 May 1979. *J. Atmos. Sci.*, **40**, 2804–2830.

Tepper, M., 1954: Pressure jump lines in Midwestern United States, January-August 1951. Res. Paper 37, United States Weather Bureau, Washington, D.C., 70 pp.

Uccellini, L. W., 1973: A case study of apparent gravity-wave initiation of severe convective storms. Rep. 73–2, Department of Meteorology, University of Wisconsin.

Whitham, G. B., 1974: *Linear and Nonlinear Waves*, Wiley, New York, 636 pp.

CHAPTER 13

Quasi-Stationary Convective Events

Charles F. Chappell

13.1. Introduction

Quasi-stationary, or very slowly moving, storm systems are of particular interest to the forecaster, because they frequently produce heavy rainfall and flash floods. These convective weather systems are composed at any moment of many individual storms, all in various stages of their life cycles. The individual storms frequently have trajectories that carry them repeatedly over the same region, producing pulsating heavy rains that quickly cause streams and rivers to overflow their banks.

Meteorological processes on several scales must work synergistically for a storm system to be in a quasi-stationary condition for a few hours. Physical processes extending from synoptic to the very small scale of cloud droplets are involved in a delicate interplay to create a quasi-stationary storm system. These physical processes are dynamical, thermodynamical, and microphysical.

The development of nearly stationary mesoscale convective systems depends upon (1) characteristics of the larger scale environment, which largely determine the type and intensity of individual storms building up the storm system, (2) the nature of the forcing by the synoptic scale, which determines the type of mesoscale configuration that convection will adopt, and (3) the location and rates of formation and dissipation of storms that become a part of the mesoscale convective system.

13.2. Structure of Convective Storms as Modeled

Early observations indicated that thunderstorms exist in a variety of forms. Recent theoretical studies and numerical experiments have aided our understanding of the relation between environmental conditions and the convective storm type. Moncrieff and Green (1972), building on the earlier work of Ludlam (1963) and Green et al. (1966), were able to demonstrate that the quasi-steady character of certain storms creates constraints on the motion of the convective system. Using a form of the Bernoulli equation, they were able to calculate the speed of storm motion in terms of the environmental wind shear and the work of buoyancy forces within the storm. This formulation was developed for a two-dimensional convective system (squall line), but Haman (1976, 1978) was able to modify this approach for application to storms with three-dimensional structure.

In recent years several numerical modeling studies have added insight on the structure of convective storms. Schlesinger (1978, 1980), Wilhelmson and Klemp (1978), and Rotunno and Klemp (1985) documented that non-hydrostatic pressure gradients can produce preferential uplift on the flanks of an original updraft. They demonstrated that storms evolving in a unidirectional, strongly sheared environment will split into mirror-image storms that move to the right and left of the mean wind shear.

Weisman and Klemp (1982), following the lead of Moncrieff and Green (1972) and Moncrieff and Miller (1976), investigated the relation between convective storm type, the available potential buoyant energy, and characteristics of the environmental wind field using the Klemp-Wilhelmson (1978) three-dimensional cloud model. A relationship between vertical wind shear and buoyancy was expressed in terms of a Bulk Richardson Number R:

$$R = PBE/(\overline{U}^2/2) \ . \tag{13.1}$$

PBE is the potential buoyant energy given by

$$PBE = \int_{LFC}^{EL} g \frac{\theta_p - \theta_e}{\theta_e} dz \ , \tag{13.2}$$

where θ_p and θ_e are the potential temperatures of the parcel and the environment, respectively. PBE is the work done per unit mass on the environment by a buoyant air parcel as it rises from its level of free convection (LFC) to its equilibrium level (EL), or the positive area on an energy-conserving thermodynamic diagram. $\overline{U}$ is the vertical wind shear given by the density-weighted mean wind speed over the lowest 6 km minus the mean wind speed taken over the lowest 1/2 km of the profile, or

$$\overline{U} = \overline{U}_{6000} - \overline{U}_{500} \ . \tag{13.3}$$

The denominator of R, $\overline{U}^2/2$ is not only a measure of the wind shear in the lower troposphere, but can also be considered a measure of the inflow kinetic energy made available to the storm by the vertical wind shear (Moncrieff and Green, 1972).

The numerator of R, Eq. (13.2), measures directly the potential strength of the updraft and indirectly the potential vigor of the downdraft and boundary-layer outflow. The measure of wind shear in the denominator represents both the strength of the surface inflow feeding the storm, and the ability of the updraft to take on rotation. R thus represents the relative balance between several factors relevant to storm type and structure.

Weisman and Klemp (1982) found that for a given amount of buoyancy, weak wind shear produces short-lived single cells. Low-to-moderate wind shears produced secondary cell development similar to multicellular storms; moderate-to-strong wind shears were associated with split or supercell-type storms. These results suggest that an optimal buoyancy/shear condition probably exists for the development of supercell-type storms. Apparently, a quasi-balance exists between the amount of low-level air being drawn into the storm and ability of the storm's main updraft to carry it up and out. Short-lived single cells, on the other hand, carry their own seeds of destruction; the development of precipitation quickly destroys the nearly vertical updraft that tends to grow in an environment of little wind shear. Between supercell and single-cell conditions, there exists a range of R where vertical wind shear is sufficient to unload precipitation particles from the updrafts, and potential buoyant energy for updraft growth exceeds the inflow kinetic energy to the storm. In this case, new and separate updraft formation apparently occurs, leading to the formation of multicellular storms.

These preliminary modeling studies address a first simple configuration of the wind field. There are shortcomings inherent in the way R is defined:

- Vertical distribution of buoyancy is not considered.
- Vertical distribution of moisture is not taken into account.
- Directional turning of the vector wind shear is not considered.
- Details of the distribution of vertical wind shear are not taken into account.

Nevertheless, the modeling results are highly encouraging and suggest that it may be possible to anticipate a spectrum of convective storm types through a limited number of observable features of the environment.

13.3. Types of Convective Storms

13.3.1. Single-Cell Storms

Single-cell storms normally occur in environments where winds are relatively light and vertical wind shear is small. These storms usually persist for less than an hour, are unsteady, and are relatively small (5–10 km). Threats of excessive rainfall from single-cell storms are not significant because of the storms' short lifetimes.

13.3.2. Supercell Storms

Supercell storms (Browning and Ludlam, 1962) are larger, intense, and persistent, and normally produce more severe weather than other types of thunderstorms. A highly organized internal circulation that reaches a nearly steady state enables the supercell to propagate continuously. Newly forming updrafts continue to feed and reinforce the main updraft of the storm over a significant period of time. Supercell storms usually occur in an environment with strong vertical shear, where the wind shear vector turns clockwise (veers) with height below 500 mb.

Since supercell storms occur in environments containing strong winds and wind shear, the precipitation efficiency of these storms is reduced by entrainment processes. The reduced precipitation efficiency has two effects:

- Combined with relatively rapid storm motion, it lowers the probability of prolonged excessive rainfall at a given location. Because newly formed updrafts continue to feed the main updraft of the storm, rather than providing for the growth of new cells, these storms do not propagate discretely.
- It reduces the threat of excessive rains, since it is the ability of a storm system to propagate discretely, in a direction opposite to the mean motion of the cells, that is vital to the formation of quasi-stationary mesoscale convective rainstorms.

13.3.3. Multicellular Storms

Multicellular storm systems consist of a series of evolving cells. Cells typically form on or near the storm periphery at 10 to 15 min intervals, and each eventually becomes the dominant cell of the storm complex, building to higher levels as it approaches and finally merges with the main storm complex. Precipitation forms in the new updraft and is held aloft temporarily while the cell matures. Finally, precipitation unloads as the cell matures, resulting in a heavy gush of rain.

At low levels, cooler air diverging from the downdraft intersects the inflowing air along a gust front, creating a region of strong low-level convergence favorable for the development of new updrafts. The result is a series of new updrafts that tend to form on the right or right-rear flank of the storm system. As each updraft reaches its maximum strength, it generally penetrates the equilibrium level (where parcel and environmental temperatures again become equal), and briefly produces an overshooting top. The top then subsides and a new cell takes over. As the top collapses, the accompanying divergence aloft produces an expansion of the storm anvil. This upper divergence results from the rapid decrease in updraft speed forced by the negative buoyancy present above the equilibrium level. Part of the updraft air mixes with downdraft and environmental air and descends in the precipitation region. Some updraft air also moves downwind in the storm anvil.

Multicellular storms occur in an environment containing significant buoyant energy; lifted indexes of $-8°C$ were documented by Marwitz (1972). Low-to-moderate vertical wind shear is also present in multicellular environments. Average shear values in the cloud layer are on the order of 2.5 m s^{-1} per kilometer of depth, as shown in Marwitz (1972). The wind shear vector normally shows little clockwise turning with height, but some counterclockwise turning has occasionally been noted. Multicellular storm systems, especially those that become quasi-stationary, are most likely to develop intense, longer-lived convective rainstorms.

13.4. Movement of Convective Storms

It is convenient to think of the motion of a storm system as the sum of two vectors; the mean velocity of the cells constituting the storm system, and the propagation velocity due to the formation of new cells on the periphery of

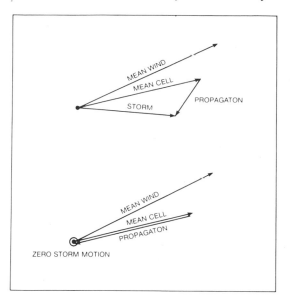

Figure 13.1. (Top) The effect of propagation on storm motion, and (bottom) the relationship between propagation and mean cell motion, for developing a quasi-stationary storm complex.

the storm (Fig. 13.1, top). The mean cell velocity may lie along, or on either side of, the vector mean wind of the cloud layer, and discrete propagation may occur anywhere along the storm periphery. If it occurs on the leading edge of the storm system, an accelerating effect results; if it occurs on a rear flank, deceleration of the system results. It is also clear from Fig. 13.1 (bottom) that not only must the propagation velocity equal the mean cell motion, but new cells must form on the storm flank opposite to the direction of cell motion if a stationary storm system is to develop.

To forecast excessive rains and flash floods, it is more important to monitor the movement of the storm flank with active convection, rather than the motion of the storm system centroid. When a storm is becoming stationary, new cell growth and discrete propagation are occurring in the rear storm flank, and it is nearly always this more active portion of the storm that poses the greatest threat of heavy rain and flooding.

13.4.1. Observations of Convective Storm Motion

Espy (1861) and Humphreys (1914, 1940) suggested that convective clouds should move with the vector mean wind in the cloud layer. Radar studies of convective storm movement by Brooks (1946) and Byers and Braham (1949) show this to be nearly true on the average, but these studies focused on relatively small storms.

In contrast, studies by Byers (1942), Newton and Katz (1958), and Marwitz (1972) indicated that larger thunderstorm clusters usually move systematically, with a substantial component toward the right of the vector mean wind of the cloud layer.

The study of radar-observed convective storms by Brooks (1946) is also consistent with these findings. Brooks found that the steering level (level where the direction of the environmental wind and the direction of storm motion are coincident) of storms moves upward as the size of the storm increases. He also introduced the concept that the movement of a group of cells differs from that of the individual cells, owing to the effect of propagation. The peculiar motion of some storms has baffled meteorologists since storm-tracking radars became available. Both right- and left-moving storms are often the result of a storm splitting process. Storms that deviate to the left of the vector mean wind have been observed by Newton and Fankhauser (1964), Hammond (1967), Charba and Sasaki (1971), Brown et al. (1973), Cotton et al. (1982), and others. Trajectories of convective storms may vary through their life cycle. S-shaped trajectories are occasionally observed as storms move more nearly with the vector mean wind early and late in their lives when they are relatively small, but deviate significantly to the right of the mean wind when they are larger and most strongly developed. Newton and Katz (1958) found that many large storms move toward the right, not only of the vector mean wind, but of the winds at all levels. Thus, the concept of a steering level is not valid.

Newton and Katz (1958) and Newton and Newton (1959) examined the movements of convective rainstorms on 24 days. Concentrated areas of rainfall 50–100 km across tended to move along tracks toward the right of the vector mean wind in the 850–500 mb layer. Average deviation was 25°. Growth of new cells on the right flank was favored by a relative flow of low-level moist air, while water-vapor starvation occurred on the left flank. Other forces are also thought to affect new cell generation on a storm's periphery:

- A deflecting force, arising from Bernoulli effects, that a rotating cylinder experiences when embedded in a moving fluid—the Magnus Force (Fujita and Grandosa, 1968; Goldman and Wilkins, 1973).
- Hydrodynamic vertical pressure gradients on the storm boundary (Newton and Newton, 1959; Hitschfeld, 1960; Schlesinger, 1978, 1980; Rotunno and Klemp, 1982).
- Low-level convergence and lifting of potentially unstable air associated with the gust front (Weaver, 1979; Wilhelmson and Chen, 1982; Knupp and Cotton, 1982).

Newton and Fankhauser (1964) showed that convective storms tend to move in such a way as to adjust the supply of water vapor to storm requirements (precipitation plus liquid water lost to evaporation aloft). Thus, smaller storms tend to move to the left of the vector mean wind direction at nearly the same speed or slightly slower, whereas large storms move to the right of and slower than the vector mean wind, with a deviation proportional to their diameter.

It is seen from Fig. 13.2 that a large rightward-deviating storm moves in a direction having an upwind component with respect to the mean low-level wind relative to the storm ($\overline{V}_{RL}$). Thus, its movement relative to the wind in the moist boundary layer is greater than that corresponding to smaller storms, where storm velocity (V_S) is usually directed to the left of the vector mean wind ($\overline{V}$).

This agrees with the concept that precipitation increases with storm size, as the square of its diameter (D). For if the storm intercepts the water vapor

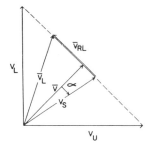

Figure 13.2. The increase in the mean low-level wind relative to the storm ($\overline{V}_{RL}$) as the rightward deviation of the storm from the vector mean wind (α) increases. (After Newton and Fankhauser, 1964.)

along a track determined by its relative motion through the moist boundary layer, then the rate of vapor supplied to the storm is proportional to $D\,\overline{V}_{RL}$. It follows that when vapor supply and storm requirements are in equilibrium, $\overline{V}_{RL}$ should be proportional to D if rainout is to be proportional to size or D^2. It has been found that the deviation of individual radar cells about the vector mean wind agrees reasonably well with this simple concept.

Newton and Fankhauser (1975), using data from the National Severe Storms Project, summarized the deviations of storm movements from the direction of the mean wind for several convective days. The vector mean wind ($\overline{V}$) was represented by the mean of the 850, 700, 500, and 300 mb winds, which approximates the pressure-averaged wind in the 900–200 mb convective layer. The component of the 850–300 mb wind shear normal and directed to the right of the vector mean wind was designated V_{sn}. The ratio $V_{sn}/\mid\overline{V}\mid$ therefore represents a measure of the wind veer through the troposphere. The results (Fig. 13.3) show a clear tendency for large storms (centered around 25 km diameter) to move farthest to the right of the vector mean wind. Medium-sized echoes (centered around 15 km diameter)

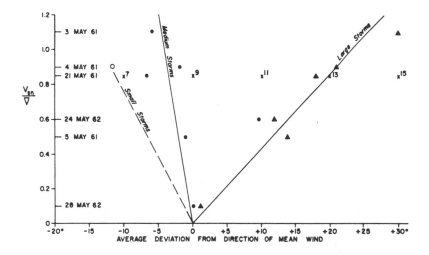

Figure 13.3. Summary of storm movements for six days. Total sample represents 334 cases. Large storms are centered around 25 km diameter, medium storms around 15 km diameter, and small storms around 5 km diameter. Standard deviation of the direction of storm motion is about 10° for any given size. (After Newton and Fankhauser, 1975.)

generally moved a little to the left of the vector mean wind, and small echoes (centered around 5 km diameter) moved up to 30° to the left. These results indicate that the deviation of storm movement from the mean wind direction is not only well correlated with storm size but is proportional to the wind veer.

Weaver (1979) postulated that a source of deviate storm motion may lie in synoptic and meso-α-scale boundary-layer features. He presented a case in which a radar echo expanded slowly northeastward with time, but the most intense radar reflectivity core remained nearly stationary on the west and southwest flanks of the storm. New convection formed frequently in the vicinity of the strong boundary-layer convergence zone, anchoring the west end of the storm. In further studies involving eight storms, he observed that about half the storms moved in the same direction and at the approximate speed of the moving mesoscale convergence feature forcing the storms. The remainder moved in a direction and at speeds in between the vectors representing the motion of the mesoscale convergence feature and the mean cloud-layer wind. Weaver summarized the results by speculating that storm motion depends upon the relative strengths of three factors: the mean cloud-layer wind vector; thunderstorm-induced convergence features; and the orientation, strength, and movement of mesoscale, boundary-layer convergence regions.

Large thunderstorm clusters, frequently referred to as Mesoscale Convective Complexes, or MCCs (Maddox, 1980), have smaller meso-β-scale regions of intense convection. These meso-β-scale elements constitute the most vigorous convective updrafts within the MCC, and most severe weather and excessive precipitation are associated with this region of the MCC. Merritt and Fritsch (1984) studied the motion of about 100 of these meso-β-scale convective portions of MCCs. They found a poor correlation between the movement of the β-scale elements and the winds at any tropospheric level. In fact, a steering level was frequently absent. However, they did find that the mean cloud-layer shear vector provided a good estimate of the direction of the intense convective- element movement for most MCCs. The motion the convective elements followed closely was in the direction of the 850–300 mb shear vector, or along the 850–300 mb thickness contours (Fig. 13.4). The speed appeared to be modulated by the strength and relative position of the low-level moisture convergence, which would also appear largely to determine the direction of the storm propagation vector.

13.4.2. Storm Propagation

Storm propagation is an additional influence on storm motion. It may speed up or slow down a mesoscale convective system (MCS) depending on whether propagation is occurring on a forward or rear flank of the system. An MCS is less restrictively defined than an MCC (Maddox, 1980). It is any multicellular storm or group of interacting storms that suggests some organization in its forcing (e.g., squall line). An MCC is also an MCS. The production of a quasi-stationary or very slowly moving MCS depends on propagation characteristics that generate new cells on the storm flank

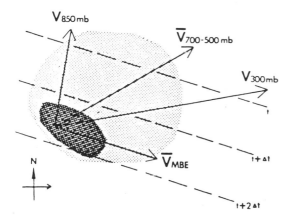

Figure 13.4. Direction of meso-β-scale element (MBE) movement relative to an 850 mb and a 300 mb wind, and a mean midtropospheric wind $\overline{V}$. Dashed lines are 850–300 mb thickness contours. Light stippling is the cold-cloud shield of MCC; dark stippling is the region of MBEs. (From Merritt and Fritsch, 1984.)

opposite to the mean cell motion, while cell dissipation occurs on the storm flank coincident with the mean cell motion.

Wilhelmson and Chen (1982) explained the propagation of multicellular storms with their numerical studies. They found that discrete propagation can significantly affect storm motion, usually causing the storm system to move to the right of individual cells. Frequently, enhanced convergence in the boundary layer due to downdrafts triggered new cells, which typically formed 4–10 km away from older cells. However, this mechanism of new cell formation does not satisfactorily explain the 10–15 min periodicity frequently observed for new cell development, since the time required for a new cloud to produce precipitation at the ground is longer than 15 min. Some of their simulations suggest that the separation speed between old cells and the outflow may be a more controlling factor on the timing of new cell generation. In this case, the primary role of individual precipitation-induced downdrafts is to reinforce and maintain the overall storm outflow, and the timing of individual downdrafts is not directly related to the frequency of new cell formation.

The magnitude of propagation depends on the rate at which new cells form, the separation distance between the newly developing cell and the storm, and the rate at which the new cells enlarge and accrete to the storm boundary. For example, if under a given set of meteorological conditions new cells form every 10 min and at an average separation distance of 6 km, the maximum effect of propagation velocity on retarding storm motion is 36 km h^{-1}, assuming that propagation velocity is opposite to the mean cell motion and newly generated cells grow rapidly to merge with the storm system.

Factors that control the separation distances and the rate at which new cells develop and grow are not well documented. It is reasonable to assume that the rate of new cell formation will depend on the magnitude of the potential buoyant energy available to feed the cell genesis region of the storm, and on the rate at which it is being replenished. Destabilization of the troposphere may occur from differential temperature advection, which warms the lower levels and cools the upper levels of the troposphere. Synoptic and mesoscale forcing, which produce low-level mass convergence and high-level mass divergence, also destabilize the environment and promote continuing thunderstorm activity, or can trigger it initially. The orientation and strength of mesoscale boundary-layer convergence (Weaver, 1979) has been found to be important in initiating new convection, as has orographic lifting in areas of complex terrain (Maddox et al., 1978).

The rapid genesis of cells in a specific region, which is required to produce a large propagation velocity, requires a focusing mechanism to repeatedly initiate convection in the same location relative to the storm. Normally this requires that the release of the potential buoyant energy be inhibited and confined to a small area. There must exist a layer of negative buoyancy, which must be overcome (through lifting of air parcels to their level of free convection) prior to release of the instability. Under these conditions, buoyant energy can be stored until triggered and focused by some lifting process. Potential lifting mechanisms include mechanical lifting along gust fronts, favorable hydrodynamic pressure distributions on the periphery of previous storm cells, mesoscale zones of boundary-layer convergence (frontal, drylines, etc.) and orographic lifting. As long as environmental conditions remain essentially unchanged, the storm can continue to feed from a reservoir of potentially unstable air, and new cell formation can be focused in a relatively small area by these lifting mechanisms.

The generation of a quasi-stationary MCS requires that the storm propagation vector be opposite to the mean cell motion. This, in turn, requires that the storm system be positioned uniquely with respect to regions of maximum potential buoyant energy, and strong low-level mass and moisture convergence. The desired configuration is to juxtapose these features with the rear flank of the MCS. This particular orientation often occurs in nature as a storm system moves east through the low-level jet and corridor of most unstable air. As this happens, the rear flank of the storm system becomes the most favorable region for new cell generation, and dissipation of earlier activity begins on the leading flank where conditions become more unfavorable. The net effect is to decelerate the storm system, and perhaps, to produce a quasi-stationary MCS.

13.5. Meteorological Environments Associated With Quasi-Stationary MCSs

Quasi-stationary or slowly moving MCSs pose the greatest threat of excessive rains and flash floods. These storms can arise a number of ways. At stronger environmental wind speeds, it becomes difficult for discrete propagation processes to offset the rapid mean cell motion within the storm. This is perhaps the main reason that quasi-stationary, or very slowly moving, mesoscale convective storms predominate during the summer months, when potential buoyant energy is plentiful but environmental winds are weaker. However, the strong baroclinic conditions of fall, winter, and spring, which often produce fast-moving squall-line convection and supercell storms, can also produce quasi-stationary MCSs. In fact, some of the larger and longer-lived systems develop under these conditions. In this latter case, thunderstorms may develop into meso-β-scale storm systems prior to accreting to the parent storm system. Thus meso-β-scale storm clusters may become the primary mechanism for discrete propagation, leading to the development of huge quasi-stationary meso-α-scale convective systems.

Merritt and Fritsch (1984) characterized the environment prior to the formation of MCCs; Maddox *et al.* (1979) described synoptic patterns associated with flash floods. Maddox *et al.* found that meteorological patterns associated with flash floods east of the Rockies could be grouped into three categories: synoptic, frontal, and mesohigh.

13.5.1. Synoptic Forcing

Flash floods may develop in association with a strong synoptic-scale cyclone or frontal system. A major upper-level trough is usually moving slowly eastward, and the associated surface front is often quasi-stationary. Convective storms repeatedly develop and move rapidly over the same general area, which is generally just ahead of the cold front in the tongue of warm moist air flowing north from the Gulf of Mexico. Often a weak warm front or a rain-cooled bubble of air helps to trigger the storms. Occasionally under these conditions, storm cells grow rapidly into meso-β-scale storm clusters which, in turn, accrete to a larger meso-α-scale cluster. Thus, discrete propagation can be accomplished on a larger scale by joining small MCSs to the larger parent storm cluster rather than by the attachment of individual storm cells. These synoptic-type events sometime affect portions of several states and may linger for two or three days. These events are frequently accompanied by more general flooding, as well as several flash flood episodes. They occur most frequently in the fall and spring.

Environmental winds associated with synoptically forced events veer about 35 degrees from the surface up to 500 mb, with less veering above. Upper winds are approximately parallel to the surface frontal zone. The composite sounding for these events is conditionally unstable, and precipitable water is almost twice the mean climatological value.

A synoptic-type pattern is characterized by a slowly moving major trough at 500 mb and an associated slowly moving or quasi-stationary cold front, usually oriented north-northeast to south-southwest (Fig. 13.5). Thunderstorms erupt in the warm, moist unstable air lying east of the stationary front and move northeast. Moisture convergence is strong ahead of the low-level jet, and storms develop repeatedly on the southern or southwestern periphery of earlier storms and move northeast. The intensity of thunderstorm activity is occasionally modulated by short-wave troughs that rotate northeastward out of the major trough. A warm frontal boundary oriented northwest to southeast ahead of the slowly moving cold front may help to concentrate and focus the thunderstorm activity.

13.5.2. Frontal Forcing

Frontal-type patterns producing flash floods are characterized by a stationary or very slowly moving synoptic-scale frontal boundary (usually ori-

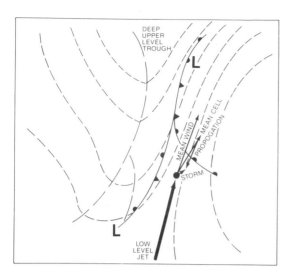

Figure 13.5.   Configuration of important synoptic features that can lead to the formation of a synoptic-type, quasi-stationary MCS. Storm cells erupt in the corridor of unstable air, enhanced by the presence of low-level mass and moisture convergence, ahead of a slowly moving or nearly stationary frontal system. As rain-cooled air is generated, zones of mesoscale convergence are created that contribute to further storm cell formation. Mean cell motion is directed north-northeast, and new cells develop and replace the old. Strong discrete propagation directed south-southwest produces a nearly stationary MCS with heavy pulsating rains, especially along the southern periphery of the system.

ented west to east) that helps to trigger thunderstorm activity and focus storm location. Heavy rains occur on the cool side of the surface front as warm unstable air flows over the frontal zone, providing the energy source for the storms. Again, upper-level winds nearly parallel the surface front, and convective storms repeatedly develop and move over the same region. These events are distinctly nocturnal. In most cases, a meso-α-scale short-wave trough approaches the threat area and stimulates convection. In some cases a weak mesolow moves along the frontal boundary, increasing convergence and inflow into the storm area. Frontal events occur mainly during the warm half of the year.

Environmental winds for frontal-type flash floods exhibit significant veering below 700 mb, and less veering above. Wind speed changes little with height. This veering favors the movement of a storm nearly parallel to the frontal zone, where conditionally unstable air continues to arrive unimpeded on the storm's right rear flank. The composite sounding for frontal events is somewhat more unstable than for synoptic events. Precipitable water also exceeds that for synoptic events although it is somewhat less in terms of the mean climatological value.

An example of the frontal configuration favorable for the development of quasi-stationary convective rainstorms is shown in Fig. 13.6. Note that the vector mean wind has a significant component directed toward the cooler air mass. This allows relatively large and intense storms, which are likely to develop in connection with significant synoptic-scale forcing and large potential buoyant energy, to be directed to the right of the vector mean wind but still not intercept the nearly stationary frontal system.

New cell formation would be enhanced along that segment of the frontal boundary where potential buoyant energy and moisture convergence are greatest. This segment might be identified by the juxtaposition of the low-level thermal axis and moist tongue, since low-level temperature and moisture largely control the amount of available potential buoyant energy. Max-

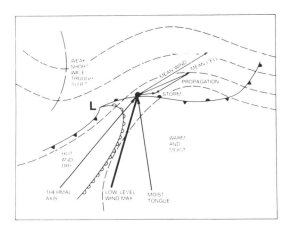

Figure 13.6. Configuration of important synoptic features that can lead to the formation of a frontal-type, quasi-stationary MCS. Storm cluster passes to the east of the low-level thermal axis, moist tongue, and maxima of mass and moisture convergence. New cell formation is therefore enhanced on the rear flank of the storm system, giving rise to strong discrete propagation directed west and southwest. The MCS stalls as new cells on the western flank assume similar trajectories and produce pulsating heavy rains. Mean cell motion is directed at some small angle away from the frontal system toward the cooler air, so that ongoing thunderstorm activity does not disrupt the frontal lifting of the potentially unstable air.

ima in low-level mass and moisture convergence would also be expected near the intersection of the low-level wind maximum and the surface frontal zone. If the locations of the maxima of potential buoyant energy and low-level mass and moisture convergence nearly coincide, periodic cell formation in this location will generate a significant propagation component that generally retards storm motion. This occurs as storm clusters move east and this favored cell formation area becomes coincident with the rear flank of the storm system.

The resulting storm motion (Fig. 13.6) is nearly parallel to the frontal system, or even directed slightly toward the cooler air mass. This ensures the persistence of a quasi-stationary lifting mechanism since storms do not move across and destroy the frontal zone. Rather, the repeated downdraft formation from new cells serves to anchor and maintain the front, providing a mechanism for overcoming any layer of negative buoyancy by lifting the potentially unstable air to its level of free convection. Thus, there is a relatively small area (meso-β-scale) adjacent to the rear of the storm complex where conditionally unstable air continues to be available and mesoscale and cloud-scale lifting processes persist, promoting the continued regeneration of new storms. The distribution of potential buoyant energy aloft (slight negative buoyancy topped by large positive buoyancy) ensures that the instability is not released until the air is lifted near the periphery of the storm complex, culminating in a sudden and concentrated release of the potential buoyant energy in new cells close to the parent storm.

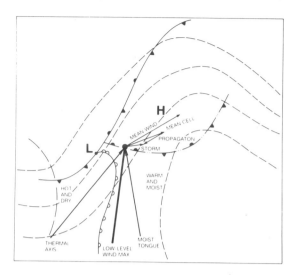

Figure 13.7. Configuration of important synoptic features that can lead to the formation of a mesohigh-type, quasi-stationary MCS. Thunderstorm activity usually develops along or near the stationary front, frequently develops a squall line structure with attendant mesohigh pressure, and moves to the east of the low-level thermal axis, moist tongue, and maxima of mass and moisture convergence. New cell formation is enhanced on the trailing portion of the original squall line, as the unstable air is lifted over the rear of the dome of cool air formed by previous rain and downdrafts. Strong discrete propagation is directed west and southwest as new cells on the trailing flank of the mesohigh and squall line assume similar trajectories and produce pulsating heavy rains. Mean cell motion is directed at some small angle away from the boundary toward the cooler air so that thunderstorm activity does not disrupt the ongoing lifting of the potentially unstable air by the dome of cool air.

13.5.3. Mesohigh Forcing

Mesohigh-type flash floods are associated with a nearly stationary thunderstorm outflow boundary, generated by previous convective activity. The heaviest rains occur on the cool side of the boundary, usually south or southwest of the mesohigh pressure center (Fig. 13.7). About half of these events occur to the east of a slowly moving large-scale frontal system; the remainder are far removed from significant frontal systems. Again, upper-level winds are parallel to the outflow boundary and storms repeatedly develop and move over the same area. These events are also distinctly nocturnal, and most are associated with a meso-α-scale short-wave trough at upper levels. These events are mainly a warm season phenomenon; almost 80% occur in May through August.

Environmental winds for mesohigh-type flash floods exhibit significant veering up to 700 mb, and less veering above. Again, wind speed changes little with height. The composite sounding shows a Showalter Index of -5, the most unstable air mass of all the types that produce flash floods. (The Showalter Index is derived by taking a parcel of air at 850 mb with a given temperature and moisture content, lifting it dry adiabatically until saturated and then moist adiabatically to 500 mb. The temperature of the lifted parcel is then subtracted from the ambient temperature at 500 mb to obtain the value of the Showalter Index.) Precipitable water is similar to that for frontal-type events, running about 162% of the mean climatological value.

The evolution of this type of excessive rain event usually begins with thunderstorm activity that develops along or just ahead of the slowly moving cold front and just to the north of a weak low-pressure center located on the front. This thunderstorm activity usually intensifies and develops into a relatively large area of storms, which may have a squall line feature on its leading edge. The squall line moves eastward through the low-level moist tongue and decays upon reaching the upper-level ridge. Normally, a mesohigh pressure system develops behind this initial squall line, because of cooling of the lowest levels by downdraft air and the continuing stratiform rain to the rear of the squall line.

Thunderstorm activity may redevelop to the southwest of the mesohigh pressure center where the maximum potential buoyant energy and greatest moisture convergence exist. The dome of cooler air near the surface, generated by the previous thunderstorm activity, frequently provides sufficient lifting to bring the warm, moist, inflowing air to its level of free convection. Thus, a prolonged and concentrated release of convection on the west or southwest flank of the initial storm complex is favored, causing it to remain nearly stationary over a period of time. It is in this region that frequent new cell generation leads to repeated storms with similar trajectories, giving pulsating heavy-to-excessive rains over their path.

These mesohigh-type events usually occur at night, and the low-level moisture convergence is enhanced by boundary-layer cooling and decoupling of the boundary layer. Frequently, the approach of a short-wave trough also strengthens or maintains the low-level wind maximum.

An important forecast problem related to both the frontal and mesohigh events is to anticipate the ending of an excessive rain event. This may result from a decrease, or interruption, of the potential buoyant energy and moisture convergence feeding the active cell generation area, or from the arrival of a short-wave trough aloft, which may interrupt the favorable larger scale forcing and alter the environmental winds. The result can be a change in intensity and/or motion of the storm complex and its active generation region. Arrival of the short-wave trough brings a decrease or reversal of the positive vertical motion field in the cell generation region, and, without the larger scale lifting to help destabilize the environment, the convection ends or translates downstream with the short-wave trough.

13.5.4. Joint Frontal and Mesohigh Forcing

Another pattern that produces quasi-stationary MCSs and excessive rains is a combination of frontal and mesohigh events (Fig. 13.8). In this case, a very slowly moving and shallow warm front is oriented west to east, or northwest to southeast, ahead of a slowly moving cold front. The air mass streaming northward ahead of the cold front is conditionally very unstable. However, substantial negative buoyancy is present below the level of free convection, and the potential buoyant energy is not released upon reaching the surface position of the stationary front or warm front. The air then rises

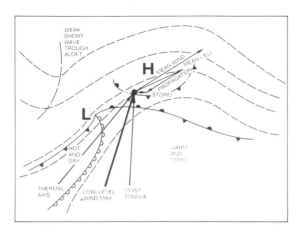

Figure 13.8. Configuration of important synoptic features that can lead to
the formation of a combination frontal-
and   mesohigh-type,   quasi-stationary
MCS. Attributes of both frontal and
mesohigh types may apply.

over the frontal surface for a considerable distance before it finally attains
its level of free convection and convection is released.

A large and nearly stationary storm complex may then form some 100–
300 km north of the surface warm front, as the cluster of thunderstorms
moves eastward through the tongue of overrunning unstable air, and new
cell formation shifts to its rear flank. Cooling from downdrafts and heavy
rains forms a bubble of colder air (mesohigh pressure center) near the ground,
which serves as an additional lifting mechanism. Frequently, the warm front
progresses slowly north toward the storm, while the storm complex remains
essentially stationary. Occasionally, the warm front reaches the storm complex and the event in its final stages resembles a mesohigh-type event. The
meteorological environment of this combination-type event closely resembles
that of a frontal-type event.

13.5.5. A Parallel Set of Patterns

Consistent with the earlier work of Maddox *et al.* (1979), Merritt and
Fritsch (1984) found that MCCs develop in three distinct regimes. These
categories were also named synoptic, frontal, and mesohigh, and although
they have much in common with the descriptions of Maddox *et al.* (1979),
there are also important differences. A study of 100 MCCs, and MCC-
like events, showed that MCC development occurred where the conditional
instability is high in the presence of low-level warm advection and moisture
convergence. Considerable variation in the hodograph curvature pattern
was noted in the pre-MCC environment but in the mean, strong veering of
the environmental winds with height was observed from the surface up to
500 mb. Above 500 mb the hodograph displayed either less veering with
height up to 300 mb, or slight backing. Vertical wind shear was relatively
strong in the case of synoptic MCCs and for most frontal type MCCs. These
MCCs occurred within significant baroclinic flow and appear to be linked
dynamically with upper-level jet streams. As winds aloft and vertical shear
become weaker, mesohigh MCCs seem to occur more often.

13.6. Examples of Quasi-Stationary MCSs

The Big Thompson flash flood of 31 July 1976 (Caracena *et al.*, 1979) is typical of flash floods that occur in the eastern foothills of the Rocky Mountains. Strong easterly flow up the high plains brought an unusually large amount of low-level moisture into the foothills of northern Colorado. Dewpoint temperatures from 60° to 65° flowed westward into the foothills behind a cool front across southern Kansas and southeast and north-central Colorado.

The air mass flowing into the foothills was very unstable (Lifted Index = −6). (The Lifted Index is derived by mixing the lowest 3000 ft of the atmosphere to obtain the mean potential temperature and mean vapor mixing ratio of an air parcel representative of this layer. This air parcel is then lifted dry adiabatically until saturated and then moist adiabatically to 500 mb. The temperature of the lifted parcel is then subtracted from the ambient temperature at 500 mb to obtain the value of the Lifted Index.) Winds aloft were easterly, 25–50 kt at and below cloud base, and winds at middle and upper levels were southerly, 10–20 kt. There was a layer of negative buoyancy just below the level of free convection, and therefore the moist air mass streaming in from the east required lifting of about 1000 m before the potential buoyant energy could be released. The lifting was provided by the terrain and resulted in an explosive and concentrated release of the buoyant energy at an elevation of about 8000 ft along the foothills. Winds aloft veered strongly below 500 mb and thunderstorms drifted slowly north, parallel to the north-south ridges.

A quasi-stationary MCS developed over the foothills. The storm system was aligned north-south along the ridges. New cells generated along the southern edge of the storm and subsequently accreted to the mesoscale storm system, producing a propagation velocity approximately equal to, but opposite in direction to, the mean cell velocity (Fig. 13.9). During its stationary period, an estimated 8 inches of rain fell in about $1\frac{1}{2}$ hours over the steep walls of the Big Thompson Canyon.

The Johnstown flash flood of 19 July 1977 (Hoxit *et al.*, 1978) was produced by a mesohigh-type, quasi-stationary MCS. A meso-α-scale wave at 500 mb was located just west of the Johnstown area the morning before the storm. During the early afternoon a squall line was oriented northeast-southwest from south-central New York to southwestern Pennsylvania. Thunderstorms were also occurring over northwestern Pennsylvania as warm moist air from Ohio flowed east and was lifted over the cool downdraft air left by the first squall line (Fig. 13.10). This thunderstorm activity became oriented northwest to southeast and moved slowly southeast during the late afternoon and night. For several hours new cells formed over northwestern Pennsylvania and moved southeast to Johnstown. The air mass was very unstable (−7 lifted index) as shown by the Pittsburgh rawinsonde taken at 2000 EDT, 19 July 1977. However, with the onset of boundary-layer cooling during the evening, the moist air required lifting over the dome of cool air to release the potential buoyant energy. Lifting provided for a contin-

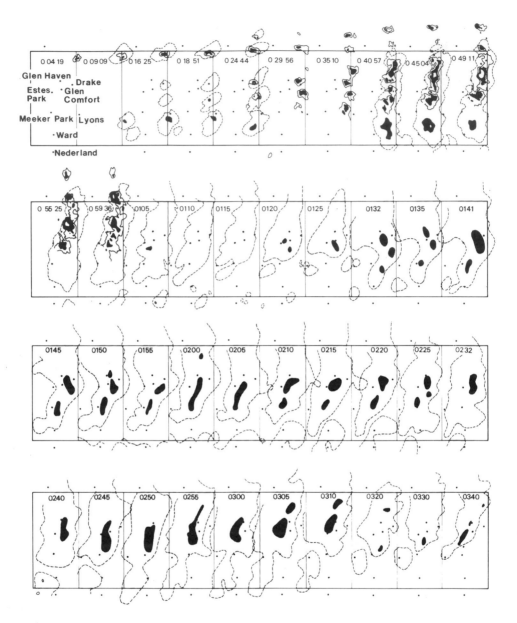

Figure 13.9. Radar echoes of the Big Thompson storm of 31 July–1 August 1976. Radar reflectivities are as observed from the NCAR National Hail Research Experiment radar (10 cm) located at Grover, Colo., for the period up to 0100 GMT, 1 August 1976, followed by the NWS Limon, Colo., radar data after 0100 GMT, 1 August 1976. Greater detail is evident from the NCAR radar at Grover because of a narrower beam width and closer location to the storm. Vertically integrated precipitable water (VIP) level one is gray; level two is white; level three is black; level four is white.

ued release of the buoyant energy and a concentration of the thunderstorm activity in a northwest-southeast zone running through Johnstown. For several hours the propagation velocity was large and directed northwestward or opposite to the mean cell motion and thus created a quasi-stationary MCS.

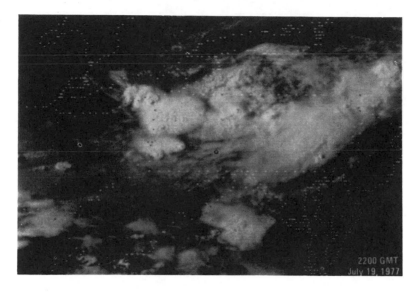

Figure 13.10. Geostationary Operational Environmental Satellite (GOES) photograph for 2200 GMT, 19 July 1977. An active squall line extends from southeastern New York to northern New Jersey, southeastern Pennsylvania to northern Maryland. A zone of thunderstorms oriented northwest-southeast has developed on a line across extreme northeastern Ohio and western Pennsylvania. Johnstown is located at the small circle between the two active thunderstorm regions. This represents the organizing stage of the MCC.

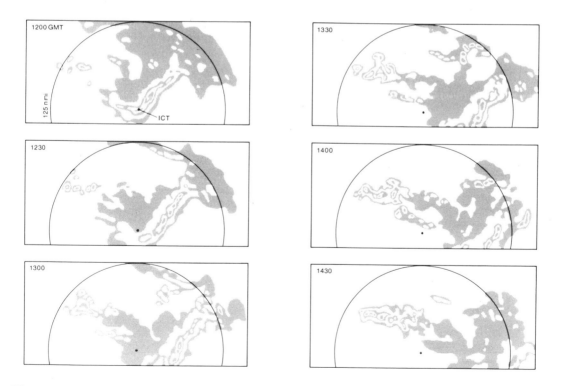

Figure 13.11. Wichita radar depictions of central Kansas quasi-stationary rainstorm, 22 June 1981. All VIP levels are shown, beginning with gray for level one and alternating white and gray for higher levels. Times are GMT.

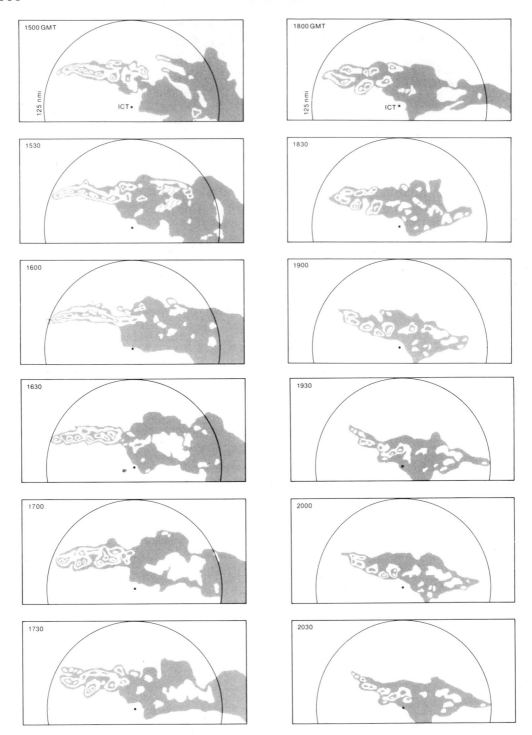

Figure 13.11. Continued.

An outstanding frontal-type case of rapid cell generation and strong propagation, leading to the formation of a stationary convective system, occurred in central Kansas on 22 June 1981. Figure 13.11 shows that at 1200 GMT a large echo with a short squall line on its leading edge was bearing down on the radar site at Wichita. The squall line passed through Wichita and was followed by a large stratiform region of light rain. A few thunderstorm cells developed to the west where very unstable air was overrunning a stationary front and the cool dome left by the previous thunderstorm activity. Rapid generation of new storms continued for several hours on the west flank of the MCS, producing an intense zone of thunderstorms. Several inches of rain resulting in flash flooding occurred as storms moved repeatedly over the same region.

REFERENCES

Brooks, H. B., 1946: A summary of some radar thunderstorm observations. *Bull. Amer. Meteor. Soc.*, **27**, 557–563.

Brown, R. A., D. W. Burgess, and K. C. Crawford, 1973: Twin tornado cyclones within a severe thunderstorm: Single Doppler radar observations. *Weatherwise*, **26**, 63–71.

Browning, K. A., and F. H. Ludlam, 1962: Airflow in convective storms. *Quart. J. Roy. Meteor. Soc.*, **88**, 117–135.

Byers, H. R., 1942: Nonfrontal thunderstorms. Dept. Meteor., Univ. of Chicago, Misc. Rep. 3, 26 pp.

Byers, H. R., and R. R. Braham, Jr., 1949: *The Thunderstorm.* U.S. Government Printing Office, Washington, D.C., 287 pp.

Caracena, F., R. A. Maddox, L. R. Hoxit, and C. F. Chappell, 1979: Mesoanalysis of the Big Thompson storm. *Mon. Wea. Rev.*, **107**, 1–17.

Charba, J., and Y. Sasaki, 1971: Structure and movement of the severe thunderstorms of 3 April 1964 as revealed from radar and surface mesonetwork data analysis. *J. Meteor. Soc. Japan*, **49**, 191–213.

Cotton, W. R., R. L. George, and K. R. Knupp, 1982: An intense, quasi-steady thunderstorm over mountainous terrain. Part I: Evolution of the storm-initiating mesoscale circulation. *J. Atmos. Sci.*, **39**, 328–342.

Espy, J. P., 1861: *The Philosophy of Storms.* Little, Brown, 341 pp.

Fujita, T., and H. Grandosa, 1968: Split of a thunderstorm into anticyclonic and cyclonic storms and their motion as determined from numerical model experiments. *J. Atmos. Sci.*, **25**, 416–439.

Goldman, J. L., and E. M. Wilkins, 1973: Drag experiments with cylinders of varying roughness related to flow around thunderstorm cells. *J. Geophys. Res.*, **78**, 913–919.

Green, J. S. A., F. H. Ludlam, and T. F. R. McIlveen, 1966: Isentropic relative flow analysis and the parcel theory. *Quart. J. Roy. Meteor. Soc.*, **92**, 210–219.

Haman, K. E., 1976: On the airflow and motion of quasi-steady convective storms. *Mon. Wea. Rev.*, **104**, 49–56.

Haman, K. E., 1978: On the motion of a three-dimensional quasi-steady convective storm in shear. *Mon. Wea. Rev.*, **106**, 1622–1626.

Hammond, G. R., 1967: Study of a left-moving thunderstorm of 23 April 1964. Tech. Memo. IERTM–NSSL 31, ESSA, Norman, Oklahoma, (NTIS#PB–174681), 75 pp.

Hitschfeld, W., 1967: The motion and erosion of convective storms in vertical wind shear. *J. Meteor.*, **17**, 270–282.

Hoxit, L. R., R. A. Maddox, C. F. Chappell, F. L. Zuckerberg, H. M. Mogil, I. Jones, D. R. Greene, R. E. Saffle, and R. A. Scofield, 1978: Meteorological Analysis of the Johnstown, Pennsylvania Flash Flood, 19–20 July 1977. NOAA Tech. Rep. ERL 401–APCL 43 (NTIS#PB–

297412/9GA), 71 pp.

Humphreys, W. J., 1914: The thunderstorm and its phenomena. *Mon. Wea. Rev.*, **42**, 348–380.

Humphreys, W. J., 1940: *Physics of the Air* (3rd ed.). McGraw Hill, New York, 676 pp.

Klemp, J.B., and R. B. Wilhelmson, 1978: The simulation of three-dimensional convective storm dynamics. *J. Atmos. Sci.*, **35**, 1070–1110.

Knupp, K. R., and W. R. Cotton, 1982: An intense, quasi-steady thunderstorm over mountainous terrain. Part II: Doppler radar observations of the storm morphological structure. *J. Atmos. Sci.*, **39**, 343–358.

Lemon, L. R., and C. A. Doswell, 1979: Severe thunderstorm evolution and mesocyclone structure as related to tornadogenesis. *Mon. Wea. Rev.*, **107**, 1184–1197.

Ludlam, F. H., 1963: Severe local storms: A review. In *Severe Local Storms*, Meteor. Monogr. American Meteorological Society, Boston, 1–28.

Maddox, R. A., 1980: Mesoscale convective complexes. *Bull. Amer. Meteor. Soc.*, **61**, 1374–1387.

Maddox, R. A., L. R. Hoxit, C. F. Chappell, and F. Caracena, 1978: Comparison of meteorological aspects of the Big Thompson and Rapid City floods. *Mon. Wea. Rev.*, **106**, 375–389.

Maddox, R. A., C. F. Chappell, and L. R. Hoxit, 1979: Synoptic and meso-α-scale aspects of flash flood events. *Bull. Amer. Meteor. Soc.*, **60**, 115–123.

Marwitz, J. D., 1972: The structure and motion of severe hailstorms. Part II: Multicell storms. *J. Appl. Meteor.*, **11**, 180–188.

Merritt, J. H., and J. M. Fritsch, 1984: On the movement of the heavy precipitation areas of mid-latitude mesoscale convective complexes. Preprints, 10th Conference on Weather Forecasting and Analysis, Clearwater. American Meteorological Society, Boston, 529–536.

Moncrieff, M. W., and J. S. A. Green, 1972: The propagation and transfer properties of steady convective overturning in shear. *Quart. J. Roy. Meteor. Soc.*, **98**, 336–353.

Moncrieff, M. W., and M. J. Miller, 1976: The dynamics and simulation of tropical cumulonimbus and squall lines. *Quart. J. Roy. Meteor. Soc.*, **102**, 373–394.

Newton, C. W., and J. C. Fankhauser, 1964: On the movements of convective storms, with emphasis on size discrimination in relation to water-budget requirements. *J. Appl. Meteor.*, **3**, 651–668.

Newton, C. W., and J. C. Fankhauser, 1975: Movement and propagation of multicellular convective storms. *Pure Appl. Geophys.*, **113**, 747–764.

Newton, C. W., and S. Katz, 1958: Movement of large convective rainstorms in relation to winds aloft. *Bull. Amer. Meteor. Soc.*, **39**, 129–136.

Newton, C. W., and H. R. Newton, 1959: Dynamical interactions between large convective clouds and environment with vertical shear. *J. Meteor.*, **16**, 483–496.

Rotunno, R., and J. B. Klemp, 1982: The influence of the shear-induced pressure gradient on thunderstorm motion. *Mon. Wea. Rev.*, **110**, 136–151.

Rotunno, R., and J. B. Klemp, 1985: On the rotation and propagation of simulated supercell thunderstorms. *J. Atmos. Sci.*, **42**, 271–292.

Schlesinger, R. E., 1978: A three-dimensional numerical model of an isolated thunderstorm. Part I: Comparative experiments for variable ambient wind shear. *J. Atmos. Sci.*, **35**, 690–713.

Schlesinger, R. E., 1980: A three-dimensional numerical model of an isolated thunderstorm. Part II: Dynamics of updraft splitting and mesovortex couplet evolution. *J. Atmos. Sci.*, **37**, 395–420.

Weaver, J. F., 1979: Storm motion as related to boundary-layer convergence. *Mon. Wea. Rev.*, **107**, 612–619.

Weisman, M. L., and J. B. Klemp, 1982: The dependence of numerically simulated convective storms on vertical wind shear and buoyancy. *Mon. Wea. Rev.*, **110**, 504–520.

Wilhelmson, R. B., and C. S. Chen, 1982: Simulations of the development of successive cells along a cold outflow boundary. *J. Atmos. Sci.*, **39**, 1456–1465.

Wilhelmson, R. B., and J. B. Klemp, 1978: A numerical study of storm splitting that leads to long-lived storms. *J. Atmos. Sci.*, **35**, 1974–1986.

CHAPTER 14

Mesoscale Structure of Hurricanes

Robert W. Burpee

14.1. Introduction

Tropical cyclones can produce widespread damage and account for the loss of many lives. Each year approximately 80 tropical cyclones with maximum sustained winds ≥ 20 m s^{-1} occur over the globe (Gray, 1979). But, because of the relative rarity of tropical cyclones at any one location, forecasters at many local offices have limited tropical cyclone experience.

In the Western Hemisphere, tropical cyclones with maximum sustained winds ≥ 18 m s^{-1} are termed tropical storms and are assigned names. When their maximum sustained winds are ≥ 33 m s^{-1}, these cyclones are called hurricanes. During the past 35 years, the average number of named storms in the Atlantic basin each year has been 10, of which 6 become hurricanes (Clark, 1983).

Background material on the nature and history of hurricanes is available. Gray (1968, 1979) determined climatological statistics of tropical cyclones in all ocean basins and also described typical environmental conditions favorable for the formation and intensification of tropical cyclones. Anthes (1982) discussed the formation, maintenance, and decay of tropical cyclones, as well as detailed their structure, numerical simulation, and forecasting. Ooyama (1982) presented a concise summary, without mathematical equations, of theoretical progress in understanding the role of organized moist convection in tropical cyclones.

Although satellites are playing an increasing role in estimating storm position (Gaby et al., 1980), intensity (Dvorak, 1975; Kidder et al., 1978), and rainfall potential (Rodgers and Adler, 1981), they are not treated here. Rather, analyses of observations from research aircraft and land-based platforms that were recorded within 150 km of a storm's eye are presented. This information may be useful in local forecasting offices that are not regularly involved in forecasting during the landfall of hurricanes.

14.2. General Storm Structure

14.2.1. Mesoscale Observations

Research aircraft and coastal radars have provided nearly all the observations of the mesoscale structure of hurricanes. The first research flights were made into hurricanes in 1956, and radar data were first collected in hurricanes in the mid-1940s (e.g., Wexler, 1947). With the instrumentation capability that was available until the early 1970s, only very general properties of the mesoscale structure of hurricanes could be determined. In the mid-1970s, advances in aircraft instrumentation made it possible to record digital radar data and to determine accurate horizontal and vertical winds on research aircraft. About the same time, mobile equipment was developed to record digital radar data at the coastal station of the National Weather Service nearest to the hurricane landfall. As a result of these and other technological advances, detailed observations of the mesoscale structure of mature hurricanes have been available since 1977. However, relatively few hurricanes occurred in the Atlantic basin from 1977 to 1984. In view of the structural differences between hurricanes that have been observed and the changes that can occur in individual hurricanes, more storms need to be observed and analyzed before many basic features of the mesoscale structure are clearly understood.

The National Oceanic and Atmospheric Administration (NOAA) has two WP–3D aircraft that fly research and reconnaissance missions in hurricanes. During research missions, the aircraft have flown at altitudes as low as 150 m and as high as 8000 m and with airspeeds of 110–135 m s^{-1}. It is not possible to obtain flight-level data over a wide range of spatial and temporal scales simultaneously with only two aircraft. The research flight patterns are typically designed to monitor the mesoscale structure of a storm within 150 km of the center for 24–36 h, by scheduling back-to-back missions at a single level, or to concentrate both aircraft on a convectively active region of a storm for at least 2–3 h.

14.2.2. Large-Scale Features

In the fully developed stage, a hurricane is a nearly circular, warm-core vortex that extends throughout the depth of the troposphere and has a typical horizontal scale of a few hundred kilometers. Maximum horizontal winds occur 10–50 km from the storm center and 500–1000 m above the ocean surface. Figure 14.1 shows two radial profiles of horizontal winds from Hurricane Alicia plotted relative to the storm's moving center. The wind data were recorded by a research aircraft that passed through the center of the hurricane at an altitude of 1500 m. The winds to the right and left of the storm track have maximum values in excess of 40 m s^{-1} 18–20 km from Alicia's center. The winds decrease from about 40 m s^{-1} to 20 m s^{-1} as the radius increases from 20 to 150 km. In the Northern Hemisphere, winds measured on the right side of a tropical cyclone tend to be stronger than those on the left because of the storm's translation speed. In Fig. 14.1, the winds computed relative to the storm motion on the right of Alicia's track are a

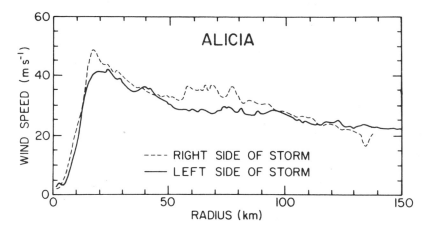

Figure 14.1. Radial profile of relative horizontal winds in Hurricane Alicia. The winds were recorded by one of the NOAA research aircraft at 1500 m at approximately 0100 GMT, 18 August 1983 (unpublished data). The relative winds were computed by subtracting the storm's motion from the winds measured by the aircraft.

few meters per second stronger than those on the left side. The difference in relative wind speed between the left and right sides occurred because the storm was moving slower than the component of air motion along the storm track at the altitude of the aircraft, and because of a convectively induced transient in the eyewall of the northeast side of the storm.

Hawkins and Rubsam (1968) determined the structural characteristics of several kinematic and thermodynamic variables in the inner core region of Hurricane Hilda. Figure 14.2 shows their analysis of horizontal wind speed relative to the moving hurricane. The analysis is based on aircraft data recorded at five pressure levels between 900 and 150 mb. At radii with the strongest low-level winds, the horizontal wind speed decreases very slowly in the vertical. Gray (1966) hypothesized that the low values of vertical shear of the horizontal wind are caused by deep cumulus convection that mixes momentum rapidly in the vertical. The radial flow in the inner core region of a hurricane is difficult to determine precisely because the calculation of the radial component of the wind is very dependent on accurate positioning of the storm center. Typically the radial flow is characterized by a layer of inflow about 200 mb thick that is located just above the ocean surface and a deeper layer of outflow in the upper troposphere. From 800 to 400 mb, the radial flow tends to be rather weak. The warm core structure of a hurricane is illustrated in Fig. 14.3. Maximum temperature anomalies are typically 15°C and occur at about 250–300 mb. Frank (1977a, b) composited radiosonde observations from many tropical cyclones in the western Pacific and determined the average kinematic and thermodynamic structure of these storms.

Radar observations reveal that the center or eye of a hurricane does not contain any significant precipitation (e.g., Wexler, 1947). The eye is surrounded by a nearly circular ring of deep cumulus convection called the eyewall. The deep convection in the eyewall is usually not symmetrically

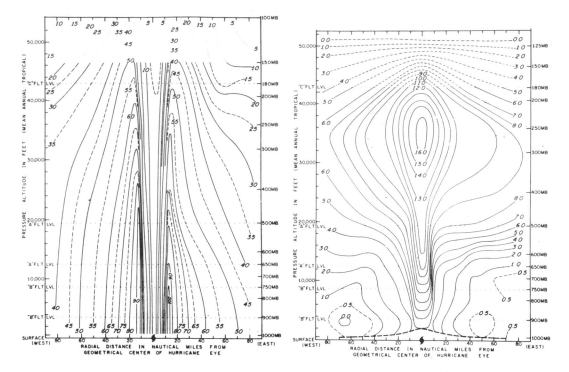

Figure 14.2. Vertical cross section of wind speed in knots (1 m s^{-1} = 1.94 kt) from Hurricane Hilda, 1 October 1964. The wind speeds were computed relative to the motion of the center of the hurricane. (From Hawkins and Rubsam, 1968.)

Figure 14.3. Vertical cross section of temperature anomalies (°C) from Hurricane Hilda, 1 October 1964. The temperature anomalies are deviations from Jordan's (1958) mean tropical sounding. (From Hawkins and Rubsam, 1968.)

distributed. The inner edge of the eyewall in some hurricanes may be < 15 km in diameter, and in other hurricanes the diameter may be >80 km. In a few intense hurricanes, the diameter of the eyewall may vary over this range in 1 or 2 days. The maximum reflectivity in the eyewall is typically 45–50 dBZ, a much lower value than is usually found in middle-latitude thunderstorms. Echoes with maximum reflectivities of 51–53 dBZ have been observed, but they occur infrequently and last only a few minutes. Outside the eyewall, there are typically several spiral rainbands that have a few isolated areas of deep cumulus convection and extensive areas of stratiform rain (e.g., Atlas et al., 1963).

The shape of the eye on radar is circular in some hurricanes (Fig. 14.4) at a time when the minimum central pressure was about 970 mb. The diameter of Alicia's eye at this time was about 20 km. In some intense hurricanes, like Allen of 1980, two concentric eyewalls have been observed on radar for several hours (Marks, 1985). In other hurricanes, such as Frederic when its minimum pressure was about 945 mb, the shape of the eyewall is more elliptical (Fig. 14.5). In this case, the major axis was about 80 km long and the minor axis was about 60 km. Figures 14.4 and 14.5 are plotted on the same scale and represent the very different precipitation structures that may

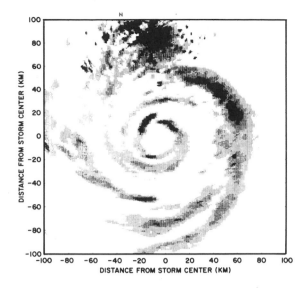

Figure 14.4. Reflectivity data recorded during Hurricane Alicia from the WSR–57 radar in Galveston, Tex., at 0014 GMT, 18 August 1983. The thresholds for the five contour levels are 18, 28, 33, 38, and 43 dBZ. The tilt angle is 0.4°, and the domain of the display is 200 km by 200 km. The domain is centered on the hurricane. North is at the top center, and the Galveston radar is at the top slightly left of center. (From Burpee and Marks, 1984.) Most of the wind observations in Fig. 14.1 were recorded within 1 h of the time of this radar sweep.

occur in fairly intense hurricanes. The precipitation in tropical storms and weak hurricanes is organized in rainbands that may not encircle the storm center completely. There is frequently a major rainband with embedded deep convection that spirals in toward the center.

The pattern of convection in the eyewall and inner rainbands does not usually remain the same throughout the lifetime of a particular storm. For example, Marks (1985) presented airborne radar observations gathered during six days in Hurricane Allen. The shape of the eyewall was circular on some occasions and noncircular on others. During this time, the diameter of the eye varied from a minimum of 25 km on 9 August 1980 to a maximum of 60 km on 6 August.

The physical processes that account for the shape of the eyewall in different hurricanes are not clearly understood. Willoughby (1984, private communication) speculated that the shape is determined in part by the storm

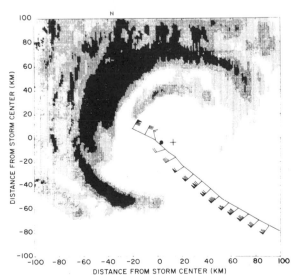

Figure 14.5. Reflectivity data during Hurricane Frederic from the WSR–57 radar in Slidell, La., at 0005 GMT, 13 September 1979. The tilt angle is 0.2°. The wind data are from one of the WP–3D research aircraft operated by NOAA's Office of Aircraft Operations. The observations were collected from 2355 to 0015 GMT as the aircraft headed northwest. The solid circle represents the wind center, and the + indicates the approximate location of the radar center. Other parameters are as in Fig. 14.4. (From Burpee and Marks, 1984.)

intensity and in part by the storm environment, such as the vertical shear
of the horizontal wind and nearby land masses. The wind center at altitudes $\leq$ 1500 m is typically located very close to the geometric center of the
eye observed on radar displays when the radar beam is below the freezing
level. On some occasions, particularly when the most intense convection is
concentrated in a relatively small sector of the eyewall, the wind center may
be displaced several kilometers from the radar center. Figure 14.5 shows an
example from Hurricane Frederic when research aircraft determined that the
wind center was about 10 km to the west of the radar center. The northwest
part of the eyewall had the most intense convection.

14.2.3. Eyewall Structure

Until research aircraft began flying in hurricanes in the middle 1950s, it
was not possible to describe the kinematic and thermodynamic structure of
the inner core region of hurricanes. Shea and Gray (1973) processed data
obtained from many flights in 1957–1969 and composited the observations
relative to the radius of maximum wind. In the eyewall region, they found
that inflow is confined to the lowest 1500 m, maximum winds occur within
the eyewall, and descent occurs in the eye. Advances in aircraft instrumentation since the middle 1970s have made possible quantitative precipitation
estimates from radar, accurate horizontal and vertical wind measurements,
and quantitative observations of cloud microphysics. Jorgensen (1984a,b)
composited data from several flights of the NOAA WP–3D aircraft in four
mature hurricanes. Although three of these hurricanes had circular eyewalls and one had an asymmetric eyewall, he found similar profiles of radial
and tangential wind, radar reflectivity, and vertical motion on several radial
passes. Jorgensen showed that the circulation in the eyewall is highly organized in a vertical plane along a radial through the storm center. Embedded
within the two-dimensional eyewall are cores of high reflectivity that are 2–5
km in diameter. Locations of features shown in Fig. 14.6 are typical. In his
analysis of the aircraft data, Jorgensen found that the 10 dBZ reflectivity
contour and the radius of maximum tangential wind slope outward with increasing height. He also showed that the strongest convective-scale vertical
motions are located 1–6 km inward from the radius of the maximum tangential wind. At low levels, the zone of maximum radar reflectivity is several
kilometers radially outward from the radius of maximum tangential wind;
in the middle troposphere, the maximum radar reflectivity and maximum
tangential wind are nearly collocated. Jorgensen determined that liquid water content and radar reflectivity in the eyewall decrease rapidly above the
freezing level (about 5 km).

Marks and Houze (1984a) found generally similar kinematic structure in
their analyses of airborne Doppler radar data collected in Hurricane Alicia.
The Doppler analyses showed updrafts along the inner edge of the eyewall
that sloped radially outward with increasing height. The maximum updrafts
were positioned above the reflectivity maximum and varied from 5 to 10 m
s^{-1}. A downdraft coincided with the reflectivity maximum in the eyewall.
Marks and Houze (1984b) also analyzed airborne Doppler radar data from

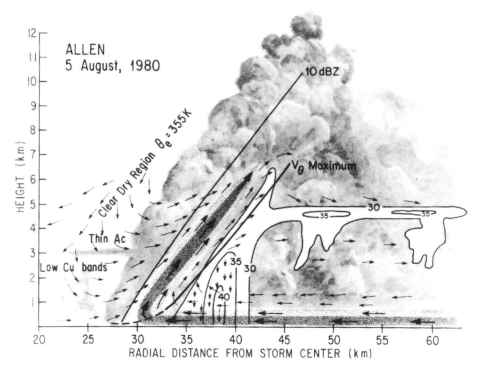

Figure 14.6. Locations of the clouds and precipitation, radius of maximum winds, and radial-vertical airflow through the eyewall of Hurricane Allen on 5 August 1980. (From Jorgensen, 1984b.)

Hurricane Debby. They showed that, at altitudes of 2–4 km in a portion of the developing eyewall, two mesoscale wind speed maxima and a mesoscale vortex were superimposed on the hurricane-scale circulation.

14.2.4. Rainband Structure

Radar reflectivity observations show that hurricane rainbands have both a stratiform and convective structure (Barnes *et al.*, 1983; Jorgensen, 1984a). Rainbands have fewer vertically oriented cores of reflectivity and fewer organized updrafts than the eyewalls have. Figure 14.7 is an example of the radar reflectivity pattern in a vertical plane in Hurricane Alicia. Radially outward from the eyewall, the rainbands are characterized by extensive horizontally homogeneous reflectivity patterns with "bright bands" of enhanced reflectivity at altitudes of 4.5–5.0 km, just below the melting level. Jorgensen (1984a) estimated that stratiform precipitation in the rainbands of Hurricanes Frederic and Allen covered areas about 10 times larger than convective precipitation.

A hurricane's rainbands seem to be more three-dimensional than its eyewall. Flight-level variables at a single level show considerable variability between successive penetrations across the same rainband. Gentry (1964) found that kinematic and thermodynamic variables measured by aircraft traverses along rainbands vary widely. As a result, characteristic features of

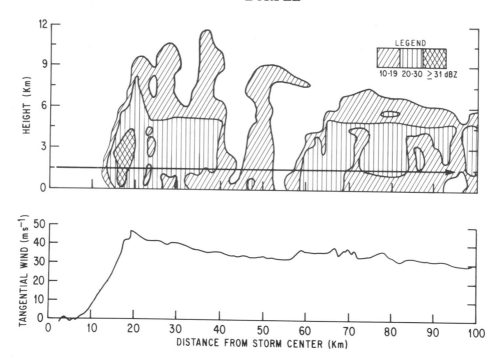

Figure 14.7. Profiles of radar reflectivity and tangential winds in Hurricane Alicia. The data were recorded from 0109 to 0128 GMT, 18 August 1983. The radar reflectivity cross section was obtained by compositing vertical rays from the tail radar at four samples per minute. The horizontal arrow indicates the altitude of the aircraft.

rainbands are much more difficult to specify than are those of the eyewall. Detailed aircraft observations of the structure of rainbands have been obtained in only two hurricanes. Barnes *et al.* (1983) composited aircraft data from repeated passes at many levels through a rainband in Hurricane Floyd, and Jorgensen and Marks (1984) deduced the structure of a rainband in Hurricane Tico from airborne Doppler radar data. In both cases, the mesoscale updraft sloped outward with height, and in the middle troposphere it was located above the highest reflectivity. At 300 m in Hurricane Floyd, the equivalent potential temperature (θ_e) was 12 K lower on the inside edge of the band than on the outside edge (Fig. 14.8). Barnes *et al.* speculated that subgrid-scale transports of θ_e by cumulus-scale cells accounted for the decrease in θ_e across the band. In the Tico rainband, however, there were no significant horizontal gradients of θ_e. The reasons for the difference in the thermodynamic structure of the two rainbands are not known.

Atlas *et al.* (1963) and Barnes *et al.* (1983) have indicated that the upwind end of some rainbands is mainly convective and that there is a transition toward less convective and more stratiform precipitation toward the downwind end. Movies of the radar reflectivity patterns of rainbands in several recent hurricanes reveal that organized mesoscale convective areas occasionally move downwind along rainbands. In such circumstances, the downwind end of a rainband may temporarily be more convective than other parts of the band.

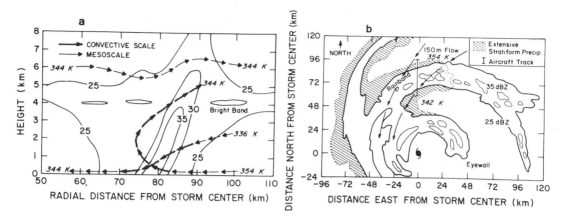

Figure 14.8. (a) Vertical view of the Hurricane Floyd rainband. Thin solid lines indicate average radar reflectivity in dBZ. Typical mesoscale and convective-scale motions are represented by arrows. The numbers along the arrows are equivalent potential temperatures. (b) Horizontal view of the rainband, showing the aircraft track, radar reflectivity, the location of cells and stratiform precipitation, the 150 m flow, and equivalent potential temperatures. (From Barnes *et al.*, 1983.)

14.3. Characteristics of Cumulus Convection in Hurricanes

The deep cumulus convection that occurs in hurricanes has many similarities with tropical cumulus clouds that have been studied in the eastern Atlantic during GATE (the GARP Atlantic Tropical Experiment) by LeMone and Zipser (1980) and in the western Pacific and Indian Ocean by Warner and McNamara (1984). LeMone and Zipser defined a convective updraft or downdraft from 1 s time series of vertical velocity measured by aircraft as a continuous positive or negative event with a horizontal distance of at least 500 m and a maximum velocity $> |0.5|$ m s^{-1}. They defined part of a draft as a core if the vertical velocity exceeded ± 1 m s^{-1} for 500 m. LeMone and Zipser found that the vertical velocities in convection during GATE were much weaker than in middle-altitude convection over land.

Jorgensen *et al.* (1985) studied the convective drafts and cores in four intense hurricanes. Updrafts were more numerous and transported more mass than downdrafts. The sizes and intensities of drafts and cores in hurricane rainbands closely resemble those from GATE. Convective cores in the eyewall tended to have vertical velocities similar to those in GATE, but they were twice as large and transported twice as much mass. As in GATE, the majority of updraft cores were weak compared with typical updrafts in middle-latitude thunderstorms over land (Fig. 14.9). The weak updrafts are consistent with the rapid decrease in radar reflectivity and cloud water above the freezing level that was described by Jorgensen (1984a). The updrafts are unable to transport large water drops up through the freezing level, where ice particles quickly induce freezing of the cloud water.

Parrish *et al.* (1984) studied radar echoes in Hurricane Frederic of 1979 and compared the results with López's (1978) radar analyses of GATE convection. They tabulated the lifetime and maximum area of cells in Hurricane

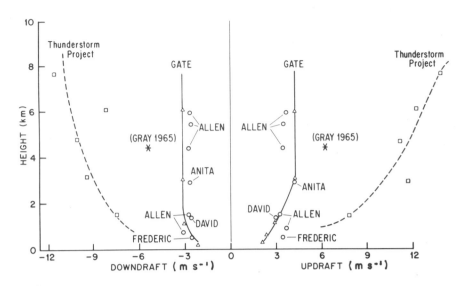

Figure 14.9. Average vertical velocity at the strongest 10% level (from log-probability plots) for updraft and downdraft cores as a function of height. Triangles are GATE values; squares are Thunderstorm Project values from Florida and Ohio; asterisks are estimates by Gray (1965) in earlier hurricanes. (Adapted from Zipser and LeMone, 1980; reproduced from Jorgensen *et al.*, 1985.)

Frederic and plotted them on log-probability paper (Figs. 14.10 and 14.11). In log-probability coordinates, cell lifetime and maximum area frequencies are nearly in a straight line. A true lognormal distribution in this coordinate system is a straight line.

Cell lifetime and maximum area distributions from the GATE results of López (1978) are also plotted in Figs. 14.10 and 14.11. The GATE cells are shown for isolated or individual cells (cells that never merge, never split, and appear to contain only one cell or reflectivity maximum during their lifetime) and cells from groups (all cells that are not individual cells). In

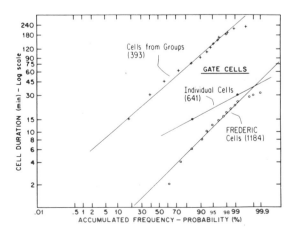

Figure 14.10. Accumulated frequency distributions of the cell duration (min) in log-probability coordinates for Hurricane Frederic (open circles) and GATE (+ and solid circles). The GATE cell frequencies are reproduced from López (1978), and the figure is from Parrish *et al.* (1984).

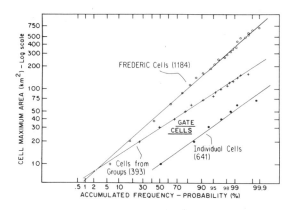

Figure 14.11. Accumulated frequency distributions of the maximum cell area (km^2) in log-probability coordinates for Hurricane Frederic (open circles) and GATE (+ and solid circles). The GATE cell frequencies are from López (1978), and the figure is from Parrish *et al.* (1984).

López's terminology, Hurricane Frederic cells are group cells because of the large echo areas within which they are observed; the hurricane cells tend to have larger areas and shorter durations. The reasons for these differences are not clearly understood, but they may be related to characteristic differences in horizontal wind speed (the average speed of the cells in Frederic was 30 m s^{-1}), thermodynamic stability, or the total area covered by radar echoes.

14.4. Motion of Hurricane Rainbands Relative to the Eyewall

The strong horizontal winds in the inner regions of hurricanes advect individual cumulus cells at about the speed of the low-level wind, counterclockwise about the storm center (Parrish *et al.*, 1984). The mesoscale rainbands, however, remain fairly stationary relative to the storm center in some hurricanes and rotate about the center in others. For example, Fig. 14.12

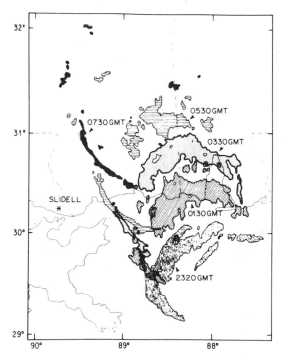

Figure 14.12. Outlines of the 41 dBZ contours from the Slidell, Pensacola, and Jackson radars at approximately 2 h intervals from 2320 GMT, 12 September, to 0730 GMT, 13 September 1979. (From Parrish *et al.*, 1982.)

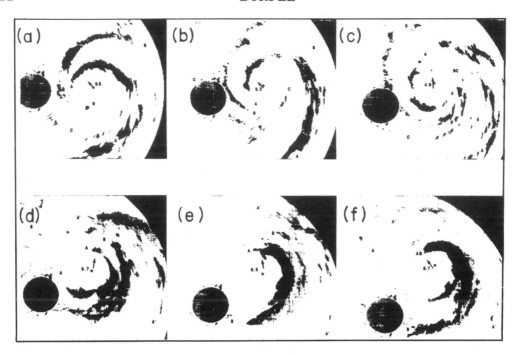

Figure 14.13. Single sweeps of the Miami WSR–57 radar at (a) 0915, (b) 0959, (c) 1045, (d) 1131, (e) 1217, and (f) 1300 GMT, 3 September 1979. Successively darker shading indicates reflectivities >25, 30, 35, and 40 dBZ. Each frame represents an area 240 km square. (From Willoughby *et al.*, 1984.)

shows that in Hurricane Frederic the most intense areas of mesoscale convection stayed in about the same storm-relative position during landfall. Land-based radar data taken before landfall (not shown) indicate that the rainbands and eyewall remained in the same storm-relative position for at least 24 h. On the other hand, it is clearly evident from Fig. 14.13 that the mesoscale rainbands in Hurricane David rotated cyclonically about the storm center while David was east of Miami. In the radar sequence, the major mesoscale convection moved completely around the center of the storm in a little more than 2 h. During that time, David had maximum sustained winds of 40 m s^{-1} and an eyewall diameter of about 70 km.

The cyclonic motion of the mesoscale convection coincided with a trochoidal oscillation in the storm track (Fig. 14.14). A trochoidal oscillation is a wave-like oscillation in the track of tropical cyclones. Lewis and Black (1977) summarized the published research on this subject and analyzed oscillations in the tracks of six hurricanes. They found that cross-track oscillations of ±20 km and periods of 7–14 h, relative to the smooth storm tracks, were typical of these storms. Willoughby *et al.* (1984) speculated that the motion of the mesoscale convection around David's eye influenced the change in the speed and direction of the eye. During the times represented in Fig. 14.13, the average storm motion was toward the northwest at about 6 m s^{-1}. When the deep convection was southwest and south of

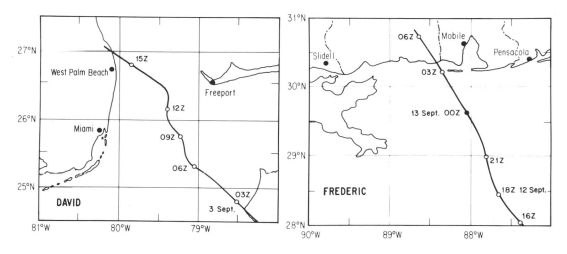

Figure 14.14. Part of the track of Hurricane David, 3 September 1979.

Figure 14.15. Part of the track of Hurricane Frederic, 12 and 13 September 1979.

the center, David's eye slowed down and nearly stalled. The eye speeded up and temporarily headed a little east of north as the deep convection passed around the east side of the storm. It is unclear to what extent those parts of the storm that were located radially outward from the eyewall were affected by the trochoidal oscillations.

Trochoidal oscillations in the tracks of hurricanes are still not well documented observationally or understood physically. From the available information, it appears that hurricanes with quasi-stationary rainbands, such as Frederic, or hurricanes that move much faster than average, tend to have smooth tracks. A portion of Hurricane Frederic's relatively smooth track is shown in Fig. 14.15; David's track in Fig. 14.14 is on the same scale. David's track oscillated more than 20 km from the mean track during the time (0915–1300 GMT) that the mesoscale convection and the rainbands rotated around the storm center. Because these oscillations occur in many storms, forecasters should be cautious about using extrapolation to estimate future storm motion.

Many intense, highly symmetric hurricanes have two concentric, nearly circular eyewalls. Willoughby *et al.* (1982) found that a horizontal wind maximum tends to be associated with both the inner and outer eyewall. The outer wind maximum and eyewall are frequently observed to constrict about the inner eyewall and ultimately to replace it. When the inner eye dissipates, the minimum central pressure of the storm often increases. Hurricane Allen of 1980 is a particularly good example of this sequence of events. Time series of Allen's minimum pressure and radius of maximum wind are illustrated in Fig. 14.16. Two complete cycles of the concentric eye contraction occurred from 4 to 8 August. Marks (1985) showed that the radius of the eyewall observed by radar decreased at the same rate as the radius of Allen's outer wind maximum.

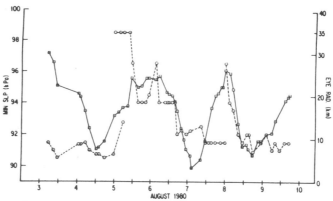

Figure 14.16. Time changes of eye diameter (dashed lines) and central pressure (solid line) for Hurricane Allen. The dates are plotted at 1200 GMT. (From Willoughby *et al.*, 1982.)

14.5. Changes in Mesoscale Convection During Landfall

Several observational and numerical studies of hurricanes during land-fall have been completed in recent years. Powell (1982) analyzed surface streamlines and isotachs based on composited aircraft, ship, and land wind observations for an 8 h period as Frederic crossed the Gulf coast (Fig. 14.17). His analyses show an abrupt discontinuity in wind speed and a change in wind direction at the coastline, such that the streamlines are oriented more toward the storm center over land. The discontinuity is clearly a result of the change in surface drag that occurs at the coast. Numerical simulations of the landfall of an idealized hurricane by Tuleya *et al.* (1984) show the same type of discontinuity at the coast. In Powell's composite analysis, it is interesting to note that the confluence of the streamlines is opposed by downstream acceleration as the streamlines cross the isotachs toward higher wind speeds. The representativeness of the composited surface wind field from Frederic is not yet well established because there have not been enough

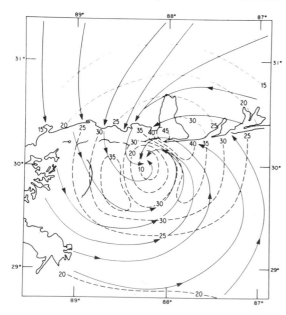

Figure 14.17. Streamline and isotach analyses for the landfall composite of Hurricane Frederic for 0000–0800 GMT, 13 September 1979. The isotachs are in meters per second, and the area of the analysis is approximately 300 km by 300 km. (After Powell, 1982.)

observations from other storms for similar analyses.

In spite of the uncertainties about the pattern of surface convergence in the right front quadrant of Frederic, the coastline appeared to enhance the initiation of major convective features. Parrish *et al.* (1982) showed that the rainfall maximum occurred slightly to the left of Frederic's track. The rainfall maximum was caused by convective bands that initially formed near the coast and to the right of the storm track. After their formation, the convective areas became better organized and increased in strength as they moved cyclonically around Frederic's center and then produced the greatest precipitation several kilometers to the left of the storm track.

Four PPI presentations from the digital radar data collected at Slidell, La., during Frederic's landfall are shown in Fig. 14.18. During the 3 h period (0100–0400 GMT), there was a tendency for intense convective cells to form near the Alabama coastline (Figs. 14.18a, b, c). As Frederic moved ashore and

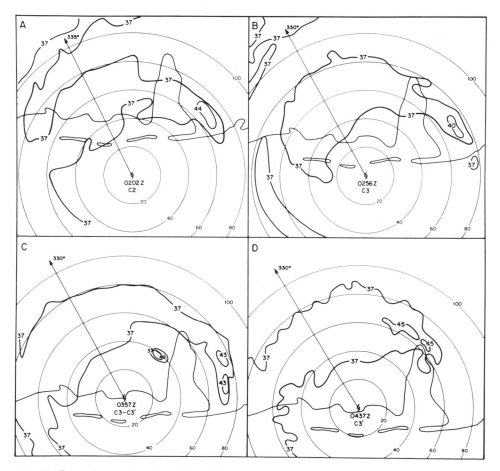

Figure 14.18. Location of the initial formation of intense convective bands near the Alabama coast along the Gulf of Mexico and Mobile Bay during the landfall of Hurricane Frederic. The intensity level of the mesoscale bands is shown by appropriate reflectivity contours ≥ 40 dBZ. The 37 dBZ contour represents the boundary of the primary rain area. The storm center is indicated by the hurricane symbol, and the range rings are at 20 km intervals. Times are in GMT on 13 September 1979. (From Burpee and Marks, 1984.)

the low-level flow developed a long fetch over Mobile Bay, intense convective cells also began to form in a second mesoscale band near the downwind end of Mobile Bay (Figs. 14.18c,d). The land-sea interface appears to have had an important effect on the initiation of organized convection in Frederic. It is not possible to corroborate this effect with the radar data from David because the rainbands tended to be parallel to the coast near Miami and David made landfall about 150 km north of the radar. The rainbands north of David's center were beyond the quantitative range of the radar. Analysis of the radar data obtained during Alicia's landfall is not yet complete.

14.6. Surface Winds and Convective Bands

It is well known that strong winds in hurricanes can cause widespread damage. For more than 10 years, Fujita (1978) and his colleagues have been analyzing aerial photographs of hurricanes that caused significant damage in the United States and mapping their estimates of the peak wind gusts. Frederic was one of these storms. Parrish *et al.* (1982) showed that most of the area with estimated peak wind gusts ≥ 45 m s^{-1} in Frederic also had radar reflectivities ≥ 41 dBZ at some time during the storm. Because of the close agreement between the areas of occurrence of high rainfall rates and significant wind damage, they suggested that convective-scale downdrafts may transport high-momentum air into the boundary layer. On the basis of his analyses of Hurricane Celia of 1971, Fujita (1978) also concluded that convective-scale downdrafts cause most of the wind damage.

If convective-scale downdrafts are responsible for the most severe wind damage in hurricanes, then the vertical shear of the horizontal wind in the lower troposphere must be relatively large over land. Evidence in support of this characteristic was provided by Novlan and Gray (1974) in a study of hurricanes that spawned numerous tornadoes at landfall. Additional support for large wind shear near the surface was reported by Bates (1977) and Powell (1982). Bates compared low-level aircraft winds with nearby surface winds and found that 10 m mean winds over land are approximately 40% of aircraft-measured winds at altitudes of 150–3000 m. Powell analyzed aircraft, ship, buoy, and land station data during the landfall of Hurricane Frederic. He estimated that 10 m winds at coastal stations are about 55% of low-level (500–1500 m) aircraft winds measured in the same position relative to the storm center. Over the ocean, Bates and Powell found 10 m mean winds to be about 70–75% of low-level aircraft winds.

It is difficult to determine the precise time that damage occurs; therefore, specific radar reflectivity patterns have not been correlated directly with wind damage. In recent years, many public and private facilities have purchased recording anemometers. The number of opportunities to compare anemometer traces with radar reflectivities is increasing. Preliminary work is under way with Alicia data to see if surface winds tend to be stronger in rainbands. Figure 14.19 compares peak wind gusts in 10 min intervals (plotted every 30 min) at a location about 50 km to the right of Alicia's track, and 30 min averaged reflectivity during the landfall of Hurricane Alicia. On

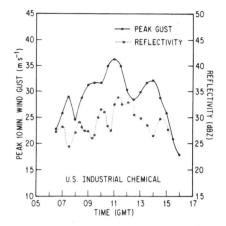

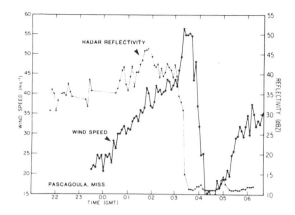

Figure 14.19. Peak wind gusts in 10 min intervals, estimated from the anemometer trace at U.S. Industrial Chemical, and the corresponding 30 min averaged reflectivities from the digital data recorded by the Texas A&M 10 cm radar interpolated at the anemometer's location, during the landfall of Hurricane Alicia. (From Burpee and Marks, 1984.)

Figure 14.20. Peak wind gusts in 5 min intervals, estimated from the anemometer trace at Pascagoula, and the corresponding 5 min averaged reflectivities from the digital data recorded by the Slidell WSR–57 radar interpolated at the anemometer's location, during the landfall of Hurricane Frederic. The minimum detectable signal of the radar at the range corresponding to Pascagoula is 11 dBZ. (From Parrish et al., 1982.)

this time scale, there is not a strong correlation between changes in peak wind gusts and changes in radar reflectivity.

In most hurricanes, radar reflectivity and peak wind gusts are not well correlated near the inner edge of the eyewall. Recent aircraft observations over the open ocean during the mature stages of Hurricanes David, Frederic, and Allen indicate that the radius of maximum winds may slope inward between standard low-level flight altitudes (500–3000 m) and the surface, and that the maximum surface winds are on some occasions on the inner edge, or slightly inside, the radar eyewall (Jones *et al.*, 1981). A similar relationship between maximum surface winds and the radar may also occur during landfall. In Hurricane Frederic, for example (Fig. 14.20), maximum winds of 50 m s^{-1} were measured on the inner edge of the eyewall, where the radar-estimated reflectivity was at or below the minimum detectable signal. Fujita (1978) also found that the maximum surface winds were inside the radar eye in Hurricane Celia.

14.7. Tornadoes Associated With Hurricanes

During 1948–1982, tornadoes were recorded in most hurricanes that crossed the U.S. coastline from Brownsville, Tex., to Long Island, N.Y. Gentry (1983) showed that the percentage of hurricanes with tornadoes increased in recent years as procedures for documenting tornado occurrence improved; from 1959 to 1960, tornadoes were observed in all tropical cyclones with at least hurricane intensity and in which the eastern quadrants of the storm's

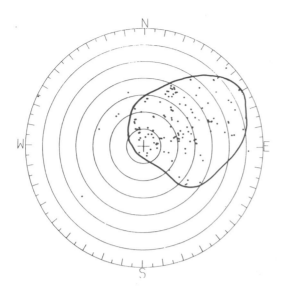

Figure 14.21. Locations of tornadoes plotted relative to the center of hurricanes that affected the United States from 1973 to 1980. The dark curve includes 95% of the cases. The range marks are in intervals of 50 km. (From Gentry, 1983.)

circulation made landfall. Most hurricane-induced tornadoes are weaker than those observed in the central and southern United States, but a few are as strong. There is a wide range in the number of tornadoes reported. About 75% of the hurricanes have fewer than 10 tornadoes. During Hurricane Beulah of 1967, however, 141 tornadoes were reported. Novlan and Gray (1974) estimated that tornadoes account for about 10% of the deaths and 0.5% of the damage in hurricanes, over the long term.

Orton (1970) and Gentry (1983) demonstrated that tornadoes typically occur either northeast and east of the storm center or in any quadrant within 50 km of the center (Fig. 14.21). Gentry noted that most tornadoes usually develop near areas of intense convection in the outer spiral rainbands, but some tornadoes also form in convectively active regions of the inner rainbands. Novlan and Gray (1974) found that nearly all hurricane-induced tornadoes occur within 200 km of the coast and in regions where the speed of the horizontal wind increases by about 20 m s^{-1} from the surface to 850 mb. In contrast with environmental conditions typical of the central plains, thermal instability is small in hurricanes. The low-level wind shear is thought to be a crucial factor in the genesis of hurricane tornadoes, but the physical processes involved in the initiation of the tornadoes are not well understood. Two mechanisms have been suggested for the development of the low-level shear. Novlan and Gray (1974) hypothesized that the shear is produced as a thermal wind response to the cold central core that develops near the surface in rapidly filling storms. More recently, Gentry (1983) argued that most tornadoes occur in areas close to the coastline where the trajectory of the low-level air has been over land for a relatively short time. In these regions, Gentry speculated that friction reduces the surface wind while the wind at slightly higher altitudes is still moving rapidly. So far, researchers have been unable to identify radar or satellite signatures that can be used to forecast tornadoes in hurricanes.

REFERENCES

Anthes, R. A., 1982: *Tropical Cyclones: Their Evolution, Structure, and Effects.* Meteor. Monogr., **19**(41), American Meteorological Society, Boston, 208 pp.

Atlas, D., K. R. Hardy, R. Wexler, and R. Boucher, 1963: The origin of hurricane spiral bands. *Geofis. Int.*, **3**, 123–132.

Barnes, G. M., E. J. Zipser, D. Jorgensen, and F. Marks, Jr., 1983: Mesoscale and convective structure of a hurricane rainband. *J. Atmos. Sci.*, **40**, 2125–2137.

Bates, J., 1977: Vertical shear of the horizontal wind speed in tropical cyclones. NOAA Tech. Memo. ERL WMPO–39 (NTIS#PB272659/AS), 19 pp.

Burpee, R. W., and F. D. Marks, Jr., 1984: Analyses of digital radar data obtained from coastal radars during Hurricanes David (1979), Frederic (1979), and Alicia (1983). Preprints, 10th Conference on Weather Forecasting and Analysis, Clearwater Beach, Fla., American Meteorological Society, Boston, 7–14.

Clark, G. B., 1983: Atlantic hurricane season of 1982. *Mon. Wea. Rev.*, **111**, 1071–1079.

Dvorak, V. F., 1975: Tropical cyclone intensity analysis and forecasting from satellite imagery. *Mon. Wea. Rev.*, **103**, 420–430.

Frank, W. M., 1977a: The structure and energetics of the tropical cyclone, I: Storm structure. *Mon. Wea. Rev.*, **105**, 1119–1135.

Frank, W. M., 1977b: The structure and energetics of the tropical cyclone, II: Dynamics and energetics. *Mon. Wea. Rev.*, **105**, 1136–1150.

Fujita, T. T., 1978: Manual of downburst identification for project NIMROD. SMRP Res. Pap. 156, Department of Geophysical Sciences, University of Chicago, 104 pp.

Gaby, D. C., J. B. Lushine, B. M. Mayfield, S. C. Pearce, and F. E. Torres, 1980: Satellite classifications of Atlantic tropical and subtropical cyclones: A review of eight years of classifications at Miami. *Mon. Wea. Rev.*, **108**, 587–595.

Gentry, R. C., 1964: A study of hurricane rainbands. National Hurricane Research Project Rep. No. 69, 85 pp. [Available from AOML Hurricane Res. Div., 4301 Rickenbacker Causeway, Miami, FL 33149.]

Gentry, R. C., 1983: Genesis of tornadoes associated with hurricanes. *Mon. Wea. Rev.*, **111**, 1793–1805.

Gray, W. M., 1965: Calculation of cumulus vertical draft velocities in hurricanes from aircraft observations. *J. Appl. Meteor.*, **4**, 463–474.

Gray, W. M., 1966: On the scales of motion and internal stress characteristics of the hurricane. *J. Atmos. Sci.*, **23**, 278–288.

Gray, W. M., 1968: Global view of the origin of tropical disturbances and storms. *Mon. Wea. Rev.*, **96**, 669–700.

Gray, W. M., 1979: Hurricanes: Their formation, structure, and likely role in the tropical circulation. In *Meteorology Over the Tropical Oceans.* D. B. Shaw (Ed.), Royal Meteorological Society, 155–218.

Hawkins, H. F., and D. T. Rubsam, 1968: Hurricane Hilda, 1964, II: Structure and budgets of the hurricane on October 1, 1964. *Mon. Wea. Rev.*, **96**, 617–636.

Jones, W. L., P. Black, V. E. Delnore, and C. T. Swift, 1981: Airborne microwave remote sensing measurements of surface winds and rain rate during Hurricane Allen. *Science*, **214**, 274–280.

Jordan, C. L., 1958: Mean soundings for the West Indies area. *J. Meteor.*, **15**, 91–97.

Jorgensen, D. P., 1984a: Mesoscale and convective-scale characteristics of mature hurricanes, Part I: General observations by research aircraft. *J. Atmos. Sci.*, **41**, 1268–1285.

Jorgensen, D. P., 1984b: Mesoscale and convective-scale characteristics of mature hurricanes, Part II: Inner core structure of Hurricane Allen (1980). *J. Atmos. Sci.*, **41**, 1287–1311.

Jorgensen, D. P., and F. D. Marks, Jr., 1984: Airborne Doppler radar study of the structure and three-dimensional airflow within a hurricane rainband. Preprints, 22nd Conference on Radar Meteorology, Zurich, Switzerland. American Meteorological Society, Boston, 572–577.

Jorgensen, D. P., E. J. Zipser, and M. A. LeMone, 1985: Vertical motions in intense hurricanes. *J. Atmos. Sci.*, **42**, 839–856.

Kidder, S. Q., W. M. Gray, and T. H. Vonder Haar, 1978: Estimating tropical cyclone central pressure and outer winds from satellite microwave data. *Mon. Wea. Rev.*, **106**, 1458–1464.

LeMone, M. A., and E. J. Zipser, 1980: Cumulonimbus vertical velocity events in GATE, Part I: Diameter, intensity, and mass flux distributions. *J. Atmos. Sci.*, **37**, 2444–2457.

Lewis, B. M., and P. G. Black, 1977: Spectral analysis of oscillations of radar-determined hurricane tracks. 11th Technical Conference on Hurricanes and Tropical Meteorology, Miami Beach, American Meteorological Society, Boston, 484–489.

López, R. E., 1978: Internal structure and development processes of C-scale aggregates of cumulus clouds. *Mon. Wea. Rev.*, **106**, 1488–1494.

Marks, F. D., Jr., 1985: Evolution of the structure of precipitation in Hurricane Allen. *Mon. Wea. Rev.*, **113**, 909–930.

Marks, F. D., Jr., and R. A. Houze, Jr., 1984a: Airborne Doppler radar observations in Hurricane Alicia. Preprints, 22nd Conference on Radar Meteorology, Zurich, Switzerland, American Meteorological Society, Boston, 578–583.

Marks, F. D., Jr., and R. A. Houze, Jr., 1984b: Airborne Doppler radar observations in Hurricane Debby. *Bull. Amer. Meteor. Soc.*, **65**, 569–582.

Novlan, D. J., and W. M. Gray, 1974: Hurricane-spawned tornadoes. *Mon. Wea. Rev.*, **102**, 476–488.

Ooyama, K. V., 1982: Conceptual evolution of the theory and modeling of the tropical cyclone. *J. Meteor. Soc. Japan*, **60**, 369–380.

Orton, R., 1970: Tornadoes associated with Hurricane Beulah on September 19–23, 1967. *Mon. Wea. Rev.*, **98**, 541–547.

Parrish, J. R., R. W. Burpee, F. D. Marks, Jr., and R. Grebe, 1982: Rainfall patterns observed by digitized radar during the landfall of Hurricane Frederic (1979). *Mon. Wea. Rev.*, **110**, 1933–1944.

Parrish, J. R., R. W. Burpee, F. D.

Marks, Jr., and C. W. Landsea, 1984: Mesoscale and convective-scale characteristics of Hurricane Frederic during landfall. Postprints, 15th Conference on Hurricanes and Tropical Meteorology, Miami, American Meteorological Society, Boston, 415–420.

Powell, M. D., 1982: The transition of the Hurricane Frederic boundary-layer wind field from the open Gulf of Mexico to landfall. *Mon. Wea. Rev.*, **110**, 1912–1932.

Rodgers, E. B., and R. F. Adler, 1981: Tropical cyclone rainfall characteristics as determined from a satellite passive microwave radiometer. *Mon. Wea. Rev.*, **109**, 506–521.

Shea, D. J., and W. M. Gray, 1973: The hurricane's inner core region, I: Symmetric and asymmetric structure. *J. Atmos. Sci.*, **30**, 1544–1564.

Tuleya, R. E., M. A. Bender, and Y. Kurihara, 1984: A simulation study of the landfall of tropical cyclones using a movable nested-mesh model. *Mon. Wea. Rev.*, **112**, 124–136.

Warner, C., and D. P. McNamara, 1984: Aircraft measurements of convective draft cores in MONEX. *J. Atmos. Sci.*, **41**, 430–438.

Wexler, H., 1947: Structure of hurricanes as determined by radar. *Ann. New York Acad. Sci.*, **48**, 821–844.

Willoughby, H. E., J. A. Clos, and M. G. Shoreibah, 1982: Concentric eyewalls, secondary wind maxima and the evolution of the hurricane vortex. *J. Atmos. Sci.*, **39**, 395–411.

Willoughby, H. E., F. D. Marks, Jr., and R. J. Feinberg, 1984: Stationary and moving convective bands in hurricanes. *J. Atmos. Sci.*, **41**, 3189–3211.

Zipser, E. J., and M. A. LeMone, 1980: Cumulonimbus vertical velocity events in GATE, Part II: Synthesis and model core structure. *J. Atmos. Sci.*, **37**, 2458–2469.

CHAPTER 15

Characteristics of Isolated Convective Storms

Morris L. Weisman
Joseph B. Klemp

15.1. Introduction

Convective storms exist under a wide variety of conditions and evolve in an equally wide variety of ways. As the understanding of convective phenomena has increased, so has appreciation of their complexity. Storm behavior is inherently dependent on the environment in which the storm grows, including thermodynamic stability, vertical wind profiles, and mesoscale forcing influences. To the extent that the important prestorm conditions can be identified (through rawinsonde ascents, surface observations, satellites, vertical profilers, etc.), current knowledge provides valuable guidance on how convection will evolve in a given environment. For example, inferences can be made about storm motion, longevity, and potential severity. Because of the complexity of the problem, however, the knowledge of storm dynamics to date is most applicable to relatively isolated convective events, i.e., individual thunderstorm cells, small groups of cells, or some very simple squall lines. To the extent that larger scale systems such as Mesoscale Convective Complexes (MCCs) are made up of individual convective cells, this knowledge of the properties of isolated convection is still very useful. But as interactions among cells, along with mesoscale and synoptic-scale influences, become important, any inferences regarding storm behavior are made with less certainty.

This discussion of the properties of isolated individual convective cells or small groups of cells is intended to serve as a basis for interpreting more complicated systems. It will be assumed that a triggering mechanism is available for the convection (e.g., diurnal heating, lifting along a frontal or mesoscale boundary), and the evolution of convection in a variety of different environments will be considered. In some instances, the triggering process may influence the resultant storm evolution (e.g., the triggering of a single cell is different from the triggering of a line of cells where each cell may interact with its neighbor), but only the features of convection that can be isolated from complicating influences are considered.

Some important topics are omitted. For example, microphysics is not discussed, and theoretical models of convection are discussed only qualitatively; how synoptic and mesoscale systems produce the convective storm environment is not considered. Further information can be found in reviews by Newton (1963), Browning (1977), Lilly (1979), Kessler (1986), and Doswell (1982).

15.2. Observed Types of Convective Storms

The concept of the convective cell is fundamental to a discussion of convective storms. The convective cell will be regarded as a region of strong updraft (at least 10 m s^{-1}) having a horizontal cross section of 10–100 km^2 and extending in the vertical through most of the troposphere. Each updraft cell has associated with it a region of precipitation that can easily be identified on radar. Research has shown that convective cells as observed on radar often evolve in identifiable, repeatable patterns. On the basis of these radar characteristics, conceptual models have been proposed for the most commonly observed storm types. These include the short-lived single cell (e.g., Byers and Braham, 1949), the multicell (Marwitz, 1972b; Newton and Fankhauser, 1975), and the supercell (Browning, 1964; Lemon and Doswell, 1979).

15.2.1. The Single-Cell Storm

The short-lived single cell is the most basic convective storm. It consists of a single updraft, which rises rapidly through the troposphere and produces large amounts of liquid water and ice. When the rain drops or ice particles become too heavy for the updraft to support, they begin to fall, creating a downdraft that quickly replaces the updraft. The downdraft is initially nearly saturated, but as it falls into the lower troposphere and mixes with drier air, strong evaporational cooling may occur. This cooling accelerates the downdraft (because of negative buoyancy), which spreads out horizontally as a cold surge (gust front) on reaching the surface. This sequence of events (Fig. 15.1) usually takes 30–50 min to complete. During this period, the storm system characteristically moves with the mean environmental wind over the lowest 5–7 km AGL. Severe weather such as high winds or hail may occur but tends to be short-lived, and tornadoes are rare.

15.2.2. The Multicell Storm

The multicell storm can be thought of as a cluster of short-lived single cells. The cold outflows from each cell, however, combine to form a large gust front, the convergence along its leading edge being generally strongest in the direction of storm motion. This convergence triggers new updraft development along and just behind the gust front, and new cells evolve as described in Sec. 15.2.1. Figure 15.2 shows this process in a vertical cross section through a multicellular hailstorm observed during the National Hail

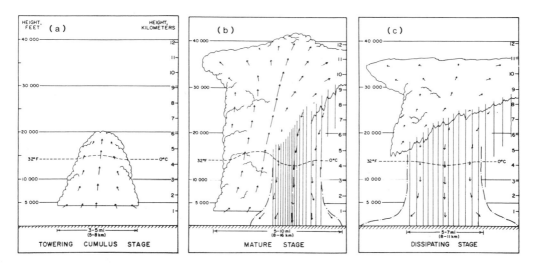

Figure 15.1. (a) The towering cumulus stage, (b) mature stage, and (c) dissipating stage of a short-lived convective cell. (Courtesy of C. A. Doswell, NOAA/ERL/WRP, Boulder, Colo.; adapted from Byers and Braham, 1949.)

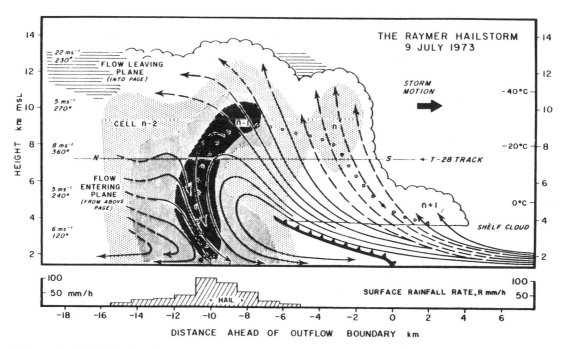

Figure 15.2. Vertical section of an ordinary multicell hailstorm, along the storm's direction of travel through a series of evolving cells ($n-2$, $n-1$, n, $n+1$). The solid lines are streamlines of flow relative to the moving system; on the left their broken ends represent flow into and out of the plane, and on the right they represent flow remaining within a plane a few kilometers closer to the reader. Light shading represents the extent of the cloud, and the three darker shades represent radar reflectivities of 35, 45, and 50 dBZ. (From Browning *et al.*, 1976.)

Research Experiment. The new cell growth often appears disorganized, but occasionally occurs on a preferred storm flank where each individual cell moves roughly with the mean wind. The storm's motion as a whole, however, may deviate substantially from the mean wind direction, owing to the discrete redevelopment of cells. Some examples of multicellular behavior are included in Figs. 15.3 and 15.4. Because of their ability to renew themselves constantly through new cell growth, multicell storms may last a long time, affecting vast areas. If the storm motion is very slow, heavy local rainfall may occur, presenting the possibility of flooding. Exceptionally strong updrafts may produce hail, and short-lived tornadoes are possible along the gust front in the vicinity of strong updraft centers.

15.2.3. The Supercell Storm

The supercell is potentially the most dangerous of the convective storm types. It may produce high winds, large hail, and long-lived tornadoes over a wide path. In its purest form it consists of a single, quasi-steady, rotating updraft, which may have a lifetime of several hours while propagating continuously to the right, or occasionally left, of the mean winds. It is often seen to evolve from multicell storm systems, and even during its quasi-steady

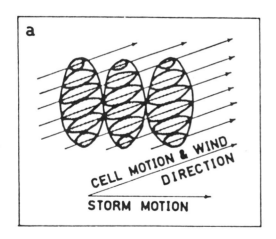

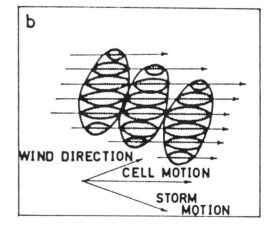

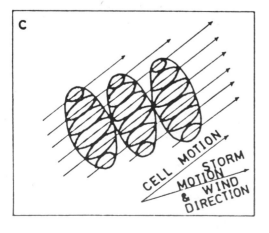

Figure 15.3. Schematic diagrams of the (a) Wokingham, (b) Alhambra, and (c) Rimbey storms, demonstrating the concept of cell motion as opposed to storm motion in an organized multicell storm system. (From Marwitz, 1972b.)

phase may comprise several rain centers, which evolve similarly to organized multicells. However, the general structure and evolution of the supercell suggest that it is dynamically different from ordinary convection.

The evolution of the archetypical supercell as it might be observed on radar is depicted in Fig. 15.5. Initial storm development (Fig. 15.5a) is essentially identical to what would be observed for a short-lived single cell or an individual cell in a multicell storm complex. The radar reflectivity pattern is vertically aligned (no weak echo regions or overhangs present), and the storm motion is generally in the direction of the mean wind (to the northeast in the figure). About 1 h into the storm's lifetime (Fig. 15.5b), the reflectivity pattern has elongated in the direction of the mean vertical wind shear vector, and the strongest reflectivity gradient is located on the southwest flank of the storm. The middle-level reflectivity field also overhangs the low-level reflectivity field on the flank of the storm by several kilometers, indicating the presence of a strong updraft. The storm has now veered to the right of the mean wind (to the east in the figure). A supercell usually reaches its mature, quasi-steady phase within 90 min (Fig. 15.5c); a hook-like appendage appears on the southwest flank of the storm, a large area of middle-level reflectivity continues to overhang the low-level echo, and often a Bounded Weak Echo

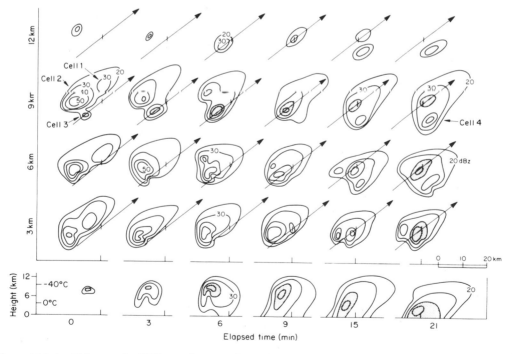

Figure 15.4. Schematic PPI sections and vertical cross sections for a multicell storm at various stages of its life cycle. PPI sections are illustrated for four elevations (3, 6, 9, and 12 km) at six different times. The arrows depict the direction of cell motion and are also geographical reference lines for the vertical cross sections shown along the bottom of the figure. Cell 3 is shaded to emphasize its history. (From Chisholm and Renick, 1972.)

Region (BWER) appears at middle levels above the edge of the low-level reflectivity gradient. A BWER usually indicates the presence of both strong updraft and strong rotation about a vertical axis in its vicinity (Weisman *et al.*, 1983).

A time series of radar reflectivity structure for a storm that occurred on 19 April 1972 near Norman, Okla. (Fig. 15.6), portrays a commonly observed trait of supercell storms. About 1 h into the storm's lifetime, the rain center appears to split into two diverging echo masses; the more intense southern storm veers to the right and slows its motion while the northern storm moves more quickly to the northeast (see Sec. 15.3).

Figure 15.7 presents the surface features commonly observed during a supercell's tornadic phase. Note the similarities to models of occluded synoptic-scale waves. The strong circulation associated with the surface mesocyclone wraps the gust front around the southern flank of the storm, overtaking the frontal boundary associated with the forward flank downdraft. The tornado usually forms at the tip of the occlusion (on the edge of the hook echo) on the gradient between updraft and downdraft (but within updraft). Fig. 15.8 illustrates the complete occlusion process. A new mesocyclone and updraft

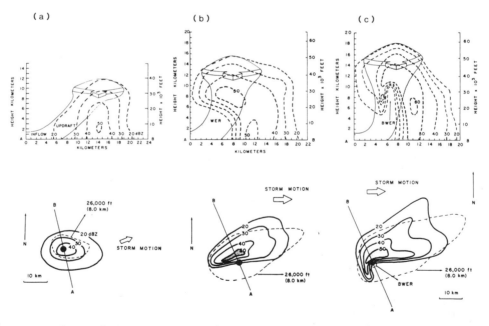

Figure 15.5. (Above) Vertical cross sections as might be observed on a radar scope during the (a) early, (b) middle, and (c) mature phases of a supercell storm. Low-level inflow, updraft, and outflow aloft (solid lines) are superimposed on the radar reflectivity (dashed lines). Reflectivities greater than 50 dBZ are stippled. The updraft becomes more intense as the storm evolves to its mature phase. WER implies a weak echo region and BWER implies a bounded weak echo region. (Below) Composite tilt sequences. Solid lines are the low-level reflectivity contours, dashed lines outline the echo >20 dBZ derived from the middle-level elevation scan, and the black dot is the location of the maximum top from the high-level scan. (Adapted from Lemon, 1980.)

may form at the triple point of the occlusion as the old storm center is cut off from its supply of warm air. Not all supercells go through this occlusion process, but some repeat the sequence several times, leading to a similar sequential development of tornadoes (see Ch. 18, this volume).

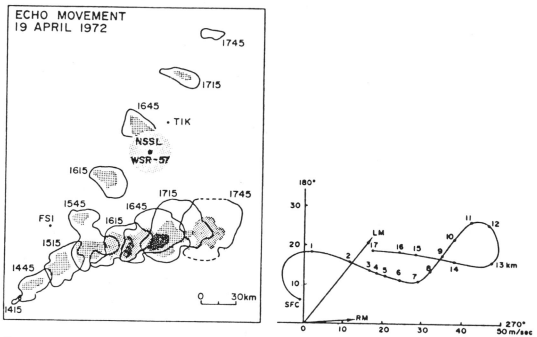

Figure 15.6. (Left) WSR-57 radar history of a splitting storm observed in south central Oklahoma. The solid contours indicate return greater than 10 dBZ, and the stippled regions indicate return greater than 40 dBZ. Times adjacent to each outline are CST. (Right) A hodograph representative of the storm's environment. RM and LM indicate the observed motion of the right-moving and left-moving cells. (Adapted from Burgess, 1974.)

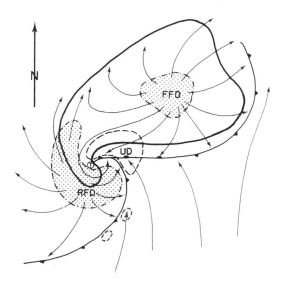

Figure 15.7. Plan view of a tornadic thunderstorm at the surface. Thick line encompasses radar echo. The thunderstorm "gust front" structure and "occluded" wave are depicted with a solid line and frontal symbols. Surface positions of the updraft (UD) are finely stippled; forward flank downdraft (FFD) and rear flank downdraft (RFD) are coarsely stippled; arrows are associated streamlines (relative to the ground). Encircled T is tornado location. (From Lemon and Doswell, 1979.)

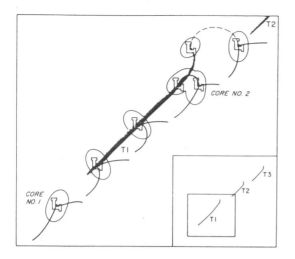

Figure 15.8. Conceptual model of mesocyclone core evolution in an occluding supercell.   Thick lines are low-level wind
discontinuities, and tornado tracks are
shaded. Insert shows the tracks of the
tornado family, and the small square is
the region expanded in the figure. The
mesocyclone occlusion can alternately be
viewed as the occlusion of the storm's
main updraft.   (From Burgess *et al.*,
1982.)

15.3. Physical Mechanisms Controlling Convective Storm Growth and Evolution

Convective storm type and severity are strongly dependent on the environmental conditions in which the storm grows, particularly thermodynamic
instability (buoyancy) and vertical wind shear. Thermodynamic instability
exerts a fundamental control on convective storm strength, since it controls
the ability of air parcels to accelerate vertically. Vertical wind shear, however,
strongly influences the form that the convection might take, i.e., whether the
convection evolves as short-lived cells, multicells, or supercells.

Some basic physical mechanisms related to the magnitude of the vertical wind shear and buoyant energy can be used to help explain (and perhaps forecast) the wide variety of convective storms observed in nature. To
demonstrate this, it is helpful to use numerical modeling studies which make
it possible to view storm evolution without the complexities and uncertainties inherent in observing storms in the real world. Recent modeling studies
have reproduced many of the observed features of the convective storm spectrum and offer new insights into the physical mechanisms underlying the
large variety of observed structures. For example, Fig. 15.9 portrays many
of the often observed features associated with tornadic storms, reproduced
in a simulation of the 20 May 1977 supercell storm. In another example,
a numerical simulation largely reproduces the more complex evolution and
motion of cells associated with a splitting storm observed on 3 April 1964
(Fig. 15.10). Model results are used with the understanding that they provide only a simplified picture of the real atmosphere. Nevertheless, they do
provide a strong foundation from which to analyze real storm behavior.

15.3.1. Thermodynamic Structure

The first task in assessing the potential for severe convection is to consider
the thermodynamic structure in the environment of the storms. The reader
is referred to the Air Force Skew T Log P manual (Air Weather Service,
1979) for a review of thermodynamic analysis and interpretation techniques.

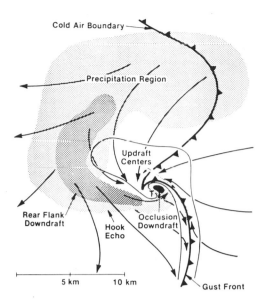

Figure 15.9. Low-level (250 m) storm features produced in a simulation of the 20 May 1977 supercell storm. The region of vertical velocity greater than 2 m s^{-1} is shaded with diagonal lines; downdrafts exceeding -2 m s^{-1} are darkly stippled. Rainwater concentrations greater than 0.5 m s^{-1} are lightly shaded. The cold frontal boundary denotes the $-1°$ C perturbation isotherm. Flow arrows represent storm relative surface streamlines. T marks the location of maximum vertical vorticity. (From Klemp and Rotunno, 1983.)

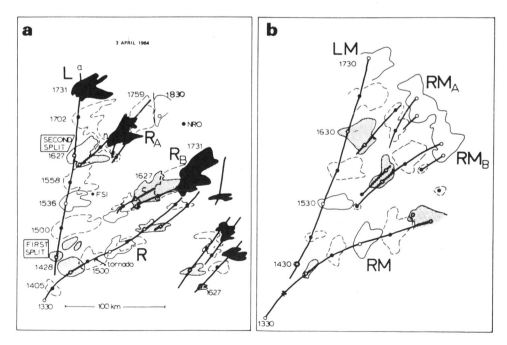

Figure 15.10. The (a) observed and (b) modeled storm development on 3 April 1964. Observed reflectivities > 12 dBZ at 0° and modeled rainwater contents > 0.5 g kg^{-1} at z =0.4 km are enclosed by alternating solid and dashed contours about every 30 min. Maxima in these fields are connected by solid lines. The storms are labeled and the contoured regions are stippled at several times for better visualization of the storm development. Labels for the modeled storms are the same as the corresponding observed storms except for the inclusion of M. The scale shown in (a) applies in (b). (From Wilhelmson and Klemp, 1981.)

The degree of instability is often measured by easily calculated indices such as the Lifted Index, Showalter Index, and Total Totals. A more accurate measure, however, can be obtained by explicitly calculating the amount of buoyant energy (B) available to a representative parcel rising vertically through an undisturbed environment.

$$B = g \int \frac{\theta(z) - \bar{\theta}(z)}{\bar{\theta}(z)} \, dz \quad , \tag{15.1}$$

where $\theta(z)$ defines the potential temperature of a representative moist adiabatically ascending surface parcel, $\bar{\theta}(z)$ defines the environmental potential temperature profile, and the integral is taken over the vertical interval where the lifted parcel is warmer than its environment. This calculation is equivalent to evaluating the positive area represented on a skew-T diagram (Fig. 15.11). If pressure perturbation effects, water loading, and mixing are ignored, B can be directly related to the maximum vertical velocity obtainable by a parcel rising vertically through the troposphere:

$$W_{\mathrm{max}} = (2B)^{1/2} \quad . \tag{15.2}$$

Magnitudes of B can be as large as 4500 m^2 s^{-2}, but generally range between 1500 and 2500 m^2 s^{-2} for moderately unstable convective days. A buoyant energy of 2500 m^2 s^{-2} would translate to a maximum possible updraft strength of 70 m s^{-1}. However, water loading, perturbed vertical pressure gradients, and mixing effects reduce these estimates by roughly 50%.

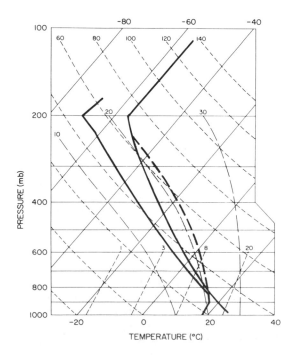

Figure 15.11. Skew-T diagram of temperature and moisture profile used in model experiments (heavy solid lines). Heavy dashed line represents a parcel ascent from the surface, based on a surface mixing ratio of 14 g kg^{-1} and a surface potential temperature of 300.7 K. The shaded region depicts the positive area defined by the parcel ascent. Tilted solid lines are isotherms; short-dashed lines are dry adiabats, and long-dashed lines are moist adiabats. (Adapted from Weisman and Klemp, 1982.)

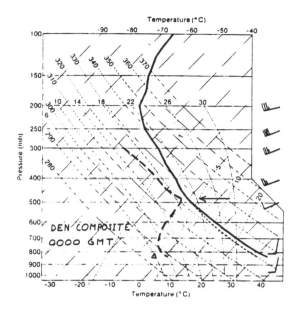

Figure 15.12. Composite sounding (temperature and dewpoint) for a series of damaging wind events in eastern Colorado, plotted on a skew-T, log-P diagram. Plotted winds (standard format) are vector means. (From Brown et al., 1982.)

Another important aspect of the thermodynamic structure is the moisture stratification. Large amounts of moisture are needed in the boundary layer to support updraft growth, but the absence of moisture above the boundary layer (2–4 km AGL) often enhances storm severity. One mechanism by which this occurs is the increased downdraft and outflow strength from the evaporational cooling that occurs as the rainy downdraft falls through or entrains the dry air. In extreme cases, convective cells with relatively weak updrafts may produce damaging winds in the form of downbursts or microbursts (Fujita, 1981). An example of a composite microburst sounding is presented in Fig. 15.12. Such cases are often characterized by small amounts of potential buoyant energy, but a very dry, deep adiabatic layer below cloud base. As rain begins to fall and evaporate in this subcloud layer, large negative buoyancy accelerates the air downward, sometimes producing damaging winds The enhanced outflow strength that accompanies increased evaporation may also indirectly enhance storm severity by strengthening the convergence along the gust front, thereby increasing the likelihood of updraft redevelopment.

15.3.2. Vertical Wind Shear

While the thermodynamic structure strongly influences the vertical accelerations in a convective storm, vertical wind shear has a strong influence on what form convection might take. The relationship between wind shear and storm type is demonstrated in Figs. 15.13 and 15.14, which display composite hodographs for observed short-lived cells, multicell, and long-lived supercell storms. In both studies the magnitude of the vertical wind shear over the lowest 6 km AGL increases as the type of convection progresses from short-lived storms to supercells.

The means by which vertical wind shear organizes convective cells has long intrigued researchers. Early observations (Byers and Braham, 1949)

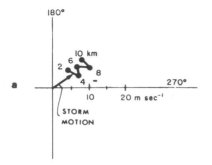

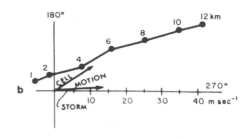

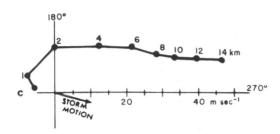

Figure 15.13. Typical wind hodographs for (a) single cell, (b) multicell,
and (c) supercell storms observed during the Alberta Hail Studies project.
(From Chisholm and Renick, 1972.)

suggested that wind shear was detrimental to the growth of small convective
elements; clouds were observed to be torn apart if the shear was too strong.
At the same time, large convective elements seemed to be enhanced by the
presence of shear.

Two physical mechanisms might help explain the organizational capacity
of vertical wind shear. The first is related to the ability of a gust front
to trigger new convective cells. The second is related to the ability of an
updraft to interact with the environmental vertical wind shear to produce
an enhanced, quasi-steady storm structure.

Consider the case of a convective cell evolving in an environment with
no wind (this can alternately be thought of as an environment having no
vertical wind shear). The convective downdraft produces a pool of cold
air which spreads horizontally at the surface equally in all directions. The
convergence at the boundary of the gust front may trigger new convection
if there is sufficient lifting to raise air parcels to the level of free convection.
Since there is no environmental wind above the surface outflow, the updraft
cell will tend to be motionless while the gust front continues to spread out at
the surface. The new cell quickly finds itself in the cold, stable environment
behind the gust front, and further updraft development is stopped.

Now consider environmental winds which increase from zero at the surface to some moderate value at higher levels (an environment of moderate
vertical wind shear). A cell developing in this environment will still produce
outflow due to the pool of cold air at the surface. Cells growing in response
to the convergence along the outflow boundary, however, will now tend to
move downshear, roughly with the mean wind over the lowest 5–7 km AGL.
This enhances cell growth along the downshear portion of the gust front by
increasing both the relative inflow into these developing cells and the time

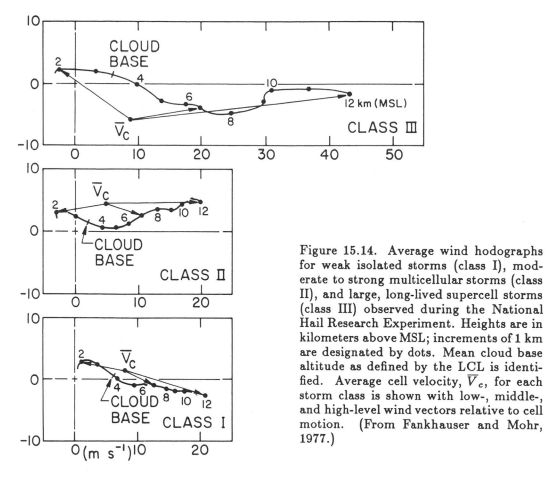

Figure 15.14. Average wind hodographs for weak isolated storms (class I), moderate to strong multicellular storms (class II), and large, long-lived supercell storms (class III) observed during the National Hail Research Experiment. Heights are in kilometers above MSL; increments of 1 km are designated by dots. Mean cloud base altitude as defined by the LCL is identified. Average cell velocity, $\overline{V}_c$, for each storm class is shown with low-, middle-, and high-level wind vectors relative to cell motion. (From Fankhauser and Mohr, 1977.)

over which the cells maintain their low-level convergence and feed on the warm air ahead of the gust front. For the proper magnitude of vertical wind shear, the cell motion and gust front motion may be the same, leading to a continual redevelopment of updrafts. To generalize this scenario to a case of non-zero surface wind, merely add the surface wind component to the wind profile at each height, and to the resultant gust front motions, cell motions, etc. Redevelopment of cells along a preferred region of an outflow boundary represents the primary physical mechanism for the sustenance of multicell storms.

As the vertical wind shear becomes stronger, the interaction of the updraft with the sheared flow becomes an important contributor to the organization and sustenance of the convection. The essential physical mechanism responsible for this is the development of rotation on the flank of the updraft (Rotunno and Klemp, 1985). This rotation originates through the tilting of horizontal vorticity inherent in the vertically sheared flow (Rotunno, 1981; Davies-Jones, 1983). If the vertical wind shear extends through the middle levels of the storm (~4–6 km), the rotation dynamically induces a pressure deficit which is strongest several kilometers above the ground and thus produces a vertical pressure gradient that accelerates surface air upward. This forcing helps both to sustain the updraft and to promote a propagation that deviates from the mean wind. Supercells owe their existence to these dynamically induced pressure forces.

In convective storms, gust front convergence and dynamic pressure forcing work together to produce the observed storm characteristics. The relative importance of each mechanism, as well as the location within the storm where each mechanism acts, is dependent on the magnitude, shape, and depth of the vertical wind shear profile. This is demonstrated in Fig. 15.15, based on recent numerical modeling results for an idealized updraft growing in a horizontally homogeneous environment (Weisman and Klemp, 1982, 1984; Klemp and Weisman, 1983). In the cases discussed below, the thermodynamic conditions are moderately unstable (buoyant energy $\sim$2200 m^2 s^{-2}). The first case considered is an environment of unidirectional wind shear; it is followed by the more commonly observed case of a curved wind shear

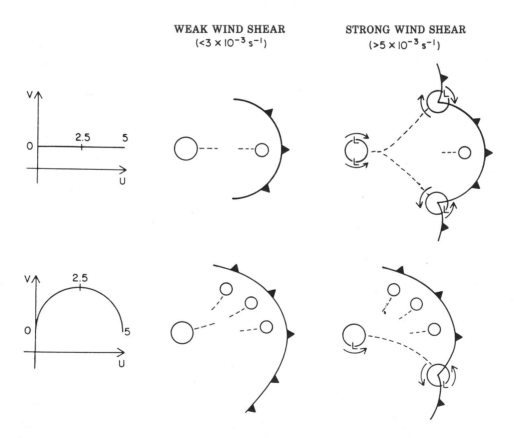

Figure 15.15. Updraft evolution in weak and strong wind shear conditions for unidirectional and clockwise-curved wind shear profiles. Hodographs on the left define the wind shear type; 0, 2.5, and 5 km levels are indicated. Large and small circles represent relatively strong and weak updrafts, respectively; the path of each updraft cell is indicated by a dotted line. Updraft structure is depicted at the early and mature phases of each storm; surface gust fronts (barbed lines) are included at the mature phase. L is the approximate position of significant middle-level mesolow features. The direction of the updraft rotation (if any) is indicated by arrows.

profile. Since Wilhelmson and Klemp (1978) showed that storm structure is most sensitive to the vertical wind shear in the lowest several kilometers AGL, shear profiles are assumed to extend only up to 5 km, with constant winds above.

Unidirectional Shear

In weak wind shear conditions, a unidirectional shear profile produces a short-lived cell whose gust front may trigger new short-lived convection. The most probable zone for cell redevelopment is directly downshear from the original cell. As the wind shear increases, low pressure begins to develop (strongest at middle levels) on both the right and left flanks of the original updraft (relative to the wind shear vector). In strong wind shear conditions, this pressure forcing is sufficient to split the updraft into two quasi-steady storms moving to the right and left of the environmental wind; the right- and left-moving updrafts rotate cyclonically and anticyclonically, respectively. Unidirectional wind shear profiles occur rather infrequently; however, when they have occurred with sufficient shear magnitude, storm splitting has been observed (e.g., Fujita and Grandoso, 1968).

Curved Shear

Much more commonly, the wind shear vector turns clockwise over the lowest couple of kilometers above the surface (see Fig. 15.13c; also, Maddox, 1976). In an idealized profile of such a situation, the wind shear vector turns clockwise throughout the depth of the profile. In weak shear conditions, short-lived cell regeneration along the gust front now occurs along both the forward and left flanks of the original storm. In strong shear conditions, the pressure forcing now occurs only on the right flank of the original updraft, producing only one quasi-steady cyclonically rotating updraft. Short-lived cells, however, still may regenerate along the left flank. If the hodograph turned counterclockwise rather than clockwise with height, storm evolution on the right flank and left flank would simply be reversed (e.g., left-moving, anticyclonic supercell with unsteady growth along the right flank). Klemp and Wilhelmson (1978b) demonstrated that the hodograph curvature in the lowest 1–2 km AGL has the most influence on favoring a particular storm flank for supercell evolution. As shown by Maddox (1976) and others, the low-level environment of severe convective storms climatologically favors a clockwise-turning hodograph. This explains the predominance of cyclonic, rather than anticyclonic, storms observed in conditions of strong wind shear.

Significance of the Wind Shear Vector

To understand the relationships between storm structure and the environmental winds, it is necessary to comprehend the role of the wind shear vector. Given a horizontally homogeneous environmental wind profile, $\boldsymbol{V}_H(z) = U(z)\boldsymbol{i} + V(z)\boldsymbol{j}$, the environmental wind shear vector is defined as

$$\frac{dV_H}{dz} = \frac{dU}{dz}\, \boldsymbol{i} + \frac{dV}{dz}\, \boldsymbol{j} \quad , \tag{15.3}$$

where $\boldsymbol{i}$ and $\boldsymbol{j}$ are unit vectors in the x (east) and y (north) directions, respectively. In the standard hodograph representation (U, V coordinates) the magnitude of the shear is proportional to the length of the hodograph curve over constant height intervals and the shear vector lies tangent to the hodograph curve at each height. Ignoring surface and Coriolis effects, developing convection depends only on the variations in the magnitude and direction of the wind shear vector with height (Klemp and Wilhelmson, 1978b; Rotunno and Klemp, 1982). The orientation of ground-relative winds has no fundamental significance!

Consider, for example, the strongly sheared environments shown by the hodographs in Figs. 15.16a and 15.16b. Both have exactly the same shear

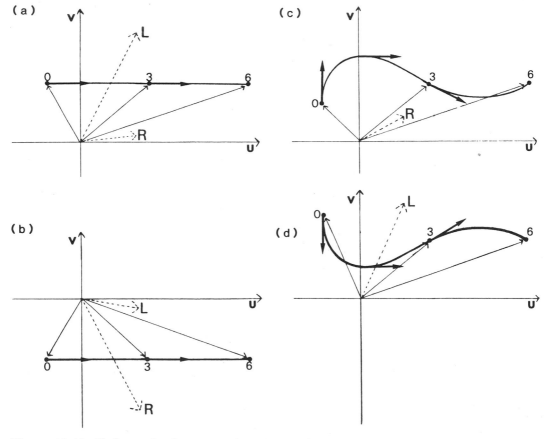

Figure 15.16. Hodographs demonstrating the relationship between wind shear, wind direction, and storm motion for (a,b) unidirectional and (c,d) curved shear profiles. Heights are in kilometers. Thin solid vectors depict the ground-relative wind directions at the three heights. Thin dashed vectors depict ground-relative storm motions for left-moving (L) and right-moving (R) supercells. Thick vectors point in the direction of the wind shear vector at the given heights.

profile; the only difference is that in (a) the ground-relative winds all have a southerly component and in (b) they have a northerly component. Numerical simulations for these two cases (again, ignoring surface and Coriolis effects) would produce identical storms. Mirror-image split storms would evolve, one propagating to the right and one to the left of the hodograph line. In (a), the ground-relative winds veer with height and the northern (left-moving) storm moves faster than the mean wind while the southern (right-moving) storm moves to the right of and more slowly than the winds at all levels. However, in (b), the situation is reversed; the environmental winds back with height, the southern storm moves faster than the mean winds, and the northern storm moves to the left of and more slowly than the winds.

Another strong shear example is illustrated in Figs. 15.16c and 15.16d. In both hodographs, the ground-relative winds veer with height. However, in (c) the low-level wind shear vector veers with height, favoring the development of right-moving, cyclonically rotating storms; in (d) the low-level shear vector backs with height, favoring the growth of left-moving, anticyclonically rotating storms.

15.4. Models of Wind Shear and Updraft Redevelopment

The dependence of updraft redevelopment and motion on environmental wind shear can explain a large variety of convective precipitation patterns. A recent modeling study by Klemp and Weisman (1983) demonstrates this for a representative series of hodograph shapes and magnitudes. The tool for this study is the Klemp-Wilhelmson (1978a) numerical cloud model. Storms are initialized by an isolated, axisymmetric warm bubble in a horizontally homogeneous environment characterized by the temperature and moisture sounding represented in Fig. 15.11 (buoyant energy ~2200 m^2 s^{-2}). Simulations use seven different wind profiles which are representative of a larger series of experiments in which the depth, shape, and magnitude of the wind shear profile were varied over a wide range of conditions observed in the vicinity of storms.

The simulations are displayed in Fig. 15.17. The important storm characteristics discussed for the six experiments are those that can be easily observed with conventional radars and surface networks.

15.4.1. Case A: Short-Lived Multicell

This experiment (Fig. 15.17, case A) is representative of most low-shear simulations that produce relatively short-lived multicellular storms. During the initial 30 minutes of simulation, the warm bubble produces an updraft that produces, in turn, a large amount of liquid water. A downdraft develops and rain begins to reach the ground, producing a pool of cold air. After 40 min, the main updraft associated with the initial cell is near the location of heaviest surface rainfall, and the cold surface outflow has begun to propagate outward from this region. After 80 min, the initial updraft has disappeared and a new weaker updraft has developed along the storm's left flank in response to the gust front convergence. Note that the gust front has begun to propagate ahead of the storm. In 120 min, the storm has weakened considerably; the second updraft has disappeared, and two even weaker updrafts have taken its place. The gust front has moved a considerable distance ahead of the rainfall and updraft region, and the system continues to decline in strength. Through its lifetime, the rain region moves roughly with the mean wind.

15.4.2. Case B: Supercell on the South End of a Multicellular Line

For this experiment (Fig. 15.17, case B) the hodograph has the same shape as in experiment A (180° turning over the lowest 5 km) but twice the shear magnitude. The magnitude and depth of the shear profile is now

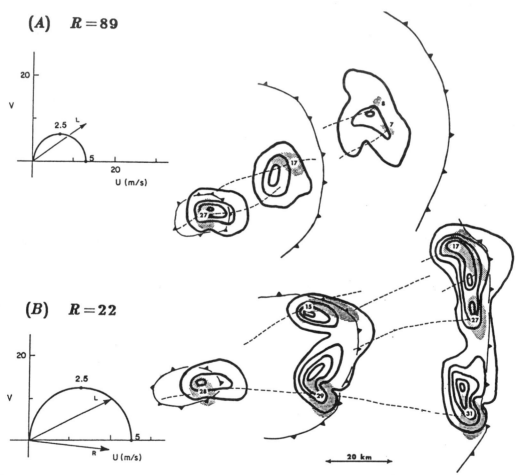

Figure 15.17. Hodograph and storm structures at 40, 80, and 120 min for cases A–F described in text. Storm positions are relative to the ground; dashed lines represent updraft cell path. Low-level (1.8 km) rainwater fields (similar to radar reflectivity) are contoured at 2 g kg^{-1} inervals. Regions in which the middle-level updraft (4.6 km) exceeds 5 m s^{-1} are shaded. Surface gust fronts are defined by the $-1°$C perturbation surface isotherm. Numbers at the updraft centers represent maximum vertical velocity (m s^{-1}) at the time. On hodographs, heights are labeled in km agl and arrows indicate the mean storm motion between 80 and 120 min. R = bulk Richardson number as discussed in Sec. 15.5.1. Cases A and B, multi-cell and supercell storms.

sufficient to produce strong positive vertical pressure gradient forces on the right flank of the initial updraft. This induces the initial cell to evolve into a quasi-steady supercell, which proceeds to move to the right of the mean winds. At 80 min, new, unsteady rain centers are developing on the left flank (in association with new updrafts) and continue to redevelop along the gust front through the 120 min of simulation. The right-flank supercell is always stronger than the left-flank storms, and hook structures in the rainwater contours are apparent at 80 and 120 min. Note that the gust front never moves far from the rain and updraft region. At 120 min the overall storm system has the appearance of a short squall line that consists of a supercell on the southern end of a multicellular line of storms.

15.4.3. Case C: Right-Flank Supercell Split from Weaker Left-Flank Storm

This experiment (Fig. 15.17, case C, top) has the same magnitude and depth of shear as experiment B, but the hodograph now forms a straight line between 2.5 and 5 km. Again, the initial storm evolves into a right-moving, quasi-steady supercell with weaker unsteady activity on the left flank. The rainwater contours for the supercell do not display a hook-type structure, but strong gradients of rainwater are found on the southwest flank of the storm in conjunction with the region of strong updraft. With less hodograph curvature than in case B, the left-flank storms are more distinctly isolated from the right-flank supercell. Consequently, the storm system now appears as two diverging echo masses, with stronger activity in the southern, right-moving storm. If the environmental hodograph forms a straight line (dashed line in case C hodograph), the initial storm splits into mirror image right- and left-moving supercells (as in Fig. 15.17, case C, bottom).

15.4.4. Case D: Right-Flank Supercell

The hodograph is the same for experiment D (Fig. 15.17, case D) as for experiment C, except that the linear portion of the shear profile is extended to 7.5 km. The initial storm again evolves into a quasi-steady, right-moving supercell, but the rain is now being drawn significantly downshear by the stronger upper-level flow. This visual structure now matches observational models presented by Lemon and Doswell (1979) and others. New development on the left flank is much weaker than in case C and almost disappears by 120 min. Note that although the supercell updrafts in cases B, C, and D are dynamically similar, the distribution of precipitation varies significantly from case to case, depending on the depth and shape of the shear profile.

15.4.5. Case E: Weak Squall Line

In this experiment (Fig. 15.17, case E), the shear profile used in case B is truncated at 2.5 km. Although this represents a strong magnitude of shear, it does not occur over a deep enough layer to produce the pressure forcing necessary to sustain a supercell storm. However, new, unsteady updrafts grow readily downshear of the original cell, along the gust front. The storm system moves roughly with the mean winds, and by 120 min has the appearance of a 50-km-long squall line. The cold surface outflow moves slightly faster than the precipitation region, but new cells growing behind the gust front are still able to reach moderate intensities.

15.4.6. Case F: Squall Line; Spearhead Echo Evolving into Bow and Comma Echoes

The shear profile in case F (Fig. 15.17, case F) is the same as in case E except that the magnitude of the shear has been increased by 50%. In addition, the lowest level moisture has been increased from 14 g kg^{-1} (as in Fig. 15.11) to 15 g kg^{-1} to help highlight the key storm features. Even though the shear is limited to the lowest 2.5 km, the extreme magnitude of the shear

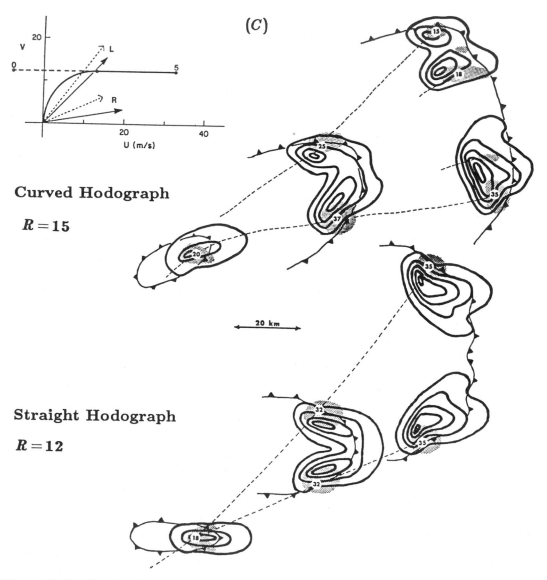

Figure 15.17. Continued. Case C, splitting supercell storms.

is now sufficient to produce a somewhat steady updraft on the right flank of the storm system. This supercell, however, remains weaker than those that developed with a deeper layer of shear. The overall storm evolution now resembles that proposed by Fujita (1981) for a severe downburst-producing thunderstorm system. At 80 min, the initial rain cell has developed a spearhead configuration, and by 120 min the storm system has evolved into a 60-km-long squall line with a rotating comma head on the northern flank. Surface winds in the vicinity of the break in intense rainwater reach 35–40 m s^{-1} between 80 and 120 min.

15.4.7. Observed Splitting Storm

With the results of the modeling experiments, it is possible to interpret the behavior of the storm depicted in Fig. 15.6. The hodograph displays strong clockwise veering of the wind shear vector below 1 km with strong unidirectional shear between 1 and 7 km. Thus, storm splitting would be expected (somewhat as in cases C and D), and the low-level hodograph curvature would favor the development of the right flank storm.

15.4.8. Discussion

In all the modeling cases, the thermodynamic stratification in the environment of the storm was kept the same. However, as noted before, the degree of instability or middle-level dryness can have a fundamental effect on updraft and downdraft strength. Such variations can feed back to the dependence of storm structure on vertical wind shear in important ways. For example, a weakly unstable environment (buoyant energy ~800 m^2 s^{-2}) may be able to support the growth of convection in weak shear conditions, but not in strong shear conditions. In the latter case, the organizational capacity of the wind shear may be overpowered by the tendency for the shear to

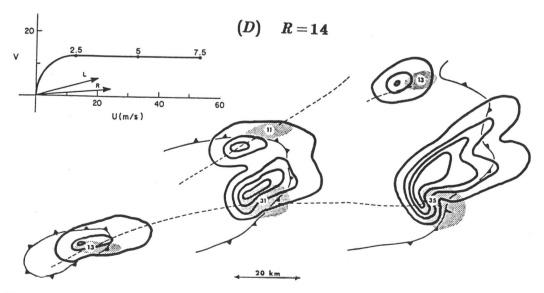

Figure 15.17. Continued. Case D, right-flank supercell.

mix out the convection; i.e., the cloud is torn apart. In a strongly unstable environment with strong wind shear, small convective elements may similarly be mixed out while large convective elements evolve into quasi-steady supercells.

In another example, variations in middle-level moisture may alter the strength of the surface outflow beneath a storm. This may feed back to storm structure by altering the speed of the gust front relative to the updraft.

Variations to the low-level thermodynamic structure may control whether a gust front can trigger new convection. If the level of free convection (LFC) is very high, the convergence along the gust front would have to occur over a much deeper layer, or affect a parcel over a longer amount of time to trigger new convection than if the LFC were very low. Thus, very moist (low LFC),

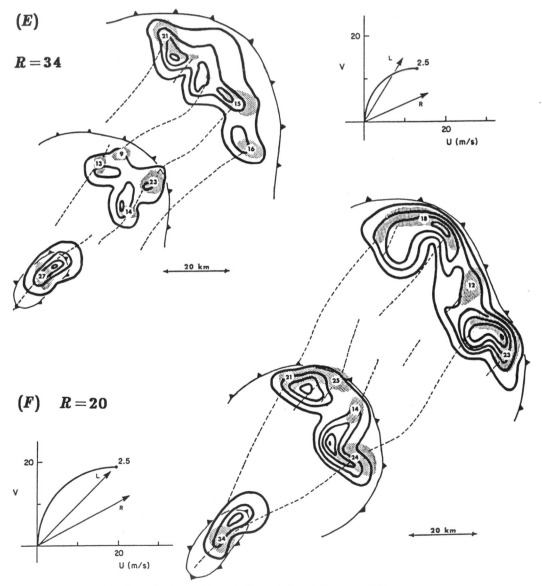

Figure 15.17. Continued. Cases E and F, multi-cellular squall lines.

tropical soundings are much more apt to have new convection triggered by gust front convergence than drier (high LFC), middle-latitude soundings.

15.5. Some Applications to Forecasting Convective Storm Type and Severity

15.5.1. Determining Storm Type

Deducing the possibility of severe convection is perhaps the most challenging job a forecaster must face during the spring and summer months. Severe weather can occur with any type of convective storm, but certain storms are more likely than others to produce severe weather. In particular, the supercell is most likely to produce long damage swaths from hail, high winds, and tornadoes. On the other hand, many serious flash floods occur with multicellular storms. Thus, knowing what type of convective storm can evolve in a given environment and by what physical mechanism it is apt to evolve or move, can be invaluable to the forecaster.

Convective storm type was shown in Sec. 15.3 to be strongly dependent on the vertical wind shear in the storm's environment (especially the lowest 5–6 km), but this dependence can be modified by thermodynamic considerations. Thus, any attempt to forecast storm type must consider both aspects of the environment. Weisman and Klemp (1982, 1984) have shown that much of the relationship between storm type, wind shear, and buoyancy can be represented in the form of a bulk Richardson number, R, defined to be

$$ R = \frac{B}{1/2\ U^2} \quad , \tag{15.4} $$

where B is the buoyant energy in the storm's environment [Eq. (15.1)] and U is a measure of the vertical wind shear. U is calculated by taking the difference between the density-weighted mean wind over the lowest 6 km of the profile and a representative surface layer wind (500-m mean wind). Similar expressions have also been used by Moncrieff and Green (1972), Ludlam (1980), and Seitter and Kuo (1983). The numerical modeling results of Weisman and Klemp (1982, 1984) and calculations of R for a series of documented storms (Fig. 15.18) both suggest that unsteady, multicellular growth occurs most readily for R>30 and that supercellular growth is confined to magnitudes of R between 10 and 40.

Calculations of R for the six hodograph cases in Sec. 15.4 are included in Fig. 15.17. The environments that produced quasi-steady supercellular updrafts (cases B, C, D, and F) have values of R between 14 and 22; the unsteady cases (A and E) have much higher values. As shown in these examples, even though a small magnitude of R is a necessary condition for the existence of supercellular convection, such environments do not preclude the simultaneous existence of nearby multicellular convection. For example, unsteady, multicellular growth occurs readily on the left flank of a supercell storm in cases B and C.

The magnitude of R indicates what type of convection might occur in a given environment, but it does not necessarily indicate the severity of that

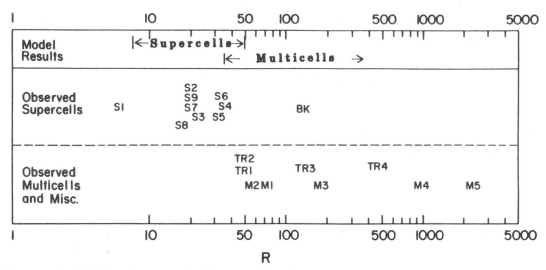

Figure 15.18. Richardson number R as calculated for a series of documented storms. Model results are summarized at the top of the figure. S1, S2,...,S9 represent supercell storms; M1, M2,...,M5 represent multicell storms; TR1, TR2,...,TR4 represent tropical cases. (Adapted from Weisman and Klemp, 1982.)

convection. For example, in an environment with only small buoyant energy ($B < 1000$ m^2 s^{-2}) and moderate wind shear (4×10^{-3} s^{-2}), R may be well within the supercell range. A forecaster would then expect some of the convective cells to have the steadiness and propagation characteristics of supercell storms. However, the lack of buoyant energy might preclude the production of severe weather (e.g., strong updrafts are still needed to produce hail or tornadic activity independent of the storm structure). Likewise, an environment with large buoyant energy (>3500 m^2 s^{-2}) and moderate wind shear may be characterized by a relatively large value of R, yet produce tornadoes or large hail within a relatively unsteady storm. For the range of B between 1500 and 3500 m^2 s^{-2}, however, the correspondence between storm type and storm severity works quite well (e.g., supercells are more severe than multicells), and R may also indicate the potential for severe weather.

A study by Rasmussen and Wilhelmson (1983) demonstrates how a forecaster might use the vertical wind shear and potential buoyant energy as depicted on the 1200 GMT sounding to assess the potential for storms that produce mesocyclones and tornadoes later in the day. A sample taken primarily from the southern Great Plains (Fig. 15.19) suggests that tornadic storms occur only in a regime of high buoyancy (> 2500 m^2 s^{-2}) and moderate-to-strong wind shear ($>3.5 \times 10^{-3}$ s^{-1}). Storms that produce mesocyclones without tornadoes occur for lower buoyant energies, but only with very strong vertical wind shear ($>5 \times 10^{-3}$ s^{-1}). Of course, buoyancy and wind shear conditions often change considerably during the day, and a forecaster must be aware of such changes. Nevertheless, this study suggests that valuable guidance can often be obtained from the 1200 GMT conditions.

15.5.2. *Estimating Storm Motion*

Once a forecaster has determined what storm type might be produced in a given environment, another important step is to estimate the storm motion. Since storm evolution responds to the characteristics of the wind shear rather than the wind at any given height, this step can be done properly only with the aid of a hodograph. Multicellular updrafts tend to move with the direction and speed of the mean wind over the lowest 5–7 km. Redevelopment of cells tends to occur downshear of existing cells. Supercells, however, have a significant component of propagation perpendicular to the mean wind shear over the lowest 5–7 km and may appear to move to the right or left of any winds on the profile. Also, the shape of the hodograph suggests which flank of a storm is most likely to produce supercellular or multicellular updrafts. The cell motions indicated on the hodographs in Fig. 15.17 provide a useful forecasting guide for a large range of wind shear conditions.

Since ground-relative cell motions are dependent on ground-relative winds as well as on shear (even though storm structure is not; see the discussion of Fig. 15.16), the examples in Fig. 15.17 can be used to portray conditions that could lead to flash flooding. To produce locally heavy rains, intense, slow moving, and relatively steady convective systems are needed. The convective cells may be either steady supercells or unsteady cells that keep redeveloping in the same general region. This occurs in all the examples in Fig. 15.17 (except case A, which has weak ambient shear), but the cells are moving rapidly relative to the ground. To change this cell speed, the

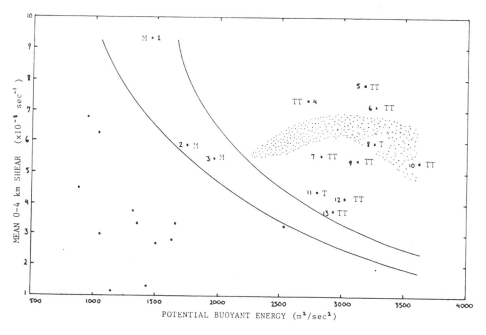

Figure 15.19. Storm occurrence as a function of 0-4 km mean shear and potential buoyant energy as estimated from 1200 GMT soundings. Each dot represents an individual case. M indicates a storm with a mesocyclone; T indicates a storm with one tornado; TT indicates a storm with more than one tornado. (Adapted from Rasmussen and Wilhelmson, 1983.)

wind profile on the hodograph must be shifted such that the storm speed relative to the ground is very slow. For example, if the wind profile in case B is shifted 20 m s^{-1} to the left so that the surface wind is strong easterly, veering to light westerly at 5 km, the supercell on the right flank of the storm system becomes virtually motionless. This potential for slow storm motion could easily be recognized by plotting the hodograph.

15.5.3. Incorporating Local Effects

Knowledge of the effects of local mesoscale variations or terrain features on storm structure and evolution is limited. However, their importance to forecasting is undeniable. Indeed, most storms are triggered in response to such features. Such features probably have the largest effect on storm evolution in conditions of low to moderate wind shear. In these conditions, the primary mechanism for cell redevelopment is low-level convergence produced along a storm's own gust front. If pre-existing boundaries provide similar low-level convergence fields in the vicinity of a storm, these features may be as effective as the storm's own outflow in guiding resultant storm growth and motion. Regions of enhanced surface moisture or instability may likewise influence the location of new updraft growth. In conditions of strong wind shear, however, the steadiness and propagation of supercell updrafts is dependent on the existence of a storm-induced middle-level mesolow which forces air up from near the surface, largely apart from the existence of the surface outflow boundary. Thus, supercell propagation would tend to be less affected by any indigenous boundaries or other inhomogeneities of the environment. Of course, if the storm's natural propagation tendencies bring it into contact with more or less favorable environmental conditions, the storm may strengthen or weaken.

A recent study by Maddox et al. (1980) highlights many observational aspects of the interaction between storms and pre-existing boundaries. It is noted that many tornadic storms move across pre-existing boundaries, even though they are moving into colder environments, which weaken the storms. On the other hand, these storms produce the most severe weather as they interact with these boundaries. Storms that naturally move along such boundaries also tend to be more severe than other nearby storms. The increase in vertical wind shear and convergence, which naturally accompanies such boundaries, was offered as a possible explanation for the enhanced severity. In another example, Szoke et al. (1984) discussed a nonsteady tornadic storm occurring in weak shear, forming and moving along a pre-existing boundary. In this case the storm motion and severity were probably strongly linked to the mesoscale boundary. Despite these examples, we must stress that supercell storms (because of their own internal dynamics) often evolve and produce severe weather totally independent of, or with a complete absence of, pre-existing boundaries.

Another problem in trying to apply simple convective models to real-time forecasts arises in determining the environmental conditions in the vicinity of the storms. Storms often occur between sparsely distributed observing stations, and at inconvenient times in between standard radiosonde observations. The forecaster is forced into extrapolating data from early morning conditions, which might be very difficult in rapidly evolving synoptic situations. In the future, vertical wind profilers may alleviate this problem by offering nearly continuous vertical wind shear conditions.

REFERENCES

Air Weather Service, 1979: The use of the Skew T, Log P Diagram in analysis and forecasting. AWS/TR–79/006 (December), Air Weather Service, Scott Air Force Base, Illinois, 8 chapters, 4 attachments (revision of AWSM, 105–124).

Brown, J. M., K. R. Knupp, and F. Caracena, 1982: Destructive winds from shallow, high based cumulonimbi. Preprints, 12th Conference on Severe Local Storms, San Antonio, Tex., American Meteorological Society, Boston, 272–275.

Browning, K. A., 1964: Airflow and precipitation trajectories within severe local storms which travel to the right of the winds. *J. Atmos. Sci.*, **21**, 634–639.

Browning, K. A., 1977: The structure and mechanism of hailstorms. In *Hail: A Review of Hail Science and Hail Suppression*, Meteor. Monogr. **16**, American Meteorological Society, Boston, 1-43.

Browning, K. A., J. C. Fankhauser, J.-P. Chalon, P. J. Eccles, R. C. Strauch, F. H. Merrem, D. J. Musil, E. L. May, and W. R. Sand, 1976: Structure of an evolving hailstorm. Part V: Synthesis and implications for hail growth and hail suppression. *Mon. Wea. Rev.*, **104**, 603–610.

Burgess, D. W., 1974: Study of a right-moving thunderstorm utilizing new single Doppler radar evidence. Master's thesis, Dept. of Meteorology, University of Oklahoma, 77 pp.

Burgess, D. W., V. T. Wood, and R. A. Brown, 1982: Mesocyclone evolution statistics. Preprints, 12th Conference on Severe Local Storms, San Antonio, Tex., American Meteorological Society, Boston, 422–424.

Byers, H. R., and R. R. Braham, Jr., 1949: *The Thunderstorm*. Supt. of Documents, U. S. Government Printing Office, Washington, D.C., 287 pp.

Chisholm, A. J., and J. H. Renick, 1972: The kinematics of multicell and supercell Alberta hailstorms. Alberta Hail Studies, Research Council of Alberta Hail Studies, Rep. 72-2, Edmonton, Canada, 24–31.

Davies-Jones, R. P., 1983: The onset of rotation in thunderstorms. Preprints, 13th Conference on Severe Local Storms, Tulsa, Okla., American Meteorological Society, Boston, 215–218.

Doswell, C. A., III, 1982: The Operational Meteorology of Convective Weather. Vol. I, Operational Mesoanalysis. NOAA Technical Memorandum NWS NSSFC-5.

Fankhauser, J. C., and C. G. Mohr, 1977: Some correlations between various sounding parameters and hailstorm characteristics in northeast Colorado. Preprints, 10th Conference on Severe Local Storms, Omaha, Neb., American Meteorological Society, Boston, 218–225.

Fujita, T., 1981: Tornadoes and downbursts in the context of generalized planetary scales. *J. Atmos. Sci.*, **38**, 1511–1524.

Fujita, T., and H. Grandoso, 1968: Split of a thunderstorm into anticyclonic and cyclonic storms and their motion as determined from numerical model experiments. *J. Atmos. Sci.*, **35**, 1070–1096.

Kessler, E., 1986: *Thunderstorms: A Social, Scientific, and Technological Documentary*. Vol. 2: *Thunderstorm Morphology and Dynamics*. Second ed. revised and enlarged, U. of Oklahoma Press, Norman, Okla., and London, 411 pp.

Klemp, J. B., and R. Rotunno, 1983: A study of the tornadic region within a supercell thunderstorm. *J. Atmos Sci.*, **40**, 359–377.

Klemp, J. B., and M. L. Weisman, 1983: The dependence of convective precipitation patterns on vertical wind shear. Preprints, 21st Conference on Radar Meteorology, Edmonton, Alberta, Canada, American Meteorological Society, Boston, 44–49.

Klemp, J. B., and R. B. Wilhelmson, 1978a: The simulation of three-dimensional convective storm dynamics. *J. Atmos. Sci.*, 35, 6, 1070–1096.

Klemp, J. B., and R. B. Wilhelmson, 1978b: Simulations of right- and left-moving storms produced through storm splitting. *J. Atmos. Sci.*, 35, 1097–1110.

Lemon, L. R., 1980: Severe Thunderstorm Radar Identification Techniques and Warning Criteria. NOAA Tech. Memo, NWS NSSFC–3, Kansas City, Mo. (NTIS #PB81–234809), 67 pp.

Lemon, L. R., and C. A. Doswell III, 1979: Severe thunderstorm evolution and mesocyclone structure as related to tornadogenesis. *Mon. Wea. Rev.*, 107, 1184–1197.

Lilly, D. K., 1979: The dynamical structure and evolution of thunderstorms and squall lines. *Annual Review of Earth and Planetary Science,* Vol. 7, Annual Reviews, Inc., Palo Alto, Calif., 117–161.

Ludlam, F. H., 1980: *Clouds and Storms.* Pennsylvania State University Press, University Park, 404 pp.

Maddox, R. A., 1976: An evaluation of tornado proximity wind and stability data. *Mon. Wea. Rev.*, 104, 133–142.

Maddox, R. A., L. R. Hoxit, and C. F. Chappell, 1980: A study of tornadic thunderstorm interactions with thermal boundaries. *Mon. Wea. Rev.*, 108, 322–336.

Marwitz, J. D., 1972: The structure and motion of severe hailstorms. Part II: Multicell storms. *J. Appl. Meteor.*, 11, 180–188.

Moncrieff, M. W., and J. S. A. Green, 1972: The propagation and transfer properties of steady convective overturning in shear. *Quart. J. Roy. Meteor. Soc.*, 98, 336–352.

Newton, C. W., 1963: Dynamics of severe convective storms. In *Severe Local Storms*, Meteor. Monogr. 5, American Meteorological Society, Boston, 33–58.

Newton, C. W., and J. C. Fankhauser, 1975: Movement and propagation of multicellular convective storms. *Pure Appl. Geophys.*, 113, 747–764.

Rasmussen, E. N., and R. B. Wilhelmson,

1983: Relationships between storm characteristics and 1200 GMT hodographs, low level shear and stability. Preprints, 13th Conference on Severe Local Storms, Tulsa, Okla., American Meteorological Society, Boston, 55–58.

Rotunno, R., 1981: On the evolution of thunderstorm rotation. *Mon. Wea. Rev.*, 109, 171–180.

Rotunno, R., and J. B. Klemp, 1982: The influence of the shear-induced pressure gradient on thunderstorm motion. *Mon. Wea. Rev.*, 110, 136–151.

Rotunno, R., and J. B. Klemp, 1985: On the rotation and propagation of simulated supercell thunderstorms. *J. Atmos. Sci.*, 42, 271–292.

Seitter, K. L., and H.-L. Kuo, 1983: The dynamic structure of squall-line type thunderstorms. *J. Atmos. Sci.*, 40, 2831–2854.

Szoke, E. J., M. L. Weisman, J. M. Brown, F. Caracena, and T. W. Schlatter, 1984: A sub-synoptic analysis of the Denver tornadoes of 3 June 1981. *Mon. Wea. Rev.*, 112, 790–808.

Weisman, M. L., and J. B. Klemp, 1982: The dependence of numerically simulated convective storms on vertical wind shear and buoyancy. *Mon. Wea. Rev.*, 110, 504–520.

Weisman, M. L., and J. B. Klemp, 1984: The structure and classification of numerically simulated convective storms in directionally varying wind shears. *Mon. Wea. Rev.*, 112, 2479–2498.

Weisman, M. L., J. B. Klemp, and J. Wilson, 1983: Dynamic interpretation of notches, WERs, and mesocyclones simulated in a numerical cloud model. Preprints, 21st Conference on Radar Meteorology, Edmonton, Alberta, Canada, American Meteorological Society, Boston, 39–43.

Wilhelmson, R. B., and J. B. Klemp, 1978: A three-dimensional numerical simulation of splitting that leads to long-lived storms. *J. Atmos. Sci.*, 35, 1037–1063.

Wilhelmson, R. B., and J. B. Klemp, 1981: A three-dimensional numerical simulation of splitting severe storms on 3 April 1964. *J. Atmos. Sci.*, 38, 1581–1600.

CHAPTER 16

Extratropical Squall Lines and Rainbands

Carl E. Hane

16.1. Introduction

Squall lines and rainbands occur over both tropical and extratropical areas of the Earth in both oceanic and land areas. They affect society in major ways because of their wide geographical range and frequency of occurrence and because they typically contain high rainfall rates. In the Great Plains of the United States, for example, a high percentage of the annual rainfall occurs during late spring and early summer. A large fraction of this rainfall comes from squall lines that propagate generally from west to east, associated with disturbances in the upper-level westerly flow, and "fueled" by copious amounts of water vapor from the low-level moist air moving north from the Gulf of Mexico. Agricultural interests in the Great Plains and the many other areas of the world are critically dependent upon this rainfall. An additional concern is the severe weather frequently accompanying squall lines, which predominantly, but not exclusively, comes in the form of high winds near the low-level outflow boundaries. A very-slow-moving large squall line can also cause flash flooding of streams and small rivers. Small hail occurs not infrequently with squall lines, and when driven horizontally by strong winds, can devastate standing crops. The very intense lightning with these systems and an occasional tornado also pose threats to society.

In this chapter an overview of a number of meteorological factors concerning extratropical squall lines and rainbands is presented. It is not the intent, nor is it possible, to review all the important research relating to these phenomena. Rather, a sampling of research results is presented in order to provide information on critical aspects of the subjects.

16.1.1. Definitions

The *Glossary of Meteorology* (1959) defines a squall line as "any non-frontal line or narrow band of active thunderstorms (with or without squalls); a mature instability line." Further, an instability line is defined as "any non-

359

frontal line or band of convective activity in the atmosphere." In general, meteorologists have not adhered strictly to these definitions in discussions and written material. The term "instability line" is now rarely used. Current usage appears to conform to the idea that any line of thunderstorms, whether associated with a front or not, is considered to be a squall line. The glossary definition does not address the question of whether the term squall line applies to all lines of thunderstorms or only those that include continuous precipitation at the ground along the line. In most cases when lines of storms contain strong convection, the structure of individual cells within the line is significantly related to whether cells are imbedded within continuous (along the line) precipitation or are isolated. For example, isolated cells in a line often have a more supercellular structure than cells imbedded within a quasi-continuous line. Here, the term "squall line" will include both continuous and broken lines of thunderstorms.

A definition for "rainband" seems even less well established. The term is here defined as the complete cloud and precipitation structure associated with an area of rainfall sufficiently elongated that an orientation can be assigned (Houze *et al.*, 1976a, b). Note that the definition is very general and in its present form includes squall lines. To differentiate between squall lines and rainbands some reference to the intensity of convection (if present) is needed. For the purposes of the following discussion, "rainband" describes banded precipitation structures that are either nonconvective or only weakly convective, and "squall line" includes all linear convective structures stronger than rainbands.

16.1.2. Examples of Squall Lines and Rainbands

Figure 16.1 shows examples of squall lines as depicted by PPI radar. Reflectivity is contoured on the display beginning with dim areas just above the reflectivity threshold, progressing to bright and then to canceled areas, as the reflectivity increases because of heavier precipitation. The sequence (dim, bright, cancel) is repeated until the highest contour value is reached. Range rings, where they are visible, are at 40 km intervals in Figs. 16.1a, b, c, d and at 100 km intervals in e and f. Figures 16.1e, f also contain rings at approximately 115, 230, and 345 km, which result from interference from the NSSL 10 cm Doppler radar.

Figure 16.1a shows a squall line that appears quasi-two-dimensional and has sharp reflectivity gradients on both the front and back (east and west in this case). The squall line shown in Fig. 16.1b, which moved generally toward the southeast before this time, includes a highly convective region on the leading edge and a mesoscale precipitation area to the northwest. Interestingly, this squall line became nearly stationary after this time and did not pass over the radar site. There is also a very distinct thin line (the bright band extending east and west of the ground clutter) with this squall line, an indication that the low-level outflow had progressed approximately 40 km ahead of the rain area north of the radar site. Figure 16.1c shows another squall line with a trailing mesoscale precipitation area. On this day (and in other cases) the cells on or near the southern or southwestern end

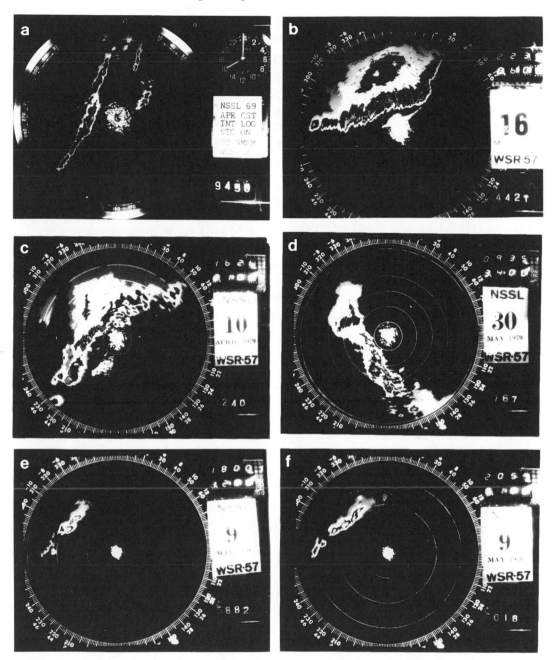

Figure 16.1. PPI displays of squall lines from the WSR–57 radar at the National Severe Storms Laboratory, Norman, Okla. Squall lines occurred on (a) 26 April 1969; (b) 16 May 1977; (c) 10 April 1979; (d) 30 May 1979; (e,f) 9 May 1979.

of the line were very severe. Specifically, cells at 160 km and 230 km range later produced strong tornadoes at Lawton, Okla., and Wichita Falls, Tex., respectively.

Squall lines that occur near extratropical cyclones are most often located in the warm sector. Sometimes, however, the line is initiated behind the cold front and propagates into the warm sector (e.g., the case discussed by Newton [1950]). Figure 16.1d shows a line that formed north of an east-west-oriented cold front and was associated with a disturbance in the upper west-southwesterly flow. Unlike the other examples, it occurred during the morning hours. This squall line was weaker than the other examples, deriving much of its energy from a layer above the surface cold air. The final example (Fig. 16.1e, f) is a squall line that moved little during a long period of its existence. The southwest end can be seen to have remained stationary over the 3 hours between photographs, and the northeast end had moved eastward only about 20 km. In this case cells formed on the southwest end of the line and moved along the line, producing the threat of severe weather over the same location for several hours.

Figure 16.2 shows examples of rainbands associated with extratropical cyclones as observed by a Doppler radar in the Pacific Northwest region of the United States. Hobbs (1981) provided descriptive terms for the various types of rainbands in that area. Figure 16.2a includes examples of warm-frontal rainbands (type 1 in Fig. 16.12). These bands are typically about 50 km wide and occur ahead of and parallel to the surface warm front in an extratropical cyclone. The orientation of the bands and warm front in this case is approximately west-northwest to east-southeast. There are two primary bands in the figure; one lies from about 100 km west of the radar to 40 km south of the radar, and the other lies 80 km north of the radar to 100 km east. Maximum reflectivity in the bands is 30–35 dBZ. The warm front is located south-southwest of the area displayed.

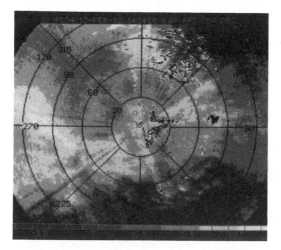

Figure 16.2a. Warm-frontal rainbands within an extratropical cyclone on 12 February 1982 as observed by a C-band Doppler radar (NCAR) on the coast of Washington. Total range is 150 km, and range marks are at 30 km intervals.

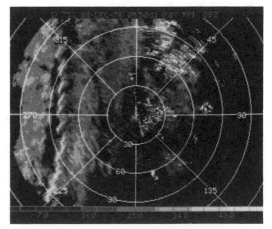

Figure 16.2b. A wide cold-frontal rainband, a narrow cold-frontal rainband, and warm sector rainbands that occurred on 8 December 1976 within an extratropical cyclone near the Washington coastline. Total range is 150 km, and range marks are at 30 km intervals.

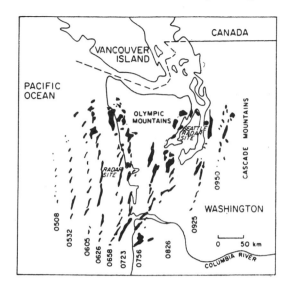

Figure 16.2c. Positions of precipitation cores at different times within a narrow cold-frontal rainband as it passed the coastline of the northwest U.S.A. Areas shown are 25 dBZ or greater, and precipitation outside the narrow cold-frontal band is omitted. The time labeled 0605 follows by 15 minutes the distribution of echo indicated in Fig. 16.2b. (After Hobbs and Biswas, 1979.)

Figure 16.2b shows rainbands in the vicinity of a cold front. The north-south band of elliptically shaped echoes about 85 km west of the radar is a narrow cold frontal rainband. This type of rainband (type 4 in Fig. 16.12) is coincident with the surface cold front. Just ahead (east) of the front and parallel to it is a wider band called a warm sector rainband (type 2 in Fig. 16.12), centered about 55 km west of the radar. Farther to the west, centered about 30 km west of the narrow cold frontal band and also parallel to the front, is a wide cold frontal rainband (type 3 in Fig. 16.12). Rainbands often persist for long periods, as Fig. 16.2c illustrates.

16.1.3. Relation to Other Mesoscale Systems

Recently, the term Mesoscale Convective System (MCS) was proposed (Zipser, 1982), to refer to all precipitation systems on spatial scales from 20 to 500 km that include deep convection during some part of their lifetimes. Examples of the most intense of these systems in middle latitudes are large isolated thunderstorms, squall lines, and Mesoscale Convective Complexes (MCCs). Rainbands, if convective and large, might also be included in the MCS category. Maddox (1980) provided a very specific definition for the MCC based upon infrared satellite imagery. Criteria relating to areal coverage, duration, and shape must be satisfied.

Squall lines with large mesoscale rain areas appear to have much in common with MCCs. The definition of squall line and our ideas about squall lines have grown over many years from surface and radar observations. The definition of MCC has resulted from relatively recent studies of convective systems using satellite information. There are undoubtedly many convective systems that could be termed either MCC or squall line. Comprehensive studies of the internal dynamics of MCSs and the conditions in which they form are needed to determine differences between squall lines and MCCs and to determine whether in many cases they are essentially the same.

16.2. Synoptic Setting For Squall Lines

16.2.1. Relation to Surface and Upper-Air Features and to Air Mass Stability

The larger scale conditions favorable for the formation of squall lines include those necessary for the formation of strong thunderstorms in general. There are also special conditions that are believed to contribute to squall line formation as opposed to isolated storms. An unstable air mass, that is, one with warm, moist air in low levels and relatively cold air aloft, is a necessity. In addition a middle-level dry layer just above the low-level moist layer can contribute to the vigor of the line if the dry air moves with a velocity such that it overtakes the line. Evaporation of rain is enhanced by the presence of this dry air, leading to cooling, increased negative buoyancy, enhanced downdrafts, and strong outflow. An environmental temperature profile with an elevated inversion (cap) can inhibit the formation of deep convection early in the day or in locations ahead of the squall line, thereby ensuring a copious supply of warm, moist air in the line, once the line forms.

An unstable environmental stratification alone is not sufficient cause for squall line formation; a mechanism for release of the instability must be present. This localized ascent can be produced in various ways. The squall line, once formed, provides its own release mechanism in the form of convergence along the gust front, which produces strong ascent into existing updraft areas or initiation of convection at new locations ahead of existing storms. Another common initiating mechanism is the convergence along a cold front or warm front. Cold fronts are frequently associated with strong, large- scale weather systems (e.g., in early spring) that include deepening surface low-pressure centers. As these synoptic-scale cyclones develop in the Central Plains of the United States (for example), warm moist air flows north from the Gulf of Mexico, contributing to destabilization of the environment. This is shown schematically in Fig. 16.3 (Newton, 1963). In Fig. 16.3a a cold upper-level trough approaches sounding location A, contributing to destabilization through cold advection at that level. In Fig. 16.3b the warm moist air (high θ_w) approaches the same site in lower layers. The effect upon the sounding at the point-A site is shown schematically in Fig. 16.3c. The environment is destabilized dramatically through steepening of the environmental lapse rate and moistening of the low levels. Figure 16.3d illustrates the effect of large-scale organized lifting on the inversion in the original sounding. Adiabatic cooling of the air initially at the inversion level results from lifting, and a deepening of the moist layer occurs at the same time.

Other mechanisms for release of instability are important in certain situations. In the Southern Plains, moist low-level air proceeds northward following the departure of surface anticyclones toward the east. The location where the top of this moist layer intersects the sloping terrain to the west is called the dryline (see Ch. 23). Convergence is sometimes present in low levels near the dryline, and at times sufficient upward motion is produced to initiate storms. The convergence is produced because westerly momentum from levels above the surface is transported downward more effectively in

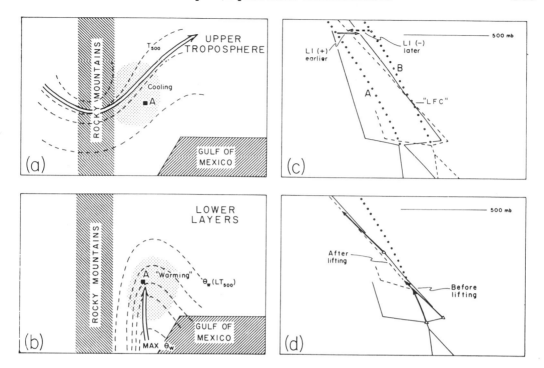

Figure 16.3. Processes in formation and modification of a potentially unstable air mass. In (a) and (b), double arrows are axes of high tropospheric and low-level jets. Dashed lines in (a) indicate 500 mb isotherms; dashed lines in (b) indicate wet-bulb potential temperature in moist layer, which when lifted reaches temperature LT_{500} at the 500 mb level. In (c), dotted lines A and B are moist adiabats corresponding to θ_w of the moist layer before and after moistening occurs. The lifted index (LI) indicates potential stability at first, changing to instability later. In (d), curves are for sounding before (solid) and after (dashed) lifting. (After Newton, 1963.)

the dry air west of the dryline. Since the initiating mechanism is in this case acting along a line, it is not uncommon for storms to form in lines along or just east of the dryline.

Another mechanism for release of instability is the dynamic lifting associated with upper-level disturbances that are associated with no apparent surface features. Figure 16.4 illustrates how a line of storms might form ahead of such an upper-level system (or along a cold front). The dashed line SS might be the west edge of a tongue of potentially unstable air. As the upper-level feature approaches, if it is stronger to the north, isolines of constant upward motion (some critical amount denoted by cold front symbol) will be oriented as shown. As the feature passes over SS, storms are first initiated near point A, later at point B, and so on. A line of storms thus forms with time between points A' and D. Line orientation is determined by the orientation of SS and the spatial distribution of the initiating mechanism.

Another mechanism, which is now frequently recognized with the help of satellite imagery, is the outflow boundary associated with earlier large areas of convection. These large areas often occur at night and begin to dissipate

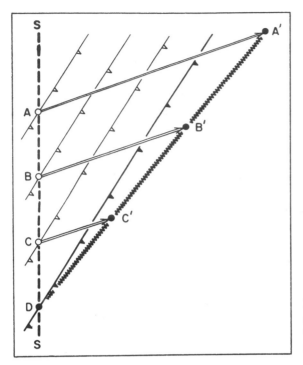

Figure 16.4. Storm initiation in a line as a surface boundary or an upper-level area of upward motion sweeps into a line source SS. (After Newton, 1963.)

during the morning hours, leaving pools of cool air and perturbed areas of pressure and wind near the ground. Sufficient convergence is often present along these boundaries to initiate convection, sometimes in lines, with the help of daytime heating. Orographic features can also act to produce storms in lines. A notable example is the "cap rock" in the Texas Panhandle, a north-south-oriented escarpment. Under moist southeasterly surface flow (upslope) and daytime heating, storms often form first in such areas. In the summer months along the eastern edge of the Rocky Mountains, convection often begins first in the mountains, because of enhanced heating at higher elevations. If storms are sufficiently numerous, deep pools of evaporatively cooled air form in the valleys of the high terrain, and with the aid of a westerly component of flow aloft, outflow rushes down the canyons toward the plains to the east. Such outflows can merge to produce a continuous eastward-advancing boundary. With a component of flow on the plains from east to west and moderate low-level moisture, storms sometimes form in lines in the convergent zone produced by the opposing flows.

Although mechanisms important for squall line initiation have been identified, much research is needed to better understand conditions under which lines of storms do or do not occur when individual mechanisms are present. Additionally, the scale of low-level convergence and associated upward motion remains unknown in many cases.

16.2.2. Initiation of Squall Lines Versus Isolated Storms

It is not clear what environmental differences produce isolated storms in some cases and storms in lines in other cases. It is even less clear why sometimes solid lines of storms develop and other times broken or scattered lines of isolated storms occur. An obvious factor is the nature of the initiating mechanism in a case where the potential instability is quasi-uniform or banded in the horizontal. For example, isolated storms occur (1) when a mesoscale wave forms on a cold front, resulting in enhanced convergence in the vicinity of the lowest pressure; (2) when two outflow boundaries intersect or when an outflow boundary and front or dryline intersect; (3) when an orographic feature such as a range of hills provides enhanced elevated heating and/or lifting over a restricted area. Effects of land versus water (discussed in Ch. 22) can produce many more examples of mechanisms that form isolated storms or lines of storms.

A distinction should be made between a squall line that contains closely spaced cells with nearly continuous precipitation throughout its length and a squall line consisting of a line of isolated thunderstorms. These two situations differ not only in appearance but in cell structure. It is not uncommon for a line of isolated cells to evolve into a solid line by a progressive filling of gaps in the line. When the initially isolated cells are tornadic, this "lining out" generally signals the end of tornadic activity for those cells, even though their identity is maintained within the line. In such a case the tornadic activity frequently shifts to a cell that develops on the south or southwest end of the squall line. Existing lines sometimes increase in length by a systematic addition on the south end of the line (in the case of north-south orientation). Such behavior was noted in squall lines investigated by Newton and Fankhauser (1964). This "building southward" can apparently occur on two scales. On the meso-α scale a feature in the upper atmosphere sometimes overtakes a surface feature first in a more northerly location (Fig. 16.4), and later toward the south. Another factor here might be the strength of a capping inversion which, if stronger in the south, might require organized upward motion or heating for a longer time before the environment is suitable for storm development. On a smaller scale, in both space and time, new development can occur on the south end of a squall line, because of updraft generation along the low-level outflow boundary south of existing storms.

Why squall line thunderstorms at times remain isolated and at other times form quasi-continuous bands is not completely known. The evolution from isolated lines of cells to quasi-continuous lines is generally understood to result when low-level outflows from separate storms move to produce a continuous band of convergence. In a classification of squall lines by Jain and Bluestein (1981) this type of evolution is termed either "broken line formation" or "broken areal formation". Looking at the empirical evidence, Miller (1972) emphasized the importance of vertical wind shear in stating that for optimal squall line formation conditions, the middle-level winds (~5 km altitude) should have a component of at least 25 kt perpendicular to the lower moisture ridge. No differentiation was made, however, between the broken and the solid line cases.

In recent three-dimensional numerical cloud-modeling experiments, Klemp and Weisman (1983) examined the dependence of convective precipitation patterns upon wind shear. Their wind shear profiles are shown in Fig. 15.17 (Ch. 15, this volume). Numerical simulations were initiated by a thermal bubble in an otherwise horizontally uniform environment. Vertical profiles of temperature and water vapor in the environment were not varied among the experiments. The low shear experiment, labeled A, produced a short-lived multicellular storm; cases D and E resulted in much stronger isolated storms. Cases B, C, and F resulted in what were characterized as various types of squall lines. The common feature in the hodograph for cases B, C, and F is the constant magnitude of shear up to a certain altitude with the shear vector rotating clockwise, followed by zero shear above that altitude. Squall lines may on occasion form in the manner associated with profiles B, C, and F, owing to the presence of more than one initialization point along a line.

Wilhelmson and Klemp (1983a) attempted to account for the effect of storm development along a line by initiating numerical experiments with equally spaced thermals and specifying periodic boundary conditions on boundaries normal to the line. In experiments employing a straight-line hodograph normal to the line, splitting storms formed and later merged to form a line of storms whose individual structure differed markedly from the structure occurring with only one initiating perturbation. However, in a case where curvature was imposed in the hodograph below 1 km, the right-most storm of the split pair developed into a single supercell while the left one dissipated and developed into a line of supercells. In other numerical experiments Wilhelmson and Klemp (1983b) initiated storms along a line of converging air containing small random temperature perturbations. The hodograph used in this case is a smoothed version of a hodograph from analysis of the squall line observed 19 May 1977 by Kessinger *et al.* (1983); it is somewhat similar to case E of Fig. 15.17. The simulation (see Sec. 16.3.3) resulted in elongated areas of precipitation containing long-lived cells that resemble those observed on this day in a number of ways. The cells that developed were internally different from the supercells that would have formed had the simulation been initiated from a single perturbation.

It is clear that vertical shear of the horizontal wind is a major factor in determining whether squall line storms are isolated or more continuous. However, the form of initiating mechanism is also very important since line initiation produces storms that interact with one another under many environmental shear conditions. Research will continue to attempt to identify combinations of shear and initiating mechanism that govern the nature and spatial distribution of thunderstorms in squall lines.

16.3. Internal Structure of Squall Lines

Research on the internal structure of squall lines has been extensive over the past 40 years. Much of the knowledge gained concerning tropical squall lines (e.g., Zipser, 1977; Houze, 1977; LeMone, 1983; Roux *et al.*, 1984) has relevance to the understanding of extratropical systems. It is clear that tropical and extratropical squall line studies have drawn upon each other interactively to advance understanding of such systems.

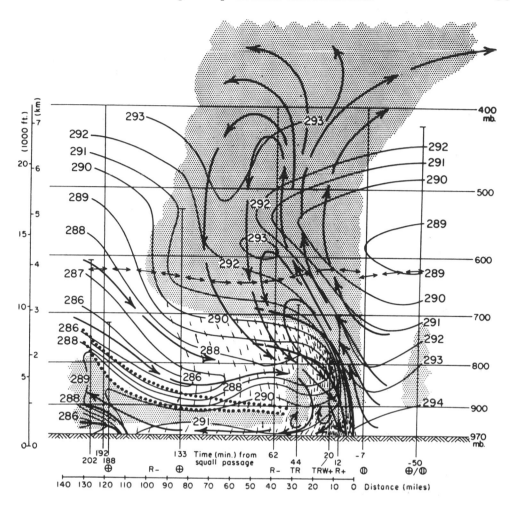

Figure 16.5a. Vertical cross section through a pre-frontal squall line that passed over the Ohio network of the Thunderstorm Project on 29 May 1947. Heavy lines indicate surfaces of stable layers (polar front at left); dotted lines are boundaries of stable layer in squall sector with relatively dry air above it; thin lines are isopleths of θ_w; cloudy area is stippled; arrows showing airflow are schematic. (After Newton and Newton, 1959.)

16.3.1. *The Intense Convection in Squall Lines and Its Relation to the Environment*

The first comprehensive data in squall lines were obtained over the Ohio network of the Thunderstorm Project (Byers and Braham, 1949) in the late 1940s. Rawinsondes were released frequently as a squall line passed over this network on 29 May 1947. The data were analyzed by Newton and Newton (1959), who produced a schematic vertical cross section normal to the line (Fig. 16.5). An earlier analysis by Newton (1950) showed that this squall line first formed west of a synoptic-scale cyclone and cold front, approximately 14 hours before the time of this cross section. Downward transfer of westerly momentum in the rainy area produced a rapid eastward movement of the low-level convergent area (outflow boundary) along which new convection

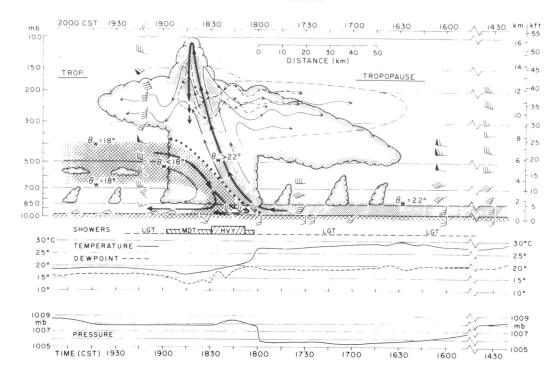

Figure 16.5b. Cross section through squall line of 21 May 1961 as it passed Oklahoma City. Hatching indicates probable extent of high θ_w air of low-level origin; cross-hatching indicates location of low θ_w air of probably middle-level origin. Heavy arrows are axes of main drafts; thin arrows are streamlines, dashed where air emanates from core of stratospheric tower. Long dashes suggest outline of air plume originating in storm; at lower altitude cloud plume consists of small precipitation particles. (After Newton, 1966.)

formed. In Fig. 16.5a the cold front is at lower left. The average circulation relative to the moving system is shown, along with the distribution of wet-bulb potential temperature and areas of cloud and rain.

From Fig. 16.5a, the importance of vertical shear in the horizontal environmental wind is clear. Low-level flow on the front (east) side of the line ascends in an upshear (westward) sloping updraft generally along a tongue of high wet-bulb potential temperature. Condensation in the updraft (whose details are unknown here, especially in upper levels) results in rain falling from the updraft in both convective and mesoscale regions of the line. Dry air (low wet-bulb potential temperature) overtakes the system from the west. Because of its dryness, this air is most effective in evaporating rain, becomes negatively buoyant, and descends in both the mesoscale and convective rain areas. Downdrafts spread at the ground and carry some of the westerly momentum from middle levels, thereby enhancing convergence along the outflow boundary and leading to maintenance of the convection.

The highly convective region in a squall line is by definition always present; however, many lines lack the more stratiform mesoscale region indicated in Fig. 16.5a toward the rear of the line. An example is the squall line investigated by Newton (1966), which passed through central Oklahoma on 21 May 1961 (Fig. 16.5b). The partly schematic cross section is based upon radar and surface data and some rawinsonde information. Again the upshear slope of the updraft is emphasized along with the middle-level source of downdraft air. From computations using this data it was concluded that no significant part of the air rising in the updraft is likely to return to low levels. The air descending in upper levels on edges of stratospheric towers is separate from the downdraft in the low-level rainy area. Additionally, the back-sheared and forward anvil regions are composed of both air that has never left the troposphere and air that has circulated through the stratosphere. The traces from the surface network in Fig. 16.5b indicate the characteristic changes that occur with passage of the storm's gust front: a sharp rise in pressure, a marked shift and strengthening of the wind, and a significant decrease in temperature and water vapor. Though much is known about the type of system shown in Fig. 16.5b, work is needed to ascertain such things as the source region of downdraft and anvil air and the importance of the third dimension in determining system structure and evolution.

16.3.2. *Mesoscale Precipitation Areas in Squall Lines*

A squall line that contains a mesoscale area of lighter precipitation to the rear of the advancing strong convection is here termed a squall line complex. Sanders and Emanuel (1977) speculated upon the evolution of such systems, in which an early stage contains only convection, a mature stage contains both convection and a mesoscale area of rain, and a dissipating stage includes only the mesoscale area. These systems often form away from fronts in more horizontally homogeneous air masses, although the pre-frontal squall line (Fig. 16.5a) also contains an associated mesoscale rain area. These systems have been recognized for many years and have gained attention because of the large amount of rainfall they often produce. Relying primarily upon synoptic surface observations and visual appearance of clouds, Bergeron (1954) postulated a structure for these systems (Fig. 16.6). The area of vigorous convection is toward the right (east) side, and a large area of precipitation trails toward the west (total horizontal extent of precipitation is ~200 km).

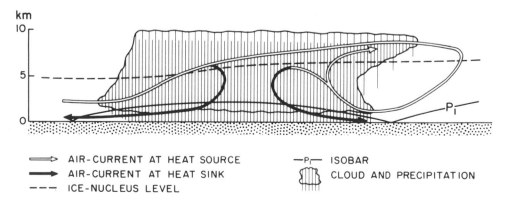

Figure 16.6. Vertical cross section through a convective system formed by the fusion of several convective cells in very humid, unstable air. (After Bergeron, 1954.)

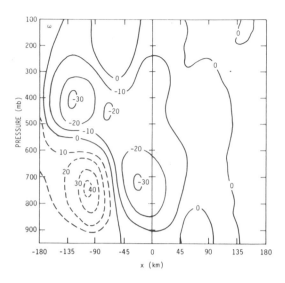

Figure 16.7. Vertical motion deduced in an Oklahoma squall line that occurred on 22 May 1976. Values are 10^{-3} mb s^{-1} and were deduced from the continuity equation, using rawinsonde network observed horizontal winds. Horizontal position (x) is distance ahead of the squall line's leading edge. (After Ogura and Liou, 1980.)

Mesoscale ascent is postulated to occur at high levels in the area trailing the convective region. Downward motion is indicated in the rainy area at middle and lower levels in the central part of the system and in clear air ahead of the system. Heat source and heat sink refer to regions of local heating and local cooling, respectively. The correspondence between the structure postulated in Fig. 16.6 and that deduced from recent Doppler radar measurements is striking.

Evidence for the presence of mesoscale ascent in the trailing rain area of squall lines has been provided by Ogura and Liou (1980). In Fig. 16.7 the vertical motion field based upon data from a network of rawinsondes is shown for the case of a squall line complex that occurred over Oklahoma on 22 May 1976. Mesoscale ascent is indicated in the area of strong convection (just west of $x = 0$) and another area in middle levels (7 km) more than 100 km west of the line's leading edge. An area of mesoscale descent is indicated below approximately 5 km centered approximately 100 km west of the leading edge. Mesoscale descent in the trailing precipitation area has also been noted in studies of tropical squall lines (e.g., Zipser, 1977; Houze, 1977; Leary and Houze, 1979).

Houze and Smull (1982) investigated the same squall line, utilizing reflectivity and radial velocity data from one of the National Severe Storms Laboratory (NSSL) Doppler radars. Their findings are summarized schematically in Fig. 16.8, which shows a vertical cross section (east on right) normal to the squall line. In this case evidence was found for an updraft that originates ahead of the eastward-directed outflow and tilts toward the back of the line over a horizontal distance of approximately 30 km until it reaches a height of about 8 km in the most intense (reflectivity) cell in the line. This updraft enters a jet of front-to-rear air at 4 to 5 km. Weaker reflectivity maxima, interpreted as dissipating cells moving toward the rear of the line in a relative sense, are located west of the intense convection. A melting layer is also indicated just below 4 km altitude in the region between −55

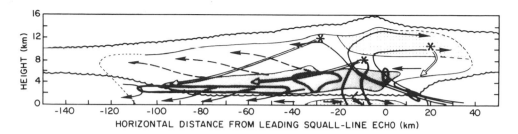

Figure 16.8. Conceptual model based upon single-Doppler radar observation of a squall line complex that passed over Oklahoma on 22 May 1976. Estimated cloud outline is scalloped; outside (mostly solid) contour outlines radar echo; heavy lines denote areas of higher reflectivity. Light shading indicates regions of system's relative wind component from left to right; heavy shading shows jet of maximum horizontal wind from right to left. Thin streamlines show two-dimensional relative flow consistent with wind and echo structure. Hypothesized ice particle trajectories are denoted by asterisks and double arrows. Dashed lines indicate speculative parts of the model. (After Houze and Smull, 1982.)

and −110 km. The melting layer is hypothesized to result from melting ice particles that form in the stronger updrafts and are advected westward as they fall, as indicated by the trajectories. The vertical velocity structure in the stratiform region above the melting layer is somewhat speculative since it is inferred from measurements from one Doppler radar; however, it is consistent with the results of Ogura and Liou (1980). Comparison with studies of squall lines in the tropics also reveals similarities between systems like this and tropical systems.

Another squall line was observed by single Doppler radar in northern Illinois (Zipser and Matejka, 1982). This line also contained a trailing stratiform region of precipitation and a bright band near 4 km altitude. A difference between this and the case reported by Houze and Smull (1982) (H-S) is that there was pronounced rear-to-front flow at 2 to 5 km altitude in the stratiform region in this case, whereas flow in this region was almost entirely from front to rear in the H-S case. Zipser and Matejka proposed that the upper layer of the stratiform region is supplied moisture from the debris of convection in the forward part of the system and that the upper stratiform region is characterized by mesoscale ascent. In this case and in a tropical case, strong middle-level shear in the wind component normal to the line resulted in a sloping mesoscale updraft over a sloping mesoscale downdraft, apparently leading to maintenance of the air circulation in the mesoscale precipitation area.

Another Oklahoma squall line complex, which occurred on 19 May 1977, was investigated by Kessinger (1983). In this case Doppler velocity information was used again, but data from four Doppler radars were combined to produce the three components of air motion over the observed volume. (See PPI radar depiction in Sec. 6.5.2, Fig. 16.17.a.) This squall line contained a convective region near the eastern edge and a meso-β-scale area of precipitation trailing to the west, as in other cases. A schematic representation of the airflow and reflectivity structure is shown in cross section in Fig. 16.9. In

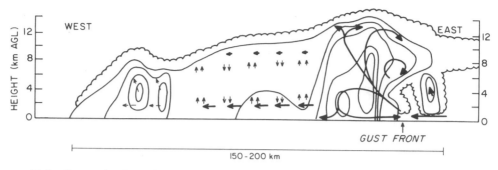

Figure 16.9. Conceptual model of the 19 May 1977 squall line based upon Doppler-observed wind and water fields. Arrows trace the airflow relative to the line. Solid isolines represent qualitative reflectivity. Scalloped lines depict estimated cloud boundary. Small vertical arrows represent perturbations superimposed upon mean flow; dashed arrows represent speculative flow. The isolated pre-line echo at right is present only at certain locations along the line. (Adapted from Kessinger, 1983.)

this particular line the updraft in the convective region sloped from front to rear at a steep angle. In the trailing meso-β region there were perturbations in both horizontal and vertical velocities (small arrows), but in the mean, line-relative flow was either very weak or toward the rear at all levels. A distinct feature was an area of enhanced reflectivity and vertical motion (details uncertain) near the western edge of the precipitation area. Many small cells ahead of the system are believed to be very important to the maintenance of the line in this particular case. Small cells ahead of a tropical squall line, which merged with the line itself, were also noted in a case studied by Houze (1977). Vertical motions (both upward and downward) in the meso-β region in this case appear to be larger than values expected with general mesoscale ascent and descent. In the mean, vertical motions are probably upward in this region in upper levels and downward in lower levels, but this remains to be investigated.

In summary, squall line complexes that have been studied in recent years have much in common but also exhibit important differences. Observations indicate that vigorous convection lies on the low-level relative inflow side of a meso-β-scale precipitation area containing low-level downdrafts. This arrangement appears to exist in all the cases described above and in other cases, including MCCs in Texas (Leary, 1983) and near the British Isles (Browning and Hill, 1984). In the meso-β-scale precipitation area, flow is generally away from the convective band at all levels, but there are exceptions. The sense of slope of the area of updrafts near the convective region appears to be the same in all cases, although the degree of slope (and therefore horizontal area affected) varies. A bright band appears to be present in the meso-β-scale precipitation area with rising motion generally above that level, but the horizontal distribution and strength of that motion is unknown and probably varies among systems. In many cases propagation of the convection by merger with and growth of new cells appears important, but the organization of these propagative mechanisms probably differs from case to case. Differences in observed structure are partly a function of the time of observation during the life history of these sytems as well as a function of what elements a study has emphasized. Clearly, much more research is needed to determine with greater certainty which features are common to all systems, how such systems evolve, and how they are affected by differing environments.

16.3.3. Numerical Simulation of Squall Lines

Convection in squall lines has been numerically simulated in both two-dimensional and three-dimensional cloud models. Modeling studies allow for the changing of conditions (e.g., environmental), to assess the effect on the structure and evolution of the resulting system. Any two-dimensional slab-symmetric simulation is, in effect, modeling an infinitely long line of convective clouds since by definition no gradients in any variables are allowed along the line. No attempt is made here to review all the two-dimensional modeling of deep convective clouds that has been carried out since Malkus and Witt (1959) performed the first two-dimensional experiments with dry thermals. Instead, examples are given of a two-dimensional and a three-dimensional application. Both models are applied to the 19 May 1977 squall line in Oklahoma.

The two-dimensional model of Hane (1973) was applied to the convection along the eastern edge of the 19 May 1977 squall line complex. The model domain was 16 km deep but only 40 km wide, so that the full width of the line could not be simulated. Results have been reported by Kessinger *et al.* (1982) and Kessinger (1983). Nearby soundings were used to define vertical profiles of wind, temperature, and moisture and to specify a convergent region in the wind field as an initial perturbation over a system of grid points spaced at 400 m in the horizontal and vertical. A 2 h simulation included an early growth stage followed by a series of maxima in vertical velocity within the convective region. Figure 16.10 shows the structure at 50 minutes into the simulation along with a vertical cross section of Doppler-derived velocity and reflectivity. A limited area of mesoscale precipitation west of the convection and a westward sloping updraft have evolved. Height of the cloud top is several kilometers less than observed, owing to lesser updraft velocities. Schlesinger (1984) in comparative two- and three-dimensional numerical simulations showed that restricting airflow to two dimensions results in larger downward-directed pressure gradients in updrafts. An interesting result, discernible owing to the fine time resolution of the model results, is the manner in which the major convective area is maintained. Pulses of upward motion are periodically initiated at the gust front (at 32.5 km in figure) and travel westward over the outflow toward the region of major convection (17–22 km), causing intermittent invigoration of the main updraft region. The flow near the surface shows areas of convergence and divergence in response to small areas of rain-induced downdrafts, which in turn interact with the impulses initiated near the gust front. Observed cross sections also contain perturbations in the east-to-west flow in the area east of the major updraft region. In spite of the restrictive nature of two-dimensional simulations, it appears

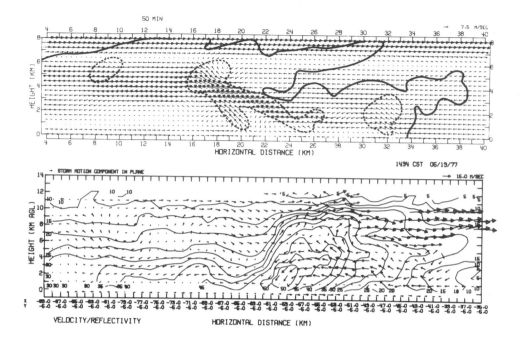

Figure 16.10. Top: Wind vectors, rainwater outline (solid), and cloud cores (dashed) from a 2-D cloud model using environmental soundings from 19 May 1977. Bottom: Vertical cross section of the 19 May squall line showing wind vectors and reflectivity (dBZ). Both model output and observed fields show westward-tilting updrafts and the existence of small areas of upward motion moving from east to west (right to left) above the westerly low-level outflow. These are located in model output at 22, 26, and 32 km and in observations at −42 and −48 km.

that at least some features are well reproduced and that the selective use of such models is justified.

Three-dimensional simulations allow for variations along the line and therefore the development of much stronger convection at certain locations. Wilhelmson and Klemp (1983b) used soundings from this same day to specify vertical profiles of wind, temperature, and moisture for a three-dimensional simulation. The line of storms is initiated by specifying a converging line of air with random thermal perturbations along it. Periodic boundary conditions are used along the north and south boundaries, but the 40 km north-south extent of the domain allows several cells to exist along the line at one time. The domain extends 60 km east to west; it is 16 km deep with 2 km grid point separation horizontally and 0.5 km separation vertically.

Figure 16.11 shows modeled evolution of the rainwater and vertical velocity fields during a 1 h period. Cell motion matches exceedingly well the observed motion on this day, and the rainwater maxima west of vertical velocity maxima are also as observed. Despite the long-lived nature of the cells (e.g., B and C) the cells within the line appear to be internally different from supercells. Simulation using the observed shear profile on this day and a single initiating perturbation would probably have resulted in a supercell storm. The solution fields were averaged along the line at the last time shown in

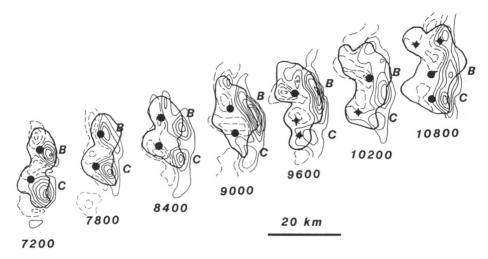

Figure 16.11. Horizontal cross sections at 2.25 km (altitude) from a 3-D cloud model showing vertical velocity contours at 2 m s^{-1} intervals and the 0.5 g kg^{-1} rain contour. Times are given in seconds. Eastward movement of the velocity field is exaggerated so that contours can be clearly seen; northward movement is exactly as simulated. Maxima in the rainwater field are indicated by solid circles. (After Wilhelmson and Klemp, 1983b.)

Fig. 16.11, to ascertain whether important features could be seen in a two-dimensional composite. The westward tilt of the updraft, displacement of rainwater maximum to the west, warm updrafts, cold outflow, and essential elements of the pressure distribution were all apparent in the composite.

Recent three-dimensional modeling efforts have begun to produce realistic results in simulating various features of observed isolated storms; however, numerical simulation that treats convection explicitly in squall line complexes is a new area of research. The relatively large time scales and areas associated with these sytems, and the difficulty of prescribing realistic boundary and initial conditions, complicate the running of numerical experiments. Larger scale (hydrostatic) modeling experiments (e.g., Fritsch and Chappell, 1980), which include much larger domains, must parameterize the effects of convection. Results of such experiments reproduce first-order flow and structure on the meso-β scale, but omit detail. Future explicit (nonhydrostatic) cloud-scale modeling must explore the effects of varying vertical wind shear along with determining the influence of different initial and boundary conditions. A significant challenge lies in incorporating the influence of larger scale temporal and spatial changes into convective-scale models, while accounting for effects of the convection itself on the larger scale.

16.3.4. Severe Weather in Squall Lines

Squall lines can exist with or without severe weather. The most common form of severe weather, when it does occur, is the very strong straight wind that flows outward ahead of the line when downdrafts spread near the ground and carry horizontal momentum from higher levels of the environment. Cloud-to-ground lightning, especially those powerful strokes that lower positive charge to ground (Rust *et al.*, 1983), are an ever-present danger in squall lines. There is a suggestion that some of the positive strokes are imbedded in the mesoscale rain areas trailing squall lines, propagating for tens of kilometers before striking the ground. Squall lines often produce hail, but on the average, stones are smaller than in strong isolated storms (Nelson and Young, 1979). However, even small hail when driven by strong winds can devastate crops and other vegetation. Flash flooding is not uncommon with squall lines, especially those that are very broad or move very slowly.

The preceding remarks refer mainly to squall lines composed of a continuous band of precipitation and outflow. Isolated storms that form lines (which may be considered in the squall line category also) have different internal structure. Tornadoes usually occur only in these isolated storms in lines or in cells that have formed or persist on the south or southwest end of a squall line. Exceptions to this are small short-lived vortices, which sometimes form in horizontal shear zones along irregular outflow boundaries ahead of vigorous squall lines. Additionally, squall lines sometimes develop wavelike patterns (line echo wave pattern), analogous to synoptic-scale frontal waves, in which severe weather frequently develops. Careful examination of a large body of squall line data is needed in order to establish more firmly the frequency of various types of severe weather in squall lines compared with other types of strong convection.

16.4. Rainbands

Rainbands in extratropical regions occur primarily in association with well-organized extratropical cyclones. The existence of rainbands in other synoptic-scale environments is acknowledged; however, the preponderance of research on rainbands has been carried out in the extratropical cyclone setting. For this reason and because space is limited here, this discussion focuses on rainbands associated with extratropical cyclones.

Much rainband research has been carried out in recent years in the Pacific Northwest of the United States (e.g., Hobbs *et al.*, 1980; Houze *et al.*, 1976). The results may to some extent be generalized to other parts of the world. Pacific Northwest findings are probably most applicable to extratropical cyclones that have crossed large expanses of ocean, rather than to systems in a genesis or slightly later stage. Specifically, such systems appear very similar to those affecting the British Isles. Earlier work in the British Isles by Browning and Harrold (1969, 1970), Harrold (1973), Browning *et al.* (1973), and Browning (1974) reveals relationships of various rainbands to frontal structures similar to those relationships found in the Pacific Northwest.

Rainband research in the northeastern United States has been more limited, but banded structure has been observed accompanying an extratropical cyclone (Kreitzburg and Brown, 1970). Nozumi and Arakawa (1968) reported banded structures, similar to those in the Pacific Northwest, associated with cyclones over the subtropical ocean near Japan. Banded precipitation structure on a much larger scale and not associated with extratropical cyclones has also been reported in the Japan area. Ninomiya and Akiyama (1974) described a large-scale band associated with the low-level jet near

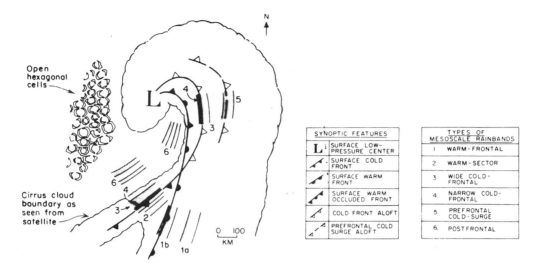

Figure 16.12. Locations of rainband types observed in extratropical cyclones of the Pacific Northwest. (After Hobbs, 1981.)

Japan during the summer. Another variety of large-scale rainband, which also occurs near Japan in summer, but on the west side of an upper-level ridge, was described by Ninomiya *et al.* (1983).

Work in the Pacific Northwest region of the United States is summarized here as an example of research in several regions of the world. It should be noted that the work in the Pacific Northwest has emphasized microphysics within rainbands, whereas work in other regions has generally stressed kinematics and forecasting.

16.4.1. *Synoptic Setting for Rainbands of the Pacific Northwest*

Houze and Hobbs (1982) classified rainbands in extratropical cyclones as six defined types. The locations of the types in relation to frontal structures within cyclones are illustrated in Fig. 16.12.

(1) Warm-frontal bands are typically about 50 km wide and occur within the forward portion of the frontal system. They are located within areas of deep warm advection and have orientations similar to that of the warm front. They may be located ahead of the warm front (1a) or be coincident with the surface warm front (1b).

(2) Warm-sector bands are typically 50 km wide or less and occur in the warm sector.

(3) Wide cold-frontal bands are about 50 km wide and either encompass or are behind the surface cold front. In the case of occluded fronts, they are associated with the cold front aloft.

(4) The narrow cold-frontal band is only about 5 km wide and lies along the surface cold front.

(5) Pre-frontal, cold-surge bands are associated with surges of cold air ahead of cold fronts.

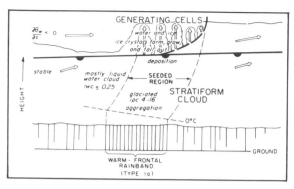

Figure 16.13. Model of a warm-frontal rainband shown in vertical cross section. Cloud structure and predominant mechanisms for precipitation growth are indicated. The motion of the rainband is from left to right. (After Hobbs, 1978.)

(6) Post-frontal bands are lines of convective clouds that form well behind and generally parallel to the cold front.

Other features sometimes present are wavelike rainbands that are superimposed upon the other rainbands, and unbanded convective clouds (sometimes roughly hexagonal cells) occurring well behind the cold front. In any particular cyclone not all types of rainbands need be present. Terminology used for Pacific Northwest rainbands appears to have at least some roots in the British work. The terms warm-frontal bands and warm-sector bands were used by Browning and Harrold (1969). Additionally, Browning and Harrold (1970) and Nozumi and Arakawa (1968) reported upon narrow cold-frontal bands, but the term was not used.

16.4.2. Internal Structure of Pacific Northwest Rainbands

Warm-frontal rainbands represent mesoscale regions of enhanced precipitation within the general area of ascent resulting from warm advection ahead of an advancing cyclone. Figure 16.13 schematically shows the structure believed to be associated with such bands. The warm-frontal intersection with the ground is far to the left, and the cross section is oriented normal to the band. It is believed that these bands are seeded from above by generating cells. These cells produce ice particles that fall through the stratiform cloud below and grow by deposition, aggregation, and riming. Streamers of ice from individual generating cells can give rise to enhanced precipitation rates over mesoscale areas within the bands. Most of the precipitation that falls is made up of water originating from the stratiform region, though the seeder clouds play a crucial role in enhancing precipitation. The air above the stable stratiform region is potentially unstable. Probable sources of this instability above the warm front are either directly from the potential instability of the tropical maritime air moving over the warm front or from differential advection in the middle troposphere. Several mechanisms have been suggested for the release of the potential instability in the seeder area. The theory that seems to agree best with the properties of the observed bands suggests that gravity waves could be ducted through the stable layer below the warm front. The vertical motions associated with these waves could promote the release of potential instability in the seeder zone or produce higher concentrations of condensed water in the stratiform region (or both).

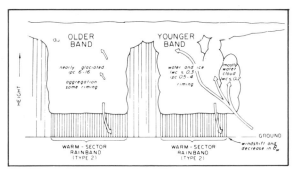

Figure 16.14. Model of two warm-sector rainbands shown in vertical cross section. The structure of the clouds and predominant mechanisms for precipitation growth are shown. Open arrows depict airflow relative to the rainbands whose ground-relative motion is from left to right. (After Hobbs, 1978.)

Warm-sector rainbands, depending upon their definition, may include pre-frontal squall lines. The weaker versions of these rainbands, which are the subject here, are indeed similar in some respects to squall lines. Figure 16.14 schematically depicts a vertical cross section through two such bands, which are oriented parallel to the cold front (not shown, off to the left). In contrast to the warm-frontal bands, the warm-sector bands can contain deep convective cells through the full depth of the band. Instability is released by vertical motion forced at a gust front as shown in Fig. 16.14 at lower right. These bands often occur in series, the newer bands being farther ahead of the cold front. It has been determined that the newer bands are more strongly convective and contain higher concentrations of supercooled water than the older bands, which are primarily composed of ice particles. The predominant particle growth mechanisms appear to be riming in the newer bands and aggregation in the older ones. It has been suggested that these rainbands develop as a result of internal gravity waves that propagate away from the cold front. Additional research is needed to determine if this or some other mechanism (a number have been suggested) is responsible for their formation.

Wide cold-frontal rainbands result from regions of enhanced lifting over the cold-frontal surface (Fig. 16.15). These bands resemble warm frontal rainbands in that generator cells above the front in unstable air produce ice crystals that grow as they fall through the clouds below the cold front.

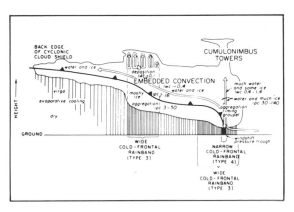

Figure 16.15. Model of the clouds associated with a cold front, showing narrow and wide cold-frontal rainbands in vertical cross section. Cloud structure and predominant precipitation growth mechanisms are shown. A strong convective updraft and downdraft above the surface front and pressure trough, and broader ascent over the elevated cold front are indicated by open arrows depicting the relative airflow. Ground-relative motion of the rainbands is from left to right. (After Hobbs, 1978.)

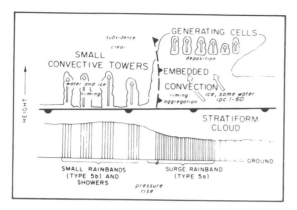

Figure 16.16. Model of rainbands associated with a pre-frontal surge of cold air aloft, ahead of an occluded front. The broken cold-frontal symbol indicates the leading edge of the surge (primary cold front is to the left of figure). Cloud structure and predominant precipitation growth mechanisms are indicated. Open arrows indicate convective ascent and airflow relative to the cold surge. Ground-relative motion of the cold surge and rainbands is from left to right. (After Hobbs, 1978.)

Symmetric instability has been suggested as a probable mechanism for the formation of these bands.

Narrow cold-frontal rainbands (Fig. 16.15) occur at the surface cold front where convergence produces a narrow updraft whose velocity may be a few meters per second. They may also occur above the cold-frontal passage aloft in a warm type occlusion (Fig. 16.12). Cloud towers associated with the updraft may occasionally penetrate the larger cloud shield. The cold air continually moves under the air in a low-level southerly jet which supplies moisture to the small-scale updraft. Liquid water rather than ice is abundant in these bands so that ice particles grow primarily by riming, resulting in graupel and hail formation on some occasions. Elliptically shaped precipitation cores, oriented at 30–35° angles to the cold front, contain the heaviest precipitation in these bands. These cores form in areas of preferred low-level convergence resulting from irregularities in the shape of the mesoscale outflow boundary. In many respects the passage of these cores at the surface is very similar to the passage of a squall line gust front. Circulations (about vertical axes) may occasionally develop in the vicinity of these cores, and in one case (Carbone, 1982) a tornado has been documented. The breakup of the narrow cold-frontal bands into cores is believed to be due to gravity currents or to instabilities produced by the strong horizontal shear across the front in the wind component parallel to the front.

Pre-frontal cold-surge bands result when, in an occlusion, the cold air advances over the warm front in a series of surges. The strongest surge is the cold front itself, but weaker surges also occur ahead of the front and are indicated at the surface by a temporary slight rise in pressure or a decrease in the rate of pressure fall. Behind the pre-frontal surge, small convective clouds with associated wavelike rainbands may be present (Fig. 16.16). These result from instability due to the cold air aloft behind the surge line. The surge rainband itself falls from a deep band of cloud that precedes or straddles the cold surge and is in many respects similar to the wide cold-frontal and warm-frontal rainbands. The deep band may contain imbedded convection and may be seeded by generating cells aloft.

Post-frontal rainbands (Fig. 16.12) are lines of convection that form in cold air masses behind regions of strong subsidence that immediately follow cold-frontal passage. These bands are more readily observable, either visually or from satellite imagery, because of their displacement to the west of the cyclone's cloud shield. Newer bands sometimes form ahead of existing lines, suggesting that at times they behave as organized convective systems. Similar to warm-sector bands, the newer bands contain primarily supercooled water; growth is by riming, and showers of graupel are common. The older bands are more glaciated, and particle growth is primarily by aggregation. Since no large-scale air mass boundaries appear to influence these bands, a mechanism such as wave-CISK (conditional instability of the second kind), may be important for band formation (Parsons and Hobbs, 1983).

Although types of extratropical cyclone rainbands are described separately, interactions can occur between the various types. The most common interaction is between narrow and wide cold-frontal rainbands. In addition, orographic features can have profound effects upon the formation and evolution of rainbands (Houze and Hobbs, 1982).

16.5. Forecasting Squall Lines and Rainbands

Forecasting the formation and evolution of squall lines and rainbands is a major problem. In many cases the data used for forecasts are available (to man or machine) on spatial and temporal scales much larger than those scales of the phenomena. Often the behavior of the smaller scale must be inferred from the larger scale observations, on the basis of both physical reasoning and experience. The experience factor is related to the storm formation climatology of a given region (terrain features, land/water locations, etc., are important), and to the process of learning how to attain and sort through all the important meteorological factors relevant in a given situation. Important observational systems currently utilized on smaller time and space scales include hourly synoptic surface observations, radar data, and meteorological satellite data. Potential sources of higher resolution data in the future are satellite sounding systems and ground-based remote (profiler) systems.

The term "forecast" in relation to these mesoscale phenomena could more properly be replaced by "nowcast" since it is probable that such events cannot be forecast in any reasonable detail beyond 12 hours (or less in many cases). The conditions in which the phenomena form (without details of timing and location) can in many cases be forecast over considerably longer periods. ("Forecast" is here used in a general sense.)

16.5.1. Factors Influencing the 4–10 Hour Forecast

Conditions favorable for squall line development are in most cases the same as those needed for the formation of vigorous thunderstorms in general. An unstable air mass including warm moist air in low levels and cold air aloft is necessary, in addition to a dry layer just above the deep moist layer (say, above 2 km AGL). The dry air contributes to the potential instability in that it is important in the evaporative cooling, downdraft formation, and downward transfer of momentum processes. Vertical shear of the horizontal wind is important in that, among other reasons, it ensures that the line of storms will propagate (either discretely or continuously) in such a way as to increase the relative inflow of moist air in low levels. The form of the vertical shear profile is probably important also in determining the structure of individual storms in lines.

A mechanism for release of the above-mentioned instability is the other major necessary ingredient. As mentioned previously, there are a variety of mechanisms for localized lifting of unstable air that must be taken into account. Air mass boundaries (fronts) are probably the most common initiating mechanisms with regard to squall lines. For reasons not completely understood, squall lines sometimes form in the warm sector ahead of cold fronts. Such developments may be related (for example) to linear regions of convergence ahead of the front or to vertical motions in association with variations in wind speed along the low-level jet. Other mechanisms (see Sec. 16.2.1) are short waves aloft with no associated surface feature, drylines, gravity waves initiated at the front, low-level outflow boundaries from earlier convection, and various orographic effects.

Useful empirical rules have been developed through experience in forecasting squall lines (e.g., Miller, 1972). A few examples:

- The amount of wind shear necessary for squall line development is such that the wind component at 14,000–16,000 ft and perpendicular to the low-level and middle-level moisture ridge is at least 25 kt.

- The angular shear between the low-level and middle-level winds is at least 30 degrees on the forward side of the upper-level trough.

- If other parameters are favorable for the development of severe thunderstorms, the most intense activity is expected about 100 miles behind the 850/500-mb-thickness ridge.

- Development along the dryline occurs only if the moist layer is at least 3,000–6,000 ft deep, the lapse rate is unstable, and insolation is at least moderate.

As with other empirical rules, exceptions must be studied carefully and understood so that limitations may be placed upon the situation to which each rule is applied.

The factors discussed apply to a 6–8 h forecast of the location and timing of squall line development. In practice, for a midmorning forecast of late afternoon activity, one would do quite well to forecast development within 2 h and 100 km of actual occurrence. In addition, most of the time, development of squall lines cannot be distinguished from development of isolated storms (say a scattered line or areal development) because of a lack of complete understanding as to what factors cause these different modes and because resolution in both time and space of upper-air data is lacking.

In the last decade or so, much progress has been made in mesoscale numerical modeling (e.g., Pielke, 1984). In the future, with the development of rapid data assimilation and processing, such models could contribute greatly toward forecasting squall lines and other mesoscale phenomena. Currently such modeling applications suffer from inadequate data for initialization of model predictive equations. In addition such models require parameterization of the effects of deep convection that are not directly resolvable on the model temporal and spatial scales. These parameterization schemes require further development in order to account for interactions between the hydrostatic mesoscale and the nonhydrostatic convective scale. Operational use of such models would provide the forecaster of mesoscale events with an additional valuable tool to help in the decision-making process.

16.5.2. Factors Influencing the 0–3 Hour Forecast

The forecast problem continues on smaller time and space scales following formation of squall lines. Questions arise concerning the speed of the line, direction of motion, increases or decreases in intensity, preferential development in particular locations along the line, and whether a squall line complex will develop (to name a few). In general it can be said that convective systems that are organized on the mesoscale are more amenable to short-term forecast than are individual thunderstorms.

Empirical rules have been developed to estimate the motion of squall lines. Miller (1972) stated that steering of squall lines is generally in the direction of the 500 mb winds at 40% of the speed. Boucher and Wexler (1961) in an analysis of lines of precipitation detected by radar over the northeastern United States and in Illinois developed an empirical relation between line motion and the 700 mb wind component normal to the line. The latter study apparently included both squall lines and extratropical cyclone rainbands. There are of course exceptions to these rules, which do not allow for accelerations or decelerations without changes in line orientation. These rules are perhaps best used prior to squall line development or during squall line formation, since extrapolation can be used effectively once motion of the line is determined.

Increases or decreases in intensity are very difficult to forecast since small-scale environmental data are needed to make such judgments. A possible exception is in the early stages of line development where knowledge of atmospheric stability may allow foretelling continued intensification. Occasionally, decreases in intensity may be forecast when the line approaches an unfavorable environment such as a region of cooler or dryer air in low levels. Preferential development in one part of the line or another may occur also. A common occurrence is new cell development on the southern or southwestern end of a line, but factors favorable for continuation or cessation of this growth mode are largely unknown. Motion of the line is complicated by new cell development and other types of growth involving up to three types of motion in some lines. Newton and Fankhauser (1975) noted that there may be a motion of individual cells near the motion of the mean wind, a cell-cluster motion somewhat more to the right, and a line motion even more to the right. The line moves to the right by "accretion" and loss of clusters on the right and left, respectively, and clusters move less to the right (of the individual cell motion) by accretion and loss of cells on the right and left, respectively. From results of 3-D cloud simulations, it may be concluded that

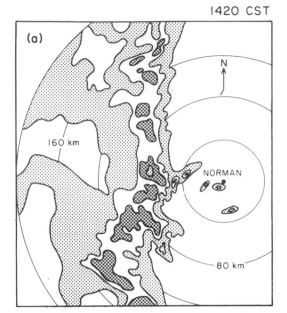

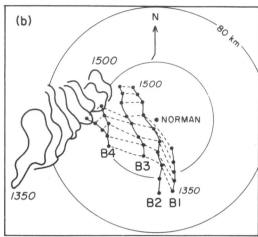

Figure 16.17. (a) The Oklahoma squall line of 19 May 1977 at 1420 CST as observed by NSSL's WSR–57 radar. Reflectivity levels (dBZ) shown are light hatching = 24 to 34, white = 34 to 46, dark hatching = 46 to 58, and light hatching = 58 to 70. The line was moving east at about 10 m s^{-1}. (b) Tracks of small cells within a line segment ahead of the squall line, and track of a major reflectivity center within the line, from 1350 to 1500 CST.

the type of storm that forms is very much a function of the environmental wind profile.

Other squall lines have included preferential growth of cells or clusters at other locations within the line. The squall line depicted in Fig. 16.17a was analyzed by Hane *et al.* (1985) to ascertain the modes by which the strong convection along the leading edge was maintained. The squall line moved east and overtook the line of small cells extending generally east-southeastward ahead of it, just south and west of the radar. The tracks of individual small cells are from the south-southeast; the track of a large area of high reflectivity within the line is from the southwest. This portion of the line showed similar behavior over a period of several hours during which the most intense activity occurred at the point where line and small cells merged. It appears that a weak mesolow was located near the merger point and a pressure trough containing horizontal convergence extended east from that point. The small cells grew in an environment of deep moisture and little wind shear, moving with the mean wind in that layer. Identification of such features, along with the environmental conditions in which they occur, should be helpful in very-short-term forecasts of preferential growth areas in squall lines.

Development of mesoscale rain areas trailing lines in an important factor to be considered in a short-term forecast. The larger area increases both the total rainfall and duration of rainfall at a point, thereby affecting both agricultural and general public interests. The evolution beginning with formation of the mesoscale area through dissipation of the squall line complex is not understood in detail, and environmental conditions contributing to this evolution have not been identified.

16.5.3. Rainband Forecasting

The forecasting of extratropical cyclone rainbands first involves forecasting movement and changes in intensity of the cyclone itself, in practice guided by numerical weather prediction models. The rainbands themselves often exist for 6 to 12 hours, so that, once identified, they can be forecast for 1 or 2 hours by extrapolation. Browning *et al.* (1982) experimented with the forecast of these systems using a radar network, and pointed out important factors that contribute to forecast errors. In general, successful forecast techniques for dealing with the individual bands have not been developed, especially for the less-organized features.

REFERENCES

Bergeron, T., 1954: The problem of tropical hurricanes. *Quart. J. Roy. Meteor. Soc.*, **80**, 131–164.

Boucher, R. J., and R. Wexler, 1961: The motion and predictability of precipitation lines. *J. Meteor.*, **18**, 160–171.

Browning, K. A., 1974: Mesoscale structure of rain systems in the British Isles. *J. Meteor. Soc. Japan*, **52**, 314–327.

Browning, K. A., and T. W. Harrold, 1969: Air motion and precipitation growth in a wave depression. *Quart. J. Roy. Meteor. Soc.*, **95**, 288–309.

Browning, K. A., and T. W. Harrold, 1970: Air motion and precipitation growth at a cold front. *Quart. J. Roy. Meteor. Soc.*, **96**, 369–389.

Browning, K. A., and F. F. Hill, 1984: Structure and evolution of a mesoscale convective system near the British Isles. *Quart. J. Roy. Meteor. Soc.*, **110**, 897–913.

Browning, K. A., M. E. Hardman, T. W. Harrold, and C. W. Pardoe, 1973: The structure of rainbands within a midlatitude depression. *Quart. J. Roy. Meteor. Soc.*, **99**, 215–231.

Browning, K. A., C. G. Collier, P. R. Larke, P. Menmuir, G. A. Monk, and R. G. Owens, 1982: On the forecasting of frontal rain using a weather radar network. *Mon. Wea. Rev.*, **110**, 534–552.

Byers, H. R., and R. R. Braham, 1949: *The Thunderstorm.* U.S. Government Printing Office, Washington, D.C., 287 pp.

Carbone, R. E., 1982: A severe frontal rainband. Part I. Stormwide hydrodynamic structure. *J. Atmos. Sci.*, **39**, 258–279.

Fritsch, J. M., and C. F. Chappell, 1980: Numerical prediction of convectively driven mesoscale pressure systems. Part II: Mesoscale model. *J. Atmos. Sci.*, **37**, 1734–1762.

Glossary of Meteorology, 1959: R. E. Huschke (Ed.), American Meteorological Society, Boston, 638 pp.

Hane, C. E., 1973: The squall line thunderstorm: Numerical experimentation. *J. Atmos. Sci.*, **30**, 1672–1690.

Harrold, T. W., 1973: Mechanisms influencing the distribution of precipitation within baroclinic disturbances. *Quart. J. Roy. Meteor. Soc.*, **99**, 232–251.

Hobbs, P. V., 1978: Organization and structure of clouds and precipitation on the mesoscale and microscale in cyclonic storms. *Rev. Geophys. Space Phys.*, **16**, 741–755.

Hobbs, P. V., 1981: Mesoscale structure in midlatitude frontal systems. Proceedings, IAMAP Symposium on Nowcasting: Mesoscale Observation and Short-Range Prediction, Eur. Space Agency Publ. SP-165, 29–36.

Hobbs, P. V., K. R. Biswas, 1979: The cellular structure of narrow cold-frontal rainbands. *Quart. J. Roy. Meteor. Soc.*, **105**, 723–727.

Hobbs, P. V., T. J. Matejka, P. H. Herzegh, J. D. Locatelli, and R. A. Houze, Jr., 1980: The mesoscale and microscale structure and organization of clouds and precipitation in midlatitude cyclones. I: A case study of a cold front. *J. Atmos. Sci.*, **37**, 568–596.

Houze, R. A., Jr., 1977: Structure and dynamics of a tropical squall line system. *Mon. Wea. Rev.*, **105**, 1540–1567.

Houze, R. A., Jr., and P. V. Hobbs, 1982: Organization and structure of precipitating cloud systems. *Adv. in Geophys.*, **24**, 225–315.

Houze, R. A., Jr., and B. F. Smull, 1982: Comparison of an Oklahoma squall line to mesoscale convective systems in the tropics. Preprints, 12th Conference on Severe Local Storms, San Antonio, Tex., American Meteorological Society, Boston, 338–341.

Houze, R. A., Jr., J. D. Locatelli, and P. V. Hobbs, 1976a: Dynamics and cloud microphysics of the rainbands in an occluded frontal system. *J. Atmos. Sci.*, **33**, 1921–1936.

Houze, R. A., Jr., P. V. Hobbs, K. R. Biswas, and W. M. Davis, 1976b: Mesoscale rainbands in extratropical cyclones. *Mon. Wea. Rev.*, **104**, 868–878.

Jain, M., and H. Bluestein, 1981: A classification of severe squall line development using WSR-57 radar data. Preprints, 20th Conference on Radar Meteorology, Boston, Mass., American Meteorological Society, Boston, 251–254.

Kessinger, C. J., 1983: An Oklahoma squall line: A multiscale observational and numerical study. M.S. thesis, University of Oklahoma, Norman, 211 pp.

Kessinger, C. J., C. E. Hane, and P. S. Ray, 1982: A Doppler analysis of squall line convection. Preprints, 12th Conference on Severe Local Storms, San Antonio, Tex., American Meteorological Society, Boston, 123–126.

Kessinger, C. J., P. S. Ray, and C. E. Hane, 1983: An Oklahoma squall line: A multiscale observational and numerical study. CIMMS Report No. 34, Univeristy of Oklahoma, Norman, 211 pp.

Klemp, J. B., and M. L. Weisman, 1983: The dependence of convective precipitation patterns on vertical wind shear. Preprints, 21st Conference on Radar Me-

teorology, Edmonton, Alta., American Meteorological Society, Boston, 44–49.

Kreitzberg, C. W., and H. A. Brown, 1970: Mesoscale weather systems within an occlusion. *J. Appl. Meteor.*, **9**, 417–432.

Leary, C. A., 1983: Internal structure of a mesoscale convective complex. Preprints, 21st Conference on Radar Meteorology, Edmonton, Alta., American Meteorological Society, Boston, 70–77.

Leary, C. A. and R. A. Houze, 1979: The structure and evolution of convection in a tropical cloud cluster. *J. Atmos. Sci.*, **36**, 437–457.

LeMone, M. A., 1983: Momentum transport by a line of cumulonimbus. *J. Atmos. Sci.*, **40**, 1815–1834.

Maddox, R. A., 1980: Mesoscale convective complexes. *Bull. Amer. Meteor. Soc.*, **61**, 1374–1387.

Malkus, J. S., and G. Witt, 1959: The evolution of a convective element: A numerical calculation. In *The Atmosphere and the Sea in Motion*, Rockefeller Inst. Press, New York, 425–439.

Miller, R. C., 1972: Notes on analysis and severe-storm forecasting procedures of the Air Force Global Weather Central. AWS Tech. Rep. 200 (Rev.).

Nelson, S. P., and S. K. Young, 1979: Characteristics of Oklahoma hailfalls and hailstorms. *J. Appl. Meteor.*, **18**, 339–347.

Newton, C. W., 1950: Structure and mechanism of the prefrontal squall line. *J. Meteor.*, **7**, 210–222.

Newton, C. W., 1963: Dynamics of severe convective storms. In *Severe Local Storms*, Meteor. Monogr., 5(27), 33–58.

Newton, C. W., 1966: Circulations in large sheared cumulonimbus. *Tellus*, **18**, 699–712.

Newton, C. W., and J. C. Fankhauser, 1964: On the movements of convective storms, with emphasis on size discrimination in relation to water-budget requirements. *J. Appl. Meteor.*, **3**, 651–668.

Newton, C. W., and J. C. Fankhauser, 1975: Movement and propagation of multicellular convective storms. *Pure Appl. Geophys.*, **113**, 747–764.

Newton, C. W., and H. R. Newton, 1959: Dynamical interactions between large convective clouds and environment with vertical shear. *J. Meteor.*, **16**, 483–496.

Ninomiya, K., and T. Akiyama, 1974: Band structure of mesoscale echo clusters associated with the low-level jet stream. *J. Meteor. Soc. Japan*, **52**, 300–313.

Ninomiya, K., R. Hasegawa, and Y. Tatsumi, 1983: Subtropical rainband along the western side of a slow moving ridge—observation and real data forecast experiments. *J. Meteor. Soc. Japan*, **61**, 606–618.

Nozumi, Y., and H. Arakawa, 1968: Prefrontal rainbands located in the warm sector of subtropical cyclones over the ocean. *J. Geophys. Res.*, **73**, 487–492.

Ogura, Y., and M. T. Liou, 1980: The structure of a midlatitude squall line: a case study. *J. Atmos. Sci.*, **37**, 553–567.

Parsons, D. B., and P. V. Hobbs, 1983: The mesoscale and microscale structure and organization of clouds and precipitation in midlatitude cyclones. XI: Comparisons between observational and theoretical aspects of rainbands. *J. Atmos. Sci.*, **40**, 2377–2397.

Pielke, R. A., 1984: *Mesoscale Meteorological Modeling*. Academic Press, London, 632 pp.

Roux, F., J. Testud, M. Payen, and B. Pinty, 1984: Pressure and temperature perturbation fields retrieved from dual-Doppler radar data: An application to the observation of a West-African squall line. *J. Atmos. Sci.*, **41**, 3104–3121.

Rust, W. D., D. R. MacGorman, and R. T. Arnold, 1983: Positive cloud-to-ground lightning in severe storms and squall lines. Preprints, 13th Conference on Severe Local Storms, Tulsa, Okla., American Meteorological Society, Boston, 211–214.

Sanders, F., and K. A. Emanuel, 1977: The momentum budget and temporal evolution of a mesoscale convective system. *J. Atmos. Sci.*, **34**, 322–330.

Schlesinger, R. E., 1984: Effects of the pressure perturbation field in numerical models of unidirectionally sheared thunderstorm convection: Two versus three dimensions. *J. Atmos. Sci.*, **41**, 1571–1587.

Wilhelmson, R. B., and J. B. Klemp, 1983a: Numerical simulation of severe storms within lines. Preprints, 13th Conference on Severe Local Storms, Tulsa, Okla., American Meteorological Society, Boston, 231–234.

Wilhelmson, R. B. and J. B. Klemp, 1983b: Comparison of Doppler and numerical model data of an Oklahoma squall line. Preprints, 21st Conference on Radar Meteorology, Edmonton, Alta., American Meteorological Society, Boston, 91–96.

Zipser, E. J., 1977: Mesoscale and convective-scale downdrafts as distinct components of squall-line circulation. *Mon. Wea. Rev.*, **105**, 1568–1589.

Zipser, E. J., 1982: Use of a conceptual model of the life-cycle of mesoscale convective systems to improve very-short-range forecasts. In *Nowcasting*, K.A. Browning (Ed.), Academic Press, New York, 191–204.

Zipser, E. J., and T. J. Matejka, 1982: Comparison of radar and wind cross-sections through a tropical and a midwestern squall line. Preprints, 12th Conference on Severe Local Storms, San Antonio, Tex., American Meteorological Society, Boston, 342–345.

CHAPTER 17

Mesoscale Convective Complexes in the Middle Latitudes

R. A. Maddox
K. W. Howard
D. L. Bartels
D. M. Rodgers

17.1. Introduction

A type of large, long-lived convective weather system that frequently occurs over middle latitudes of the United States was identified, and discussed by Maddox (1980), Fritsch and Maddox (1981), Fritsch *et al.* (1981), Bosart and Sanders (1981), Wetzel *et al.* (1983), and Maddox (1983). It is termed Mesoscale Convective Complex (MCC). The satellite images in Fig. 17.1 show the infrared enhancement temperatures from two MCCs over central portions of the United States. The definition employed by Maddox (1980) to identify an MCC used characteristics observable in satellite imagery (Table 17.1); internal structural characteristics are not addressed by the definition. The criteria were selected to identify very large and long-lived MCCs that could be studied by utilizing routine synoptic upper-air and surface observations. An extensive climatology of convective systems (Bartels *et al.*, 1984) indicates that smaller mesoscale systems occur much more frequently than the larger MCCs discussed here.

Fritsch *et al.*(1981) showed that MCCs produce widespread rainfall and that they probably account for a significant portion of growing-season rainfall over much of the United States corn and wheat belts. Maddox (1980) and Wetzel *et al.* (1983) found that many MCCs produced locally intense rainfalls and flash flooding in addition to widespread beneficial rains (see also Bosart and Saunders, 1981). Some MCCs also produce a variety of other severe convective phenomena, including tornadoes, hail, wind, and intense electrical storms. Almost one of every four MCCs results in injuries

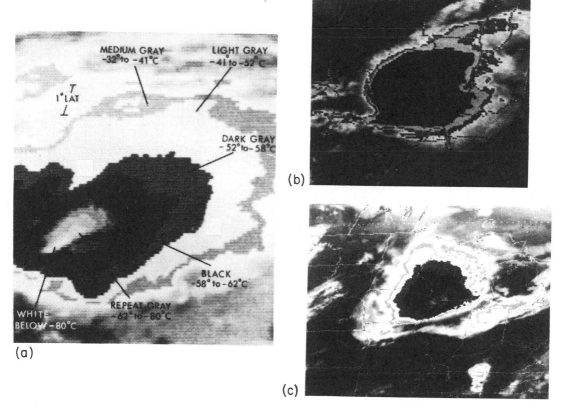

Figure 17.1. Satellite views of MCCs. (a) Key to infrared enhancement levels and corresponding temperature ranges. Enhanced IR satellite images showing MCCs over the central United States: (b) 7 June 1982, 1430 GMT, and (c) 1 July 1983, 1300 GMT.

or deaths, indicating that MCCs are significant weather events. MCCs of recent years are reviewed in Maddox (1980), Maddox *et al.* (1982), Rodgers *et al.* (1983), and Rodgers *et al.* (1985).

17.2. Climatological Attributes

Enhanced infrared (IR) satellite images (operational data routinely available from the Kansas City Satellite Field Service Station at intervals of approximately 30 min) have been examined for 1978–1982 to develop an MCC climatology (see Bartels *et al.*, 1984). Although the image sets are far from complete, because of machine trouble, power outages, wrong sectors available, etc., the life cycles of ∼160 MCCs have been documented. Significant weather that each MCC system produced during the period between initiation and termination has also been determined, using *Storm Data* (published by and available from the National Environmental Satellite, Data, and Information Service, NCDC, Federal Bldg., Asheville, N.C. 28801; volumes 20–25 were used in this work).

Paths that most of these MCCs followed (excluding August and September) are shown in Fig. 17.2. The tracks are for the centroid of the T ≤ −32°C

Table 17.1. Definition of mesoscale convective complex (MCC) based on analyses of enhanced IR satellite imagery*

Criterion	Physical characteristics
Size[†]	A: Cloud shield with continuously low IR temperature $\leq -32°C$ must have an area $\geq 100,000$ km^2.
	B: Interior cold cloud region with temperature $\leq -52°C$ must have an area $\geq 50,000$ km^2.
Duration	Size definitions A and B must be met for a period ≥ 6 h.
Maximum extent	Contiguous cold-cloud shield (IR temperature $\leq -32°C$) reaches maximum size.
Shape	Eccentricity (minor axis/major axis) ≥ 0.7 at time of maximum extent.

*From Maddox (1980).
[†]Initiation occurs when size definitions A and B are first satisfied. Termination occurs when size definitions A and B are no longer satisfied.

cloud shield. Termination, as used here, denotes the end of intense convection; however, weaker convection, showers, and cloudiness persist for hours (see Wetzel *et al.*, 1983). Once the MCCs formed, movement tended to be with the mean flow in the 700–500 mb layer. Merritt and Fritsch (1984) found generally similar results but also discussed the meteorological settings that can lead to anomalous movement.

Figure 17.3 plots the number of MCCs in terms of cloud-top area (colder than $-52°C$) at maximum extent. The great majority of MCCs had maxi-

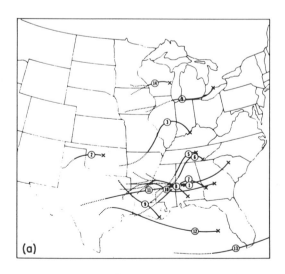

Figure 17.2. Tracks of MCCs documented during 1978–1982 . Start of the heavy solid line corresponds with "initiation," circled number with "maximum extent," and x with "termination," as defined in Table 17.1. Dotted line indicates movement of a developing thunderstorm prior to MCC initiation. (a) MCCs in April.

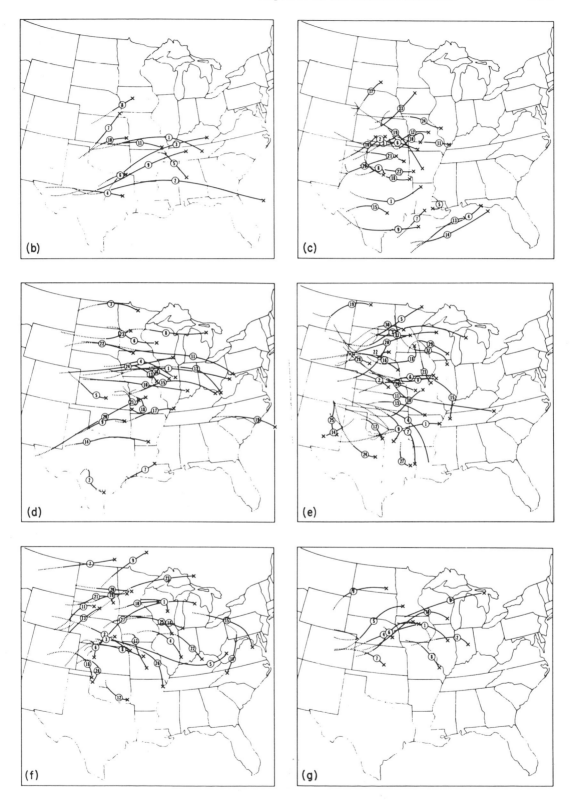

Figure 17.2. Continued. (b) MCCs, 1–15 May; (c) MCCs, 16–31 May; (d) MCCs, 1–15 June; (e) MCCs, 16–30 June; (f) MCCs, 1–15 July; (g) MCCs, 16–31 July.

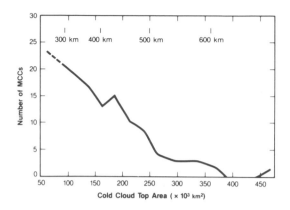

Figure 17.3. Number of MCCs versus area of cold-cloud top ($\leq -52°C$) at "maximum extent." Selected square roots of the area are shown above the curve to convey the length scales involved.

mum areas in the range 75,000–200,000 km^2. Since the mature MCCs usually produce rainfall beneath much of the region of T $\leq -52°C$, the large area of precipitation produced by MCCs is suggested. Although larger MCCs occur infrequently, the lower end of the distribution reflects the arbitrary nature of the MCC definition (i.e., the requirement for $\geq$ 50,000 km^2 cloud-top area of temperature $\leq -52°C$ for at least 6 h).

The 6 yr overview of satellite imagery has served to illustrate a number of interesting aspects of MCCs. Many complexes result from mergers and interactions between groups of storms that develop in different locations. Some of the MCCs are initially squall lines that gradually acquire MCC characteristics as they persist and grow in size. This often happens on the southern, or trailing portion, of squall lines where the associated (or triggering) cold fronts often weaken and become slow moving. As the warm season progresses, the favored region of MCC occurrence shifts slowly northward so that by July and August MCCs primarily affect the north-central states. Regardless of the time of year, MCCs are relatively rare along the East Coast and west of the Rocky Mountains.

Dirks (1969) and Wetzel et al. (1983) studied the eastward propagation of convective complexes from the Rocky Mountains to the plains of Colorado. They noted that many storms that develop over the mountains (termed "orogenic" complexes by Wetzel et al.) eventually grow into large nocturnal storms over the plains. About a quarter of the complexes documented to date had their origins in thunderstorm activity that was first detected (in satellite images) over the Rocky Mountains or their eastern slopes. Of the sample considered, about half the storms grew from initial storm developments west of longitude 100°W and about half initiated between 90°and 100°W (and only seven initiated east of 90°W). Thus, although thunderstorms that build west of 100°W often play a key role in the development of central United States MCCs, their involvement is obviously not a necessary ingredient.

Figure 17.4 shows the number of MCCs that reached certain life cycle stages during 2 h periods of the day. Regardless of where they originate, MCCs usually have their roots in thunderstorm activity that first develops during the late afternoon. It is important to realize that many convective complexes develop and begin to organize during the evening but do not

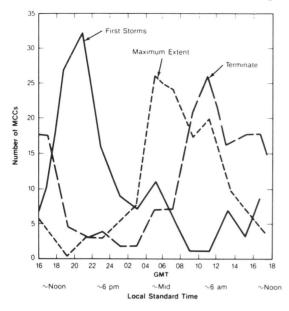

Figure 17.4. Diurnal variation in the number of MCCs, from 1600 GMT (1000 local time) to 1800 GMT (1200 local time) the next day.

persist and blossom into MCCs. However, under favorable meteorological conditions, MCCs persist and reach maximum size around local midnight—about the time that the nocturnal low-level jet stream reaches maximum strength, has its strongest westerly component, and is located nearest the surface (Hoxit, 1975). The MCCs examined tended to weaken during the morning hours when the low-level jet should be backing and weakening.

The tracks and the data shown in Figs. 17.2 and 17.4 indicate that MCCs are probably responsible, in large part, for the well-known nocturnal maxima in precipitation and thunderstorm frequency over the central United States (Kincer, 1916; Wallace, 1975). The nocturnal maximum of thunderstorm incidence reported by Wallace extends from Oklahoma northward to Minnesota, Wisconsin, and northern Michigan and is strongly peaked around 0100 local time (about 0700 GMT), the time that MCCs typically reach their greatest extent (see Fig. 17.4). It is important to note that, even though many of the MCCs studied moved into or across the nocturnal precipitation region, they had their roots in thunderstorm activity that developed during late afternoon, often far to the west.

More than half the MCCs examined produced heavy rains and flooding. A climatology by Crysler *et al.* (1982) shows that heavy precipitation events (>4.0 inches [101.6 mm] in <8 h) in Missouri and Illinois have a maximum frequency near local midnight—just the time of night that these regions are often affected by large, mature MCCs. Maddox (1980) presented a hypothetical life cycle for MCCs in which it was stressed that severe weather was most likely early in the cycle. Figure 17.5 is a composite of *Storm Data* reports logged during 12 nocturnal MCCs, and indeed the bulk of the severe weather occurred prior to initiation (defined in Table 17.1). However, some summertime MCCs produce high winds during much of their lifetime (Johns and Hirt, 1983). Heavy rainfalls become more likely as MCCs mature, especially if they are slow moving.

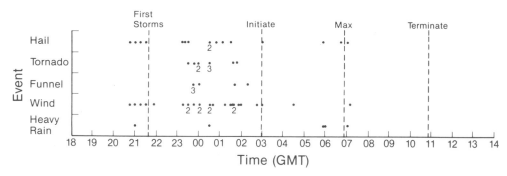

Figure 17.5. Severe weather occurrences relative to the life cycles of 12 nocturnal MCCs. (From Watson, 1984, personal communication.)

17.3. Synoptic Settings

17.3.1. Characteristic Large-Scale Patterns

Maddox (1983) presented composite analyses of the large-scale settings attending 10 MCCs. The important aspect of the analyses is that only the repetitive features of the large-scale environment survive the compositing and smoothing applied to the data.

The typical MCC develops and organizes within a precursor mesoscale region of upward vertical motion (e.g., 5–10 μb s^{-1} at 500 mb) over the Great Plains ahead of a middle-level short-wave trough. This upward motion appears to reflect pronounced low-level warm advection rather than differential vorticity advection. The warm advection pattern shifts eastward across the Great Plains as low-level winds increase and veer during the night. However, the mesoscale environment reflects the intense convection, and a deep, warm-core, net upward circulation develops. This circulation is overlain by a distinct cold core in the upper troposphere and lower stratosphere that is attended by intense outflow. Other than the strong coupling of low-level warm advection and upward motion, the events that transpire during the night cannot be interpreted in a quasi-geostrophic sense. This agrees with Wetzel *et al.* (1983), who found that the MCCs they studied did not act in the classic baroclinic sense to transport sensible heat northward. Once the MCC decays, the upper-tropospheric outflow rapidly weakens; however, the thermal perturbations appear more persistent. Intensification of the middle- and upper-level short-wave trough and ridge (to the east) often precedes a "comma cloud" in satellite images.

For the average case, the MCC interacts with its environment by moistening a deep tropospheric layer and by creating an upper-tropospheric temperature perturbation. Apparently the amplification of the short-wave trough after 1200 GMT is a manifestation of the large-scale adjustment to the temperature perturbation. Because the time scale of development of this temperature perturbation is typically less than a half-pendulum day $(2\pi/f)$, the mesoscale flow remains imperfectly balanced in the quasi-geostrophic sense. This trough may persist and move eastward for several days. The lower troposphere remains essentially decoupled from the adjustment process; however, in certain cases, MCC interactions and feedbacks may be self-perpetuating, resulting in a persistent, deep tropospheric wave and an extended MCC existence (see Bosart and Sanders, 1981; Wetzel *et al.*, 1983).

Other results from Maddox (1983) are briefly summarized below:

- Composite analyses of upper-air data show that MCCs interact with and modify their larger scale environment and, thereby, may affect the evolution of meteorological features over much of the Eastern United States.

- The large-scale setting provides a conditionally unstable thermodynamic structure over a large region ahead of, and to the right of, the advancing MCC.

- The nocturnal increase in speed and significant veering of the low-level winds enhances both the warm advection and influx of moist unstable air; radiative cooling decouples the entire MCC from the near-surface layer.

- Convective elements occur within a mesoscale environment that is, with time, becoming moist through a large depth and in which the vertical wind shear is decreasing. Thus, during the mature phase, downdrafts may be less intense and precipitation efficiency higher than in the development and growth phases.

- The MCC eventually moves into a more stable, and convectively less favorable, environment, apparently initiating its demise.

17.3.2. Examples of MCCs

MCCs can develop in a wide variety of large-scale environments and can produce very severe weather. Satellite images of four MCCs that produced widespread damaging winds are shown in Fig. 17.6. The large-scale settings that produced these events are illustrated in Fig. 17.7, which shows the cold cloud shield (T $\leq -52°C$), environmental relative winds and streamlines, and contours of θ_e. Note that all these MCCs feed on high θ_e air located to the front and right-front (Fig. 17.7a). At 700 mb (Fig. 17.7b) relative flow enters the MCCs from the front and is characterized by low θ_e, which presumably helps drive the strong downdrafts and mesoscale outflows. The 500 mb patterns (Fig. 17.7c) are less similar and generally characterized by weaker gradients in θ_e. Two of the MCCs still have distinct inflow along their leading edges. At 200 mb (Fig. 17.7d) the relative flow is strongly diffluent and is from front to back for two MCCs and more or less from back to front for the other two. Figures 17.7d and 17.6 show that the appearance of the cirrus shields reflects strongly the 200 mb relative flows. These high-wind MCCs, in addition to having very low θ_e air available at middle levels, tend to occur along slow-moving frontal zones and in environments exhibiting substantial vertical wind shear (contrast Figs. 17.7a and 17.7d), and tend to move fairly rapidly (mean speed 21 m s^{-1}).

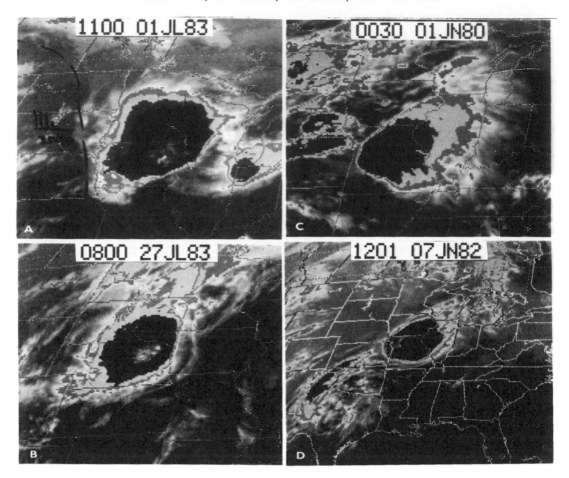

Figure 17.6. IR satellite images showing four MCCs that produced widespread high-wind gusts.

Other MCCs often produce very heavy nighttime rains during the summer. Four MCCs, each of which produced 6–15 inch (150–375 mm) rains, are illustrated in Figs. 17.8 and 17.9. At low levels (Fig. 17.9a) very high θ_e air again feeds the front to right-side quadrants of the MCCs. At 700 mb (Fig. 17.9b) θ_e is generally much higher than for the strong-wind MCCs, and relative inflow of low θ_e air seems much less pronounced. The MCC with the distinct cyclonic relative circulation (13 August 1982) produced rains exceeding 15 inches (375 mm) in less than 12 h.

The analyses at 500 mb (Fig. 17.9c) show an apparent relative cyclonic circulation for three of the MCCs, indicating a "protected" (e.g., little inflow of environmental air) characteristic, and also illustrate weak θ_e gradients. In upper levels (Fig. 17.9d) the heavy rain events are characterized by very distinct, anticyclonic relative outflow. Not surprisingly, these events occurred within environments characterized by weaker vertical wind shear (contrast Figs. 17.7 and 17.9) and moved more slowly (average speed 10 m s^{-1}) than the high-wind MCCs.

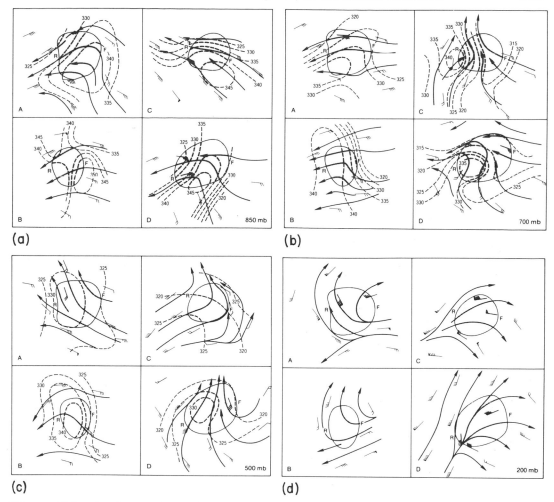

Figure 17.7. Analyses of conditions attending high-wind MCCs at (a) 850 mb, (b) 700 mb, (c) 500 mb, and (d) 200 mb. Streamlines are relative wind flow, and dashed lines are contours of θ_e. The $-52°C$ cloud shields are outlined. F and R mark the front and rear of the traveling MCC. Storm-relative winds are in knots. A, B, C, and D represent the frames respectively located in Fig. 17.6. The spacing of the upper-air observing sites (i.e., 300–400 km) is an indicator of the size of the MCCs.

During the early springtime months, MCCs along the Gulf Coast sometimes produce both widespread and extended severe weather and very heavy rainfalls. Figure 17.10 shows a satellite image of such a complex. MCCs tend to occur well to the east of a major trough (see Fig. 17.11) but in conjunction with rapidly moving, weak short waves and strong, persistent low-level warm advection in the vicinity of an east-west stationary or warm frontal zone. These complexes tend to move slowly to the right of the strong middle-level flow, owing to propagation effects as new cells repeatedly develop along the southern and southwestern flanks. In these MCCs individual thunderstorm cells are moving very rapidly with the mean flow, and it is the mesoscale complex that often remains surprisingly stationary (see Ch. 13) while producing locally heavy precipitation.

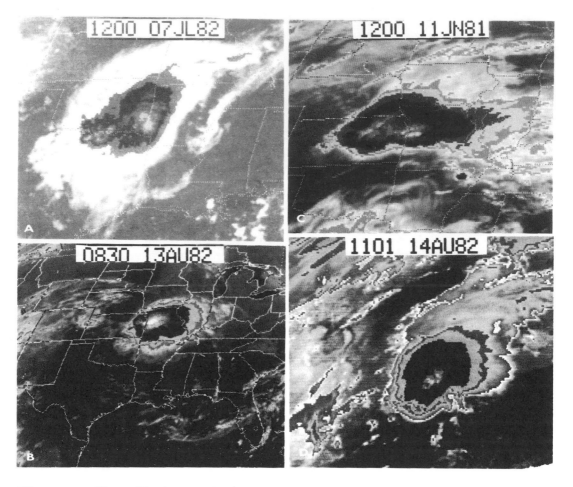

Figure 17.8. IR satellite images showing four MCCs that produced heavy rainfall (6 to more than 15 inches).

17.4. Internal Precipitation Structure of MCCs

The physical reasoning that led to the MCC definition reflects the following hypothesis: Large, long-lived canopies of cold cloud probably signal persistent mesoscale lifting in the upper half of the troposphere. It seems likely that the longer these cirrus shields persist, and the more circular they become, the greater the relative strength and influence of mesoscale (versus synoptic-scale) vertical motion fields. It is apparent that relationships between satellite-observed cloud-top temperatures and the convective structure beneath the anvil vary dramatically during the life of convective complexes. For example, an intense, individual thunderstorm may be characterized by a large, cold anvil that is associated only with a relatively small radar echo. Later, if it develops into an MCC, the entire cold-cloud shield (i.e., T $\leq$ −52°C) may lie over a region of more stratiform radar echo and precipitation. The two events discussed here as examples were studied with films of low-level reflectivity data from National Weather Service 10 cm

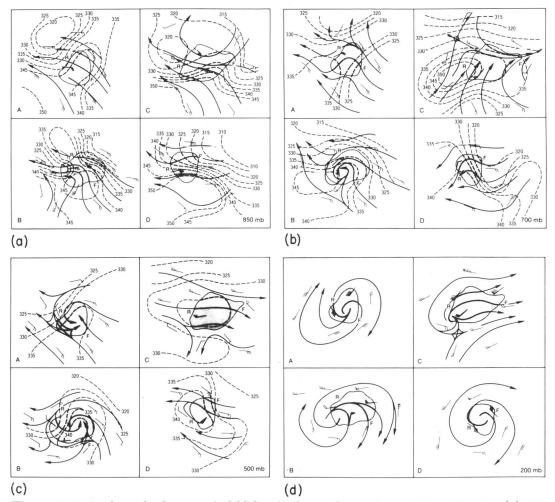

(a)　　　　　　　　　　　(b)

(c)　　　　　　　　　　　(d)

Figure 17.9. Analyses for heavy-rain MCCs, similar to the analyses in Fig. 17.7, for (a) 850 mb, (b) 700 mb, (c) 500 mb, (d) 200 mb. A, B, C, and D represent the frames respectively located in Fig. 17.8.

Figure 17.10. Enhanced IR satellite image for 0600 GMT, 12 April 1980, showing an MCC along the Gulf Coast.

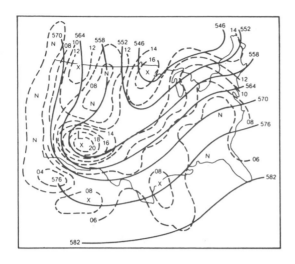

Figure 17.11. Twelve-hour Limited-area Fine Mesh (LFM) forecast of 500 mb heights (solid lines) and vorticity (dashed lines) valid at 0000 GMT, 12 April 1980.

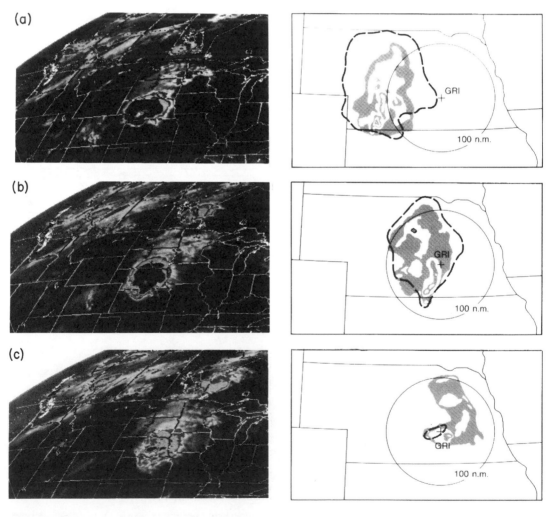

Figure 17.12. Simultaneous enhanced IR satellite images and radar composites for (a) 0900 GMT, (b) 1200 GMT, and (c) 1500 GMT on 14 August 1982.

radars. The radar data were examined in conjunction with hourly surface and precipitation data.

17.4.1. Case Study I: 13–14 August 1982

A slow-moving but long-lived MCC occurred during the night of 13–14 August 1982. The MCC formed as a number of individual thunderstorms merged over the Nebraska Panhandle and then moved slowly eastward (~18 kt), reaching eastern Nebraska about 15 h later. Figure 17.12 depicts the latter half of the life cycle of this MCC as viewed by both satellite and radar. At 0900 and 1200 GMT the satellite depiction displays a large, nearly circular shield of cloud with coldest tops in the south-central portion of the MCC. The radar depiction at 0900 GMT indicates two lines of stronger echo (generally oriented N-S) west and west-southwest of Grand Island and a large area of weaker echo of <40 dBZ north of the lines. The most intense storm reaches >56 dBZ at the southern end of the trailing, stronger echo line. By 1200 GMT the leading line has strengthened and is now trailed by a large area of 15 dBZ stratiform precipitation; in addition, widespread 15 dBZ echo is occurring beneath the satellite-viewed cold-cloud shield from north-northeast to northwest of the echo line. The strongest thunderstorms, now only 41–45 dBZ, are still located at the southern end of the line. By 1500 GMT the MCC has weakened rapidly (note the almost total lack of satellite-observed very cold cloud, although large areas of eastern Nebraska are still covered by 15 dBZ radar echo.

Certain portions of the life cycle of this MCC appear very similar to convective systems of the tropics described by Zipser (1977) and Houze (1977) and the middle-latitude complexes studied by Leary and Rappaport (1983); most notably, the radar depiction of the southern half of the MCC at 1200 GMT is quite like the systems discussed by these authors. However, the internal structure of this MCC changed very rapidly with time (see Figs. 17.12a, b, c), as did apparent relationships of satellite-depicted IR cloud-top temperature structure to radar echo. The area of stratiform precipitation (considered to be the 15 dBZ echo region) gradually expanded so that within the mature MCC (1200 GMT) the very cold cloud shield and region of radar echo were nearly coincident. Early in its life cycle, cold cirrus coverage was much greater than that of echo, but late in the cycle, broad areas of echo existed and little cold cloud was apparent in the satellite images.

17.4.2. Case Study II: 22 June 1981

During the early morning hours of 22 June 1981 an MCC developed rapidly over western Kansas, and several hours later two distinct IR cold cores developed within the complex over central Kansas. This complex produced >7 inches (175 mm) of rain over a relatively small area of Kansas. The MCC subsequently split into two separate MCCs. Hourly surface observations were again used, in conjunction with radar and satellite data, to diagnose the internal precipitation structure of both complexes.

The life cycle of the two MCCs, as depicted by enhanced IR satellite imagery, is illustrated in Fig. 17.13. Initial convection developed within a

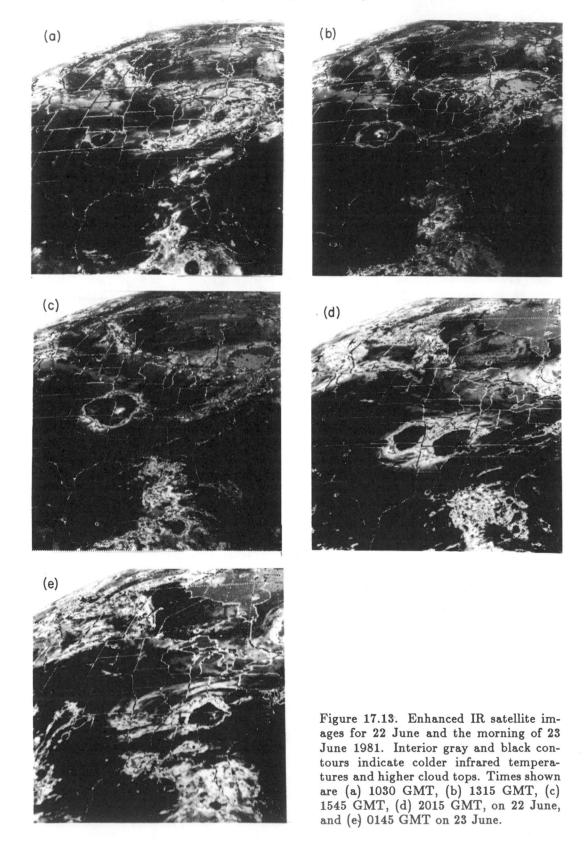

Figure 17.13. Enhanced IR satellite images for 22 June and the morning of 23 June 1981. Interior gray and black contours indicate colder infrared temperatures and higher cloud tops. Times shown are (a) 1030 GMT, (b) 1315 GMT, (c) 1545 GMT, (d) 2015 GMT, on 22 June, and (e) 0145 GMT on 23 June.

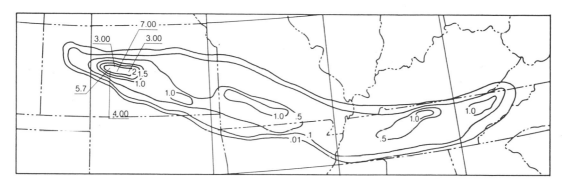

Figure 17.14. Accumulated rainfall (inches) produced by MCCs on 22 and 23 June (see Fig. 17.13).

surface trough located to the lee of the Colorado Rockies. As mentioned earlier, Wetzel *et al.* (1983) found that afternoon thunderstorms that develop along the foothills and Front Range of the Rockies often move eastward and form the roots of significant Great Plains mesoscale complexes. But in this case the initial storms appeared to develop over the high Plains late at night. From 0600 to 1000 GMT, individual thunderstorm cells moved eastward to the north of an east-west stationary front. By 1030 GMT (Fig. 17.13a) a developing MCC was depicted over central Kansas in satellite imagery. A second region of cold temperatures had developed within the cirrus shield to the west of the main system by 1315 GMT (Fig. 17.13b). High winds, heavy rainfall, and flash flooding were reported with the intense thunderstorms associated with this second area of cold tops. The leading convective complex (Figs. 17.13c, d, e) moved rapidly eastward in concert with a weak midtropospheric (500 mb) short wave. The trailing complex occurred in an unusual location to the west of the short wave as a persistent low-level jet continued to feed unstable air into central Kansas. The western system (which technically did not meet the MCC criteria) finally decayed during the afternoon as subsidence behind the short wave probably became strong enough to suppress convection; the eastern MCC moved all the way to the Appalachians before finally dissipating at about 0200 GMT on 23 June.

Figure 17.14 shows the accumulated rainfall produced by both storms. It is apparent that MCCs can result in substantial precipitation events for the central United States. Measurable precipitation accumulated along a swath approximately 1500 km long and 300 km wide. Widespread areas within this swath received more than an inch (25 mm) of precipitation. Heaviest amounts occurred beneath the western convective complex, exceeding 3 inches (75 mm) in only 6 h at several locations. These rains produced flash flooding in Hoisington, Kans., forcing evacuation of hundreds of residents.

Detailed examination of the precipitation structure of these convective complexes centered on the period during which the western MCC developed and split from the initial MCC. Figure 17.15 illustrates this period (1000 to 1600 GMT), at approximately 2 h intervals, using enhanced IR satellite

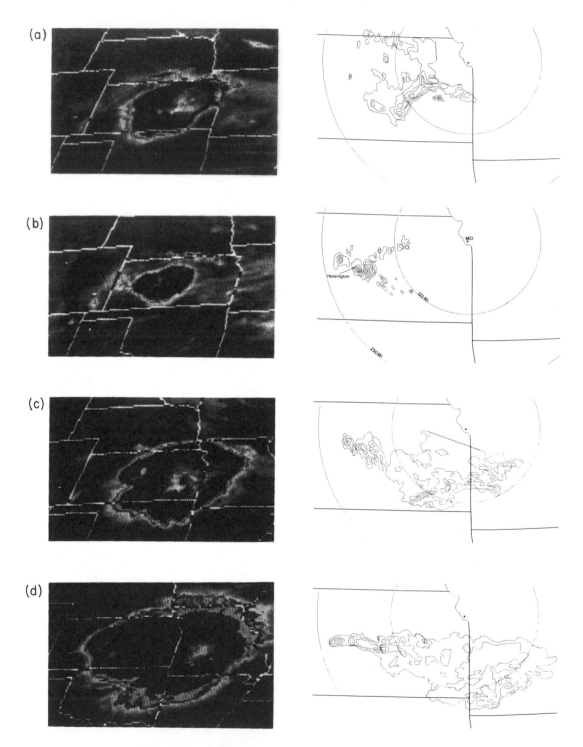

Figure 17.15. Enhanced IR satellite images and WSR-57 radar composites for (a) 1000 GMT, (b) 1200 GMT, (c) 1400 GMT, and (d) 1600 GMT on 22 June 1981. Intensity contour levels, in dBZ, are (1) <30, (2) 30–40, (3) 41–45, (4) 46–49, (5) 50–56, and (6) ≥ 57.

imagery and composite radar maps constructed from the NWS 10 cm radar data (low-level PPI films).

At 1000 GMT the radar composite (Fig. 17.15a) indicated two lines of individual cells that culminated at the west in three very large, nearly circular, level 5 (50–56 dBZ) and 6 (>56 dBZ) thunderstorms. Even though individual storms were dominant at this time, a single, large cirrus canopy was already present. By 1200 GMT remarkable changes had occurred in the structure of the complex. Radar (Fig. 17.15b) indicated a very strong squall line (oriented NNE-SSW) over east-central Kansas. This line was moving rapidly east-southeast with the flow at middle and upper levels and was trailed by a region of more stratiform precipitation. Additionally, the northern portion of the MCC contained two lines (oriented WNW-ENE) of distinctly convective echoes and another region of stratiform echo over northeast Kansas.

Two hours later (Fig. 17.15c), internal changes were again apparent. The line of echoes leading the complex had weakened, become less well defined, and assumed a WSW-ENE orientation. A large area of stratiform echo was now present west and north of the leading edge. New and intense thunderstorms had developed to the rear of the original system where the low-level flow apparently continued to advect moist, unstable air into the region of the outflow boundary left behind by the first convection. By 1600 GMT (Fig. 17.15d) the eastern mesoscale complex was dominated by stratiform reflectivities less than 40 dBZ, and some stronger cells (41–45 dBZ) were distributed rather randomly within the leading third of the MCC. The western complex was now a very distinct east-west line of intense storms with little associated stratiform echo.

17.4.3. Characteristics From Case Studies

Although the two case studies of MCC structure are based on operational data sets, several important characteristics of the complexes are apparent. All complexes were distinctly three dimensional and certainly could not be termed "quasi-steady." The complexes in the two studies exhibited both similarities to and marked differences from tropical convective systems that have been widely discussed in the scientific literature. Clearly, we need better quality radar data, for the entire life cycles of a number of middle-latitude systems. However, relatively simple, or generally applicable, observational paradigms for middle-latitude convective complexes will not be developed easily.

It is also apparent that the radar signatures (and thereby the rainfall patterns) present within these MCCs and their relationship to the satellite-depicted cloud-top temperature changed dramatically during the lifetimes of the complexes. The implication is that it is probably not reasonable to apply a single, or static, satellite rain estimation technique to satellite data throughout the life cycle of MCCs.

Some important, but unanswered, questions about MCCs include the following:

- Are there small-scale controls that govern initial thunderstorm formation in the MCC precursor environment?

- What is the role of thunderstorm-scale downdrafts and downdraft interactions within a nascent MCC?

- Precisely how do individual convective storms interact to initiate and drive large mesoscale circulations?

- What are the small-scale characteristics and structure of the diagnosed region of mesoscale upward circulation?

- What are the relative contributions of convective and stratiform precipitation?

- What are the dynamic and thermodynamic constraints that determine the size of a given MCC?

- Are there other controls on dissipation above and beyond the MCC's movement into a convectively unfavorable regime?

- Can the location, timing, and intensity of MCC development be forecast more accurately?

Detailed study of synoptic network radar data, hourly precipitation data, hourly and special surface observations, etc., is helping elucidate smaller scale aspects of MCC structure and evolution (e.g., precipitation intensities and distribution, spatial distribution, and temporal evolution of intense thunderstorm outflows), thus allowing more detailed comparison and contrast between tropical and middle-latitude convective systems. However, the eventual specification of the energetics and detailed structure of MCCs requires special data sets that can be gathered only during intensive field efforts designed to probe the life cycles of a number of MCCs.

17.5. Forecasting MCCs

One of the first, and probably easiest, ways that operational forecasting of MCCs can be improved is through recognition (by use of radar and satellite) of the type of convective organization occurring (see Zipser's [1982] discussion on very-short-range forecasting). Repetitive weather characteristics relating to MCCs could be reflected in the forecast or update.

Since MCCs are prolific precipitation producers, the forecast should reflect higher probabilities of precipitation (POPs) and the operational emphasis should shift to assessment of the heavy rain and flooding threats. This assumes that satellite imagery shows an MCC developing over and/or approaching the forecast area. Figures 17.16–17.19 illustrate several aspects of the precipitation forecast problems. Two MCCs (Fig. 17.16) traveled eastward, one trailing the other, across Kansas and Missouri. The rainfall produced during the mature and dissipating stages of the western complex was significant (Figs. 17.17 and 17.18); both the area and amounts of rainfall dramatically illustrate the obvious difference between precipitation of

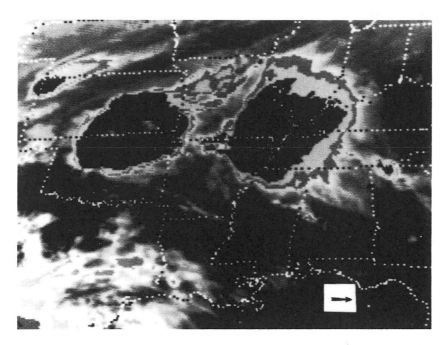

Figure 17.16. Enhanced IR satellite image for 0645 GMT, 20 May 1979, showing two MCCs.

MCCs (Fig. 17.19) and the hit-or-miss rain often associated with individual thundershowers.

Weather conditions at Chanute, Kans., as the two MCCs passed over the station (Fig. 17.20), indicate that MCCs affect all aspects of sensible weather (i.e., those elements most important to the forecast) from temperature to ceiling and visibility. Indeed, the cool outflows from MCCs often have

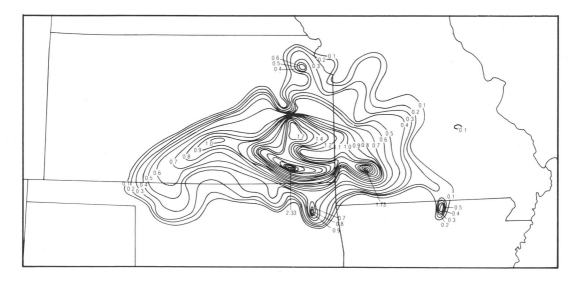

Figure 17.17. Accumulated rainfall (inches) between 0600 and 1200 GMT, 19–20 May 1979, in the area affected by the mature western MCC. (From Rockwood *et al.*, 1984.)

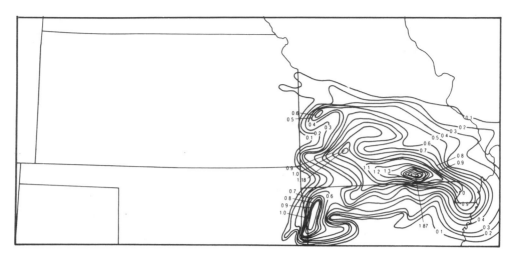

Figure 17.18. Accumulated rainfall (inches) between 1200 and 1800 GMT, 19–20 May 1979, in the area affected by the dissipating western MCC. (From Rockwood *et al.*, 1984.)

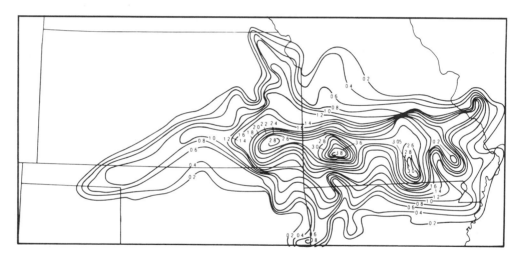

Figure 17.19. Twenty-five-hour total storm precipitation (inches) for MCCs on 19–20 May 1979. (From Rockwood *et al.*, 1984.)

dramatic effects on the temperature after a long-lived nighttime occurrence. Furthermore, large MCCs are visible over the northern United States in the satellite image of Fig. 17.21. The differences between observed maximum temperatures and model output statistical (MOS) guidance for the ensuing day are plotted in Fig. 17.22. The guidance forecasts were much too warm; in effect, the MCCs provided the forecasters in the central part of the country the opportunity to make significant improvements to computer-based guidance.

Accurate medium-range (6–12 h) prediction of MCCs remains a very difficult problem, owing largely to the spatial and temporal distribution of data as well as scanty understanding of mesoscale processes. Nevertheless, application of subjective forecast techniques, and careful re-analysis of routinely

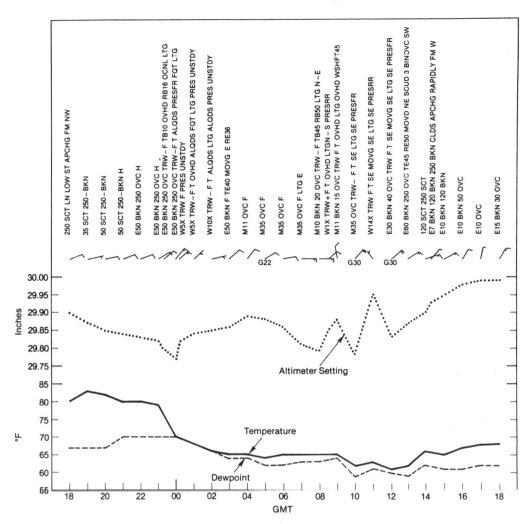

Figure 17.20. Weather conditions at Chanute, Kans., from 1800 GMT, 19 May 1979, to 1800 GMT, 20 May 1979. (From Rockwood *et al.*, 1984.)

Figure 17.21. Enhanced IR image for 0700 GMT, 4 June 1980, showing two large MCCs.

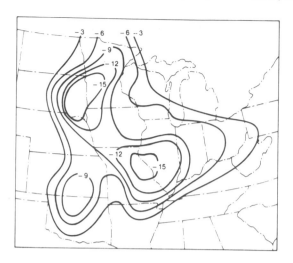

Figure 17.22. MOS maximum temperature errors (°F) (observed maximum temperature minus MOS forecast) on 4 June 1980. (After Maddox and Heckman, 1982.)

available data (Doswell, 1982), have resulted in some success in predicting the development of MCCs (Rodgers *et al.*, 1984). Results thus far have strongly suggested that immediate improvements in medium-range forecasting of MCCs could be achieved if forecasters in the field had access to the detailed information available in the numerical forecasts' gridded data. Furthermore, forecasts being produced by "state-of-the-art" mesoscale research models (see, e.g., Anthes *et al.*, 1982; Fritsch and Brown, 1982; Kaplan *et al.*, 1984; Perkey and Maddox, 1985) indicate great potential for explicit numerical forecasts of mesoscale convective weather. It will, however, be a number of years before the forecaster has such models available to support real-time, operational meteorology.

REFERENCES

Anthes, R. A., Y. Kuo, S. G. Benjamin, and Y. F. Li, 1982: The evolution of the mesoscale environment of severe local storms: Preliminary modeling results. *Mon. Wea. Rev.*, 110, 1187–1213.

Bartels, D. L., J. M. Skradski, and R. D. Menard, 1984: Mesoscale convective systems: A satellite-data-based climatology. NOAA Tech. Memo. ERL ESG–6, Environmental Research Laboratories (NTIS#PB85–187862), 58 pp.

Bosart, L. R., and F. Sanders, 1981: The Johnstown flood of July 1977: A long-lived convective storm. *J. Atmos. Sci.*, 38, 1616–1642.

Crysler, K. A., R. A. Maddox, L. R. Hoxit, and B. M. Muller, 1982: Diurnal distribution of very heavy precipitation over the central and eastern United States. *Nat. Wea. Digest*, 7(1), 33–37.

Dirks, R. A., 1969: A climatology of central Great Plains mesoscale convective systems. Final Report, ESSA Grant E–10–68G, Atmospheric Science Dept., Colorado State Univ., Fort Collins, Colo., 60 pp.

Doswell, C. A., III, 1982: The operational meteorology of convective weather. Volume 1: Operational mesoanalysis. NOAA Tech. Memo. NWS NSSFC–5, National Weather Service (NTIS-#PB83–162321), 131 pp.

Fritsch, J. M., and J. M. Brown, 1982: On the generation of convectively driven mesohighs aloft. *Mon. Wea. Rev.*, 110, 1554–1563.

Fritsch, J. M., and R. A. Maddox, 1981: Convectively driven mesoscale weather systems aloft, Part I: Observations. *J. Appl. Meteor.*, 20, 9–19.

Fritsch, J. M., R. A. Maddox, and A. G. Barnston, 1981: The character of mesoscale convective complex precipitation and its contribution to warm season rainfall in the United States. Preprints, 4th Conference on Hydrometeorology, Reno, Nev., American Meteorological Society, Boston, 94–99.

Houze, R. A., Jr., 1977: Structure and dynamics of a tropical squall-line system. *Mon. Wea. Rev.*, **105**, 1541–1567.

Hoxit, L. R., 1975: Diurnal variations in planetary boundary-layer winds over land. *Bound.-Layer Meteor.*, **8**, 21–38.

Johns, R. H., and W. D. Hirt, 1983: The derecho—A severe weather producing convective system. Preprints, 13th Conference on Severe Local Storms, Tulsa, Okla., American Meteorological Society, Boston, 178–181.

Kaplan, M. L., J. W. Zack, V. C. Wong, and G. D. Coats, 1984: The interactive role of subsynoptic scale jet streak and planetary boundary layer processes in organizing an isolated convective complex. *Mon. Wea. Rev.*, **112**, 2213–2238.

Kincer, J. B., 1916: Daytime and nighttime precipitations and their economic significance. *Mon. Wea. Rev.*, **44**, 628–633.

Leary, C. A., and E. N. Rappaport, 1983: Internal structure of a mesoscale convective complex. Preprints, 21st Conference on Radar Meteorology, Edmonton, Alta., Can., American Meteorological Society, Boston, 70–77.

Maddox, R. A., 1980: Mesoscale convective complexes. *Bull. Amer. Meteor. Soc.*, **61**, 1374–1387.

Maddox, R. A., 1983: Large-scale meteorological conditions associated with mid-latitude mesoscale convective complexes. *Mon. Wea. Rev.*, **111**, 1475–1493.

Maddox, R. A., and B. E. Heckman, 1982: The impact of mesoscale convective weather systems upon MOS temperature guidance. Preprints, 9th Conference on Weather Forecasting and Analysis, Seattle, Wash., American Meteorological Society, Boston, 214–218.

Maddox, R. A., D. M. Rodgers, and K. W. Howard, 1982: Mesoscale convective complexes over the United States during 1981—Annual summary. *Mon. Wea. Rev.*, **110**, 1501–1514.

Merritt, J. H., and J. M. Fritsch, 1984: On the movement of the heavy precipitation areas of mid-latitude mesoscale convective complexes. Preprints, 10th Conference on Weather Forecasting and Analysis, Tampa, Fla., American Meteorological Society, Boston, 520–536.

Perkey, D. J., and R. A. Maddox, 1985: A numerical investigation of a mesoscale convective system. *Mon. Wea. Rev.*, **113**, 553–566.

Rockwood, A. A., D. L. Bartels, and R. A. Maddox, 1984: Precipitation characteristics of a dual mesoscale convective complex. NOAA Tech. Memo., ERL ESG–6, Environmental Research Laboratories (NTIS#PB84-226711), 50 pp.

Rodgers, D. M., K. W. Howard, and E. C. Johnston, 1983: Mesoscale convective complexes over the United States during 1982—Annual summary. *Mon. Wea. Rev.*, **111**, 2363–2369.

Rodgers, D. M., D. L. Bartels, R. D. Menard, and J. H. Arns, 1984: Experiments in forecasting mesoscale convective weather systems. Preprints, 10th Conference on Weather Forecasting and Analysis, Tampa, Fla., American Meteorological Society, Boston, 486–491.

Rodgers, D. M., M. J. Magnano, and J. H. Arns, 1985: Mesoscale convective complexes over the United States during 1983. *Mon. Wea. Rev.*, **113**, 889–901.

Wallace, J. M., 1975: Diurnal variations in precipitation and thunderstorm frequency over the conterminous United States. *Mon. Wea. Rev.*, **103**, 406–419.

Wetzel, P. J., W. R. Cotton, and R. L. McAnelly, 1983: A long-lived mesoscale convective complex, Part II: Evolution and structure of the mature complex. *Mon. Wea. Rev.*, **111**, 1919–1937.

Zipser, E. J., 1977: Mesoscale and convective-scale downdrafts as distinct components of squall-line structure. *Mon. Wea. Rev.*, **105**, 1568–1589.

Zipser, E. J., 1982: Use of a conceptual model of the life-cycle of mesoscale convective systems to improve very-short-range forecasts. In *Nowcasting*. K. Browning (Ed.), Academic Press, New York, 191–204.

CHAPTER 18

Tornadoes and Tornadogenesis

Richard Rotunno

18.1. Introduction

Its extreme violence and dramatic appearance make the tornado the object of awe, fear, and wonder. The tornado and its parent thunderstorm (usually a "supercell") have stimulated much research, both theoretical and applied. Over the past decade or so, several new developments have allowed for great advances in our understanding of the tornado. However, there is no single theory that explains all the commonly observed features. Our knowledge is fragmentary, consisting of a collection of solutions to the governing equations, each applying to a specific situation but none to the entire milieu of the tornado. Recently there have been several comprehensive reviews (the general review by Davies-Jones, 1986; the review of laboratory work by Snow, 1982; the popular article by Snow, 1984; the review of [axisymmetric] numerical simulations by Lewellen, 1976). The focus here is on work that leads directly to an explanation of the tornado.

Until our image of the tornado becomes sharper, the following popular definition will suffice:

> **tornado,** *n.* A rotating column of air usually accompanied by a funnel-shaped downward extension of a cumulonimbus cloud and having...[winds] whirling destructively at speeds of up to 300 miles per hour. (*American Heritage Dictionary of the English Language*)

This definition reflects a fundamental property of the tornado: it occurs in association with a thunderstorm. Naturally, this was one of the first observations made. Ferrel (1889), noting the concurrence of tornadoes and thunderstorms, proffered a theory that tornadoes form when the thunderstorm updraft encounters a "gyratory" wind field. The reasoning behind this remark is obvious enough, but the premises of Ferrel's theory raise (at least) two further fundamental questions: What is the origin and nature of this "gyratory" wind field? And, granting the gyratory wind field, can the essence of the tornado mechanism be described simply by the imposition of an updraft on a rotating wind field?

The first question specifically relates to thunderstorm motion on the scale 10–50 km. We have quantitative knowledge of the flow on this scale because it has been observed by aircraft and multi-Doppler radar with a resolution of a few kilometers. The second question concerns the way the larger-scale thunderstorm produces the smaller-scale vortex. Here our knowledge is largely qualitative, but impressive nonetheless. Over the past decade, progress in this area has been made by systematic scientific eyewitness observation coordinated with radar surveillance of tornadic thunderstorms. There are consistent cloud features, radar reflectivity patterns, and signatures from single-Doppler radar from which certain indirect inferences may be made about the thunderstorm flow on a scale of 1–5 km. Some quantitative knowledge also exists on the airflow on the tornado scale (10–500 m) because air motion may be inferred from the motion of small debris and cloud material, which can be filmed.

The following summary of observations—starting with the tornado's parent thunderstorm, moving down to smaller scales, and ending with those of the tornado itself—considers only the most intense variety of tornado, which occurs within the so-called supercell thunderstorm; it is the most thoroughly documented and, from a practical point of view, the most important. Knowledge gained from three-dimensional numerical simulations of supercell-like convection can be combined with the considerable body of knowledge concerning vortices formed in cylindrical domains where suction (thunderstorm updraft) is imposed at the top of the cylinder and rotation ("gyratory" wind field) is imposed at the sides, to patch together a theory that explains the tornado phenomenon on scales from the parent storm on down to the tornado itself.

18.2. Observations

18.2.1. Thunderstorm-Scale Motions

Tornadoes form within supercell thunderstorms that develop in environments of strong vertical wind shear and large latent instability (Fawbush and Miller, 1954). The synoptic patterns that produce such a combination of wind shear and instability are fully discussed elsewhere (e.g., Newton, 1963).

The application of Doppler-radar technology to study of the supercell thunderstorm has led to significant new discoveries concerning its structure and life cycle. An important precursor of tornadogenesis is the mesocyclone, an organized rotation about a vertical axis on a scale of 5–10 km. Its two-dimensional flow pattern in a horizontal plane is discernible when two or more Doppler radars are trained on the supercell (e.g., Ray, 1976). Once a mesocyclone has developed, another feature termed the Tornado Vortex Signature (TVS) may occur. The TVS is a strong small-scale anomaly in the horizontal wind shear, detected at middle levels within a thunderstorm; it may precede the tornado by as much as 20 minutes (Brown *et al.*, 1978).

The other significant development in recent years is the systematic eye-witness observation of the tornado and supercell thunderstorms by scientific observers (Golden and Morgan, 1972; Bluestein, 1980; Lemon and Doswell, 1979; Davies-Jones, 1982). Features such as the special position of the tornado within the storm, the relation of thunderstorm outflow boundary to the tornado, and distinctive cloud formations have been repeatedly documented.

Discovery of tornado precursors, and improved observations, made possible a delineation of the life cycle of the supercell. According to Burgess *et al.* (1977) and Lemon and Doswell (1979), there are, broadly speaking, three stages in the evolution of a tornadic supercell.

Organizing Stage

In the beginning, there is a protean group of growing and decaying cells. Then, suddenly, one of the cells grows to such a large size that it dominates the others. This "supercell" begins to propagate to the right of the environmental winds at all levels, and develops a mesocyclone at middle levels. The radar reflectivity, generally most intense in the downdraft region, evinces, in a range-height cross section, an upper-level extension toward the region of updraft. This strong echo overhang is thought to be the result of the rapid storm udpraft suspending the precipitation (Browning and Ludlam, 1962). The region below this overhang has since received the acronym WER (Weak Echo Region) (Chisholm, 1973).

Mature Stage

The mesocyclone builds down to lower levels, and downdrafts intensify at middle and low levels. The echo overhang is now so pronounced that the weak echo region becomes bounded (BWER) when viewed in horizontal cross section. This may mean that updrafts are intensifying. A hook-shape appendage, which is also indicative of storm rotation (Browning and Donaldson, 1963), forms on the right (when the view is in the direction of storm motion) rear side of the echo mass. The TVS, funnel clouds, and weak tornadoes (if any) usually form at this time. At the surface there is a pool of cold air, produced by the evaporation of precipitation, which spreads out and forms the "gust front," an interface with the warm, moist inflow.

Collapse Stage

The BWER begins to fill, and downdrafts intensify. The gust front surges strongly and become highly contorted in the vicinity of the main updraft. In association with this development, a tornado forms at full strength and may last from a few to several tens of minutes.

The distinctive airflow pattern of the mature supercell thunderstorm was discovered by Browning (1964). Figure 18.1 shows the two main branches of the circulation in a frame of reference that moves the thunderstorm. The updraft is predominantly fed by the warm, moist air from the right front quadrant of the system; the major portion of the downdraft, located on the left rear quadrant, is due to middle-level air from the right flank, which

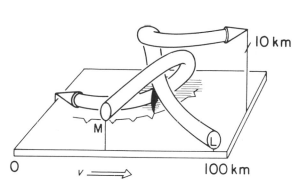

Figure 18.1. Browning's 3-D model of the airflow within a supercell thunderstorm that travels to the right of the mean wind. Updraft and downdraft circulations are depicted in a frame of reference moving with the storm. L (low) and M (middle) refer to the predominant levels of origin of the updraft and downdraft, respectively. Also shown are the approximate extent of the precipitation at the surface (hatched area), and positions of the surface gust front, and the tornado (when present). (From Browning, 1964.)

crosses ahead of the updraft. This arrangement apparently lets the supercell persist in a near-steady state.

Later corroboration of the Browning model came from a combination of radar and aircraft measurements by Fankhauser (1971; see his Fig. 19) and by analysis of Doppler radar observations by Klemp *et al.* (1981). Figure 18.2 shows the flow in the supercell thunderstorm as derived from the analysis of data taken simultaneously by four Doppler radars. Figures 18.2c and 18.2d depict the flow at the mature stage of the storm's life. In a frame of reference moving with the storm, the low-level air approaches from the southeast and rises through the updraft. The middle-level air approaches from the south-southwest and passes around the east side over to the north side where it descends in downdrafts that extend to the surface and then flows southwestward.

These data also show the intense rotation in the supercell. At $z = 1$ km (Fig. 18.2c) the principal flow feature is the mesocyclone; at $z = 4$ km (Fig. 18.2d) there is anticyclonic rotation to the north of the mesocyclone. Several other multi-Doppler radar studies of supercell thunderstorms show these same general features (e.g., Ray, 1976; Eagleman and Lin, 1977; Brandes, 1978; Heymsfield, 1978).

18.2.2. Tornado-Scale Motions

Figures 18.2e and 18.2f show the storm as it enters its collapse stage. Here the updraft becomes highly contorted into a "horseshoe" shape at $z = 1$ km and actually divides in two at $z = 4$ km. It was reported by ground-based observers that the tornado made contact with the ground just before and reached full intensity during this period. This horseshoe-shaped updraft appears to be a common characteristic of thunderstorms as they enter their tornadic phase. From ground-based observation of the direction of blowing dust, rapid temperature changes, raising or lowering of cloud bases, and many other detailed signatures, schematics of surface weather features for a tornadic thunderstorm at this stage were constructed by Golden and Purcell (1978a) and by Lemon and Doswell (1979).

Lemon and Doswell's schematic is shown in Fig. 15.7 (Ch. 15, this volume). The environmental flow is generally from the south while the storm

Model

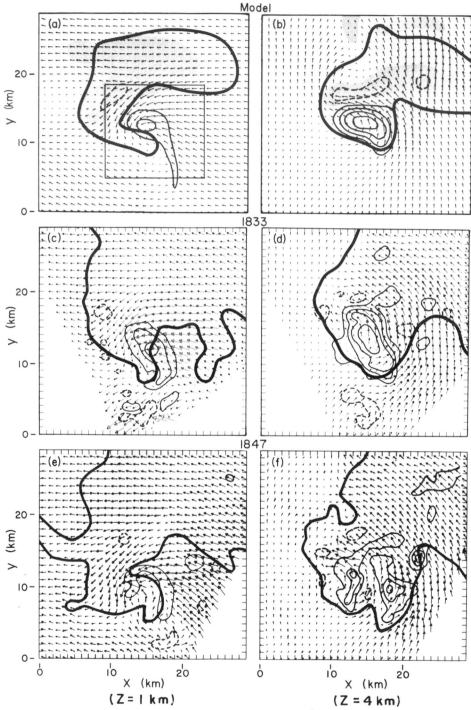

Figure 18.2. Cross sections at (left) $z = 1$ km and (right) $z = 4$ km, for a 3-D model simulation (a, b) and the Del City storm Doppler radar synthesized winds at 1833 CST (c, d) and 1847 CST (e, f). The x-axis points eastward. Vertical velocity is contoured in 5 m s^{-1} intervals, except for 1847 which is in 10 m s^{-1} intervals. Shaded areas designate downdraft regions. The heavy solid line denotes the 0.5 g kg^{-1} rainwater contour in the model and the 30 dBZ reflectivity contour in the observed storm. Horizontal flow vectors are scaled such that one grid interval represents 20 m s^{-1}. (From Klemp *et al.*, 1981.)

outflow is directed more or less radially outward from the rear-flank down-draft and the forward-flank downdraft. The gust front and hook-shape in the reflectivity field are apparent. According to the diagram, the updraft is located in the area partially bounded by the hook. The tornado at its mature stage is located between the updraft and the rear-flank downdraft. The latter is inferred from the "clear slot," an area through which the sun may shine, produced by rapid erosion of cloud material just southwest of the tornado and seen in many late-afternoon observations of tornadoes. Golden and Purcell's (1978a) map, shown in Fig. 18.3, is qualitatively similar and includes somewhat more detail concerning cloud and precipitation features.

Figure 18.4 pictures the Union City, Okla., tornado during its mature stage from the vantage point indicated in Fig. 18.3 by NSSL TEAM. The view is to the northwest; note the general erosion of the cloud base to the southwest of the tornado and the low cloud bases and precipitation to the northeast. Lemon and Doswell (1979) point out that this general erosion

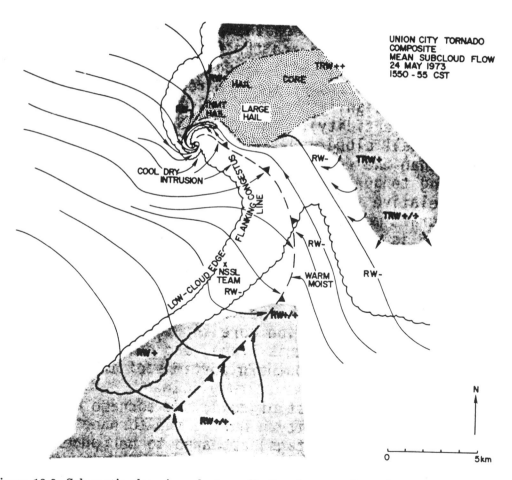

Figure 18.3. Schematic plan view of a tornadic thunderstorm from the surface observations of Golden and Purcell (1978a), showing the location of the surface gust front, major low-level cloud boundaries, and precipitation types and intensities. Note the flanking-line cloud base relative to surface gust front, stream flow, and tornado.

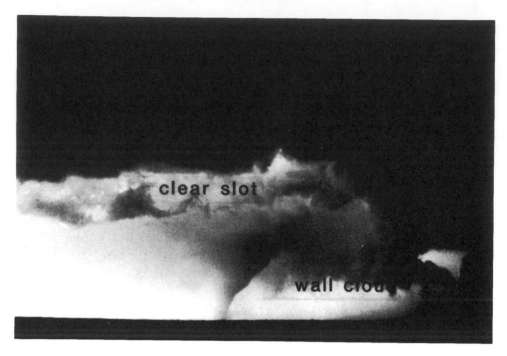

Figure 18.4. The Union City tornado funnel at the beginning of its mature phase. View is to the northwest from the vicinity of the NSSL TEAM site in Fig. 18.3 (photo from Moller *et al.*, 1974). Two features that commonly attend the tornado at its mature phase are labeled.

on the southwest side, together with the appearance of outward blowing dust just below, strongly suggests downdraft on that side. Furthermore, the low cloud base and rapidly rising cloud tags on the northeast side strongly suggest updraft on that side. The ineluctable conclusion is that the tornado lies betweeen the storm's updraft and downdraft. This was observed many years before by Ward (1961) in the first scientific storm chase on record; the placement of the tornado between the updraft and the downdraft in Browning's model (Fig. 18.1) was based on Ward's observations. Golden and Purcell's (1978b) subjective estimate of the streamflow around the Union City tornado, inferred from debris trajectories and damage analysis, also places the tornado between the updraft and the downdraft.

In many cases new tornadoes form from lowered rotating cloud bases, which form on the gust front as it surges eastward (Burgess *et al.*, 1982; Jensen *et al.*1983). These are probably the same circulations that Agee *et al.* (1976) inferred from damage patterns (which are smaller than the mesocyclone but bigger than the tornado) and which they called the tornado cyclone.

Although the tornado is located very close to the storm's downdraft, close inspection of movies of tornadoes reveals that on and just outside the funnel there is clearly very intense rising motion. From detailed observations and analyses of the motion of dust, debris and cloud tags, deductions of the wind field may be made (see the review by Golden, 1976). The most prominent of these studies are the ones by Hoecker (1960), Golden and Purcell (1977,

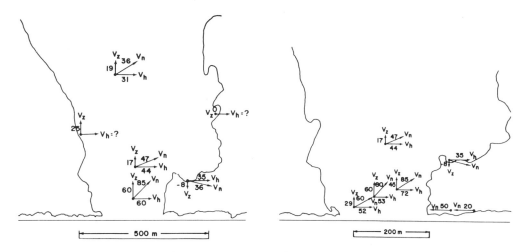

Figure 18.5. Scaled outlines of the Great Bend tornado's dust column; representative net, horizontal, and vertical velocity vectors (m s^{-1}) are superimposed. Velocities were derived photogrammetrically from movie loops over a 5 min interval. (From Golden and Purcell, 1977.)

1978b), Forbes (1978), and Campbell *et al.* (1983). Although there is some question about the more detailed conclusions drawn from these analyses, there are common basic features. First, the most intense winds occur very close (within 100 m) to the ground. Second, the vertical velocity is comparable in magnitude with the horizontal velocity, and it too reaches very strong values close to the ground. Figure 18.5 shows these features for the Great Bend, Kans., tornado. Here the view is to the south; there is very intense swirling and rising motion near the ground; there is a strong inward jet along the ground approaching from the southwest and weak subsiding flow away from the tornado on the west side at approximately 200 m above ground level at a larger radius. Hence, there is an updraft on a much smaller scale and much more intense than would be likely anywhere else within the thunderstorm (see Fig. 18.2d). And, observations similar to those in Figs. 18.3–18.5 indicate that it is located adjacent to the thunderstorm downdraft.

Hoecker's (1960) study of the Dallas tornado is less precise than later analyses, for reasons discussed by Morton (1966) and Golden (1976). However, it is probably more than mere coincidence that the flow pattern obtained by Hoecker has been approximately reproduced (in some parameter range) by almost every axisymmetric vortex model, numerical or laboratory, that has ever been run. (See Hoecker's analysis in Fig. 18.6; asymmetries have been neglected but note the strong swirl and vertical motion very close to the ground. See also Sec. 18.3.)

18.3. Theory

18.3.1. Thunderstorm-Scale Motions

Figures 18.2a and 18.2b show the results of a three-dimensional numerical simulation of the Del City supercell thunderstorm, the Doppler analysis for

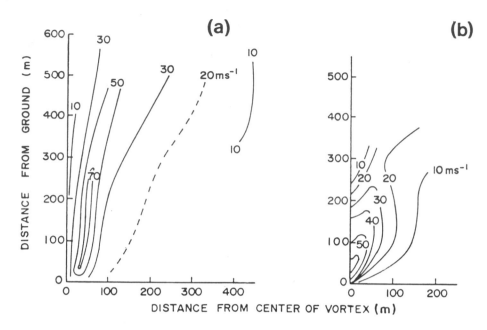

Figure 18.6. Hoecker's photogrammetric analyses of the 1957 Dallas tornado showing (a) tangential and (b) vertical velocities. (From Davies-Jones, 1986.)

which is shown in Figs. 18.2c, d (mature stage) and 18.2e, f (collapse stage). The simulation exhibits a mesocyclone, hook echo, and qualitatively correct arrangement of major downdrafts, outflows, and inflows. The explanation of the supercells rotational properties is the beginning point for the explanation of the tornado; it relies heavily on the model results and, to the extent they agree with the data, an explanation of the former is an explanation of the latter, *mutatis mutandis*.

The following discussion assumes a good working knowledge of fluid dynamics, in particular, vorticity dynamics (e.g., Morton, 1966), Ertel's theorem for the conservation of potential vorticity (e.g., Dutton, 1976, p. 382) and what is meant by "conservation" of a particular quantity.

It was suggested by Browning and Landry (1963) and more fully developed by Barnes (1970) that rotation within the supercell derives from the upward tilting of the horizontal vorticity associated with the environmental shear as the updraft propagates to the right of the mean wind. All three-dimensional convection models are initiated with zero vertical vorticity in the environment (if Coriolis effects are neglected). Since all the air in the updraft or downdraft has vertical vorticity and all that air enters from the environment where all the vorticity is horizontal, the vertical vorticity must develop through tilting. Rotunno (1981), Lilly (1982), and Davies-Jones (1984) developed some simple relations to illustrate how a propagating updraft/downdraft couplet acquires vertical vorticity as it propagates within the environmental shear flow. The following is a paraphrase of Davies-Jones's (1984) description based on linear theory.

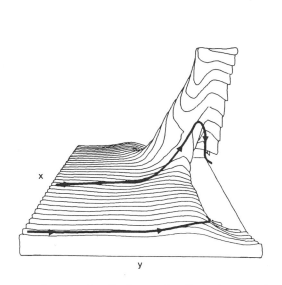

Figure 18.7. A perspective (the view is westward) of the 336 K equivalent potential temperature surface from a 3-D numerical simulation of a rightward-propagating supercell thunderstorm. The updraft is evidenced by the column of high-equivalent-potential-temperature air. The vortex lines, by the conservation of the initially zero equivalent potential vorticity, must lie on this surface. Therefore they tilt upward on the updraft's south flank, so vertical vorticity is positive. Low-equivalent-potential-temperature air descends on the north flank in the rainy downdraft; the vortex lines must tilt downward, so vertical vorticity is negative on the north flank. (Adapted from Rotunno and Klemp, 1985.)

An updraft-downdraft pair can be represented by a hill in the displacement field of the isentropic surfaces. In the case shown in Fig. 18.2 the flow relative to the hill at $z = 4$ km is from the south, and the environmental vorticity points toward the north-northwest (see the environmental wind hodograph in Fig. 3 of Klemp *et al.*, 1981). Thus, the vortex lines of the mean shear flow lie on the surface of the hill and tilt upward on the hill's south side and downward on its north side. Since the vertical velocity should also be positive on the south side of the hill and negative on the north, it is clear that, under the stated conditions, the linear theory predicts that an updraft will have positive vertical vorticity and a downdraft, negative, if the storm-relative winds approach from the south (or, alternatively, as the system propagates to the right of the shear vector).

A more comprehensive theory proposed by Davies-Jones (1984) and Rotunno and Klemp (1985) is based on the conservation of "equivalent potential vorticity," which is the scalar product of the vorticity vector with the gradient of equivalent potential temperature. Because, in the model's initial state, all the vorticity is horizontal and all the variation of equivalent potential temperature is vertical, the initial equivalent potential vorticity is zero and, if conserved, must remain so. Geometrically, this means that all the vortex lines must always lie in surfaces of constant equivalent potential temperature. And, since equivalent potential temperature is nearly conserved, analysis of the topology of these surfaces instantly yields the desired information on how vortex lines must run through the storm. Figure 18.7 shows a three-dimensional perspective of a surface of equivalent potential temperature of a simulated supercell that is propagating southward (to the left). The vortex lines associated with the environmental shear point northward and adhere to the surface that rises up on the south flank and sinks down on the north flank, owing to the intrusion of low-equivalent-potential-temperature air from middle levels. Thus, there is a cyclonically rotating updraft and

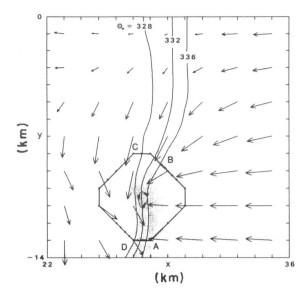

Figure 18.8. Cross section at $z = 250$ m of the flow in an idealized 3-D supercell simulation. The updraft is shaded in increments of 1 m s^{-1} beginning with the 1 m s^{-1} values. The wind vectors are plotted every 2 km; a vector of a length equal to one grid interval (1 km) signifies a wind of 10 m s^{-1}. The position of maximum vertical vorticity is indicated by the closed contour with the arrows. Lines of constant equivalent potential temperature (K) are indicated. (From Rotunno and Klemp, 1985; they studied the evolution of the circulation round the curve ABCD, which is not discussed here.)

an anti-cyclonically rotating downdraft propagating to the right of the mean wind.

In the foreground there is another vortex line, which turns toward the southwest, disappears from view, and then passes near the low-level vertical vorticity maximum. This suggests that the mechanism of low-level vorticity generation is more complicated than the simple tilting of the horizontal vorticity associated with the environmental shear. Klemp and Rotunno (1983) noted that a good portion of the flow reaching the location of maximum low-level vertical vorticity approaches from the northeast (see Fig. 18.2a), where there is a boundary between the warm environmental air approaching from the southeast and the rain-cooled air flowing from the north. Along this boundary, strong horizontal vorticity is generated baroclinically in the direction of flow. As the flow approaches the updraft, the tilting of this horizontal vorticity produces cyclonic vorticity, which is subsequently stretched to produce the observed large values of vertical vorticity. Figure 18.8 illustrates these ideas. It represents the supercell flow in the mature stage just before collapse; clearly, the maximum vertical vorticity is located in updraft and is between the cold air and the warm air.

That the vertical vorticity should be non-zero in the gradient zone of equivalent potential temperature follows from the conservation of the (zero) equivalent potential vorticity. The latter implies that the circulation around any closed material curve lying wholly within a surface of constant equivalent potential temperature must be zero. However, any horizontal material curve drawn around the location of the vertical vorticity maximum will have a non-zero (positive) circulation and so must cut through surfaces of constant equivalent potential temperature. Hence, the rotation at low levels depends in an essential way on the juxtaposition of the storm's cold outflow and warm inflow airstreams.

It is now possible to answer the question, "What is the origin and nature of the gyratory wind field?" The middle-level rotation is due to the upward tilting of environmental horizontal vorticity, and it transports potentially cold air to the forward and left flanks, where, being evaporatively chilled by rain from the northward sloping updraft, it descends and forms a cold pool at the surface on the left and forward flanks. This is just the right place for baroclinically generating horizontal vorticity of the proper sign to produce cyclonic vorticity at low levels when this air subsequently encounters the updraft. The second (and more difficult) question concerns how a flow like the one shown in Fig. 18.8 can give birth to a tornado.

18.3.2. Tornado-Scale Motions

Klemp *et al.*'s (1981) simulation (Figs. 18.2a, b) appears to capture the basic features of the mature supercell but does not reach a collapse stage. Klemp and Rotunno (1983), noting that the collapsed updraft involves smaller scales, interpolated Klemp *et al.*'s solution to a finer mesh and ran the 3-D cloud model on the domain indicated in Figure 18.2a for several minutes. Figure 18.9 shows that the updraft becomes contorted into the characteristic horseshoe shape, and the vertical vorticity field takes the shape of an annular ring having two local maxima diametrically opposite. Klemp and Rotunno (1983) argued that these low-level vorticity maxima correspond to the tornado cyclones alluded to in Sec. 18.2.2.

Klemp and Rotunno (1983) found that the low-level rotation increased dramatically when the horizontal resolution was increased and that this preceded the development of the horseshoe-shaped updraft. They argued that the increasing low-level rotation induces cyclostrophically low perturbation pressure near the ground, which produces an unfavorable vertical pressure gradient for the updraft. Thus, the updraft decreases at its center and the horseshoe shape and general updraft weakening are produced. This argument is a variant on the "vortex-valve" invoked by Lemon *et al.* (1975) to explain the decline in the Union City, Okla., storm as low-level rotation increased. Similar behavior was found by Brandes (1978) in his analysis of the tornadic storm that struck near Harrah, Okla., on 8 June 1974. A composite picture of this flow is in Fig. 15.9 (Ch. 15, this volume).

Why does the low-level rotation increase dramatically above its middle-level values? Possibly the combination of the strong baroclinic production of horizontal vorticity, and its tilting and subsequent amplification through strong convergence near the surface, sooner or later produces more vorticity there than can be transported or diffused away.

And why is the consequent horseshoe-shaped updraft more conducive to tornado cyclone formation and how do tornado cyclones produce tornadoes? The reason is not clear, but cylindrical-domain models offer possible guidance.

Figure 18.10 represents the archetypal vortex chamber. Fluid is drawn through the chamber by suction applied at the exhaust hole, and rotation is added as the fluid passes through the sides of the chamber. Almost all numerical models pertaining to tornadoes are designed as some variation on the vortex-chamber theme. Thus, these remarks on the vortex-chamber flow

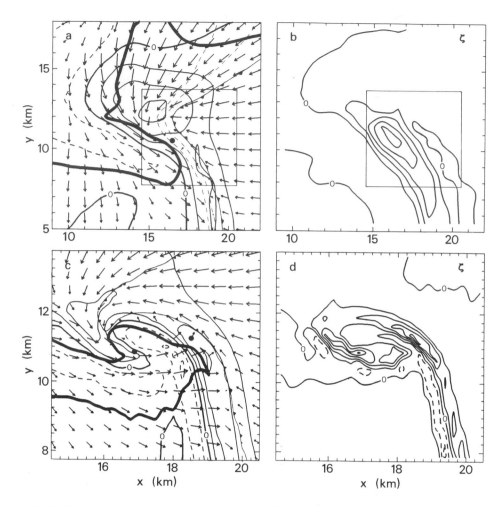

Figure 18.9. Cross section at $z = 250$ m for the (a) flow field from the storm-scale simulation displayed in the location of a high-resolution simulation indicated by the box in Figure 18.2a; (b) vertical vorticity associated with the flow field in (a) contoured in 0.005 s^{-1} intervals; (c) flow field at 6 min in high-resolution simulation displayed in the region indicated by the box in (a); (d) vertical vorticity at 6 min displayed in the same region as (c) with a 0.02 s^{-1} contour interval. In (a), the vertical velocity is contoured in 1 m s^{-1} intervals and the heavy solid line represents the 0.5 g kg^{-1} rainwater contour. One grid interval represents 20 m s^{-1} for the horizontal flow vectors. The $-1°$C perturbation potential temperature is denoted by the cold frontal boundary; the two thin dashed lines behind this boundary locate the $-2°$ and $-3°$C isotherms. The location of the maximum vertical vorticity is marked with a solid black circle. In (c) the notation is as in (a) except that the vertical velocity is contoured in 2 m s^{-1} intervals and one grid interval represents 10 m s^{-1} for horizontal flow vectors, which are plotted only at every other grid point. (From Klemp and Rotunno, 1983.)

apply as well to the numerical vortex models (which, except for Rotunno's [1984] are all axisymmetric).

In light of the foregoing analysis of the three-dimensional-cloud-model results, which are in qualitative agreement with what is generally observed, the relation of the chamber results to the tornado phenomenon is not as straightforward as was once thought. Consider that the flow in the chamber

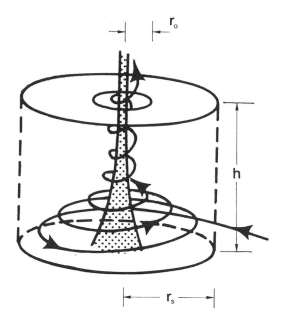

Figure 18.10. Flow in a vortex chamber. Fluid enters with a tangential velocity component, spirals inward, and exits axially at smaller radius. (Adapted from Davies-Jones, 1976.)

is homogeneous and axisymmetric (as it enters through the sides), while the flow around the vorticity maximum shown in Fig. 18.8 manifestly and fundamentally is not. Work in this area has proceeded on the hope that, in spite of the fact that the flow on the storm scale is asymmetric and heterogeneous, some of the results of the vortex-chamber-type work are relevant. Given the present state of computers, it is impossible for the three-dimensional cloud models, even with nested grids, to simulate both the supercell on a domain large enough to include the undisturbed environment and the tornado with a resolution fine enough to capture the significant dynamical processes. However, even if this were possible, the results of such a calculation might be too difficult to decipher. Therefore, there really is, at present, nowhere else to turn for guidance toward understanding the fluid dynamics on the tornado scale.

Thus we arrive at the second question posed in the Introduction: "Can the essence of the tornado mechanism be described simply by imposing an updraft on a rotating wind field?"

According to the review of laboratory modeling by Davies-Jones (1976), only the model parameters used in the experiments by Ward (1972) resemble those occurring naturally. Ward (1972) thought of the vortex chamber as the laboratory-analog of the mesocyclone. The rotating screen represents the observed low-level rotation source, the exhaust radius represents the updraft radius, and the height of the chamber represents the layer of moist, potentially unstable air feeding the mesocyclone updraft.

Following Ward, and in view of Fig. 18.8, suppose that the depth of the chamber h corresponds to the distance between the ground and cloud base ($\sim$1 km), and the exhaust radius r_o corresponds to the radial extent of the most intense part of the low-level updraft (say, 2 km). Let the vortex chamber represent the region shown in Fig. 18.8, centered on the location of maximum updraft with the outer radius $r_s \sim 3$ km. Before Ward's work, the radius of the exhaust was taken to be much smaller than the height of the inflow layer (that is, the aspect ratio h/r_o was large) whereas in Ward's model it is of order 1. Thus, Ward's chamber is geometrically more similar to the mesocyclone than were earlier models.

Further analysis of Ward's data (Davies-Jones, 1973), further experiments (Church et al., 1979), and numerical simulations of Ward's experiment (Harlow and Stein, 1974; Rotunno, 1977, 1979, 1984; Gall, 1982) all show that the dynamics of the vortex obtained in the chamber depend strongly on the swirl ratio, $S = r_o\Gamma/2Q$, where Γ is the circulation about the central axis at $r = r_o$, and Q is the rate of volume flow through the chamber. Since Q and Γ may be controlled independently, values of S close to those for the mesocyclone may be used so that the experimental vortex is dynamically similar to the natural vortex.

On the basis of laboratory experiments and numerical simulations, Davies-Jones (1986) described the different kinds of vortex motion produced in the chamber as S is increased from small to large values. Figure 18.11a shows that at small values of S the flow is basically a nonrotating updraft. At any given level above the boundary layer, the pressure rises as the flow approaches the center axis because it must decelerate. Thus the boundary layer separates under the influence of this adverse pressure gradient, and any rotation that exists in the entering stream is carried aloft before reaching the axis. At slightly larger values of S (Fig. 18.11b), the rotation in the outer flow makes the pressure gradient in the direction of flow favorable so that the boundary layer remains attached and a "one-celled" (updraft throughout) vortex is formed. When S is increased further (Figs. 18.11c, d), a "two-celled" (a central downdraft surrounded by updraft) vortex is formed. To understand this behavior, consider the two-celled flow shown in Fig. 18.11d where the central downdraft penetrates all the way to the surface.

Rotunno (1977), using an axisymmetric numerical model of the Ward experiment, showed that the two-celled vortex is the only one obtainable when a zero-stress lower boundary condition is used (see also Harlow and Stein, 1974; Smith and Howells, 1983); in such a case, there is no boundary layer. Consider the initial-value problem where at $t = 0$ there is an irrotational, inwardly and upwardly directed flow with no swirl. Since there is no boundary layer, there is, in this case, no separation like that shown in Fig. 18.11a. For $t > 0$, angular momentum with respect to the central axis is introduced over the depth of the inflow layer. As this air is transported inward, the nearly conserved angular momentum implies that the air must swirl about the center axis ever more quickly as it gets closer and closer to the axis. As this proceeds, the pressure on the center axis falls, and the initial updraft experiences an adverse pressure gradient on the center axis, which causes the vortex to "fill in" from above the chamber exhaust, where the rotation has not yet penetrated. Since there is only a finite force drawing air through the chamber, the flow stops its inward progress at some radius determined by the amount of convergence imposed and the amount of angular momentum it has (i.e., its swirl ratio) and thereafter travels upward. This process results in

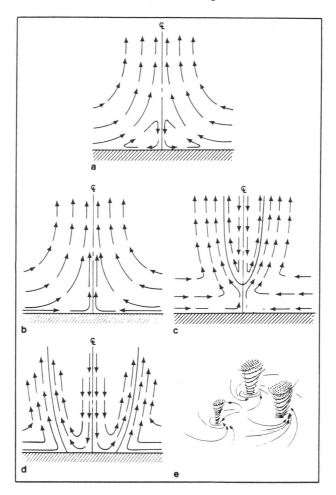

Figure 18.11. Types of vortices produced in the vortex chamber as swirl ratio increases: (a) weak swirl—boundary layer separates and passes around lower corner; (b) one-celled vortex; (c) one-celled vortex over lower portion, two-celled vortex over upper portion, separated by vortex breakdown; (d) two-celled vortex; and (e) multiple vortices. In (a)–(d) only meridional motion is shown. (From Davies-Jones, 1986.)

the two-celled flow shown in Fig. 18.11d and is a commonly observed feature of engineering-vortex-chamber flows (see the review by Lewellen, 1971).

When a no-slip (radial and azimuthal velocity = 0) lower boundary condition is used (Rotunno, 1977, 1979; Lewellen and Teske, 1977; Smith and Leslie, 1978, 1979; Lewellen and Sheng, 1981; Howells and Smith, 1983; see Lewellen, 1976 for references before 1976), there is a boundary layer which, as in other rotating flows, has a profound influence (see the review by Rott and Lewellen, 1966). Above the boundary layer, the radial pressure gradient is balanced by the centrifugal force of the air parcels as they swirl around the central axis. However, within the boundary layer, the centrifugal force decreases because the swirling velocity must decrease to zero by the no-slip condition while the radial pressure gradient remains virtually unchanged (a feature of boundary layers, in general), except very close to the axis. Thus there is an unbalanced inward force, which implies that radial velocity should be inward in the boundary layer. The radial dependence of the radial velocity within the boundary layer determines (by mass continuity) whether the vertical velocity is positive or negative at the top of the layer. This dependence, in turn, is a function of the radial dependence of the swirl velocity in the flow above the boundary layer.

Bodewadt (1940) considered a semi-infinite flow in solid-body rotation (swirl proportional to radius) above an infinite flat plate. He found that there is uniform upward motion at the top of the boundary layer, the radial flow being supplied to the boundary layer at infinite radius. At the other extreme, the boundary-layer equations for the flow of a potential vortex (swirl inversely proportional to radius) over a disc of finite radius were solved numerically by Burgraff *et al.* (1971). They found that the vertical velocity is negative at all radii over the finite disk, and this implies that mass is flowing radially inward in the boundary layer at all radii. These solutions imply catastrophic behavior near the origin where the inflowing boundary-layer air must turn and flow upward. This "eruption" of the boundary-layer flow near the axis forms the one-celled vortex shown in Fig. 18.11b. (Other important contributions to this idea are found in Turner, 1966; Barcilon, 1967; and Maxworthy, 1972).

When the swirl ratio has an intermediate value, the one- and two-celled vortices coexist as shown in Fig. 18.11c. The transition from the lower, one-celled, boundary-layer-induced to the upper, two-celled vortex is an example of "vortex breakdown," which is a phenomenon observed in other flow situations (see the review of vortex breakdown by Leibovich, 1978). As S is increased, the height of the breakdown lowers to the ground, and the two-celled vortex dominates the flow. With increasing S, the radial extent of the central downdraft increases, and the shear layer between the weakly swirling downdraft and the rapidly swirling updraft becomes sharper. This layer becomes unstable and multiple vortices are formed (Ward, 1972; Davies-Jones and Kessler, 1974; Snow, 1978; Rotunno, 1978, 1984; Staley and Gall, 1979; Gall, 1983), as illustrated in Fig. 18.11e.

How does all this relate to the observations that were presented in Sec. 18.2? Start with the collapse phase of the supercell, where the relatively broad rotating updraft undergoes a transition to the horseshoe-shaped updraft apparent in observations (Figs. 18.2 and 18.3) and simulations (Fig. 18.9). Klemp and Rotunno (1983) explained this transition as a consequence of the increasing low-level rotation, which induces cyclostrophically low pressure near the ground and so sets up an adverse pressure gradient that retards the updraft at its center. This explanation is in essence the same as the one given for the two-celled vortex (Fig. 18.11d), but because of the basic east-west asymmetry in the mesocyclone, there is a horseshoe-shaped updraft instead of an annular ring of updraft.

As it happens, for larger values of S, the two-celled vortex is unstable to three-dimensional perturbations. Using a three-dimensional numerical model in cylindrical coordinates, Rotunno (1984) simulated the instability and the ensuing multiple vortices that ultimately form and persist in a quasi-

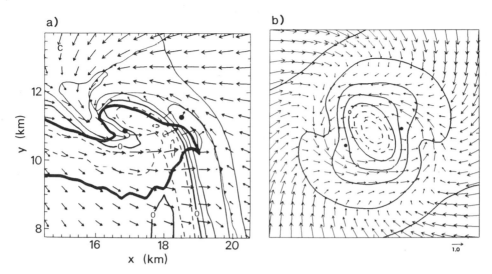

Figure 18.12. Comparison of (a) the high-resolution simulation shown in Fig. 18.9c with (b) a 3-D simulation of the multiple vortices produced in the Ward-type vortex chamber. In (b), the cross section is taken at one grid level above the ground. Within the larger chamber-scale vortex, the contours of vertical velocity show there is a two-celled vortex with asymmetric vortices located between the updraft and downdraft. (From Rotunno, 1984.)

steady equilibrium with a modified two-celled vortex. Figure 18.12 shows this flow (with the mean rotation rate subtracted) in comparison with Klemp and Rotunno's fine-resolution simulation of the mesocyclone. Here the two smaller circulations imbedded within the larger circulation are in evidence; they are located in the annular region between the updraft and downdraft but are centered in the updraft.

For these larger values of S, the two-celled vortex is approximately half the size (as determined by the radius of the maximum tangential velocity) of the exhaust radius. Therefore, if the exhaust radius is meant to represent the mesocyclone updraft, then the two-celled vortex is also of the scale of the mesocyclone. Likewise, the smaller centers of rotation are too large to be directly identified with the tornado. Further, the photogrammetric studies all indicate that there is intense rising motion in the tornado funnel, whereas in both the Rotunno (1984) and the Klemp-Rotunno (1983) simulations the centers of rotation are in weak updraft (zero-stress conditions were used in those studies, owing to limited vertical resolution). For these reasons, it appears that the smaller vortices shown in Fig. 18.12 are more analogous to tornado cyclones than to the tornado, which has intense rising motion at very low levels.

How does a tornado cyclone foster tornadogenesis? Looking again to the vortex-chamber studies, suppose that the vortex chamber is a model of the tornado cyclone. Here the reasoning leading to the conclusion that only Ward's model is geometrically similar to the natural phenomenon is less certain. As is evident in Figure 18.2e, the storm-scale updrafts become narrower, while the inflow depth remains approximately the same, implying

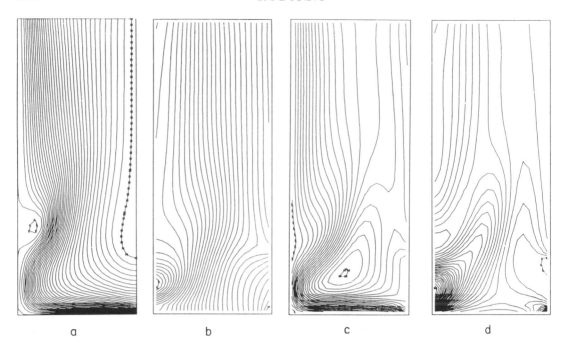

a   b   c   d

Figure 18.13. A typical axisymmetric vortex simulation using a no-slip lower boundary condition: (a) streamlines; (b) pressure contours (note the low on the axis of symmetry); (c) tangential velocity contours; and (d) vertical velocity contours. Note the resemblance between this flow and that of the Hoecker vortex shown in Fig. 18.6. (From Harlow and Stein, 1974.)

a larger local aspect ratio. However, this may not be important. In spite of differing aspect ratios, the kind of intense vortex with strong vertical and tangential velocities very close to the lower surface observed through photogrammetry is obtained, not only in the Ward chamber, but in other vortex-chamber studies where the swirl ratio is in the proper range (Barcilon, 1967; Ying and Chang, 1970; Maxworthy, 1972; Wan and Chang, 1972; Chi, 1977). Furthermore, all axisymmetric numerical models produce this kind of vortex when the parameters are in the proper range (Figs. 5–7 of Chi, 1977; Figs. 5.1–5.3 of Lewellen and Sheng, 1980; Figs. 4.4 and 4.6 of Rotunno, 1979; Fig. 5 of Smith and Leslie, 1979; Figs. 2–3 of Howells and Smith, 1983; Fig. 3 of Wilson and Rotunno, 1982).

An example of this is shown in Fig. 18.13, taken from Harlow and Stein's simulation of the Ward-chamber flow. These findings suggest that the final amplification that produces the tornado is the interaction of the tornado cyclone with the ground. Thus, the interaction of the tornado cyclone with the lower boundary is required to explain the final, intense, end-wall vortex that is the tornado. One might fairly question this argument in light of the fact that the TVS (when present) originates aloft. However, the dynamical connection between the TVS and the visible funnel in "contact" with the ground is not known. That is to say, the appearance of a TVS aloft may not be relevant. Indeed, the tornadic storm analysed by Klemp *et al.* (1981) did not even have a TVS (see Fig. 18.2).

If the swirl ratio at the tornado cyclone is larger, multiple vortices (see Fujita, 1971) are produced. Here the Ward device appears to be superior to the others because the two-celled vortex from which the multiple vortices form does depend on the geometry of the chamber. When multiple vortices are observed, the aspect ratio and the swirl ratio of the tornado cyclone are both order 1. On the other hand, when there is a single funnel, Hoecker-type vortices are observed, the swirl ratio is in the low-to-moderate range, and the aspect ratio is not as important.

It is clear that the tornado does not fit a simple model like the spin-up that skaters experience when they pull in their arms. (conservation of angular momentum). Tornadoes occur near the place where the storm's cold outflow meets the warm inflow; this requires baroclinic generation of horizontal vorticity as a source of the tornado's rotation. Observations of tornado-associated updrafts of 80 m s^{-1} at 100 m AGL suggest that complex boundary-layer interactions are important.

REFERENCES

Agee, E. M., J. T. Snow, and P. R. Clare, 1976: Multiple vortex features in the tornado cyclone and the occurrence of tornado families. *Mon. Wea. Rev.*, **104**, 552–563.

Barcilon, A., 1967: Vortex decay above a stationary boundary. *J. Fluid Mech.*, **27**, 155–175.

Barnes, S. L., 1970: Some aspects of a severe, right-moving thunderstorm deduced from mesonetwork rawinsonde observations. *J. Atmos. Sci.*, **27**, 634–648.

Bluestein, H., 1980: The University of Oklahoma Severe Storms Intercept Project—1979. *Bull. Amer. Meteor. Soc.*, **61**, 560–567.

Brandes, E. A., 1978: Mesocyclone evolution and tornadogenesis: Some observations. *Mon. Wea. Rev.*, **106**, 995–1011.

Brown, R. A., L. R. Lemon, and D. W. Burgess, 1978: Tornado detection by pulse Doppler radar. *Mon. Wea. Rev.*, **106**, 29–38.

Browning, K. A., 1964: Airflow and precipitation trajectories within severe local storms which travel to the right of the mean wind. *J. Atmos. Sci.*, **21**, 634–639.

Browning, K. A., and R. J. Donaldson, Jr., 1963: Airflow and structure of a tornadic storm. *J. Atmos. Sci.*, **20**, 533–545.

Browning, K. A., and C. R. Landry, 1963: Airflow within a tornadic storm. Preprints, 10th Weather Radar Conference, Washington, D.C., American Meteorological Society, Boston, 116–122.

Browning, K. A., and F. H. Ludlam, 1962: Airflow in convective storms. *Quart. J. Roy. Meteor. Soc.*, **88**, 117–135.

Bodewadt, V. T., 1940: Die Drehstromung uber festem Grunde. *Z. Angew. Math. Mech.*, **20**, 241–253.

Burgess, D. W., R. A. Brown, L. R. Lemon, and C. R. Safford, 1977: Evolution of a tornadic thunderstorm. Preprints, 10th Conference on Severe Local Storms, Omaha, Nebr., American Meteorological Society, Boston, 84–89.

Burgess, D. W., V. T. Wood, and R. A. Brown, 1982: Mesocyclone evolution statistics. Preprints, 12th Conference on Severe Local Storms, San Antonio, Tex., American Meteorological Society, Boston, 422–424.

Burgraff, O. R., K. Stewartson, and R. Belcher, 1971: Boundary layer induced by a potential vortex. *Phys. Fluids*, **14**, 1821–1833.

Campbell, B. D., E. N. Rasmussen, and R. E. Peterson, 1983: Kinematic analysis of the Lakeview, Texas Tornado. Preprints, 13th Conference on Severe Local Storms, Tulsa, Okla., American Meteorological

Society, Boston, 62–65.

Chi, J., 1977: Numerical analysis of turbulent end-wall boundary layers of intense vortices. *J. Fluid Mech.*, **82**, 209–222.

Chisholm, A. J., 1973: Radar case studies and airflow models. In *Alberta Hailstorms*, Meteor. Monogr., **36**, American Meteorological Society, Boston, 1–36.

Church, C. R., J. T. Snow, G. L. Baker, and E. M. Agee, 1979: Characteristics of tornado-like vortices as a function of swirl ratio: A laboratory investigation. *J. Atmos. Sci.*, **36**, 1755–1766.

Davies-Jones, R. P., 1973: The dependence of core radius on swirl ratio in a tornado simulator. *J. Atmos. Sci.*, **30**, 1427–1430.

Davies-Jones, R. P., 1976: Laboratory simulations of tornadoes. Proceedings, Symposium on Tornadoes: Assessment of Knowledge and Implications for Man, Lubbock, Tex., Texas Tech University, 151–174.

Davies-Jones, R. P., 1982: Tornado interception with mobile teams. In *Thunderstorms: A Social and Technological Documentary*, Vol. III. E. Kessler (Ed.), U.S. Govt. Printing Office, Washington, D.C., 33–46.

Davies-Jones, R. P., 1984: Streamwise vorticity: The origin of updraft rotation in supercell storms. *J. Atmos. Sci.*, **41**, 2991–3006.

Davies-Jones, R. P., 1986: Tornado dynamics. In *Thunderstorms: A Social and Technological Documentary*, Vol. II. E. Kessler, (Ed)., Second edition revised and enlarged, University of Oklahoma Press, Norman, Okla., and London, 197–236.

Davies-Jones, R. P., and E. Kessler, 1974: Tornadoes. In *Weather and Climate Modification*. W. N. Hess (Ed.), John Wiley and Sons, New York, 552–595.

Dutton, J. A., 1976: *The Ceaseless Wind.* McGraw-Hill, New York, 579 pp.

Eagleman, J. R., and W. C. Lin, 1977: Severe thunderstorm internal structure from dual-Doppler radar measurements. *J. Appl. Meteor.*, **16**, 1036–1048.

Fankhauser, J. C., 1971: Thunderstorm-environment interactions determined from aircraft and radar observations. *Mon. Wea. Rev.*, **99**, 171–192.

Fawbush, E. J., and R. C. Miller, 1954: The types of air masses in which North American tornadoes form. *Bull. Amer. Meteor. Soc.*, **35**, 154–165.

Ferrel, W., 1889: *A Popular Treatise on the Winds.* Wiley, New York, 505 pp.

Forbes, G. S., 1978: Three scales of motion associated with tornadoes. Final Report to the U.S. Nuclear Regulatory Commission, NUREG/CR–0363 (NTIS#PB–288 291), 359 pp.

Fujita, T. T., 1971: Proposed mechanism of suction spots accompanied by tornadoes. Preprints, 7th Conference on Severe Local Storms, Kansas City, Mo., American Meteorological Society, Boston, 208–213.

Gall, R. L., 1982: Internal dynamics of tornado-like vortices. *J. Atmos. Sci.*, **39**, 2721–2736.

Gall, R. L., 1983: A linear analysis of the multiple vortex phenomenon in simulated tornadoes. *J. Atmos. Sci.*, **40**, 2010–2024.

Golden, J. H., 1976: An assessment of wind speeds in tornadoes. Proceedings, Symposium on Tornadoes: Assessment of Knowledge and Implications for Man, Lubbock, Tex., Texas Tech University, 5–42.

Golden, J. H., and B. B. Morgan, 1972: The NSSL-Notre Dame tornado intercept program, Spring 1972. *Bull. Amer. Meteor. Soc.*, **53**, 1178–1180.

Golden, J. H., and D. Purcell, 1977: Photogrammetric velocities for the Great Bend, Kansas, tornado of 30 August 1974: Accelerations and asymmetries. *Mon. Wea. Rev.*, **105**, 485–492.

Golden, J. H., and D. Purcell, 1978a: Life cycle of the Union City, Oklahoma, tornado and comparison with waterspouts. *Mon. Wea. Rev.*, **106**, 3–11.

Golden, J. H., and D. Purcell, 1978b: Airflow characteristics around the Union City tornado. *Mon. Wea. Rev.*, **106**, 22–28.

Harlow, F. H., and L. R. Stein, 1974: Structural analysis of tornado-like vortices. *J. Atmos. Sci.*, **31**, 2081–2098.

Heymsfield, G. M., 1978: Kinematic and dynamic aspects of the Harrah tornadic storm from dual-Doppler radar data. *Mon. Wea. Rev.*, **106**, 233–254.

Hoecker, W. H., 1960: Wind speed and airflow patterns in the Dallas tornado of April 2, 1957. *Mon. Wea. Rev.*, **89**,

533–542.

Howells, P., and R. K. Smith, 1983: Numerical simulations of tornado-like vortices. Part I: Vortex evolution. *Geophys. Astrophys. Fluid Dyn.*, **27**, 253–284.

Jensen, B., E. N. Rasmussen, T. P. Marshall, and M. A. Mabey, 1983: Storm scale structure of the Pampa storm. Preprints, 13th Conference on Severe Local Storms, Tulsa, Okla., American Meteorological Society, Boston, 85–88.

Klemp, J. B., and R. Rotunno, 1983: A study of the tornadic region within a supercell thunderstorm. *J. Atmos. Sci.*, **40**, 359–377.

Klemp, J. B., R. B. Wilhelmson, and P. S. Ray, 1981: Observed and numerically simulated structure of a mature supercell thunderstorm. *J. Atmos. Sci.*, **38**, 1558–1580.

Lemon, L. R., and C. A. Doswell, III, 1979: Severe thunderstorm evolution and mesocyclone structure as related to tornadogenesis. *Mon. Wea. Rev.*, **107**, 1184–1197.

Lemon, L. R., D. W. Burgess, and R. A. Brown, 1975: Tornado production and storm sustenance. Preprints, 9th Conference on Severe Local Storms, Norman, Okla., American Meteorological Society, Boston, 100–104.

Lewellen, W. S., 1971: A review of confined vortex flows. NASA Report CR-1772, 219 pp.

Lewellen, W. S., 1976: Theoretical models of the tornado vortex. Proceedings, Symposium on Tornadoes: Assessment of Knowledge and Implications for Man, Lubbock, Tex., Texas Tech University, 107–143.

Lewellen, W. S., and Y. P. Sheng, 1981: Modeling tornado dynamics. Final Report to the U.S. Nuclear Regulatory Commission (NTIS#NUREG/CR-1585), 277 pp.

Lewellen, W. S., and M. E. Teske, 1977: Turbulent transport model of low level winds in a tornado. Preprints, 10th Conference on Severe Local Storms, Omaha, Nebr., American Meteorological Society, Boston, 291–298.

Leibovich, S., 1978: The structure of vortex breakdown. *Ann. Rev. Fluid Mech.*, **10**, 221–246.

Lilly, D. K., 1982: The development and maintenance of rotation in convective storms. In *Intense Atmospheric Vortices*, L. Bengtsson and M. J. Lighthill (Eds.), Springer-Verlag, Berlin, 149–160.

Maxworthy, T., 1972: On the structure of concentrated columnar vortices. *Astronaut. Acta.*, **17**, 363–374.

Moller, A., C. Doswell, J. McGinley, S. Tegtmeier, and R. Zipser, 1974: Field observations of the Union City tornado in Oklahoma. *Weatherwise*, **27**, 68–77.

Morton, B. R., 1966: Geophysical vortices. In *Progress in Aeronautical Sciences*, **7**, Pergamon Press, 145–192.

Newton, C. W., 1963: Dynamics of severe convective storms. In *Severe Local Storms*, Meteor. Monogr., **5**(27), 33–58.

Ray, P. S., 1976: Vorticity and divergence within tornadic storms from dual Doppler radar. *J. Appl. Meteor.*, **15**, 879–890.

Rott, N., and W. S. Lewellen, 1966: Boundary layers in rotating flow. In *Progress in Aeronautical Sciences*, **7**, Pergamon Press, 111–144.

Rotunno, R., 1977: Numerical simulation of a laboratory vortex. *J. Atmos. Sci.*, **34**, 1942–1956.

Rotunno, R., 1978: A note on the stability of a cylindrical vortex sheet. *J. Fluid Mech.*, **87**, 761–771.

Rotunno, R., 1979: A study in tornado-like vortex dynamics. *J. Atmos. Sci.*, **36**, 140–155.

Rotunno, R., 1981: On the evolution of thunderstorm rotation. *Mon. Wea. Rev.*, **109**(5), 577–586.

Rotunno, R., 1984: An investigation of a three-dimensional asymmetric vortex. *J. Atmos. Sci.*, **41**, 283–298.

Rotunno, R., and J. B. Klemp, 1985: On the rotation and propagation of simulated supercell thunderstorms. *J. Atmos. Sci.*, **42**, 271–292.

Smith, R. K., and P. Howells, 1983: Numerical simulations of tornado-like vortices. Part II: Two-cell vortices. *Geophys. Astrophys. Fluid Dyn.*, **27**, 285–298.

Smith, R. K., and L. M. Leslie, 1978: Tornadogenesis. *Quart. J. Roy. Meteor. Soc.*, **104**, 189–199.

Smith, R. K., and L. M. Leslie, 1979: A numerical study of tornadogenesis in a rotating thunderstorm. *Quart. J. Roy. Meteor. Soc.*, **105**, 107–127.

Snow, J. T., 1978: On inertial instability as related to the multiple vortex phenomenon. *J. Atmos. Sci.*, **35**, 1660–1671.

Snow, J. T., 1982: A review of recent advances in tornado vortex dynamics. *Rev. Geophys. Space Phys.*, **20**, 953–964.

Snow, J. T., 1984: The tornado. *Sci. Amer.*, **250**, 86–97.

Staley, D. O., and R. L. Gall, 1979: Barotropic instability in a tornado vortex. *J. Atmos. Sci.*, **36**, 973–981.

Turner, J. S., 1966: The constraints imposed on tornado-like vortices by the top and bottom boundary conditions. *J. Fluid Mech.*, **25**, 377–400.

Wan, C. A., and C. C. Chang, 1972: Measurements of the velocity field in a simulated tornado-like vortex using a three-dimensional velocity probe. *J. Atmos.*

Sci., **29**, 116–127.

Ward, N. B., 1961: Radar and surface observations of tornadoes of May 4, 1961. Preprints, 9th Weather Radar Conference, Kansas City, Mo., American Meteorological Society, Boston, 175–180.

Ward, N. B., 1972: The exploration of certain features of tornado dynamics using a laboratory model. *J. Atmos. Sci.*, **29**, 1194–1204.

Wilson, T., and R. Rotunno, 1982: Numerical simulation of a laminar vortex flow. Proceedings, Int. Conference on Computational Methods and Experimental Measurements, Washington, D.C., Springer-Verlag, 203–215.

Ying, S. J., and C. C. Chang, 1970: Exploratory model study of tornado-like vortex dynamics. *J. Atmos. Sci.*, **27**, 3–14.

CHAPTER 19

Upslope Precipitation Events

Roger F. Reinking
Joe F. Boatman

19.1. Introduction

Plains on the lee side of mountain ranges in the middle latitudes commonly receive their winter precipitation from circulations that are counter to the climatologically prevailing westerlies; moisture-bearing easterly currents are driven up the slopes of the lee-side plains, toward the mountains, by synoptic circulations. "Upslope storms" have been so named under the presumption that advection over the generally rising terrain induces lifting, cloudiness, and precipitation. This term is inexact; the lifting from low to high plains is more subtle than lifting by mountain ranges, and must be considered relative to superimposed lifting over frontal surfaces or by cyclonic convergence. "Upslope," by popular usage, nonetheless suffices to categorize the cloud systems described in this chapter.

Two extremely different driving circulations determine the characteristics of upslope events: the fully developed extratropical cyclone and the shallow arctic anticyclone. Mesoscale forecasting is made difficult by a number of variable and often contradictory factors that include blocking or cyclogenesis induced by the mountain range, depth and vorticity of the synoptic circulations, overriding countercurrents, oceanic or regional sources of moisture, degree and irregularities of topographic lifting, and dynamic and thermodynamic support for various microphysical precipitation-forming processes. Observational studies of the dynamical and microphysical events and processes that cause winter upslope cloudiness and precipitation are reviewed, with the focus on upslope precipitation in the rain shadow of the Rocky Mountains, where the full range of interactive processes and scales comes into play. Appendix A lists the general characteristics of middle-latitude upslope precipitation events.

19.2. Climatological Controls and Impacts of Upslope Events

Climatologically, mountains in the prevailing middle-latitude westerlies shelter their eastern foothills and plains from clouds and precipitation, especially in winter. Orographic lifting of the westerlies amplifies precipitation in the mountains, but spillover contributes relatively little to the eastern slopes. A lee-side moisture deficit is created, the magnitude of which is determined by fundamental climatological controls including elevation, extent, and orientation of the mountain range, location with respect to long waves and air mass sources, proximity to major sources of moisture, and continentality. The lee-side cloud and precipitation deficit can be reduced only by moisture-bearing circulations from the east.

The fundamental climate controls in concert set the stage for upslope precipitation events in the rain shadows of middle-latitude mountain ranges. In the conterminous United States, these controls come together, with varied collective effects, on the slopes east of the Cascade Range, the Sierra Nevada, the Rocky Mountains, the Appalachian Mountains, and many minor ranges. The lee side of the high Sierra Nevada is the most arid; here upslope precipitation does occur but is minimized by the sheltering effect of the complex topography of the Great Basin. The Appalachians offer an opposite extreme. Abundant moisture and a less continental clime are provided by the Gulf of Mexico and the Atlantic Ocean. The eastern terrain differential from the sea coasts to the Appalachian foothills provides a lift of about 1 km (~90 mb) in a horizontal distance of some 400–650 km (~0.17 mb km^{-1}). The northeast-southwest orientation of the Appalachians is often parallel to winter storm tracks, so the east side is commonly exposed to the counter-clockwise circulations of passing cyclones. Consequently, a dry side of the Appalachians does not exist; indeed, the southeastern slopes normally receive more precipitation than the northwestern ones.

The processes that cause upslope events are epitomized over the high, semi-arid plains east of the Rocky Mountains. The season for high plains upslope weather is September to April, and the most common time of occurrence is December to March. The northern plains have an extended season, and relatively more events occur in the early spring (Whiteman, 1973a).

On one extreme, the high plains winter storm driven by a fully developed extratropical cyclone can produce extremely heavy snows and blizzard conditions. On the other, the arctic anticyclone with upslope flow often does not qualify as a storm but merely generates layers of stratus. About 55% of the anticyclonic upslope circulations result in only light or very light precipitation rates and quantities. [Light: Trace to 0.3 cm h^{-1}; maximum 0.003 cm (6 min)$^{-1}$. Very light: Scattered drops or flakes that do not completely wet or cover an exposed surface, regardless of duration.] A few of these circulations do produce moderate to heavy precipitation (Whiteman, 1973a). The weather analyst and forecaster must deal with a continuum of upslope cloud and precipitation events bounded only by these extremes.

The plains within the United States rise from ~500 m elevation in the Upper Missouri River Basin westward ~700 km to ~1500 m elevation at the foot of the Rockies (~0.16 mb km^{-1}), and rise 1500 m over the ~1400 km from the Gulf Coast to the Rockies (~0.11 mb km^{-1}); the gradients are steeper nearer the mountains (Fig. 19.1a). Another 1500–2500 m lifting can be induced by the Rockies themselves. These terrain effects are not sufficient to overcome the precipitation shadows induced by the Rockies; contrary to the gradual east-to-west steepening of the high plains terrain (Fig. 19.1a), the

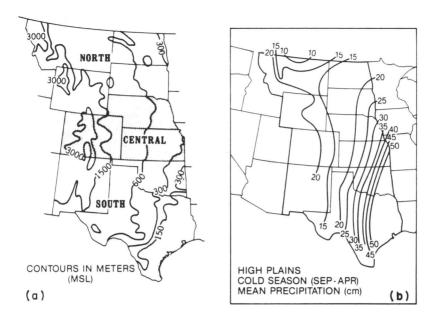

Figure 19.1. (a) High plains topography. (After Whiteman, 1973a.) (b) Water-equivalent cold season (September–April) mean precipitation for the high plains. (Based on data from Whiteman 1973b.)

broad mesoscale precipitation gradient lessens in that direction, at least until the mountains are encountered (Fig. 19.1b). Upslope-induced precipitation does, nevertheless, account for one-third to one-half of the regional annual precipitation. The east-west precipitation gradient in winter is about twice that in summer, and reflects frontal rather than thermal convective processes.

Broad mesoscale features of snowfall are evident in Fig. 19.2. Annually, about 1 m of snowfall is normal for the northern high plains. The transition to rain is evident in the decreasing mean snowfall southward from Kansas. This snowfall, primarily from upslope storms, is regulated by much more than the terrain-induced lifting. A good example can be made of feedbacks from the snow cover itself, to illustrate that many factors are intertwined: Walsh (1984) noted that some of the largest changes that occur in the Earth's surface over a period of several days to several months are due to the variations in snow cover. There is a regular winter snow cover over the northern high plains states (Fig. 19.1a), and occasional snow cover between these states and southern Texas. Snow increases the surface albedo, effectively insulates the ground, and while melting or subliming, is an effective sink of latent heat. The net effect is that air over snow cover is generally cold and incapable of holding large amounts of moisture. Overriding, relatively warm air will be cooled, perhaps to condensation. Treidl (1970) calculated that warm, moist air moving northward over a ~15 cm snow cover will cool 4–5°C per day by heat conduction to the surface. Thus, an encroaching storm may be weakened by loss of sensible heat to the snow cover. On longer time scales, a cool mean surface temperature, reduced by reflection of solar radiation from extensive snow cover, can delay spring warming (Walsh, 1984) and

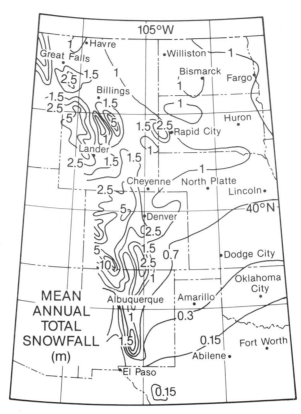

Figure 19.2. Mean annual total snowfall for the high plains. (After Whiteman, 1973b.)

may lengthen the season of upslope storms. Intermittent bare or even dry ground and snow-covered ground, temporally or spatially, is common in the high plains. Such complex but important factors are not easily accounted for by the forecaster.

The impacts of upslope events are not to be belittled. Seasonally, these weather systems impact water supply and air quality. Mountain runoff and soil moisture on the plains are regulated directly by the precipitation and indirectly by shading of the land. Successive years with little upslope moisture reduce east slope forests and high plains grasslands. Spring germination of wheat depends on upslope moisture; even winter survival of urban vegetation is greatly aided by upslope precipitation. Interacting with atmospheric constituents, the storms cleanse the air and determine the chemical quality of precipitation (Chaumerliac *et al.*, 1983; Nagamoto *et al.*, 1983); air pollution photochemistry in the now sprawling Front Range urban corridor of Colorado can be slowed by prolonged upslope cloudiness; and the aerosols of pollution can prolong upslope cloudiness and haze, which in turn reduce the transfer of solar and thermal radiation.

On a storm-by-storm basis, upslope events affect people in different ways. A contribution to the water supply is normally a positive factor. However, cattle on the open range may freeze or starve. Air traffic to high plains airports is often closed or slowed by snowfall and low visibility for hours and even days. Aircraft have iced and crashed. Highway travelers have been stranded, and the less fortunate have died in their vehicles. Urban and rural roadway departments have fallen far behind the challenge of snow removal, and the economically massive ski industry has on occasion been unable to get its customers from the plains to the mountains (or back).

The challenge to, and the need for, the forecaster of upslope events is evident. Not only does the forecaster need an acute awareness of the subtle mesoscale meteorological and topographic features that influence the nature of these storms; the forecaster also needs an understanding of the precipitation physics. Appendix B summarizes the mesoscale factors of upslope storms; Appendix C lists their synoptic extremes.

19.3. The Upslope Winter Anticyclone— Kinematics and Dynamics

The concepts of slope-induced weather are most clearly derived from examination of upslope winter anticyclones. These systems normally produce a shallow cloud mass within a moist layer that has a distinct top limited by an inversion, a feature that distinguishes these storms from the cyclonic systems. According to a 10 year record compiled by Whiteman (1973a), the average upslope cloud thickness over the high plains (Texas to Montana) is about 2 km, and the associated mean surface east wind component is 1–2 m s^{-1}; mean cloud top temperatures are between $-8°C$ and $-11°C$, the colder mean temperatures occurring farther upslope; thermodynamic stability is prevalent—mean lapse rates are approximately 2–4°C km^{-1}, and inversions within the cloud layer are not uncommon; only about 55% of the 10 to 14 such systems that occur each year precipitate. The gradations in the possible form and intensity of anticyclonic upslope cloud systems are schematically shown in Fig. 19.3.

19.3.1. Frontal Processes

Typically, a cold front leads continental polar or arctic air in anti-cyclonic flow southward over the high plains. The cold-air outbreak normally follows the passage of a surface cyclone across the northern plains states or southern Canada. The passing surface cyclone and the ensuing cold air mass are associated with a 500 mb trough that protrudes southward from a closed polar low, which may be as cold as $-40°C$ or $-50°C$ at its core (Boatman and Reinking, 1984). Any cloud system associated with the shallow surface anticyclone is not supported by upward motions induced by the polar front jet; indeed, the anticyclonic circulation may be accompanied by subsidence in the left rear quadrant of the jet. The upslope storm or cloudiness occurs while surface pressure over the high plains rises (Boatman and Reinking, 1984).

These anticyclonic circulations can induce the purest of upslope motion, that caused solely by flow over rising terrain. The condensation and stability from the forced lifting can be described by the simple principles of vertical pressure layer displacement and horizontal divergence (e.g., Hess,

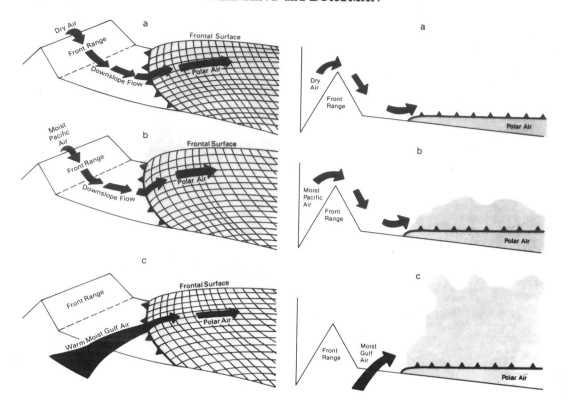

Figure 19.3. Schematic models of upslope anticyclonic storms from two viewpoints. (a–a) Stratus clouds are formed by upslope lifting from meager moisture within the polar air mass; convergence with the downslope flow is weak and/or the downslope flow is dry; the clouds are confined under the frontal surface. (b–b) Frontal lifting of relatively moist *in situ* air leads to cloud development above the polar air mass; these clouds may supplement upslope stratus formed within the polar air. (c–c) Moist air imported from the Gulf of Mexico is lifted by the air mass to generate a deep cloud system possibly with convection; the frontal surface rather than the terrain becomes the primary lifting slope; stratus formed within the arctic air is of secondary importance to the intensity of the storm. (From Boatman and Reinking, 1984.)

1959). Complexities arise in defining the effective slope, however, when consideration must be given not only to the slope of varied terrain, but also to the slopes of the leading front, the top of the arctic air mass, and the isentropic surfaces in overriding air. Complexities of the thermodynamics and dynamics increase accordingly.

In the simplest sense, an orographic component of lifting is induced as the cold air mass spreads viscously, westward and upward over the terrain toward the Rockies. The vertical displacement tends to lessen the thermodynamic stability of the air mass, while the divergence and vertical shrinking of the anticyclone tend to enhance it. The net result most often is to maintain stable air (Whiteman, 1973a). The upslope lifting may generate stratus clouds within the cold air mass from the limited moisture it carries or picks up by evaporation from previously wetted ground (Fig. 19.3a).

Note: 7 ⊾─○ = 270°/15 m·s⁻¹

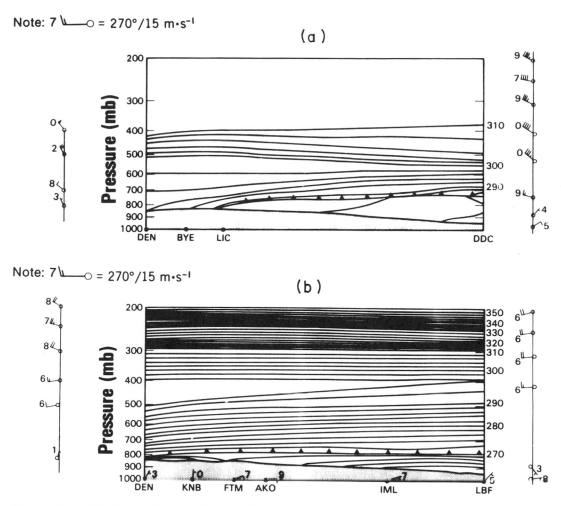

Figure 19.4. Wind soundings and atmospheric cross sections of contoured potential temperature (K) for (a) Denver, Colo., to Dodge City, Kans., on 16 January 1982, and (b) Denver to North Platte, Nebr., on 4 February 1982. (From Boatman and Reinking, 1984.)

Case studies of two upslope events that presented these features were made by Boatman and Reinking (1984). The two arctic anticyclones produced small or trace amounts of precipitation and overcast skies that persisted for 12 to 18 h in each case. The air masses were only 100 mb (1.5 km) thick or less (Fig. 19.4). The upper boundary of the air mass in each case possessed little or no significant slope, except in the vicinity of the surface front. The air was potentially stable above the arctic air masses in each case, and potentially stable to neutral within. As indicated by the strong stabilities and low mixing ratios, the observed upslope clouds were stratiform and confined within the arctic air masses. The cloudiness was induced when northeast- to southwest-oriented horizontal pressure gradient forces produced upslope vertical velocities of the order of 1–2 cm s⁻¹. These velocities were similar to those found in simulated circulations of a cold northeasterly airflow across the Canadian western plains, derived from the mesoscale boundary layer model of Raddatz and Khandekar (1977).

Given situations such as these, the basic shape of the cloud is that of a wedge, very shallow at the leading edge, and increasing in depth because of lowering cloud bases and sometimes rising tops to the east, where the cold dome is deeper (Fig. 19.4). Both the slope of the terrain and the conservation of absolute potential vorticity in the anticyclonic flow contribute to this wedging characteristic (Boatman and Reinking, 1984; Palmén and Newton, 1969).

The upslope cold fronts are often preceded by chinooks or more gentle, but still very dry, westerly, downslope flow along the east flank of the mountains. Convergence and lifting of the dry air by the cold front is most often insufficient to form clouds above the arctic air mass. Time-lapse photographs by the late Doyne Sartor show shallow clouds within the easterly flow of an arctic air mass lapping against the foothills, and being sheared off and evaporated as they are forced upward into a clear, downslope, overriding countercurrent (film library, National Center for Atmospheric Research, Boulder, Colo.). Shapiro (1984) described a case in which west-northwesterly downslope flow prevailed ahead of an upslope front; near-adiabatic descent from the Continental Divide and midday solar heating produce surface temperatures as warm at 15°C, with a temperature-dewpoint spread of $\sim 11°C$, in the lee. The temperature dropped to about $-5°C$ in 2.5 h with passage of the cold front. The sharp front generated a violent dust storm. Forced lifting over the 1.5 km depth of the cold dome was just sufficient to form a line of shallow, broken stratocumuli; these propagated with the front, but produced no precipitation.

Without such preceding or overriding, dry, downslope flow, the upslope cloud systems may be reinforced by frontal lifting of *in situ* moisture present in the warmer air encountered in the polar flow. A deeper cloud mass, with layers above the cold dome, may be expected in this situation (Fig. 19.3b). A cloud mass in arctic air with some enhancement from overriding air within the frontal zone was analyzed in a case study by Walsh (1977). Snow was produced over a wide region of Wyoming by this arctic air mass as it advanced southward over gradually rising terrain. The characteristic wedge form of the upslope cloud was evident from aircraft measurements: the cloud was shallow at the leading edge, but the depth increased in the downslope direction. Cloud base at the leading edge was 2.6 km m.s.l. and lowered downslope; the corresponding cloud tops increased from $\sim$3 km to a maximum of $\sim$4 km m.s.l. in a distance of $\sim$140 km (Fig. 19.5). Winds within the arctic air mass were from the northeast at 8 m s^{-1}; those in the converging air were southwesterly and increased with altitude from 5 to 15 m s^{-1}. Walsh deduced that air ahead of and within the arctic front (the stable zone of concentrated isentropes) moved upward and veered somewhat, out of the plane of Fig. 19.5; the air behind the front also veered into the plane of the figure. Walsh observed undulations in the cloud tops and suggested that these were generated by gravity waves traveling along the interface of the air mass; gravity waves at this interface, and particularly at the nose of

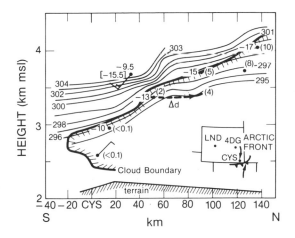

Figure 19.5. Vertical cross section of equivalent potential temperature (K) for an upslope cloud system formed in a continental arctic air mass. The section is perpendicular to the surface front and passes through Cheyenne (CYS) and Douglas (4DG), Wyo. (inset). Winds, temperatures (°C), a clear-air dewpoint (°C, in brackets), and ice particle concentrations (L^{-1}, in parentheses) are noted; Δd and arrow mark a frontal region of transition in the droplet size-concentration spectrum. (After Walsh, 1977.)

the front, have been directly observed with an FM-CW radar and a profiling radiometer (Fig. 19.6). The terrain and upslope wind indicated average upslope lifting at 1 or 2 cm s^{-1}. The warmer air upgliding over the advancing cold dome produced vertical velocities of ~10 cm s^{-1}, as estimated from the southwest winds and the general slope of the θ_e surfaces (Fig. 19.5). Walsh concluded from this analysis and correlated cloud physics evidence (Sec. 19.4) that the lifting of the air within the shallow, stable frontal zone was the major mechanism supplying water to the cloud system.

The atmospheric motions and cloud systems associated with these upslope anticyclones are not necessarily restricted to the arctic air mass and the air in or immediately above the frontal zone. Mountain ranges are known to induce vertical motions in overriding air through depths of the troposphere exceeding by many times the topographic relief (e.g., Ludlam, 1980, p. 372). Similarly, dense arctic air masses may act as barriers to induce perturba-

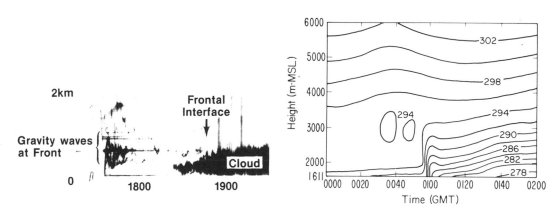

Figure 19.6. Time-height cross section (left) from FM-CW radar, and (right) of potential temperature (K) through an arctic front that passed through Denver, Colo., on 16 January 1982. Front-generated gravity waves are evident from the vertical fluctuations in the radar return (Gossard *et al.*, 1984) and from the upward bulge in the isentropes calculated from microwave radiometer measurements with an atmospheric Profiler (Decker, 1984). The formation of the upslope cloud within the arctic air mass is also shown by the dark area in the FM-CW radar return.

tions at higher altitudes in the overriding flow. The perturbations due to the mountains and the anticyclone are not easily separated in the upslope situation, but in the systems documented by Boatman and Reinking (1984), precipitating cloud layers formed in rising westerlies well above the upslope, easterly anticyclonic flow. These stratiform clouds formed as middle-level moisture was forced upward to condensation, as evidenced by the sloping isentropic surfaces in the middle troposphere (Fig. 19.4).

The frontal cloud-forming processes in shallow upslope circulations may be more dramatically enhanced if a southerly flow of moist Gulf air encounters and overrides the polar air (Fig. 19.3c). In this situation, the leading edge of the cold front and the upper boundary of the cold dome can become the primary lifting slope, and the primary direction of "upslope" flow will be toward the north and east. The cloud system may then become considerably more complex. The features of this type of system begin to blend with those of the upslope cyclone.

19.3.2. Case Study: 24 March 1982

The dynamic processes within the narrow zone of the aforementioned cold front that initially produced a dust storm were analyzed by Shapiro (1984). One cannot say whether Shapiro's case is characteristic of upslope cold fronts, but the high resolution of the observations and the analysis provide a unique look at the frontal processes.

The front advanced southwestward over the Colorado high plains, passed through the PROFS mesonet (a mesoscale, surface, real-time observing network on Colorado's high plains, operated by the Program for Regional Observing and Forecasting Services [Reynolds, 1983]) and past the 300 m instrumented tower of the Boulder Atmospheric Observatory (BAO). The advance

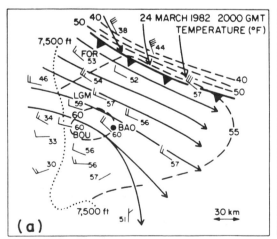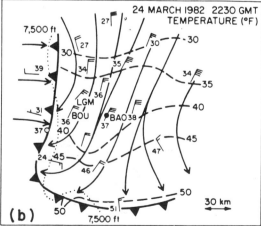

Figure 19.7. Streamlines and surface temperature analysis for (a) 2000 GMT and (b) 2230 GMT, during passage of an arctic front through the PROFS mesonetwork. Black dot marks the location of the BAO tower. Dotted line is 7500 ft (2300 m) elevation. Each long wind vector barb equals 10 kt; short barb equals 5 kt. (From Shapiro, 1984.)

of the front over a 2.5 h period is shown in Fig. 19.7. The confluence of the warm downslope flow and the advancing cold air is evident, as are the strong winds behind the front. This front propagated past the BAO at 17 m s^{-1}. Such a rapid frontal advance would be sufficient to generate temporary blizzard conditions if accompanied by snow.

Shapiro examined the lowest 300 m of the frontal zone (Fig. 19.8). The width of the surface frontal zone was less than 200 m, and passage occurred in about 10 s. The pre-frontal solar heating of the ground produced a superadiabatic temperature stratification in the lowest 50 m where the front nosed over in response to frictional drag (Fig. 19.8a). In the 10 s passage, front-normal wind speeds increased by 6 m s^{-1} as the frontal potential temperature dropped 6 K. The net horizontal convergence generated ascending motion that increased to 6 m s^{-1} between the ground and 300 m in the leading edge of the frontal zone; this was followed by sinking in the trailing edge of -1 to -2 m s^{-1} (Fig. 19.8b). Shapiro found these vertical circulations to be typical in pattern but some 10 times stronger than those observed or simulated by Sanders (1955), Williams (1974), or Keyser and Anthes (1982). Sanders, e.g., reported a minimum frontal width of 25 km and calculated a maximum ascent rate of 0.25 m s^{-1}. However, Browning and Harrold (1970) used Doppler radar to observe passage of a sharp, North Atlantic cold front of ~3 km depth, with horizontal dimensions of only 2 km and vertical motions as strong as $+8$ m s^{-1} within the frontal zone. Theoretically, only turbulent mixing of the frontal circulations will limit contraction of the surface frontal gradients (Shapiro, 1984; Williams, 1974), and such mixing

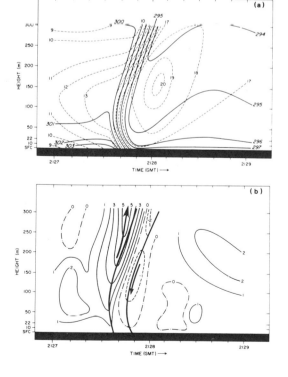

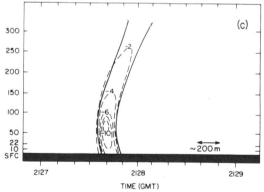

Figure 19.8. The 2 min time-height cross-section analyses for the case shown in Fig. 19.7. (a) Front-normal wind speed (m s^{-1}) and potential temperature (K); (b) vertical motion (m s^{-1}) in vicinity of frontal boundaries (heavy solid lines); (c) total temperature gradient frontogenesis (10^{-4} K m^{-1} s^{-1}) occurring within frontal boundaries. (From Shapiro, 1984.)

was evidently limited in this case by the potential stability that prevailed everywhere except near the ground at the front (Fig. 19.8a).

Cross-front confluence strengthened the thermal gradient throughout the front. The differential vertical motion, or tilting, as influenced by the potential stability, was frontogenetic below ~ 100 m and frontolytic above that level. Shearing deformation was negligible. The summed effect, net frontogenesis, is given by

$$\frac{d}{dt}\left(\frac{\partial \theta}{\partial y}\right) = \frac{\partial v}{\partial y}\frac{\partial \theta}{\partial y} - \frac{\partial w}{\partial y}\frac{\partial \theta}{\partial z} \,. \qquad (19.1)$$

It was observed throughout the 300 m depth of the frontal zone, especially below ~ 100 m where air parcels were being ingested into the frontal zone from the warm side (Fig. 19.8c); the processes were the same but the frontogenetic tendencies in this case were 10^2 times greater than those calculated for the case study by Sanders (1955). Browning and Harrold (1970) did not analyze frontogenesis but noted substantial horizontal convergence and deformation associated with the front they observed (convergence of $\sim 10^{-2}$ s^{-1} over 500 m vertically and horizontally, deformation up to 80×10^{-5} s^{-1} over a zone 10 km wide).

19.3.3. Mesoscale Terrain Effects

The advance of these fronts over terrain is not always as clear-cut as shown in Fig. 19.5. The coarse, synoptic-scale features of a surface front and temperature field are shown at 6 h intervals in Fig. 19.9a, for the 3–4 February 1982 case studied by Boatman and Reinking (1984); this scale of analysis does show how the advance of the arctic air was blocked by the Rocky Mountains. Figure 19.9b shows the mesoscale features of the temperature field and front over the PROFS mesonet; the corresponding terrain is depicted in Fig. 19.10. In Fig. 19.9b, areas influenced by winds with an easterly, upslope component are shaded. Westerly winds were prevalent over the mountains (at the left of each figure) at all times. Over the plains, the surface winds and the advancing front responded to local terrain to the north of the depicted mesoscale area. The arctic air first climbed southward toward Cheyenne (note the terrain in Fig. 19.5). It then spilled southward into the depicted mesonet, and in some portions of this downslope flow it developed the winds with westerly components indicated by the unshaded areas behind the contorted mesoscale cold front. The cold air over the mesonet subsequently deepened and climbed over the Palmer Divide south of Denver (DEN, Figs. 19.9 and 19.10). Only then did the cold air mass completely cover eastern Colorado, such that the patterns with synoptic and mesoscale resolution reached similarity. The mesoscale contortions of the airflow can significantly affect the precipitation pattern.

19.4. The Upslope Winter Anticyclone—Precipitation Physics

The nature of even the shallow upslope stratiform cloud layer is influenced by complex microphysical variables and processes. These processes act

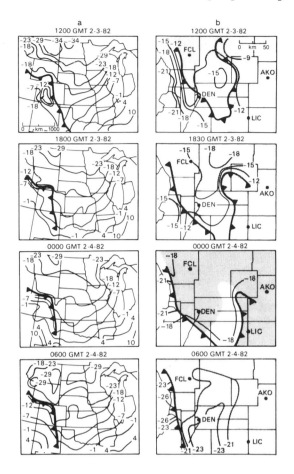

Figure 19.9. (a) Sequence of surface temperatures and movement of an arctic front, as depicted by synoptic-scale data. (b) Approximately the same time sequence, depicted by mesoscale data from the PROFS mesonetwork; shading represents areas where wind had an easterly, upslope component. (From Boatman and Reinking, 1984.)

in concert with the dynamics to determine the cloud microstructure and colloidal stability, and the precipitation rate, amount, duration, and efficiency. The variables of primary microphysical importance are, of course, the lifting, the available moisture (or moisture flux), and the temperatures within the layers where the clouds form. The evolution of cloud and precipitation is then determined by how much liquid water is provided by condensation, how the cloud droplets become distributed by size and concentration, whether the cloud layer is warm or supercooled, what mechanisms are active for conver-

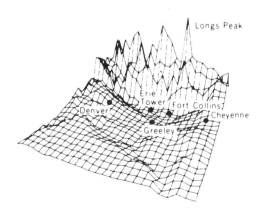

Figure 19.10. Perspective of the high plains terrain between Cheyenne, Wyo., and Denver, Colo., backed by the Rocky Mountains; horizontal grid spacing is ~10 km; total vertical relief is ~3 km. (From Chaumerliac *et al.*, 1983.)

sion of water vapor and liquid to ice, what concentrations and types of ice particles are generated, and how much growth these ice particles attain in their fall through the cloud and to the ground.

19.4.1. Temperature Climatology

Climatologically, and in most individual cases, thermal stability prevails within the cloud, so the precipitation is little affected by convection (Sec. 19.2); processes at the cold front, and springtime surface heating may provide exceptions. The mean upslope cloud base temperatures range from about +4°C in the southern high plains to about −8°C in the northern high plains; the 0°C cloud-base isotherm runs east-west through the Texas Panhandle (Whiteman, 1973a). Thus, in the mean, clouds north of Texas will be entirely supercooled, but may be fed with cloud droplets that have had time to grow in the warmer-than-freezing downslope reaches of the circulations. Cloud top temperatures in the nonprecipitating systems average about −8°C to −10°C; only slightly more cloud top cooling (≥ 1.5°C) provides a climatological transition to precipitating systems (Whiteman, 1973a). This might be explicable in terms of the temperature-dependent process of heterogeneous ice crystal nucleation (i.e., nucleation on particulate matter); however, there are wide, case-by-case, warm and cold deviations from the mean in the cloud top temperatures of both the nonprecipitating and precipitating upslope clouds (Whiteman, 1973a; Weickmann, 1981).

At the nonprecipitating extreme, a cloud only ∼1 km deep extending from Denver to North Platte was observed to be isothermal through its depth. Despite supercooling, it contained no ice particles but was composed entirely of cloud drops ∼10 μm in diameter and smaller (Reinking, flight log book, 19 December 1980. A forward scattering laser spectrometer probe was used to detect and measure the droplets; Knollenberg, 1981). The moisture flux and dynamic forcing were evidently insufficient to unsettle the observed colloidal stability.

19.4.2. Snow Crystal Generation and Growth in Seeder-Feeder Couplets

The mechanisms that inhibit or induce precipitation from the clouds in anticyclonic flow are well illustrated by some case studies. The two shallow systems sampled by Boatman and Reinking (1984) were producing light to very light precipitation. In both cases, totally supercooled upslope stratus formed within the arctic air mass. In both cases a second supercooled layer of cloud formed well above the upslope cloud, in the overriding countercurrent that brought its moisture from a totally different source (Figs. 19.4a, b; Sec. 19.3; and figures in Boatman and Reinking). None of the cloud layers exceeded 1 km in depth. The primary determinants of similarities and differences in the microphysical processes and the consequent precipitation in these cases were (1) the available moisture in the cloud layers, (2) the cloud temperatures, and (3) the saturation levels with respect to both water and ice between the upper and lower cloud layers.

The presence of liquid water in each of the layers was documented by very light but persistent aircraft icing, by collection of rime ice for cloud

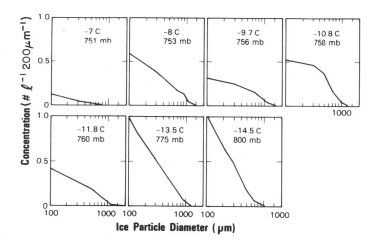

Figure 19.11. Ice particle generation in the lower, upslope cloud in the case of 16 January 1982 (Boatman and Reinking, 1984.) Note the increase in ice particle concentration downward, from cloud top (−7°C) to cloud base (−14.5°C). The maximum total concentration was about 4 L^{-1}.

chemistry measurements (Nagamoto et al., 1983), and with a forward scattering laser spectrometer. The water contents of the lower, upslope clouds were each consistent with the small amounts of water vapor confined within the respective arctic air masses.

The temperatures within each cloud layer were found to be sufficiently cold to generate ice crystals by primary, heterogeneous nucleation, which was indeed the most likely source of the ice, given (1) the characteristic temperature dependence of ice particle concentrations, and (2) the lack of moisture for the droplet growth required for some mechanisms of ice particle multiplication. The measurements clearly showed, e.g., the dependence of the ice generation on temperature through a −14.5°C to −7°C inversion within the warmer upslope cloud in one of the arctic air masses (Fig. 19.11). In Case A (Fig. 19.4a) the upper cloud was dramatically colder (about −25°C to −30°C), whereas in Case B (Fig. 19.4b), the temperature was −15°C ± 3°C in both the upslope and overriding cloud layers. All the cloud layers were saturated with respect to water. The air between the cloud layers in Case A was ice-supersaturated and nearly water-saturated. The air between the layers was far from saturation in Case B, but the air beneath the upslope cloud was nearly water-saturated and completely ice-supersaturated. The resulting patterns in ice particle growth illustrate the significance of this information with regard to the physical principles of ice crystal growth (Hobbs, 1974).

Once nucleated, the ice particles in each case grew predominantly by vapor deposition. The cloud liquid water contents were insufficient to promote significant growth by riming. In Case B, the smallest ice particle sizes were observed at the top of each cloud layer, crystal growth was evident in both the upper and the lower layers, and beneath the lower cloud. Depositional growth habits and rates of ice crystals are temperature dependent, and growth is most rapid in the $-15°C \pm 3°C$ temperature regime (Magono and Lee, 1966; Ryan *et al.*, 1976). Thus crystals of stellar or dendritic growth habits, some as large as 2000 μm, formed in the very shallow upper cloud deck; the total concentration at the base was 0.1 L^{-1}.

These crystals began to sublime as they fell into the unsaturated clear air between the cloud layers, but crystals with the largest initial diameters survived settling through the 1-km-deep clear layer where ice saturation ratios were only ~0.1–0.6. Nucleation produced new crystals beginning at the upslope cloud top; small crystals were present throughout this cloud and concentrations increased downward from ~0.2 L^{-1} to 0.8 L^{-1}. The habits of the new crystals were platelike, consistent with the slightly colder temperature regime in the lower cloud ($-16°C$ to $-18°C$). Crystal aggregation was minimal, and in accord with a snowflake collection efficiency of less than 0.01 for temperatures colder than $-13°C$ (Pruppacher and Klett, 1978). The observed water vapor levels permitted depositional growth to continue between the lower cloud and the ground. The storm produced a total of 2.0 to 2.5 cm of snow in 18 h, in the area of observation (Boatman and Reinking, 1984).

The atmospheric temperature structure and the consequent ice crystal generation were slightly more complex in Case A. Only very small ice particles, indicative of very recent nucleation, were detected at the top of the very cold upper cloud. The growth habits of crystals observed falling from this layer, bullet or columnar combinations, were again consistent with the formation temperature regime. Growth occurred both within the upper cloud and in the clear but ice-saturated air below.

With the ice saturation ratio as high as 1.3 between the cloud layers, additional nucleation was possible even in the clear air. Many crystals nucleated here by deposition and developed platelike and stellar growth habits consistent with the $-10°C$ to $-20°C$ region. However, the liquid water content in the cold upper cloud, and the saturation levels between the clouds, were evidently insufficient to produce very many large crystals with fall speeds great enough to traverse the intercloud distance. Competition for the limited available water resulted mainly in small crystals with small fallspeeds, albeit a few 800 μm crystals reached and seeded the upslope cloud layer (Fig. 19.11, 751 mb).

Aggregates as large as 2000 μm formed between $-6.8°C$ and $-8°C$ within this upslope cloud; this was the warmest part of the cloud system, and was within the temperature regime that appears to create relatively high snowflake collection efficiencies (Pruppacher and Klett, 1978). At lower but colder altitudes, the concentration remained the same, suggesting no further aggregation. The aggregation evidently occurred mainly in the top of the cloud layer (Fig. 19.11, 753 mb), and nucleation of new (initially small) crystals increased downward (Fig. 19.11, 756–775 mb); growth of these new crystals was limited by the small amounts of available liquid water, despite temperatures appropriate for rapid depositional growth. This combination of aggregation and nucleation evidently stablized the mean particle sizes within the upslope cloud at ~400–500 μm; it also generated a broader mean size spectrum than did the upslope cloud of more uniform and colder temperatures in Case B. Again growth continued below cloud base; ice particles observed near the ground in Case A occurred as 1–2 mm single crystals and aggregates. Despite the large sizes, the crystal generation rate was slow, and a snowfall of "only several mm" was produced by this storm (Boatman and Reinking, 1984).

Of major importance from these studies is the illustration of the "seeder-feeder" phenomenon, by which the lower cloud is microphysically stimulated by and feeds moisture to the natural ice "seeds" supplied by the upper cloud, to produce precipitation. Given these two cases with very similar dynamics and outwardly similar middle-level and upslope cloud layers, the differences in the moisture and temperature structure in these weakest of precipitating anticyclonic systems were sufficient to cause some significant variations in ice nucleation rates and crystal growth processes in and between the cloud layers. The seeder-feeder process evidently did not significantly affect precipitation duration or amount in these cases; however, in systems with stronger sources of moisture, this process becomes more important to depositional, riming, and aggregational modes of crystal growth and to the resulting precipitation. From analyses of varied cloud systems, Raddatz and Khandekar (1979), Matejka *et al.* (1980), and Bader and Roach (1977) all specified that substantial and replenished liquid water in the underlying cloud was critical for seeder crystal riming and an effective seeder-feeder couplet.

A quite different, dynamically induced seeder-feeder couplet was observed by Walsh (1977). This coupling is between the physically merged cloud volumes formed in the upslope flow and in flow overriding the cold front. Differing nucleation and growth of the cloud droplets resulted from the two airflows in the case study (Fig. 19.5). The slow lift from the upslope flow produced very low supersaturations, small droplet concentrations ($\sim$55 cm^{-3}), and narrow droplet spectra, well within the cold air mass and the upslope cloud. Here spectral broadening was limited by the low available moisture, despite the long fetch and time available for droplet growth. The liquid water content within the upslope cloud was less than 0.02 g m^{-3}. The measured droplets were still fewer in number ($\sim$30 cm^{-3}) and smaller, and the spectra were narrower (mean diameter and standard deviation, $\bar{d} \pm \sigma \simeq 6\mu\text{m} \pm 2\mu\text{m}$) at the outer edge of the saturated, overriding air in the frontal zone. However, penetrating deeper into the cloud (along the dashed arrow in Fig. 19.5) but remaining within the frontal zone, progressively larger droplets, droplet concentrations, and liquid water contents were measured. (Along the arrow, the concentrations increased steadily from $\sim$30 cm^{-3} to 140 cm^{-3}, $\bar{d} \pm \sigma$ increased from approximately $6\mu\text{m} \pm 2\mu\text{m}$ to $10\mu\text{m} \pm 7\mu\text{m}$, and the cloud liquid water content increased from $\sim$0.005 to 0.102 g m^{-3}.) Thus, the overriding air provided lifting and a flux of moisture that was more significant to the system than that carried by the upslope flow.

Ice crystals were observed in concentrations about an order of magnitude larger than expected from nominal background nucleation 10 L^{-1} at $-17°$C, 2 L^{-1} at $-13°$C, < 0.1 L^{-1} at $-10°$C (Cooper and Vali, 1976; and Fig. 19.5). Elevated supersaturations may have been caused by the frontal lifting and assisted by the gravity waves noted in Sec. 19.3. The supersaturations would have promoted efficient ice nucleation; also, mixing between the frontal zone and the quieter upslope air may have led to some in-cloud accumulation of the ice crystals, as suggested by Cooper and Vali.

The $-13°C$ to $-17°C$ temperature regime along the frontal zone (Fig. 19.5) in combination with the frontal condensation and liquid water flux, was ideal for promoting rapid dendritic ice crystal growth. These crystals grew to 3000–4000 μm diameters and then settled through, and collected additional moisture from the upslope cloud, moisture that was not likely to be efficiently removed by processes within the upslope cloud mass alone. The gentle lifting of the upslope air provided little resistance to the crystals, which settled to the ground.

19.4.3. Formation of Graupel in Shallow Stratus

Overall, the temperatures of clouds in upslope systems, and the habits of the ice crystals formed, vary throughout the possibilities for arctic, polar, and less frigid air masses; the water that the crystals precipitate by depositional growth can vary accordingly. Another mechanism for water extraction from these clouds is ice particle riming. Crystals or large frozen drops that not only collect vapor by deposition but also collect cloud droplets by accretion can significantly increase the efficiency of precipitation.

Graupel formation in deep cloud systems with ample liquid water and substantial supercoolings is quite feasible. However, small graupel ("snow pellets," "snow grains") is commonly observed precipitating from the shallow clouds of upslope anticyclones, and particularly from relatively warm clouds spanning a temperature range no wider or cooler than $-6°C$ to $-10°C$ (Weickmann, 1981). A cloud with a sufficient water vapor flux is required to make the droplets grow enough to become available for accretion. This condition can generally be met, even though the liquid water contents have been low in some of the sampled upslope clouds. Formation of fast-falling and droplet-collecting graupel embryos by the freezing of large drops is unlikely; large vapor-grown drops are not formed in the anticyclonic upslope systems, and as for the relatively warm clouds, Pruppacher and Klett (1978) presented evidence that drops smaller than about 2000 μm diameter will not regularly freeze as bulk water at $-10°C$ or warmer. However, crystals with growth habits indicating vapor deposition are regularly nucleated in shallow clouds with tops around $-10°C \pm 3°C$, although the rates of nucleation may be small. This is evident, for example, from cloud temperatures and corresponding forms and fluxes of snow crystals in upslope precipitation described by Weickmann (1981). The graupel produced by these clouds is often small (~ 1 mm), and the ice particle kernels that initiate this graupel are so small that Weickmann suggested that these graupel form without such an embryo. While debate as to the mechanism of graupel formation has ensued, the works of Fukuta et al. (1982, 1984) offer a most convincing explanation.

In their supercooled-cloud tunnel, these investigators grew crystals isothermally between $-4.5°C$ and $-18°C$ (growth habits agreed with previous work of Ryan et al., 1976, and others). They observed that the isometric

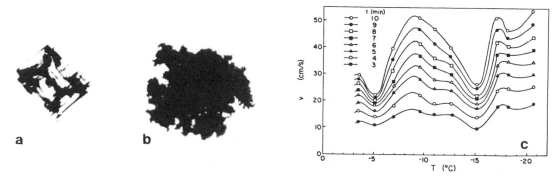

Figure 19.12. (a) Isometric ice crystal, formed at −9.5°C, after 10 min growth, 0.14 mm long. (b) 1.05 mm graupel, formed from crystal of the isometric type, after 25 min growth. (c) Ice crystal fall velocity plotted as a function of temperature for increasing growth times; the peak between −8°C and −11°C is for the isometric crystals of the type shown in (a).

crystals (*a*-axis = *c*-axis, short columns or thick plates) that form near −10°C have a tendency to collect droplets and become graupel (Figs. 19.12a, b), whereas the depositional growth of more asymmetric ($a \gg c$) crystals such as dendrites grown at −15°C outpaces the riming, so the depositional crystal identity is retained. Further, the terminal fallspeeds of the isometric crystals were found to be the fastest of the various types. Although the fall velocities of crystals of all habits increased with depositional growth time, the rate of this increase was found to be greater and more sustained for the isometric crystals in the −10°C regime than for crystals growing in regimes some 3°C warmer or colder (Fig. 19.12c). The shape and relatively high density of the isometric crystals contribute to this difference.

Finally, Fukuta *et al.* (1984) found that the enhanced fallspeed increases riming efficiency, such that the growth mode switchover from the depositional growth (where mass $m \propto t^{3/2}$) to riming growth of "spherical" particles (where $m \propto t^6$) happens in the shortest period of time (∼1.7 min), the dense isometric crystals growing at −10°C. These crystals grow much more slowly by deposition than crystals in regimes either 5°C warmer or 5°C colder (Ryan *et al.*, 1976), but more rapidly by riming. Thus the earliest graupel development, and on the smallest of kernels, is expected with crystal growth at and near −10°C.

Given an upslope cloud 1 km deep with a top temperature of about −10°C, an isometric crystal formed at the top and falling at the relatively rapid rate of 50 cm s^{-1} (Fig. 19.12c) will have ample time (∼33 min, or more with lifting) to collect cloud droplets and become graupel. This hypothesis for the frequent occurrence of graupel in precipitation from shallow upslope clouds with a −6°C to −10°C regime appears to be quite satisfactory.

19.4.4. Snow Crystal Flux and Precipitation Rate

The precipitation rates of upslope systems vary widely. Whiteman (1973a) found that, in the mean, the precipitation intensity from upslope systems increases with cloud thickness (Fig. 19.13). The precipitation rate

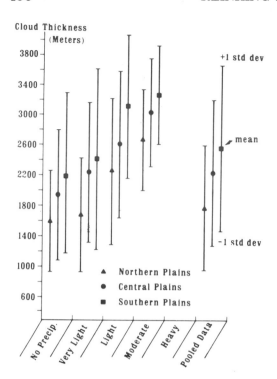

Figure 19.13. Precipitation intensity as a function of upslope cloud thickness for three regions of the high plains. (From Whiteman, 1973a.)

must equal or be less than the condensation rate, and will be influenced by the rates of nucleation and growth of the ice particles. Weickmann (1981) adapted a dew scale to measure the mass rate of snowfall from upslope storms; he also measured the crystal concentrations (and habits) precipitated per unit time by collecting and photographing them on black velvet. Thus, he was able to correlate short-term variations in the precipitation rate with the snow crystal flux [number/(10 s)(100 cm^2)] and the average crystal mass. He found that the time-variations of precipitation rate and snow crystal flux were parallel in some storms (Fig. 19.14a); in others, these two parameters were not directly correlated (Fig. 19.14b).

Upper-air data for these cases are incomplete, but the suggestion from Weickmann's data is that the precipitation rate and ice particle flux are parallel for the shallow upslope systems and nonparallel for the deep systems of those with seeder-feeder couplets. For example, the cloud in the parallel case in Fig. 19.14a was some 1600 m deep, and the top was marked by a definite inversion with dry air above. In contrast, the cloud system for the nonparallel case in Fig. 19.14b was at least 4 km deep, and cloud top above that level was indefinite. Weickmann also mentioned a seeder-feeder couplet with nonparallel crystal flux and precipitation rate (10 February 1981). The necessary deduction here is that, given a certain ice nucleation rate, the snow crystal flux and precipitation rates are driven by the condensation rate and therefore by the mesoscale flow and lifting. The less complex this is, the better the evidence of parallelism.

Thus, for example, the systems with single sources of moisture and lifting, and a small range of temperature through the cloud depth, would show the

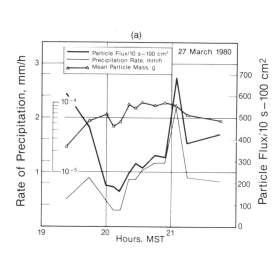
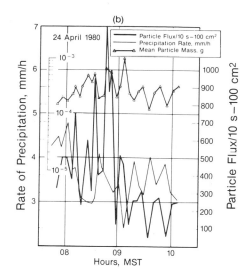

Figure 19.14. Precipitation rates and snow crystal fluxes (a) varying in parallel and (b) showing uncorrelated variation, during two upslope storms of differing character. The mean ice particle mass is also shown. (From Weickmann, 1981.)

parallel feature of precipitation rate and snow crystal flux. This would be most common in the anticyclonic systems. Seeder-feeder couplets, however, can develop precipitation from quite independent air flows and sources of moisture. Although the precipitation rate and crystal flux might vary in parallel for the feeder cloud, that relationship would be obscured at the ground by the mix of the couplet. Within deep cyclonic storms, the wind fields, moisture sources, and lifting processes vary considerably, so again the parallel relationship would become obscured. The effects of a wider temperature regime for ice generation is added to this, so that if the cloud mass is not uniformly deep and moisture is not uniformly fed to the whole layer, the crystal fluxes from the different temperature regimes will not vary in unison.

19.5. The Upslope Winter Cyclone—Kinematics and Dynamics

19.5.1. Synoptic Features

The high plains upslope storm driven by a winter cyclone is characterized by a deep, cold-core trough or closed low at 500 mb with strong dynamic support throughout the troposphere. The core of the polar front jet stream dips over New Mexico and Arizona, or even into Mexico. This dynamic support may be enhanced by effects of the high plains and Rocky Mountain topography on the flow, and by latent heat released as substantial moisture is drawn northward from the Baja California region and/or the Gulf of Mexico. Cyclogenesis occurs as the center of the surface cyclone tracks eastward near the Utah-Arizona and Colorado-New Mexico borders, leaves the mountainous terrain, and enters the lee of the Rockies (Fawcctt and Saylor, 1965). The circulation of the surface cyclone in these storms draws polar air southward and westward (upslope) over the high plains to meet moist Gulf

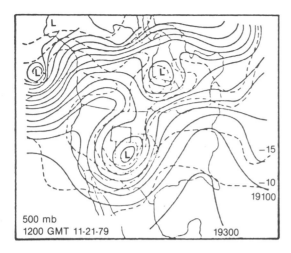

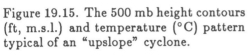

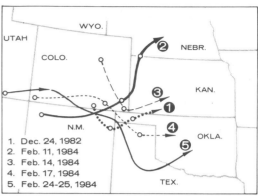

Figure 19.15. The 500 mb height contours (ft, m.s.l.) and temperature (°C) pattern typical of an "upslope" cyclone.

Figure 19.16.   Tracks of some recent storms that produced either heavy snowfall or blizzard conditions, or both, on the high plains. (After Henz, 1984.)

air. The clash of these air masses can set the stage for heavy precipitation events. But even in these situations, the location, amount, and intensity of the precipitation, and the driving winds in relation to the high plains topography depend substantially on the location of formation of the surface low-pressure center, its track and moisture supply, and its rate of movement in conjunction with the upper-air trough. Figure 19.15 shows an example of the 500 mb circulation of a storm that produced 25–66 cm of snow along the east flank of the Rockies (Auer and White, 1982). The movement of this storm slowed for some time as the closed-off circulation became nearly vertical through 300 mb, thus contributing to the heavy snowfall. On Christmas Eve in 1982, a blizzard in eastern Colorado was generated by a very similar trough (Schlatter *et al.*, 1983); that storm set a new record for 24 h snow accumulation in Denver—60 cm—breaking the old record of 58.4 cm set on 25 April 1885. The snow contained 5.1 cm of water. A comparable amount of water (4–5 cm) was produced from less snow (30–45 cm) in a synoptically similar storm along Colorado's Front Range on 20–21 April 1984.

Fawcett and Saylor (1965) provided a climatology of synoptic circulations and storm tracks. They noted that cyclogenesis is typical of the late winter/early spring season; however, occurrences throughout the winter are common. The tracks of a few recent surface cyclones are illustrated in Fig. 19.16. The localized effects of these storms were as varied as their tracks. The 1982 Christmas Eve blizzard generated winds in excess of 25 m s^{-1}, along with the heavy snows, in northeastern Colorado. The 11 February 1984 storm with a more northerly track also generated high winds (>20 m s^{-1}) and drifting snow in northeastern Colorado, but more moderate snowfall in Denver (5–15 cm). Comparable snowfall but no blizzard came with the storm of 24–25 February 1984. Not infrequently, a cyclone will leave the Denver area

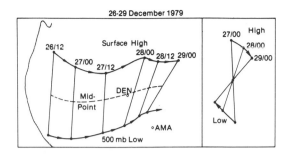

26-29 December 1979

Figure 19.17. Tracks of surface anticyclone and upper-level cyclone (left) and movement of their centers relative to the midpoint, in a 2 day period. Denver, Colo. (DEN), and Amarillo, Tex. (AMA), are noted. (From Riehl and Reinking, 1981.)

relatively unaffected, but will impact the high plains some greater distance from the mountains.

A regional forecaster suggests that lows with surface intensification beginning in the Four Corners area most commonly bring moist easterly winds and heavy wet snows, but not blizzards, to eastern Colorado (Henz, 1984). The snowfall can continue in the high plains for 2 or 3 days; the wind gradually shifts to more northerly directions, and the heaviest precipitation area shifts somewhat southward as the storm center tracks eastward. Prediction may be enhanced because the upper-air trough moves slowly, spending some 48–72 h over the Utah-Arizona area. The rate of progression of the aforementioned storm of this type, shown in Fig. 19.15, was about 22 km h^{-1}.

In storms that generate blizzards, the upper-level trough is often initially fast moving, and the surface lows more often develop rapidly (in 12–18 h) in the southeast Colorado/Oklahoma-Texas Panhandle area (Henz, 1984). In the case of the Christmas Eve blizzard, surface pressures fell rapidly [4–5 mb (3h)$^{-1}$] and a surface low (990 mb) formed in southeast Colorado in ∼ 13 h (Schlatter *et al.*, 1983); within just a few hours, a surge of cold air from Montana, generating 4–5 mb h^{-1} pressure rises, merged with the circulation of the southeast Colorado low, and weather conditions over the Colorado plains deteriorated rapidly. Once formed, the surface cyclones that bring blizzards tend to track more slowly eastward (Henz, 1984).

Easterly or northeasterly winds are generated when these surface lows form or track relatively far south and west. Upslope flow with the easterly components develops ahead of the surface low and continues as the low pressure center tracks eastward. A Denver sounding taken during the Christmas Eve storm was saturated to ∼350 mb and showed a very deep upslope flow, with winds veering from northerly to easterly between the surface and ∼275 mb (Schlatter *et al.*, 1983). As in the Christmas Eve case, the onset of precipitation following the time of lowest pressure is typical of upslope events. Also, the beginning of snowfall in the upslope region during the Christmas Eve blizzard, and in the case studied by Auer and White (1982), coincided with closing of the 700 mb circulation, which signaled strong convergence and lifting at the surface.

Another synoptic feature that offers potential as a forecasting tool is the coupled eastward movement of the 500 mb low and the dense surface high that accounts for the lee-side cold surge. Such a couplet is tracked in Fig. 19.17. After 1200 GMT on the middle day of observation, the anticyclone accelerated relative to the upper low; this relative motion (Fig. 19.17, right) led to a persistent upslope flow in the lee of the Rockies (Riehl and Reinking, 1981).

19.5.2. Mesoscale Features and Terrain Effects

During the events represented in Fig. 19.17, mesoscale features of the temperature field and turning of the wind field to upslope directions in the upper cold low were identified, using an aircraft and standard synoptic information (Fig. 19.18). It appears that a stable, shallow, mesoscale wave was traveling along the polar front or at least perpendicular to the temperature gradient. This wave apparently passed between rawinsondes unnoticed. The aircraft encountered turbulence at 700 mb, within the core of the wave; measured vertical gusts were about ± 1 m s^{-1} ($w'_{rms} = \pm 0.5$ m s^{-1}). This may have been a slightly unstable area, where mesoscale mixing and precipitation processes were relatively more stimulated than in the wide area of greater stability (Riehl and Reinking, 1981). Surface observations indicated that the wave marked the forward edge of the upslope snowfall.

The estimated area-average, water-equivalent precipitation for this entire storm of 24 h duration was about 1.8 cm, of which about 0.6 cm fell in the first 6 h of snowfall. Conversion of the aircraft-measured 0.25 g m^{-3} cloud liquid water content to ice every 30 min would have yielded the 0.1 cm h^{-1} precipitation rate. A crude estimate of the precipitation-generating vertical motion in this system can be made from this rate, the observed cloud depth of 200 mb (~ 2 km) early in the storm, and the saturation specific humidity (q_s) from standard soundings.

The condensate C depends on the time rate of change of q_s. Assuming that

$$\frac{dq_s}{dt} = w\frac{\partial q_s}{\partial z} \, , \tag{19.2}$$

where w is the vertical motion and z is height,

$$C = \int \rho w \frac{\partial q_s}{\partial t} \delta z \tag{19.3}$$

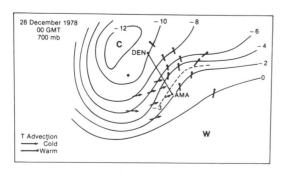

Figure 19.18. Cyclone 700 mb isotherms and their intercepts with streamlines. The line between Denver and Amarillo indicates the path of a research flight. (From Riehl and Reinking, 1981.)

per unit surface area. The change in q_s over the depth of the cloud was $\sim 3 \times 10^{-3}$ g g^{-1}, which is verified by the approximately equal, measured cloud liquid water content. Thus, integrating over the cloud depth gives,

$$C \simeq 3 \times 10^{-6}\overline{w}E(\text{g cm}^{-2}\text{s}^{-1}) , \qquad (19.4)$$

where E is the precipitation efficiency and $\overline{w}$ is the vertical average of w. Equating condensate to precipitation P, gives $\overline{w} \simeq 10$ cm s$^{-1}/E$. When $P = 0.1$ cm h^{-1}, $\overline{w} \simeq 10$ cm s^{-1} for $E = 1 (20$ cm s^{-1} for $E = 0.5$). Simple calculations indicate that this motion could be generated by low-level convergence of 1 or 2×10^{-4} s^{-1}. The deduced ascent rate is much faster than the 1 cm s^{-1} forced by the slope of the terrain and a 10 m s^{-1} upslope wind. Thus the terrain-forced upslope ascent over the high plains appears to be negligible in the setting of this cyclonic precipitation.

A different deduction and a useful computation technique are offered by Auer *et al.*(1980) and Auer and White (1982) for the cyclonic case of 20–21 November 1979 (Fig. 19.15). They first computed the vertical motion on the storm scale from interpolated rawinsonde data. An area for computation was established by centering a 1110 km (600 n mi) square box on the storm, and dividing it to contain four storm quadrants. The box was oriented and propagated with the storm. Vertical mass transport at several pressure levels between the surface and 300 mb was determined by following tubes of the conserved property, equivalent potential temperature (θ_e), through intersections with each pressure surface, and considering the mass accumulations due to the horizontal mass fluxes normal (M_n) and parallel (M_s) to the propagating box. Surface topography was accounted for. The average horizontal and vertical mass flows and vertical velocities over 24 h are shown in Fig. 19.19.

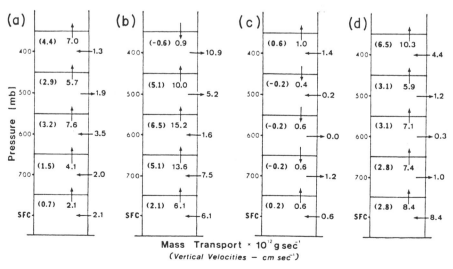

Figure 19.19. The 24 h average horizontal and vertical mass flow, and corresponding vertical motions, in the four quadrants of the cyclonic storm of 20–21 November 1979: (a) left rear, (b) left front, (c) right rear, (d) right front. (From Auer and White, 1982.)

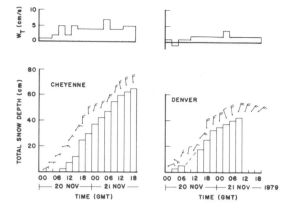

Figure 19.20. Snowfall accumulation with time and prevailing 3 h surface winds at Cheyenne and Denver, and corresponding terrain-induced vertical velocities W_t for the cyclonic storm of 20–21 November 1979. (After Auer *et al.*, 1980.)

The heaviest snowfall, with respect to the cyclone center, fell from the northeast (leading left) quadrant of the storm. In this quadrant, area average vertical velocities due to the cyclone dynamics were 2.1 to 6.5 cm s^{-1} below the 400 mb level. The maximum upward mass transport and vertical motion in the storm took place here at the level of nondivergence, 550 mb. The mean 550 mb temperature in this left front quadrant was -16°C; the coincidence of this temperature with the maximum vertical motion was a key to the heavy snowfall (see Sec. 19.6).

Auer *et al.*(1980) estimated the vertical velocities forced by mesoscale terrain during this storm, and compared them with the rates of accumulation of snowfall at Cheyenne and Denver (Fig. 19.20). Both cities received the most snowfall while winds were predominantly northerly, rather than easterly. The terrain north of Cheyenne (Fig. 19.5) lends more upslope inducement to northerly winds than does that around Denver (Fig. 19.10); this is evident in Fig. 19.20. The upslope winds generated maximum vertical velocities of about 2 cm s^{-1} near Denver but 4–7 cm s^{-1} near Cheyenne. The upslope-induced motions in the vicinity of Denver were of little consequence relative to the stronger ascents due to the cyclone dynamics. However, the horizontal winds of 15–20 kt (8–10 m s^{-1}) produced meaningful vertical motions around Cheyenne, where more total snowfall accumulated at greater rates than around Denver. Near Cheyenne, at least, it is likely that the upslope component of vertical motion contributed significantly to cloudiness and precipitation within the cold surge behind the surface front and became relatively more important as the upper-air support diminished. This was suggested by Auer *et al.* (1980) who noted that the precipitation rates late in the storm were about one-third those early in the storm at both Cheyenne and Denver; the later precipitation rate of ~0.8 cm h^{-1} is comparable with that not uncommonly observed during upslope anticyclonic flow. Indeed, Auer *et al.* suggested that the snowfall late in this cyclonic system was due to upslope components alone.

Raddatz and Khandekar (1977, 1979) examined upslope-enhanced storms over the Canadian western plains. They compared the results of a mesoscale numerical model of a "typical warm-front-type occluded cyclone" with actual

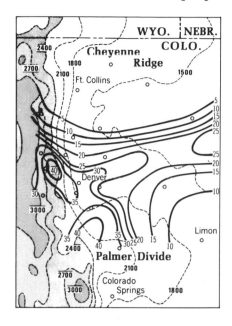

Figure 19.21. Snow depth (inches) from the 1982 Christmas Eve blizzard measured within the PROFS mesonetwork of northeast Colorado. (After Schlatter *et al.*, 1983.)

precipitation, for terrain that is somewhat more varied and locally steeper than the high plains of Colorado. The terrain-induced vertical velocities, enhanced by the mesoscale heat and momentum fluxes of a September storm, were estimated to be about 15 cm s^{-1} over the steeper terrain, and considerably stronger than vertical motions associated with the synoptic-scale system. Here the situation of conditional instability came into play, thus differentiating this case from those typically much more stable cases of mid-winter.

Overall, the potential effects of topography on the winter cyclone should not be belittled. If this point is not clear already, it is made lucid by the Schlatter *et al.* (1983) mesoscale analysis of Colorado's Christmas Eve blizzard. Denver was in the northwest quadrant of the storm, and synoptic-scale lifting was considerable. However, satellite imagery shows bare ground in southeast Wyoming where winds, after blowing across the Cheyenne Ridge, produced a downslope flow (Fig. 19.10, right side; and Fig. 19.21); only 2–5 cm of snow fell there and subsequently sublimed in the downslope winds. Only light snow and some blowing dust were observed slightly farther south. Also, the south side of the Palmer Divide (Fig. 19.10, left side; and Fig. 19.21) received much less snow than its north slope, where the wind was forced to blow uphill. The windward slopes received up to 112 cm (44 inches) of snow. A band of the heaviest snowfall was enhanced by the west-southwestward topographic rise toward Denver, upslope along the Platte River Valley (Fig. 19.21; and Fig. 19.10, diagonally across foreground).

19.6. The Upslope Winter Cyclone—Precipitation Physics

The microphysical processes within the upslope cyclonic systems, just as in the anticyclonic systems, are unique only inasmuch as cloud development from terrain-forced lifting is important. The importance of that has already been demonstrated. Given the normally strong moisture supply, the varied sources of moisture in the low- and high-level circulations, and the potentially large cloud depth and temperature spread at the crest of these storms, the microphysical precipitation-generating mechanisms should be expected to be as complex as the cyclone dynamics.

19.6.1. Cloud Layering and Washout

A cyclonic storm may initially duplicate the characteristics of an anticyclonic storm, if observed in the cold surge that is drawn into the cyclogenetic area. Gossard *et al.* (1984), in a cyclonic case of 11 May 1982, used the FM-CW radar to observe first cloud formation within a Canadian air mass behind a cold front. This occurred before any overriding clouds were observed. The shallow upslope cloud was very similar to that observed in the purely anticyclonic case shown in Fig. 19.6.

In the outer reaches of a cyclonic system, stacked seeder-feeder cloud layers are likely to occur. These normally merge into a cloud mass that is continuous with depth nearer the cyclone core. Riehl and Reinking (1981) observed a deep precipitating stratus cloud mass overlain by spatially intermittent altostratus and cirrus clouds; the greatest quantities of ice particles (30–40 L^{-1}) were observed in the portion of the cloud system farthest upslope (near Denver) where cloud layers had merged and nearly glaciated. Outlying portions of the cloud were laden with 0.2–0.4 g m^{-3} of liquid water, especially in the mesoscale wave discussed in Sec. 19.5. These water contents exceeded the norm for the anticyclonic systems. The water was found in supercooled cloud layers (where heavy aircraft icing was encountered), and in the lowest cloud mass, which was warmer than $0°C$ in the downslope, inflow region. Measurements with airborne laser spectrometer probes revealed ice particles (up to ~ 7 L^{-1}) falling in clear air between layers, so the seeder-feeder process was operative (Fig. 19.22). Vapor from the subliming crystals served to moisten the clear air and encourage the merger of stratus layers farther upslope.

The mesoscale numerical model of upslope cyclonic flow by Raddatz and Khandekar (1979) includes the assumption that clouds in the mixed planetary boundary layer contribute most of the total surface precipitation. The low-level, colloidally stable "feeder" clouds form in response to the orographic forcing; large-scale ascent generates high-level "seeder" clouds, which produce hydrometeors large enough to fall and efficiently wash out the water contents from the low-level clouds. The cyclone observations presented by Raddatz and Khandekar support the model. This seeder-feeder concept was also supported by Bader and Roach (1977), in their model of orographic rainfall produced in warm sectors of depressions. They concluded that the "washout" of hill-enhanced, low-level clouds by "seeder" particles from above is an effective mechanism for warm-sector precipitation. Such washout was in progress in the case studied by Riehl and Reinking. Some ice particles from high layers were large enough to form 500–600 μm raindrops upon melting, and to induce further drop growth by collision and coalescence in the

Figure 19.22. A seeder-feeder couplet: precipitation falling between cloud layers merging in the circulation of an upslope cyclone. (From Riehl and Reinking, 1981.)

warmer-than-freezing cloud volume, thus enhancing water extraction from the lower cloud layers.

Graupel and other ice particles were observed in significant concentrations (up to 20 L^{-1}) falling through the $-5°C$ level in some volumes of the cloud in the latter case. Cloud droplet size distributions were broad enough (diameters up to $\sim$30 μm) to be conducive to at least a marginally active process of secondary ice generation by rime splintering, which is most active between $-3°C$ and $-8°C$ (Hallett and Mossop, 1974). The cloud there also appeared to have tops as cold as $-8°C$ to $-10°C$, which would promote isometric crystal growth and subsequent graupel formation as postulated earlier on the basis of the laboratory studies of Fukuta and colleagues.

19.6.2. Imbedded Convection

Imbedded convection has received almost no attention so far in studies of upslope storms, but it does occur. Banded storm structures are suggested by variations in precipitation rate (e.g., Fig. 19.14). Also, cyclogenesis on 3 March 1984 produced deep convection in the Texas Panhandle (as documented by satellite photos), and an associated broken overcast with embedded convection in the Denver area where, between 1800 and 2100 GMT, Reinking observed a brief shower that produced only conical graupel of low density (slow fallspeed), which averaged 1.25 cm in diameter! These particles (Fig. 19.23) were undoubtedly products of convection, as well as significant cloud water extractors. Detailed examinations of extratropical cyclones, as influenced by the orography of Washington state, have revealed similar and additional complexities in deep-storm microphysics (Houze *et al.*, 1976; Hobbs, 1978; Matejka *et al.*, 1980): Imbedded cumulus convection, in mesoscale rainbands fed by low-level moisture, appeared to enhance ice parti-

Figure 19.23. Conical, low-density 1.25 cm graupel produced by an upslope convective shower near Boulder, Colo., 1300 MST, 3 March 1984.

cle growth; riming and aggregation (more of the latter) occurred just above the melting layer. Older rainbands glaciated, but ice crystal growth continued by aggregation. Precipitation "cores" also originated in higher-level generating cells that were produced by the lifting of potentially unstable air. The cores produced "seed" ice particles, which also grew by riming and aggregation as they fell through lower cloud layers. Appreciable mesoscale precipitation in these cyclonic storms was found to be "invariably associated with high concentrations ($\sim$1–100 L^{-1}) of ice particles" (Hobbs, 1978). However, on localized scales, cores of updrafts in cold frontal rainbands carried relatively high liquid water contents (0.45–1.4 g m^{-3}, 50–100% of computed adiabatic values) and ice particle concentrations that were only one-half to one-tenth the concentrations in the surrounding clouds. The cyclone precipitation was concentrated within rainbands 5–50 km wide, and within still smaller elements 10–50 km^2 within the rainbands. It was suggested that small hills played a role in triggering or enhancing the rainbands when the air was unstable.

19.6.3. The Importance of $-15^\circ C$

Of great importance to upslope precipitation is the presence, within at least part of the cloud mass, of the dendritic ice crystal regime (about -15°C $\pm$ 3°C) where conversion of water to ice by vapor deposition is most rapid. This regime was present in the upper reaches of the deepest part of the storm studied by Riehl and Reinking (1981); there most of the water had been extracted by dendrite crystal growth and precipitated, and only a nearly glaciated cloud mass remained. Auer and White (1982) estimated that 27% of the condensate was released in the dendritic growth regime, in the case they studied. Efficient collection of rime by dendritic crystals, during their growth by deposition, is necessarily included in this effect. Auer and White concluded that cyclonic upslope storms that carry heavy moisture fluxes through temperature regimes conducive to dendritic crystal growth are the heaviest snow producers.

Appendix A: General characteristics of middle-latitude upslope precipitation events

Location

- Sloping terrain on climatological lee of mountain ranges

Season

- Winter (September–April, Northern Hemisphere)

Types of circulations, modified by terrain

- Polar or arctic anticyclone
- Extratropical cyclone
- Anticyclone merging into cyclone

Surface pressure tendency

- *Rising* at onset of precipitation

Key climate controls

- Elevation, extent, and orientation of mountains influencing blocking and cyclogenesis
- Slope of lee-side terrain influencing lifting
- Location relative to long (Rossby) waves
- Air mass and moisture sources and proximity
- General continentality

Examples of effects

- Cloudiness, snow, blizzards, reduction in solar and thermal radiation exchange
- Beneficial:

    Water supply enhancement
    Pollutant photochemistry reduction
    Pollutant washout

- Detrimental:

    Impeded ground and air transportation, reduced safety
    Livestock loss

Heating fuel/energy consumption

- Mixed:

    Ski industry economics (snow vs. access)

Appendix B: Synoptic extremes of middle-latitude
upslope precipitation events

Anticyclone

- Cold air mass spreading upslope with conservation of absolute potential vorticity

- Subsidence in the left rear quadrant of polar jet

- Low-to-moderate moisture in polar air, varied moisture in overriding air

- Generally thermodynamically stable, extensive stratus clouds

- Temperature, moisture fields of shallow air mass and overriding layers setting stage for precipitation form, rate, and amount

Cyclone

- Deep cold-core trough or cold low at 500 mb

- Cyclogenesis, initiated to south and upstream of upslope impact area

- Forced topographic lifting superimposed on cyclone vertical motion

- Variations in regional precipitation with variations in path of low center

- Heaviest snow with slow-moving trough, blizzards with rapid cyclogenesis, and initially fast-moving upper trough

- Temperature and moisture fields over much of depth of troposphere, setting stage for precipitation form, rate, and amount

Appendix C: Mesoscale factors of middle-latitude
upslope precipitation events

Terrain

- Slope and its orientation to moisture flux regulate forced lifting to condensation, destabilization.

- Irregularities and drag block, channel, and contort advection and fronts, influence frontogenesis and lifting.

- Relative importance may vary directly or inversely with synoptic dynamic support, in anticyclones and cyclones, respectively.

- Effect is enhanced by destabilization or formation of seeder-feeder couplets.

Kinematics/dynamics

- Daily, seasonal surface albedo and sensible/latent heat exchange precondition air to be lifted.

- Low-level upslope winds and fetch, moisture, and thermodynamic stability regulate terrain effects.

- Slope/lifting by front/air mass must be weighed relative to slope/lifting by terrain.

- Low- and midtroposphere isentropic structure provides evidence of lifting and potential seeder cloud-feeder cloud couplets.

- Mesoscale waves, rainbands, imbedded convection, and frontal gravity waves may be enhanced by terrain.

Microphysics

- Upslope moisture advection and induced vertical motion, superimposed on storm-driven lifting, determine condensation rate, cloud form, and depth.

- Singular or converging moisture fields and stable or convective lifting influence range/uniformity of upslope storm liquid water content, drop sizes, and cloud layering or vertical continuity.

- Predominant ice crystal generation rate, form, density, growth rate and mode, and resulting precipitation rate all depend on temperature regimes reached by greatest water vapor and liquid water fluxes.

- Level of saturation with respect to ice regulates survival of ice crystals precipitating between and below cloud layers.

- Frontal or midtropospheric seeder clouds generally increase precipitation efficiency of upslope feeder clouds.

REFERENCES

Auer, A. H., Jr., and J. M. White, 1982: The combined role of kinematics, thermodynamics and cloud physics associated with heavy snowfall episodes. *J. Meteor. Soc. Japan*, **60**, 500–507.

Auer, A. H., Jr., T. Karacostas, T. Krauss, J. White, B. Boe, and M. Bradford, 1980: An examination of the synoptic events, atmosphere water budget, and mesoscale influences associated with the Front Range blizzard of 20–21 November 1979. Preprints, 8th Conference on Weather Forecasting and Analysis, Denver, Colo., American Meteorological Society, Boston, 29–30.

Bader, M. J., and W. T. Roach, 1977: Orographic rainfall in warm sectors of depressions. *Quart. J. Roy. Meteor. Soc.*, **103**, 269–280.

Boatman, J. F., and R. F. Reinking, 1984: Synoptic and mesoscale circulations and precipitation mechanisms in shallow upslope storms over the western high plains. *Mon. Wea. Rev.*, **112**, 1725–1744.

Browning, K. A., and T. W. Harrold, 1970: Air motion and precipitation growth at a cold front. *Quart. J. Roy. Meteor. Soc.*, **96**, 369–389.

Chaumerliac, N., E. Nickerson, and R. Rosset, 1983: A three-dimensional numerical simulation of atmospheric cleansing during the 1982 Boulder Upslope Cloud Observation Experiment (BUCOE). In *Precipitation Scavenging, Dry Deposition, and Resuspension*, Vol. 1., H. Pruppacher et al. (Eds.), Elsevier, New York, 627–648.

Cooper, W. A., and G. Vali, 1976: Ice crystal concentrations in wintertime clouds. Preprints, International Cloud Physics Conference, Boulder, Colo., American Meteorological Society, Boston, 91–96.

Decker, M. T., 1984: Observation of low-level frontal passages with a multifrequency microwave radiometer. Analysis of some cloud and frontal events, E. E. Gossard (Ed.), NOAA Environmental Research Laboratories (NTIS#PB84–179662), 27–32.

Fawcett, E. B., and H. K. Saylor, 1965: A study of the distribution of weather accompanying Colorado cyclogenesis. *Mon. Wea. Rev.*, **93**, 359–367.

Fukuta, N., M. Kowa, and N.-H. Gong, 1982: Determination of ice crystal growth parameters in a new supercooled cloud tunnel. Preprints, Conference on Cloud Physics, Chicago, Ill., American Meteorological Society, Boston, 325–328.

Fukuta, N., N.-H. Gong, and A.-S. Wang, 1984: A microphysical origin of graupel and hail. Proceedings, 9th International Cloud Physics Conference (3 vols.), Tallinn, USSR, International Association of Meteorology and Atmospheric Physics, 257–260.

Gossard, E. E., T. Detman, J. B. Snider, and R. Zamora, 1984: Coordinated observation of cloud events by radar and microwave radiometer. Analysis of some cloud and frontal events, E. E. Gossard (Ed.), NOAA Environmental Research Laboratories (NTIS#PB84–179662), 1–26.

Hallett, J., and S. C. Mossop, 1974: Production of secondary ice crystals during the riming process. *Nature*, **249**, 26–28.

Henz, J., 1984: Predicting State weather no piece of cake. *Rocky Mountain News*, 1 March 1984, p. 160.

Hess, S. L., 1959: *Introduction to Theoretical Meteorology.* Holt, Rinehart and Winston, New York, 362 pp.

Hobbs, P. V., 1974: *Ice Physics.* Clarendon Press, Oxford, 837 pp.

Hobbs, P. V., 1978: Organization and structure of clouds and precipitation on the mesoscale and microscale in cyclonic storms. *Rev. Geophys. Space Phys.*, **16**, 741–755.

Houze, R. A., Jr., J. D. Locatelli, and P. V. Hobbs, 1976: Dynamics and cloud microphysics of the rainbands in an occluded frontal system. *J. Atmos. Sci.*, **33**, 1921–1936.

Keyser, D., and R. A. Anthes, 1982: The influence of planetary boundary layer physics on frontal structure in the Hoskins-Bretherton horizontal shear model. *J. Atmos. Sci.*, **39**, 1783–1802.

Knollenberg, R. G., 1981: Techniques for probing cloud structure. In *Clouds: Their Formation, Optical Properties, and Effects*, P. V. Hobbs and A. Deepak (Eds.), Academic Press, New York, 15–92.

Ludlam, F. H., 1980: *Clouds and Storms: The Behavior and Effect of Water on the Atmosphere.* Pennsylvania State Univ. Press, 372 pp.

Magono, C., and C. W. Lee, 1966: Meteorological classification of natural snow crystals. *J. Fac. Sci. Hokkaido Univ.*, Ser. 7, **2**, 321–335.

Matejka, T. J., R. A. Houze, Jr., and P. V. Hobbs, 1980: Microphysics and dynamics of clouds associated with mesoscale rainbands in extratropical cyclones. *Quart. J. Roy. Meteor. Soc.*, **106**, 29–56.

Nagamoto, C. T., F. Parungo, R. Reinking, R. Pueschel, and T. Gerish, 1983: Acid clouds and precipitation in eastern Colorado. *Atmos. Environ.*, **17**, 1073–1082.

Palmén, E., and C. W. Newton, 1969: *Atmospheric Circulation Systems: Their Structure and Physical Interpretation.* Academic Press, New York, 603 pp.

Pruppacher, H. R., and J. D. Klett, 1978: *Microphysics of Clouds and Precipitation.* D. Reidel Publ. Co., Boston, 714 pp.

Raddatz, R. L., and M. L. Khandekar, 1977: Numerical simulation of cold easterly circulations over the Canadian western plains using a mesoscale boundary-layer model. *Bound.-Layer Meteor.*, **11**, 307–327.

Raddatz, R. L., and M. L. Khandekar, 1979: Upslope enhanced extreme rainfall events over the Canadian western plains: A mesoscale numerical simulation. *Mon. Wea. Rev.*, **107**, 650–661.

Reynolds, D. W., 1983: Prototype workstation for mesoscale forecasting. *Bull. Amer. Meteor. Soc.*, **64**, 264–273.

Riehl, H., and R. F. Reinking, 1981: Ice crystal processes in Colorado upslope storms. Case study of an upslope storm. NOAA Tech. Memo. ERL WMPO-44, (available from the authors, NOAA Environmental Research Laboratories), (NTIS#PB82–109547), 30 pp.

Ryan, B. F., E. R. Wishart, and D. E. Shaw, 1976: The growth rates and densities of ice crystals between −3°C and −21°C. *J. Atmos. Sci.*, **33**, 842–850.

Sanders, F., 1955: An investigation of the structure and dynamics of an intense surface frontal zone. *J. Meteor.*, **12**, 542–552.

Schlatter, T. W., D. V. Baker, and J. F. Henz, 1983: Profiling Colorado's Christmas Eve blizzard. *Weatherwise*, **36**, 60–66.

Shapiro, M. A., 1984: Structure and physical processes within meteorological tower measurements from a surface cold front. Analysis of some cloud and frontal events, E. E. Gossard (Ed.), NOAA Environmental Research Laboratories (NTIS#PB84–179662), 63–78.

Treidl, R. A., 1970: A case study of warm air advection over a melting snow surface. *Bound.-Layer Meteor.*, **1**, 155–168.

Walsh, J. E., 1984: Snow cover and atmospheric variability. *Amer. Sci.*, **72**, 50–57.

Walsh, P. A., 1977: Cloud droplet measurements in wintertime clouds. M. S. thesis, Dept. of Atmospheric Science, Univ. of Wyoming, Laramie, 170 pp.

Weickmann, H., 1981: Mechanism of shallow winter-type stratiform cloud systems. NOAA Environmental Research Laboratories (NTIS#PB82–170176), 61 pp.

Whiteman, C. D., 1973a: Some climatological characteristics of seedable upslope cloud systems in the high plains. NOAA Tech. Rep. 268–APCL–27 (NTIS-#COM–73–50924/2GI), 43 pp.

Whiteman, C. D., 1973b: Variability of high plains precipitation. NOAA Tech. Rep. 287–APCL–31 (NTIS#COM–74–533510), 43 pp.

Williams, R. T., 1974: Numerical simulations of steady state fronts. *J. Atmos. Sci.*, **31**, 1286–1296.

CHAPTER 20

Mountain Waves

Dale R. Durran

20.1. Introduction

The basic flow pattern across a long ridge of mountains is determined by the mountain width. If the ridge is wide enough that the time required for air to cross it is greater than order $1/f$ (where f is the Coriolis parameter), rotational effects generate a disturbance with large displacements in the horizontal x-y plane. As the width decreases to less than 100 km, the perturbations in the horizontal plane disappear and waves in the vertical x-z plane develop. When the wind blows over such a ridge, air parcels are displaced vertically and, if the atmosphere is stably stratified, they descend and may oscillate about their equilibrium levels. The gravity waves that result, called mountain waves or lee waves, have been observed in mountainous regions all over the world.

The presence of mountain waves is frequently revealed by distinctive orographic clouds that form in the wave crests. Various types of mountain waves produce the different types of wave clouds. An identification of a particular type of wave cloud can be used to make some qualitative deductions about the vertical variation in wind speed and stability over the mountains.

Large-amplitude mountain waves can produce several weather phenomena that significantly affect human activity and therefore require the attention of the weather forecaster. The strong downslope winds observed along the lee slopes of mountain barriers are usually associated with large-amplitude waves. Dangerous regions of clear-air turbulence are also produced by these waves.

20.2. Theory of Linear Waves Forced by Sinusoidal Mountain Ridges

The most fundamental properties of mountain waves can be profitably examined by considering the steady-state, two-dimensional airflow over "small-amplitude" mountains (so that linear theory can be used). The two-dimensional assumption is appropriate if the mountains are assumed to extend indefinitely in the direction parallel to the ridge and be sufficiently narrow that the Rossby number governing the flow is large (so Coriolis forces may be neglected). Consider the equations for an inviscid Boussinesq fluid, the simplest case that contains the essential physics governing the flow, and (consistent with our small-amplitude assumption) linearize them about a horizontally uniform basic state with a mean wind U, a reference potential temperature θ_0, and a mean potential temperature gradient $d\bar{\theta}/dz$. The result may be written

$$U\frac{\partial u}{\partial x} + w\frac{dU}{dz} + \frac{\partial P}{\partial x} = 0 \,, \tag{20.1}$$

$$U\frac{\partial w}{\partial x} + \frac{\partial P}{\partial z} = b \,, \tag{20.2}$$

$$U\frac{\partial b}{\partial x} + N^2 w = 0 \,, \tag{20.3}$$

$$\frac{\partial u}{\partial x} + \frac{\partial w}{\partial z} = 0 \,, \tag{20.4}$$

where $b = g\theta/\theta_0$, $P = c_p\theta_0\pi$, and $N^2 = (g/\theta_0)(d\bar{\theta}/dz)$. Here (u,w) represent the perturbation velocity components in the Cartesian (x,z) coordinate system, θ is the perturbation potential temperature, and π the perturbation Exner function, $(p/p_0)^{R/c_p}$. The Brunt-Väisälä frequency is given by N. The remaining constants have their conventional meanings.

Equations (20.1)–(20.4) may be combined to form a single equation for w:

$$\frac{\partial^2 w}{\partial z^2} + \frac{\partial^2 w}{\partial x^2} + \ell^2\, w = 0 \,, \tag{20.5}$$

where

$$\ell^2 = \frac{N^2}{U^2} - \frac{1}{U}\frac{d^2 U}{dz^2} \tag{20.6}$$

is the Scorer parameter.

As a further simplification, let the terrain profile be defined as an infinite set of periodic ridges,

$$h(x) = h_0 \cos kx \,, \tag{20.7}$$

and let N and U be constant with height.

Since the Earth's surface is fixed, the normal component of the velocity must vanish at the lower boundary. Thus,

$$w[x, h(x)] = (U + u)\frac{\partial h}{\partial x} \,, \tag{20.8}$$

which can be approximated, to the same order of accuracy as the linearized equations, as

$$w(x, 0) = U\frac{\partial h}{\partial x} = -Uh_0 k \sin kx \,. \tag{20.9}$$

Solutions to (20.5) can be obtained in the form

$$w(x, z) = \hat{w}_1(z) \cos kx + \hat{w}_2(z) \sin kx . \qquad (20.10)$$

Substituting (20.10) into (20.5) yields an equation that determines the vertical structure of the perturbation velocity field. Both $\hat{w}_1$ and $\hat{w}_2$ satisfy

$$\frac{\partial^2 \hat{w}_i}{\partial z^2} + (\ell^2 - k^2)\hat{w}_i = 0 \qquad i = 1, 2 . \qquad (20.11)$$

Since N and U are assumed to be constant with height, $\ell^2 - k^2 = m^2$ is a constant and the solution of (20.11) may be written

$$\hat{w}_i(z) = \begin{cases} A_i e^{\mu z} + B_i e^{-\mu z} & k > \ell \\ A_i' \cos mz + B_i' \sin mz & k < \ell, \end{cases} \qquad (20.12)$$

where $\mu^2 = -m^2$. Note that the vertical structure of the wave depends on the relative magnitudes of the Scorer parameter and the horizontal wavenumber. When $k > \ell$ the wave amplitude decreases (or increases) exponentially with height; when $k < \ell$ the vertical velocity above a fixed point on the ground oscillates between regions of upward and downward motion as z increases.

The coefficients A_i, B_i, A_i', and B_i' in (20.12) are determined by boundary conditions imposed at $z = 0$ and in the limit as z approaches infinity. When $k > \ell$, the first term in the general solution corresponds to a wave whose amplitude grows exponentially without bound as z increases. This behavior is not physically reasonable since the mountain is the energy source for the disturbance. Thus, the boundary condition at infinity requires that $A_i = 0$. Then $B_1 = 0$, $B_2 = -Uh_0 k$, in order to satisfy the "tangential flow" condition (20.9).

In the case $k < \ell$, trigonometric identities can be used to write the general solution as

$$\begin{aligned} w(x, z) = &C_1 \sin(kx + mz) + C_2 \sin(kx - mz) \\ &+ C_3 \cos(kx + mz) + C_4 \cos(kx - mz) , \end{aligned} \qquad (20.13)$$

where both $m > 0$, $k > 0$. The lower boundary condition requires $C_1 + C_2 = -Uh_0 k$, and $C_3 + C_4 = 0$. These coefficients are uniquely specified by applying a boundary condition in the limit as z approaches infinity. The nature of this upper boundary condition is not as obvious as it was in the case $k > \ell$, a circumstance that has led to some confusion in the past. The terms with coefficients C_1 and C_3 correspond to waves in which lines of constant phase $(kx + mz = \text{constant})$ tilt upstream. It can be shown that these waves transport energy upward and momentum downward. The opposite is true for the remaining two terms. Since the mountain acts as the energy source for the wave disturbance, the correct choice of coefficients is $C_1 = -Uh_0 k$, $C_2 = C_3 = C_4 = 0$, which requires that all energy transport by the waves be directed upward as z approaches infinity.

The same choice of the upstream-tilting wave can be obtained by a variety of other arguments (see Smith, 1979). It should be noted, in particular, that the upstream-tilting wave is invariably obtained as the steady-state numerical solution to initial-value problems involving vertically propagating mountain waves. Unfortunately, some authors continue to suggest that the choice of the upper boundary condition is somehow still an open question. Atkinson (1981) follows Scorer's incorrect development and does not adequately alert the reader that the result for propagating waves, Atkinson's Eq. (34), is wrong.

Perhaps it is easier to understand why the upstream-tilting wave is the correct solution by considering the perturbation pressure field. In this simple situation, the mean wind is constant with height, and the horizontal momentum and continuity equations can be combined to yield

$$\frac{\partial P}{\partial x} = U\frac{\partial w}{\partial z} . \qquad (20.14)$$

In the case $k < \ell$, the solutions for the perturbation pressure fields in the upstream- and downstream-tilting waves are

$$P = \begin{cases} -U^2 h_0 m \sin(kx + mz) & \text{upstream tilting} \\ U^2 h_0 m \sin(kx - mz) & \text{downstream tilting} . \end{cases} \qquad (20.15)$$

Note that at $z = 0$, the extrema in the perturbation pressure field are shifted 90° relative to the location of the troughs and ridges in topography. The wave that tilts upstream with height $[-\sin(kx+mz)]$ produces high pressure upwind of the ridge crest and low pressure downwind. The wave that tilts downstream produces the opposite pressure pattern. The asymmetry in the pressure distribution across the ridge gives rise to a net pressure force on the topography. The upstream-tilting wave exerts a force on the mountain in the direction of the mean flow, and therefore an equal and opposite force is exerted by the terrain, which acts to decelerate the mean flow. In the case of downstream-tilting waves, the resulting pressure forces are directed upstream and simultaneously accelerate the mean flow in the direction it is already going. This last situation is clearly contrary to physical intuition, suggesting that the correct solution is the upstream-tilting wave.

In summary, the perturbation vertical velocity field in the waves forced by the sinusoidal terrain profile (20.7) can be written

$$w(x,z) = \begin{cases} -U h_0 k e^{-\mu z} \sin kx & k > \ell \\ -U h_0 k \sin(kx + mz) & k < \ell . \end{cases} \qquad (20.16)$$

The two types of wave structures are illustrated in Fig. 20.1. The waves in the case $k > \ell$ (Fig. 20.1a) decay exponentially with height (evanescent waves), whereas in the $k < \ell$ case (Fig. 20.1b), the waves propagate vertically, without loss of amplitude.

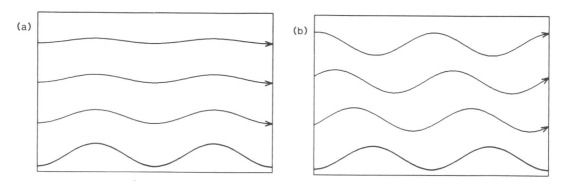

Figure 20.1. Streamlines in the steady airflow over an infinite series of sinusoidal ridges when (a) the wavenumber of the topography exceeds the Scorer parameter (narrow ridges) or (b) the wavenumber of the topography is less than the Scorer parameter (wide ridges).

The speed of the air flowing through these waves can also be determined from Fig. 20.1. It follows from the change in spacing between streamlines, which in this case are actually contours of a stream function, that in the case of the vertically propagating wave there will be weak winds over the windward slope and strong winds in the lee. This amplification of the lee-side wind is believed to be associated with the development of strong downslope winds like the Colorado chinook. In the case of the evanescent wave, the distribution of wind speed is symmetric about the mountain peak, the strongest winds being directly above the crest.

The physical reason for the difference between the two types of waves may be described qualitatively as follows. The requirement $k < \ell$ is equivalent, in the absence of curvature in the mean wind profile, to $Uk < N$, which states that the intrinsic frequency of the forcing by the terrain must be less than the Brunt-Väisälä frequency. The Brunt-Väisälä frequency is the highest frequency at which buoyancy forces can support periodic motion in a stably stratified fluid; parcels of air oscillating at the Brunt-Väisälä frequency move vertically, straight up and down. Lower frequency oscillations are obtained when the parcel paths are tilted at some angle off the vertical. It can be shown that if ϕ is the angle between the slanted parcel trajectories and the vertical, the frequency of the oscillation will be $N \cos \phi$ (Gill, 1982, p. 132). The troughs and crests in vertically propagating gravity waves tilt in order to match the natural frequency of the atmospheric oscillations to the intrinsic frequency forced by the airflow over the terrain. As a result, the slope of the wave crests (which is identical to the slope of a parcel trajectory) satisfies

$$Uk = N \cos \phi \ . \qquad (20.17)$$

However, if the intrinsic frequency of the forcing exceeds N, no real angle ϕ will satisfy (20.17), and no buoyancy-driven oscillation can be established. In this case, buoyancy forces act to dampen the oscillation, like a spring forced at a nonresonant frequency. The wave are damped more rapidly as the difference between Uk and N increases.

20.3. Theory of Linear Waves in More Realistic Situations

The mountain wave solutions (20.16) apply only to an air mass with constant stability, flowing at a uniform mean speed, across an endless series of sinusoidal ridges. If more realistic terrain profiles and atmospheric structures are considered, other solutions to (20.5) can be obtained which bear a strong resemblance to observed mountain waves. The following is a description of how the wave response is influenced by isolated ridges and vertical variation in the Scorer parameter, the mathematical derivations are often omitted; they may be found in Smith (1979) and Queney *et al.* (1960).

20.3.1. Influence of Shape of Terrain

Suppose the mountain contour consists of a single ridge so that the terrain elevation eventually drops to some reference level at all distances sufficiently far upstream and downstream. Just as Fourier series can be used to represent a wide variety of periodic functions as an infinite sum of sines and cosines, the isolated mountain can, under rather general conditions, be constructed from periodic functions by the use of Fourier transforms. The Fourier transform (F) of a real function ϕ and its inverse (F^{-1}) may be defined:

$$\hat{\phi}(k) = F[\phi(x)] - \frac{1}{\pi} \int_{-\infty}^{\infty} \phi(x)e^{-ikx}dx;$$
$$\phi(x) = F^{-1}[\hat{\phi}(k)] = \text{Re} \int_{0}^{\infty} \hat{\phi}(k)e^{ikx}dk . \tag{20.18}$$

The Fourier transform is particularly useful in this application because it has the property that $F(\partial^n\phi/\partial x^n) = (ik)^n\hat{\phi}$.

If the fundamental equation for linear gravity waves (20.5) is Fourier transformed in the horizontal, the result is (20.11), where $\hat{w}(k,z) = F[w(x,z)]$. Thus, the behavior of each component $\hat{w}(k_0,z)$ of the transformed vertical velocity field is identical to that obtained by forcing the atmosphere with sinusoidal topography having the wavenumber k_0. Thus, the results obtained in Sec. 20.2 are also applicable to the case of isolated topography. The only complication arises from the requirement that after $\hat{w}(k,z)$ is determined, the actual vertical velocity field must be recovered by application of the inverse transform.

Consider again the case in which N and U (and hence ℓ) are constant with height. The function $\hat{w}(k,z)$ is the complex analog of (20.12), which, after the free slip and radiation boundary conditions are evaluated, is

$$\hat{w}(k,z) = ikU\hat{h}(k,z) \, \exp[i(\ell^2 - k^2)^{1/2}z] \qquad k > 0 . \tag{20.19}$$

Here $\hat{h}$ is the Fourier transform of the terrain profile; it determines the relative weight accorded to each wavenumber in the final solution $w(x,z)$. If the mountain is very narrow, the dominant weighting will be at wavenumbers greater than ℓ so the solution will consist primarily of evanescent waves. On the other hand, if the mountain is sufficiently wide, the dominant weighting

478   *DURRAN*

is at wavenumbers less than ℓ and the solution consists primarily of vertically propagating waves.

Figure 20.2 above shows how the waves generated by a bell-shaped ridge,

$$h(x) = \frac{h_0 a^2}{a^2 + x^2} \, , \tag{20.20}$$

vary with the difference between a^{-1} and ℓ (N and U are again constant with height). This mountain has a maximum height of h_0 at $x = 0$, and falls to $\frac{1}{2}h_0$ at $x = \pm a$; thus a^{-1} represents a scale characteristic of the wavenumbers forced by the mountain. For very narrow mountains, where $a^{-1} >> \ell$ (Fig. 20.2a), the wave pattern is symmetric with respect to the ridge crest, and the perturbations decay with height, just like the evanescent waves in Sec. 20.2. For a wide mountain, where $a^{-1} << \ell$ (Fig. 20.2c), the waves propagate vertically, and lines of constant phase tilt upstream. Assuming the condition that $a^{-1} << \ell$ is equivalent to taking the hydrostatic limit, in which case $k << \ell$ and (20.19) reduces to

$$\hat{w}(k,z) = ik \, U\hat{h}(k,z) \, e^{i\ell z} \qquad k > 0 \, , \tag{20.21}$$

eliminating the dependence of vertical wavelength on horizontal wavenumber. As a result, the mountain profile is reproduced at every level that is an integral multiple of $2\pi/\ell$. (This result is independent of the shape of the mountain.)

In the third case, $a^{-1} = \ell$ (Fig. 20.2b), the solution is dominated by vertically propagating nonhydrostatic waves (i.e., $k < \ell$ without $k << \ell$) and the situation is more complicated. The phase lines still tilt upstream, and energy is transported upward. However, unlike hydrostatic waves, in which the transport occurs directly over the mountain (because the horizontal group velocity of a stationary hydrostatic wave is zero), nonhydrostatic waves transport energy both upward and downstream (because the horizontal

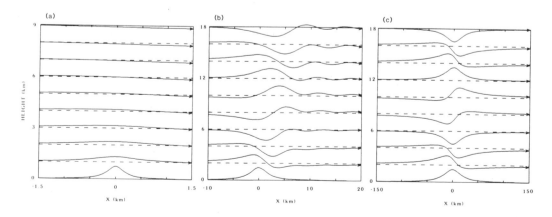

Figure 20.2. Streamlines in the steady airflow over an isolated bell-shaped ridge. (a) $a^{-1} >> \ell$, narrow ridge; (b) $a^{-1} = \ell$, width of the ridge comparable with the Scorer parameter; (c) $a^{-1} << \ell$, wide ridge, but not so wide that rotational effects become significant.

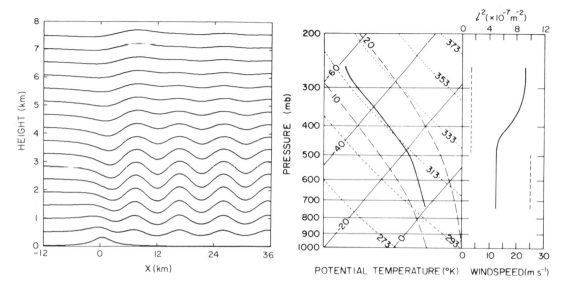

Figure 20.3. Streamlines in the steady airflow over an isolated bell-shaped ridge when the vertical variation of the Scorer parameter has the two-layer structure shown in Fig. 20.4.

Figure 20.4. A vertical distribution of temperature and wind speed (solid lines), which implies a discontinuous two-layer structure for the Scorer parameter (dashed line on right panel).

group velocity of a stationary nonhydrostatic wave is directed downstream). As a result, the waves in Fig. 20.2b appear in a wedge-shaped region downstream of the ridge.

It is important to appreciate that if a mountain is sufficiently wide to produce essentially hydrostatic waves (but not so wide that Coriolis forces become significant), there will be only one wave crest in the air flowing over the mountain. Additional crests do not appear downstream from the mountain unless nonhydrostatic effects are significant. Even in the case shown in Fig. 20.2b, the wave amplitude decays rapidly downstream from the mountain.

20.3.2. Influence of Atmospheric Structure

Trapped lee waves are a different type of mountain wave. The airflow in a system of trapped waves (also known as resonant lee waves) is shown in Fig. 20.3. Note that most of the wave activity is confined to the lower troposphere on the lee side of the mountain. As demonstrated by Scorer (1949), this type of long wave train occurs only when ℓ^2 decreases with height. Reference to (20.6) reveals that a decrease in ℓ^2 will result from an increase in wind speed, a decrease in stability, or an increase in the curvature of the wind speed profile. Scorer examined the behavior of linear waves in a two-layer atmosphere in which ℓ^2 was constant in each layer. An example of this kind of two-layer structure is shown in Fig. 20.4, which describes the upstream conditions for the solution shown in Fig. 20.3. Note that a discontinuity in ℓ^2 can occur without discontinuities in θ and U. Scorer

showed that the wavenumber of the trapped waves must satisfy the following
resonance condition:

$$(\ell_L^2 - k^2)^{1/2} \cot[(\ell_L^2 - k^2)^{1/2}H] = -(k^2 - \ell_U^2)^{1/2} , \qquad (20.22)$$

where ℓ_U and ℓ_L are the Scorer parameters in the upper and lower layers,
and H is the depth of the lower layer. A necessary condition for the existence
of a solution to (20.22) is

$$\ell_L^2 - \ell_U^2 > \frac{\pi^2}{4H^2} , \qquad (20.23)$$

which states that the difference in wave propagation characteristics in the two
layers must exceed a certain threshold before the waves can be "trapped." If
(20.23) is satisfied by a sufficient margin, there may be multiple solutions to
(20.22), in which case the mode with the longest horizontal wavelength will
usually dominate.

The wavenumber of the resonant wave solution to (20.22) must satisfy
the inequality

$$\ell_L > k > \ell_U , \qquad (20.24)$$

which implies that the wave propagates vertically in the lower layer, and
decays exponentially in the upper layer. One might expect the waves to tilt
upstream with height throughout the region in which they are vertically propagating; however, as shown in Fig. 20.3, trapped waves have no tilt. This is
because, when the upward-propagating waves (which are originally triggered
by the mountain) reach the upper layer, they cannot continue to propagate
upward; they are reflected as downward-propagating waves. The downward-
propagating waves subsequently are reflected as upward-propagating waves
when they strike the ground. As this process continues, an infinite number of reflections take place. There is no loss of amplitude when the waves
are reflected because there are no energy exchanges with the upper layer or
the flat ground downstream of the mountain. As a result, the disturbances
that appear downstream are the superposition of equal-amplitude upward-
and downward-propagating waves, and thus have no tilt. To illustrate this
mathematically, the trapped waves in Fig. 20.3 have the functional form

$$w(x, z) = \beta \sin \alpha z \cos kx , \qquad (20.25)$$

in the lower layer. Here β and α are constants (see Queney *et al.* [1960]
for the details). By using trigonometric identities, (20.25) can be written as
the sum of two equal amplitude waves that tilt upstream and downstream,
respectively:

$$w(x, z) = \frac{\beta}{2} \sin(\alpha z + kx) + \frac{\beta}{2} \sin(\alpha z - kx) . \qquad (20.26)$$

Referring again to Fig. 20.3, note that, in addition to the trapped waves,
a weak vertically propagating wave also appears over the mountain. This

develops because the mountain produces some forcing at wavenumbers less than ℓ_U, thereby generating waves that can propagate through the upper layer. The mountain shown in Fig. 20.3 is specified by (20.20) with $a = 2.5$ km, $h_0 = 300$ m. If the mountain width is increased, the amplitude of the vertically propagating waves would increase and the amplitude of the trapped waves would decrease.

This example illustrates the important conclusion that the behavior of linear mountain waves depends entirely on the shape of the terrain profile and the mean atmospheric structure. The range of possible wave motions (i.e., which wavelengths will be vertically propagating, which will be evanescent, and which, if any, will be resonant and trapped) is determined by the atmospheric structure, particularly the mean horizontal wind speed and static stability. The shape of the terrain determines the strength of the forcing applied to each wavelength. In most atmospheric situations associated with mountain waves, vertically averaged values of N and U typically lie in the ranges 0.008 to 0.02 s^{-1} and 10 to 40 m s^{-1}, implying that most wavelengths less than 3 km ($k > N_{\max}/U_{\min}$) will be evanescent, and most wavelengths greater than 30 km ($k < N_{\min}/U_{\max}$) will propagate vertically. Trapped waves generally occur at intermediate wavelengths between 5 and 25 km.

20.4. Interpretation of Clouds in Satellite Photos

20.4.1. Types of Clouds

Wave-induced orographic clouds are often evident in satellite photographs. However, when interpreting these photographs, operational meteorologists tend to assume that all cloud-producing mountain waves propagate horizontally (Fig. 20.3). In fact, vertically propagating waves (Fig. 20.2c) commonly generate clouds as well, and the two types of clouds can usually be distinguished. Once the type of wave has been indentified, certain general conclusions can be drawn about the atmospheric conditions in the vicinity of the waves.

Clouds From Trapped Waves

The clouds in Fig. 20.5 show an extensive region of lee wave activity covering most of Nevada and portions of Oregon, Idaho, and Utah on 2 May 1984 at 1715 GMT. These are trapped waves of the type shown in Fig. 20.3, and can be recognized as such by these characteristics:

- There are multiple wave crests downstream from the initial disturbance.
- The wavelength, which varies between 10 km in northwestern Nevada and 23 km in southeastern Nevada, lies in the range at which waves are likely to be trapped.

The first waves appear to be generated by the Sierra Nevada and the Cascades, although the continued excitation of trapped waves at distances far downstream is probably produced by the mountains in the basin and range region of Nevada.

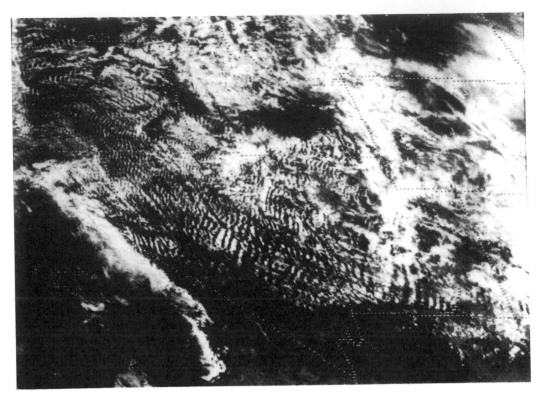

Figure 20.5. Visible satellite imagery for 1715 GMT, 2 May 1984.

Given that the waves revealed in Fig. 20.5 are trapped waves, one can draw general conclusions about the state of the atmosphere in their vicinity. In order for mountain waves of any type to exist, a sufficiently strong wind must be directed across the mountain at the ridge-top level. The minimum wind speed required for waves will vary with the size and shape of the mountain, but seems to lie in the range from 7 to 15 m s^{-1} (Queney *et al.* 1960). Since the waves are trapped, ℓ^2 values in the upper troposphere should be significantly smaller than those in the lower troposphere. This requirement is usually satisfied by a large increase in the wind with height, and the presence of one or more stable layers in the lower troposphere. Fig. 20.6 shows the 700 and 200 mb analyses for this case. Note that the winds are oriented almost perpendicular to the northern Sierras and to the Nevada ranges at both 700 and 200 mb, and there is a dramatic increase in wind speed with height. The thermodynamic properties of this air mass are illustrated in Fig. 20.7 by the Winnemucca sounding taken at 0000 GMT on 3 May. A pronounced stable layer is evident in the lower troposphere; the moist layer that probably contained the wave clouds is also apparent.

Clouds From Vertically Propagating Waves

A different type of situation is shown in Fig. 20.8, in which clouds reveal the presence of mountain waves along the eastern edge of the Front Range of

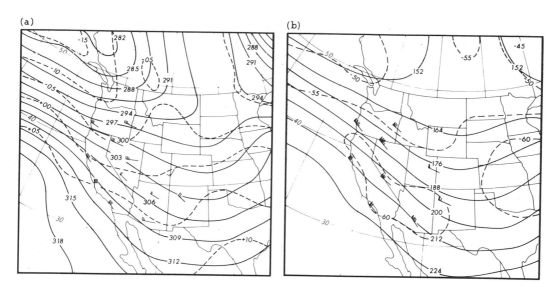

Figure 20.6. Analysis for 1200 GMT, 2 May 1984. (a) 700 mb; (b) 200 mb.

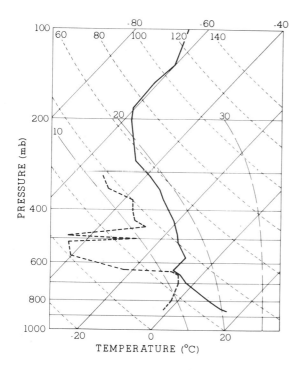

Figure 20.7. Winnemucca sounding for 0000 GMT, 3 May 1984. The 1200 GMT sounding for 2 May is similar, but shows two distinct inversions in the layer between 610 and 460 mb.

Colorado's Rocky Mountains at 1815 GMT on 7 November 1983. One can conclude that the waves are vertically propagating for the following reasons:

- Only one wave crest is visible.

- The horizontal wavelength, though hard to determine, exceeds the width of the wave cloud, which is roughly 100 km. This is far too long a wavelength to be trapped.

Figure 20.8. Visible satellite imagery for 1815 GMT, 7 November 1983.

Given that the waves are vertically propagating, one should again expect a significant flow across the mountains, but in this case, little can be definitely concluded about the vertical distribution of ℓ^2. As shown in Figs. 20.9 and 20.10, the increase in wind speed with height is less pronounced and the low-level stable layers are weaker than in the previous example. This seems to suggest that the difference between the waves in Figs. 20.5 and 20.8 can be accounted for by differences in the atmospheric conditions (i.e., insufficient decrease in ℓ^2 to trap the waves). However, the most significant difference is probably due to the topography. The Front Range is a wide mountain range so it tends to force long waves, which can propagate vertically even when ℓ^2 is very small. Reliable conclusions about the vertical distribution of wind speed and stability can generally be obtained only when trapped waves are present.

20.4.2. Distribution of Clouds

Returning to Fig. 20.8, note that although the winds over the Sierra Nevada and the Rockies are quite similar, the cloud patterns are just opposite. The clouds are largely restricted to the upwind side of the Sierras, whereas they appear on the downstream side of the Rockies.

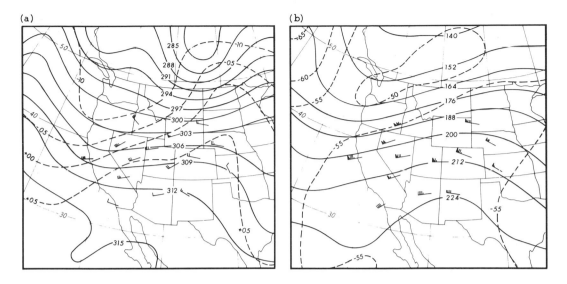

Figure 20.9. Analysis for 1200 GMT, 7 November 1983. (a) 700 mb; (b) 200 mb. The poor agreement between the wind and height fields at 700 mb can probably be attributed to orographic effects.

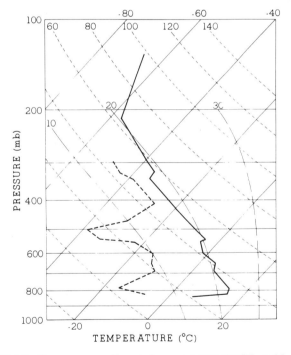

Figure 20.10. Denver sounding taken at 1200 GMT, 7 November 1983.

Middle- and Low-Level Clouds

The difference betweeen the middle- and low-level cloud distributions can be easily accounted for by differences in humidity. The moist maritime air mass flowing inland from the Pacific must be lifted 3 to 4 km to traverse

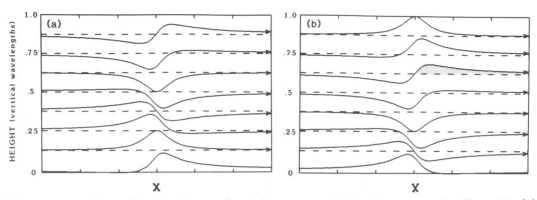

Figure 20.11. Streamlines in the steady airflow over an isolated asymmetric ridge with (a) a steep windward slope; (b) a steep leeward slope. Probable location of wave-induced cirrus is shaded.

the Sierras. In fact, most of the low-level flow is blocked and never passes directly over the main ridge, but there is still sufficient lifting to produce extensive regions of middle- and low-level clouds on the windward side of the mountain. At 1800 GMT (the approximate time of the satellite picture) surface stations upstream of the mountain in the Central Valley report multiple layers of stratiform and cumuliform clouds. On the other hand, no clouds appear upstream of the Colorado Front Range because the air approaching the crest, having already traversed several mountain ranges, is drier, and because less lifting is required to clear the crest of the Front Range. (Less lifting is required to traverse the Rockies because the air upstream starts at the elevation of the high intermountain plateau, not at sea level as in the case of the Sierra Nevada.)

The distribution of low- and middle-level cloudiness is similar in the lee of both mountain ranges. The cloud-free area in the lee of the Sierras, sometimes referred to as a foehn gap, is produced by descending air, which experiences net subsidence since the upstream flow is blocked. An additional contribution to the lee-side clearing occurs as moisture is removed by precipitation over the mountains. Bishop, a surface station on the lee side, reports precipitation in sight but distant from the station, presumably from the foehn (or cap) cloud along the ridge crest. Although it is not obvious from the satellite photo, the lee side of the Rockies is almost clear of middle- and low-level clouds. As in the case of the Sierras, subsidence over the lee slopes produces a reduction of relative humidity. Dry air is apparent below 400 mb in the Denver sounding (Fig. 20.10).

High-Level Clouds

The difference in the distribution of high cloud is harder to explain. One can hypothesize that it is due to differences in the terrain profiles or to moisture effects, or to some combination of the two.

The effect of differences in the terrain shape is illustrated in Fig. 20.11, which shows the linear hydrostatic waves that develop in a Boussinesq fluid flowing with uniform wind speed and stability over asymmetric mountains.

In Figs. 20.11a,b the mountain contours correspond to the fifth and first streamlines above the surface in Fig. 20.2c; the only difference in the three cases is the phase of the disturbance at the ground. Fig. 20.11a shows the streamline pattern forced by a mountain with a steep windward slope and a gentle lee slope. This corresponds to the large-scale character of the Sierras, which rise from sea level on the windward side, but descend only to the level of the intermountain plateau in the lee. In contrast, the Colorado Front Range corresponds to the situation shown in Fig. 20.11b, where the elevation upstream exceeds the elevation in the lee. It must be emphasized that this description applies only to the long wavelength characteristics of the mountain; on a smaller scale, the lee slope of the Sierras descending into Owens Valley is actually steeper than the windward slope. The vertical coordinate in Fig. 20.11 is labeled in vertical wavelengths ($L = 2\pi U/N$). For representative values of N and U, the heights of the high clouds in Fig. 20.8 would lie between $0.5L$ and $0.75L$. At these heights there would be a net downward displacement in the lee of the Sierras (Fig. 20.11a) and a net upward displacement in the lee of the Rockies (Fig. 20.11b), which would produce the observed distribution of high clouds.

A second possible explanation involves the influence of moisture on the flow upstream of the Sierras. Suppose that both ranges are represented by symmetric profiles and that the orographic cloud is forming at a physical height which corresponds to a level of $0.75L$ over the Rockies as illustrated in Fig. 20.12b. Durran and Klemp (1983) have demonstrated that a region of deep cloudiness on the upstream side of a mountain can increase the wavelength in vertically propagating waves by 30%, because the local stability is reduced in the regions where the air is saturated. If the deep clouds upstream of the Sierras were to increase the vertical wavelength by 30% relative to the (dry) value over the Rockies, the same physical height at which orographic cloud appears downstream of the Rockies would correspond to just $0.57L$ over the Sierras, which (as shown in Fig. 20.12a) would be an unfavor-

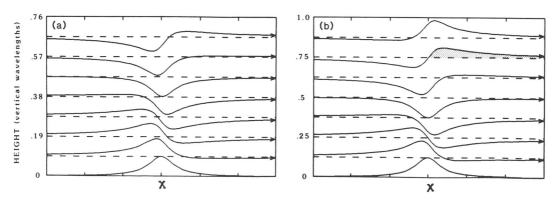

Figure 20.12. Streamlines in the steady airflow over an isolated bell-shaped mountain when the vertical wavelength in (a) exceeds the wavelength in (b) because of a reduction in the stability due to cloudiness over the windward slopes. Probable location of wave-induced cirrus is shaded.

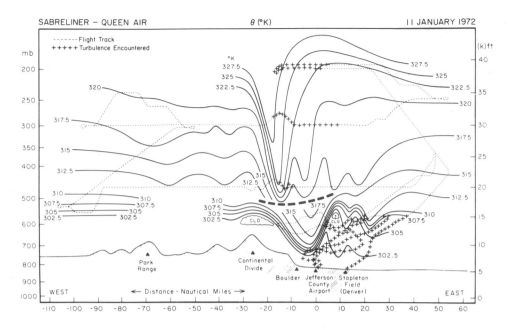

Figure 20.13. Cross section of the potential temperature field observed in a very strong mountain wave over Boulder, Colo., on 11 January 1972. The heavy dashed lines separate observations taken at different times; the dotted lines show the aircraft flight tracks; the crosses indicate regions of turbulence. (From Lilly and Zipser, 1972.)

able region for lee-side cloud development because there is a net downward displacement throughout most of the region.

If the previous two hypotheses are combined, so that the streamlines in Fig. 20.11a are adjusted to reflect an increase in vertical wavelength, the difference between the net vertical displacements in the upper troposphere becomes even more pronounced. A more complete analysis of this case is hampered by the lack of detailed information about the actual cloud levels, and uncertainties about nonlinear effects.

20.5. Development of Large-Amplitude Mountain Waves

The concepts that rely on linear theory (Secs. 20.2–20.3) are implicitly limited to relatively low-amplitude waves. Although they produce impressive cloud formations, these low-amplitude waves have little direct impact on human activities. In contrast, large-amplitude mountain waves can produce damaging downslope winds and dangerous regions of clear-air turbulence. An example of one such large-amplitude wave is shown in Fig. 20.13, which is an analysis of data collected over Boulder, Colo., on 11 January 1972 by Lilly and Zipser (1972). Understanding of the dynamics governing the behavior of large-amplitude mountain waves is, unfortunately, incomplete. Three possible mechanisms have been suggested to explain the development of large-amplitude waves such as the one shown in Fig. 20.13.

20.5.1. Hydraulic Jump

Long (1953) and several subsequent investigators examined hydraulic analogs to the atmosphere consisting of two or more immiscible fluids flowing over a barrier. Long established a set of conditions under which a hydraulic jump will form downstream of the barrier, and suggested that the mechanism that produces the hydraulic jump may be similar to one that produces strong waves and downslope winds in the atmosphere. This approach has the merit of explicitly accounting for the nonlinearities in the system; however, unlike the atmosphere, all the models considered have an upper boundary consisting of either a free surface or a rigid lid. Since no upward energy transport can occur across the upper boundary, vertically propagating atmospheric waves cannot be properly represented in hydraulic models. At any rate, since most of the cases examined have been limited to a fluid with no more than three homogeneous layers (i.e., $N = 0$ in each layer), any comparison of results with waves in the continuously stratified atmosphere must be primarily qualitative.

20.5.2. Reflection of Upward-Propagating Waves

A second mechanism for the generation of large-amplitude waves has been suggested by Klemp and Lilly (1975), who examined the behavior of linear waves in a multilayer atmosphere with constant stability and wind shear in each layer. Unlike Scorer, they limited their investigation to wavelengths that would be long enough to propagate vertically throughout every layer, even those with the smallest values of ℓ (in fact they assumed the waves were hydrostatic). When an upward-propagating wave encounters a region in which the Scorer parameter changes rapidly, part of its energy is reflected back into a downward-propagating wave. The wave amplitude below the reflecting layer is thus determined by the superposition of upward- and downward-propagating waves. Klemp and Lilly found that in the case of a three-layer atmosphere flowing over sinusoidal topography, the optimal superposition occurs when each of the lower layers is one-fourth of a vertical wavelength deep, in which case the downslope wind speed exceeds that which develops in the presence of a single uniform layer by the factor $N_1 N_3 / N_2^2$, where N_1, N_2, and N_3 are the Brunt-Väisälä frequencies in the lower, middle, and upper layers. If the upper layer is assumed to represent the stratosphere, this result suggests that the development of strong waves will be favored when the lower troposphere is relatively stable and the lapse rate in the remainder of the troposphere is nearly dry adiabatic.

The multilayer approach can be extended to more layers, and has the advantage that, as the number of layers is increased, the mean state used in the calculations can be configured to closely match an actual sounding. Klemp and Lilly (1975) used a many-layered model, with reasonable success, to predict downslope winds in Boulder, Colo., from soundings taken upstream. However, since this is a linear model, it must be applied with caution to the case of large-amplitude waves because the perturbation fields in these waves are comparable in magnitude to the mean flow. Recent numerical experiments suggest that, in the case of some large-amplitude waves, the predictions from linear multilayer models can be rather misleading.

20.5.3. Self-Induced Critical Layer

A third mechanism for the generation of large-amplitude waves has been suggested by Clark and Peltier (Clark and Peltier, 1977; Peltier and Clark, 1979). They examined the behavior of large-amplitude mountain waves with a nonlinear nonhydrostatic model, obtaining good agreement between their calculations and the flow observed on 11 January 1972 over Boulder, Colo. (Fig. 20.13). Their numerical approach has the advantage that it explicitly accounts for nonlinear and nonhydrostatic effects, and it allows a detailed representation of the atmospheric flow structure. They observed that a substantial increase in the strength of the wave occurs once the streamlines overturn and the wave "breaks." The wave-breaking region is characterized by strong mixing and a local reversal of the horizontal wind. Clark and Peltier suggest that the energy in the upward-propagating wave is trapped below this self-induced critical layer (the region where the wind reverses direction), thereby producing large-amplitude waves such as the one in Fig. 20.13. A disadvantage to the numerical approach is that the results reflect the highly complex interaction of several factors, and therefore can be rather difficult to interpret. At present, there is still some controversy about the role played by self-induced critical layers on the subsequent amplification of the wave (Lilly and Klemp 1980; Peltier and Clark, 1980).

20.6. Forecasting Mountain Waves

In spite of the advances in our understanding in the last 25 years, it is still not possible to make detailed forecasts of the strength, duration, and precise location of the mountain wave phenomena that occur in the real atmosphere. As a result, little improvement can be made to the forecasting advice offered by Queney *et al.* (1960) and Colson (1954), which was based on a combination of results from linear theory and observational experience. Note, for example, that the work of Peltier and Clark (1979) on the importance of wave overturning provides little guidance to the forecaster trying to assess the potential for wave development from an upstream sounding. One reasonable approach would be to estimate the wave response according to linear theory, and assume that if overturning is predicted it will occur in the actual nonlinear flow as well. In fact, it has not yet been clearly established whether the optimal upstream conditions for linear waves closely coincide with those for large-amplitude waves.

Queney *et al.* (1960) described the conditions in which strong waves are likely to develop:

- The mountain barrier in question has a steep lee slope. Theoretical support for this once largely empirical criterion was obtained by Smith (1977) and Lilly and Klemp (1979), who confirmed that asymmetric mountains with steep leeward and gentle windward slopes are the most effective generators of large-amplitude mountain waves.

- The wind is directed across the mountain (roughly within 30° of perpendicular to the ridge line) throughout a deep layer of the troposphere. The wind speed at the level of the crest should exceed a terrain-dependent value of 7 to 15 m s^{-1}, and should increase with height.

- The upstream temperature profile exhibits an inversion or a layer of strong stability near mountain top height, with weaker stability at higher levels. Colson (1954) also suggested that weak stability below the inversion favors the development of waves in the lee of the Sierra Nevada.

These conditions were originally established in order to predict trapped waves, so that the second and the third imply a decrease in the Scorer parameter with height, between the inversion and the tropopause. The evaluation of an approximation to the Scorer parameter,

$$\ell^* = \frac{N}{U} , \qquad (20.27)$$

is also recommended. (The U_{zz}/U term is not always negligible, but is omitted because it is difficult to evaluate from the radiosonde reports available in an operational environment.) Waves will be favored whenever the ℓ^* profile decreases significantly with height.

The importance of vertically propagating waves and the role played by the stratosphere have been recognized only in the last decade. The guidelines for forecasting vertically propagating waves, based again on linear theory and observations, are identical to conditions described by Queney *et al.* (1969) and listed above, except for the role played by wind speed (the wind direction and minimum speed criteria are unchanged). In the case of trapped waves, an increase in the wind speed with height produces a decrease in ℓ^*, enhancing the potential for wave trapping, so that roughly speaking, the stronger the upper level winds, the greater the chance for lee waves. Klemp and Lilly (1975) showed that the amplitude of vertically propagating waves reaches a maximum when the phase shift between the ground and the tropopause is one-half vertical wavelength. This implies that for a given stability profile, the optimal wind speed distribution should satisfy

$$\pi = \int_{z_s}^{z_t} \ell^* ds , \qquad (20.28)$$

where z_t is the height of the tropopause and z_s is the height of the topography or the top of a blocked layer of air upstream of the mountain, and $\pi = 3.1416$. If the stability distribution is suitable for intense waves, strong upper tropospheric winds will generally be required to satisfy (20.28), but the presence of very strong winds aloft (as in the case of the jet stream) will not necessarily favor the development of large-amplitude vertically propagating waves.

If strong waves of any type are forecast, clear-air turbulence and downslope winds are likely to develop as well. When the potential for downslope winds is estimated, the synoptic-scale pressure gradient across the mountains should also be considered. As discussed in Sec. 20.2, mountain waves generate a mesoscale pressure distribution with high pressure upstream of the crest and low pressure in the lee. Strong downslope winds are more likely to develop when the synoptic-scale pressure gradient is in phase with the wave-induced pressure gradient. For example, the strongest Colorado chinooks occur during wave events when there is a large region of high pressure upstream of the mountains to the west, and a rapidly developing lee-side trough or low pressure center in the high plains to the east or northeast.

REFERENCES

Atkinson, B. W., 1981: *Meso-Scale Atmospheric Circulations.* Academic Press, New York, 496 pp.

Clark, T. L., and W. R. Peltier, 1977: On the evolution and stability of finite-amplitude mountain waves. *J. Atmos. Sci.,* **34,** 1715–1730.

Colson, D., 1954: Meteorological problems in forecasting mountain waves. *Bull. Amer. Meteor. Soc.,* **35,** 363–371.

Durran, D. R., and J. B. Klemp, 1983: A compressible model for the simulation of moist mountain waves. *Mon. Wea. Rev.,* **111,** 2341–2361.

Gill, A. E., 1982: *Atmosphere-Ocean Dynamics.* Academic Press, New York, 662 pp.

Klemp, J. B., and D. K. Lilly, 1975: The dynamics of wave-induced downslope winds. *J. Atmos. Sci.,* **32,** 320–339.

Lilly, D. K., and J. B. Klemp, 1979: The effects of terrain shape on non-linear hydrostatic mountain waves. *J. Fluid Mech.,* **95,** 241–261.

Lilly, D. K., and J. B. Klemp, 1980: Comments on "The evolution and stability of finite-amplitude mountain waves. Part II: Surface wave drag and severe downslope windstorms." *J. Atmos. Sci.,* **37,** 2119–2121.

Lilly, D. K., and E. J. Zipser, 1972: The front range windstorm of 11 January 1972—a meteorological narrative. *Weatherwise,* **25,** 56–63.

Long, R. R., 1953: A laboratory model resembling the "Bishop-wave" phenomenon. *Bull. Amer. Meteor. Soc.,* **34,** 205–211.

Peltier, W. R., and T. L. Clark, 1979: The evolution and stability of finite-amplitude mountain waves. Part II: Surface wave drag and severe downslope windstorms. *J. Atmos. Sci.,* **36,** 1498–1529.

Peltier, W. R., and T. L. Clark, 1980: Reply. *J. Atmos. Sci.,* **37,** 2122–2125.

Queney, P., G. Corby, N. Gerbier, H. Koschmieder, and J. Zierep, 1960: The airflow over mountains. WMO Tech. Note 34, 135 pp.

Scorer, R., 1949: Theory of waves in the lee of mountains. *Quart. J. Roy. Meteor. Soc.,* **75,** 41–56.

Smith, R. B., 1977: The steepening of hydrostatic mountain waves. *J. Atmos. Sci.,* **34,** 1634–1654.

Smith, R. B., 1979: The influence of mountains on the atmosphere. In *Advances in Geophysics,* **21,** Academic Press, New York, 87–230.

CHAPTER 21

Lee Cyclogenesis

R. T. Pierrehumbert

21.1. Introduction

Determining the processes governing the geographical distribution of cyclogenesis has long been a problem of great theoretical and practical importance in dynamic meteorology. The pioneering work of Petterssen (1956) on the climatology of cyclogenesis established mountainous terrain as a major contributor to cyclogenesis. Through a study of Northern Hemisphere sea level pressure data from 1899 to 1939, he found distinct maxima in cyclogenesis and cyclone occurrence in the lee of most of the major mountain ranges of the hemisphere (notably the Rocky Mountains and the Alps). In this context, "cyclogenesis" was defined as a fall in pressure of greater than a certain magnitude, and a "cyclone" was defined as a region of low pressure. Chung and Reinelt (1973) and Chung et al. (1976) applied similar techniques to subjectively analyzed surface pressure maps for 1958, and essentially confirmed Petterssen's conclusions over North America. In such studies the identification of cyclones with low pressure regions is based on an assumption of approximate geostrophic balance; this can lead to very misleading results in mountainous terrain. Nevertheless, Petterssen's work has played a key role in most subsequent research on lee cyclogenesis, and has served as a major stimulus to the field.

21.1.1. Climatology of Cyclones

Recent work on the climatology of cyclones is rather more ambiguous with regard to the association of cyclone occurrence with mountain ranges. Notably, Blackmon et al. (1977) studied the geographical pattern of time variance of surface pressure and 300 mb geopotential height. They isolated the part of the transience attributable to synoptic-scale disturbances by first filtering the data to allow only motions with a time scale of ~2.5–10 days. In contrast with Petterssen's results, they found that, in the wintertime, the synoptic transient activity was primarily concentrated in two storm tracks: one over the Pacific Ocean and the other over the Atlantic. A slight secondary maximum of surface pressure variance was found in the lee of the Rockies, but no appreciable maximum was seen in the lee of the Alps. This study suggests that cyclogenesis in the lee of the Rockies is a shallow phenomenon, as its signal is apparent in surface and 850 mb data, but not at all in the 300 mb data. Murakami and Ho (1981) studied objectively analyzed data for 1979, and also failed to find distinct variance maxima in the immediate lee of the Rockies.

The reasons for the discrepancy are not clear. Because cyclones grow as they move downstream, maps of variance indicate regions of maximum cyclone intensity rather than regions of cyclogenesis. This distinction has no bearing on Petterssen's maps of cyclone occurrence though. Blackmon's technique has the virtue of eliminating quasi-stationary and very-short-lived disturbances; on the other hand, Chung and Reinelt also attempted to distinguish between stationary and moving disturbances. Perhaps the main reason for the discrepancy is that Petterssen and his followers defined a cyclone as a region of closed isobars. In contrast, large variance of pressure can occur without the appearance of any closed isobars at all, provided the mean westerly wind is strong enough; conversely, closed isobars can appear in a region of weak westerlies even if the variance is small.

Although the climatology of lee cyclogenesis is uncertain, there is considerable evidence from individual case studies, particularly in the Alps, that intense cyclogenesis does indeed occur from time to time in the lee of mountain ranges (see, e.g., Buzzi and Tibaldi, 1978). In fact, three moderate-to-strong cases were documented during the Alpine Experiment (ALPEX) carried out during March and April 1982. It is possible that strong lee cyclogenesis occurs too infrequently to yield an appreciable signal in the analysis of Blackmon *et al.* (1977), though it is no doubt a source of severe weather in specific instances.

In one sense, the patterns found by Blackmon *et al.* represent a type of lee cyclogenesis. The storm tracks are associated with regions of high baroclinicity, in a process that is now fairly well understood from a theoretical standpoint (see Frederiksen, 1983; Pierrehumbert, 1984a). These baroclinic zones are associated with climatological stationary troughs over the oceans. To a certain extent, the Pacific trough is due to a stationary, lee, Rossby wave train emanating from the Himalayas, and the Atlantic trough is due to a wave train emanating from the Rockies (Held, 1983). In this regard, the topography helps determine the existence and location of the storm tracks. However, topography is only one of the sources of the stationary planetary wave pattern, the other being atmospheric heating. The following discussion deals primarily with rapidly intensifying disturbances in the immediate lee of mountains, rather than with longer range orographic effects.

21.1.2. Dynamics of Lee Cyclogenesis

Research on the dynamics of lee cyclogenesis began concurrently with efforts to understand cyclogenesis in general; thus, much of the early work on cyclogenesis in the lee of the Rockies was directed at the general problem of development rather than the problem of lee cyclogenesis *per se*. There is a certain amount of confusion as to which results are specific to orographically induced cyclones. In addition, the pioneering work on lee cyclogenesis was carried out at a time before the importance of quasi-geostrophic theory and

Charney's baroclinic instability theory of cyclogenesis was widely appreciated. Consequently, a number of somewhat dated approaches to the problem of development have survived in the lee cyclogenesis literature, though they have largely disappeared from the rest of dynamic meteorology.

In the following, some of these points of confusion are clarified. However, there is at present no satisfactory theory of lee cyclogenesis. Moreover, observations indicate that lee cyclogenesis can occur in many different ways, manifested as a group of loosely related phenomena. Thus, a single theoretical explanation of lee cyclogenesis may not be possible. The intent is not to provide a theory of lee cyclogenesis, but rather to discuss the fundamental physical processes involved in the effect of topography on developing cyclones.

21.2. Observational Background

21.2.1. Cyclogenesis in the Lee of the European Alps

A large body of literature deals with observations and numerical modeling of Alpine cyclogenesis; Buzzi and Speranza (1983) present an overview of this material. In addition, a rich data set on Alpine cyclogenesis was recently collected during the Alpine Experiment (ALPEX). The discussion here is based on objective analysis of the ALPEX Level IIA data, performed at the Geophysical Fluid Dynamics Laboratory.*

Figure 21.1 shows a 12-hourly sequence of 850 mb temperature/geopotential maps for the 4–5 March 1982 ALPEX cyclogenesis case. Since the Alps extend to roughly 700 mb, this level is well below the height of the mountain barrier. The event begins at 0000 GMT on 4 March; a predominantly westerly current is flowing largely parallel to the mountains. A weak, broad trough is beginning to form to the west of the Alps, advecting cold air southward. The pattern is consistent with what one would expect in connection with developing baroclinic instability. During the next 12 hours, the trough deepens and moves east, whereupon it begins to interact with the mountains. Cold air penetrates into the Mediterranean east of the Alps but is blocked immediately north of the Alps, as evidenced by the shape of the 286 K isotherm. At this time, there is little airflow directed against the northern face of the mountains. Rapid cyclogenesis occurs during the next 12 hours, and a cutoff low has appeared by 0000 on 5 March. At this time, a northwesterly jet is directed against the mountain, and a ridge begins to form to the north of the barrier. The low-level cyclone deepens no further. At 1200 on 5 March, the occlusion process is well under way. Note that much cold air has been blocked north of the mountain, leading to a strengthening of the ridge.

The corresponding 500 mb maps are shown in Fig. 21.2. Again, the early development has many of the characteristics of baroclinic instability. In contrast with the low-level charts, there is little distortion of the thermal field

---

* The ALPEX analysis system at the Geophysical Fluid Dynamics Laboratory was developed in collaboration with Bruce Wyman. He also assisted in producing Figs. 21.1 and 21.2.

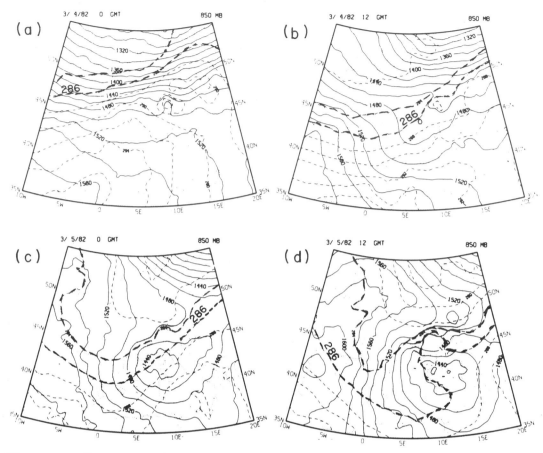

Figure 21.1. Geopotential (850 mb height) and temperature maps of Alpine cyclogenesis on 4–5 March 1982. Plain contours are height in meters; dashed contours are temperature in kelvins. (a) 4 March at 0000 GMT; (b) 4 March at 1200 GMT; (c) 5 March at 0000 GMT; (d) 5 March at 1200 GMT.

above the mountains, and the trough remains broad until 0000 on 5 March. Although the low-level development ceases at this time, the upper-level development continues, leading to the emergence of a closed low at 1200 on 5 March.

It must be emphasized that, even if attention is restricted to the alpine case, lee cyclogenesis can occur in many different ways, depending primarily on the orientation of the large-scale wind field with respect to the mountain. The classic case studied by Buzzi and Tibaldi (1978) occurred in an environment of predominantly northwesterly wind. Consequently, the interaction of a rapidly moving cold front with the mountain was much more prominent during the early stages of the cyclogenesis. The weak ALPEX cyclogenesis case of 29–30 April was similar to the case studied by Buzzi and Tibaldi. On the other hand, the ALPEX case of 24–25 April was dominated by the interaction of a pool of cold, high-vorticity air with the mountain, and bore no resemblance whatever to any instance of cyclogenesis that had been previously studied.

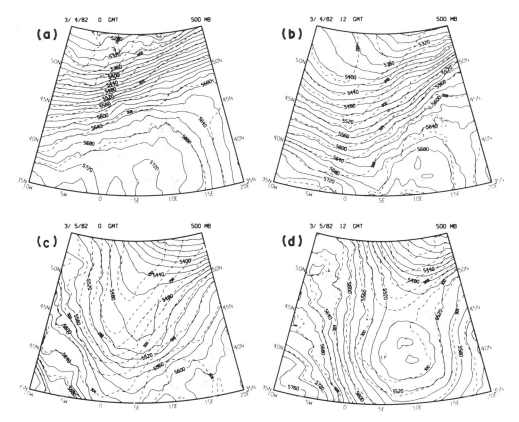

Figure 21.2. Geopotential (500 mb height) and temperature maps of Alpine cyclogenesis on 4–5 March 1982, as in Fig. 21.1.

21.2.2. Cyclogenesis in the Lee of the Rocky Mountains

Figure 21.3 reproduces the sequence of 12-hourly 700 mb and 850 mb geopotential height maps from the case study of the 21–22 January 1951 cyclone, carried out by McLain (1960). (It is difficult to find more recent case studies of Rocky Mountain cyclogenesis.) As is typical, this case begins with a primary cyclone wave propagating into the mountains from the Pacific. At 0300 on 21 January, the low-level primary cyclone (which is unfortunately off the western edge of the map in this analysis) has reached the coast; at this time the upper-level geostrophic wind is largely westerly, but the lower-level geostrophic wind is mostly parallel to the mountain. Over the next 12 hours, the system moves eastward and strong low-level westerlies cross the mountain barrier south of the primary cyclone. The geostrophic vorticity of the primary low weakens as it crosses the mountain. At 850 mb a pronounced ridge forms over the mountain and a trough forms in the lee. These features are nearly absent at 700 mb. By 0300 on 22 January a cutoff low has formed over the lee slope near the Canadian border. It is evident at 850 and 700 mb, but absent at 500 mb and above (charts not shown). This low intensifies markedly during the next 12 hours, as it moves down the lee slope. The intensification is strongest at low levels. In short, the parent

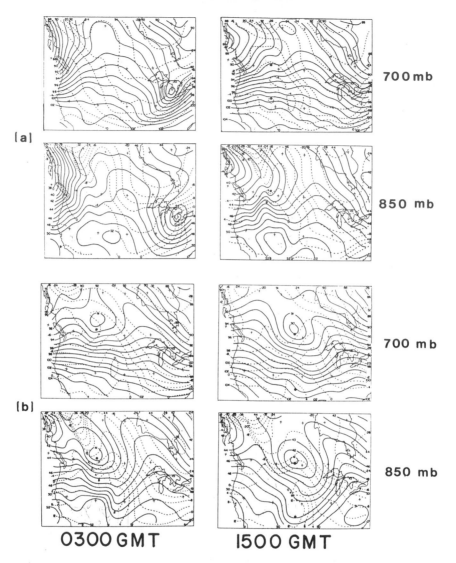

Figure 21.3. Geopotential (700 and 850 mb heights) and temperature maps of Rocky Mountain cyclogenesis on 21–22 January 1951. Plain contours are height, and dashed contours are temperature. (a) 21 January at 0300 and 1500 GMT; (b) 22 January at 0300 and 1500 GMT.

cyclone disappears, then reappears in the lee somewhat to the south and at a somewhat smaller scale.

In both the Alps and the Rocky Mountains, the formation of lee cyclones involves the physics of cyclogenesis, of airflow over mountains, and of interactions between the two.

21.3. Mechanisms of Cyclone Development

A cyclone is a concentrated area of cyclonic vorticity. Therefore, to understand cyclogenesis, one must look to the sources of cyclonic vorticity. Consider motions of a thin layer of fluid on a sphere rotating with angular velocity Ω. Introduce local Cartesian coordinates with x pointing east and y north, and choose pressure p as the vertical coordinate; further, let u, v, and ω be the velocities in the x, y, and p directions. Assume that the depth scale of the motion is much less than the horizontal scale of the motion, so that the hydrostatic approximation is valid and the contribution of the vertical velocity to the horizontal Coriolis force is negligible. Then the horizontal momentum equations become

$$\frac{du}{dt} - fv = -\partial_x \phi \tag{21.1}$$

$$\frac{dv}{dt} + fu = -\partial_y \phi , \tag{21.2}$$

where the material derivative is given by

$$\frac{d}{dt} = \partial_t + u\partial_x + v\partial_y + \omega\partial_p , \tag{21.3}$$

and all horizontal derivatives are to be taken on constant-pressure surfaces. In these equations, ϕ is the geopotential and $f = 2\Omega\sin\theta$, where θ is the latitude. The equation for the vertical component of the absolute vorticity is formed by subtracting the y derivative of (21.1) from the x derivative of (21.2). The result is

$$\frac{dQ}{dt} = Q\partial_p\omega + (\partial_y\omega\partial_p u - \partial_x\omega\partial_p v) , \tag{21.4}$$

where

$$Q = (\partial_x v - \partial_y u) + f . \tag{21.5}$$

The left-hand side of (21.4) is the change of Q following a fluid parcel. The right-hand side represents absolute vorticity production by vortex stretching (first term) and tilting (second term). Historically, most attention has centered on the stretching term. In particular, if the vertical velocity increases with height in a region of cyclonic absolute vorticity, the stretching term is a source of cyclonic absolute vorticity; since the Earth's vorticity is cyclonic, horizontal convergence can create cyclonic relative vorticity from a state of no relative motion. For mesoscale motions, the tilting term could in principle be equally important. In any event, the sources of vorticity are intimately linked with the pattern of vertical velocity. Hence, all theories of cyclogenesis amount to the delineation of the circumstances under which vertical velocity fields giving rise to cyclonic vorticity sources can be created and maintained.

21.3.1. The Vertical Motion Field

The problem is to relate the vertical motion field to meteorological quantities (such as vorticity and temperature advection) that can be easily observed and physically interpreted. This was the goal of the work of Sutcliffe and Petterssen. The objective can be consistently attained only for large-scale motions (i.e., motions characterized by a small Rossby number).

Petterssen's work on cyclogenesis was inspired by the work of Sutcliffe (1947), who made use of a selective application of the assumption of geostrophic balance in order to derive a diagnostic relation between vertical velocity and the horizontal advections of potential temperature and vorticity. In doing so, Sutcliffe came very close to discovering quasi-geostrophic theory. Sutcliffe's approach was not based on a systematic scale analysis of the equations of motion; in consequence, he retained a number of terms in his equation that could have been discarded, as they are generally no larger than terms he neglected in applying the geostrophic approximation. The resulting formulation, although not strictly incorrect, was more cumbersome than it needed to be.

Within the context of the quasi-geostrophic theory derived by Charney (1948), a more elegant diagnostic equation for vertical velocity can be derived. This relation is now known as the "omega equation." It is an interesting historical note that the connection between the omega equation and the work of Petterssen and Sutcliffe is not made in Petterssen's 1956 textbook, even though this book included a chapter by Eliassen on the quasi-geostrophic theory.

Derivations of the quasi-geostrophic equations are found in Charney (1948) and Pedlosky (1979). A statement of the result follows. (For the sake of simplicity, attention is restricted to scales of motion sufficiently small that f may be regarded as constant.) Let U, L, and D be the velocity, length, and depth scales of the motion, whence the Rossby number $\mathrm{Ro} = U/(fL)$ is defined. When Ro is small the left-hand side of (21.4) is dominated by the advection of the geostrophic vorticity by the geostrophic wind. Small Ro also implies that the tilting term on the right-hand side of (21.4) is negligible and that the relative vorticity is negligible compared with f in the stretching term. The vorticity equation becomes

$$\frac{d_0 \varsigma_g}{dt} = f(\partial_p \omega) , \tag{21.6}$$

where

$$\frac{d_0}{dt} = \partial_t + u_g \partial_x + v_g \partial_y$$

$$(u_g, v_g) = (-\partial_y \psi, \partial_x \psi)$$

$$\varsigma_g = \partial_x v_g - \partial_y u_g ,$$

and where $\psi = \phi/f$ is the geostrophic stream function. An important feature of (21.6) is that within the quasi-geostrophic approximation, cyclogenesis results from stretching of the Earth's vorticity alone, as the stretching of the relative vorticity is negligible. Let N be the Brunt-Väisälä frequency, and define the Burger number $S = ND/(fL)$. If S is of order unity or more, the horizontal potential temperature perturbations are small, and the thermodynamic equation can be written

$$\frac{d_0}{dt}(\partial_p \psi) + \frac{\omega}{\bar{\rho}^2 g^2} \frac{N^2}{f} = 0 \qquad (21.7)$$

through systematic use of the hydrostatic equation and the equation of state. In (21.7) $\bar{\rho}(p)$ is the background density. Equations (21.6) and (21.7), together with suitable boundary conditions, form a closed system.

With hindsight, Petterssen's ideas on cyclogenesis can best be expressed in terms of the quasi-geostrophic omega equation. To form this equation we eliminate the time derivative between (21.6) and (21.7) by subtracting the horizontal Laplacian of (21.7) from the vertical derivative of (21.6). This yields

$$\nabla^2 \omega + \frac{f^2 \bar{\rho}^2 g^2}{N^2} \partial_{pp} \omega = \frac{f \bar{\rho}^2 g^2}{N^2} [\partial_p(\mathbf{v}_g \cdot \nabla \varsigma_g) - \nabla^2(\mathbf{v}_g \cdot \nabla \partial_p \psi)] , \qquad (21.8)$$

which is an elliptic equation for ω. The forcing function on the right-hand side is the difference between the vertical derivative of the vorticity advection and the horizontal Laplacian of the temperature advection. Petterssen reasoned that a vertical motion field conducive to low-level development occurs in connection with certain specific configurations of vorticity and temperature advection, and particularly when an upper-level rapidly moving trough overtakes a slowly moving surface trough. As an illustration of this process, consider the action of a uniformly sheared geostrophic current $U = U_z z$ on a geopotential perturbation field dependent on x and p alone. The total stream function is

$$\psi(x, y, p) = -U_z yz + \psi'(x, p) . \qquad (21.9)$$

Upon substituting (21.9) into (21.8), we find

$$\nabla^2 \omega + \frac{f^2 \bar{\rho}^2 g^2}{N^2} \partial_{pp} \omega = -\frac{2U_z f \bar{\rho} g}{N^2} \partial_{xx}(\partial_x \psi') . \qquad (21.10)$$

To simplify the solution, attention is restricted to a fluid with constant N^2 and constant background density $\bar{\rho}$. In addition, all density variations are neglected except in the buoyancy term (this assumption is known as the Boussinesq approximation). In this case, $p = p_0 - \bar{\rho} g z$, where p_0 is the mean surface pressure, and (21.10) may be written

$$\nabla^2 \omega + \frac{f^2}{N^2} \partial_{zz} \omega = -\frac{2U_z f \bar{\rho} g}{N^2} \partial_{xx}(\partial_x \psi') \qquad (21.11)$$

in z coordinates. Suppose that the perturbation wind has the form $v_g(x, z) = \partial_x \psi' = \cos(kx + mz)$, representing a sequence of ridges and troughs, which tilt westward with height if k and m are positive. Equation (21.11) then reduces to

$$\partial_{xx} \omega + \frac{f^2}{N^2} \partial_{zz} \omega = \frac{2U_z k^2 f \bar{\rho} g}{N^2} \cos(kx + mz) . \qquad (21.12)$$

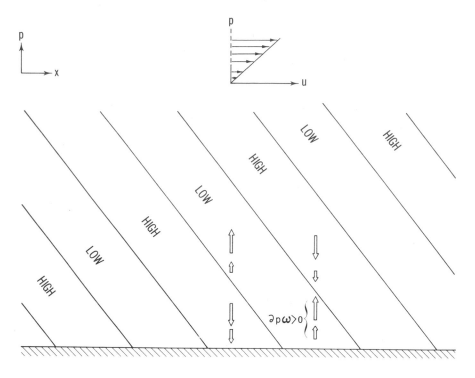

Figure 21.4. Vertical velocity pattern associated with a tilted trough-ridge system in the process of being distorted by westerly shear.

If we require that the vertical motion vanish at the ground, the solution of (21.12) is

$$\omega = \frac{-k^2}{k^2 + (m^2 f^2/N^2)} \frac{2U_z f \bar{\rho} g}{N^2} \left[\cos(kx + mz) - \cos(kx) e^{-k\frac{N}{f}z} \right] . \quad (21.13)$$

The ground-level stretching associated with this field is then

$$\partial_p \omega \mid_{z=0} = \frac{-k^2 m}{k^2 + (m^2 f^2/N^2)} \frac{2U_z f}{N^2} \left[\sin(kx) - \frac{Nk}{fm} \cos(kx) \right] . \quad (21.14)$$

The vertical velocity pattern relative to the pressure pattern is sketched in Fig. 21.4. From the fact that the geopotential perturbation near the ground is $\sin(kx)/k$, so the lows appear where $kx = -\pi/2$, $3\pi/2$, etc., it follows that the low-level stretching acts to intensify the low-level cyclones, provided that the wind shear U_z is positive and the perturbation has a steep westward tilt (specifically $m/k \gg N/f$). Barotropic perturbations, even in the presence of strong basic state baroclinicity, do not lead to initial development. This illustrates the general principle that a cyclone develops when a rapidly moving upper trough (with its associated vorticity advection) overtakes a low-level slowly moving trough (whose cold advection is greatest on its western flank). The same association between phase tilt and development can be deduced through energetics. It can be shown that, in the presence of westerly shear, only eddies with a westward phase tilt can release the potential energy of the basic state (see, e.g., Pedlosky, 1979).

21.3.2. Relation to Baroclinic Instability

Petterssen's result shows when a given vorticity configuration will initially lead to low-level cyclonic development. In time, the vorticity configuration evolves, and the cyclonic development may cease. Baroclinic instability theory, on the other hand, shows that some configurations can maintain their pattern indefinitely (within the limitations of linear theory), continually releasing potential energy in a process that leads to exponential growth of cyclonic vorticity. The relevance of Petterssen's mechanism in a given situation depends on how long the given initial vorticity pattern persists. Important recent work by Farrel (1984) on the initial value problem for baroclinic instability indicates that there are circumstances under which Petterssen's mechanism can lead to appreciable cyclogenesis, even if the underlying system is not baroclinically unstable in the exponential sense. (In fact, the example described above amounts to a reanalysis of Farrel's example in terms of vertical velocity.) In the absence of mountains, both mechanisms of baroclinic development achieve low-level cyclogenesis by means of vortex stretching, by maintaining ascending motion somewhat above the ground (where vertical velocity vanishes).

21.3.3 Connection with Lee Cyclogenesis

The basic processes responsible for middle-latitude cyclogenesis away from mountains have been described. It remains to understand the way in which mountains affect these cyclogenetic circulations. The treatment given in Sec. 21.3.1 is based on large-scale quasi-geostrophic dynamics, and is not valid for the small-scale distortions expected in the vicinity of mountains such as the Alps and the Sierra Nevada. It is often said that mountains enhance lee cyclogenesis by blocking low-level advection, whereby the vertical gradient of vorticity advection is enhanced, leading to enhanced cyclogenesis. This argument has little or no merit; as shown in Pierrehumbert and Wyman (1985), the blocking of flow by mountains is a fundamentally ageostrophic process, so that the associated circulations cannot be described in terms of the omega equation. On the other hand, the effects on cyclogenesis of geostrophically balanced orographic distortions (such as the standing anticyclone discussed in Sec. 21.4) can be consistently analyzed in terms of the omega equation.

In general, we anticipate two fundamentally different lee cyclogenesis mechanisms; the first is appropriate to large-scale mountains like the Rockies for which quasi-geostrophic ideas have some utility, and the second is appropriate to small-scale mountains like the Alps for which dynamics is dominated by ageostrophic diversion of low-level flow around the mountain.

21.4. Pressure and Wind Patterns Associated with Steady Flow Over Mountains

To understand the environment in which lee cyclogenesis takes place it is necessary to understand the patterns of pressure, wind, and vorticity that occur during steady flow over a mountain. Consider a mountain ridge oriented in the north-south direction, with height $h(x)$. Let h_m be the characteristic height of the mountain and L be its characteristic width. Suppose that the wind impinging on the mountain has a characteristic magnitude U, and the Brunt-Väisälä frequency has a typical magnitude N. Assume further that the density scale height has the constant value $H = -\bar{\rho}/\bar{\rho}_z$, and that the Coriolis parameter has a constant value f_0. The key nondimensional parameters are then (1) the Rossby number $\text{Ro} = U/(fL)$, (2) the internal inverse Froude number $\text{Fr} = Nh_m/U$, and (3) the nondimensional scale height $C = NH/(fL)$. Ro measures the importance of the Coriolis force, Fr measures the extent to which stratification causes fluid parcels to slow down as they ascend the upwind slope, and C measures the importance of compressibility effects.

The steady pattern varies as a function of these parameters. In the following description of the variation, attention is confined to motions that are small perturbations of a basic state with uniform zonal wind U and Brunt-Väisälä frequency N. The linear system was thoroughly explored in the pioneering work by Queney (1947). An accessible review of Queney's work may be found in Smith (1979) and Gill (1982), and additional properties of the solutions are discussed in Pierrehumbert (1984b). Note that in linear theory, the mountain height enters the problem only as a proportionality constant affecting the overall strength of the forcing. Hence, in the linear regime Fr influences the strength of the response, but not its fundamental character.

21.4.1. Case of Infinite Ro

The case of infinite Ro corresponds to a very narrow mountain or a very strong wind. In this circumstance, the Coriolis effect can be neglected. Additionally, if L is much larger than the characteristic vertical scale U/N, the hydrostatic approximation is valid and the equations of motion can be solved analytically for a bell-shaped mountain $h(x) = h_m/[1 + (x/L)^2]$. In this case the perturbation cross-mountain wind is given by

$$u' = \frac{e^{z/(2H)} N h_m [\sin(k_s z) + (x/L)\cos(k_s z)]}{1 + (x/L)^2}, \qquad (21.15)$$

where $k_s = N/U$. [Although it is not widely recognized, this formula is strictly valid only when $(k_s H)^{-1}$ is small.] At the surface, this solution yields upslope decelerations and downslope enhancement of the wind impinging on the barrier. Aloft, the motion takes the form of a vertically propagating gravity wave that oscillates with vertical wavelength $2\pi/k_s$ (see Fig. 21.5a). The oscillation extends to infinite altitude without diminution of energy.

Because the Coriolis force has been neglected, the along-mountain wind v' vanishes and there is no vertical vorticity associated with the motion. In the absence of Coriolis force the pressure is given by $p' = -\bar{\rho}(z)Uu'$. At $z = 0$, u' is negative on the upwind side and positive on the lee side; hence the surface pressure exhibits a high on the upwind side and a low to the lee, as illustrated in Fig. 21.5a. This lee trough is caused by adiabatic descent of air in the lee. There is no cyclonic vorticity associated with the trough. Hence, pressure fall in the lee of steep mountains does not imply cyclogenesis, and a climatology of pressure fluctuations is not a reliable indicator of cyclone occurrence in mountainous terrain.

21.4.2. Case of Small Ro

Small Ro corresponds to a very weak wind or a very broad mountain. In this limit, the dominant balance is between the Coriolis force and the pressure gradient force, and the quasi-geostrophic approximation is valid. Eliminating the vertical velocity between (21.6) and (21.7) yields the quasi-geostrophic potential vorticity equation. For y-independent motions, this equation becomes

$$\frac{d_0 q_g}{dt} = 0 , \tag{21.16}$$

where

$$q_g = \partial_{xx}\psi + \bar{\rho}^{-1}\partial_z \left[\left(\frac{f^2\bar{\rho}}{N^2} \right) \partial_z\psi \right] \tag{21.17}$$

in z coordinates; f has been assumed constant in (21.16). The bottom boundary condition is obtained from (21.7), using the requirement $w = d_0 h/dt$ at the ground, and becomes

$$\frac{d_0}{dt} \left(\partial_z\psi + \frac{N^2}{f}h \right) = 0 \quad \text{at} \quad z = 0 . \tag{21.18}$$

If the wind is initially uniform, q_g vanishes initially, and (21.16) implies that $q_g = 0$ for all time. The equation for the perturbation stream function $\psi'(x, z)$ is then

$$0 = \partial_{xx}\psi' + \bar{\rho}^{-1}\partial_z \left[\left(\frac{f^2\bar{\rho}}{N^2} \right) \partial_z\psi' \right]$$

$$= \partial_{xx}\psi' + \frac{f^2}{N^2} \left(\frac{\bar{\rho}_z}{\bar{\rho}} \partial_z\psi' + \partial_{zz}\psi' \right) , \tag{21.19}$$

assuming N^2 is constant. This is a linear elliptic equation for ψ'. The bottom boundary condition for (21.19) is obtained by linearizing (21.18) about a uniform zonal wind. The result is

$$\partial_z\psi' + \frac{N^2}{f}h = 0 \quad \text{at} \quad z = 0 . \tag{21.20}$$

As the upper boundary condition, the energy density is required to remain bounded at infinite z. When the density scale height is constant, (21.19) is a constant-coefficient equation, and can easily be solved by Fourier transform. If $h(x)$ is written

$$h(x) = (2\pi)^{-1/2} \int_{-\infty}^{\infty} \hat{h}(k)e^{ikx}dk \,, \tag{21.21}$$

then ψ' is readily found to be

$$\psi'(x,z) =$$

$$(2\pi)^{-1/2}\frac{N^2}{f}e^{z/(2H)} \int_{-\infty}^{\infty} \frac{2H\exp\left[-\frac{z}{2H}\left(1+\frac{4k^2N^2H^2}{f^2}\right)^{1/2}\right]\hat{h}(k)e^{ikx}}{\left(1+\frac{4k^2N^2H^2}{f^2}\right)^{1/2}-1}dk \,. \tag{21.22}$$

The vertical relative vorticity is $\varsigma_g' = \partial_{xx}\psi'$. From (21.22), then, the vorticity pattern is given by

$$\varsigma_g' =$$

$$(2\pi)^{-1/2}\frac{N^2}{f}e^{z/(2H)} \int_{-\infty}^{\infty} \frac{-2k^2H\exp\left[-\frac{z}{2H}\left(1+\frac{4k^2N^2H^2}{f^2}\right)^{1/2}\right]\hat{h}(k)e^{ikx}}{\left(1+\frac{4k^2N^2H^2}{f^2}\right)^{1/2}-1}dk \,. \tag{21.23}$$

The net vorticity generated by the mountain is obtained by integrating (21.23) over all x. This can be analytically integrated using the fact that the integral of e^{ikx} over all x is $(2\pi)\delta(k)$ (where δ is the Dirac delta-function), so that the only contribution to the integrated vorticity comes from the $k=0$ component in (21.23). Thus,

$$\int_{-\infty}^{\infty} \varsigma_g'(x,z)dx = v(\infty,z) - v(-\infty,z) = -\frac{(2\pi)^{1/2}f\hat{h}(0)}{H} = -\frac{fA}{H} \,, \tag{21.24}$$

where A is the cross-sectional area of the mountain. The net circulation is independent of height, and is anticyclonic; the non-zero circulation results in a permanent turning of the wind crossing the barrier. Note that (21.24) is precisely the same as the expression for the circulation generated by a homogeneous fluid of finite depth H crossing a barrier. However, the circulation in the stratified case is generated purely by compressibility effects, and vanishes in the incompressible limit ($H \to \infty$). The importance of compressibility in generating net anticyclonic circulation in stratified flow over obstacles was first made explicit by Smith (1979), who used a somewhat different argument than that used here.

The vorticity pattern predicted by (21.24) can be obtained analytically in two special cases. Consider first a broad mountain, with $L \gg NH/f$ (i.e., a width much greater than the Rossby radius of deformation based on the scale height). Then $\hat{h}(k)$ is nearly zero except when $|k| \ll f/(NH)$, and the integrand in (21.24) can be replaced by its small-k form. This yields

$$\varsigma_g' = -(2\pi)^{-1/2} \int \frac{f\hat{h}(k)e^{ikx}}{H} dk = -\frac{fh(x)}{H} . \qquad (21.25)$$

Thus, for a broad mountain, the vorticity is independent of height, and is everywhere anticyclonic.

Consider next a narrow mountain, with $L \ll NH/f$. Then, most of the energy in the spectrum of h is contained in wavenumbers much larger than $f/(NH)$, and the integrand in (21.24) can be replaced by its large-k limit, provided x is not too large. [The latter restriction arises because the large-x behavior of (21.24) is always dominated by the $k = 0$ contribution.] Then, near the mountain (21.24) can be approximated as

$$\varsigma_g'(x, z) = -(2\pi)^{-1/2} N \int |k|\hat{h}(k) \exp\left(ikx - \frac{N}{f}|k|z\right) dk . \qquad (21.26)$$

This can be integrated analytically for the bell-shaped mountain, as was first done by Queney:

$$\varsigma_g'(x, z) = -\frac{Nh_m}{L} \frac{(1 + \frac{z}{D})^2 - (\frac{x}{L})^2}{\left[(1 + \frac{z}{D})^2 + (\frac{x}{L})^2\right]^2} . \qquad (21.27)$$

In this case, the vorticity decays with height rather rapidly, and the depth of penetration is the Rossby "depth of deformation" based on the mountain width, $D = fL/N$. The maximum anticyclonic vorticity occurs at the mountain peak ($x = 0$) at ground level, and has a value of Nh_m/L. At each height there is a region of anticyclonic vorticity centered on the mountain crest, and weak tails of cyclonic vorticity upstream and downstream of this region. Within the narrow-mountain approximation, the effects of compressibility are negligible, and the net cyclonic vorticity in the tails exactly cancels the net anticyclonic vorticity above the mountain. Consequently, there is no permanent turning of the flow.

Between these two extremes of $fL/(NH)$, the flow exhibits some characteristics of each limit. At all levels, streamlines bend slightly cyclonically as they approach the mountain, turn sharply anticyclonically as they cross the crest, and turn back cyclonically in the lee (Fig. 21.5c). The 1000 km breadth of the Rocky Mountains puts them in the intermediate regime.

An important feature of the quasi-geostrophic solution for uniform flow over topography is that the topographically induced motion is independent

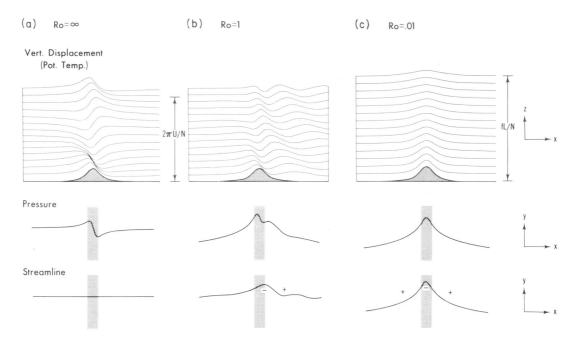

Figure 21.5. Steady flow over a ridge for Ro = ∞, Ro = 1, and Ro = 0.1. The first row gives the vertical displacement field in the x-z plane; these may also be regarded as contours of potential temperature. The second row shows the surface pressure perturbation, displayed as the pattern of a typical isobar. The bottom row gives the path taken by an air parcel near the ground as it crosses the mountain barrier. Stippling indicates position of the mountain.

of the wind speed U. This refutes a common fallacy in discussions of the effect of intensifying the wind impinging on the mountain. It is often said that intensified flow down the lee slope of the mountain intensifies the cyclonic vorticity by enhancing the orographically induced stretching. It is true that the enhanced winds increase the cyclonic vorticity source in the lee; however, this effect is offset by the fact that air parcels then spend less time in the source region, and so pick up less vorticity. In the uniform flow case discussed above, the two effects cancel exactly. Section 21.5 describes how a particularly extreme form of transience can mitigate this cancellation.

21.4.3. Case of Intermediate Ro

The mesoscale case, Ro = 0[1], does not permit an analytic solution. However, the following picture can be pieced together from numerical and asymptotic results given in Queney (1947) and Pierrehumbert (1984b). At low levels, the trajectory curves cyclonically as it approaches the mountain, and begins to curve anticyclonically near the crest of the mountain. The flow in the lee takes the form of an inertial wave with wavenumber $k_f = f/U$, which decays like $x^{-1/2}$ with downstream distance. In contrast with the quasi-geostrophic case, the inertial wave causes the maximum anticyclonic curvature to occur at approximately one-fourth the inertial wavelength downstream of the crest, and this maximum is followed by a weaker maximum of cyclonic vorticity one-half inertial wavelength farther downstream. The oscillation is strongest when the mountain is on the order of one-half inertial wavelength wide. Because the lee wave is inertial, it has little signature in the pressure pattern. Thus, we can also have cyclonic vorticity in the lee of a mountain without a low pressure center.

Aloft, the x-z cross section of the isentropes (and hence streamlines) looks much like the infinite Ro cases above the mountain, with the important difference that the dispersive effects of the Coriolis force cause the disturbance to decay to zero with height. The depth scale of the decay may be estimated as follows: The vertical group velocity of hydrostatic gravity waves with vertical wavenumber k_s and horizontal wavenumber k_x is Nk_x/k_s^2 (see Gill, 1982), and the time it takes for the Coriolis force to assert itself is on the order of $1/f$. The height attained by time $1/f$ is thus on the order of $D = Nk_x/(fk_s^2)$; when k_x is estimated as $1/L$, D is estimated as Ro/k_s. Above this height, geostrophic adjustment causes the isentropes to flatten rather quickly. As expected, the penetration depth becomes infinite as Ro becomes infinite.

Another important feature of the mesoscale case is that low-level anticyclonic vorticity is generated above and somewhat downstream of the mountain crest, where the isentropes draw together. However, the pattern reverses aloft at an altitude on the order of π/k_s, leading to the generation of predominantly cyclonic vorticity above the mountain at this altitude. It is possible that this supply of cyclonic vorticity plays a role in upper-level cyclogenesis for mesoscale mountains. The mesoscale pattern is summarized in Fig. 21.5b.

Nonlinear effects are discussed in Pierrehumbert (1985a) and Pierrehumbert and Wyman (1985). When Ro > 1, the oncoming flow below mountain top level is strongly blocked when Fr > 2; the blocked region extends upstream for a distance approximated by the radius of deformation, Nh_m/f. Under typical circumstances, the Alps are expected to produce marked blocking.

21.5. Transient Generation of Orographic Cyclones

Transient effects typically generate cyclonic vorticity in the lee. The basic idea, originated by Huppert and Bryan (1976), is that when flow over a mountain is impulsively started, quiescent air residing initially at the top of the mountain is forced into the lee. This air experiences the lee-side cyclonic vorticity generation without ever having been subjected to the upwind anticyclonic source, and hence acquires cyclonic relative vorticity. The orographic "starting vortex" moves downstream and evolves with time.

This mechanism can be illustrated within the context of a simple quasigeostrophic model. As in Sec. 21.4, a two-dimensional geometry with mountain profile $h(x)$ is assumed. To allow for the effects of baroclinicity (which were not considered by Huppert and Bryan), the equations of motion are linearized about a sheared profile $U = U_0 + zU_z$, in which U_z is constant. Further, making the Boussinesq approximation [$\bar{\rho} = \text{constant in (21.17)}$] causes the basic-state potential vorticity gradient to vanish. Thus, the perturbation potential vorticity vanishes for all time if it vanishes initially. The vertical structure is then governed by (21.19) with H taken to be infinite, and the transient effects are limited to the boundary condition (21.18). Transience in this system was studied by Smith (1984) in relation to stationary baroclinic lee waves when U_z is negative. Here, attention is focused instead on the characteristics of the transient lee cyclone that is generated when U_z and U_0 are positive.

Linearization of the bottom boundary condition about the sheared basic state yields

$$(\partial_t + U_0\partial_x)\partial_z\psi' - (\partial_x\psi')U_z = -\frac{N^2}{f}U_0\partial_x h , \qquad (21.28)$$

at $z = 0$. The solution may be represented as the sum of a steady part ψ_s and a transient part ψ_t, satisfying

$$U_0\partial_x \left(\partial_z\psi_s - \frac{\psi_s U_z}{U_0} + \frac{N^2}{f}h \right) = 0 \qquad (21.29a)$$

$$(\partial_t + U_0\partial_x)\partial_z\psi_t - (\partial_x\psi_t)U_z = 0 . \qquad (21.29b)$$

[The boundary condition (21.20) used for the unsheared case can be recovered from (21.29a) by integrating the latter once with respect to x and substituting $U_z = 0$.] If the flow is started impulsively from rest at $t = 0$, with initial condition $\psi_t(x, z, t) = -\psi_s(x, z)$ at $t = 0$, then the task is to find the subsequent time evolution of ψ_t.

The steady state is obtained by solving (21.19) subject to the lower boundary condition (21.29a). Equation (21.19) implies that the motion decays exponentially in z without any phase shift [as in (21.22), and in contrast with (21.12)]. When $U_z = 0$, the solution reduces to (21.27). In fact, (21.27) will be an approximate solution to (21.29a) whenever the length scale of the mountain is sufficiently small; because the depth scale of (21.27) is fL/N, the first term in (21.29a) will dominate the shear term whenever $fL/N << U_0/U_z$. For simplicity, attention is restricted to this case. The approximate steady state then has high pressure and anticyclonic vorticity at its core, surrounded by weak tails of cyclonic vorticity. Correspondingly, the transient part ψ_t has low pressure and cyclonic vorticity at its core initially.

It is straightforward to solve (21.19) subject to (21.29a) by means of the Fourier transform. If $\hat{\psi}_s(k, z)$ is the Fourier transform of $\psi_s(x, z)$, then the result is

$$\psi_t(x, z, t) |_{z=0} =$$

$$- (2\pi)^{-1/2} \int_{-\infty}^{\infty} \hat{\psi}_s(k, 0) \exp[ik(x - U_0 t)] \exp\left(-i\frac{k}{|k|}\frac{U_z f}{N}t\right) dk . \quad (21.30)$$

Consider first the unsheared case $U_z = 0$. Equation (21.30) reduces to $\psi_t(x, 0, t) = -\psi_s(x - U_0 t, 0)$. Hence, a transient low forms in the lee and is advected downstream at speed U_0 without change of shape. The vertical structure is identical to that of the steady state, apart from a reversal in sign. The cyclonic vorticity is strongest at the surface, and decays over a characteristic depth scale fL/N. Because the cyclonic vorticity decays toward zero with height, the structure has a warm core. In the unsheared case, inclusion of compressibility effects (finite H) is straightforward, and it can be shown that the mobile starting vortex in this case also has the same form as the steady state, apart from a sign reversal. An important difference is that, according to (21.25), for sufficiently broad mountains the mobile lee cyclone loses its warm core structure and becomes barotropic.

When U_z is non-zero, the starting vortex still moves downstream at speed U_0, but undergoes an oscillation of shape with angular frequency fU_z/N. As before, the shed vortex is initially cyclonic, but at a time $\pi N/(fU_z)$ later, the oscillatory factor in (21.30) becomes -1 for all wavenumbers, and the cyclone turns into an anticyclone. Hence, baroclinic effects act to destroy the cyclonic vorticity of the starting vortex. This might have been expected in light of the results of Sec. 21.3, as the shed vortex has no phase tilt with height, and therefore is not configured to release energy from the zonal flow. It must be emphasized that the addition of compressibility results in nontrivial modifications to the sheared case by introducing a non-zero meridional gradient of basic state potential vorticity. This makes the basic state baroclinically unstable. It is not yet clear whether the shed vortex in the compressible, sheared case will initially grow or decay.

When the Rossby number is order unity or greater, the picture is considerably modified by ageostrophic effects. In the ageostrophic regime, the formation of the starting vortex is expected to be a two-stage process. In the first stage, adiabatic descent of air parcels in the lee of the mountain creates a pool of warm air and low pressure in the lee, which begins to move downstream. Initially, this warm pool has little or no associated vertical vorticity, because the Coriolis force has had insufficient time to act. In time, the warm pool adjusts to geostrophic equilibrium by radiating inertial-gravity waves. Geostrophic adjustment involves adjustment of both the mass field and the wind field; hence, in the course of adjustment, the pressure drop at the center of the low weakens, the circulation becomes stronger, and the horizontal scale of the low increases. The proportion of energy that is lost to gravity wave radiation depends on the depth and length scales of the initial motion, and on deviation of the initial motion from geostrophy.

21.6. Theories of Lee Cyclogenesis

21.6.1. Lee Cyclogenesis With Broad Mountains (Rockies)

The classical theory of cyclogenesis in the lee of broad mountains like the Rockies assumes a flow pattern like that shown in Fig. 21.5c. Cyclogenesis is ascribed to the cyclonic vorticity source over the lee slope. The source arises because vertical velocity is negative at the ground and decays to zero with height, leading to positive $\partial \omega/\partial p$. This argument ignores the fact that air must first pass through an equally strong anticyclonic vorticity source on the upwind slope; indeed, Sec. 21.4 showed that the steady vorticity over broad mountains is everywhere anticyclonic. However, there is still some merit in the argument. With a 10 m s^{-1} wind, the overall massif of the Rockies is characterized by Ro = 0.1 and $C = 1$; the steady flow indeed looks like Fig. 21.5c, and is dominated by a standing anticyclone. Vorticities on the order of $0.5f$ are expected. The simplest hypothesis is that the cyclogenetic circulation described in Sec. 21.3 becomes linearly superposed with the standing mountain-induced flow. Thus, as the developing cyclone crosses the mountain, its vorticity is largely canceled by the standing anticyclone; it appears to grow rapidly as it emerges on the other side, though the strength in the end is the same as if the mountain had been absent. In this picture, the mountain temporarily masks the cyclone, but does not play any important dynamical role. There is little documented observational or numerical evidence as yet to indicate that anything besides masking is present. Indeed, considerable support for the masking hypothesis can be found in the numerical experiments of Hayes (1985).

The scales of the Rockies also permit the formation of an appreciable warm-core cyclonic starting vortex, like that discussed in Sec. 21.5. This could happen if the low-level flow west of the mountains shifted suddenly from southerly to westerly. There are some indications of this in Mclain's case study, in which the wind shifts are associated with the arrival of the primary cyclone. The starting vortex should emerge south of the primary cyclone (where the low-level flow is westerly) simultaneously with the formation of the standing anticyclone. An intensified lee development could result if the starting vortex became superposed with the primary cyclone. Taking ageostrophic effects into account (which are certainly present when the relative vorticity is as large as $0.5f$), the vortex stretching associated with the primary cyclone may intensify the vorticity of the starting vortex, leading to an intensity greater than predicted by linear superposition. The drying of moist maritime air through rainfall as the primary cyclone crosses the mountain may also lead to intensified development, though this effect has never been quantified.

Although the overall profile of the Rockies is characterized by small Ro, mountains like the Sierra Nevada have many local features small enough that Ro = 0[1]. These features can cause appreciable blocking of the oncoming flow, though on a scale small compared with that of the overall massif. The consequence of such blocking is uncertain.

Hess and Wagner (1948) noted that the variation of Coriolis force with latitude (the "beta effect") allows a parcel to acquire net cyclonic vorticity as it crosses the mountain, leading to the formation of a standing lee trough. A common fallacy is the identification of this lee trough with lee cyclogenesis. In modern terms, the lee trough is part of the stationary Rossby wave train emanating from the mountain. Such waves, with wavelengths of several thousand kilometers, dominate the climatological stationary wave pattern but are much larger than the scales of interest in lee cyclogenesis.

21.6.2. Lee Cyclogenesis With Narrow Mountains (Alps)

Under typical conditions the Alps are characterized by $Ro = 0[1]$, $Fr = 2$, and $NH/(fL) = 10$. The masking hypothesis is untenable for the Alps, because Ro is not small. Even if Ro were small (e.g., for weaker winds) the depth fL/N of the standing anticyclone would be insignificant and its breadth would be too small to mask the primary cyclone. In the indicated parameter range, blocking of low-level flow around the mountain is both expected and observed. There is no complete theory describing the modifications of baroclinic instability by this type of blocking. Radinovic recognized the importance of blocking, though his theory relied in a rather complicated way on the shape of the cold front in the lee, and moreover depended heavily on quasi-geostrophic reasoning (see Buzzi and Speranza, 1983, for a summary). More simply, one expects the mountain to hold back the advection of cold air, leading to anomalously high pressures on the upwind side and anomalously low pressures in the lee (compared with what would have occurred without mountains). This prediction is consistent with numerical studies of the effects of the Alps on cyclogenesis (see, e.g., Tosi *et al.*, 1983). Low pressure need not imply cyclogenesis, but under a variety of circumstances geostrophic adjustment can create a cyclonic circulation from a low pressure region.

There has been much theoretical work on the Alpine problem, though the results are far from conclusive. Pierrehumbert (1985b) solved the problem of baroclinic instability in the presence of a barrier that blocks low-level flow. The theory employed reproduces a number of aspects of the numerical experiments on orographic modification of baroclinic instability, though it does not account for the high-low pattern described above. Smith (1984) proposed a linear theory involving forced stationary waves, which under special conditions is able to produce considerable lee pressure falls. Speranza *et al.* (1984) proposed a theory of orographically modified instability that does yield a high-low pattern. The relevance of the last two theories is rendered somewhat uncertain by their reliance on properties of quasi-geostrophic flow over smooth obstacles; the relevance of the theory in Pierrehumbert (1985) may be questioned on account of the assumed highly idealized geometry.

The key process in all these scenarios is the modification of baroclinic instability by the presence of a mountain. Although first documented in the vicinity of the Rockies, the vorticity advection patterns discussed in Sec. 21.3 are part and parcel of baroclinic instability in general and are not specific to lee cyclogenesis. The traditional understanding of lee cyclogenesis amounts to the linear superposition of this pattern with quasi-geostrophic orographically induced flow. This picture has some merit for the relatively broad Rockies, but in the narrower Alps ageostrophic effects dominate and in consequence the interaction is more difficult to analyze. It must be recognized, however, that much of the intricacy of the Alpine problem was brought to light only through the extensive body of theoretical, numerical, and observational work carried out in connection with the Alpine Experiment. The problem of Rocky Mountain cyclogenesis is, in comparison, practically unexplored. It is not unlikely that a more extensive examination of the problem will reveal considerable subtleties.

REFERENCES

Blackmon, M., J. M. Wallace, N. C. Lau, and S. Mullen, 1977: An observational study of the Northern Hemisphere wintertime circulation. *J. Atmos. Sci.*, **34**, 1040–1053.

Buzzi, A., and A. Speranza, 1983: Cyclogenesis in the lee of the Alps. In *Mesoscale Meteorology—Theories, Observations and Models.* D. K. Lilly and T. Gal-Chen (Eds.), D. Reidel, New York, 55–142.

Buzzi, A., and S. Tibaldi, 1978: A case study of Alpine lee cyclogenesis. *Quart. J. Roy. Meteor. Soc.*, **104**, 271–287.

Charney, J. G., 1948: On the scale of atmospheric motions. *Geophys. Publ.*, **17(2)**, 17 pp.

Chung, Y. S., and E. R. Reinelt, 1973: On cyclogenesis in the lee of the Canadian Rocky Mountains. *Arch. Meteor. Geophys. Bioklim., Ser. A*, **22**, 205–226.

Chung, Y. S., K. D. Hage, and E. R. Reinelt, 1976: On lee cyclogenesis and airflow in the Canadian Rocky Mountains and the East Asian Mountains. *Mon. Wea. Rev.*, **104**, 879–891.

Farrel, B., 1984: Modal and non-modal baroclinic waves. *J. Atmos. Sci.*, **41**, 668–673.

Frederiksen, J., 1983: Disturbances and eddy fluxes in Northern Hemisphere flows: Instability of three-dimensional January and July flows. *J. Atmos. Sci.*, **40**, 836–855.

Gill, A. E., 1982: *Atmosphere-Ocean Dynamics.* Academic Press, New York, 662 pp.

Hayes, J., L., 1985: A numerical and analytical investigation of lee cyclogenesis. Ph.D. dissertation, Naval Postgraduate School, Monterey, California, 138 pp.

Held, I. M., 1983: Stationary and quasi-stationary eddies in the extratropical troposphere: Theory. In *Large-Scale Dynamical Processes in the Atmosphere*, B. Hoskins and F. Pearce (Eds.). Academic Press, New York, 127–167.

Hess, S. L., and H. Wagner, 1948: Atmospheric waves in the northwestern United States. *J. Meteor.*, **5**, 1–19.

Huppert, H., and K. Bryan, 1976: Topographically generated eddies. *Deep-Sea Res.*, **8**, 655–679.

Mclain, E. P., 1960: Some effects of the Western cordillera of North America on cyclonic activity. *J. Meteor.*, **17**, 105–115.

Murakami, T., and L. Y. C. Ho, 1981: Orographic influence of the Rocky Mountains on the winter circulation over the contiguous United States: Part II, Synoptic-scale (short period) disturbances. *J. Meteor. Soc. Japan*, **59**, 683–708.

Pedlosky, J., 1979: *Geophysical Fluid Dynamics.* Springer-Verlag, New York, 624 pp.

Petterssen, S., 1956: *Weather Analysis and Forecasting.* McGraw-Hill, New York.

Pierrehumbert, R. T., 1984a: Local and global baroclinic instability of zonally varying flow. *J. Atmos. Sci.*, **41**, 2141–2162.

Pierrehumbert, R. T., 1984b: Linear results on the barrier effects of mesoscale mountains. *J. Atmos. Sci.*, **41**, 1356–1367.

Pierrehumbert, R. T., 1985a: Formation of shear layers upstream of the Alps. *Revista di Meteorologia Aeronautica*, Special Issue on Proceedings of the 5th Course on Meteorology of the Mediterranean, Erice, Italy, 1983.

Pierrehumbert, R. T., 1985b: A theoretical model of orographically modified cyclogenesis. *J. Atmos. Sci.*, **42**, 1244–1258.

Pierrehumbert, R. T., and B. Wyman, 1985: Upstream effects of mesoscale mountains. *J. Atmos. Sci.*, **42**, 977–1003.

Queney, P., 1947: Theory of perturbations of stratified currents with applications to airflow over mountain barriers. Misc. Tech. Rep. 23., Dept. of Meteorology, University of Chicago.

Smith, R. B., 1979: The influence of mountains on the atmosphere. *Adv. Geophys.*, **21**, 87–220.

Smith, R. B., 1984: A theory of lee cyclogenesis. *J. Atmos. Sci.*, **41**, 1159–1168.

Speranza, A., A. Buzzi, A. Trevisan, and P. Malguzzi, 1985: A theory of deep cyclogenesis in the lee of the Alps: Part I, Modifications of baroclinic instability by localized topography. *J. Atmos. Sci.*, **42**, 1521–1535.

Sutcliffe, R. C., 1947: A contribution to the problem of development. *Quart. J. Roy. Meteor. Soc.*, **73**, 370–383.

Tosi, E., M. Fantini, and A. Trevisan, 1983: Numerical experiments on orographic cyclogenesis: Relationship between the development of the lee cyclone and the basic flow characteristics. *Mon. Wea. Rev.*, **111**, 799–814.

CHAPTER 22

Mesoscale Circulations Forced by Differential Terrain Heating

R. A. Pielke
M. Segal

22.1. Introduction

The forcing of mesoscale systems by horizontal gradients of surface heating is conceptually one of the most straightforward physical processes in the atmosphere. Therefore, it is surprising that thermally forced mesoscale systems are not exploited more by weather forecasters to provide improved predictions of local weather.

The existing forecasting methodologies traditionally emphasize large-scale forcing. However, in many situations locally induced meteorological systems are more important in determining local weather conditions. Thus, improvement of local weather forecasting is linked to improved understanding of mesoscale circulations through the use of both observations and models.

The classic thermally forced mesoscale systems include the land and sea breezes, mountain-valley winds, and urban heat island circulations. Less commonly considered types of thermally generated mesoscale flows include those due to horizontal contrasts in ground wetness, to snow-covered-soil/bare-soil contrasts, to spatial variations in cloud cover, and to vegetation/bare-soil contrasts.

22.2. Basic Equations

The derivation of the equations used to described mesoscale flow is reported in detail in Pielke (1984, Ch. 4). Therefore, only the results of direct relevance to thermally forced mesoscale circulations are presented here. Only circulations that are forced by terrain gradients in heating and cooling are discussed.

22.2.1. The Circulation Equation

From Pielke (1984, Ch. 4), the equation of motion can be written in the context of a mesoscale model as

$$\frac{\partial \overline{V}}{\partial t} = -\overline{V}\cdot\nabla\overline{V} - \alpha_0\nabla p' - \alpha_0\nabla_H p_0 + \frac{\alpha'}{\alpha_0}g\mathbf{k} - 2\mathbf{\Omega}\times\overline{V} - \left(\frac{1}{\rho_0}\right)\nabla\cdot\rho_0\overline{V''V''}\,, \quad (22.1)$$

where $\overline{V}$ is the vector velocity, $\mathbf{\Omega}$ is the angular velocity of the Earth, g is the gravitational acceleration, α is specific volume, p is pressure, and ρ is density $(\rho = 1/\alpha)$. The subscript $(\ )_0$, indicates a domain-averaged (i.e., synoptic-scale) quantity; a prime denotes a mesoscale perturbation from the synoptic average. The operator ∇ is three-dimensional; ∇_H is applied only on the horizontal. The dot and cross products are denoted by $\cdot$ and $\times$, respectively. The overbar on a variable indicates the sum of the synoptic average and the mesoscale perturbation. A double prime represents subgrid-scale deviations from the grid-volume-averaged quantities; $\rho_0\overline{V''V''}$ indicates the subgridscale flux of momentum. The individual terms in (22.1) have the following meanings:

$\partial\overline{V}/\partial t$	Local change of velocity
$\overline{V}\cdot\nabla\overline{V}$	Advection of velocity. As shown, for example, by Martin and Pielke (1983), this term concentrates or spreads out the temperature gradient through nonlinear effects as well as translates the atmospheric system.
$\alpha_0\nabla p'$	The mesoscale pressure gradient (forcing term for the mesoscale velocity accelerations).
$\alpha_0\nabla_H p_0$	The horizontal synoptic-scale gradient of pressure (assumed to be zero in this analysis).
$\dfrac{\alpha'}{\alpha_0}g\mathbf{k}$	Buoyancy term.
$2\mathbf{\Omega}\times\overline{V}$	The Coriolis term. The term is needed to represent the effect of a rotating coordinate system (i.e., the Earth) on motion.

$\dfrac{1}{\rho_0}\nabla \cdot \rho_0 \overline{V''V''}$ Subgrid-scale flux of velocity. This term represents the mechanism by which velocity is decelerated through dissipation into the ground, as well as accelerated or decelerated through turbulent transfers of momentum in all three spatial directions.

When the Stokes theorem is used (e.g., Kaplan, 1952, p. 275), circulation C on an arbitrary surface S can be expressed as

$$C = \oint \rho_0 \overline{u}_T \, dl = \iint_S \mathrm{curl}_n \rho_0 \overline{V} \, dS \ , \qquad (22.2)$$

where $\rho_0 \overline{u}_T$ is the momentum component tangent to the path of integration l and the integrand on the right side of (22.2) represents the density-weighted vorticity normal (i.e., in the direction n) to the surface S. The surface S is bounded by the path represented by l.

By using the definition of vorticity and the anelastic continuity equation (i.e., $\nabla \cdot (\rho_0 \overline{V}) = 0$), the change in circulation with time can be obtained from (22.1), through the curl operation ($\nabla \times$), which yields

$$\frac{\partial \overline{\xi}}{\partial t} = - \nabla \times (\nabla \cdot \rho_0 \overline{V}\ \overline{V}) + g \nabla \times \boldsymbol{k} \left(\frac{\alpha'}{\alpha_0^2} \right)$$
$$- 2\nabla \times (\Omega \times \rho_0 \overline{V}) - \nabla \times \nabla \cdot \rho_0 \overline{V''V''} \ , \qquad (22.3)$$

where $\overline{\xi} = (\nabla \times \rho_0 \overline{V})$.

The individual terms in (22.3) denote the following:

$\partial \overline{\xi}/\partial t$ The local tendency of vorticity.

$-\nabla \times (\nabla \cdot \rho_0 \overline{V}\ \overline{V})$ The gradient of the resolvable flux of vorticity and the tilting term whereby resolvable vorticity is transferred between the three spatial components as a result of velocity shear.

$\nabla \times \nabla \cdot \rho_0 \overline{V''V''}$ Include the subgrid-scale term in which vorticity is created or destroyed by small-scale motions.

$g\nabla \times \boldsymbol{k}(\alpha'/\alpha_0^2)$ The solenoidal term in which vorticity is created or removed as a result of gradients in density. For shallow atmospheric circulations (as discussed in the next paragraph) this term becomes $g\nabla \times \boldsymbol{k}(\rho_0 \theta'/\theta_0)$ so that differential heating is one mechanism to change the vorticity.

$$2\nabla \times (\Omega \times \rho_0 \overline{V})$$

The solid-body rotation term in which vorticity is created or destroyed as a result of motion on the rotating Earth.

Thermally forced circulations result from the second term on the right of (22.3), since

$$\frac{\alpha'}{\alpha_0} \simeq \frac{T_v'}{T_0} - \frac{p'}{p_0} \simeq \frac{\theta'}{\theta_0} - \frac{c_v}{c_p}\frac{p'}{p_0} . \tag{22.4}$$

These relations are obtained from the linearized form of the ideal gas law and definition of potential temperature $[\theta = T_v(1000/p)^{R_d/c_p}]$ where c_v and c_p are the heat capacities for air at constant volume and pressure, and R_d is the gas constant for dry air; p is given in millibars. T_v is the virtual temperature $[T_v = T(1 + 0.61q); q$ is the specific humidity].

The pressure perturbation term in (22.4) can be ignored (Dutton and Fichtl, 1969) for shallow atmospheric systems (i.e., where the vertical scale of the circulation, L_z, is much less than the density scale height of the atmosphere, H_α, where $[(1/\alpha_0)(\partial\alpha_0/\partial z)]^{-1} = H_\alpha$). For this case

$$\frac{\alpha'}{\alpha_0} \simeq \frac{T_v'}{T_0} \simeq \frac{\theta'}{\theta_0}$$

so that the solenoidal term in (22.3) becomes

$$\text{solenoidal term} \simeq g\nabla \times \boldsymbol{k}\left(\frac{T_v'}{\alpha_0 T_0}\right) \simeq g\nabla \times \boldsymbol{k}\left(\frac{\theta'}{\alpha_0\theta_0}\right) . \tag{22.5a}$$

From (22.2) and (22.5a), therefore, thermally forced atmospheric circulations develop and change with time when a spatial gradient in temperature exists on a surface. That is,

$$\frac{\partial C}{\partial t} = \oint \rho_0 \frac{\partial \overline{u}_T}{\partial t}dl = \int\int_S \frac{\partial \overline{\xi}}{\partial t}ds , \tag{22.5b}$$

where $\overline{\xi}$ is the magnitude of the density-weighted vorticity normal to the surface S.

22.2.2. Equation of Motion for Hydrostatic Flow

On the mesoscale, it is convenient to impose a condition on (22.1), namely, that the flow be in hydrostatic equilibrium. For this case (22.1) reduces to

$$\frac{\partial \overline{V}}{\partial t} = -\overline{V}\cdot\nabla\overline{V} - \frac{1}{\rho_0}\nabla\cdot\rho_0\overline{V''V''} - \alpha_0\nabla_H p' - 2\Omega\times\overline{V} \tag{22.6}$$

$$\frac{\partial p'}{\partial z} = \frac{\alpha'}{\alpha_0^2}g = \frac{\theta'}{\alpha_0\theta_0}g \quad \text{(for shallow atmospheric circulations)}$$

$$= \frac{g}{\alpha_0}\left(\frac{\theta'}{\theta_0} - \frac{c_v}{c_p}\frac{p'}{p_0}\right) \text{(for deep atmospheric circulations)}. \tag{22.7}$$

In (22.6) the pressure gradient is the forcing term for the horizontal accelerations. Because of the hydrostatic assumption there are no explicit vertical
accelerations. Differentiating (22.7) with respect to x and y (i.e., ∇_H), re-
versing the order of the differential operations, and invoking the conditions
that α_0 and θ_0 be functions only of z yields

$$\frac{\partial}{\partial z}(\nabla_H p') = \frac{g}{\alpha_0}\nabla_H\left(\frac{\alpha'}{\alpha_0}\right) = \frac{g}{\alpha_0\theta_0}\nabla_H\theta' \text{ (for } L_z \ll H_\alpha)$$

$$= \frac{g}{\alpha_0}\left(\frac{1}{\theta_0}\nabla_H\theta' - \frac{c_v}{c_p p_0}\nabla_H p'\right) \text{ (for } L_z \cong H_\alpha) . \tag{22.8}$$

Integrating (22.8) from the surface to an arbitrary level z_1 results in

$$\nabla_H p'|_{z_1} = \nabla_H p'|_{z=0} + g\int_0^{z_1}\frac{1}{\alpha_0}\nabla_H\left(\frac{\alpha'}{\alpha_0}\right) . \tag{22.9}$$

The level to which (22.9) is integrated depends on the height to which
significant fluctuations in specific volume α' occur. Nonzero values of α' occur because of diabatic heat inputs, and because of adiabatic motions that
result in response to diabatic heating. For thermally forced mesoscale systems this height is expected to be proportional to the height of the boundary
layer. Hence, except for extremely deep boundary layers, the associated atmospheric system would, in general, be expected to be shallow with respect
to H_α.

Therefore, to illustrate the expected order of magnitude and the terms of
influence on the horizontal pressure gradient, $z_1 = z_I$ can be used where z_I is
the height of the boundary layer. Thus for shallow systems, (22.9) becomes

$$\nabla_H p'|_{z_1} = \nabla_H p'|_{z=0} + g\int_0^{z_I}\frac{\nabla_H\theta'}{\alpha_0\theta_0}\,dz . \tag{22.10}$$

At the initial time of the development of a thermally forced mesoscale system,
$\nabla_H p'$ at $z = 0$ is presumed zero so that the horizontal pressure gradient force
is directly proportional to the vertically integrated temperature gradient.
Once this thermal circulation develops, temperature perturbations can occur
above z_I (i.e., because of vertical motion). However, at the initial time, if
the temperature fluctuations are caused by surface heating, θ' variations will
be initially confined to heights lower than z_I.

To illustrate the utility of (22.10) in explaining thermally forced systems, assume that the boundary layer is well mixed (i.e., a convective boundary layer). Equation (22.10) then reduces to

$$\nabla_H p'|_{z_I} = \nabla_H p'|_{z=0} + \frac{g z_I}{\alpha_0 \theta_0} \nabla_H \hat{\theta}' \, , \tag{22.11}$$

where $\hat{\theta}'$ is the average value of θ' between the surface and z_I. For this case, the horizontal pressure gradient and therefore the horizontal acceleration (from 22.6) is directly proportional to (1) the magnitude of the horizontal temperature gradient, and (2) the depth to which this gradient in temperature extends. The expression given by (22.11) is equivalent to the statement that horizontal acceleration is proportional to the horizontal gradient of heat content. This conclusion is similar to that found by Pearson (1973). Pielke *et al.* (1982) discussed this type of scale analysis in more detail.

If local density variations are ignored in the conservation-of-mass equation, which is the common practice for mesoscale flows (e.g., see Pielke, 1984, pp. 23–29), vertical velocity can be obtained from

$$\frac{\partial \rho_0 w}{\partial z} = -\rho_0 \left(\frac{\partial u}{\partial x} + \frac{\partial v}{\partial y} \right) \tag{22.12a}$$

or (for shallow atmospheric systems) from

$$\frac{\partial w}{\partial z} = - \left(\frac{\partial u}{\partial x} + \frac{\partial v}{\partial y} \right) \, . \tag{22.12b}$$

Therefore, if a horizontal temperature gradient results in a horizontal gradient in the horizontal acceleration, vertical velocity results.

Equations (22.12) and (22.6) are equivalent to the circulation equation given by (22.5b) except that the hydrostatic equation has been assumed. Because of this assumption, whereas vertical accelerations directly result from (22.5b), vertical accelerations are only implicit in the hydrostatic form of (22.1). These implicit vertical accelerations in the hydrostatic system are a result of horizontal accelerations, along with the requirement for mass conservation as described by (22.12).

The major conclusion of this analysis is that for atmospheric systems that are thermally forced at the surface, the intensity of the resultant circulation is directly proportional to the magnitude of the temperature gradient and the depth in the atmosphere to which the temperature perturbation extends.

As stated previously, if the spatial and temporal characteristics of the temperature are known, solution of (22.6) and (22.12) is possible. Otherwise, as is common in model simulations, the thermodynamic equation is needed. The thermodynamic equation can be written as (Pielke, 1984, p. 49)

$$\frac{\partial \overline{\theta}}{\partial t} = -\overline{\boldsymbol{V}} \cdot \nabla \overline{\theta} - \frac{1}{\rho_0} \nabla \cdot \rho_0 \overline{\boldsymbol{V}'' \theta''} + \overline{S}_\theta \, , \tag{22.13}$$

where $\overline{S}_\theta$ is the source/sink term for heat and includes average radiative flux divergence and phase changes of water. When significant water is present to influence $\overline{S}_\theta$, a conservation equation is also needed for water substance. The appendix to Chapter 4 of Pielke (1984) presents three sets of consistent equations to describe atmospheric flow.

A qualitative description of land-water thermally driven circulations in the absence of synoptic flow is presented in Pielke (1984, p. 456–457). As evident directly from (22.6) or from Pielke (184, p. 456) the wind field, at least initially, is perpendicular to the pressure gradient at all levels, even above the boundary layer. Therefore, in the terminology of synoptic models, these surface-forced, thermally driven circulations are not near gradient balance (i.e., a substantial ageostrophic wind exists). The divergent component of the wind is initially much larger than the rotational component. After a period of time (a substantial portion of the inertial period), the Coriolis effect, represented by the fourth term on the right of (22.6), will cause the wind to turn toward the right in the Northern Hemisphere and to become more parallel to the isobars above the level at which ground frictional effects [i.e., the second term on the right of (22.6)] are large.

Thus the following conditions characterize a surface, thermally forced system:

- The system initially consists of only a divergent horizontal wind component.
- If the thermal gradient persists long enough, Coriolis turning will become significant and a rotational component of the wind will develop.
- The wind will not monotonically approach a steady-state rotational component in the absence of ground friction (e.g., the gradient wind), even in the presence of a constant thermal field, but will overshoot its equilibrium value. This overshoot, caused by the development of a mesoscale thermal gradient, is one source of low-level jets, as discussed by McNider and Pielke (1981), and Mizzi and Pielke (1984).

Note that the basic concepts that describe thermally forced mesoscale flows apply equally well in irregular or flat terrain.

22.3. Surface Heat Balance

Specific physical mechanisms create surface-generated thermal gradients. The mechanisms include horizontal differential heating having these causes:

- Land-water contrasts.
- Elevated terrain.
- Urban-rural contrasts.
- Gradients in soil moisture content.
- Gradients in snow cover.
- Variations in cloud shadowing.
- Contrasts in ground albedo and vegetation.

At the surface, if no heat storage is permitted, differential heating results from horizontal gradients in one or more of the following:

Q_G :   Ground heat conduction.

Q_{SH}; Q_{LE} :   Surface sensible and latent turbulent heat flux.

Q_{LW} :   Net long-wave irradiance.

Q_{SW} :   Net short-wave irradiance.

Q_A :   Anthropogenic heat input.

A surface heat budget can be written for this situation as

$$Q_{SW} + Q_{LW} + Q_{SH} + Q_{LE} + Q_G + Q_A = 0 \,, \qquad (22.14)$$

where the surface is assumed to be infinitesimally thin so that no heat storage is permitted. An equilibrium surface temperature is required for (22.14) to balance. Spatial gradients in this equilibrium temperature, in conjunction with the overlying thermodynamic and moisture stratification, will dominate the upward or downward flux of heat for thermally forced mesoscale systems. This will result in the horizontal temperature gradients required to drive the mesoscale circulations.

22.4. Sea/Land Breeze

The predictive characteristics of the sea breeze (SB) and the land breeze in the eastern Mediterranean were documented in ancient times: "Southward goes the wind, then turns to the north; it turns and turns again; back then to its circling goes the wind" (Ecclesiastes 1:6). The Ashdod wind rose (Fig. 22.1) verifies this biblical observation. Termination of the daylight SB is followed by an accelerated veering of the flow southward; during the day the SB direction veers from south to the north. This veering characteristic of the sea/land breeze along the coast in the Northern Hemisphere is attributed to the Coriolis effect as interpreted in modern studies (e.g., Haurwitz, 1947; Neumann, 1977). The other characteristic of this mesoscale feature, its daily recurrence, is typical in the eastern Mediterranean (Skibin and Hod, 1979).

The sea/land breeze affects a large portion of the populated coastal areas of the world and therefore, compared with other mesoscale circulations, it has been extensively studied. This thermally induced flow dominates the local weather, particularly in the subtropical latitudes (Edinger and Helvey, 1961; Hsu, 1970; Skibin and Hod, 1979; Gentilli, 1971; Segal *et al.*, 1982a); thus it is a major forecasting factor there during the warm season. The SB has been indicated to exist as far north as the arctic coast of Alaska (Moritz, 1977; Kozo, 1982). In the subtropical regions, in the absence of strong synoptic-scale flows, the SB circulation may penetrate onshore more than 100 km, and peak surface layer wind speeds due to the SB may be 5–10 m s^{-1}. The typical vertical depth of the lower level flow is usually less than 1000 m (Hsu, 1970). The distance of inland penetration, the depth of onshore flow, and the SB intensity become smaller at higher latitudes. When the SB interacts with the synoptic flow, a significant modification in those characteristics is expected (Estoque, 1962). Additionally, other characteristics of the SB, for example, the SB front (which defines the edge of the inland-penetrating marine air mass) and the return flow aloft, in general tend to be less noticeable as the synoptic flow increases.

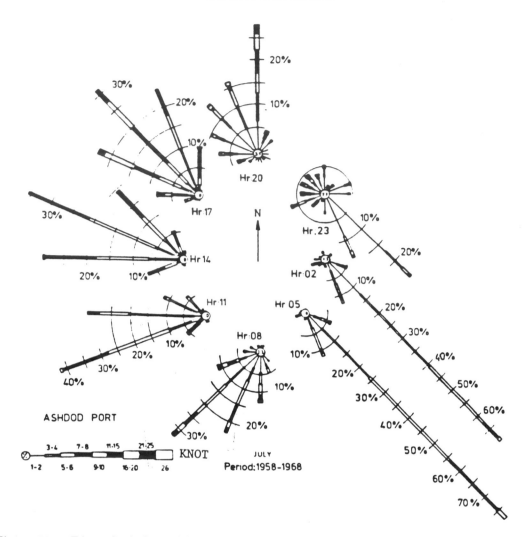

Figure 22.1. Diurnal wind roses for Ashdod Port based on July 1958–1968. The lines point in the direction from which the wind comes; speed is denoted by the width and by the shading of the wind rose lines. Ashdod Port is located on Israel's Mediterranean coast ~ 25 km south of Tel Aviv. In that area in summer, the winds are primarily sea and land breezes. (From Neumann, 1977.)

The land breeze circulation, as observed and modeled in the warm season, is generally significantly less intense than that of the SB. The numerical model study by Pearson (1975) and the theoretical study by Mak and Walsh (1976), for example, provide comparative evaluations of the intensity of both circulations.

In the 1940s and 1950s, analytical studies of the sea/land breeze circulations based on the solution of linear models provided the first substantial general theoretical insight into these circulations (e.g., studies by Schmidt, 1947; Defant, 1950; Smith, 1955, 1957). When such models are used, however, it is impossible to resolve, for example, nonlinear processes such as the SB front.

With the emergence of the computer era, numerical models for simulating the sea/land breeze circulations have gradually developed. The first models were two-dimensional and represented a vertical cross section normal to the coast (e.g., Estoque, 1961; Fisher, 1961; Neumann and Mahrer, 1971). They were followed by three-dimensional models with more accurate and complete parameterizations of the physical processes that force the land and sea breeze (Pielke, 1974; Tapp and White, 1976; Anthes and Warner, 1978; Carpenter, 1979). With the new generation of numerical models, quantitative three-dimensional features of the SB have been simulated and applied to a number of case studies.

22.4.1. Sea-Breeze-Induced Cloudiness

Formation of deep cumulus convection in coastal areas during the warm season appears to be strongly influenced by the sea breeze in many subtropical and tropical locations. This effect has probably been studied more intensively for the south Florida area during the summer (Frank *et al.*, 1967; Pielke, 1974; Ulanski and Garstang, 1978; Blanchard and Lopez, 1984). The subsidence generated by the presence of deep convective clouds often clears the sky of nearby small cumulus. Offshore during the day, the sinking portion of the SB similarly results in clear skies. At night the pattern tends to be reversed, displaying preferential cumulus convection offshore in the land breeze convergence areas (see Sec. 22.4.2).

Over relatively narrow islands or peninsulas, daytime convergence of the SB at the center is likely to enhance upward motion and therefore cloudiness as compared with a non-island case (Neumann and Mahrer, 1975; Abe and Yoshida, 1982). Abe and Yoshida investigated the relation between peninsula width and peak values of the vertical velocities at the latitude of Japan during the summer. Figure 22.2 provides an actual example of cloud formation over Cuba involved with SB convergence, including enhancement where the convergence lines from the two coasts merge.

It is worth noting that an alternate situation occurs over lakes. Owing to the enhanced subsidence involved with the return lake breeze circulation aloft and the suppression of sensible heat fluxes over the lake, reduced daytime cloudiness is expected there. This situation is also illustrated in Fig. 22.2; Lake Okeechobee is cumulus-cloud free, which is typical during the day.

22.4.2. Land-Breeze-Induced Cloudiness

During the night the land breeze circulation generates offshore upward vertical velocities. Although these vertical velocities are generally smaller than those involved with the SB, they can still trigger cloud formation. Such formation of clouds is made more likely by (1) a high moisture content of the marine atmosphere, (2) an unstable marine boundary layer (caused, for example, by the offshore advection of cold air over relatively warmer water), or (3) light onshore synoptic flow that enhances offshore convergence.

Neumann (1951) studied the relatively high frequency of nocturnal thunderstorms compared with those recorded in the daytime, along the south-

Figure 22.2. Geostationary satellite image of the Caribbean and south Florida for 1913 GMT, 25 May 1974.

eastern Mediterranean coasts during the winter. He suggested that the land breeze effect was enhanced by the curvature of the shore, which provided the opportunity for an intensification of cloud development in the offshore environment during the night. Houze *et al.* (1981) and Johnson and Priegnitz (1981) suggested that the early morning convective activity observed along the coasts of north Borneo during the winter monsoon is significantly affected by the land breeze.

When very cold air advects over relatively warm lake water, the sensible and latent heat fluxes to the atmosphere can be very large. The resultant deepening and moistening may lead to stratocumulus and cumulus cloud formation, and afterward to heavy snowfall. These situations are well known along the windward shores of the Great Lakes during winter arctic-flow outbreaks. These lake-effect snowstorms can occur downwind of any large water body that is traversed by cold, arctic air. Two feet of snow, for instance, fell in Salt Lake City on 17–18 October 1984, as a result of cold air advection across the Great Salt Lake. The thermal differences generated between the lake and the land can also result in land-breeze-type flows, which when opposing the large-scale flow can enhance snowfall in the land breeze convergence zone (Hjelmfelt and Braham, 1983).

Passarelli and Braham (1981) presented radar and aircraft data for three winter cases of shoreline-parallel snow bands that occurred over Lake Michigan. They concluded that in these cases a winter land breeze from one or both shores had an important role in organizing the low-level convergence and convective motions. Passarelli and Braham compared these cases with earlier studies of lake-effect snow bands near and over Lakes Erie and Ontario, emphasizing the role of the land breeze in the generation of these bands. Using mesoscale numerical models, Alpert and Neumann (1983) and Hjelmfelt and Braham (1983) simulated cases reported by Passarelli and Braham, and Braham (1983), and provided additional evidence as to the importance of the land breeze in the generation of snow bands.

22.4.3. Heat Load Conditions Associated With Sea Breezes

Heat load conditions are one of the most important forecasting characteristics involved with the SB during the warm seasons over many parts of the world. Heat load conditions in the outdoor environment are affected by the air temperature, air moisture, wind speed, and solar radiation. Indices considering all or part of these parameters in the quantitative evaluation of the head load are described in the literature (e.g., Munn, 1970). The SB, by advecting cooler marine air onshore, and by increasing ventilation (when the synoptic flow is absent or light) contributes in general to a moderation of the heat load conditions.

The most significant role of the SB in moderating heat load conditions is likely to be in the following situations (which are most frequent in the subtropical latitudes):

(1) Synoptic subsidence associated with a strong, stagnant subtropical anticyclone generally causes an increase of air temperature near the ground, along with a reduction in the daytime depth of the planetary boundary layer. Reduced wind speeds, typical of this synoptic condition, together with a shallow planetary boundary layer (which is probably capped by a strong subsidence inversion) often produce very moist air within this layer. Altogether this pattern results in adverse heat load conditions. The development of the SB in this situation results in a reduction of air temperature, and an increase in wind speed, moderating the heat load (although it also involves an increase in the relative humidity).

(2) Offshore synoptic flows are involved in many occasions with increased temperatures along coastal areas (e.g., the mid-Atlantic coasts of the United States during the summer, which are often under the influence of southwesterly low-level flow associated with the Bermuda High; the Santa Ana winds in California; the Khamsin winds [see Winstanley, 1972] in the southeastern Mediterranean). In such synoptic situations the SB circulation is shifted closer to the coast (Estoque, 1962) so that the coastal area may be affected for some period with calm winds resulting from the convergence of the SB and synoptic flows. If the synoptic flow is not too intense, some inland penetration of the SB in the afternoon is likely, resulting in an improvement of the heat load condition.

Numerical mesoscale model studies of the coastal area of Israel (Segal and Mahrer, 1979) and the Delmarva peninsula and Chesapeake Bay region (Segal and Pielke, 1981) evaluated the regional heat load under conditions similar to those described above. It is likely that when numerical mesoscale models are applied for local weather prediction in coastal areas during the summer, a major product will be the evaluation of heat load conditions, using one of the existing heat load indices. Such information will also be useful to electric utilities in planning demand loads.

22.4.4. Air Quality in Coastal Areas Associated With Sea Breezes

Qualitative forecasting of daytime air quality may be provided through a ventilation index that considers the product of the planetary boundary layer depth and the average wind within this layer. In the coastal areas, however, forecasts of air quality may require considering the effects of the daytime veering of the SB; of an internal thermal boundary layer near the shoreline (Lyons, 1975); of substantial vertical velocities; and of possible return flows aloft. In the recent decade, increasing research attention has been given to the study of air-pollution meteorology along coastal areas, using both observational and modeling methodologies (e.g., Lyons, 1975; Keen et al., 1979; Blumenthal et al., 1978; Young and Winchester, 1980; Bornstein and Thompson, 1981; Segal et al., 1982b; Schultz and Warner, 1982; Shair et al., 1982).

The existence of a strong high-pressure system or the offshore synoptic flow mentioned in Sec. 22.4.3, has the potential for producing high air pollution concentrations. The first situation is well known in the summer, for example, in the Los Angeles Basin. The development of the SB there, however, while increasing advection, may also be involved with reducing the onshore mixing layer as a result of the onshore advection of the marine boundary layer (e.g., Edinger, 1959; Anthes, 1978; Schultz and Warner, 1982). Schultz and Warner, nevertheless, found in a summer-period model study of the Los Angeles Basin, that although the planetary boundary layer height is not particularly deep after the passage of the SB front, strong SB inflow maintains relatively high ventilation. However, a gradual decrease in the modeled ventilation toward the end of the afternoon was due to a general decrease in the intensity of the circulation. For the second situation, the possible creation of calm wind zones in the coastal area where the synoptic flow and SB oppose each other will probably be associated with an increase of pollutant concentrations. Additionally, an offshore advection during the first half of the day may be reversed in the afternoon if the SB flow dominates the offshore synoptic flow by that time. In such a case, pollutants accumulated within the stable marine atmosphere may cause an increase in the onshore concentrations, when being advected onshore (Segal et al., 1981; 1982b). Similarly, the change from a land breeze to a SB circulation in the morning is possibly involved with an onshore advection of pollutants that were advected offshore during the night (Lyons, 1975; Shair et al., 1982).

22.4.5. Air-Sea Interactions Associated With Sea Breezes

Studies of the offshore segment of the SB and the land breeze circulation are few. They include, for example, the intensive observational study along the coasts of Oregon during the Coastal Upwelling Experiment (CUE) (Elliott and O'Brien, 1977) and the study along the Texas coast in the 1960s (Hsu, 1970).

The attention given to the study of air-sea interactions along the coast of Oregon is a result of the major economic importance of the upwelling region off the coast. A large portion of the world's fish catch is from upwelling regions. Conceptually there should be a feedback interaction between the upwelling and SB intensity; however, the two modeling studies that have looked at this relation along the Oregon coast suggest that the feedbacks are small. Mizzi and Pielke (1984) related the weak feedback to the relatively mild effect of the cooled surface water on sensible heat fluxes into the atmosphere. Clancy *et al.* (1979) attributed it to the narrow spatial coverage of the upwelling zone.

Pielke (1981) listed various aspects related to offshore forecasting. These include, for example, the estimation of sea state along the coasts. Intensification of the SB in the afternoon is generally involved with an increase in surface wave height and a tendency for wave breaking. Finally, the onset of the SB may be involved with onshore advection of marine fog and stratus. Such situations are observed, for example, along the west coast of the United States (Leipper, 1968). They may also occur on the coast along the eastern Mediterranean during fall and spring (Levi, 1967).

22.5. Slope/Valley Flows

In a region with irregular terrain, local wind patterns can develop because of the differential heating between the ground surface and the free atmosphere at the same elevation some distance away. A larger diurnal temperature variation usually occurs at the ground, so that during the day the higher terrain becomes an elevated heat source, whereas at night it is an elevated heat sink. Two categories of winds are generally recognized: slope flow, and valley winds. These types are easiest to recognize when the prevailing large-scale flow is light.

Slope flow refers to (1) cool, dense air flowing down elevated terrain at night, or (2) warm, less dense air moving toward higher elevations during the day. Such air movements are often referred to, respectively, as nocturnal drainage flow and the daytime upslope. The nocturnal drainage flow is also called a katabatic wind, and the daytime upslope is also referred to as an anabatic wind.

Valley winds are up- and down-valley circulations that develop from along-valley horizontal pressure gradients in one segment of a valley. Such gradients occur because of the input, by the slope flow, of air having a different temperature structure than the air in adjacent segments.

Slope/valley thermally induced flows may have great importance in determining local weather when major synoptic disturbances do not occur. Compared wth the sea/land breeze, however, they have received less observational and modeling attention, at least until the last few years.

Defant (1951) reviewed the earlier studies of these flows. Recent studies of these flows include, for example, that by Mahrer and Pielke (1977a) who modeled a diurnal cycle around a two-dimensional ridge. Banta and Cotton (1981), and Banta (1984) provided an observational evaluation of the development of the daytime planetary boundary layer in a broad mountain valley and its interaction with the synoptic flow aloft. Bader and McKee (1983) simulated the development of upslope flow following sunrise.

Small-scale valley nocturnal drainage flows have received intensive research attention in recent years through ASCOT (Atmospheric Studies in Complex Terrain). ASCOT is devoted to an improvement in the understanding of these flows as they affect air quality in the western United States. Observational results from this program were reported, for example, by Doran and Horst (1981) and Clements and Nappo (1983). Modeling efforts in ASCOT were reported, for example, by Doran and Horst (1983), Garrett (1983), and Yamada (1983).

Unlike the sea and land breezes, these circulations are not easy to characterize quantitatively. The coupling of the slope/valley flow with existing synoptic flows is more involved than in the sea/land breeze case. On the basis of the studies mentioned, it is suggested that for light synoptic flows the daytime upslope flows are about 3–6 m s^{-1}, and the depth of the layer involved is several hundred meters or more. The nighttime valley flows in the ASCOT studies have a similar speed, although the drainage layer was typically several tens of meters deep. Observations in ASCOT indicated that the depth of the nocturnal drainage flow increases linearly with downslope distance.

22.5.1. Convective Cloudiness Associated With Slope/Valley Flows

One of the main forecasting aspects involved with the daytime development of the upslope flows is its relation to convective cloud formation. The upward velocities associated with these flows, and frequent interaction with the synoptic flow, provide in many mountainous areas a triggering mechanism for cloud formation and afternoon showers.

Orville (1965) modeled the effect of a daytime upslope circulation on cumulus initiation along very steep mountains. Other studies have focused on the investigation of the summer daytime moist convective activity in mountainous regions and adjacent areas of Colorado and nearby states. These studies (e.g., Maddox, 1981; Klitch et al., 1985; Holroyd, 1982) have illustrated the importance of the daytime, thermally forced flows to local weather.

A special surface network was established in the Front Range area of eastern Colorado as part of the Program for Regional Observing and Forecasting Services (PROFS), to provide input for the forecasting of thermally forced and other types of mesoscale flows. Toth and Johnson (1985) used wind data from these stations to generate a climatology of July 1981 surface flow over the region. They found that the daytime flow was dominated by a thermally induced upslope flow (see Fig. 22.3). During that month a local confluence was found during midday along major east-west ridges of the region, and consequently, these ridges were preferred locations for afternoon thunderstorm activity.

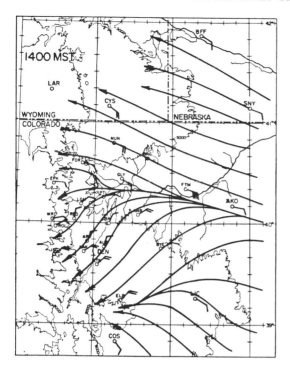

Figure 22.3. Surface streamline analysis for 1400 Mountain Standard Time (MST) over northeastern Colorado in the summer. One full wind barb $= 1$ m s^{-1}. (From Toth and Johnson, 1985.)

22.5.2. Nocturnal Valley Flows

During the nocturnal period when the sky is clear and the synoptic flow is light, the basin of a valley is characterized by the accumulation of relatively cold, stable, and stagnant air. Although some turbulence exists within the flow along the slopes, it often is substantially reduced or even eliminated as the flows converge at lower elevations. Hence, in lower elevations, the vertical mixing by turbulence at night is often negligible so that the cold air accumulates in the bottom of valleys and basins. Several forecasting aspects may be related to this situation:

(1) The air quality at night in such topographical locations is expected to be relatively poor when the valley is subject to low-level pollutant emissions. Similarly, haze or fog formation is likely to occur at night if the air is moist (Pilié et al., 1975).

(2) The breakup of the nocturnal inversion, following sunrise, and the development of the daytime planetary boundary layer, generally cause the elimination of this stagnant valley air mass, as described, for example, by Whiteman and McKee (1982). When the valley is snow covered, however, the stagnant flow may continue until later in the morning, or, occasionally may not be eliminated at all, resulting in a long-term buildup of pollution. This accumulation of pollution becomes pronounced when a valley has few or no outlets that would permit a secondary out-valley flow.

(3) During the cold season, the accumulation of cold air in valleys may result in frost or freeze. In areas where agricultural activity exists during this

season (e.g., southern California and the southern areas of the United States), frost and freeze forecasts have economic significance. The establishment of a topoclimatic agrometeorological network and complementary modeling of the valley flow to provide a frost climatology for a valley can be useful for daily forecasting in these agricultural areas (e.g., Zemel and Lomas, 1977; Bagdonas *et al.*, 1978).

22.5.3. Low-Level Jet Associated With Heating/Cooling of Elevated Terrain

Although alternate hypotheses are described in the literature (e.g., Uccellini, 1980), one mechanism for the formation of the low-level jet (LLJ) is generally regarded as a response of the atmosphere boundary layer to the diurnal thermal forcing over sloping terrain (see the analytical studies by Holton, 1967; Bonner and Paegle, 1970). The horizontal temperature gradient that is created by the diurnal heating and cooling cycle can result in a LLJ by creating a larger thermal wind component (McNider and Pielke, 1981).

The LLJ pattern seen in Fig. 22.4 is based on data from a special pibal network from Amarillo Tex., to Little Rock, Ark. Model simulations by McNider and Pielke (1981) for typical conditions in a two-dimensional cross section of the Great Plains reproduced a similar structure. The model results indicate that the diurnal boundary heating over the sloping terrain of the Great Plains is directly associated with the formation of the LLJ.

Pitchford and London (1962) concluded that the high frequency of nocturnal convective activity over the Great Plains is linked to the convergence associated with the nocturnal LLJ, and Wallace (1975) provided further discussion. Maddox (1983) suggested a linkage between development of Mesoscale Convective Complexes (MCCs) and the warm advection involved with the LLJ over the Great Plains.

At present, there is considerable controversy concerning the genesis mechanism of these features. A suggested hypothesis states that initiation of MCCs near the Front Range of the Rocky Mountains is caused primarily by mesoscale thermal forcings as described in the following.

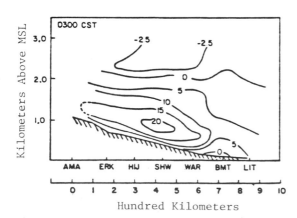

Figure 22.4. Contours of the southerly velocity component over sloping terrain, at night (0300 CST), as reported by Hoecker (1963) from a special pibal network between Amarillo, Tex., and Little Rock, Ark.

A flow with an easterly component develops over the Great Plains during the day because of the heating of the sloping terrain. This upslope flow also occurs along the slopes of the east-west ridges that extend out into the Plains (e.g., the Palmer Divide and the Cheyenne Ridge).

Westerly flow develops over the Rockies and moves out over the Great Plains as the cumulonimbus activity and deep boundary layer entrain westerly upper-level flow and mix it down to the surface (e.g., Banta, 1984; Toth and Johnson, 1985). Downslope flow also develops north and south of the east-west ridges, in response to the cooler air transferred downward by the cumulus convection and its negative buoyancy. The region of convergence of the upslope and downslope flows is a region where subsequent cumulus convection is expected if the atmosphere is sufficiently unstable and moist. These regions of convergence would tend to be focused in the areas between the east-west ridges in much the same way a convex curvature enhances sea breeze convergence as described in Pielke (1974).

In the late afternoon the boundary layer tends to become decoupled from the surface, resulting in an acceleration of the winds, and therefore enhanced convergence as long as upslope and downslope flows persist. Thus as LLJs begin to evolve from these upslope and downslope flows, convergence is enhanced until a substantial portion of the inertial period passes. The upslope flow also provides moisture from the east.

These localized, more-or-less circular regions of convergence are those hypothesized to be the genesis locations of MCCs. Maddox (1981) concluded that such preferred regions occur, and the annual summaries of MCC activities published in the *Monthly Weather Review* support the suggestion (e.g., see Rodgers *et al.*, 1983). However, additional quantitative analyses are needed for confirmation.

22.5.4. Mountainous Coastal Area Circulations

The presence of mountains next to water bodies results in a circulation that blends the sea/land breeze and mountain/valley flows (Mahrer and Pielke, 1977; Ookouchi *et al.*, 1978). Daytime SB penetration uphill interacting with the thermally induced upslope flow would be expected to enhance the flow. However, the relatively cool air advected inland by the SB could reduce the forcing for the induced upslope flow. Segal *et al.* (1983), examining this aspect through numerical model simulations and scale analysis, evaluated the influence of terrain steepness and the water-body cooling effect on the resultant flows.

The mesoscale flows involved with such a terrain-coastal configuration can be rather strong, particularly when the flow crosses to the lee side of the mountains (Olsson *et al.*, 1973; Bitan, 1977, 1981; Alpert *et al.*, 1982). In the lee of mountains under this type of flow, a noticeable increase of the air temperature similar to that associated with a chinook can be observed (Ashbel, 1939). Such warming is thought to be due in this case to the entrainment of potentially warmer air aloft into the boundary layer, caused by the enhanced turbulence associated with the air flowing over the rough terrain (e.g., Pielke, 1984, p. 476).

The development of the SB in such coastal areas may be involved with orographic precipitation, such as that frequently observed in the mountainous islands of the tropics (Riehl, 1954).

The intensification of offshore synoptic flows through a topographically forced internal gravity wave as they cross mountainous coastal areas can reduce or even eliminate the onshore penetration of the SB. The prediction of the inland penetration of the SB can be very important from a fire weather point of view along the Pacific Coast during Santa Ana conditions (Sommers, 1981). Mahrer and Segal (1979), for a similar flow pattern along the coastal area of Israel, concluded that the SB can penetrate rapidly in the late afternoon toward an inland mountain ridge (about 50 km onshore) as the SB becomes strong enough to overcome the opposing synoptic winds, although earlier in the day it was kept along the shore line by the synoptic flow.

With respect to the nocturnal period, Doron and Neumann (1977) and Mahrer and Pielke (1977a) investigated combined drainage flows and land breezes. They found that with mountain heights of several hundred meters or so, thermally induced nocturnal offshore flows in middle latitudes are dominated by the drainage flow.

22.6. Urban Heat Island and Urban Circulation

The difference in surface properties of urban and rural areas leads to differences in the thermal fluxes involved with (22.14). Consequently higher air temperatures in the urban area are generated in many situations, particularly during the nocturnal period, resulting in the so-called urban heat island. Factors such as larger heat capacity of the urban surfaces and lower daytime evaporation were found by Myrup (1969) to be the prime causes of the urban heat island. Sisterson and Dirks (1978) found a daytime reduction of 10–20% in the specific humidity during the Metropolitan Meteorological Experiment (METROMEX) (Changnon, 1981). (St. Louis was chosen for this experiment because of its relatively flat terrain and the comparatively clear distinction between urban and rural zones. Urban areas near coastlines and in irregular terrain are often strongly influenced by sea/land breezes and slope/valley winds.) Other contributing factors for the onset of the urban heat island may be attributed, for example, to the differences in the surface albedo, and to anthropogenic heat release in the urban area. The urban heat island is most pronounced under clear skies and light synoptic wind situations (Oke, 1973). Advection involved with relatively strong synoptic flow tends to eliminate the thermal differentiation between the rural and urban areas.

Conceptually, there is a similarity between the expected circulation involved with the nocturnal urban heat island and that involved with the generally light lake-land breeze circulation. However, in the urban case the significant frictional drag, due to buildings, may be a major retardant of wind flow.

Numerical model studies (Delage and Taylor, 1970; Bornstein, 1975) demonstrated that relatively light horizontal flows are associated with the urban heat island. Wong and Dirks (1978) found that for St. Louis when the heat island was strong and extended some distance aloft, and the synoptic winds were light, the thermally induced pressure perturbation from the city was the dominant force on the local winds. The result was an acceleration of the airflow as it converged into the heat island. With strong winds and a weak thermal field, the frictional drag appeared to become a significant force and decreased the wind speed over the city. Vukovich and Dunn (1978) used a three-dimensional primitive equation model for sensitivity tests of the effect of the St. Louis area on developing a mesoscale flow perturbation and found that the heat island intensity and the boundary layer stability have dominant roles in the development of heat island circulations. Other aspects of the urban atmosphere have been discussed and reviewed comprehensively by Oke (1979) and Landsberg (1981).

Probably the most significant forecasting parameter involved with the urban heat island is its nocturnal temperature departure from the rural areas. In many cases it was observed to be higher by several degrees than the temperature in the surrounding rural area (see examples in Landsberg, 1981). Comparison of climatological temperature data from adjacent urban/rural stations provides an estimate of the expected difference for forecasting purposes.

The urban heat island is expected to reduce the intensity of the nocturnal surface inversion. Bornstein (1968), for example, found that in the New York urban area, surface inversions were less intense and far less frequent than in the surrounding non-urban area. This characteristic contributes to improved nocturnal pollutant dispersion and to lessening the probability for formation of radiative fog over the urban area.

An increase in precipitation amounts downwind of urban areas has been reported in several locations. Explanations for this trend relate to the higher surface roughness of the urban area, which results in low-level convergence and therefore upward vertical velocities, as well as upward motions caused by the enhanced surface heating in the urban area (Changnon, 1981). Hjelmfelt (1982) simulated the urban heat island of St. Louis and produced positive vertical velocities downwind of the city. Such a pattern was obtained because of the urban surface roughness convergence effect and the downwind shifting of the heat island circulation by the synoptic flow. Hjelmfelt's results are consistent with the hypothesis that some of the increased precipitation in the St. Louis area during the summer (Huff and Vogel, 1978) may result from the deepening of the boundary layer caused by the enhanced sensible heat fluxes and increased surface roughness of the urban area. Several studies (e.g., Landsberg, 1981, Ch. 8) attribute increase in precipitation (when supercooled clouds exist) to a larger number of ice nuclei due to the anthropogenic pollution.

22.7. Nonclassical Mesoscale Circulations

In recent years growing attention has been given in the research literature to the evaluation of nonclassical thermally induced mesoscale circulations. These systems are established when horizontal gradients in the sensible heat fluxes, resulting from differential soil wetness, snow cover, cloud cover, etc., cause a horizontal temperature gradient within the planetary boundary layer. In addition, research attention has been given to the evaluation of the impact of major changes in the sensible heat fluxes (due to one or more of the conditions listed in Sec. 22.3) on the more classical circulations discussed in Secs. 22.4–22.6.

Horizontal gradients in sensible heat flux are generally the prime forcing of dry atmospheric mesoscale circulations; studies such as those by Carlson and Boland (1978), McCumber and Pielke (1981), Garrett (1982), and Zhang and Anthes (1982) provide some insight into the relation between surface characteristics and sensible heat fluxes. Zhang and Anthes, for example, suggested the following hierarchy of surface properties as modifiers of sensible heat fluxes: (1) soil moisture, (2) roughness of the surface, (3) albedo, (4) soil thermal capacity. A normal range of variations of these properties was considered in their analysis.

Mesoscale circulations that result from differential heating rates over land, as well as modifications in land and sea breezes, and slope/valley flows due to major alterations in sensible heat fluxes, are evaluated in the following subsections. With respect to weather forecasts in such situations, however, it is worth noting the following:

- The thermal forcing related to conditions of soil moisture, snow, cloudiness, and other causes may be temporary (time scale of days), seasonal, or permanent.
- Modeling (or other types of evaluation) of those conditions for purposes of prediction may be complicated because of large spatial irregularities in the domain of interest, the transience of contrast lines, and the need to quantify the forcing characteristics. Therefore, classification of those conditions is needed, as well as research, to provide better insight as to their importance and to develop procedures to incorporate quantitative effects into very-short-range weather prediction.

22.7.1. Soil Moisture

Modifications of soil moisture can last for only a few days following sporadic heavy precipitation, or as long as a season in an irrigated agricultural area. Numerical model studies by McCumber (1980) and Physick (1980) illustrate the reduction in the intensity of sea breeze circulations with increasing soil wetness. Ookouchi et al. (1984) showed a similar impact while considering the daytime development of thermal upslope flows. Figure 22.5, which is from the last study, illustrates the effect of soil moisture along a mountain slope on the intensity of the induced flows. Wet slopes significantly reduced the peak values of the predicted vertical velocities.

A contrast in soil wetness along flat terrain may produce, for the extreme case, surface flows similar in magnitude to those obtained in a sea breeze equivalent case as illustrated in Fig. 22.6.

1600 LST

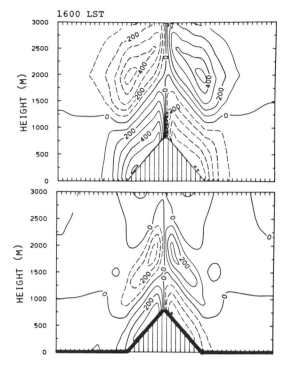

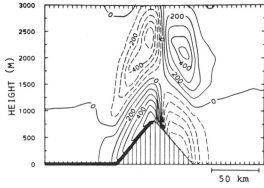

Figure 22.5. Vertical cross section of a simulated ridge presenting the cross-ridge component (cm s^{-1}) in the afternoon for various soil moisture situations. Dark line indicates high soil moisture availability (0.5); otherwise, the soil is regarded as having a low moisture availability (0.05). (From Ookouchi *et al.*, 1984.)

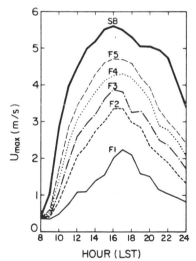

Figure 22.6. The maximum wind speed at 5 m height along a flat-domain cross section, where a relatively dry region (soil moisture availability = 0.05) is contrasted by a region of soil moisture availability m = 0.1 (F1); m = 0.2 (F2); m = 0.3 (F3); m = 0.5 (F4); m = 1 (F5); and by a sea surface (SB).

22.7.2. Snow

Horizontal contrasts in snow cover (e.g., between bare soil and adjacent snow cover) can also result in mesoscale circulations. Snow cover effects are parameterized (in a simplified manner) in large-scale models. However, this effect has not been treated in mesoscale models. That is surprising, considering the relatively large area and the long duration of snow cover in many populated areas of the world.

Snow is known to increase the albedo, reduce aerodynamic roughness, and alter surface heat and moisture conductivity into the ground. These effects, in addition to melting, and possible penetration of solar radiation into the snow layer, markedly modify sensible heat fluxes and the resultant heat input into the atmosphere.

Model and observational studies that evaluate the interaction of snow layers with the air above have generally been confined in the atmosphere to the surface layer (e.g., Schlatter, 1972; Halberstam and Schieldge, 1981). Modeling of the PBL over a snow-covered surface by Halberstam and Melendez (1979) (1–D simulation) indicates peculiar thermal characteristics in the planetary boundary layer such as the generation of a highly stable, thin surface layer with irregular turbulence bursts.

Contrasts between snow and bare soil in mesoscale domains are often widespread. They can be detected most easily through surface observations or from satellite images. Such contrasts should result in significant thermal differences leading to mesoscale circulations (Johnson et al., 1984). In a winter case study over central Oklahoma, Bluestein (1982) found evidence that diabatic cooling over snow cover, contrasted by heating and vertical mixing over adjacent bare soil, played an important part in generating a strong surface horizontal temperature gradient. Schlatter et al. (1983) illustrated the horizontal snow cover variability in a mesoscale region following the Christmas Eve 1982 snow storm in eastern Colorado. On the day following the storm, a difference of about 5°C was observed in the air temperature at meteorological shelter height between the snow-covered and bare areas.

22.7.3. Cloudiness (Nonconvective)

The reduction of the incoming solar radiation by cloud shielding may produce a significant decrease in the sensible heat fluxes. Like soil moisture and snow, cloudiness can affect sea/land breezes (Gannon, 1978) and slope/valley flow, as well as generate mesoscale circulations over land when there are spatial gradients in the cloud cover.

Several studies have addressed such situations. Purdom (1982), for instance, provided observational evidence from satellite imagery that implied the occurrence of thermal circulations between cloud-covered and clear mesoscale areas. Bailey et al. (1981) applied a three-dimensional numerical model to help explain the occurrence of an intense stationary multicell thunderstorm over London. In that study, shading by an observed layer of medium-level clouds, which was included in the forecast model, had an important effect on the prediction. A severe rainstorm appeared to be related to a mesoscale circulation that was induced by the contrast between adjacent clear and cloudy areas. Further discussion of this case is provided in Carpenter (1982). Recently, McNider et al. (1984) evaluated the initiation of a line of thunderstorms as a result of a mesoscale circulation caused by a horizontal gradient in morning cloud cover over Texas.

22.7.4. Other Causes of Surface Forcing

Any region with a nonhomogeneous surface is a candidate for surface-induced mesoscale circulations. Mahrer and Pielke (1977b) simulated noticeable, daytime, thermally induced mesoscale circulations while considering albedo contrasts of 0.2 (0.2 to 0.4). Studies by McCumber (1980), Garrett (1982), and Blanchard and López (1984) imply that vegetation cover has an influence on the development of thermally induced mesoscale circulations. Anthes (1984) provided a comprehensive review and evaluation of various aspects of this type of circulation. Contrasts between vegetative and bare-soil regions should be investigated for their ability to generate significant, thermally induced mesoscale circulations.

22.8. Modeling

In Sec. 22.2 it was concluded that the intensity of surface-forced thermally driven mesoscale systems is directly related to the depth and magnitude of diabatic effects (cooling or heating of the air). Diabatic heating, for the systems discussed in this chapter, has its source at or close to the Earth's surface (22.14). It was also determined that these atmospheric systems have a substantial divergent component to the wind.

Therefore, to determine observationally the existing structure of a thermally forced system we need to monitor at least (1) the current spatial distribution of surface temperature and (2) the current spatial distribution of the divergent and rotational components of the horizontal wind. Satellite imagery of the surface in the infrared can provide evaluations of the first. Doppler radars and lidars permit accurate estimation of the second.

In the absence of a monitoring system for the horizontal winds, (22.6) and (22.12) can be used to estimate the horizontal and vertical winds if the time history and current distribution of the temperature field are known. This follows from (22.6); if the horizontal pressure gradient is known, e.g., by using a more general form of (22.10), the other variables in (22.6) are directly expressed in terms of the velocities (i.e., the advection and Coriolis effects) or are parameterized as a function of these velocities and the observed temperature distribution (i.e., the subgrid scale flux). Although a three-dimensional distribution of temperature would be ideal, even the surface distribution of temperature would be useful if a reasonable estimate of its vertical structure could be obtained from a standard source such as the synoptic radiosonde network, and an equation for conservation of heat and water substance could be used (e.g., Eq. 22.13; see also Eqs. 4.23 and 4.24 in Pielke, 1984).

Even in the absence of extensive observational data, however, realistic simulations are possible since the primary forcing of surface, thermally forced mesoscale systems is directly related to the surface characteristics. This is in contrast to other types of mesoscale systems such as squall lines or tropical cyclones, which propagate into a region. For example, the configuration of a land-water boundary is known, so for the land and sea breeze, the characterization of the surface forcing is obvious without any meteorological

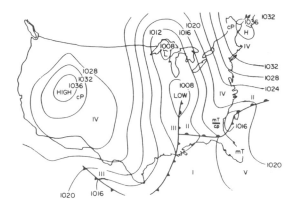

Figure 22.7. Categories I–V (see Table 22.1) applied to a surface analysis chart for 9 January 1964 over the United States. (From Pielke, 1982.)

data. Land-use patterns, the previous rainfall distribution, etc., can provide spatial variations of surface characteristics inland. Although the thermodynamic structure and the synoptic flow direction and speed directly influence the form of these systems, the distribution of surface heating is most directly related to the development of these thermally driven systems.

For quantitative prediction of the mesoscale atmospheric flows described in Secs. 22.3–22.7, numerical models must be constructed. These models describe meteorological systems by using mathematical descriptions of the conservation relations for heat, momentum, mass, and other atmospheric constituents as described, for example, in Pielke (1984).

There are two distinct approaches to modeling of mesoscale systems for forecasting purposes:

- Modeling in real time on a day-by-day basis.
- Modeling representative synoptic situations that have a high degree of persistence and/or repetitiveness.

At present, simulations of thermally forced mesoscale systems on a continuous day-by-day basis for one specific location are expensive. For instance, at the time of writing, about 1 h of calculation time on the NCAR CRAY or a CYBER 205 is needed for a 24 h simulation of the sea- and land-breeze circulations over south Florida, with a 33×36 horizontal grid mesh of 11 km intervals, 16 vertical levels, and a time step of 90 s. Thus, to simulate a year of this phenomenon for a portion of the U.S. coast, assuming that the model is integrated each day, would require about 365 hours or $\gtrsim 15$ days of commited computer time.

Modeling representative situations is an alternative to this procedure, at least for those mesoscale systems in which the surface forcing is similar day after day. Pielke (1982) summarized a procedure to classify the large-scale flow regime into five categories. As illustrated in Fig. 22.7 and explained in Table 22.1, there are sound physical reasons for this separation.

With respect to surface-forced mesoscale systems, lake effect storms are best defined when strong, cold, synoptic flow advects over a warm ocean body. Therefore they occur almost exclusively in category III (i.e., behind a cold front with cyclonic curvature to the isobars and cold advection). Other surface-forced thermal flows, however, are best developed under lighter synoptic flow and with clear skies so that the diurnal variation of radiative flux divergence is most pronounced. Therefore, sea and land breezes, slope and valley flows, urban circulations, etc., would be expected to occur predominantly in categories IV and V.

Thus to use a mesoscale model to describe the local variations of meteorology due to these circulations, it may be sufficient to integrate only representative (i.e., climatologically most expected) conditions. Snow (1981) used this approach to estimate the climatologically expected wind energy for three coastal areas in the eastern United States and found that only three mesoscale model runs were necessary to characterize the mesoscale wind energy in at least two of the three regions of study.

The procedure to combine climatological synoptic data and mesoscale model simulations can be summarized as follows:

- Determine the frequency of each category within a mesoscale region on a monthly basis over a long-term period (e.g., at least several years).
- Within each category, use the surface pressure analysis to determine the frequency of geostrophic winds of discrete classes of speed and direction (e.g., less than and greater than 4 m s^{-1}; SSE through ESE). For these discrete classes evaluate the average and standard deviation of the thermodynamic profile.
- Use the mesoscale model to simulate the local conditions expected during the most frequent categories and geostrophic wind classes for each month. A limited number of simulations for the most common conditions in each month (or even for each season) will permit the compilation of a climatological mesoscale atlas, which should be of considerable use to forecasters and in environmental studies.

Because of the general month-to-month similarity within seasons, the number of climatological mesoscale simulations can probably be minimized. Assuming, on the basis of experience, that no more than eight simulations per season are required, the one-time computer requirements, using current model capabilities, are about 32 h of CRAY or CYBER 205 time. By the end of 1986, as a result of computer code vectorization, much smaller resources will be required. Only if the mesoscale model were substantially improved in its physical representations, or if finer spatial resolution were required, would the climatologically most frequent mesoscale calculations have to be redone.

22.9. Use of Satellite Observations

Satellite technology has produced new techniques to identify situations in which nonhomogeneous conditions exist over mesoscale areas. Carlson *et al.* (1981) and Wetzel *et al.* (1984) discussed procedures to estimate soil moisture content through satellite sensing. Snow cover and cloudy areas can also be identified through synoptic observations and satellite sensing. Satellite systems can now provide reasonable temporal and spatial resolution within mesoscale domains. Significant surface properties that are permanent

or seasonal can be mapped and used forcing to estimate the climatologically
expected mesoscale response.

Table 22.1. Classification of mesoscale environments,
based on large-scale flow*

Category	Air mass	Location of flow regime / Characteristics†
I	mT	*In the warm sector of an extratopical cyclone.* Thickness and vorticity advection weak, with little curvature to the surface isobars; low-level convergence is limited, and an upper-level ridge tends to produce subsidence. Southerly low-level winds are typical.
II	mT/cP, mT/cA, mP/cA	*Ahead of the warm front in the region of cyclonic curvature to the surface isobars.* Warm air advecting upslope over the cold air stabilizes the thermal stratification; positive vorticity advection and low-level frictional convergence add to the vertical lifting. Because of the warm advection, the geostrophic winds veer with height. Low-level winds are generally northeasterly through southeasterly.
III	cP; cA	*Behind the cold front in the region of cyclonic curvature to the surface isobars.* Positive vorticity advection and negative thermal advection dominate; resulting cooling causes strong boundary layer mixing. The resulting thermal stratification in the lower troposphere is neutral, or even slightly superadiabatic. Gusty winds are usually associated with this sector of an extratropical cyclone. Because of the cold advection, the geostrophic winds back with height. Low-level winds are generally northwesterly through southwesterly.
IV	cP; cA	*Under a polar high in a region of anticyclonic curvature to the surface isobars.* Negative vorticity, weak negative thermal advection, and low-level frictional divergence usually occur, producing boundary layer subsidence. Because of relatively cool air aloft, the thermal stratification is only slightly stabilized during the day, despite the subsidence. At night, however, the relatively weak surface pressure gradient causes very stable layers near the ground on clear nights, owing to long-wave radiational cooling. The low-level geostrophic winds are usually light to moderate, varying slowly from northwesterly to southeasterly as the ridge moves eastward.
V	mT	*In the vicinity of a subtropical ridge.* Vorticity and thickness advection, and the horizontal pressure gradient at all levels, are weak. The large upper-level ridge, along with the anticyclonically curved low-level pressure field, produces weak but persistent subsidence. This sinking causes a stabilization of the atmosphere throughout the troposphere. Low-level winds over the eastern United States associated with these systems tend to be southeasterly through southwesterly.

*Adapted from Pielke (1982).
†This discussion applies to the Northern Hemisphere.

REFERENCES

Abe, S., and T. Yoshida, 1982: The effect of the width of peninsula on the sea breeze. *J. Meteor. Res. Japan*, **60**, 1074–1084.

Alpert, P., and J. Neumann, 1983: A simulation of Lake Michigan's winter land breeze on 7 November 1978. *Mon. Wea. Rev.*, **111**, 1873–1881.

Alpert, P., A. Cohen, E. Doron, and J. Neumann, 1982: A model simulation of the summer circulation from the Eastern Mediterranean past Lake Kinneret in the Jordan Valley. *Mon. Wea. Rev.*, **110**, 994–1006.

Anthes, R. A., 1978: The height of the planetary boundary layer and the production of circulation in a sea breeze model. *J. Atmos. Sci.*, **35**, 1231–1239.

Anthes, R. A., 1984: Enhancement of convective precipitation by mesoscale variations in vegetative covering in semiarid regions. *J. Climate Appl. Meteor.*, **23**, 541–554.

Anthes, R. A., and T. T. Warner, 1978: Development of hydrodynamic models suitable for air pollution and other mesometeorological studies. *Mon. Wea. Rev.*, **106**, 1045–1078.

Ashbel, D., 1939: The influence of the Dead Sea on the climate of its neighbourhood. *Quart. J. Roy. Meteor. Soc.*, **65**, 185–194.

Bader, D.C., and T.B. McKee, 1983: Dynamical model simulation of the morning boundary layer development in deep mountain valleys. *J. Climate Appl. Meteor.*, **22**, 341–351.

Bagdonas, A., J. C. George, and J. F. Gerber, 1978: Techniques of frost prediction and methods of frost and cold protection. WMO Technical Note No. 157, Geneva, Switzerland, 160 pp.

Bailey, M. J., K. M. Carpenter, L. R. Lowther, and C. W. Passant, 1981: A mesoscale forecast for 14 August 1975— The Hampstead storm. *Meteor. Mag.*, **110**, 147–161.

Banta, R. M., 1984: Daytime boundary-layer evolution over mountainous terrain. Part I: Observations of the dry circulations. *Mon. Wea. Rev.*, **112**, 340–356.

Banta, R., and W. R. Cotton, 1981: An analysis of the structure of local wind systems in a broad mountain basin. *J.*

Appl. Meteor., **20**, 1255–1266.

Bitan, A., 1977: The influence of the special shape of the Dead Sea and its environment on the local wind system. *Arch. Meteor. Geophys. Bioklim.*, Ser. B, **24**, 283–301.

Bitan, A., 1981: Lake Kinneret (Sea of Galilee) and its exceptional wind system. *Bound.-Layer Meteor.*, **21**, 477–487.

Blanchard, D. O., and R. E. López, 1984: Variability of the convective field pattern in south Florida and its relationship to synoptic flow. NOAA Tech. Memo. ERL ESG-4, Environmental Research Laboratories (NTIS#PB84-177393), 77 pp.

Bluestein, H. B., 1982: A wintertime mesoscale cold front in the southern plains. *Bull. Amer. Meteor. Soc.*, **63**, 178–185.

Blumenthal, D. L., W. H. White, and T. B. Smith, 1978: Anatomy of a Los Angeles smog episode: Pollutant transport in the daytime sea breeze regime. *Atmos. Environ.*, **15**, 893–907.

Bonner, W. D., and J. Paegle, 1970: Diurnal variations in boundary layer winds over the south-central United States in summer. *Mon. Wea. Rev.*, **98**, 735–744.

Bornstein, R. D., 1968: Observations of the urban heat island effect in New York City. *J. Appl. Meteor.*, **18**, 1118–1129.

Bornstein, R. D., 1975: The two-dimensional URBMET urban boundary layer model. *J. Appl. Meteor.*, **14**, 1459–1477.

Bornstein, R. D., and W. T. Thompson, 1981: Effects of frictionally retarded sea breeze and synoptic frontal passages on sulfur dioxide concentration in New York City. *J. Appl. Meteor.*, **20**, 843–858.

Braham, R. R., 1983: The midwest snow storm of 8–11 December 1977. *Mon. Wea. Rev.*, **111**, 253–272.

Carlson, T. B., and F. E. Boland, 1978: Analysis of urban-rural canopy using a surface heat flux/temperature model. *J. Appl. Meteor.*, **17**, 998–1013.

Carlson, T. N., J. N. Dodd, S. G. Benjamin, and J. N. Cooper, 1981: Satellite estimation of the surface energy balance, moisture availability and thermal inertia. *J. Appl. Meteor.*, **20**, 67–87.

Carpenter, K. M., 1979: An experimental forecast using a nonhydrostatic mesoscale model. *Quart. J. Roy. Meteor. Soc.*, **105**, 629–655.

Carpenter, K. M., 1982: Model forecasts for locally forced mesoscale systems. In *Nowcasting* (K. A. Browning, Ed.), Academic Press, New York, 223–234.

Changnon, S. A. (Ed.), 1981: *METRO-MEX: A Review and Summary.* Meteor. Monogr. 40, American Meteorological Society, Boston, 181 pp.

Clancy, R. M., J. D. Thompson, J. D. Lee, and H. E. Hurlburt, 1979: A model of mesoscale air-sea interaction in a sea breeze-coastal upwelling regime. *Mon. Wea. Rev.*, 1476–1505.

Clements, W. E., and C. J. Nappo, 1983: Observations of a drainage flow event on a high-altitude simple slope. *J. Climate Appl. Meteor.* **22**, 331–335.

Defant, F., 1950: Theorie der land- und seewind. *Arch. Meteor. Geophys. Bioklim.* Ser. A., **2**, 404–425.

Defant, F., 1951: Local winds. In *Compendium of Meteorology* (T.F. Malone, Ed.), American Meteorological Society, Boston, 655–672.

Delage, Y., and P. A. Taylor, 1970: Numerical studies of heat island circulations. *Bound.-Layer Meteor.*, **1**, 201–226.

Doran, J. C., and T. W. Horst, 1981: Velocity and temperature oscillations in drainage winds. *J. Appl. Meteor.*, **20**, 361–364.

Doran, J. C., and T. W. Horst, 1983: Observations and models of simple nocturnal slope flows. *J. Atmos. Sci.*, **40**, 708–717.

Doron, E., and J. Neumann, 1977: Land and mountain breezes with special attention to Israel's Mediterranean coastal plain. *Israel Meteor. Res. Pap.*, **1**, 109–122.

Dutton, J. A., and G. H. Fichtl, 1969: Approximate equations of motion for gases and liquids. *J. Atmos. Sci.*, **26**, 241–254.

Edinger, J. G., 1959: Changes in the depth of the marine layer over the Los Angeles Basin. *J. Meteor.*, **16**, 219–226.

Edinger, J. G., and R. A. Helvey, 1961: The San Fernando convergence zone. *Bull. Amer. Meteor. Soc.*, **42**, 626–635.

Elliott, D. L., and J. J. O'Brien, 1977: Observational studies of the marine boundary layer over an upwelling region. *Mon. Wea. Rev.*, **105**, 86–98.

Estoque, M. A., 1961: A theoretical investi-

gation of the sea breeze. *Quart. J. Roy. Meteor. Soc.*, **87**, 136–146.

Estoque, M. A., 1962: The sea breeze as a function of the prevailing synoptic situation. *J. Atmos. Sci.*, **19**, 244–250.

Fisher, E. L., 1961: A theoretical study of the sea breeze. *J. Meteor.*, **18**, 216–233.

Frank, N. L., P. L. Moore, and G. E. Fisher, 1967: Summer shower distribution over the Florida Peninsula as deduced from digitized radar data. *J. Appl. Meteor.*, **6**, 309–316.

Gannon, P. T., Sr., 1978: Influence of Earth Surface and Cloud Properties on the South Florida Sea Breeze. NOAA Tech. Rep. ERL–402 NHEML–2, Environmental Research Laboratories (NTIS-#PB297398), 91 pp.

Garrett A. J., 1982: A parameter study of interaction between convective clouds, the convective boundary layer, and forested surface. *Mon. Wea. Rev.*, 1041–1059.

Garrett, A. J., 1983: Drainage flow prediction with a one-dimensional model including canopy, soil and radiation parameterization. *J. Climate Appl. Meteor.*, **22**, 79–91.

Gentilli, J., 1971: Dynamics of the Australian troposphere. In *World Survey of Climatology*, **13**, Elsevier, New York, 53–117.

Halberstam, I. M., and R. Melendez, 1979: A model of the planetary boundary layer over a snow surface. *Bound.-Layer Meteor.*, **16**, 431–452.

Halberstam, I. M., and J. P. Schieldge, 1981: Anomalous behavior of the atmosphere over melting snowpack. *J. Appl. Meteor.*, **20**, 255–265.

Haurwitz, B., 1947: Comments on the sea breeze circulation. *J. Meteor.*, **4**, 1–8.

Hjelmfelt, M. R., 1982: Numerical simulation of the effects of St. Louis on mesoscale boundary layer airflow and vertical air motion: Simulations of urban vs. non-urban effects. *J. Appl. Meteor.*, **21**, 1239–1257.

Hjelmfelt, M. R., and R. R. Braham, Jr., 1983: Numerical simulation of the airflow over Lake Michigan for a major-effect snow event. *Mon. Wea. Rev.*, **111**, 205–219.

Hoecker, W. H., 1963: Three southerly low-level jet systems delineated by the

Weather Bureau special pibal network of 1961. *Mon. Wea. Rev.*, **91**, 573–582.

Holroyd, E. W., 1982: Some observations on mountain-generated cumulonimbus rainfall on the northern Great Plains. *J. Appl. Meteor.*, **21**, 560–565.

Holton, J. R., 1967: The diurnal boundary layer wind oscillation above sloping terrain. *Tellus*, **19**, 199–205.

Houze, R. A., Jr., S. G. Geotis, F. D. Marks, Jr., and A. K. West, 1981: Winter monsoon convection in the vicinity of North Borneo. Part I: Structure and time variation of the clouds and precipitation. *Mon. Wea. Rev.*, **109**, 1595–1614.

Hsu, S.-A., 1970: Coastal air circulation system: Observations and empirical model. *Mon. Wea. Rev.*, **98**, 487–509.

Huff, F. A., and J. L. Vogel, 1978: Urban, topographic and diurnal effects on rainfall in the St. Louis region. *J. Appl. Meteor.*, **17**, 565–577.

Johnson, R. H., and D. L. Priegnitz, 1981: Winter monsoon convection in the vicinity of North Borneo. Part II: Effects on large-scale fields. *Mon. Wea. Rev.*, **109**, 1615–1628.

Johnson, R. H., G. S. Young, J. J. Toth, and R. M. Zehr, 1984: Mesoscale weather effects of variable snow cover over northeast Colorado. *Mon. Wea. Rev.*, **112**, 1141–1152.

Kaplan, W., 1952: *Advanced Calculus*. Addison-Wesley, Reading, Mass., 275 pp.

Keen, C. S., W. A. Lyons, and J. A. Schuh, 1979: Air pollution transport studies in a coastal zone using kinematic diagnostic analysis. *J. Appl. Meteor.*, **18**, 606–615.

Klitch, M. A., J. F. Weaver, F. P. Kelly, and T. H. VonderHaar, 1985: Convective cloud climatologies constructed from satellite imagery. *Mon. Wea. Rev.* **113**, 326–337.

Kozo, T. L., 1982: An observational study of sea breezes along the Alaskan Beaufort Sea coast. Part I. *J. Appl. Meteor.*, **21**, 891–905.

Landsberg, H. E., 1981: *The Urban Climate*. Academic Press, New York, 275 pp.

Leipper, D. F., 1968: The sharp smog bank and California fog development. *Bull. Amer. Meteor. Soc.*, **49**, 354–358.

Levi, M., 1967: Fog in Israel. *Israel J. Earth Sci.*, **16**, 2–21.

Lyons, W. A., 1975: Turbulent diffusion and pollutant transport in shoreline environment. In *Lectures on Air Pollution and Environmental Impact Analysis* (D. A. Haugen, Ed.), American Meteorological Society, Boston, 136–208.

Maddox, R. A., 1981: The structure and life-cycle of midlatitude mesoscale convective complexes. Ph.D. dissertation, Colorado State Univ., Fort Collins, 311 pp.

Maddox, R. A., 1983: Large scale meteorological conditions associated with midlatitude, mesoscale convective complexes. *Mon. Wea. Rev.*, **111**, 1475–1493.

Mahrer, Y., and R. A. Pielke, 1977a: The effects of topography on the sea and land breezes in a two-dimensional numerical model. *Mon. Wea. Rev.*, **105**, 1151–1162.

Mahrer, Y., and R. A. Pielke, 1977b: The meteorological effect of the changes in surface albedo and moisture. *Israel Meteor. Res. Pap.*, **2**, 55–70.

Mahrer, Y., and M. Segal, 1979: Simulation of advective sharav conditions over Israel. *Israel J. Earth Sci.*, **28**, 103–106.

Mak, K-M, and J. E. Walsh, 1976: On the relative intensities of sea and land breezes. *J. Atmos. Sci.*, **33**, 242–251.

Martin, C. L., and R. A. Pielke, 1983: The adequacy of the hydrostatic assumption in sea breeze modeling over flat terrain. *J. Atmos. Sci.*, **40**, 1472–1481.

McCumber, M. C., 1980: A numerical simulation of the influence of heat and moisture fluxes upon mesoscale circulations. Ph.D. dissertation, University of Virginia, Charlottesville, 255 pp.

McCumber, M. C., and R. A. Pielke, 1981: Simulation of the effects of surface fluxes of heat and moisture in a mesoscale numerical model. Part I: Soil layer. *J. Geophys. Res.*, **86**, 9929–9938.

McNider, R. T., and R. A. Pielke, 1981: Diurnal boundary layer development over sloping terrain. *J. Atmos. Sci.*, **38**, 2198–2212.

McNider, R. T., G. J. Jedlovec, and G. S. Wilson, 1984: Data analysis and model simulation of the initiation of convection on April 24, 1982. Preprints, 10th Conference on Weather Forecasting and Analysis, Clearwater Beach, Fla., June 1984, American Meteorological Society, Boston, 543–549.

Mizzi, A. P., and R. A. Pielke, 1984: A numerical study of the mesoscale atmospheric circulation observed during a coastal upwelling event on August 23, 1972. Part I: Sensitivity studies. *Mon. Wea. Rev.*, **112**, 76–90..

Moritz, R. E., 1977: On a possible sea breeze circulation near Barrow, Alaska. *Arctic Alp. Res.*, **9**, 427–431.

Munn, R. E., 1970: *Biometeorological Methods*. Academic Press, New York, 336 pp.

Myrup, L. O., 1969: A numerical model of the urban heat island. *J. Appl. Meteor.*, **8**, 908–918.

Neumann, J., 1951: Land breezes and nocturnal thunderstorms. *J. Meteor.* **8**, 60–67.

Neumann, J., 1977: On the rotation rate of the direction of sea and land breezes. *J. Atmos. Sci.*, **34**, 1913–1917.

Neumann, J., and Y. Mahrer, 1971: A theoretical study of the land and sea breeze circulation. *J. Atmos. Sci.*, **28**, 532–542.

Neumann, J., and Y. Mahrer, 1975: A theoretical study of the lake and land breezes of circular lakes. *Mon. Wea. Rev.*, **103**, 474–485.

Oke, T. R., 1973: City size and the urban heat island. *Atmos. Environ.*, **7**, 769–779.

Oke, T. R., 1979: Review of urban climatology, 1973–1976. WMO Technical Note No. 163, Geneva, Switzerland, 100 pp.

Olsson, L. E., W. P. Elliot, and S. I. Hsu, 1973: Marine air penetration in western Oregon: An observational study. *Mon. Wea. Rev.*, **101**, 356–362.

Ookouchi, Y., M. Uryu, and R. Sawada, 1978: A numerical study of the effects of a mountain on the land and sea breezes. *J. Meteor. Soc. Japan.*, **56**, 368–385.

Ookouchi, Y., M. Segal, R. C. Kessler, and R. A. Pielke, 1984: Evaluation of soil moisture effects on the generation and modification of mesoscale circulations. *Mon. Wea. Rev.*, **112**, 2281–2292.

Orville, H. D., 1965: A numerical study of the initiation of cumulus clouds over mountainous terrain. *J. Atmos. Sci.*, **22**, 684–699.

Passarelli, R. E., Jr., and R. R. Braham, 1981: The role of the winter land breeze in the formation of Great Lakes snow storms. *Bull. Amer. Meteor. Soc.*, **62**, 482–491.

Pearson, R. A., 1973; Properties of the sea breeze front as shown by a numerical model. *J. Atmos. Sci.*, **30**, 1050–1060.

Pearson, R. A., 1975: On the asymmetry of the land-breeze sea-breeze circulation. *Quart. J. Roy. Meteor. Soc.*, **101**, 529–536.

Physick, W. L., 1980: Numerical experiments of the inland penetration of the sea breeze. *Quart. J. Roy. Meteor. Soc.*, **106**, 735–746.

Pielke, R. A., 1974: A three-dimensional numerical model of the sea breezes over south Florida. *Mon. Wea. Rev.*, **102**, 115–139.

Pielke, R. A., 1981: An overview of our current understanding of the physical interactions between the sea- and land-breeze and the coastal waters. *Ocean Management*, **6**, 87–100.

Pielke, R. A., 1982: The role of mesoscale numerical models in very-short-range forecasting. In *Nowcasting* (K. A. Browning, Ed.), Academic Press, New York, 207–221.

Pielke, R. A., 1984: *Mesoscale Meteorological Modeling*. Academic Press, Orlando, Fla., 612 pp.

Pielke, R. A., R. T. McNider, C. L. Martin, and M. Segal, 1982: Thermal driven effects and the parameterization of boundary layer diffusion. Proceedings, Workshop on the Parameterization of Mixed Layer Diffusion, October 20–23, 1981, Las Cruces, New Mexico, U. S. Army, 193–207.

Pilié, R. J., E. J. Mack, W. C. Kocmond, C. W. Roger, and W. J. Eadie, 1975: The life cycle of valley fog. Part I: Micrometeorological characteristics. *J. Appl. Meteor.*, **14**, 347–363.

Pitchford, K. L., and J. London, 1962: The low-level jet as related to nocturnal thunderstorms over midwest United States. *J. Appl. Meteor.*, **1**, 43–47.

Purdom, J. F. W., 1982: Subjective interpretation of geostationary satellite data for nowcasting. In *Nowcasting* (K. A. Browning, Ed.), Academic Press, New York, 149–166.

Riehl, H., 1954: *Tropical Meteorology*. McGraw-Hill, New York, 392 pp.

Rodgers, D. M., K. W. Howard, and E. C. Johnston, 1983: Mesoscale convective complexes over the United States during

1982. *Mon. Wea. Rev.*, **111**, 2363–2369.

Schlatter, T. W., 1972: The local surface energy balance and subsurface temperature regime in Antarctica. *J. Appl. Meteor.*, **11**, 1048–1062.

Schlatter, T. W., V. D. Barker, and J. F Henz, 1983: Profiling Colorado's Christmas Eve blizzard. *Weatherwise*, **36**, 60–66.

Schmidt, F. H., 1947: An elementary theory of the land- and sea-breeze circulation. *J. Meteor.*, **4**, 9–15.

Schultz, P., and T. T. Warner, 1982: Characteristics of summer-time circulations and pollutant ventilation in the Los Angeles Basin. *J. Appl. Meteor.*, **31**, 672–682.

Segal, M., and Y. Mahrer, 1979: Heat load conditions in Israel — A numerical mesoscale model study. *Int. J. Biometeor.*, **23**, 279–284.

Segal, M., and R. A. Pielke, 1981: Numerical model simulation of biometeorological heat load condition—Summer day case study for the Chesapeake Bay area. *J. Appl. Meteor.*, **20**, 735–749.

Segal, M., R. A. Pielke, and Y. Mahrer, 1981: Evaluation of onshore pollutant recirculation over the Mediterranean coastal area of central Israel. *Israel J. Earth Sci.*, **30**, 39–46.

Segal, M., Y. Mahrer, and R. A. Pielke, 1982a: Application of a numerical mesoscale model for the evaluation of seasonal persistent regional climatological patterns. *J. Appl. Meteor.*, **21**, 1754–1762.

Segal, M., R. T. McNider, R. A. Pielke, and D. S. McDougal, 1982b: A numerical model simulation of the regional air pollution meteorology of the Greater Chesapeake Bay area—Summer day case study. *Atmos. Environ.*, **16**, 1381–1397.

Segal, M., Y. Mahrer, and R. A. Pielke, 1983: A study of meteorological patterns associated with a lake confined by mountains — The Dead Sea case. *Quart. J. Roy. Meteor. Soc.*, **109**, 549–564.

Shair, F. H., E. J. Sasaki, D. E. Carlan, G. R. Cass, and W. R. Goodin, 1982: Transport and dispersion of airborne pollutants associated with the land breeze-sea breeze system. *Atmos. Environ.*, **16**, 2043–2053.

Sisterson, D. L., and R. A. Dirks, 1978: Structure of the daytime urban moisture field. *Atmos. Environ.*, **12**, 1943–1949.

Skibin, D., and A. Hod, 1979: Subjective analysis of mesoscale flow patterns in northern Israel. *J. Appl. Meteor.*, **18**, 329–337.

Smith, R. C., 1955: Theory of airflow over a heated land mass. *Quart. J. Roy. Meteor. Soc.*, **81**, 382–395.

Smith, R. C., 1957: Air motion over a heated land mass, II. *Quart. J. Roy. Meteor. Soc.*, **83**, 248–256.

Snow, J. W., 1981: Wind power assessment along the Atlantic and Gulf Coast of the United States. Ph.D. dissertation, Department of Environmental Sciences, University of Virginia, Charlottesville, 244 pp.

Sommers, W. T., 1981: Waves on a marine inversion undergoing mountain leeside wind shear. *J. Appl. Meteor.*, **20**, 626-636.

Tapp, M. C., and P. W. White, 1976: A nonhydrostatic mesoscale model. *Quart. J. Roy. Meteor. Soc.*, **102**, 277–296.

Toth, J. J., and R. H. Johnson, 1985: Summer surface flow characteristics over northeast Colorado. *Mon. Wea. Rev.*, **113**, 1458–1469.

Uccellini, L. W., 1980: On the role of upper tropospheric jet streaks and leeside cyclogenesis in the development of low-level jets in the Great Plains. *Mon. Wea. Rev.*, **108**, 1689–1696.

Ulanski, S. L., and M. Garstang, 1978: The role of surface divergence and vorticity in the life cycle of convective rainfall. Part I. Observations and analysis. *J. Atmos. Sci.* **35**, 1047–1062.

Vukovich, F. M., and J. W. Dunn, 1978: A theoretical study of the St. Louis heat island: Some parameter variations. *J. Appl. Meteor.*, **17**, 1585–1594.

Wallace, J. M., 1975: Diurnal variations in precipitation and thunderstorm frequency over the conterminous United States. *Mon. Wea. Rev.*, **103**, 406–419.

Wetzel, P. J., D. Atlas, and R. H. Woodward, 1984: Determining soil moisture from geosynchronous infrared data: A feasibility study. *J. Climate Appl. Meteor.*, **23**, 375–391.

Whiteman, C. D., and T. B. McKee, 1982: Breakup of temperature inversions in deep mountain valleys: Part II. Thermodynamic model. *J. Appl. Meteor.*, **21**, 290–302.

Winstanley, D., 1972: Sharav. *Weather*, **27**, 146–160.

Wong, K. K., and R. A. Dirks, 1978: Mesoscale perturbations on airflow in the urban mixing layer. *J. Appl. Meteor.*, **17**, 677–688.

Yamada, T., 1983: Simulations of nocturnal drainage flows by a $q^2\ell$ turbulence closure model. *J. Atmos. Sci.*, **40**, 91–106.

Young, G. S., and J. W. Winchester, 1980: Association of non-marine sulfate aerosol with sea breeze circulation in Tampa Bay. *J. Appl. Meteor.*, **19**, 419–425.

Zemel, Z., and J. Lomas, 1977: An objective method for assessing representativeness of a station network measuring minimum temperature near the surface. *Bound.-Layer Meteor.*, **10**, 3-14.

Zhang. D., and R. A. Anthes, 1982: A high-resolution model of the planetary boundary layer—Sensitivity tests and comparisons with SESAME-79 data. *J. Appl. Meteor.*, **21**, 1594–1609.

CHAPTER 23

The Dryline

Joseph T. Schaefer

23.1. Introduction

Over the Great Plains area of the United States, a narrow zone of extremely sharp moisture gradient often parallels the Rocky Mountains during the spring and early summer. This zone separates moist air flowing off the Gulf of Mexico from dry air flowing off the semi-arid high plateau regions of Mexico and the southwest. In the literature, it is given various, virtually interchangeable names, including dewpoint front (Beebe, 1958), dry front (Fujita, 1958), Marfa front (Matteson, 1969), and dryline (McGuire, 1960).

Thermal considerations preclude classifying the dryline as either a warm or a cold front. Because of water vapor absorption (Brunt, 1944, pp. 124–146), and the complex nature of the surface energy budget in regions of rapid variations in soil moisture content (Lanicci and Carlson, 1983), there is a distinct diurnal variation in the sign of the temperature gradient across the dryline. During the night and early morning hours, the dry air west of the dryline is typically cooler than the moist air (Fig. 23.1). However, during the heat of the day, the opposite holds and the dry air becomes markedly warmer than the moist air (Fig. 23.2). Centralized analyses often totally ignore the dryline.

The sharpness of the moisture gradient is affected by orographic elevation. Regions at elevations greater than 500 m above sea level experience much more intense drylines than lower areas do. A diurnal variation is present in the strength of the moisture gradient (Rhea, 1966); during the midafternoon hours, the dryline is typically much better defined than it is during the early morning.

The synoptic-scale motion of the dryline exhibits a diurnal trend, moving eastward during the day and westward at night (Figs. 23.1–23.3). Daytime eastward velocities are greatest in the morning hours. Similarly, the fastest nocturnal return occurs before local midnight. However, mesoscale analysis indicates that these generalities about motion do not necessarily hold for individual segments of a dryline. Different sections of the dryline not only can move at different speeds, but often even move in different directions. Also, it must be noted that a migrating cyclone can override this motion pattern.

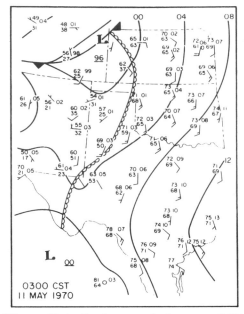

Figure 23.1. Surface analysis, 0300 CST, 11 May 1970.

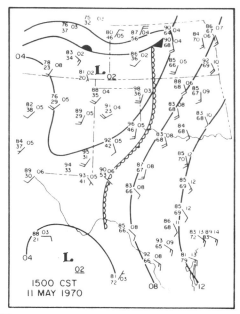

Figure 23.2. Surface analysis, 1500 CST, 11 May 1970.

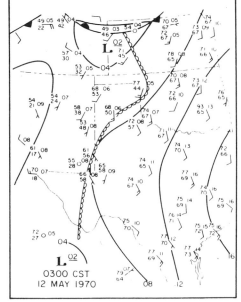

Figure 23.3. Surface analysis, 0300 CST, 12 May 1970.

23.2. Characteristics of Drylines

Although the sharp moisture gradient of the dryline is often associated with a pressure trough and/or windshift line, neither feature is necessary for dryline existence. As seen in Fig. 23.2, the dryline does not necessarily lie in a pressure trough. Furthermore, the dryline does not even lie in a trough of perturbation pressure, i.e., an area of maximum negative deviation of pressure from a linear time trend (Koch and McCarthy, 1982). Similarly, a line of organized veering surface winds, which is often located in the vicinity of the dryline, can often be found far into the moist or the dry air (Matteson, 1969).

Analysis of the kinematic properties of the wind field show that the mass convergence maximum is typically to the west and that along the dryline significant convergence does not necessarily occur (Ogura *et al.*, 1982). Even if the moisture gradient is in an area of rapid directional changes, the convergence is often dominated by the strong speed changes upstream from the dryline in the dry air (Doswell, 1976). Relative vorticity patterns are weak and unorganized, bearing no obvious relationship to the dryline. This tends to eliminate Ekman pumping (Holton, 1972) as a source of vertical motion.

It is necessary to use a moisture variable to locate the dryline. Mixing ratio, dewpoint temperature, or equivalent potential temperature (θ_e) can be used to identify the transition zone between the moist and the dry air. Since NWS observations are taken in terms of dewpoint temperature, this is by far the most convenient for operational use. On the other hand, mixing ratio is a conservative measure of only the moisture content of the atmosphere. This simplicity gives mixing ratio a strong theoretical appeal for studies of the preconvective dryline.

The convention in frontal analysis is to locate fronts on the warm side of the gradient, but no such convention exists for the dryline. Observations in the areas prone to dryline occurrence are so sparse that in places the moisture field appears to be discontinuous. At other times, weak moisture gradients within both the moist and dry air act in concert with the coarse observational grid to smear out the region of intense gradient. Accordingly some arbitrary isopleth is chosen with the observed gradient, the wind structure, and other observational clues. It must be emphasized that the winds can be very deceptive in indicating dryline position.

For analysis of the dryline, the 9 g kg^{-1} isohume or the 55°F isodrosotherm is recommended as the first guess over the Plains. This is approximately the lowest value that seems able to support tornadic thunderstorms (Williams, 1976). At times multiple zones of enhanced gradient are present. This is especially true after a single dryline episode has lasted several days. When such cases are being analyzed, secondary drylines should be indicated, since significant thunderstorm activity will frequently develop along a line of veering surface winds associated with a "secondary" dryline, even though the strongest moisture gradient is well to the west (Tegtmeier, 1974).

Satellite imagery can sometimes be used to help locate the dryline. Visible images often depict a thin line of convective clouds above the dryline. Infrared imagery often shows a boundary of "black stratus" at the dryline. These pseudo-clouds are noted nocturnally east of the dryline where the moisture content has kept temperatures warmer than those in the dry air to the west. Also, a radar-observed thin line is an excellent indicator of dryline position (Fujita, 1970).

23.2.1. Pseudo-Drylines

A zone of rather intense drying sometimes develops behind thunderstorms (Ninomiya, 1971). Such a "dryline" is in fact a rear flank downdraft (Lemon and Doswell, 1979). It is similar to the warm, dry tongue observed by Newton (1950) in the 900 to 700 mb layer, but strong enough to reach the surface. Care must be taken to distinguish between such transient storm-induced phenomena and the true dryline, which has relatively long life and is often associated with storm development.

23.2.2. Dryline Climatology

Although a dryline can occur any time, it is essentially a late-spring/early-summer phenomenon. It is typically oriented in a north-south direction and tends to parallel topographic contours. Although it is oserved as far north as Nebraska and the Dakotas, and as far east as the Texas-Louisiana border, it is more frequently observed in the Texas Panhandle and the western portions of Oklahoma and Kansas.

Rhea (1966) perused 3-hourly surface charts for April through June, 1959 to 1962, and found that an organized dewpoint discontinuity zone of at least 10°F existed between neighboring points on 45% of the days examined. Schaefer (1973) refined the identification criteria to require, in addition, a 6 h duration, a fairly uniform dewpoint field having a mean value of 50°F east of the dryline, and a diurnal change in the sign of the corresponding temperature gradient. Twenty-two distinct dryline cases were found over Texas and Oklahoma during April, May, and June 1966 through 1968. These 22 cases took place over all or part of 114 days (41% of the total).

More recently, Peterson (1983) studied the West Texas region from Oklahoma to the Rio Grande and 100°W longitude to the New Mexico border. He found an annual average of 30 dryline days during April, May, and June. Individual years had 26 to 58 dryline days. Lubbock, Tex., approximately the center of the study area, reported an average of 12 (between 6 and 21) annual dryline passages. Peterson further noted that there is a broad diversity in dryline configurations, contrasts, and temporal behavior.

23.2.3. Dryline Significance

The dryline plays a very significant role in the thunderstorm climatology of the Southern Plains region. On 70% of the days when a dryline is present, new radar echoes occur within 400 km of the dryline. Rhea (1966) compared surface charts with radar composite maps and found the frequency of new echo development relative to the line of confluent surface wind that is located near the dryline (Fig. 23.4). Entries were weighted so that if thunderstorms developed randomly, the plotted frequency would be uniform. Obviously, this is not the case; approximately one-third of the storms developed within 20 km of the dryline (windshift line), and one-half of them developed from 20 km west to 100 km east. It is emphasized that storms were found "west" of the dryline because Rhea defined the dryline as the wind confluence line rather than as the discontinuity in the moisture gradient.

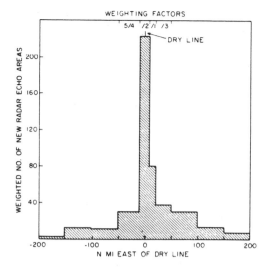

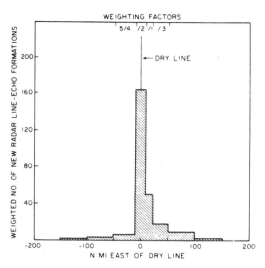

Figure 23.4. Frequency of new radar echo area development relative to surface position of the dryline for April, May, and June, 1959 through 1962. (After Rhea, 1966.)

Figure 23.5. Frequency of new radar line-echo formation relative to surface position of the dryline for April, May, and June, 1959 through 1962. (After Rhea, 1966.)

The fact that storms tend to develop near the dryline does not necessarily make the dryline the convective trigger (Ogura *et al.*, 1982). However, as Rhea (1966) noted, storms that form in the environs of the dryline tend to become well-organized, forming lines of moderate to strong intensity (Fig. 23.5). Interestingly, 60% of the new echoes that evolved into lines remained essentially stationary during the organizing process.

Temperatures are very dependent upon the dryline position. Because of the reduced diurnal variation in the moist air, 10°F forecast errors can be caused by the unexpected passage of a dryline. The dryline also plays a role in the agricultural meteorology of the West Texas region. With dewpoint differences of >10°F across a narrow zone, slight variations in the position of the dryline can dramatically affect crop drying, spraying efficiency, and the effectiveness of irrigation. The hot, dry winds west of the dryline frequently wither the corn crop (Cline, 1893). The dryline also affects flying conditions, as the winds to the west of it lift dry dirt and sand to produce significant dust storms.

23.3. Morphology of the Dryline Environment

Low-level aircraft penetrations of the immediate dryline zone indicate that the density remains essentially the same in both the moist and the dry air (McGuire, 1960). This is consistent with mesoscale surface data. Ogura and Chen (1977) analyzed the pre-storm dryline environment using the 1966 National Severe Storms Laboratory (NSSL) mesonetwork, which had a mean station spacing of about 25 km. They found a gradual monotonic change in virtual potential temperature (θ_v) of about 2 K in 170 km along a path that intersected the dryline. The NSSL network was drawn in, to a mean spacing

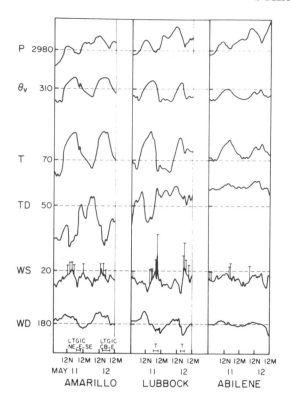

Figure 23.6. Traces of surface parameters constructed from hour observations
for 11–12 May 1970.

of 13.5 km in 1974. A study of the evolution of the dryline with those data
(McCarthy and Koch, 1982) also found that the pre-convective dryline was
located in a weak density gradient area. However, as convective activity
increased, a zone of minimum density developed that was about 20 km wide
and roughly centered on the dryline.

23.3.1. Horizontal Structure

Hourly traces of surface weather parameters observed near the dryline
(Fig. 23.6) illustrate the basic features of the dryline environment. Abilene,
Tex. (ABI), remained in the moist air east of the dryline for the period considered. (The position of the dryline in Fig. 23.2 is its closest position to
Abilene during this episode.) Weather conditions (except for the thunderstorm at the end of the trace) were stagnant. Winds were generally from
the south at 10–20 kt (5–10 m s^{-1}), and gusts were few. Dewpoint temperatures varied by only 6°F; dry bulb temperatures varied as much as 17°F.
Because of the lack of dewpoint variation, the θ_v trace closely resembles the
temperature curve.

Except for a nocturnal visit by the moist air, Amarillo, Tex. (AMA),
remained dry throughout the period. Even though there were large fluctuations in dewpoint temperatures, mixing ratios were low and remained low.
For instance, the observed dewpoint change from 35°F to 18°F corresponded
to only a 2.5 g kg^{-1} drop in mixing ratio. In contrast, a much smaller dewpoint variation from 41°F to 50°F also represents a 2.5 g kg^{-1} mixing ratio

change. Dewpoint changes alone are not a good indicator of moisture content change; a threshold value must also be used.

The AMA and ABI temperature curves illustrate that stations in the dry air experience a greater diurnal temperature change than do those in the moist air. Further, a rapid deceleration in the cooling rate is noted upon the advent of the higher dewpoint in the late evening. During the daytime hours, the winds at AMA were from the southwest, but they backed to southerly as the surface started to cool in the early evening. Rising dewpoint temperatures followed the shift of the winds to a more southerly direction. Winds in the dry air at AMA were much gustier than in the moist air at ABI. Lightning was observed to the east of AMA on both evenings.

Marked daytime and evening dryline passages are evident at Lubbock, Tex. (LBB). The winds veered as the station dried, and a backing preceded the moisture return. Gusty surface winds occurred during the time LBB was in the dry air. Pressure traces at all three stations were quite similar. Afternoon and evening thunderstorms occurred both days at Lubbock.

23.3.2. Vertical Characteristics

Data from a few aircraft penetrations at midday, after any nocturnal inversion had dissipated, provide the only detailed measurements of the vertical structure of the dryline. From flights at the 800 and 700 mb levels, Fujita (1958) estimated that the dryline "surface is tilted eastward by as much as 1:30." Three cases of "stairstep" penetrations of the dryline were analyzed by McGuire (1960). A nearly vertical boundary between the moist and dry air was found in the lowest 100 mb. Above this level only a slight slope to the east existed. Near the ground, the boundary was displaced slightly eastward and was more diffuse than it was immediately above the contact layer. This agrees with both Fujita's observation of 1:30 slope below 700 mb and with the virtually infinite slope required by applying Margules' equation to a front that exhibits no density contrast (Hess, 1959, pp. 230–234).

There is a marked difference in the midday stratification of the air on either side of the dryline. To the west, a nearly adiabatic ("well-mixed") layer, at times extending to altitudes higher than 500 mb, is observed immediately off the surface. In contrast, the moist air east of the dryline is capped by a low-level inversion typically located between 700 and 800 mb. Above this lid, a well-mixed layer is observed through the midtroposphere (Carlson and Ludlam, 1968). In general, the midday vertical profile of the dryline resembles a step. An essentially vertical boundary off the surface rapidly arcs through 90°, becoming quasi-horizontal eastward from the surface dryline.

There is considerable fine-scale detail in the inversion surface. Aircraft data (NSSP Staff, 1963) reveal upward-bulging moist tongues of lower potential temperature. These variations are not random since they have continuity. Furthermore, they bear no obvious relationship to local heat sources or orographic features.

Rawinsonde data and surface observations were used to construct an east-west cross section at 0600 CST (3 h later than Fig. 23.1) along a slice from Tucson, Ariz., to Shreveport, La. (Fig. 23.7). This section intersected the

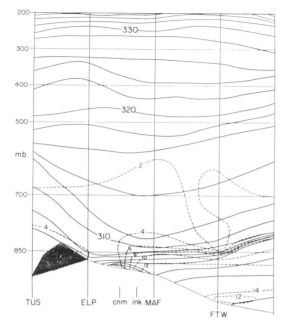

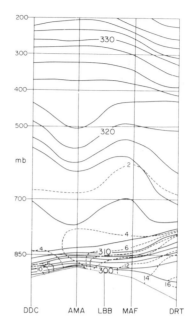

Figure 23.7. Cross section from Tucson, Ariz., to Shreveport, La., for 0600 CST, 11 May 1970. Solid lines are isentropes; dashed lines are isohumes (mixing ratio in g kg^{-1}).

Figure 23.8. Cross section from Dodge City, Kans., to Del Rio, Tex., for 0600 CST, 11 May 1970.

dryline between Carlsbad, N. Mex. (dewpoint $T_d = 25°$F), and Wink, Tex. ($T_d = 57°$F). Since these are early morning data, the potential temperature and mixing ratio patterns are different than at midday. A surface-based nocturnal radiational inversion is present across the entire section. To the east, a second inversion with potential temperatures ranging between 302 K and 308 K caps the moisture. In the dryline zone, these two distinct inversions merge, so that at El Paso, Tex. (the first sounding west of the dryline), the 308 K isentrope caps the surface layer. There is no boundary in potential temperature field separating the moist air from the dry air.

A northeast-to-southwest cross section (from Dodge City, Kans., to Del Rio, Tex.) was constructed from these same data (Fig. 23.8). It shows the same basic structure as Fig. 23.7. This cross section intersects the dryline twice; Amarillo, Tex., is the only dry station. Again, the capping inversion is visible. The temperature of the actual lid varies; it is between the 310 K and 312 K isentropes at Del Rio and between the 306 K and 310 K isentropes at Midland, Tex.. This shows that the inversion is not a material surface. Diabatic flow can and does occur in the lower layers of the atmosphere. The merger of the lid and the nocturnal inversion is seen not only in the dry air but also in regions recently reinvaded by moisture. Note that at Dodge City the T_d was 37°F at 0300 (Fig. 23.1) but is 52°F at 0600 CST.

In both cross sections, there are layers of decreased stability above the low-level inversion. However, the lack of detail in the pattern makes it impossible to distinguish any spatially continuous well-mixed layer aloft, east

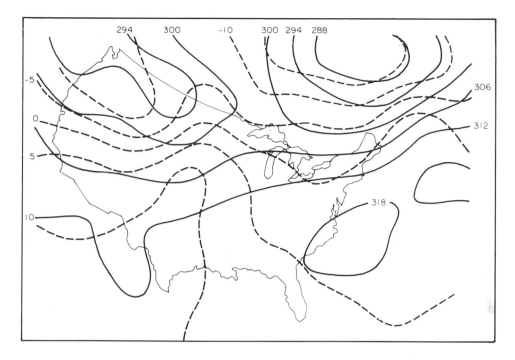

Figure 23.9. 700 mb chart, 0600 CST, 11 May 1970.

of the dryline. While midtropospheric flow is generally from the southwest with a long overland fetch before arrival in the West Texas area, the dryline region is under a weak trough immediately adjacent to the well-developed Bermuda high (Fig. 23.9).

During the day the nocturnal inversion dissipated, and 12 h later the expected potential temperature and moisture configurations were present. The Tucson-Shreveport cross section (Fig. 23.10) indicates that to the east of the dryline, a lid still exists between 304 K and 308 K. The upward bowing of the moisture in the dryline region is a reflection of the convective clouds along it. In the dry air to the west of the dryline, a surface-based mixed layer extends upward to midtropospheric levels. A second east-west cross section, about 200 mi north of this one, running from Winslow, Ariz., to Little Rock, Ark., is also presented (Fig. 23.11). The lid here is also quite evident. Since the rawinsonde stations are rather far from the dryline, there was less cloud contamination and bowing of moisture along the dryline. Again the surface-based well-mixed layer extends to midtroposphere. The southerly middle-level flow is bringing the air that is immediately west of the dryline in the more southern cross section (Fig. 23.10) to areas east of the dryline but above the lid in this one.

The northeast-southwest cross section (Fig. 23.12) is in agreement with this interpretation. Again, the clouds in the Midland, Tex., area give a bowing of the moisture above the dryline location. Cooling surface temperatures indicate that the nocturnal inversion is already starting to appear in the dryline region. Further, the dryline has been retrograding for about an hour

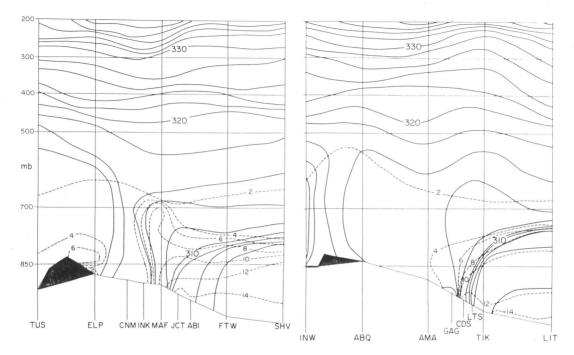

Figure 23.10. Cross section from Tucson, Ariz., to Shreveport, La., for 1800 CST, 11 May 1970.

Figure 23.11. Cross section from Winslow, Ariz., to Little Rock, Ark., for 1800 CST, 11 May 1970.

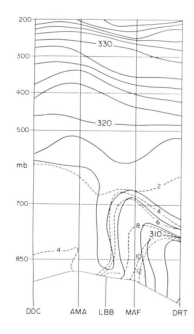

Figure 23.12. Cross section from Dodge City, Kans., to Del Rio, Tex., for 1800 CST, 11 May 1970.

at this time. Above the nascent radiational inversion there is free flow across the dryline from west to east.

In this case, the air in the well-mixed surface layer west of the dryline has been advected over the moist air. The air to the west of the dryline has the same characteristics as the air above the lid to the east. This agrees with

Schaefer's (1973) findings that the pressure-weighted mean wet bulb potential temperature in the air above the lid east of the dryline was typically the same as that in the surface-based mixed layer to the west. It is statistically impossible to use the thermal stratification to reject the concept that all the dry air belongs to the same air mass.

However, at times the low- to middle-level flow is such that the associated air on the dry side is in contact with the high, arid plateau regions of northern Mexico long enough to become excessively warm and identifiable as a new type of air mass (i.e., the Mexican plume) (Carlson *et al.*, 1983). If this air is subsequently advected northeastward over the moist air, an intense lid forms. Numerical simulations demonstrate that the generation of such hot air enhances the lee trough to the east of the Rockies and thus increases the confluence between the air flowing off the plateau and the less-modified air to the east and north (Benjamin and Carlson, 1985). This confluence is frontogenetic, creating an elevated boundary between the Mexican plume air and that of the mixed layer of the dryline.

23.4. Development and Propagation

A discussion of the origin of the dryline must consider the creation of the lid inversion, the moistening of the eastern air mass, and finally the creation of the dryline itself. Even though rapid drying across an inversion surface usually indicates subsidence, a study of the synoptic patterns that evolved into drylines during 1966 indicated that each and every capping inversion could be traced back to the previous frontal passage and the associated post-frontal inversion (Schaefer, 1974a). Vertical and horizontal differential advection by the ageostrophic wind maintain this inversion.

23.4.1. Dryline Origin

The moist air east of the dryline can usually be traced back to a continental origin. This air has been highly modified by evapotranspiration. The moisture is in the return flow portion of the air mass near its western periphery. Typically, it takes a couple of days for the moisture to return, but there is so much variation caused by the soil conditions, air mass, ground temperature difference, etc., that close monitoring is necessary. Since moisture is often advected to the High Plains in the air above the contact layer, surface data alone cannot reliably indicate the moisture return. The coarse spacing (spatial and temporal) of rawinsonde data makes satellite imagery essential for monitoring the moisture return.

The dryline itself forms through a frontogenetic process. Inhomogeneities in the wind field (intimately related to the features of the sloping terrain) interact with the ambient mixing ratio gradient to cause dryline formation. Numerical simulations show that both confluence and horizontal shear in the boundary layer winds are effective agents in dryline genesis (Anthes *et al.*, 1982).

23.4.2. Motion During Quiescent Conditions

During synoptically quiescent conditions, the daytime eastward dryline movement is generally much faster than advection will allow (Schaefer, 1973, 1974b). One mechanism that can account for the excessive daytime movement and embodies the observed step-shaped moisture profile is mixing of moist surface air with dry air aloft. Mixing would cause dryline motion in the following manner. After dawn, boundary layer mixing starts as surface temperatures rise in response to insolation. West of the dryline, the nocturnal radiational inversion is rapidly replaced by a surface layer with an adiabatic lapse rate. Any moisture trapped beneath the inversion would be freely mixed into the dry air aloft, causing surface desiccation. Since heat is mixed more efficiently than momentum, the surface convergence maximum is positioned west of the moisture discontinuity.

Since the general terrain of the Southern Plains slopes downward to the east, the depth of the moist layer also increases eastward from the dryline location. Thus, the heat input required to erase the capping inversion increases in that direction with a corresponding delay of inversion breakdown. Parallel to the dryline, the surface elevation varies gradually (as does the height of the inversion), so that the amount of heating required to break the inversion is nearly uniform. When the needed amount of heat has been absorbed, the low-level moist air mixes with the dry air aloft. The surface dewpoint temperature drops rapidly, and the dryline "leaps" eastward to a position where no appreciable mixing between air masses has occurred.

In the late afternoon and evening, the dry air cools rapidly. A nocturnal inversion forms west of the dryline. The inhibition of the vertical mixing of momentum leads to a decrease of the low-level winds in the dry air. East of the dryline, the flow features a strong, diurnally varying, ageostrophic easterly component, which varies along with the lee trough and reaches a maximum during the early evening (Benjamin and Carlson, 1985). Thus, across the dryline there is a net easterly wind. The dryline is advected westward by these winds. As cooling continues during the night, the dry air becomes denser than the moist air and is no longer easily displaced. From this time until insolation again induces eastward motion, the dryline remains nearly stationary.

Daytime dryline displacement under quiet conditions can be estimated by examining rawinsonde data and determining whether surface heating will be sufficient to break the low-level inversion. Temporal variations in inversion strength are a complicating factor; it is emphasized that quasi-geostrophic concepts are inadequate for estimating them (Watson, 1971). A more difficult forecast problem is nocturnal dryline displacement. Since it is advective, the midafternoon surface geostrophic winds (Sangster, 1960) can be used to gauge potential windspeeds. Evening surface winds will blow at perhaps 25–50% of this velocity.

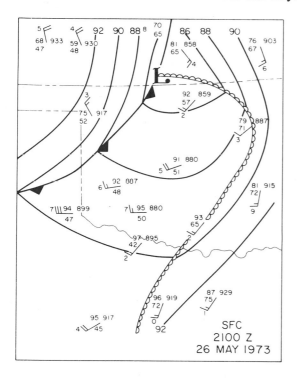

Figure 23.13. Dryline bulge from 1500 CST, 26 May 1973. (After Tegtmeier, 1974.)

23.4.3. Dryline Bulges

Relatively large-scale (100–800 km and 6–12 h duration) eastward bulges (Fig. 23.13) are often apparent on the dryline (Tegtmeier, 1974). Typically, a middle- to upper-level jet is in the vicinity (Fig. 23.14) when dryline bulges form. There are several plausible explanations (Tidwell, 1975) for this phenomenon. The immediate agent in all of them is enhanced dry air advection driven by excessive winds in the dry air, but the momentum source differs.

Convergence (as well as divergence) is associated with mid- to upper-tropospheric perturbations. Any such disturbance west of the dryline will feature an area of forced subsidence aloft (Bjerknes and Holmboe, 1944). Any high-momentum middle-level air that is present is thus injected into the mixed layer and transported to the surface by turbulent mixing. Observations that a minimum of gradient Richardson number (surface to 700 mb) and a maximum of eddy viscosity occur immediately west of the dryline lend credence to convergence as a mechanism of dryline bulges.

Standing mountain waves have also been proposed as a mechanism for transporting high-momentum middle-level air to the surface (Atkinson, 1981, pp. 25–108). Essentially, such lee waves require an elevated stable region featuring increased wind velocities over the ridge. The phase shift with height of perturbations in the elevation of the isentropic levels characteristic of mountain waves is commonly observed over the dryline. Numerical simulations suggest that standing mountain waves generated by the Mexican plateau can affect areas as far as 500 km to the east (Anthes *et al.*, 1982).

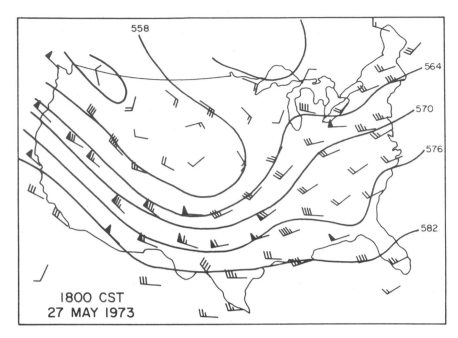

Figure 23.14. 500 mb chart, 1800 CST, 26 May 1973.

In a basic horizontal shear flow characterized by gradient Richardson numbers between 0.25 and 0.95, convective rolls, parallel to the shear, can develop as a result of baroclinic-symmetric instability. McGinley and Sasaki (1975) postulated that these symmetric roll vortices transport upper-level momentum to the surface. Baroclinic instability occurs when turbulence or diabatic processes force local regions of negative potential vorticity. Energy is derived from the kinetic energy of the base flow (Emanuel, 1983). For dryline conditions, wavelengths of about 500 km or less will amplify. This mechanism transports high-momentum air downward in the circulation of the roll, causing bulges with a wavelength compatible with the roll spacing (Fig. 23.15).

Whatever the mechanism, developing dryline bulges exhibit common characteristics. Surface winds tend to deviate toward the direction of the flow aloft, and strong gustiness appears. Highly ageostrophic conditions develop (Tegtmeier, 1974). Often, dust streaks caused by these winds can be observed in satellite imagery (Anthony, 1978). Forecasters should closely monitor surface observations to the west of the dryline for these trends. This is critical if a jet stream is crossing a portion of the dryline.

23.4.4. Fine-Scale Motion

Analyses of surface mesonetwork data reveal that fine-scale perturbations also propagate along the dryline. These disturbances propagate much faster than the winds. Koch (1979) observed that such perturbations in the dryline were typically collocated within 15 km upwind from a disturbance in the streamline pattern. The apparent waves could be detected in the dry air, but in the moist air they were found only within 20 km of the dryline.

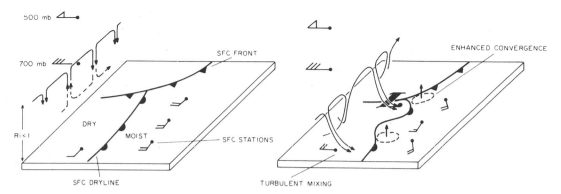

Figure 23.15. Effect of symmetric instability on drylines: (Left) Morning. (Right) After-noon. (After McGinley and Sasaki, 1975.)

These waves had a phase velocity of 22 m s^{-1}, a wavelength of 22 km, and a period of 17 min. Perturbations in the component of the wind parallel to the mean flow in the dry air were approximately 75° out of phase with perturbations in the normal component. This phase relationship is close to the 90° displacement that would be present if the wave were an internal gravity wave.

Theory shows that gravity waves can be maintained only in an environment that features a layer with a Richardson number less than 0.5 which is at or near an altitude where wind speed equals the wave phase velocity, i.e., a critical level (Gossard and Hooke, 1975). Koch (1979) examined rawinsonde data taken on both sides of the dryline to see if these conditions were met and if these perturbations were possibly gravity waves. He found a layer from 2.9 to 3.2 km (AGL) meeting the necessary conditions. Even though gravity waves are evanescent, and not ducted, so that their amplitude decreases exponentially with distance away from the critical level, significant amounts of energy can propagate to the surface. Koch further speculated that wave reflections off the lid inversion cause destructive interference in the moist air and thus limit the waves to areas near the dryline.

The gravity waves do not directly affect the moisture field, but cause perturbations in the wind field. The perturbed winds advect dry air eastward and are the immediate agent in dryline motion. Thus, gravity waves advance the dryline rather than just rippling along it like waves on a string (McCarthy and Koch, 1982).

23.5. The Dryline as a Focus of Convection

Often, the dryline is either directly involved with convective development or is an integral part of the severe thunderstorm environment. Because of this, mechanisms that serve as a convective trigger must exist in association with the dryline. Ogura *et al.* (1982) found several mechanisms that seemed to trigger thunderstorms in the moist air east of the dryline. However, on the day studied, "The dryline had no direct bearing with the initiation of the severe storm." In contrast, Koch and McCarthy (1982) examined an extremely active thunderstorm day on which mesoscale convective systems repeatedly developed over central Oklahoma. The first series of storms occurred during the early morning hours (0400–0800 CST) along a warm front but a considerable distance east of the dryline. However, beginning at about noon (1210 CST) thunderstorms started developing near the dryline. During the afternoon, three distinct systems occurred. All of them developed within 20 km of the dryline. Furthermore, it appears that each system was triggered by a different mechanism.

23.5.1. Convection as a Response to Upper Perturbations

The dryline is the westernmost boundary of moist, convectively unstable air. Because of this, the area along it and immediately east is the first region susceptible to thunderstorm development encountered by the vertical motion field associated with traveling disturbances in the westerlies. Rhea's dryline study (1966) notes that 71% of the active drylines studied were associated with a discrete feature in either the flow pattern or temperature field at 500 mb. The most frequently observed feature was a "short wave trough" 100 to 200 mi upstream from the point of first radar echo development. Diffluence at 500 mb was noted only 4% of the time.

The presence of a distinct Mexican plume, whose western boundary lies somewhat east of the dryline position, increases the effectiveness of the dryline at focusing convection induced by upper perturbations. Recall that a Mexican plume consists of air that has been excessively heated by contact with the high plateau regions and then advected northeastward over the moist, relatively cool surface air east of the dryline. Beneath this plume, the inversion that caps the low-level moisture is intensified. At the western boundary of this plume there is a baroclinic zone aloft, across which the air temperature lowers to values more attuned to those measured in the well-mixed layer west of the dryline. A distinct weakening in the strength of the inversion occurs beneath the elevated baroclinic zone.

When a plume exists, the ageostrophic circulations associated with upper-level perturbation carry the moist surface air east of the dryline north-northwestward. Initially, convection is inhibited by the strong lid. When the moist air flows out from beneath this excessive inversion (i.e., underruns the baroclinic zone aloft), especially violent convection occurs as the instability is suddenly released (Carlson *et al.*, 1983).

23.5.2. Forced Convection Along the Dryline

In a study of hourly surface charts created by compositing analyses from 23 cases in a tornado relative coordinate system, Livingston (1983) found a bulge to be a recognizable entity even in the mean presentation. Further, his analysis of perturbation pressure revealed that an area of low pressure developed along the bulge. This low is not symmetric around the bulge, but is elongated into the moist air. Tornado activity typically occurred about 3 h after and 120 km northeast of the low.

The observed convergent zone along the apex of the bulge can be related to a combination of increased westerlies by downward momentum transport

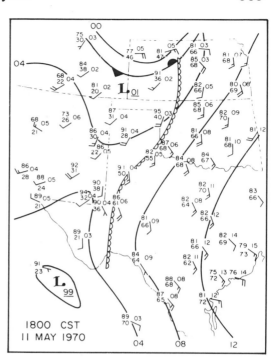

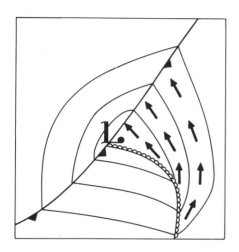

Figure 23.16. Dryline wave favorable for thunderstorm formation. (After Tegtmeier, 1974.)

Figure 23.17. Surface analysis, 1800 CST, 11 May 1970.

and by isallobaric flow. Thunderstorms are most likely to form along the portion of a dryline bulge (Fig. 23.16) that is oriented from northwest to southeast (Tegtmeier, 1974). This configuration tends to maximize the convergence as the winds in the moist air wrap around the dry air.

During the nocturnal retreat of the dryline, forced convergence is also important. During the early evening, as surface temperatures cool, winds west of the dryline die out as a radiational inversion forms and decouples the boundary layer from the free atmosphere. However, east of the dryline, the wind speeds are steady, and wind directions show a tendency to back (i.e., become normal to the dryline). Convergence and moisture convergence maximize along the retrogressing dryline. Underrunning becomes important as air flows out from beneath the lid into regions under the nascent inversion. For example, at Lubbock, Tex., on 11 May 1970, a major tornado (F5) occurred as the dryline passed on its way west. This was one of two tornadoes that developed between 1930 and 2045 CST in the Lubbock area that evening (Fujita, 1970). (The synoptic situation for that day is seen in Figs. 23.1–23.3, 23.6–23.11.) The surface chart from $1\frac{1}{2}$ h before these tornadoes (Fig. 23.17) shows the strong winds in the moist air blowing nearly perpendicular to the dryline.

23.5.3. Gravity Waves as a Trigger

In the dryline environment, gravity waves have the potential of triggering convection. In Koch and McCarthy's (1982) case study, propagating mesoscale convergence areas were noted. These areas occurred in conjunction with the gravity waves discussed in the section on fine-scale dryline motion (Sec. 23.4.4). The first radar echo associated with six individual thunderstorms developed within 10 km to the southwest of such a convergence zone. Convergence developed an average of 35 min before the echo first appeared.

By analyzing the equation defining internal gravity waves, Koch (1979) estimated that for the observed phase speed, period, and ambient conditions gravity waves originated at a location 2.9 km AGL. These waves would produce a maximum vertical displacement of 2.1 km. A profile of vertical displacement was constructed by demanding the exponential decay with height intrinsic to such evanescent waves. At the top of the low-level inversion a displacement of 380 m could be expected. This value is reasonable and compares quite favorably with the approximately 300-m-amplitude waves noted on the inversion surface during National Severe Storms Project aircraft flights (NSSP, 1963). Applying the 380 m displacement to the rawinsonde data for Koch and McCarthy's case shows that the vertical motion from the gravity wave would be sufficient to break the inversion and release the potential instability, and thus could trigger deep convection.

23.5.4. Nonlinear Biconstituent Diffusion

As previously noted, a band of cumuliform clouds is often located along the dryline. Often, some of these clouds develop into strong thunderstorms that propagate eastward ahead of the dryline. Even after this occurs, the cloud band frequently is left behind (or redevelops) along the dryline (Fig. 23.18). Those observations argue for a convective mechanism that is intrinsic to the dryline itself.

Since the density of air is a function of both its temperature and its moisture content, two independent mechanisms (warming and moistening) can lead to the development of buoyant elements. Air density is directly related to the virtual potential temperature (θ_v) with a defining equation:

$$\theta_v = \theta\left(\frac{1 + \frac{q}{0.622}}{1 + q}\right) \simeq \theta(1 + 0.608q) , \qquad (23.1)$$

where θ is the potential temperature and q is the mixing ratio. If diffusion is the only physical process considered and eddy coefficient closure is assumed, the first law of thermodynamics and the conservation of moisture are simply

$$\frac{\partial \theta}{\partial t} = K\nabla^2\theta$$

$$\frac{\partial q}{\partial t} = K\nabla^2 q , \qquad (23.2)$$

where t is time, and K the eddy diffusivity (assumed constant for simplicity). The Lewis number (ratio of eddy diffusivity to eddy conductivity) is assumed to be unity in order to exclude nonlinear effects. Combining the two parts of (23.2) gives the time rate of change of the virtual potential temperature as

$$\frac{\partial \theta_v}{\partial t} = K \nabla^2 \theta_v - 1.216 K \nabla \theta \cdot \nabla q \ . \tag{23.3}$$

Even though diffusion, which yields an exponential decay of any initial nonlinearities, controls the individual constituents, the virtual potential temperature equation contains an additional contribution that is a direct result of the nonlinear density relation. Thus, each component configuration changes at a rate determined by its own distribution and is independent of the other component, and virtual potential temperature (density) changes at a rate proportional to the product of the constituent gradients. So, even if the virtual potential temperature field is initially uniform, it cannot maintain its spatial uniformity except for the trivial case when both constituents have a uniform distribution. In a nonlinear biconstituent system, diffusion acts to create density irregularities (Schaefer, 1975).

To see this physically, consider the type of convective circulation that the temperature field across a dryline would create by itself. The air to the west (the warm, dry side) would rise relative to that in the east (the cool, moist side), forcing a direct circulation along the dryline. On the other hand, the moisture by itself would drive an indirect circulation, and the moist air

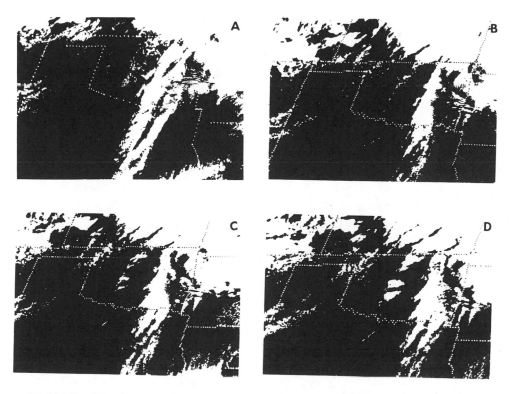

Figure 23.18. Satellite imagery from 17 May 1981 showing dryline and storm development. (Courtesy of R. Rabin.) (A) 1300 CST; (B) 1430 CST; (C) 1500 CST; (D) 1530 CST.

would rise relative to the dry air. Unless these two opposing circulations are of exactly equal magnitude, the dryline must exhibit vertical motion!

To see this mathematically, consider a dryline situation in which there is no virtual temperature difference between the moist and dry air, so that

$$\theta_v = \theta_1(1 + kq_1) = \theta_2(1 + kq_2) \,, \tag{23.4}$$

where the subscripts 1 and 2 denote conditions in the moist and dry air respectively and k is a constant equal to 0.608. If the two air masses are mixed, the resulting virtual potential temperature becomes

$$\overline{\theta_v}^m = \overline{\theta}(1 + k\overline{q}) \,, \tag{23.5}$$

where $\overline{()}^m$ denotes the average of the mixed air masses,

$$\overline{\theta} = \frac{\theta_v}{2}\left(\frac{1}{1 + kq_1} + \frac{1}{1 + kq_2}\right) , \tag{23.6}$$

and

$$\overline{q} = \frac{(q_1 + q_2)}{2} \,. \tag{23.7}$$

After much algebra it can be shown that

$$\overline{\theta_v}^m = \theta_v\left[\frac{(1 + k\overline{q})^2}{(1 + kq_1)(1 + kq_2)}\right] . \tag{23.8}$$

The density perturbation (percent change in virtual potential temperature) due to mixing alone is easily shown to be

$$\frac{\delta\theta_v}{\theta_v} = \frac{\overline{\theta_v}^m - \theta_v}{\theta_v} = \frac{k^2(\overline{q}^2 - q_1q_2)}{1 + 2k\overline{q} + kq_1q_2} \,, \tag{23.9}$$

but by the definition of θ_v, we can simplify this to

$$\frac{\delta\theta_v}{\theta_v} = \frac{k^2}{4}\left[\frac{(q_1 - q_2)^2}{1 + k(q_1 - q_2) + k^2q_1q_2}\right] . \tag{23.10}$$

Because the mixing ratios are small, this can be approximated by

$$\frac{\delta\theta_v}{\theta_v} = \frac{k^2}{4}(q_1 - q_2)^2 \,. \tag{23.11}$$

Since for a Boussinesq fluid the density perturbation times the acceleration of gravity equals vertical acceleration, a 10 g kg^{-1} mixing ratio difference will force a vertical acceleration of about $9 \cdot 10^{-5}$ s^{-2}. After 1 h, vertical velocities of 33 cm s^{-1} will develop from still air (assuming frictionless flow and no condensation). Although this process operates on a very small spatial scale, it produces uplift sufficient to account for the observed convection.

23.5.5. *Frontogenetic Circulations*

Thunderstorms frequently are initiated along a dryline when a frontal zone moving eastward overtakes (or merges) with it. The final mesosystem during Koch and McCarthy's (1982) case study serves as an excellent example. The dryline ceased to be active at about 1600 CST. However, a weak cold front that was actually undergoing frontolysis was approaching the stalled dryline from the west. At 1730, when two features were within 10 km of each other, explosive thunderstorm development began.

Frontogenesis will occur if an existing virtual potential temperature gradient and a confluent flow are superposed (Miller, 1948). Convergence is not necessary; deformation will suffice. An innate feature of the frontogenetic process is a vertical circulation. This circulation gains in significance as the magnitude of the relative vorticity approaches that of the Coriolis parameter (Hoskins and Bretherton, 1972).

These requirements for an ageostrophic frontogenetic circulation cannot be related to the general dryline environment. Although the typical dryline features confluent flow, there is little accompanying virtual potential temperature gradient and the relative vorticity field is rather featureless. However, in the vicinity of a dryline-front intersection, a density gradient (along the front) is present and frontogenesis can be a factor. Berry and Bluestein (1982) examined such intersections and found only a weak relationship between the amount of frontogenesis and the intensity of subsequent convection.

However, when a front overtakes a dryline, the wind shift line becomes more closely aligned with moisture gradient and a strong vorticity field is generated. This allows a significant frontogenetic circulation to develop. The associated vertical motion can be an effective agent for releasing potential instability. For the frontogenetic circulation to be effective, a high-vorticity, confluent flow near the dryline-front intersection is needed.

23.5.6. *The Inland Sea Breeze*

The weak zone of monotonically increasing virtual potential temperature across the dryline can force a solenoidal circulation similar to a sea breeze (Ogura and Chen, 1977). However, for such weak thermal contrasts, the background flow plays a significant role. If the geostrophic flow opposes the low-level solenoidal flow, a convergence zone is established. Not only does this directly increase the upward branch of the circulation, but the associated advective tightening of the temperature gradient also increases the circulation. In contrast, a background geostrophic flow parallel to the low-level circulation causes more widespread but weaker upward vertical velocity.

Sun and Ogura (1979) numerically simulated this effect. An initial thermal gradient of 4°C per 10 km generated vertical velocities on the order of 10–100 cm s^{-1}. For westerly geostrophic flow (favorable background flow in the dryline environment), it was noted that the circulation can cause a significant perturbation in the height of the capping inversion. The western side rises, while the eastern lowers (Fig. 23.19). This distortion of the inversion yields solenoids that intensify the upper return portion of the inland sea

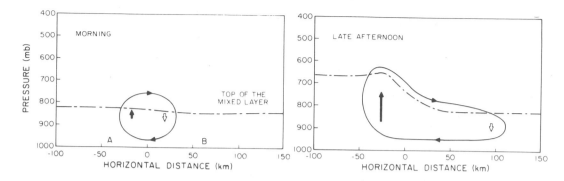

Figure 23.19. Vertical cross sections across dryline, showing the influence of background flow on the height of inversion.

breeze. A feedback circuit is thus initiated; the circulation disturbs the inversion that increases the strength of circulation. Eventually, vertical displacements become sufficiently strong to break the cap. For easterly geostrophic flows the simulated circulation was too weak to initiate this feedback. Since an inverted trough features easterly geostrophic flow, this mechanism could explain the observation that the presence of an inverted trough inhibits the development of severe convection in an otherwise favorable environment.

23.6. Drylines Worldwide

This discussion has considered only the dryline of the Great Plains of the United States. However, other parts of the globe also have dryline phenomena. Over India the dryline is a significant flow feature during the pre-monsoon months; days that have intense convection are characterized by its presence (Weston, 1972). Drylines are also regularly observed in eastern China; typically, heavy showers and hail storms are found to occur at the junction of the dryline and front (Golden, 1980). In central West Africa, squall lines (disturbance lines) account for most of the rainfall received (Hamilton and Archbold, 1945). The origin of West African storms is intimately related to the "intertropical front" (now called the inter-tropical convergence zone). Although this front runs approximately east-west, it is a dryline in that it separates dry desert air from moist monsoonal air and has the typical step-shaped profile (Eldridge, 1957). These disturbance lines differ from the North American dryline in that squall lines with the former are oriented roughly normal rather than parallel to their fronts. However, the generation of these squall lines is extremely dependent upon the position of the "front."

These features are similar to their United States counterpart in surface representation, vertical profile, and effect upon convection. Thus, knowledge of the genesis and the controlling forces for the United States dryline should help in forecasting for other regions of the world.

REFERENCES

Anthes, R. A., Y. H. Kuo, S. G. Benjamin, and Y.-F. Li, 1982: The evolution of the mesoscale environment of severe local storms: Preliminary modeling results. *Mon. Wea. Rev.*, **110**, 1187–1213.

Anthony, R. W., 1978: Dust storm—Severe storm characteristics of 10 March 1977. *Mon. Wea. Rev.*, **106**, 1219–1223.

Atkinson, B. W., 1981: *Meso-scale Atmospheric Circulations*. Academic Press, London, 495 pp.

Beebe, R. G., 1958: An instability line development as observed by the tornado research airplane. *J. Meteor.*, **15**, 278–282.

Benjamin, S. G., and T. N. Carlson, 1986: Some effects of surface heating and topography on the regional severe storms environment, Part I: 3-D simulations. *Mon. Wea. Rev.*, **114**, 307–329.

Berry, E., and H. B. Bluestein, 1982: The formation of severe thunderstorms at the intersection of a dryline and a front: The role of frontogenesis. Preprints, 12th Conference on Severe Local Storms, San Antonio, Tex., American Meteorological Society, Boston, 597–602.

Bjerknes, J., and J. Holmboe, 1944: On the theory of cyclones. *J. Meteor.*, **1**, 1–22.

Brunt, D., 1939: *Physical and Dynamical Meteorology*. Second Edition, Cambridge University Press, London, 428 pp.

Carlson, T. N., and F. H. Ludlam, 1968: Conditions for the occurrence of severe local storms. *Tellus*, **20**, 203–226.

Carlson, T. N., S. G. Benjamin, G. S. Forbes, and Y.-F. Li, 1983: Elevated mixed layers in the regional severe storm environment: Conceptual model and case studies. *Mon. Wea. Rev.*, **111**, 1453–1473.

Cline, I. M., 1893: Hot winds in Texas, May 29 and 30, 1892. *Amer. Meteor. J.*, **9**, 437–443.

Doswell, C. A., III, 1976: Subsynoptic scale dynamics as revealed by use of filtered surface data. NOAA Tech. Memo. ERL NSSL–79 (NTIS#PB–265433/AS), 40 pp.

Eldridge, R. H., 1957: A synoptic study of West African disturbance lines. *Quart. J. Roy. Meteor. Soc.*, **83**, 303–314.

Emanuel, K.A., 1983: The Lagrangian parcel dynamics of moist symmetric instability. *J. Atmos. Sci.*, **40**, 2368–2376.

Fujita, T. T., 1958: Structure and movement of a dry front. *Bull. Amer. Meteor. Soc.*, **39**, 574–582.

Fujita, T. T., 1970: The Lubbock tornadoes: A study of suction spots. *Weatherwise*, **23**, 160–173.

Golden, J. H., 1980: Forecasting and research on severe storms in China: A summary of two seminars. *Bull. Amer. Meteor. Soc.*, **61**, 7–21.

Gossard, E. E., and W. H. Hooke, 1975: *Waves in the Atmosphere*. Elsevier Scientific, Amsterdam, 456 pp.

Hamilton, R. A., and J. W. Archbold, 1945: Meteorology of Nigeria and adjacent territory. *Quart. J. Roy. Meteor. Soc.*, **71**, 231–264.

Hess, S. L., 1959: *Introduction to Theoretical Meteorology*. Holt, New York, 362 pp.

Holton, J. R., 1972: *An Introduction to Dynamic Meteorology*. Academic Press, New York, 319 pp.

Hoskins, B. J., and F. P. Bretherton, 1972: Atmospheric frontogenesis models: Mathematical formulation and solution. *J. Atmos. Sci.*, **29**, 11–37.

Koch, S. E., 1979: Mesoscale gravity waves as a possible trigger of severe convection along a dryline. Ph.D. dissertation, University of Oklahoma, Norman, 195 pp.

Koch, S. E., and J. McCarthy, 1982: The evolution of an Oklahoma dryline. Part II: boundary-layer forcing of mesoconvective systems. *J. Atmos. Sci.*, **39**, 237–257.

Lanicci, J. M., and O. N. Carlson, 1983: Sensitivity of the planetary boundary layer to changes in soil moisture availability: Some numerical experiments from SESAME IV. Preprints, 6th Conference on Numerical Weather Prediction, Omaha, Nebr., American Meteorological Society, Boston, 313–319.

Lemon, L. R., and C. A. Doswell, III, 1979: Severe thunderstorm evolution and mesocyclone structure as related to tornado genesis. *Mon. Wea. Rev.*, **107**, 1184–1197.

Livingston, R. L., 1983: On the subsynoptic pre-tornado surface environment. Ph.D. dissertation, University of Missouri, Columbia, 70 pp.

Matteson, G. T., 1969: The west Texas dry

front of June, 1967. M. S. thesis, University of Oklahoma, Norman, 63 pp.

McCarthy, J., and S. E. Koch, 1982: The evolution of an Oklahoma dryline. Part I: A meso- and subsynoptic-scale analysis. *J. Atmos. Sci.*, **39**, 225–236.

McGinley, J. A., and Y. K. Sasaki, 1975: The role of symmetric instabilities in thunderstorm development on drylines. Preprints, 9th Conference on Severe Local Storms, Norman, Okla., American Meteorological Society, Boston, 173–180.

McGuire, E. L., 1960: The vertical structure of three drylines as revealed by aircraft traverses. Paper presented at AMS Conference on Severe Local Storms (1st); also published in 1962 as National Severe Storms Project Report, No. 7, Kansas City, Mo., 11 pp.

Miller, J. E., 1948: On the concept of frontogenesis. *J. Meteor.*, **5**, 169–171.

Newton, C. W., 1950: Structure and mechanism of the pre-frontal squall line. *J. Meteor.*, **7**, 210–222.

Ninomiya, K., 1971: Mesoscale modification of synoptic situations from thunderstorm development as revealed by ATS III and aerological data. *J. Appl. Meteor.*, **10**, 1103–1121.

NSSP (National Severe Storms Project) Staff Members, 1963: Environmental and thunderstorm structures as shown by National Severe Storms Project observations in spring 1960 and 1961. *Mon. Wea. Rev.*, **91**, 271–292.

Ogura, Y., and Y. Chen, 1977: A life history of an intense mesoscale convective storm in Oklahoma. *J. Atmos. Sci.*, **34**, 1458–1476.

Ogura, Y., H. Juang, K. Zhang, and S. Soong, 1982: Possible triggering mechanisms for severe storms in SESAME-AVE IV (9–10 May 1979). *Bull. Amer. Meteor. Soc.*, **63**, 503–515.

Peterson, R. E., 1983: The west Texas dryline: Occurrence and behavior. Preprints, 13th Conference on Severe Local Storms, Tulsa, Okla., American Meteorological Society, Boston, J9–J11.

Rhea, J. O., 1966: A study of thunderstorm formation along drylines. *J. Appl. Meteor.*, **5**, 58–63.

Sangster, W. E., 1960: A method of representing the horizontal pressure force without reduction of station pressures to sea level. *J. Meteor.*, **17**, 166–176.

Schaefer, J. T., 1973: The motion and morphology of the dryline. NOAA Tech. Memo ERL NSSL-66 (NTIS#COM-74-10043), 81 pp.

Schaefer, J.T., 1974a: The life cycle of the dryline. *J. Appl. Meteor.*, **13**, 444–449.

Schaefer, J. T., 1974b: A simulative model of dryline motion. *J. Atmos. Sci.*, **31**, 956–964.

Schaefer, J. T., 1975: Nonlinear biconstituent diffusion: A possible trigger of convection. *J. Atmos. Sci.*, **32**, 2278–2284.

Sun, W. Y., and Y. Ogura, 1979: Boundary layer forcing as a possible trigger to a squall line formation. *J. Atmos. Sci.*, **36**, 235–254.

Tegtmeier, S. A., 1974: The role of the surface, sub-synoptic, low pressure system in severe weather forecasting. M. S. thesis, University of Oklahoma, Norman, 66 pp.

Tidwell, C. G., 1975: A synoptic and subsynoptic study of the June 8, 1974 severe thunderstorm and tornado outbreak in Oklahoma. M.S. thesis, University of Oklahoma, Norman, 65 pp.

Watson, G. F., 1971: A diagnostic study of the kinematical and physical processes maintaining a strong low-level subsidence inversion over land. Ph.D. dissertation, Florida State University, Tallahassee, 149 pp.

Weston, K. J., 1972: The dryline of northern India and its role in cumulonimbus convection. *Quart. J. Roy. Meteor. Soc.*, **98**, 519–531.

Williams, R. J., 1976: Surface parameters associated with tornadoes. *Mon. Wea. Rev.*, **104**, 540–545.

CHAPTER 24

Formulation of Mesoscale Numerical Models

Donald J. Perkey

24.1. Introduction

Most realistic numerical models of atmospheric circulations today are complex computer codes developed to run on the largest and fastest of today's computers. This complexity is necessary if the model is to reproduce the complicated physical processes controlling atmospheric flow.

However, more than a complicated numerical model is necessary to provide either increased understanding to the researcher or reliable guidance to the forecaster. In addition to a numerical model that replicates the physics of the atmosphere, a complete system, including pre- and post-model analysis techniques, is necessary. Prior to running of a model, the initial data must be (1) checked for errors and consistency (data validation), (2) interpolated and analyzed onto a regular grid (objective analysis), and (3) prepared for the numerical model equations (initialization). After the model has been run, the model results must be processed to provide diagnostic information and graphical displays before they can be interpreted. Thus, even the best numerical model may fail because of weaknesses in other components of the system.

Review articles on regional- and meso-scale modeling and numerical weather prediction have been written by Kreitzberg (1978, 1979), Pielke (1981), and most recently by Anthes (1983). Also texts by Haltiner and Williams (1980) and Pielke (1984) are valuable sources of information concerning numerical modeling. For the history and techniques of numerical weather prediction, with emphasis on large-scale global and climate models, see Kasahara (1977).

In this discussion, the atmospheric scale being considered is assumed to be smaller than that resolved by the conventional rawinsonde network (spacing $\sim$300 km), but much larger than individual convective clouds. The scale of particular interest is the meso-α scale (Orlanski, 1975) which has a characteristic horizontal scale of 250 to 2500 km. Models that resolve these scales are often called regional or mesoscale models and typically have grid intervals of 25 to 150 km. Attention is concentrated on methodologies most prevalent in today's regional- and meso-scale numerical models.

24.2. The Primitive Equations

The "primitive" (i.e., derived from conservation principles of the basic physical variables such as momentum, thermodynamic energy, and mass) form of the dynamic and thermodynamic equations describing the atmosphere can be found in most introductory meteorology textbooks (for example, Dutton, 1976; Holton, 1972; Wallace and Hobbs, 1977). This form of the equations is in contrast to the "balanced" (the wind and mass fields are in quasi-geostrophic balance) or "filtered" (gravity waves are removed or filtered from the solution) form of the governing equations used in the first Numerical Weather Prediction (NWP) integrations (Charney *et al.*, 1950). These integrations used the "equivalent-barotropic vorticity" equation, which can be derived from the vorticity and divergence equations (Haltiner and Williams, 1980, pp. 208–209).

In spite of the familiarity of the primitive equations, they are restated here in a form that lends itself to discussing how to formulate an atmospheric numerical model. For most primitive equation models, this implies that the equations should be written in Eulerian form; i.e., the time change in a quantity at a given location or grid point should be set equal to the appropriate agents of change.

Also, because the "resolution" (size or scale of motions that can be properly described by the model) is limited by the model's grid spacing, effects of motions smaller than this limit must be included in an "average" or "statistical" sense. For example, the $\partial/\partial t_{conv}$ terms in the equations below represent the transport of quantities by moist and dry convective elements that are inherently too small to be resolved by the model. As such, their effects are "parameterized" (see Sec. 24.3). Parameterization techniques are also used to include the effects of surface fluxes and turbulent transports by eddies, the $\partial/\partial t_s$ terms, which are also too small to be explicitly calculated by the model.

During the 1970s researchers discovered that for shorter range, smaller-grid-interval forecasts (in contrast to general circulation climate forecasts), atmospheric models needed more detailed treatment of moist processes. An example of the equations representing these processes for the moist variables of water vapor, cloud water, and precipitation (both liquid and solid) can be found in Kessler (1969). It should be noted that these equations must be highly parameterized because cloud physical processes occur on quite small scales.

Most of the notation has its traditional meteorological meaning; however, the Appendix contains definitions of all the symbols.

Conservation of horizontal momentum:

$$\underbrace{\frac{\partial u}{\partial t}}_{\substack{\text{local time} \\ \text{change}}} = \underbrace{-u\frac{\partial u}{\partial x} - v\frac{\partial u}{\partial y} - w\frac{\partial u}{\partial z}}_{\text{advection}} - \underbrace{\frac{1}{\rho}\frac{\partial p}{\partial x}}_{\substack{\text{pressure} \\ \text{gradient} \\ \text{force}}} + \underbrace{fv}_{\substack{\text{Coriolis} \\ \text{force}}}$$

$$+ \underbrace{\frac{uv}{r_e}\tan(\alpha)}_{\substack{\text{curvature} \\ \text{term}}} + \underbrace{\frac{\partial u}{\partial t_{\text{conv}}}}_{\substack{\text{convective} \\ \text{transport}}} + \underbrace{\frac{\partial u}{\partial t_s}}_{\substack{\text{eddy} \\ \text{transport}}} \tag{24.1}$$

$$\underbrace{\frac{\partial v}{\partial t}}_{\substack{\text{local time} \\ \text{change}}} = \underbrace{-u\frac{\partial v}{\partial x} - v\frac{\partial v}{\partial y} - w\frac{\partial v}{\partial z}}_{\text{advection}} - \underbrace{\frac{1}{\rho}\frac{\partial p}{\partial y}}_{\substack{\text{pressure} \\ \text{gradient} \\ \text{force}}} - \underbrace{fu}_{\substack{\text{Coriolis} \\ \text{force}}}$$

$$+ \underbrace{\frac{u^2}{r_e}\tan(\alpha)}_{\substack{\text{curvature} \\ \text{term}}} + \underbrace{\frac{\partial v}{\partial t_{\text{conv}}}}_{\substack{\text{convective} \\ \text{transport}}} + \underbrace{\frac{\partial v}{\partial t_s}}_{\substack{\text{eddy} \\ \text{transport}}} \tag{24.2}$$

Conservation of vertical momentum:

$$\underbrace{\frac{\partial w}{\partial t}}_{\substack{\text{local time} \\ \text{change}}} = \underbrace{-u\frac{\partial w}{\partial x} - v\frac{\partial w}{\partial y} - w\frac{\partial w}{\partial z}}_{\text{advection}} - \underbrace{\frac{1}{\rho}\frac{\partial p}{\partial z}}_{\substack{\text{pressure} \\ \text{gradient} \\ \text{force}}} - \underbrace{g}_{\substack{\text{acceleration} \\ \text{due to} \\ \text{gravity}}}$$

$$+ \underbrace{\frac{u^2 + v^2}{r_e}}_{\substack{\text{curvature} \\ \text{term}}} + \underbrace{\frac{\partial w}{\partial t_{\text{conv}}}}_{\substack{\text{convective} \\ \text{transport}}} + \underbrace{\frac{\partial w}{\partial t_s}}_{\substack{\text{eddy} \\ \text{transport}}} \tag{24.3}$$

Conservation of thermal energy:

$$\underbrace{\frac{\partial T}{\partial t}}_{\substack{\text{local time} \\ \text{change}}} = \underbrace{-u\frac{\partial T}{\partial x} - v\frac{\partial T}{\partial y} - w\frac{\partial T}{\partial z}}_{\text{advection}} + \underbrace{\frac{1}{c_p}Q}_{\substack{\text{latent heat of} \\ \text{condensation/} \\ \text{evaporation}}}$$

$$+ \frac{1}{\rho c_p}\frac{\partial p}{\partial t} + \frac{1}{\rho c_p}u\frac{\partial p}{\partial x} + \frac{1}{\rho c_p}v\frac{\partial p}{\partial y} + \frac{1}{\rho c_p}w\frac{\partial p}{\partial z}$$

$$\underbrace{\hspace{8cm}}_{\text{compressional warming}}$$

$$+ \underbrace{\frac{\partial T}{\partial t_{\text{conv}}}}_{\substack{\text{convective} \\ \text{transport}}} + \underbrace{\frac{\partial T}{\partial t_{\text{rad}}}}_{\substack{\text{radiational} \\ \text{heating/} \\ \text{cooling}}} + \underbrace{\frac{\partial T}{\partial t_s}}_{\substack{\text{eddy} \\ \text{transport}}} \qquad (24.4)$$

Conservation of vapor:

$$\underbrace{\frac{\partial q}{\partial t}}_{\substack{\text{local time} \\ \text{change}}} = \underbrace{-u\frac{\partial q}{\partial x} - v\frac{\partial q}{\partial y} - w\frac{\partial q}{\partial z}}_{\text{advection}} - \underbrace{\frac{\partial q}{\partial t_{\text{cond } q=>c}}}_{\substack{\text{condensation} \\ \text{of water} \\ \text{vapor}}}$$

$$+ \underbrace{\frac{\partial q}{\partial t_{\text{evap } c=>q}}}_{\substack{\text{evaporation} \\ \text{of cloud} \\ \text{water}}} + \underbrace{\frac{\partial q}{\partial t_{\text{evap } r=>q}}}_{\substack{\text{evaporation of} \\ \text{precipitation}}} + \underbrace{\frac{\partial q}{\partial t_{\text{conv}}}}_{\substack{\text{convective} \\ \text{transport}}} + \underbrace{\frac{\partial q}{\partial t_s}}_{\substack{\text{eddy} \\ \text{transport}}} \qquad (24.5)$$

Conservation of cloud water (both liquid and solid):

$$\underbrace{\frac{\partial c}{\partial t}}_{\substack{\text{local time} \\ \text{change}}} = \underbrace{-u\frac{\partial c}{\partial x} - v\frac{\partial c}{\partial y} - w\frac{\partial c}{\partial z}}_{\text{advection}} + \underbrace{\frac{\partial c}{\partial t_{\text{cond } q=>c}}}_{\substack{\text{condensation} \\ \text{of water} \\ \text{vapor}}} - \underbrace{\frac{\partial c}{\partial t_{\text{evap } c=>q}}}_{\substack{\text{evaporation} \\ \text{of cloud} \\ \text{water}}}$$

$$- \underbrace{\frac{\partial c}{\partial t_{\text{conv } c=>r}}}_{\substack{\text{conversion to} \\ \text{precipitation}}} - \underbrace{\frac{\partial c}{\partial t_{\text{coll } c=>r}}}_{\substack{\text{collection by} \\ \text{precipitation}}} + \underbrace{\frac{\partial c}{\partial t_{\text{conv}}}}_{\substack{\text{convective} \\ \text{transport}}} \qquad (24.6)$$

Conservation of precipitation (both liquid and solid):

$$\underbrace{\frac{\partial r}{\partial t}}_{\substack{\text{local time} \\ \text{change}}} = \underbrace{-u\frac{\partial r}{\partial x} - v\frac{\partial r}{\partial y} - w\frac{\partial r}{\partial z}}_{\text{advection}} - \underbrace{\frac{\partial r}{\partial t_{\text{evap } r=>q}}}_{\substack{\text{evaporation of} \\ \text{precipitation}}} + \underbrace{\frac{\partial r}{\partial t_{\text{conv} c=>r}}}_{\substack{\text{conversion to} \\ \text{precipitation}}}$$

$$+ \underbrace{\frac{\partial r}{\partial t}\bigg|_{\text{coll } c=>r}} + \underbrace{\frac{\partial r}{\partial t}\bigg|_{\text{rain}}} + \underbrace{\frac{\partial r}{\partial t}\bigg|_{\text{conv}}} \qquad (24.7)$$

collection by   rainout   convective
precipitation   precipitation

Conservation of mass:

$$\underbrace{\frac{\partial \rho}{\partial t}}_{} = \underbrace{- u\frac{\partial \rho}{\partial x} - v\frac{\partial \rho}{\partial y} - w\frac{\partial \rho}{\partial z}}_{} \underbrace{- \frac{\partial u}{\partial x} - \frac{\partial v}{\partial y} - \frac{\partial w}{\partial z}}_{} \qquad (24.8)$$

local time   advection   divergence
change

Ideal gas law:

$$p = \rho\, R\, T \qquad (24.9)$$

Equations (24.1)–(24.9) are not in their complete form; i.e., approximations and assumptions have already been made. For example, the Coriolis term has already been dropped from the vertical momentum equation (24.3), the argument being that it is small compared with the terms retained. The equations have been written with height as the vertical coordinate for simplicity, but it should be noted that most current mesoscale models employ some form of vertical coordinate that incorporates the terrain (see Sec. 24.3.1 for a discussion of vertical coordinate systems).

The solutions to the equations include horizontally and vertically propagating gravity waves, inertial waves, and sound waves, as well as the meteorological waves and circulations in which we are primarily interested (Haltiner and Williams, 1980, Ch. 2). This plethora of propagating waves can lead to numerical problems (see Sec. 24.3.5).

In addition to the approximations that have already been made to the governing equations, most numerical models that focus on meso-α-scale features involve an assumption that the vertical acceleration of a parcel $[(\partial w/\partial t) + (u\partial w/\partial x) + (v\partial w/\partial y) + (w\partial w/\partial z)$, i.e., $d\dot{w}/dt]$, the curvature term $[(u^2 + v^2)/r_e]$, the vertical friction force $(\partial w/\partial t_s)$, and the convective transport $(\partial w/\partial t_{\text{conv}})$ are small compared with the vertical pressure gradient $(\rho^{-1}\partial p/\partial z)$ and gravitational (g) forces; i.e., they assume hydrostatic balance between the vertical pressure gradient and the acceleration due to gravity (see Pielke, 1981, Sec. 3.3.2, for a formal scale analysis to justify this approximation). The resulting hydrostatic equation, $\partial(\ln p)/\partial z = -g/RT$, is used to evaluate pressure, usually by integrating down from the model top. (The pressure at the model top is calculated by using a specified top boundary condition such as $w_{\text{top}} = 0$ or $\omega_{\text{top}} = 0$.) This, however, leaves no equation for calculating the vertical velocity.

To remedy this, the vertical velocity is typically calculated by using a form of the equation for conservation of mass (24.8). This is accomplished by solving (24.8) for $\partial w / \partial z$ and integrating up from a specified ($w_{\text{surface}} = 0$) or calculated (w_{surface} is determined by the slope of the terrain) lower boundary condition. Density is diagnosed using (24.9).

In summary, a typical forecast algorithm might proceed in the following steps:

(1) Starting from an initial state in which all variables are known at all grid locations, tendencies for u, v, T, q, c, and r are calculated by using their respective conservation equations.

(2) New values of these dependent variables are calculated or stepped forward by using their initial values and their calculated tendencies.

(3) New values of p_{top} are calculated from the top boundary condition.

(4) New values of p are diagnosed by integrating the hydrostatic equation down from the model top.

(5) New values of ρ are diagnosed by substituting the already calculated new values of p and T into the Ideal Gas Law.

(6) Finally, new values of w are diagnosed by integrating the conservation of mass equation up from the surface.

(7) Steps 1 through 6 are repeated as often as required.

24.3. Numerical Approximations

Because the primitive equations are nonlinear, they cannot be solved analytically in their complete form. Therefore, numerical techniques must be used to obtain solutions. Although the atmosphere and the equations used to represent the atmosphere are continuous, to solve these partial differential equations numerically requires approximating the equations by rewriting them in discrete or finite-difference form. Richardson (1922) described the methods and techniques that he used in trying to solve the governing equations numerically. He also noted that his methods yielded unacceptable solutions. However, with the development of the electronic computer in 1945, simplification of the atmospheric equations to the equivalent-barotropic vorticity equation, and a better understanding of numerical analysis (Courant *et al.*, 1928), it became possible for Charney *et al.*(1950) to make the first successful numerical weather forecast.

Errors are still introduced when continuous equations are approximated by their finite-difference equivalents, and those errors must be considered when a numerical model is being constructed or used. For a more complete treatment of the numerical methods employed by numerical models, see Haltiner and Williams (1980) and Pielke (1984).

24.3.1. Coordinate Systems

The vertical coordinate variable can be height, pressure, or potential temperature, or any other monotonic function. The choice determines the exact form of the primitive equations. Kasahara (1974) derived the primitive equations using a generalized vertical coordinate ς, which is related monotonically to height. He also indicated how to transform the equations from this system to another. Most present-day mesoscale models use a vertical coordinate that is a function of either pressure or vertical distance (Anthes, 1983). These coordinate variables are based on the so-called σ-coordinate system introduced by Phillips (1957). In Phillips' system, σ is defined as the pressure at the level of interest divided by the pressure at the model's lower boundary (the Earth's surface) and, because it depends on pressure, is often referred to as a σ_p system. Among other useful properties, σ_p surfaces follow the lower boundary of the model; i.e., they follow the terrain. This effect decreases as pressure decreases upward.

A system with similar properties, but based on height, is the σ_z system where σ_z is defined as follows:

$$\sigma_z = \left(\frac{z - E}{H - E}\right) H, \tag{24.10}$$

where z is the height above sea level, H is the height at which the σ-surfaces reduce to z surfaces (often H is set equal to the model's top level), and E is the terrain elevation. Pielke and Martin (1981) discussed the transformation of the primitive equations into the σ_z system. Note that this system also follows the terrain at levels below $z = H$. If the model extends above $z = H$, the vertical coordinate variable typically reverts to z.

The primary advantage of these terrain-following systems is that they allow simplified treatment of the lower boundary conditions in models that contain orography. With these systems, model levels do not intersect the Earth's surface. However, as with most things, simplification in one place introduces complications elsewhere. In this case the complication emerges in the evaluation of the pressure gradient force. During the transformation, the pressure gradient term is split into two terms that have opposite signs and are of almost equal magnitude over steep terrain. To determine the difference between these two large terms and thus determine the resulting pressure gradient force, the terms must be calculated accurately and consistently (Simmons and Burridge, 1981).

Horizontally, the equations are often cast on an x-y grid transformed with appropriate map projection of the Earth to account for the spherical shape of the Earth (see Sec. 1.8 of Haltiner and Williams, 1980), or they are cast on a latitude-longitude grid with appropriate corrections to account for the convergence of lines of longitude as they approach the north and south poles. There is no apparent advantage or disadvantage to either method.

24.3.2. Grid Structures

The simplest and most straightforward system to arrange the dependent variables in the vertical is depicted in Fig. 24.1a. In this system all dependent variables are evaluated at each model level. In the scheme depicted by Fig. 24.1b, the mass variable p and the vertical velocity w are evaluated on the k levels, and the horizontal velocity u and v and temperature T are evaluated between the k levels, at the $k \pm 1/2$ levels. This type of vertical staggering of variables has advantages for evaluating the hydrostatic equation (24.3) as simplified according to the discussion in Sec. 24.2; i.e.,

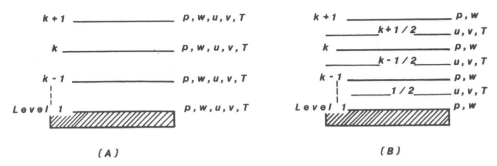

Figure 24.1. Two possible vertical grid structures.

$\Delta(\ln p)/\Delta z = -g/R\overline{T}$ where $\overline{T}$ is the layer-average temperature over the layer defined by Δz). If pressure and temperature are staggered, the average temperature is exactly what is evaluated at the $k + 1/2$ locations and thus, no calculation such as $\overline{T} = (T_{k+1} + T_k)/2$ is necessary to integrate downward from p_{k+1} to obtain p_k.

Also, as discussed in Sec. 24.2, some form of the equation for the conservation of mass (24.8) is normally used to evaluate the vertical velocity by integrating $\partial w/\partial z$ upward from the surface. Thus, the vertical staggering of the horizontal velocity components and the vertical velocity has the same advantage as staggering temperature and pressure.

Arakawa and Lamb (1977) and Haltiner and Williams (1980) discussed the advantages and disadvantages of various horizontal grid structures. Figure 24.2 shows four options. Grid (a) is an unstaggered grid in which all variables are evaluated at each node of the grid. The other grid structures are staggered, and some of the variables are evaluated at locations between the grid nodes. Notice that the height gradient $\partial h/\partial x$, used to calculate the mass-driven acceleration (i.e., the pressure gradient force, but in pressure coordinates instead of height coordinates) in the u and v tendency equations (24.1) and (24.2), can be evaluated by using a centered difference (see Sec. 24.3.3) over a distance of one grid interval instead of two grid intervals if grid (c) or grid (d) is used. Thus, these staggered grids yield twice the resolution in evaluating the gradient terms for the same number of grid

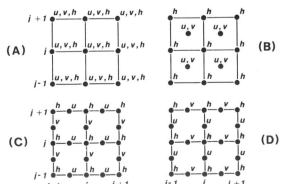

Figure 24.2. Four of Arakawa and Lamb's (1977) grid configurations.

points as in grid (a). Because the interaction between the mass and flow fields is critical to geostrophic balance, grids (c) and (d) yield more realistic approximations to this process.

24.3.3. Spatial Discretization

The advection terms (e.g., $u\, \partial u/\partial x$) and the pressure gradient terms (e.g., $\partial p/\partial x$) in the primitive equations both contain partial derivatives with respect to space. These continuous derivatives can be reformulated into their finite-difference form.

The partial derivative $\partial/\partial x$ represents a change in the variables as one proceeds in the x-direction, i.e., a change in the variable in space. A Taylor expansion of a function f about x yields the following expression:

$$f(x \pm \Delta x) = f(x) \pm \Delta x \frac{\partial f(x)}{\partial x} + \frac{(\Delta x)^2}{2!} \frac{\partial^2 f(x)}{\partial x^2} \pm \frac{(\Delta x)^3}{3!} \frac{\partial^3 f(x)}{\partial x^3} + \cdots$$

$$+ \frac{(\Delta x)^n}{n!} \frac{\partial^n f(x)}{\partial x^n} . \qquad (24.11)$$

Choosing the positive Δx and solving for $\partial f(x)/\partial x$, while neglecting the terms containing powers of Δx greater than or equal to 2, yields

$$\frac{\partial f(x)}{\partial x} = \frac{f(x + \Delta x) - f(x)}{\Delta x} = \frac{\Delta f(x)}{\Delta x} . \qquad (24.12)$$

Thus, the first derivative or gradient of a variable η can be approximated as follows:

$$\frac{\partial \eta}{\partial x} \approx \frac{\Delta \eta}{\Delta x} = \frac{\eta_{i+1} - \eta_i}{x_{i+1} - x_i} , \qquad (24.13)$$

where Δx denotes the distance between one grid point and the next, and the subscript $i+1$ denotes the value of the variable at the $i+1$ grid point or at $i\Delta x$ from the left-hand edge of the grid. Note that this is similar to approximating the derivative by a straight line passing through points at x_i and x_{i+1} (see Fig. 24.3a). Also note that this approximation can be considered to have first-order accuracy because all terms in Δx with power greater than 1 have been neglected. Note that, in this case, the approximation is not very good.

A better approximation can be found by combining the approximations of the derivative for the positive and the negative direction. Thus, expanding (24.11) for $+\Delta x$ and $-\Delta x$ and subtracting the two resulting equations yields the following approximation to $\partial \eta/\partial x$:

$$\frac{\partial \eta}{\partial x} \approx \frac{\Delta \eta}{\Delta x} = \frac{\eta_{i+1} - \eta_{i-1}}{x_{i+1} - x_{i-1}} . \qquad (24.14)$$

This approximation surrounds the actual point where the derivative is being evaluated (see Fig. 24.3b) and is therefore often referred to as a centered

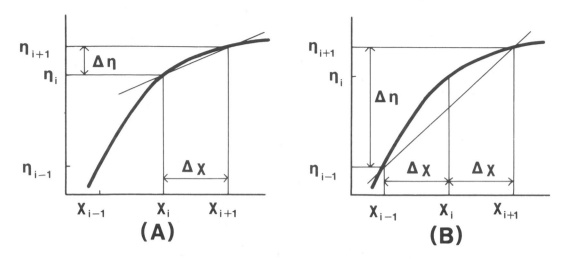

Figure 24.3. Example of a one-sided and a centered finite-difference approximation to the derivative of η with respect to x at x_i.

difference (the first approximation is called a one-sided difference). When the derivation is performed, the first term neglected is a term in $(\Delta x)^3$; thus, this approximation is second-order accurate. As demonstrated in introductory calculus, both these approximations improve as Δx decreases in distance and in the limit as $\Delta x \to 0, \Delta n/\Delta x \to \partial \eta/\partial x$.

It is also apparent that if the derivative were approximated by a higher-order polynomial, i.e., by a curve rather than a straight line, then the approximation to the derivative would be improved. The following is a five-point centered or fourth-order accurate approximation which, like the one-sided and the centered second-order accurate approximations, can be derived from the Taylor expansion:

$$\frac{\Delta \eta}{\Delta x} = \frac{2}{3\Delta x}(\eta_{i+1} - \eta_{i-1}) - \frac{1}{12\Delta x}(\eta_{i+2} - \eta_{i-2}) \ . \tag{24.15}$$

Table 24.1 compares the accuracy of the three-point versus the five-point representations for various wavelengths. Note that the five-point formulation is considerably more accurate, especially for waves with lengths between $3\Delta x$ and $8\Delta x$. But it still greatly misrepresents shorter waves. Also note that the difference between the two approximations decreases as the wavelength increases.

Although the five-point approximation is more accurate for midrange wavelengths, five-point approximations introduce some added difficulties; for example, the five-point difference cannot be calculated near the boundary of a domain and thus, boundary conditions may be more complicated. Also, five-point approximations may require more computer time and space. However, in spite of these difficulties, higher order approximations are used by several current research models (Anthes, 1983).

Table 24.1. Truncation errors for three- and five-point finite-difference approximations of $\partial/\partial x$ for waves of various lengths

Length	Fractional Error	
	Three-point	Five-point
$2\Delta x$	1.00	1.00
$3\Delta x$	0.68	0.37
$4\Delta x$	0.36	0.15
$5\Delta x$	0.24	0.07
$6\Delta x$	0.17	0.04
$7\Delta x$	0.13	0.02
$8\Delta x$	0.10	0.02
$9\Delta x$	0.08	0.01
$10\Delta x$	0.06	0.004

24.3.4. Temporal Discretization

Two common approximations for the local time derivative, $\partial/\partial t$, are similar to the approximations for the spatial derivatives discussed. $\partial/\partial t$ can be approximated by either a one-sided difference or a centered difference (leapfrog time difference). Thus, approximating $\partial \eta/\partial t$ as a one-sided or forward-in-time difference yields

$$\frac{\partial \eta}{\partial t} \approx \frac{\Delta \eta}{\Delta t} = \frac{\eta^{n+1} - \eta^n}{t^{n+1} - t^n} = f(t)^n \qquad (24.16)$$

whereas approximating the time derivative as a leapfrog difference yields

$$\frac{\partial \eta}{\partial t} \approx \frac{\Delta \eta}{\Delta t} = \frac{\eta^{n+1} - \eta^{n-1}}{t^{n+1} - t^{n-1}} - f(t)^n, \qquad (24.17)$$

where $f(t)$ is the right-hand-side or the forcing terms in the primitive equations evaluated at time n. Note that the approximation in (24.17) requires that an additional time level be available before the approximation can be evaluated. Therefore, the leapfrog difference scheme requires more computer memory or storage than does the forward-in-time scheme. Again, the higher the order of the approximation, the more accurate its representation of $\partial/\partial t$.

24.3.5. Combining Spatial and Temporal Discretizations

From the discussions in Secs. 24.3.3–24.3.4 it would seem straightforward to choose a space difference and then a time difference approximation. However, combining space and time differences to convert a continuous equation into its discrete analog has effects that must be considered. The linear advection equation can be used to illustrate such effects.

$$\frac{\partial \eta}{\partial t} + c \frac{\partial \eta}{\partial x} = 0; \qquad c = \text{constant.} \qquad (24.18)$$

This highly simplified equation predicts the behavior of a material η when it is being carried or moved along by a constant, uniform current or wind.

The solution of this equation, which can be obtained exactly by analytical methods, indicates that the field of η should remain unchanged in spatial distribution and maximum concentration as it is advected by the uniform current c. In our discussion we will think of the spatial distribution and concentration of the material as being represented by the combination of a series of waves, each with an amplitude and wavelength, and of the current c as the wave propagation speed..

Eq. (24.18) can be rewritten in finite-difference form:

$$\frac{\Delta \eta}{\Delta t} + c\frac{\Delta \eta}{\Delta x} = 0 \qquad (24.19)$$

or, when leapfrog-in-time and second-order centered-in-space approximations are employed,

$$\frac{\eta_i^{n+1} - \eta_i^{n-1}}{t^{n+1} - t^{n-1}} + c\frac{\eta_{i+1}^n - \eta_{i-1}^n}{x_{i+1} - x_{i-1}} = 0 \ . \qquad (24.20)$$

Note that if the gradient $\partial\eta/\partial x$ is underestimated, then the forcing term $-c\partial\eta/\partial x$ is also underestimated, which, in turn, causes the local change with respect to time to be underestimated.

Errors like these lead to errors both in the concentration of the material and in the speed of its propagation, i.e., in the wave amplitudes and wave speeds. For a more complete and rigorous discussion of these errors refer to Haltiner and Williams (1980, Ch. 5). Suffice it to say here that these errors are dependent on the wavelength of waves representing the concentration and distribution of the material. As stated in Sec. 24.3.3 , as Δx and Δt approach zero the accuracy of the prediction increases. But the accuracy does not depend only on Δx and Δt; it also depends on the wavelength in relation to Δx. That is, if the wavelength is large compared with Δx, then the finite-difference approximation of the gradients along that wave will be better than if the wavelength were small in relation to Δx.

Waves whose speed depends on wavelength are called dispersive. Note that the continuous form of the linear advection equation, Eq. (24.19), is nondispersive; all waves are advected with speed c, independent of their wavelength. However, as discussed above, the finite-difference form is dispersive; i.e., the transformation has altered the basic nature of the physical process.

In Richardson's time, knowledge of numerical analysis as applied to solving partial differential equations using finite-difference approximations was meager. Courant et al.(1928) discovered that one could not choose Δt and Δx independently. Indeed, for a second-order, centered-in-space, leapfrog-in-time differencing scheme, the conditions set by the Courant-Friedrichs-Lewy (CFL) relationship, Eq. (24.21), must be followed:

$$\frac{|c\Delta t|}{|\Delta x|} \leq 1 \ , \qquad (24.21)$$

where c is the speed of the most rapidly moving wave predicted by the model. Physically, this is equivalent to stating that information should not move more than one grid interval Δx in one time step Δt. Although meteorological waves do not move very fast, the equations support gravity and sound waves that may move as fast as 300 m s^{-1}. Thus, for a grid interval of 50 km, the time step should not exceed 60 s, and in practice the time step should be closer to 45 s than 60 s. If Δx is reduced by a factor of 2, then Δt also must be reduced by a factor of 2. So a doubling of the horizontal resolution over a given area requires four times more grid points and a time step of half the original time step; this amounts to an eightfold increase in computing time.

24.4. Energy Sources and Sinks

There is but one fundamental energy source that drives the atmosphere; that is solar radiation. It is uneven solar heating between equator and the poles, land and water, mountains and valleys, cities and suburbs, etc., that makes the wind blow and drives the weather. However, as simple as this may sound, before the complexities of atmospheric flow can be understood the details of how this energy is transformed from one form to another, how it is transported from one place to another, and how these energy changes affect atmospheric motions and the resulting weather must be considered. This means that to simulate a particular atmospheric circulation in a numerical model, the energy transformations and transports that are relevant to it must be included.

The primitive equations (Sec. 24.2) can be used for energy transformations that are important to atmospheric processes:

- Absorption of solar radiation by the atmosphere and the Earth's surface.

- Emission and absorption of terrestrial radiation by the atmosphere and the Earth's surface.

- Storage and release of latent heat caused by phase changes of water.

- Generation of potential energy by heating.

- Generation of kinetic energy by pressure gradient forces.

- Dissipation of kinetic energy by frictional forces.

The transport mechanisms as stated include

- Large-scale advection.

- Convective-scale transports.

- Boundary layer and turbulent mixing.

An important consideration for numerical modeling is the relationship between the scales of motion that are resolved by the model and those that are too small to be properly resolved. As stated in Sec. 24.3, the determination of those scales of motion that are resolvable, and those that are not is a function of the model grid interval. The smaller the grid interval the smaller the scales of motion that are properly simulated by the finite-difference approximations. However, the CFL criteria (Sec. 24.3.5) and limitations on computer size and speed restrain the construction of models with grid resolutions small enough to resolve scales of atmospheric motion from the synoptic scale (wavelength $\geq$ 3000 km) to the turbulence scale (wavelength $\geq$ 10 cm). Thus, the scales to be simulated must be selected first and then a grid interval and time step sufficient to yield the desired model accuracy at the selected scales. The effects of motions on scales that are too small to be properly resolved by this grid, but are essential for the proper energy changes, must be parameterized; i.e., their statistical or net effects must be included even though the details of their circulations are neglected. This implies that the scales of motions being parameterized should be much smaller than the motions of interest.

As the primary driving term in the momentum equation, the pressure or height gradient force is of great interest and should be accurately represented by the model finite-difference formulation. If the grid and time differencing has been judiciously chosen, then the pressure gradient $\partial p/\partial x$ will be resolved with sufficient accuracy, at least for the scales of primary interest. The advection of variables by the model-resolved circulations will also be represented with sufficient accuracy. However, because the primary circulations and processes involved with convection, latent heat effects, boundary layer mixing, and radiation occur on scales much too small to be resolved by a model whose primary purpose is to investigate or predict mesoscale (wavelength $\geq$ 100 km) circulations, these processes also must be parameterized.

24.4.1. Radiation Physics

For a review of radiation parameterization methods employed in large-scale NWP and climate models, see Stephens (1984). Although variations in incoming solar radiation and its resultant heating are the driving force for most atmospheric motion, most mesoscale models do not incorporate the direct atmospheric effects of either solar (shortwave) or terrestrial (longwave) radiation (Anthes, 1983). That is, most models set $\partial T/\partial t_{\mathrm{rad}} = 0$ in (24.4). It is generally argued that these effects are not important on the 6–24 h time scale. Warming of the atmosphere by solar radiation is on the order of $1°C$ per day, and atmospheric cooling by longwave emission is about $2°C$ per day. For comparison, latent heating rates can be as large as $80°C$ per day for periods up to 3–6 h (Chang *et al.*, 1982). Neglecting the longwave cooling can cause a model to grow warmer each day that it is run, but because mesoscale models are not typically run for longer than 36 h this warming is not critical. However, even though direct warming and cooling of the atmosphere by radiation may be neglected, the effect of the balance between incoming and outgoing radiation at the Earth's surface cannot be neglected.

24.4.2. Surface Energetics

On the average, approximately 50% of the solar radiation that reaches the top of the Earth's atmosphere reaches the Earth's surface, and about

25% of the incident radiation is absorbed by the atmosphere and clouds. Therefore, the atmosphere can be considered as a fluid being warmed from below more than a fluid being warmed throughout by direct absorption of solar radiation. Because of this, it is important to include the surface and its effects either directly or indirectly in a model if the diurnal effects of radiation are to be included.

Various physical processes occurring at the Earth's surface affect the surface energy balance. For example, solar energy that reaches the surface can be either absorbed or reflected by the surface, depending on the albedo of the surface. A snow-covered surface (with albedo of 0.40–0.95) reflects more energy and absorbs less energy than does a bare soil surface (with albedo of 0.05–0.40) (Oke, 1978, p. 15). Reflection by a water-covered surface depends on the solar zenith angle; albedo varies from 0.03–0.10 for small zenith angles to 0.40–1.00 for large zenith angles.

Energy that is absorbed by the surface may be used either to warm the surface or to evaporate moisture from the surface. Partitioning the energy between sensible and latent heat depends on the type of surface and its moisture characteristics. Also, before the temperature of the surface can be determined, the efficiency with which the surface changes temperature (warms or cools) as sensible energy changes (gains or loses) must be considered. The effects of differential heating between land and water surfaces is an extreme example of the effect of surface type; however, different soil and vegetation types also use energy more or less efficiently.

Because of the complexities of surface interactions like these, some models do not consider any surface interactions; they simply prescribe the fluxes of heat and moisture into the atmosphere (Perkey, 1976) or they prescribe the surface temperature and moisture as a function of time (Pielke, 1974) and then calculate the fluxes.

However, other models consider at least part of the surface interactions and use a surface energy budget formulation to calculate the surface temperature. There are several methods for accomplishing this parameterization, e.g., Gadd and Keers (1970), Anthes and Warner (1978), and Chang *et al.*(1981). More recent studies have begun to investigate the effect of including model layers that extend into the soil or water surface so that the flux into the soil and water can be better calculated (McCumber and Pielke, 1981).

24.4.3. Turbulent Energy Transfer

As the Earth's surface warms during the day because of excess incoming solar radiation and cools at night because of outgoing terrestrial radiation, differences in temperature develop between the Earth's surface and the atmosphere. Recall that the flow of thermal energy or heat is driven by temperature differences. Therefore, because the Earth's surface and the atmosphere are in contact and are at different temperatures, energy is transferred by molecular conduction. However, because molecules can move in the atmosphere (i.e., the atmosphere is a fluid), transfer is more efficient in the atmosphere than in the Earth, where molecules are bound by strong intermolecular forces. In other words, in the atmosphere energy is transferred by convective or mixing processes. As stated earlier, these processes, represented by the $\partial/\partial t_s$ terms in the primitive equations for temperature, and specific humidity, occur on scales too small to be resolved by mesoscale models and therefore must be parameterized. Frictional effects, represented by the $\partial/\partial t_s$ terms in the primitive equations for horizontal momentum, on atmospheric motion also occur on these small scales.

In contrast to global models, which typically use drag coefficient or bulk aerodynamic formulations (Haltiner and Williams, 1980) for calculating the transports of temperature, moisture, and momentum to and from the surface, most mesoscale models use parameterization techniques that are more complicated and physically more realistic. A discussion of these techniques can be found in Haltiner and Williams (1980, Ch. 8). Models using these techniques are described by Pielke (1974), Perkey (1976), and Kaplan et al.(1982).

24.4.4. Latent Heating Associated with Model-Resolvable Circulations

Under certain circumstances much can be learned from "dry" models. These models contain no moisture conservation equations, and the term $(c_p^{-1}Q)$ in (24.4) is neglected (i.e., set equal to 0). In spite of this simplification these models can provide useful information about the dynamical aspects of mesoscale circulations in which latent heat effects are not the dominant energy source. Pielke (1974), for instance, used the conservation of specific humidity equation as a tracer; he did not allow changes of phase or the thermodynamic feedback of latent heat release.

Many studies, both observational and modeling, have indicated the importance of latent heating to the growth and development of both synoptic- and meso-scale circulations (Aubert, 1957; Danard, 1964; Tracton, 1973; Perkey, 1976; Anthes and Keyser, 1979; Maddox et al., 1981; Fritsch and Maddox, 1981a, 1981b; Chang et al., 1981, 1982, 1984). Even if circulations are resolved by the model grid structure, the mechanisms controlling the detailed evolution of processes such as evaporation, condensation, cloud droplet growth, and collection of cloud drops by raindrops occur on scales far too small to be treated explicitly. Thus, their effects must be approximated by some formulation based on parameters that are available from the resolvable scale.

One approach is to have all condensed water material fall out instantaneously and be considered rain on the ground (Anthes and Warner, 1978; Mathur, 1983). This parameterization scheme is possibly the simplest, but it neglects the storage of liquid water in the form of cloud drops that never grow to raindrop size and therefore never rain out but instead re-evaporate at some later time. A somewhat more complicated scheme carries cloud water as an explicit model variable, but then makes the simple assumption that some predefined fraction of the cloud water falls out as rain. The most complicated schemes carry rainwater as a model variable and parameterize the cloud microphysical processes of conversion and collection following Kessler (1969) in terms of model parameters (Perkey, 1976).

24.4.5. *Convective-Scale Latent Heating and Mixing*

Convective-scale latent heating is associated with circulations too small to be resolved on a typical mesoscale mesh. These circulations are usually characterized by developing in a conditionally or absolutely unstable environment. Transports by such convective circulations must be parameterized. These transports are represented by the $\partial/\partial t_{conv}$ terms in the primitive equations. Frank (1983) reviewed the current state of cumulus parameterization. The form of the parameterization scheme is somewhat dependent on the scale of the atmospheric circulation being studied. The statistical effects influencing the dynamics of interest probably are considerably different for a global climate model than for a regional or mesoscale model.

Convective parameterization schemes can be divided into three categories on the basis of the "closure" assumption, i.e., the assumption used to determine the amount and rate of convective transfer:

- The convective transfer is defined as the amount required to stabilize the sounding (Manabe *et al.*, 1965). This scheme is usually referred to as convective adjustment. In its simplest form, the temperature and moisture lapse rates are adjusted to the moist adiabatic lapse rate in such a way as to conserve total energy. The resulting condensed water falls out instantaneously.

- The amount of convective transfer is related to the instantaneous large-scale convergence of mass or moisture (Kuo, 1965, 1974; Anthes, 1977). The large-scale "convective energy" is released as soon as it is generated.

- The convective transfer is proportional to local conditions (Kreitzberg and Perkey, 1976; Fritsch and Chappell, 1980). The large scale is allowed to build "convective energy," which is released according to local stability considerations; i.e., the large scale and convective scale do not have to balance on a short time scale (a few hours). They must balance over longer time scales.

The first two categories were developed primarily for use in global climate models or hurricane models. With circulations more related to the mesoscale, Fritsch *et al.*(1976) observed that the instantaneous synoptic-scale moisture convergence could not account for the observed precipitation amounts over an Oklahoma squall line; in fact, the moisture convergence could account for only about 20% of the precipitation. They suggested that the solution to this contradiction was that the mesoscale-convective system was using "energy" that had been generated during the previous hours by the large scale. In other words, the meso-convective scale was not in balance with the large scale on the time scale of a few hours. A similar conclusion is found in Kreitzberg and Perkey's (1976) discussion of the release of potential instability: "There is no reason to presuppose that potential instability is consumed at the rate it is supplied unless one is dealing with time scales of a day or more."

24.5. Numerical Artifices

In addition to the source-sink terms discussed in Sec. 24.4, mesoscale models require several numerical artifices before they yield reasonable results. For example, the atmosphere is a layer of air wrapped around a sphere much like the peel on an orange. However, in contrast to the orange peel, this layer of air does not have a definite surface; rather it gradually fades away. Mesoscale models, however, must be limited in vertical extent or they would be too large for today's computers. Therefore, most mesoscale models concentrate their grid levels within the troposphere with only a few, if any, levels in the stratosphere. Also, in order to provide resolution adequate to represent mesoscale circulations within present computers, mesoscale models must be limited in horizontal extent. Thus, artificial boundaries must be placed in somewhat arbitrary locations in models.

In addition to the imposition of such artificial boundaries, other artifices are necessitated by the numerics and not the physics of the model.

24.5.1. Horizontal and Vertical Smoothing

Horizontal diffusion and vertical diffusion are physical processes that occur on the microscale, not on the mesoscale. Their physical effect is to remove or smooth small-scale gradients that are in general fed downscale from larger atmospheric motions. In numerical models, a numerical process is required to remove small-scale (with respect to the grid interval) gradients generated by model inaccuracies and machine roundoff errors. Thus, a process similar to horizontal diffusion is required to keep model noise from building to the extent that it overwhelms the model signal. In addition to the boundary layer vertical transfers discussed in Sec. 24.4.3, many models employ some diffusion in the vertical, even above the boundary layer, to control noise. Because the purpose of these numerical schemes is to control shortwave numerical noise in the model, they are designed not as physical parameterization schemes, but rather to function as efficiently as possible, independent of the physical interpretation that might be applied to the scheme. An additional condition on these "smoothers" is that they do little harm to the larger scale, i.e., the scales of interest that are best resolved by the model.

An example of a wavelength-dependent filter in common use by mesoscale models for filtering in the horizontal is the fourth-order function

$$\frac{\partial \eta}{\partial t} = -K \frac{\partial^4 \eta}{\partial x^4} . \tag{24.22}$$

When this function is approximated by centered-in-space and leapfrog-in-time finite differencing, $\partial \eta^4 / \partial x^4$ must be evaluated at the $n-1$ time step if the advective terms are evaluated at time step n, and, for stability considerations, K must be less than $(\Delta x)^4 / 32 \Delta t$ (Perkey and Kreitzberg, 1976).

In the vertical, a second-order filter is commonly used:

$$\frac{\partial \eta}{\partial t} = K \frac{\partial^2 \eta}{\partial z^2} , \tag{24.23}$$

where, again for a centered-in-space approximation, the derivative must be evaluated at the $n-1$ time step, and in this case $K \le (\Delta z)^2/8\Delta t$ for stability. The wavelength response of this second-order filter is not as good as the response of the fourth-order filter; in other words, the second-order filter removes more energy from the middle wavelengths (6Δ to 12Δ). However, the second-order filter is simpler to apply at the boundaries of the model domain. Because there are fewer vertical grid points than horizontal grid points, the number of boundary points represents a larger percentage of the total points in the vertical than in the horizontal. Thus, the second-order filter is less complicated to implement in the vertical than the fourth-order filter.

24.5.2. Time Filtering

Depending on the exact time-differencing scheme chosen for the model, truncation errors may occur in time because of the approximations inherent in computer arithmetic. The majority of time schemes do not require an explicit time-filtering scheme. However, because of the formulation of the leapfrog time-differencing scheme, the solutions at odd and even time steps tend to split or diverge. To help overcome this tendency, Robert (1966) proposed a time filter or averaging technique to tie the odd and even solutions together:

$$\eta^\eta_{\text{new}} = \eta^\eta + \frac{\mu}{2}\left(\eta^{\eta-1} + \eta^{\eta+1} - 2\eta^\eta\right) , \tag{24.24}$$

where typically μ is between 0.3 and 0.1. Haltiner and Williams (1980, Sec. 5.7.3), Asselin (1972), and Schlesinger *et al.*(1983) discussed the use of this time filter.

24.5.3. Lateral and Vertical Boundary Conditions

Spatial boundaries are a necessity imposed by computer size and speed limitations; nature does not impose such confines. Historically, lateral boundaries have been more of a problem for models of synoptic-scale to mesoscale circulations than vertical boundaries. To quote Davies (1983), "The treatment of lateral boundaries is an intrinsic and distinctive problem associated with the formulation of regional weather prediction models. It is a problem that has bedevilled modelers from the earliest days of numerical weather prediction." Mathematically, the problem of how to construct a practical set of well-posed boundary conditions for the equations governing fluid flows has not been satisfactorily resolved. Nevertheless, in spite of the basic mathematical problems associated with specifying lateral boundary conditions, many limited-area models have been able to provide useful solutions to fluid flow problems. Baumhefner and Perkey (1982) showed that, for synoptically difficult-to-forecast cases, errors other than those caused by the lateral boundary scheme were responsible for most of the total error in the forecast. However, in order to minimize these errors, the rule of thumb should be to locate the model boundaries as far away as possible from the area of interest. Baumhefner and Perkey indicated that errors generated by the lateral boundary conditions propagated into the model domain at a rate of 20° to 30° of latitude per day.

Two strategies have been employed to address the problem of lateral boundaries:

- Nest the finer-mesh domain within a coarser-mesh domain and then integrate the prognostic equations for both domains simultaneously (e.g., Jones, 1977; Phillips, 1978).

- Nest the finer-mesh domain within a coarser-mesh domain in which the prognostic equations have been previously integrated (e.g., Perkey and Kreitzberg, 1976; Orlanski, 1976).

The first strategy has the advantage of allowing information to flow from the fine- to coarse-mesh domain as well as from the coarse- to the fine-mesh domain; i.e., information flows in both direction through the boundary. This strategy is often referred to as a two-way interacting system. The second strategy allows information to flow in only one direction, from the coarse- to the fine-mesh domain, and is therefore sometimes referred to as a one-way interacting system or as parasitic nesting.

Both strategies require care to minimize the reflection of information from the boundary back into the model domain. Light waves behave much the same way at the interface between air and water as atmospheric motion does at the interface between meshes of differing grid intervals. Waves bend or refract at the interface because of the different propagation speeds of light in air and water, and in differing mesh sizes. In Sec. 24.3.4 it was noted that the propagation speed of a wave in a finite-difference model depends on the accuracy with which the model's mesh size can reproduce the wave. Thus, because the fine mesh can reproduce the wave more accurately than the coarse mesh can, the wave travels more nearly at its true speed in the fine-mesh region than in the coarse-mesh region. Therefore, the wave changes speed as it passes through the interface between the two mesh regions and is therefore redirected or refracted.

In the vertical, almost all regional and mesoscale models have employed a rigid-lid upper boundary condition. This condition has a form similar to requiring $\omega = dp/dt = 0$ at the model top. The model top is usually at or somewhat above the tropopause (10 to 16 km). This condition has the undesirable effect of reflecting vertically propagating waves. Reflection does not allow the wave's energy to exit the model domain; reflection traps the waves in the domain where they can erroneously interact with other waves within the domain. Klemp and Durran (1983) proposed a radiation boundary condition that permits internal gravity waves to exit the domain.

24.6. Problems Still To Be Solved

Until the middle 1970's, computers were not large and fast enough and physical parameterization techniques were not sophisticated enough for operational use of mesoscale models to be feasible. But with increases in computer size and speed along with improved parameterization methods, numerical modeling of mesoscale circulations has become possible. However, in spite of the improvements, mesoscale modeling still has many problems to be solved.

Faster and more accurate numerical approximations to the governing equations are being investigated by several researchers. These include spectral and finite-element representations of the equations instead of finite-difference formulations. More accurate approximations of the important energy sources and sinks are being tested and evaluated. Because of the increased computing power of today's machines, these include parameterization schemes with more complete physics and fewer approximations. New and better filters and boundary conditions are being developed and tested.

Possibly the largest and most difficult problem for mesoscale modeling is not a problem with the models themselves but with the lack of good mesoscale meteorological data for testing the models. Meteorological data on scales smaller than the current rawinsonde network are required both for initialization and model evalution, before the maximum value can be achieved from mesoscale models. In addition to the difficulty in acquiring the data, considerable research is still needed to convert the often asynoptic and incomplete data into gridded meteorologically consistent data sets. For example, the multitudinous surface measurements taken each day do not enter into the initial data used by current operational models. Also, the immense quantity of data available from satellites has been integrated into the initial data stream with only mixed success.

Appendix: Definition of Symbols

c	specific cloud water (both liquid and solid); advection velocity
c_p	specific heat at constant pressure
f	Coriolis parameter
g	acceleration due to gravity
h	height of pressure surface
i	x-direction index
j	y-direction index
k	z-direction index
n	time index
p	pressure
q	specific humidity
r	specific precipitation (both liquid and solid)
r_e	radius of the Earth
t	time coordinate
u	x-component of velocity
v	y-component of velocity
w	z-component of velocity
x	west-east space coordinate (positive to the east)

y	south-north space coordinate (positive to the north)
z	vertical space coordinate (positive up)
E	terrain elevation
H	height at which σ-surfaces reduce to z-surfaces
K	diffusion coefficient, coefficient of smoothing
Q	latent heating per unit mass per unit time, caused by grid-resolved condensation and evaporation
R	gas constant for air
T	temperature
α	latitude
Δ	finite difference change operator
ς	generalized vertical coordinate (Kasahara, 1974)
η	arbitrary variable
μ	time filter coefficient
ω	vertical component of velocity in pressure coordinates, dp/dt
ρ	density
σ	terrain-following vertical coordinate
σ_p	terrain-following vertical coordinate based on pressure
σ_z	terrain-following vertical coordinate based on height
$\partial/\partial t_{\text{coll } c=>r}$	tendency caused by collection of cloud water by rainwater
$\partial/\partial t_{\text{cond } q=>c}$	tendency caused by condensation of vapor
$\partial/\partial t_{\text{conv}}$	tendency caused by dry and moist subgrid-scale transport
$\partial/\partial t_{\text{conv } c=>r}$	tendency caused by conversion of cloud water to rainwater
$\partial/\partial t_{\text{evap } c=>q}$	tendency caused by evaporation of cloud water
$\partial/\partial t_{\text{evap } r=>q}$	tendency caused by evaporation of rainwater
$\partial/\partial t_{\text{rad}}$	tendency caused by radiation processes
$\partial/\partial t_{\text{rain}}$	tendency caused by rain falling out of the grid volume
$\partial/\partial t_s$	tendency caused by surface fluxes and turbulent transport

REFERENCES

Anthes, R. A., 1977: A cumulus parameterization scheme utilizing a one-dimensional cloud model. *Mon. Wea. Rev.*, **105**, 270–286.

Anthes, R. A., 1983: Regional models of the atmosphere in middle latitudes. *Mon. Wea. Rev.*, **111**, 1306–1335.

Anthes, R. A., and D. Keyser, 1979: Tests of a fine-mesh model over Europe and the United States. *Mon. Wea. Rev.*, **107**, 963–984.

Anthes, R. A., and T. T. Warner, 1978: De-velopment of hydrostatic models suitable for air pollution and other mesometeorological studies. *Mon. Wea. Rev.*, **106**, 1045–1078.

Arakawa, A., and V. R. Lamb, 1977: Computational design of the basic dynamical processes in the UCLA general circulation model. *Methods in Computational Physics*, **17**, J. Chang (Ed.), Academic Press, New York, 317 pp.

Asselin, R. A., 1972: Frequency filter for time integrations. *Mon. Wea. Rev.*, **100**,

487–490.

Aubert, E. J., 1957: On the release of latent heat as a factor in large-scale atmospheric motions. *J. Meteor.*, **14**, 527–542.

Baumhefner, D. P., and D. J. Perkey, 1982: Evaluation of lateral boundary errors in a limited-domain model. *Tellus*, **34**, 409–428.

Chang, C.-B., D. J. Perkey, and C. W. Kreitzberg, 1981: A numerical case study of the squall line of 6 May 1975. *J. Atmos. Sci.*, **38**, 1601–1615.

Chang, C.-B., D. J. Perkey, and C. W. Kreitzberg, 1982: A numerical case study of the effects of latent heating on a developing wave cyclone. *J. Atmos. Sci.*, **39**, 1555–1570.

Chang, C.-B., D. J. Perkey, and C. W. Kreitzberg, 1984: Latent heat induced energy transformations during cyclogenesis. *Mon. Wea. Rev.*, **112**, 357–267.

Charney, J.G., R. Fjortoft, and J. von Neumann, 1950: Numerical integration of the barotropic vorticity equation. *Tellus*, **2**, 237–254.

Courant, R., K. Friedreichs, and H. Lewy, 1928: Uber die partiellen differenzengleichungen der mathematischen physik. *Math. Annalen*, **100**, 32–74.

Danard, M. B., 1964: On the influence of released latent heat on cyclone development. *J. Appl. Meteor.*, **3**, 27–37.

Davies, H. C., 1983: Limitations of some common lateral boundary schemes used in regional NWP models. *Mon. Wea. Rev.*, **111**, 1002–1012.

Dutton, J. A., 1976: *The Ceaseless Wind.* McGraw-Hill, New York, 579 pp.

Frank, W. M., 1983: The cumulus parameterization problem. *Mon. Wea. Rev.*, **111**, 1859–1871.

Fritsch, J. M., and C. F. Chappell, 1980: Numerical prediction of convectively driven mesoscale pressure systems. Part I. Convective parameterization. *J. Atmos. Sci.*, **37**, 1722–1733.

Fritsch, J. M., C. F. Chappell, and L. K. Hoxit, 1976: The use of large scale budgets for convective parameterization. *Mon. Wea. Rev.*, **104**, 1408–1418.

Fritsch, J. M., and R. A. Maddox, 1981a: Convectively driven mesoscale weather systems aloft. Part I. Observation. *J. Appl. Meteor.*, **20**, 9–19.

Fritsch, J. M., and R. A. Maddox, 1981b: Convectively driven mesoscale weather systems aloft. Part II: Numerical simulations. *J. Appl. Meteor.*, **20**, 20–26.

Gadd, A., and J. F. Keers, 1970: Surface exchange of sensible and latent heat in a 10-level model atmosphere. *Quart. J. Roy. Meteor. Soc.*, **96**, 297–306.

Haltiner, G. J., and R. T. Williams, 1980: *Numerical Prediction and Dynamic Meteorology.* John Wiley and Sons, New York, 477 pp.

Holton, J., 1972: *An Introduction to Dynamic Meteorology.* Academic Press, New York, 319 pp.

Jones, R. W., 1977: A nested grid for a three-dimensional model of a tropical cyclone. *J. Atmos. Sci.*, **34**, 1528–1553.

Kaplan, M. L., J. W. Zack, V. C. Wong, and J. J. Tuccillo, 1982: Initial results from a mesoscale atmospheric circulation system and comparisons with the AVE-SESAME I data set. *Mon. Wea. Rev.*, **110**, 1564–1590.

Kasahara, A., 1974: Various vertical coordinate systems used for numerical weather prediction. *Mon. Wea. Rev.*, **102**, 509–522.

Kasahara, A., 1977: Computational aspects of numerical models for weather prediction and climate simulation. *Methods in Computational Physics*, **17**, J. Chang (Ed.), Academic Press, New York, 317 pp.

Kessler, E., 1969: *On the Distribution and Continuity of Water Substance in Atmospheric Circulations.* Meteor. Monogr., **32**, American Meteorological Society, Boston, 84 pp.

Klemp, J. B., and D. R. Durran, 1983: An upper boundary condition permitting internal gravity wave radiation in numerical mesoscale models. *Mon. Wea. Rev.*, **111**, 430–444.

Kreitzberg, C. W., 1978: Progress and problems in regional numerical weather prediction. *Proc. SIAM*, **11**, Amer. Math. Soc.,32–58.

Kreitzberg, C. W., 1979: Observing, analyzing, and modeling mesoscale weather phenomena. *Rev. Geophys. Space Phys.*, **17**, 1852–1871.

Kreitzberg, C. W., and D. J. Perkey, 1976: Release of potential instability: Part 1— A sequential plume model within a hydrostatic primitive equation model. *J. Atmos. Sci.*, **33**, 456–475.

Kuo, H. L., 1965: On the formation and intensification of tropical cyclones through latent heat release by cumulus convection. *J. Atmos. Sci.*, **22**, 40–63.

Kuo, H. L., 1974: Further studies of the parameterization of the influence of cumulus convection on large-scale flow. *J. Atmos. Sci.*, **31**, 1232–1240.

McCumber, M. C., and R. A. Pielke, 1981: Simulation of the effects of surface fluxes of heat and moisture in a mesoscale numerical model; 1. Soil layer. *J. Geophys. Res.*, **86**, 9929–9938.

Maddox, R. A., D. J. Perkey, and J. M. Fritsch, 1981: Evolution of upper tropospheric features during the development of a mesoscale convective complex. *Mon. Wea. Rev.*, **38**, 1664–1674.

Manabe, S., J. Smagorinsky, and R. R. Strickler, 1965: Simulated climatology of a general circulation model with hydrologic cycle. *Mon. Wea. Rev.*, **93**, 769–798.

Mathur, M. B., 1983: A quasi-Lagrangian regional model designed for operational weather prediction. *Mon. Wea. Rev.*, **111**, 2088–2098.

Oke, T. R., 1978: *Boundary Layer Climates*, John Wiley and Sons, New York, 372 pp.

Orlanski, I., 1975: A rational subdivision of scales for atmospheric processes. *Bull. Amer. Meteor. Soc.*, **56**, 527–530.

Orlanski, I., 1976: A simple boundary condition for unbounded hyperbolic flows. *J. Comput. Phys.*, **21**, 251–269.

Perkey, D. J., 1976: A description and preliminary results from a fine-mesh model for forecasting quantitative precipitation. *Mon. Wea. Rev.*, **104**, 1513–1526.

Perkey, D. J., and C. W. Kreitzberg, 1976: A time-dependent lateral boundary scheme for limited-area primitive equation models. *Mon. Wea. Rev.*, **104**, 745–755.

Phillips, N. A., 1957: A coordinate system having some special advantages for numerical forecasting. *J. Meteor.*, **14**, 184–185.

Phillips, N. A., 1978: A Test of Finer Resolution. Office Note 171. [Unpublished manuscript available from National Meteorological Center, National Weather Service, W32, Washington, DC 20233.]

Pielke, R. A., 1974: A three-dimensional numerical model of the sea breezes over south Florida. *Mon. Wea. Rev.*, **102**, 115–139.

Pielke, R. A., 1981: Mesoscale numerical modeling. *Advances in Geophysics*, **23**, 185–344.

Pielke, R. A. 1984: *Mesoscale Meteorological Modeling*, Academic Press, New York, 612 pp.

Pielke, R. A., and C. L. Martin, 1981: The derivation of a terrain-following coordinate system for use in a hydrostatic model. *J. Atmos. Sci.*, **8**, 1707–1713.

Richardson, L. F., 1922: *Weather Prediction by Numerical Process*. Cambridge University Press (reprinted: Dover, 1965), 236 pp.

Robert, A. J., 1966: The integration of a low order spectral form of the primitive meteorological equations. *J. Meteor. Soc. Japan*, **44**, 237–244.

Schlesinger, R. E., L. W. Uccellini, and D. R. Johnson, 1983: The effects of the Asselin time filter on numerical solutions to the linearized shallow-water wave equations. *Mon. Wea. Rev.*, **111**, 455–467.

Simmons, A. J., and D. M. Burridge, 1981: An energy and angular-momentum conserving vertical finite-difference scheme and hybrid vertical coordinates. *Mon. Wea. Rev.* **109**, 758–766.

Stephens, G. L., 1984: The parameterization of radiation for numerical weather prediction and climate models. *Mon. Wea. Rev.*, **112**, 826–867.

Tracton, M. S., 1973: The role of cumulus convection in the development of extratropical cyclones. *Mon. Wea. Rev.*, **101**, 573–593.

Wallace, J. M., and P. V. Hobbs, 1977: *Atmospheric Science, An Introductory Survey*. Academic Press, New York, 467 pp.

CHAPTER 25

Assimilation and Initialization of Atmospheric Data Into Numerical Prediction Models

John B. Hovermale

25.1. Introduction

The assimilation of atmospheric data into numerical models to predict the weather is an evolving technology. A developing base of theory exists to guide general thinking, but as the many components of an assimilation system are blended many compromises must be made. A specific component, seemingly optimal in its own right, may prove inconsistent with, or preclude the use of, other powerful components. Thus the optimal assimilation system is the one not only built of powerful individual components, but also having well-matched components. The goal is to translate atmospheric physics to numbers, given an adequate data base. The definitive optimal data assimilation cannot be given here. It is possible, however, to cover the more salient prerequisites for a good data assimilation scheme, and describe briefly the problems that are most difficult to solve.

25.2. Basic Principles in Translation of Atmospheric Measurements to Model Variables

The first step in moving from the relatively exact differential equations describing model physics involves the finite approximation of the derivatives. In limited-area models this essentially means that a fluid continuum is replaced by a number of finite air volumes that are shaped like flat pancakes 10 to 100 times larger horizontally than vertically. The equations still can be expressed in this framework to maintain fundamental physical principles (e.g., mass can be exactly conserved, and total energy can be conserved in a closed system), but the variables predicted in numerical models must be defined in terms of mass-averaged values over a grid box rather than point values. It follows, of course, that one should attempt to determine these mass-averaged values from an ample supply of atmospheric data. The approach should converge toward the relation

$$\overline{\phi} = \int \phi \, dm \,, \tag{25.1}$$

where

$$\phi = \overline{\phi} + \phi' \,, \tag{25.2}$$

and

$$\int \phi' \, dm = 0 \,, \tag{25.3}$$

where ϕ is any variable and m is the total mass in the "pancake" of air.

It would be a simple matter to satisfy the above relations if numerous observations were available in each model grid box. Of course, they are not. Even as more observations become available from the broad array of composite meteorological observing systems, meteorologists in their quest to learn more about the atmosphere attempt to analyze (interpolate data to grid boxes in an attempt to estimate mass averages) on smaller and smaller scales, and may push beyond its limits the theory of sound statistical sampling. Most analysis theory is presented on the basis of determining smoothed interpolations to grid points rather than averaged values within grid boxes. And there is good reason for this most of the time since there is rarely more than one observation in a grid square. Fortunately there is a tool to provide more information within the air pancakes and supplement the meager supply of information supplied by observations at any given time.

25.3. The Relative Values of Observations in Determining Mass Means

Objective analysis is also treated in Chs. 8 and 14. However, it may be useful to re-emphasize a fundamental precept of objective analysis: The importance of data in determining a volume-averaged mean should be inversely weighted with regard to their "expected error variance." Such variance is related to the ability of the instrument to "see" what it is trying to measure. However, an observing instrument may have other aspects that compensate for observational error. (1) The system whose parameters correlate most closely with the variable of interest, $\overline{\phi}$ in (25.1), is the one whose data should receive most weight. The radiosonde gathers information at a point in horizontal space and gives averages over vertical layers that are shallow with regard to temperature and dewpoint, and somewhat deeper with regard to the wind; satellite systems average temperatures over several tens of kilometers in the horizontal and several kilometers in the vertical; temporally continuous wind-profiling instruments average over a kilometer or so in depth and less than an hour in time. In a given situation, one of these systems would be the most suitable to represent, economically, the volume-averaged atmosphere. (2) An instrument's ability to cover the entire domain under consideration should be valued as well. No one instrument is uniquely satisfactory in this respect. To establish the best composite observing system requires a blending of technologies; several factors will determine the mix.

Also of value is the overall representativeness of an observing system. Accuracy is only part of representativeness; the second, and equally important, part is the ability of the observing system to filter out noise [represented by ϕ' in (25.2)]. This variable is most closely related to actual subgrid-scale variations in the meteorological data. The ability to reduce the magnitude of this term without decreasing the amount of information obtained from observations in data-sparse grid boxes depends on how well information is carried through time. If it is done well, a few observations per box may suffice. If it is done poorly, larger grid volumes must be analyzed in order to include more current observations per grid box; it will take longer to tackle the smaller scale prediction problems on a sound scientific basis.

25.4. Supplementing a Mass-Averaged Data Base

The value of time-extrapolated (forecast) data relative to the contemporary, and more highly valued, observations, can be explained in terms of a basic logic presented by Phillips (1982). The time extrapolation method is built upon classical theory of optimal weighting of data to achieve the maximum reduction of random error in an analysis.

The analyzed value A at a grid point in an objective data assimilation scheme is obtained from a weighted average of a model-predicted value F and a value O interpolated from observing sites to the grid point; i.e.,

$$A = \alpha O + (1 - \alpha)F = F + \alpha(O - F) , \qquad (25.4)$$

where α is the fractional weight given to interpolated observations. Subtracting the true volume mean value from this expression gives

$$a = f + \alpha(\epsilon - f) , \qquad (25.5)$$

an expression of the analysis error a in terms of the first-guess error (forecast error) f and the observational (grid point) error ϵ.

Squaring each side of this statement, summing over each grid point in the domain, and dividing by the number of points, gives

$$\overline{a^2} = (1 - \alpha)^2 \overline{f^2} + \alpha^2 \overline{\epsilon^2} . \qquad (25.6)$$

This assumes that the errors ϵ and f are independent of each other.

This expression states that the analysis error variance $\overline{a^2}$, averaged over the domain is equal to the weighted sum of observational and first-guess error variances $\overline{\epsilon^2}$ and $\overline{f^2}$.

The "best analysis" is conventionally defined as the one that minimizes the analysis error variance $\overline{a^2}$. The lowest possible value of $\overline{a^2}$ can be found by determining a zero point on a curve in $(\overline{a^2}, \alpha)$ space defined by the expression

$$\frac{\partial \overline{a^2}}{\partial \alpha} = 0 . \qquad (25.7)$$

This condition leads to the optimal weighting function for blending the temporally extrapolated and spatially extrapolated information; $\overline{a^2}$ is at a minimum when

$$\alpha = \frac{\overline{f^2}}{\overline{f^2} + \overline{\epsilon^2}} \, . \tag{25.8}$$

Unfortunately the value of $\overline{f^2}$ is not always known. It varies with time and space, and even when computers are used, estimates of $\overline{f^2}$ are imperfect.
    Thus we always compute an estimate of the optimal data weight function

$$\alpha' = \frac{\overline{g^2}}{\overline{g^2} + \overline{\epsilon^2}} \, , \tag{25.9}$$

where α' is our "best" estimate of α, and g is our "best" estimate of f.
    The normalized analysis error with this inexact α' is

$$\overline{a^2} = \frac{\overline{f^2} + \overline{g^2}}{(1 + \overline{g^2})^2} \, , \tag{25.10}$$

where f, a, and g have been nondimensionalized by ϵ. If a noise-free model were used, and if it were possible to assimilate data with a very high frequency of data insertion (e.g., 10–20 min), then

$$\overline{f^2} \to \overline{g^2} \to \overline{a^2} \, . \tag{25.11}$$

The expression for $\overline{a^2}$ at any analysis time would reduce to

$$\overline{a^2}^{\tau+1} = \left(\frac{\overline{a^2}}{1 + \overline{a^2}} \right)^{\tau} \, , \tag{25.12}$$

where τ represents the time step at which new observations are assimilated. As $\tau \to \infty$, $\overline{a^2} \to 0$; that is, eventually all the observational error variance is eliminated. This ideal reduction of error variance can never be achieved because both $\overline{f^2}$ and $\overline{g^2}$ increase rapidly from $\overline{a^2}$ at time zero to some significantly higher values at $\tau + 1$ when it is again possible to introduce new observations.
    For current observational assimilation systems, $\overline{g^2} \cong 2$ to 3 times $\overline{a^2}$. If $\overline{g^2} = \overline{f^2} = 2\overline{a^2}$, then the best possible reduction in observational error variance is 50%, but if estimates $\overline{g^2}$ of the actual first-guess error variance $\overline{f^2}$ are poor, say in error by a factor of 2 or 3, the reduction of observational error variance is still better than 45%.
    The most promising approach to reduction in observational error in the objective analysis is possible through simultaneous reduction of $\overline{f^2}$ and $\overline{g^2}$. Two general methods might be employed to do this:

- Improve numerical models, thereby reducing the rate of error growth in the time extrapolation data (i.e., in the forecasts).

- Reduce the time between assimilation steps so the model forecast error has little time to grow.

The second approach appears to hold promise of most immediate payoff because current models have error-doubling times of only a day or two. Short-term model extrapolations could be substantially improved if initial shocks, created when data are inserted into the modeled physics, could be better controlled. If reduced model error is achieved where

$$\overline{f^2} << 2\overline{a^2} \; ,$$

say

$$\overline{f^2} \cong 1.2\overline{a^2} \; ,$$

then 80% of the observational error variance can be eliminated in analysis; this contrasts with the less-than-50% reduction of observational error variance in our current systems.

25.5. Aspects of More Rapid Updating of Meteorological Analyses

Forecasters have found high-frequency (1 h) surface observations to be an invaluable component of the composite meteorological observing system designed to monitor explosive weather events. Employed in concert with radar and satellite imagery, these data help provide a modicum of temporal continuity in the 12 h gap between radiosonde reports. But even when careful sectional analyses are performed with this information to gain a consistent picture of surface weather events, the value of the analyses drops off rapidly in time. Extrapolation of the current surface events cannot usefully be pushed beyond an hour or so. This is because the changes at the surface are dictated by events aloft, which go unseen between the 12 h observing times and often slip through the large spatial gaps of the upper air net. This rapid perishability in the value of surface data, as well as radar data and satellite imagery, is well known and represents the most frustrating practical problem to forecasters. Although with hindsight they often are able to realize qualitatively what set of events preceded a significant weather event, they are unable to apply their knowledge with scientifically based foresight because a complete dynamical picture is lacking.

Models tend to reject or misinterpret incomplete meteorological data sets for some of the same reasons that forecasters are misled by them. Even though extremely fine scale detail can be introduced into initial conditions of a model forecast, both directly to surface observations and by inference from radar and satellite imagery, the smoother, larger scale circulations analyzed aloft will quickly destroy or rearrange them in the forecast.

A prerequisite to improved time continuity of smaller "weather" phenomena in models is the establishment of upper-air initial conditions consistent with finely resolved weather features that are now well monitored at the surface. Field research programs have attempted to move in this direction for some years, but such data coverage has not been achieved in operations. With the advent of VAS temperature sounders and wind profilers, a consistent three-dimensional dynamical picture will be much closer. Supplemental observations derived from aircraft inertial guidance systems and Doppler radars will enhance the quality of the upper-air, regional, composite observing system of the 1990s. The use of a composite, heterogeneous observing system also aids in quality control of individual observing systems by allowing for more precise definition of biases.

But can models and their supporting analysis-initialization systems digest the one- or two-orders-of-magnitude increase in physical information that will result from the new technologies? They can if, while being constantly subjected to shocks of data transplants, they can perform a multifaceted balancing act, a feat that correctly balances dynamical forces while faithfully reproducing wave mode and scale interactions. Beyond this, energy sources and sinks that "drive" the models must be introduced with good spatial and temporal definition and should be maintained in rough balance over the entire domain.

Most of the "shock" experienced when data are introduced into the model component of a data assimilation system is manifested in terms of dynamical imbalances and the model's inability to correct or disperse such imbalances on its own. Some extra help must be given to the primitive forecast equations to reduce the shock syndrome that increases the tendency to reject valid information.

25.6. Concepts of Model Initialization

The concept of initialization of data in meteorological models originally arose from the challenge of preparing dynamically consistent data bases for time integrations with the primitive equations. In the decades following Richardson's abortive attempt to produce meteorologically realistic predictions from integrated solutions of the primitive equations, it became widely accepted that independent measurements of winds and pressures created erroneous high-amplitude and high-frequency signals that were sufficiently intense to mask the true evolution of motions related to significant weather events. In dynamical terms the cause of the problem was traced to an incorrect portrayal of the approximate (within 10% for synoptic scales) balance that is always maintained between the Coriolis and horizontal pressure gradient forces in the atmosphere (so called quasi-geostrophic balance). Uncompensated errors in wind and pressure-temperature observations and in interpolation-extrapolation of observations to model grids are the primary sources of this dynamical inconsistency.

Since in the atmosphere the net imbalanced forces were believed to be at least an order of magnitude smaller than could be reflected by observational networks, the actual accelerations in the atmosphere were expected to be ten times smaller than in Richardson's experiment, leading to actual amplitudes of "unbalanced" wave motions of insignificant size. In retrospect, this reasoning turned out to be in the right direction but perhaps led to an overreaction. It was a major factor leading to the conclusion that the meteorological forecast equations should be purged of such waves (gravity-inertia waves) in order to make the solutions insensitive to the practical evils producing erroneous dynamical imbalance.

Avoiding the direct treatment of gravity wave dynamics, although temporarily slowing progress in operational prediction with the primitive equations for a half decade or so, helped reveal some interesting and subtle aspects of atmospheric dynamics that provide insight to the behavior of current, more complete, prediction systems.

In the 1950s, the first significant decade of NWP advancement, the problems of model initialization and data assimilation were limited in scope. Geopotential heights were employed as the primary input, wind observations being used only to infer horizontal pressure gradient forces in analyses. Thereafter, nondivergent winds were specified from a "balance equation" that maintained a gradient-type relationship between winds and geopotential heights. Diagnostic divergence (or vertical velocity) could be obtained from derived winds and heights. Temperatures were inferred from the hydrostatic equation, and moisture was not a factor in operational predictions. Thus the initial state was built primarily on a foundation of analyzed pressure gradient forces; other variables were obtained primarily from this input through diagnostic relationships that maintained a quasi-geostrophic balance before and during the prediction process.

From the realization that quasi-geostrophic systems were limited in the degree of physical integrity that they could efficiently bring to bear on atmospheric problems, meteorologists began to reconsider the more basic primitive equations as the best tool for prediction. Concurrently, the removal of gravity waves (considered to be meteorological "noise" as late as the early 1960s) became a problem worthy of further consideration. The need became more apparent as efforts increased to use other types of reports: moisture, temperature, and winds as well as geopotential heights in multivariate analysis schemes using relative accuracy to weigh the importance of each type of information. The first approach in data initialization for the primitive forecast equations employed the former quasi-geostrophic equations, but this approach was found to lack both physical and mathematical accuracy.

Since the primitive forecast equations apply equally well at smaller scales, much of the initialization technology applied to global scales is relevant to regional or limited-area forecast problems. Thus it is useful to explore the classical theory that reveals the mutual relationship between slow-moving Rossby waves and gravity-inertia waves, which contain most (but not all) of their energy in fast-moving signals.

If the described frequency separation can be made, the application of linear mathematical theory to the geostrophic adjustment and initialization problems is particularly informative. It should be kept in mind, however, that the nonlinear aspects are important refinements, and are crucial to the mesoscale problems where the interaction of large- and small-scale motions plays a major, and often dominant, role.

25.7. Fundamental Types of Atmospheric Wave Motions, and Their Relationship to Weather Phenomena

It is customary to deal with shallow fluid systems to describe general principles of hydrodynamical behavior of the atmosphere. The scepticism of the student using such a simplified system is quickly reduced when the major physical characteristics are found to apply well for more complex systems. If we were to analyze the types of wave motions associated with more complex forms of the primitive atmospheric equations (e.g., with vertical stratification, compressibility, etc.) we would find essentially the same types, namely, gravity-inertia waves and mixed gravity-Rossby waves; only deep and shallow motions would exist simultaneously. However, theoretical treatments have provided considerable insight without undue complexity by simply dealing with shallow water systems and exploring a range of depths and stratifications.

For example, consider the simple two-layer system in Fig. 25.1. The layer depth h and uniform potential temperature θ in each layer are treated as parameters to distinguish deep tropospheric motions in contrast to shallow ones. The salient points can be made by considering only the effects of velocity changes in the lower layer and ignoring those in the top. The system is assumed to be rotating at a rate equivalent to that of the Earth at a given latitude. The primitive equations applicable to this system are

$$\frac{dD}{dt} + D^2 + 2\left(\frac{\partial v}{\partial x}\frac{\partial u}{\partial y} - \frac{\partial u}{\partial x}\frac{\partial v}{\partial y}\right) - \beta v + f\varsigma + g^*\nabla^2 h = 0 \qquad (25.13)$$

$$\frac{d(\varsigma + f)}{dt} + (\varsigma + f)D = 0 \qquad (25.14)$$

$$\frac{dh}{dt} + hD = 0 . \qquad (25.15)$$

The solutions pertaining to gravity modes can be eliminated from the shallow water equations simply by removing the terms

$$\frac{dD}{dt} + D^2$$

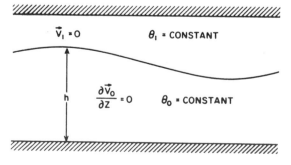

$\vec{V}_1 = 0$ θ_1 = CONSTANT

$\frac{\partial \vec{V}_0}{\partial z} = 0$ θ_0 = CONSTANT

Figure 25.1. Schematic of a simple two-layer shallow fluid system. Each layer is characterized by a constant potential temperature θ. The bottom layer is of depth h.

from the divergence equation in the full system. This amounts to removal, from the divergence equation, of all terms involving horizontal divergence. Solutions of the equations are obtained differently, and certain adjustments are made to achieve energy budget integrity, but the removal is the essential change that transforms primitive equations to quasi-geostrophic prediction equations.

The quasi-geostrophic equations provide surprising realism in atmospheric simulations, considering the fact that a recognizable portion of the atmosphere (that portion ultimately most related to "weather") is highly divergent and rapidly changing. Beyond frictionally generated circulations, one can always find, over data-rich regions in active weather situations, ageostrophic perturbations that seem to be beyond those normally associated with quasi-geostrophic theory. Such situations are normally short-lived and are not usually recognizable on the next upper-air chart 12 h later, but could still be influential weather producers.

The reasons for the transience of such situations become more apparent from classical studies of oscillatory motions created by ageostrophic imbalances. When small-amplitude periodic waveforms are tested as possible solutions to the primitive equations for the two-layer system, several possible constraints must be maintained. The assumed wave motions have an infinitude of phase velocities C, related to horizontal wave length L, mean lower layer depth H, Coriolis parameter f, variation of Coriolis parameter with latitude, β, and reduced gravity $g* = g(\theta_1 - \theta_0/\theta_1)$.

For more rapidly moving waves $(C >> V)$ the equation for wave speeds may be approximated by

$$C = \overline{V} \pm \sqrt{g^*H + (f^2/k^2)} \, , \qquad (25.16)$$

where $\overline{V}$ is the speed of surrounding fluid, and $k^2 = (2\pi/L)^2$. In the radical, g^*H represents the square of the speed of pure shallow water gravity waves and f^2/k^2 is the squared speed of a pure (horizontally oscillating) inertia wave. The combined gravity-inertia waves can move ahead of the simple advective or transport signal of the fluid parcels themselves (+ sign) or against the flow if $|\sqrt{g^*H + (f^2/k^2)}| > |\overline{V}|$ (− sign). Table 25.1 lists some typical

Table 25.1. Atmospheric phenomena and associated gravity wave parameters

Wave disturbance	Speed	Scale height	$\frac{(\theta_1-\theta_0)}{\theta_1}$	Gravity-inertia wave speed
Planetary wave	10 m s^{-1}	30 km	1/3	300 m s^{-1}
Deep tropospheric cyclone	10 m s^{-1}	20 km	1/10	150 m s^{-1}
Frontal wave disturbance	20 m s^{-1}	5 km	1/50	35 m s^{-1}

gravity-inertia wave speeds for some typical scale heights associated with key atmospheric phenomena.

Note that deep gravity-inertia waves move much faster than deep cyclones, but shallow gravity-inertia waves begin to approach the speeds associated with shallow wave frontal cyclones. For these and slower speeds, a more involved equation blending β effects with those of the gravity-inertia waves must be considered. A good approximation for the slower moving waves is given by

$$C_R = \overline{V} - \frac{\beta + [f^2 \overline{V}/(g^* H)]}{k^2 + [f^2/(g^* H)]} \ . \tag{25.17}$$

These so-called mixed gravity-Rossby waves generally move more slowly than the mean flow. Exceptions appear for relatively weak west-to-east mean flows and/or for the longer planetary waves.

Note that (25.17) is made up of a pure Rossby wave portion and two modifiers, one that enhances the β-effect (see numerator in trailing quotient term), and one that slows the strong westward propagation tendency of long (k^2 very small) waves (see denominator in trailing quotient term). The apparent enhancement of the β-effect arises through conservation of potential vorticity and results in increases of the relative vorticity of parcels as an air column moves south. The brake on westward movement occurs through a process of mass advection by the operation of ageostrophic wind components against the westward impulse produced by meridional advection of absolute vorticity.

Note that the β-effect by southward (northward) vortex tube stretching (shrinking) and the westward propagation breaking effect are both enhanced as layers become more shallow. Note (Table 25.1) that this is simultaneously related to a reduced g^* and $\overline{H}$. Both effects operate by virtue of ageostrophic wind components, which can be shown to be enhanced for shallower gravity-Rossby waves. Recall that gravity-inertia (and hence ageostrophic) waves propagate significantly more slowly in shallow layers than in the deep troposphere. The extent to which ageostrophic motions can develop and remain spatially undispersed is a key element.

The Rossby modes are of overwhelming interest when one is dealing with forecasts beyond a day, but all modes and their interactions become critical in any attempt to understand mesoscale weather on time scales shorter than about one day.

25.8. Mutual Adjustment of Pressure and Wind Patterns Toward Geostrophic Flow

The basic principles behind the mutual adjustment process can be demonstrated in a simple example posed by Rossby (1938). Building on the shallow-water dynamics, he imagined a deep jet stream of infinite east-west length and latitudinal width y, flowing in the absence of any horizontal pressure gradient. (In other words, the wind in this case is completely ageostrophic.) He then solved the equation describing the temporal evolution of wind and height fields. For the case where g^*H is large $(300 \text{ m s}^{-1})^2$, a fast-moving gravity wave is generated while mass is rapidly redistributed across the jet, bringing the height field into geostrophic balance with the wind. The wind speed is altered very little in the process, and the jet moves slightly southward. All this occurs in the vicinity of the jet within an hour, the gravity wave energy being dispersed and presumably dissipated at some distant region.

Thus when geostrophic imbalances associated with deep troposphere waves are formed, the return to balance is rapid and several orders of magnitude faster than the Rossby waves themselves. As visualized by Rossby, the process is linear; the gravity waves, once generated, act independently of the slower moving waves and simply add to the slowly evolving circulations only when the energy has not been dispersed. Thus, the assumption that these waves always maintain balance or return to balance instantly is a good approximation for the time period of the Rossby wave (3–4 days).

Although it was not thought of in terms of a physical parameterization, the quasi-geostrophic approximation was one of the first approximations introduced in NWP. It parameterized the effect of deep gravity waves by assuming that they moved at infinite speeds rather than at speeds of 150–300 m s^{-1}. Judged from the perspective of the slow speed (10 m s^{-1}) of a Rossby wave, a 300 m s^{-1} gravity wave might as well move with the speed of light. Just like other types of parameterization methods (convection, eddy viscosity, etc.) this approximation works best when the time scale separation between the two types of phenomena is broad.

Basic linear theory of wave motion and mutual adjustment theory describe the way in which the atmosphere and its models attempt to maintain balance while permitting long-term continuity of organized disturbances. The modeling problem was simplified in several significant ways. (1) The examples assumed some initially unspecified force that created an imbalance, and thereafter the atmosphere was free to oscillate under the natural constraints of gravity on a rotating Earth. (2) Although in some cases a general disturbance made up of a number of Fourier (discrete) components was assumed, these discrete waves were never allowed to interact. Thus, the problems were simplified by so-called linear conditions.

Table 25.1 shows that a time scale separation is not present for Rossby-type waves and gravity waves on frontal scales. Small frontal waves contained completely within jet zones move with speeds ≥ 20 m s^{-1}, and the shallow gravity-inertia waves move with similar speeds. Thus, once created through an imbalance of forces, these shallow (also called internal) gravity waves disperse energy at a rate one-tenth that of their deeper (external) counterparts.

The speed of the adjustment process has another important consequence, that of affecting whether wind or pressure gradient is dominant in determining the direction of adjustment. At one extreme, discussed previously, the adjustment proceeds with a speed equivalent to the maximum rate possible in the atmosphere. In this case the wind changes are almost imperceptible and the pressure gradient almost completely conforms to the original wind distribution according to geostrophic balance.

Drawing again from shallow water dynamics and the same linear theory employed in Rossby's example, Washington (1964) derived a formula that determines the relative importance of wind and pressure gradients in the adjustment process. Recognizing that potential vorticity must be conserved in the adjustment process, he derived a relationship between wind and geostrophic wind at the start and the end of a mutual adjustment process as follows:

$$\Psi_F = \frac{k^2 \Psi_i + [f^2/(g^*\overline{H})]\Psi_g}{k^2 + [f^2/(g^*\overline{H})]} , \qquad (25.18)$$

where
Ψ_i = stream function of initial wind
Ψ_g = stream function of initial geostrophic wind
Ψ_F = stream function of both wind and geostrophic wind (balanced state)

and k^2, g^*, f are the same parameters as previously defined.

Equation (25.18) shows that the final balanced flow field is a weighted mean of the initial actual and geostrophic wind fields. Outside the deep tropics, for deep disturbances (g^*H large), $f^2/(g^*H) \cong 10^{-13}$ m^{-1} and for horizontal scales of the order of 10^6 m, 99% of the adjustment is dictated by the wind field, as in the Rossby example. Even for horizontal scales of 5×10^6 m, the deep wind field plays a more effective role than the pressure gradient in determining the direction of adjustment.

The opposite behavior occurs for relatively shallow disturbances where $f^2/(g^*\overline{H}) \gg k^2$. In the middle of the meso-α scale, $L = 10^3$ km ($k^2 = 40 \times 10^{-11}$), the pressure gradient is approximately 4 times as influential as the wind in guiding the evolution to a balanced state. Recall from earlier discussion that slow-moving gravity energy dispersion slows the adjustment process. If the factors leading to imbalances operate continually, shallow systems are continually trying to adjust geostrophically, but their relatively low reaction ensures that the ageostrophic motion will remain rather large. The processes creating the ageostrophic motions can occur externally (e.g., at the surface of the Earth) and internally (e.g., by nonlinear interaction of different scales of motion). The return to balance on shallow mesoscales is made more difficult by the increased vigor of forces that create the imbalances, which often occur explosively. Sources and sinks of energy can become so important that nearly steady diabatic motion, another form of balance, can be achieved virtually exclusive of geostrophic aspects.

Numerical approximation of derivatives of the primitive equations on grids leads to some further complications in the adjustment process. On grids with conventional second-order finite differences, the energy-carrying velocity for long ($8\Delta x$ and greater) gravity-inertia waves is nearly equal to their phase speeds; the situation is significantly different for shorter waves. For waves about $4\Delta x$ in length, no energy propagation in gravity waves occurs, thus precluding the mutual adjustment process. Imbalances created around these scales will remain unless dissipated on location. One should remove any disturbances on these scales and deal only with longer waves that disperse energy correctly. Even though the dispersion process can proceed on scales larger than $4\Delta x$, it is slowed and the dominant relationship of any one variable over the other is altered. The full implications with regard to prediction accuracy of the alteration of the adjustment process on mesoscale grids have yet to be investigated in sufficient detail.

25.9 Static and Dynamical Initialization Methods Compared

Contrary to the linear theoretical examples shown earlier, models and the atmosphere are constantly being disturbed from their balanced state and waves interact significantly with one another. The actual states of the well-balanced atmosphere, and therefore accurate portrayals of it, are determined not only from relationships applying at a particular time but from a whole set of events before that time. Under the geostrophic assumption, where accelerations of parcels are zero, the relationships between hydro-dynamical variables have no direct relationship to the past or the current conditions in their immediate vicinity. These limitations are lifted substantially where quasi-geostrophic constraints are applied because the effects of flow curvature and deformation can be included and some parcel accelerations are implied. Vertical motions associated with large-scale divergence also may be determined. Compared with pure geostrophic initialization, quasi-geostrophic balancing represents a big step toward creating noise-free initial conditions for primitive equation models. For some purposes, the quasi-geostrophic systems may be adequate for balancing, but they cannot realistically simulate the geostrophic adjustment process. Also they lack the ability to incorporate accurately a number of important forcing functions that influence the relationships between variables for balanced atmospheric and/or model flows.

Foremost among the nongeostrophic factors is the influence of mountains on flow fields that can act on all scales. The forced gravity waves that can exist in nearly steady state around mountain chains and large-scale upstream blocking can be adequately handled only by the full set of primitive equations. Of equal importance in short-range prediction problems is the inclusion of energy sources and sinks estimated by model algorithms. If the initial model state does not reflect, spatially and temporally, the forces driving and damping the prediction model, significant and rapid readjustments can occur to influence early (especially precipitation) behavior.

More complete systems for initialization of the primitive equations have also been considered. The most general type, so-called dynamical initialization, proposes that the full set of prediction equations be employed to obtain a noise-free initial state. The models may be started from a quasi-geostrophic state and marched forward and then backward around the initial time until the remaining imbalances (due to mountains, latent heating, numerical approximations, etc.) have been dispersed or dissipated. An extension of this approach, which can include data insertion, involves only forward time-stepping from some time (approximately 12 h) before the present. In this way, earlier information is assimilated while the model tries simultaneously to adjust its variables to its own physics as well as to the new information in the data.

Whether the initialization problem is treated through static methods or through dynamic methods, it is desirable to achieve a balance of terms in the basic finite difference equations. In the static case, certain time derivatives are identified as vanishing, and in the dynamic case the equations are integrated through time under controls that encourage time derivatives to become small compared with spatial derivatives. The balancing is interpreted according to specific forces, accelerations, or physical processes that are approximated in the prediction model.

A new perspective in initialization has been gained in the past decade by using spectral representations of the prediction equations. This so-called normal mode treatment of the initialization problem is both instructive and highly practical (Kasahara, 1982).

In this approach the primitive equations are linearized and the solutions for Rossby-type waves and gravity waves are defined and categorized according to their vertical- and horizontal-scale characteristics. The time scale of each wave mode is determined by its vertical and horizontal scales, by static stability, and by latitude. The approach parallels that given for solitary waves in the simple model described in Sec. 25.7, except that there is a spectrum of waves, and numerical truncation can alter the behavior of the waves.

Once the relationship between time and space behavior for all waves is established for the linear equations, a linear balance can be established between the hydrodynamic variables.

If the linear solutions related to gravity-inertia waves were eliminated and only those slow-moving Rossby solutions accepted, we would find that Rossby modes create reasonably balanced solutions for forecasts with the primitive equations. The reason for this is the same as that found by Washington (1964): the wind fields (actually the nondivergent wind field) dominate the geostrophic adjustment process, and the winds in the Rossby waves contain most of the atmosphere's nondivergent flow. However, some noise (new gravity waves) is created if a nonlinear model is integrated with these linearized initial conditions.

Much of the remaining noise can be eliminated by invoking a concept of wave mode balancing first suggested by Machenhauer (1977). This concept may be understood by treating Rossby-type and gravity-type waves simultaneously. Again, interest is focused first on solitary waveform solutions. Consider a variable G that represents the amplitude $|G|$ of one (sine or cosine type) gravity wave and its phase (position of troughs and ridges in space). A variable R represents the same information for a single Rossby wave. Assume that these waves exist within a spectrum of other waves.

The equations that describe the temporal behavior of each wave can be written in a general context as follows:

$$\frac{dG}{dt} = \begin{bmatrix} \text{Linear terms describing} \\ \text{periodic oscillation} \\ \text{of gravity waves} \end{bmatrix} + \begin{bmatrix} \text{Nonlinear terms involving} \\ \text{interactions of all gravity} \\ \text{waves and Rossby waves} \end{bmatrix}$$

$$\frac{dR}{dt} = \begin{bmatrix} \text{Linear terms describing} \\ \text{periodic oscillations} \\ \text{of Rossby waves} \end{bmatrix} + \begin{bmatrix} \text{Nonlinear terms involving} \\ \text{interactions of all gravity} \\ \text{waves and Rossby waves} \end{bmatrix}$$

$$(25.19)$$

Although these equations are expressed for single waves, the total solutions must depend on both types (gravity and Rossby) of waves at all other scales. The first term on the right-hand side of each equation is related to the pure wave solutions discussed in Sec. 25.7. If only these first terms were present, the amplitudes ($|G|$ and $|R|$) would be constant in time, and G and R would represent simple sine or cosine oscillations in time, independent of each other. But when all wave interactions are included (second terms in each equation), G and R become dependent on each other as well as on the Gs and Rs representing all other wave scales.

The solutions expressed by (25.19) are the most general imaginable, where energy can be exchanged between all scales of motion and all fundamental physical types of motion. Machenhauer noted, from observations of disturbed solutions of the primitive equation expressed in the forms of (25.19), that the Gs could, to a close approximation, be represented by a constant value and a purely periodic value. He found he could remove essentially all the periodic portion by stating that

$$\frac{dG}{dt} = 0 \, . \tag{25.20}$$

This allowed him to solve for G on the assumption that its linear terms and all its interactions with other waves were in balance. When initial conditions for all gravity modes were obtained under this constraint, no periodic gravity wave motions reappeared. In other words, the initial conditions were in essentially perfect balance.

It may be recalled from earlier sections that modern weather-prediction/data-assimilation systems must place an extremely high reliance on the information that is carried through time by forecast models. This is because the data coverage, relative to the small space scales that must be predicted (and therefore analyzed), is marginally adequate to satisfy proper statistical sampling approaches. The removal of spurious gravitational modes brings temporally extrapolated information a significant step closer to reality. Therefore, it results in a more reliable input with which to judge new information as it is inserted into a data assimilation system.

By reducing the primitive equations into their fundamental (normal mode) forms and thereby distinguishing between waves that owe their existence to a rotating, round Earth and waves that are due primarily to gravity, meteorologists have found the most precise way, thus far, of establishing hydrodynamic balance in models of the atmosphere. Global centers apply this method to initialize analyses for operational forecasts. Even regional limited-area models can achieve a smoother, noise-free initialization when input fields are adjusted by this globally defined mode method.

Problems remain, but the precise mathematical formulation of the method helps define them. The method does not distinguish well between slow-moving gravity-inertia waves and Rossby waves, and it does not allow for rigorous treatment of energy sources and sinks. These shortcomings can have an effect on both global scales and mesoscales and call for more refinement beyond the basic theory to attain the ultimate model balance.

The spectral (normal mode) representation demonstrates aspects of wave behavior:

- Large-scale Rossby-type waves interact with gravity waves to create a slowly evolving mixed-mode state.

- Some energy input to the total system can slowly drive both classes of waves while maintaining a balanced state.

- In addition, some energy input can have rapid impact to create strongly forced gravity modes that behave more independently of the slow-moving Rossby modes.

It is this third class of persistent motion, in which balance is maintained between gravity mode and energy input, that must also be analyzed, initialized, and predicted if deterministic mesoscale prediction is to improve.

In order to treat a broader range of phenomena and maintain time continuity in numerical forecasts, more detailed attention will have to be given to the analyses and/or the initialization of gravity (strongly divergent) modes. The approach will probably be an iterative one, one that first establishes the amplitudes and phase relationships of gravity modes that owe their existence to Rossby modes. Then the perturbation Gs beyond this relatively steady base will have to be established.

Several sources of information are of potential value for this refinement. First, latent heating fields implied from radar or satellite observation might be interpreted through thermodynamic processes in models. Second, wind observations from temporally high resolution wind-measuring radars might be interpreted hydrodynamically with model equations. Each data base shows strong signals on time and space scales much shorter than those of the Rossby modes, and yet each is strongly identified with significant weather-producing circulations showing good continuity in time. Moreover, such information resides in the more dominant terms in the primitive equations for mesoscale motions. Thus the mathematical methods employed to introduce the data with models should prove more straightforward and effective.

REFERENCES

Kasahara, K., 1982: Nonlinear normal mode initialization and the bounded derivation method. *Rev. Geophys. Space Phys.*, **20**, 385–397.

Machenhauer, B. A., 1977: On the dynamics of gravity oscillations in a shallow water model with applications to normal mode initialization. *Beitr. Phys. Atmos.*, **50**, 253–271.

Phillips, N. A., 1982: A very simple application of Kalman filtering to meteorological data assimilation. NMC Office Note 258, National Weather Service, Washington, D.C., 27 pp.

Rossby, C.-G., 1938: On the mutual adjustment of pressure and velocity distribution in certain simple current systems, II. *J. Mar. Res.*, **1**, 239–263.

Washington, W. A., 1964: A note on the adjustment towards geostrophic equilibrium in a simple fluid system. *Tellus*, **16**, 530–534.

CHAPTER 26

Averaging and the Parameterization of Physical Processes in Mesoscale Models

William R. Cotton

26.1. Introduction

The fundamental equations of motion, and continuity equations for heat and water substance, are basically nonlinear. They contain information concerning atmospheric motion and transport over scales ranging from the largest eddies on the globe down to the very smallest eddies contributing to molecular dissipation. At the present time there is no known mathematical technique for exactly integrating this set of equations. Moreover, feasible observational systems are incapable of resolving or defining all scales of atmospheric motion contributing to the kinetic energy and transport in the atmosphere. Thus, meteorologists have been forced to distinguish between those eddies that are in a sense resolvable, either by observation systems or by some form of finite difference representation of the atmosphere (included are finite element and truncated spectral representations of the atmosphere), and the eddies that are not fully resolved either observationally or computationally.

Such unresolved eddies are defined as turbulence. Unfortunately, when only statistical properties can be used to describe or predict turbulent eddies, and to deal with the statistical properties of a system, it is necessary to introduce an averaging operator. Meteorologists have not agreed upon a single averaging operator that can be generally applied to all forms of meteorological modeling and observation. However, any averaging operator used in meteorology must satisfy certain criteria:

- The operator should provide a formal mechanism for distinguishing between resolvable and unrealistic eddies.

- The operator should produce a set of equations more amenable to integration (either analytically or numerically) than the unaveraged system of equations.

- The averaged set of atmospheric variables should be measurable by current or anticipated atmospheric sensing systems.

Anthes (1977) discussed several methods of averaging a set of equations describing a large-scale flow field with an embedded cloud layer. The first method uses the classical Reynolds averaging method by considering the convective clouds as eddies superimposed on a large-scale flow. The second method divides an area into a mean cloud and environment region and obtains the effect of cumulus clouds on this area by considering the equations for the regions separately. The latter method is most suitable for large-scale models or diagnostic studies, but as the averaging domain is decreased in size such that clouds compose 50–100% coverage over the area, the Reynolds averaging method becomes more suitable.

26.2. Ensemble Averages

When the Reynolds averaging method is employed, several different averaging operators are possible. If the fluctuating field or turbulence field is relatively stationary in time, a suitable averaging operator for a random variable ϕ measured by a sensor located at a fixed coordinate point (x_1, y_1, z_1) is

$$\overline{\phi_t} = \lim_{T \to \infty} \frac{1}{2T} \int_{-t}^{+T} \phi(x_1, y_1, z_1, t) dt . \tag{26.1}$$

Alternately, if the turbulence is homogeneous in space, one could sample a population of eddies at a given time t_1 to obtain

$$\overline{\phi_s} - \lim_{X \to \infty} \frac{1}{2X} \int_{-X}^{+X} \phi(x, t_1) dx . \tag{26.2}$$

If the atmosphere is sampled discretely, rather than continuously, an ensemble averaging operator can be defined in which a set of discrete observations k (realizations of a random variable ϕ) is made under superficially look-alike conditions at a specific location and time:

$$\overline{\phi_e} = \lim_{N \to \infty} \frac{1}{N} \sum_{k=1}^{N} \phi_k(x_1, y_1, z_1, t_1) . \tag{26.3}$$

In surface layer applications, superficially look-alike conditions are defined with respect to certain nondimensional similarity parameters such as z/L where L represents the Monin-Obukhov length scale (Monin and Obukhov, 1954).

According to the ergodic hypothesis, if the turbulence is stationary and homogeneous, the three averaging processes will be equivalent, i.e., $\overline{\phi_t} = \overline{\phi_s} = \overline{\phi_e}$. Here, ϕ could represent the three wind components (u, v, w) or temperature (T), pressure (p), total water mixing ratios (r), etc.

In practice, the flow is neither stationary nor homogeneous, so that alternate averaging operators must be sought that are suitable for typical cloud and mesoscale fields. In the case of numerical weather prediction (NWP) models, it has become common to define an averaging operator that is related to the grid scale.

26.2.1. Grid-Volume Average

In large mesoscale models or Limited-area Fine Mesh (LFM) models for which the mean flow is largely two-dimensional, the grid-volume average can be defined as simply a horizontal average. Thus Anthes (1977) and others have defined it as

$$\overline{\phi_A}(x,y) = \frac{1}{\Delta x \Delta y} \int_{x-\frac{\Delta x}{2}}^{x+\frac{\Delta x}{2}} \int_{y-\frac{\Delta y}{2}}^{y+\frac{\Delta y}{2}} \phi(x',y') dy' dx' , \qquad (26.4)$$

where Δx and Δy are the averaging intervals in the x and y directions and are generally taken to be the mesh size in the model. Deardorff (1970) and others generalized the operator to a three-dimensional grid-volume, and applied it to modeling small-scale planetary boundary layer (PBL) flow. Deardorff's version of the operator is

$$\overline{\phi_v}(x,y,z) = \frac{1}{\Delta x \Delta y \Delta z} \int_{z-\frac{\Delta z}{2}}^{z+\frac{\Delta z}{2}} \int_{y-\frac{\Delta y}{2}}^{y+\frac{\Delta y}{2}} \int_{x-\frac{\Delta x}{2}}^{x+\frac{\Delta x}{2}} \phi(x',y',z') dx' dy' dz' , \quad (26.5)$$

where the average $\overline{\phi}$ represents a running mean in space and varies continuously from point to point.

As long as the meteorological system and the spatial averaging scales, Δ, are judiciously selected such that the energy-containing eddies of the system have wavelengths considerably greater than Δ, and energy flows through Δ in a spectrally continuous (not sporadic) manner, the grid-volume average is a well-behaved function. However, in many mesoscale meteorological systems (e.g., hurricanes, convective mesoscale systems) the grid scale is in the midst of the energy-containing scales. Cumulus clouds and cumulonimbi may be converting moist enthalpy into kinetic energy on scales close to Δ. In such circumstances, $\overline{\phi}$ exhibits strong spatial variability on scales of Δ. In other words, $\overline{\phi}$ behaves as a turbulent fluctuating field in which only the smaller scale turbulence fluctuations are filtered by the averaging operator. This is true not only of the variables $\overline{\theta}$, $\overline{u}$, $\overline{v}$, $\overline{w}$, $\overline{q}$ but also of their covariances $\overline{w'\theta'}$, $\overline{u'\theta'}$, etc. In some instances the variability of the grid-volume-averaged fields is large enough to make it impractical to model such systems using finite difference techniques. The normal technique used in large-eddy simulation (LES) models is to choose a grid scale that lies in the inertial subrange, so that kinetic energy is produced on scales larger than Δ and is dissipated on scales smaller than Δ. Thus, in principle, kinetic energy simply spectrally cascades through the truncation scale. The most energetic of the unresolved eddies will then have scales close to Δ. (For a description of atmospheric power spectra and spectral analysis the reader is referred to Panofsky and Dutton, 1984.)

Because the grid-volume average is defined over a finite volume at an instant in time, much like a snapshot of the local atmosphere, it is not measurable in any practical sense. Since there are no known techniques for simultaneously measuring u, v, w, θ, r, etc., over a finite volume, models based on the grid-volume average cannot be conveniently tested.

26.2.2. The Generalized Ensemble Average

A compromise averaging operator is the generalized ensemble average:

$$\bar{\phi} = \frac{1}{\tau L_x L_y L_z}$$

$$\lim_{N \to \infty} \frac{1}{N} \sum_{k=1}^{N} \int_{z-\frac{L_z}{2}}^{z+\frac{L_z}{2}} \int_{y-\frac{L_y}{2}}^{y+\frac{L_y}{2}} \int_{x-\frac{L_x}{2}}^{x+\frac{L_x}{2}} \int_{t-\frac{\tau}{2}}^{t+\frac{\tau}{2}} \phi_k(x', y', z', t') dx' dy' dz' dt'$$

(26.6)

where L_x, L_y, L_z, and τ represent the length and time intervals of the running mean or some other appropriate length/time scale and need not be associated with model-defined grid intervals Δt, Δx, etc. In fact, when the averaged equations are integrated by finite difference techniques, L_x, L_y, L_z, and τ should be selected such that $\bar{\phi}$ varies smoothly in time and in space. Most numerical schemes better represent $\bar{\phi}$ if the grid scales are defined as some fraction of L (i.e., $\Delta x = 0.2 L_x$, $\Delta t = 0.1\tau$, etc.).

As in (26.3), the index k represents a realization of the random process. Thus in addition to being a space/time average, this operator represents the average of an ensemble of observations of superficially identical conditions occurring within the domain defined by L_x, L_y, L_z, and τ. The primary difficulty is to identify suitable superficially identical states of the atmosphere such as identical Richardson numbers, lapse rates, convergence fields.

The primary advantage of the generalized ensemble averaging operator is that it filters out all turbulence. This has definite implications for the Reynolds averaging process (Sec. 26.3). The operator has other advantages as well. Consider first the application to general-circulation-scale systems. The L_x, L_y are stipulated in accordance with a latitudinal and longitudinal belt of interest. Perhaps a simulation, forecast, or analysis with time resolution $\tau \sim$ 1 h is desired. To determine the mean vertical moisture flux through a layer L_z (i.e., subcloud layer, cloud layer, upper troposphere) where a horizontally homogeneous cloud field exists over the region, only a few realizations (e.g., aircraft cross sections or acoustic and microwave radar observations) may be needed to obtain a meaningful estimate of $\overline{w'q'}$. Perhaps even $N = 1$ is sufficient, in which case the average is similar to the ensemble average over a cloud field, such as that used by Arakawa and Schubert (1974) in their cumulus parameterization theory. This is particularly true if τ is quite small. Also for small τ and a relatively homogeneous mean field, $L_x = \Delta x$, and $L_y = \Delta y$. Then the operator converges to a grid-volume average.

Consider, however, a diurnally forced mesoscale system containing active cumulonimbi. The problem is to forecast precipitation with a time resolution

$\tau \simeq 30$ min, over a specific location having dimensions $L_x = L_y \simeq 10$ km and vertical levels $L_z \simeq 1.0$ km. It is likely that during a given 30 min diurnal period only one active cumulonimbus exists in the forecast domain. Clearly a statistically significant average cannot be established from a single realization. Certainly an ensemble of cumulonimbus eddies does not exist in the region at any one time. Furthermore, $\overline{\phi}$ and $\overline{\phi'^2}$ will vary significantly in the horizontal or vertical from a single realization. Thus by increasing N, the number of observations or realizations of look-alike cumulonimbi, the mean will become a better behaved function. However, the meaning of the predicted average must be clearly understood. The eddy fluxes $\overline{w'q'}$, or rainfall rates, may be considered to be forecasts of cloud climatology over the region for a given period and under a given large-scale forcing. When a generalized ensemble averaging operator is used, a forecast at a given time will contain the uncertainty or variance of the climatological sample. For example, under conditions for which superficially look-alike conditions occur over a given domain (L_x, L_y, L_z, τ), the forecast mean rainfall $\overline{R}$ will also be accompanied by a variance in expected rainfall $\overline{R'^2}$. The variance is due not only to variability of rainfall within the defined domain (L_x, L_y, L_z) at a given instant, but also to the variability because superficially look-alike conditions cannot be precisely defined. Use of this operator recognizes that initial conditions or boundary conditions over a finite domain (L_x, L_y, L_z, τ) cannot be observed or predicted with sufficient precision to predict rainfall at any given point (x, y, z, t) or even over a finite domain (L_x, L_y, L_z, τ). Thus, this operator introduces "inherent uncertainty" into any model forecast.

26.2.3. Average Equations by the "Top-Hat" Method

In the "top-hat" scheme, clouds are assumed to be characterized by constant values of the dynamic and thermodynamic variables, while the environment takes on other constant values. The fraction of a given area covered by cumulus clouds, for example, is designated by a, and the environment occupies a larger fraction $(1 - a)$; a may vary as a function of pressure (Fig. 26.1).

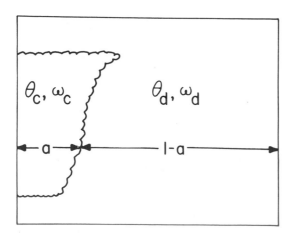

Figure 26.1. Schematic diagram of a deep cumulus cloud and the fractional area covered by the cumulus cloud. (From Anthes, 1977.)

Derivation of the equations governing the time rate of change of ϕ can take several forms (see Anthes, 1977). In one method, ϕ is assumed constant within the cloudy and environmental regions. Consequently $\overline{\phi}$ does not vary continuously in space as do the Reynolds averages, but is obtained as

$$\overline{\phi}(x) = \frac{1}{\Delta x} \int_{x_0 - \frac{\Delta x}{2}}^{x_0 + \frac{\Delta x}{2}} \phi(x')dx' = a\phi_c + (1-a)\phi_d \; . \tag{26.7}$$

For this definition $\overline{\phi}$ and a are constant over the entire interval, and $\overline{\phi'} \equiv 0$. Since $\overline{\phi}$ is not continuous over the domain this definition is not useful for deriving equations for the large-scale flow.

A second variation of the top-hat scheme assumes that ϕ_c and ϕ_d are constant but the horizontal average is a running average, similar to the Reynolds averaging technique.

$$\overline{\phi}(x) = \frac{1}{\Delta x} \int_{x - \frac{\Delta x}{2}}^{x + \frac{\Delta x}{2}} \phi(x')dx' = a\phi_c + (1-a)\phi_d \; , \tag{26.8}$$

where a represents the fractional area covered by clouds over the region centered about x. Note that the limits of integration are the only difference between (26.7) and (26.8). In (26.7), the limits correspond to fixed values x_0; in (26.8), x varies continuously. Since $\overline{\phi}$ varies continuously, this definition is suitable for deriving large-scale equations for NWP models. Then it is possible to write

$$\frac{\partial \phi_c}{\partial t} + \omega_c \frac{\partial \phi_c}{\partial p} = S_c \tag{26.9}$$

and

$$\frac{\partial \phi_d}{\partial t} + \omega_d \frac{\partial \phi_d}{\partial p} = S_d \; , \tag{26.10}$$

where S_c and S_d represent cloud scale and environmental sources and sinks of ϕ, respectively.

Strict adherence to the top-hat assumption eliminates horizontal advection. Multiplying $\partial \phi_c / \partial t$ by a and $\partial \phi_d / \partial t$ by $(1-a)$, and adding, gives

$$\frac{\partial \overline{\phi}}{\partial t} + \omega \frac{\partial \overline{\phi}}{\partial p} = \overline{S} + (\phi_c - \phi_d)\frac{\partial a}{\partial t} \; . \tag{26.11}$$

This equation is useful for truly horizontally homogeneous conditions, but is not very useful for the general case where horizontal gradients exist in the environment.

A third variation on the top-hat scheme is obtained by allowing ϕ_c and ϕ_d to vary continuously in space. Thus,

$$\overline{\phi}(x) = \frac{1}{\Delta x} \int_{x - \frac{\Delta x}{2}}^{x + \frac{\Delta x}{2}} \phi dx' = a\overline{\phi_0}^c + (1-a)\overline{\phi_d}^d \; , \tag{26.12}$$

where $\overline{()}^c$ and $\overline{()}^d$ denote horizontal averages over the cloud and environmental regions. With this definition all variables are continuous across cloud boundaries. Differentiating (26.12) yields

$$\frac{\partial \overline{\phi}}{\partial t} = \sigma \frac{\overline{\partial \phi_c}^d}{\partial t} + (1-a) \frac{\overline{\partial \phi_d}^d}{\partial t} + (\overline{\phi_c}^c - \overline{\phi_d}^d) \frac{\partial a}{\partial t} . \qquad (26.13)$$

Expressions for $\overline{\partial \phi_c}^c / \partial t$ and $\overline{\partial \phi_d}^d / \partial t$ will include horizontal advection terms. This expression for $\overline{\phi}$ is equivalent to the Reynolds averaged equation for $\overline{\phi}$.

Equations governing the interactions between cumulus clouds and the large scale were derived by Ogura and Cho (1973), Yanai *et al.* (1973), Betts (1975), Arakawa and Schubert (1974). Solution of these equations is fundamentally the parameterization problem and requires these determinations:

- Total condensation rate in an averaging volume.

- Vertical distribution of cloud-scale heating or cooling.

- Mean properties of temperature, moisture, and momentum in clouds.

- Fractional area covered by clouds.

Simple cloud models have been employed to make these determinations. Convective parameterization schemes such as Arakawa and Schubert's (1974) compute the contribution of each cloud size in a cloud ensemble to the condensation and eddy flux terms.

It should be emphasized again that, strictly speaking, the top-hat approach is valid only if there is a distinct scale separation between the cloud and large scales, and/or if the large scale is free of clouds. For convective systems (e.g., tropical disturbances, tropical cyclones, middle-latitude mesoscale convective systems) in which the averaging scale is on the order of tens of kilometers, $a \rightarrow 1.0$, and the validity of the procedure must be questioned.

26.3. The Reynolds Averaging Procedure and the Averaging Operator

The Reynolds averaging technique can be applied to the equations of motion to illustrate the basic procedure and the impact of the averaging operator. Begin with the equations of motion in the form

$$\frac{\partial}{\partial t}(\rho_0 U_i) = -\frac{\partial}{\partial x_j}(\rho_0 U_j U_i) - \frac{\partial p'}{\partial x_i} + \rho_0 \left(\frac{\alpha'_m}{\alpha_0} - r'_w\right) g \delta_{i3}$$

$$+ \rho_0 \epsilon_{ijk} f_k U_j + \rho_0 \gamma_0 \frac{\partial^2 U_i}{\partial x_j^2} , \qquad (26.14)$$

where viscous forces are assumed to behave incompressibly. Basic Cartesian tensor notation is used for convenient, abbreviated notation. The indices

$i = 1, 2, 3$ corresponding to Cartesian vector coordinates x, y, z, respectively. The following simple rules apply:

- Repeated indices are summed (e.g., $a_{ii} = a_{11} + a_{22} + a_{33}$ in three-dimensional space).

- Single indices in a term are called free indices and refer to the order of a tensor. The maximum value a free index can attain in a three-dimensional system is 3.

- Only tensors of the same order can be added.

- Multiplication of tensors can be performed as for scalars (they are commutative with respect to addition and multiplication).

- The delta function parameter is defined to simplify writing the gravitational acceleration term; thus

$$\delta_{ij} = \begin{cases} 1 & \text{for } i = j \\ 0 & \text{for } i \neq j \end{cases}.$$

Similarly the permutation symbol ϵ_{ijk} is defined to express vector cross products. Thus,

$$\epsilon_{ijk} = \begin{cases} 0 & \text{if i=j, or j=k, or i=k} \\ 1 & \text{if i,j,k are an even permutation of 1,2,3} \\ -1 & \text{if i,j,k are an odd permutation of 1,2,3} \end{cases}.$$

Hence,

$$\epsilon_{1,2,3} = 1, \ \epsilon_{1,3,2} - -1, \ \epsilon_{3,1,2} = 1, \quad \epsilon_{3,2,1} = \ 1, \ \epsilon_{3,3,1} = \epsilon_{2,1,2} - 0 \ .$$

Cartesian tensor notation yields the following equivalents with Cartesian vector notation:

$$\text{Divergence}: \ \nabla \cdot \boldsymbol{V} = \frac{\partial u_j}{\partial x_j}$$

$$\text{Advection}: \ -\boldsymbol{V} \cdot \nabla \phi = -u_j \frac{\partial \phi}{\partial x_j}$$

$$\text{Coriolis}: \ -\boldsymbol{f} \times \boldsymbol{V} = -\epsilon_{ijk} f_k U_j \ .$$

We now decompose each variable into a mean and a turbulent fluctuating component:

$$U_i = \overline{U_i} + U_i'',$$

$$p = \bar{p} + p'' = p_0 + p' + p'' \ , \tag{26.15}$$

$$\alpha_m = \overline{\alpha_m} + \alpha_m'' = \alpha_0 + \alpha_m' + \alpha_m'' \ ,$$

where p' and α'_m represent mean departures from reference state values p_0 and α_0. In accordance with scaling analysis, the turbulent fluctuations are assumed to behave incompressibly. Substitution of (26.15) into (26.14) gives

$$\frac{\partial}{\partial t}\left[\rho_0(\overline{U_i} + u''_i)\right] = -\frac{\partial}{\partial x_j}\left[\rho_0\left(\overline{U_j} + U''_j\right)\left(\overline{U_i} + U''_i\right)\right]$$

$$-\frac{\partial}{\partial x_i}(p' + p'') + \rho_0\left(\frac{\alpha'_m}{\alpha_0} + \frac{\alpha''_m}{\alpha_0} - r'_w - r''_w\right)g\delta_{i3}$$

$$+ \rho_0\epsilon_{ijk}f_k\left(\overline{U_j} + u''_j\right) + \rho_0\gamma_0\frac{\partial}{\partial x_j^2}\left(\overline{U_i} + u''_i\right),$$

$$(26.16)$$

which can be written after averaging as

$$\overline{\frac{\partial}{\partial t}(\rho_0\overline{U_i})} + \overline{\frac{\partial}{\partial t}(\rho_0 U''_i)} = -\overline{\frac{\partial}{\partial x_j}(\rho_0\overline{U_j}\,\overline{U_i})} - \overline{\frac{\partial}{\partial x_j}(\rho_0 U''_j\overline{U_i})} - \overline{\frac{\partial}{\partial x_i}(\rho_0 U''_j U''_i)}$$

$$- \overline{\frac{\partial}{\partial x_j}(\rho_0 U''_j U''_i)} - \overline{\frac{\partial p'}{\partial x_i}} - \overline{\frac{\partial p''}{\partial x_i}}$$

$$+ \overline{\rho_0\left(\frac{\alpha'_m}{\alpha_0} - r'_m\right)g\delta_{i3}} + \overline{\left(\rho_0\frac{\alpha''_m}{\alpha_0} - r''_w\right)g\delta_{i3}}$$

$$+ \overline{\rho_0\epsilon_{ijk}r_k\overline{U_j}} + \overline{\rho_0\epsilon_{ijk}f_k U''_j} + \overline{\rho_0\gamma_0\frac{\partial^2\overline{U_i}}{\partial x_j}}$$

$$+ \overline{\rho_0\gamma_0\frac{\partial^2 U''_i}{\partial x_j^2}}, \qquad (26.17)$$

where the averaging operator is one of the operators (26.1) to (26.5). The usual procedure is to assume for any variable ϕ,

$$\overline{\overline{\phi}} = \overline{\phi},$$

$$\overline{\phi''} = 0, \qquad (26.18)$$

$$\overline{\phi'' U_j} = 0 = \overline{U''_j\overline{\phi}}.$$

However, in the case of the grid-volume average, where the mean field may contain turbulent fluctuations, it is not necessarily true that

$$\overline{\frac{\partial \phi''}{\partial t}} = 0 \, ,$$

$$\overline{\frac{\partial}{\partial x_j} \left(U_j'' \overline{\phi} \right)} = 0 \, , \qquad (26.19)$$

$$\overline{\frac{\partial}{\partial x_j} \left(\phi'' \overline{U_j} \right)} = 0 \, ,$$

or even that

$$\overline{\frac{\partial}{\partial x_j} \left(\overline{U_i}\, \overline{U_j} \right)} = \frac{\partial}{\partial x_j} (\overline{U_i}\, \overline{U_j}) \, .$$

That is, correlations can exist between the shorter wavelength fluctuating variables ϕ'' and the longer wavelength mean variables $\overline{U_j}$, which still behave as turbulent fluctuating variables. Leonard (1974) discussed the inequality of the final equation in LES models and suggested models for more accurately estimating $\frac{\partial}{\partial x_j} \overline{U_i}\, \overline{U_j}$.

In the case of the generalized ensemble averaging operator (26.6), all turbulence in the mean field is removed as long as the number of realizations $\overline{N}$ is large enough. Thus, mean and fluctuating variables are uncorrelated, and (26.18) and (26.19) are exactly satisfied. Typically one assumes that (26.18) and (26.19) are satisfied even for the grid-volume averaging operation. In general, however, the magnitude of error that is introduced by this assumption is not known. Thus, for either operator, (26.17) becomes

$$\frac{\partial \overline{U_i}}{\partial t} = - \frac{1}{\rho_0} \frac{\partial}{\partial x_j} (\rho_0 \overline{U_j}\, \overline{U_i}) - \frac{1}{\rho_0} \frac{\partial}{\partial x_j} (\rho_0 \overline{U_j'' U_i''})$$

$$- \frac{1}{\rho_0} \frac{\partial \overline{p'}}{\partial x_i} + \left(\frac{\overline{\alpha_m'}}{\alpha_0} - r_w' \right) g \delta_{i3} + \epsilon_{ijk} \overline{U_j} \, , \qquad (26.20)$$

where for large Reynolds number flow we neglect the effects of molecular diffusion on the mean dependent variables. The second term on the right-hand side (RHS) of (26.20) is called the Reynolds stress term. It represents the horizontal and vertical transport of momentum by turbulent velocity fluctuations. It should be noted that had averaging not commenced with the equations of motion linearized with the thermodynamic variables, triple correlations among density and velocity fluctuations would have resulted.

Similarly, averaging the thermodynamic energy equation in terms of the variables $\theta_{i\ell}$ and continuity equations for scalar quantities such as total water mixing ratio r_T leads to

$$\frac{\partial \overline{\theta_{il}}}{\partial t} = - \frac{1}{\rho_0} \frac{\partial}{\partial x_j} (\rho_0 \overline{U_j}\, \overline{\theta_i}) - \frac{1}{\rho_0} \frac{\partial}{\partial x_j} (\rho_0 \overline{U_j'' \theta_{il}''}) + \overline{p(\theta_{il})} \, , \qquad (26.21)$$

$$\frac{\partial}{\partial x_i}(\rho_0 \overline{U}_i) = 0 \; , \tag{26.22}$$

$$\frac{\partial \overline{r_T}}{\partial t} = \frac{1}{\rho_0} \frac{\partial}{\partial x_j}(\rho_0 \overline{U}_j \overline{r_T}) - \frac{1}{\rho_0} \frac{\partial}{\partial x_j}(\rho_0 \overline{U_j'' r_T''}) + \overline{p(r_T)} \; , \tag{26.23}$$

and so on for any scalar variable. Equations (26.20), (26.21), and (26.23) are not closed, however. The terms $\overline{p(\theta_{i\ell})}$ and $\overline{p(r_T)}$ represent sources and sinks of $\theta_{i\ell}$ and r_T, respectively, due to precipitation processes. Some means of defining the correlations between velocity fluctuations and between velocity fluctuations and scalar fluctuations must be found.

26.3.1. First-Order Closure Theory

The simplest approach to closing the averaged equations is to assume that the Reynolds stresses and eddy transport terms can be expressed in terms of gradients of the mean variables. Using the concept of eddy viscosity as introduced by Boussinesq (1877), it is often assumed that the Reynolds stresses can be approximated by

$$-\overline{U_i'' U_j''} = K_m D_{ij} \; , \tag{26.24}$$

where K_m is the eddy viscosity or eddy exchange coefficient, and D_{ij} is the mean rate of deformation tensor,

$$D_{ij} = \frac{\partial \overline{U}_i}{\partial x_j} + \frac{\partial \overline{U}_j}{\partial x_i} \; . \tag{26.25}$$

Hinze (1975, pp. 522–523) argued that (26.24) is physically inconsistent because when the term is contracted for an incompressible fluid, the RHS is identically zero, whereas the LHS is zero only when there is no turbulence. A more consistent approximation would be

$$-\overline{U_i'' U_j''} \simeq -\frac{1}{3}\delta_{ij}\overline{U_\ell'' U_\ell''} + K_m D_{ij} \; , \tag{26.26}$$

where the first term on the RHS of (26.26) is proportional to the average turbulent energy. On the basis of Lilly's (1967) work, Deardorff (1970) assumed that the turbulent energy is given by

$$\frac{1}{2}\overline{U_\ell'' U_\ell''} = \frac{K_m^2}{(c_1 \Delta)^2} \; , \tag{26.27}$$

where $c_1 = 0.094$ and represents a numerical weather prediction (NWP) model grid length. Following a similar line of reasoning, it is often assumed that the turbulent transport of any scalar property A (i.e., $\theta_{i\ell}$, r_T) can be expressed in terms of the scalar eddy viscosity (K_A) as

$$-\overline{U_j'' A''} \simeq K_A \frac{\partial \overline{A}}{\partial x_j} \,. \tag{26.28}$$

Early cloud and mesoscale models contained values of K_m and K_A that were invariant in both time and space. Because the eddy viscosity must depend on the flow variables, and, unlike molecular viscosity, is not a property of the fluid itself, such an assumption is unjustified.

Smagorinsky (1963) defined a nonlinear eddy viscosity that was a function of the mean stress or deformation of the mean flow. Lilly (1967) and later Hill (1974) generalized the Smagorinsky concept by including the effects of static stability.

Klemp and Wilhelmson (1978) expressed the eddy exchange coefficient for momentum in terms of the intensity of turbulence as

$$K_m = c_m \overline{i} L \,, \tag{26.29}$$

where

$$\overline{i} = \left[\overline{(U_i'')^2} \right]^{\frac{1}{2}} \,,$$

$c_m = 0.2$, and L is a turbulence length-scale that they defined as a function of model grid dimensions:

$$L = (\Delta x \Delta y \Delta z)^{\frac{1}{3}} \,. \tag{26.30}$$

They then formulated a prognostic equation for turbulent kinetic energy. An advantage in their approach is that the eddy exchange rate K_m is not only a function of mean variables (i.e., $\overline{U_i}$, $\overline{\theta_{i\ell}}$, $\overline{r_T}$, etc.) but also a function of the locally predicted turbulent kinetic energy. Thus turbulent diffusion will occur only in those regions where active turbulence is forecast.

Alternatively, an eddy exchange coefficient can be expressed in terms of a turbulence time scale τ; i.e.,

$$K_m = c' \overline{i^2} \tau \,. \tag{26.31}$$

The time scale must be diagnosed by empirical or theoretical models. The form of (26.29) or (26.31) is obtained from dimensional arguments.

One disadvantage of the eddy viscosity closure approach in a cloud system is that eddy exchange coefficients K_A, for all mean prognostic variables, must be specified or predicted. Normally the specification of K_A is done with little basis from observation or theory for moist cloud systems.

A fundamental weakness of eddy exchange theory is its underlying assumption that turbulence always acts to diffuse down the mean gradients. There are many examples of counter-gradient transport in planetary boundary layer studies and planetary-scale eddy transport analyses. Although there exist few, if any, documented cases of counter-gradient turbulent transport associated with deep convection, most researchers would not be surprised to find that it exists, or even that it prevails at times. As a consequence, a number of investigators are seeking alternate means of closing the averaged equations.

26.3.2. Higher Order Closure Theory

Rather than make a simple first-order approximation to the correlations $\overline{U_i'' U_j''}$ or $\overline{U_i'' A''}$ such as the down-gradient approximations [Eqs. (26.24) and (26.28)], higher order closure theory involves the formulation of prognostic equations on those covariances. If a simple diagnostic model [e.g., see Eq. (26.33)] is used to evaluate the resulting triple correlation terms $\overline{U_i'' U_j'' U_k''}$ or $\overline{U_i'' U_j'' A''}$, the model is referred to as a second-order closure model; if a predictive equation is formed on the triple-correlation terms and a diagnostic equation is formed on the resulting quadruple correlation terms, the model is referred to as a third-order closure model, and so forth. For a detailed discussion of the derivation of higher order closure models see Cotton and Anthes (1987).

To employ higher order closure, equations predicting, for example, $\frac{\partial}{\partial t}(\overline{u''w''})$, $\frac{\partial}{\partial t}(\overline{w''\theta_{i\ell}''})$, $\frac{\partial}{\partial t}(\overline{w''r_t''})$, $\frac{\partial \bar{e}}{\partial t}$, are formulated, where $\bar{e}$ represents the average turbulent kinetic energy $\bar{e} = \frac{1}{2}\rho_0 \overline{U_i''}^2$. For example, the equation for the average turbulent kinetic energy is

$$
\frac{d\bar{e}}{dt} = \overset{(a)}{\frac{\partial \bar{e}}{\partial t}} + \overline{U_j}\,\frac{\partial \bar{e}}{\partial x_j} = -\rho_0\,\overline{U_i'' U_j''}\,\frac{\partial \overline{U_i}}{\partial x_j} + \overset{(b)}{\left(\frac{\overline{U_i'' \alpha_m''}}{\alpha_0} - \overline{U_i'' r_w''}\right)} g\delta_{i3}
$$

$$
\overset{(c)}{-\frac{\partial}{\partial x_j}\left(\overline{e U_j''}\right)} \overset{(d)}{-\frac{\partial}{\partial x_j}\left(\overline{U_i'' p''}\right)} \overset{(e)}{-\rho_0\gamma_0\left(\overline{\frac{\partial U_i''}{\partial x_j}}\right)^2}. \quad (26.32)
$$

Equation (26.32) illustrates that turbulent kinetic energy is a species that can be advected by the mean flow in space. There are two sources of $\bar{e}$: term (a) mechanical production, and term (b) buoyant production. Term (e) represents dissipation of $\bar{e}$ by viscous forces. Turbulence can also act to redistribute itself in space by term (c) eddy transport (velocity diffusion), and through term (d) the interaction of velocity fluctuations and pressure fluctuations (pressure-diffusion).

As with the average equations of motion, the thermodynamic energy equation, and the water continuity equations, models must be devised for terms such as (c), (d), and (e) in the turbulent kinetic energy equation.

The earliest higher order closure models approximated the velocity diffusion term (c) in analogy to modeling the Reynolds stress by an eddy viscosity. Thus,

$$-\overline{U_i'' U_j'' U_k''} \simeq K \frac{\partial}{\partial x_j} \overline{U_i'' U_k''} \,, \tag{26.33}$$

where K could be modeled in term of a length or time scale. However, several authors (Andre *et al.*, 1976a, b; Zeman and Lumley, 1976) have noted that such gradient-type diffusion models cannot represent vertical convective transport of turbulence. According to Zeman and Lumley, the gradient model often represents the wrong shape of the $\overline{U_i'' U_j'' U_k''}$ profiles and the wrong direction of eddy transport of turbulence. This can lead to serious misrepresentations of the entrainment process. Thus, to close (26.32), rate equations or diagnostic equations must be formulated for the third-order moments (Zeman and Lumley, 1976; Andre *et al.*, 1976a, b).

Similar rate or diagnostic equations must be formulated on the higher order closure terms for equations predicting the variances $\overline{\theta_{i\ell}''^2}$, $\overline{r_T''^2}$ and covariances among the thermodynamic, dynamic, and water species variables in a cloud model. At first glance the number of prognostic equations required to formulate and close a higher order cloud model is overwhelming. Unfortunately the success of such closure models is highly dependent upon the details of the closure assumption. For example, poor models of third-order closure terms could cause a model to estimate incorrectly the rate of entrainment at the tops of a population of boundary layer cumuli or stratocumuli. This error, in turn, could result in poor predictions of the top of the cloud layer, the fractional coverage of clouds, the average liquid water content, and hence, the precipitation from the cloud field. These are precisely the products that the forecaster must rely upon for guidance. Fortunately, technology is improving for closing higher order closure models. However, no generalized closure scheme yet exists that can work objectively in an unstable stratus environment, a stable or weakly stable stratus environment, and a fair weather cumulus field.

Although closure models are very complex in their formulation, they are able to produce ensemble-averaged forecast products using relatively little computer time. Bougeault (1981), for example, used a higher order closure model of boundary-layer-confined cumuli that produced ensemble-averaged liquid water contents, cloud fraction coverage, flux profiles, etc., that agreed quite well with LES model predictions at a small fraction of the computer cost of an LES model. Likewise Chen and Cotton (1983) and Chen (1984) found that their higher order closure model of marine stratocumuli compared favorably with LES model predictions and with observed cloud data. The U.S. Navy is evaluating a closure model for operational prediction of the optical properties of the marine boundary layer (Burk, 1981). Thus, although closure models are complex in their formulation, they are computationally fast enough to be useful for specialized operational forecasting or diagnostic applications.

26.4. Implications for the Forecaster or Operational Meteorologist

Meteorologists have been rather careless in defining the type of averaging operator used in the formulation of their models. This is often because of uncertainty about which form of operator is most suitable and the naive impression that it does not make a difference. However, it does make a difference, which becomes clear when applications of the averaging operators are considered.

26.4.1. Grid-Volume-Averaged Models

As noted in Sec. 26.2.1, the grid-volume-averaging concept is most appropriate for application to LES models. Since these models require very fine resolution, they are computationally very expensive. Moreover, owing to their high resolution, these models cover a very small horizontal domain, generally a few kilometers. Thus they are not suitable for operational forecast guidance or operational diagnostic applications. They are most useful for basic exploratory studies in the dynamics and physics of cloud systems. The data predicted by LES models can be averaged over a number of simulations, or realizations, to obtain ensemble-averaged data. The ensemble-averaged data, in turn, can then be used as guidance in formulating simpler, operational models or in testing the efficacy of a simpler model. Examples of LES models that have been used for this purpose are Deardorff's (1980) and Sommeria's (Sommeria, 1976; Beniston and Sommeria, 1981; Bougeault, 1981).

The grid-volume-averaging concept has also been applied to three-dimensional models of deep convective clouds (Schlesinger, 1980; Klemp and Wilhelmson, 1978; Cotton and Tripoli, 1978; and Miller and Pearce, 1974). In a sense these models could be considered LES models, although, because of their coarse resolution, they are more dependent upon closure assumptions and thereby do not necessarily meet the strict standards of an LES model.

It is quite likely that in the near future grid-volume-averaged thunderstorm models will be one-way or two-way nested to larger, regional-scale models to provide unique simulations of the larger, better organized thunderstorms, such as supercell or mesocyclone storm systems. Since such storm systems appear to concentrate most of their energy in the organized updraft and downdraft of a single, large, dominant cell, a simple subgrid-scale closure may be adequate. That is, the subgrid-scale contributions may be parameterized by a simple dissipative model under the assumption that the grid scale lies in the inertial subrange. The predictability of the model will depend primarily on the predictability of the regional-scale model in defining boundary conditions that uniquely support supercell-type storms. In a two-way nested scheme, the supercell storm could interact with the regional-scale circulations. Clearly if the regional-scale model incorrectly forecasts the environment supporting supercells, so will the supercell model, but far more dramatically. Unfortunately, it is unlikely that routine operational data will have sufficient resolution (or access speed) to be useful for directly initializing an operational supercell model in the near future.

In the case of ordinary thunderstorms or even multicellular severe thunderstorms, it is less likely that grid-volume-averaged thunderstorm models will become useful operational prediction models in the near future. This is because ordinary thunderstorms and multicellular severe storms are composed of numerous cumulus towers and cumulonimbus cells, which compete for the available moist enthalpy in the environment. To be operationally useful, a grid-volume-averaged thunderstorm model must have rather coarse resolution (i.e., 2–5 km). Unfortunately, ordinary and multicell thunderstorms often cover a rather large area, requiring the activation of fine thunderstorm meshes over large portions of a regional-scale model. Frequently the smaller flanking line cells become the dominant cell of the multicellular storm system. Thus, simple dissipative, subgrid-scale closure models are not likely to be suitable. Clearly, it will be some time before it is possible to predict ordinary and multicellular thunderstorm precipitation and winds.

It is nearly possible to use subgrid-scale closure models nested in regional-scale models for supercell storms, but questions remain: Is it possible to recognize the conditions suitable for activating such a model? If the model is activated, but gives an incorrect forecast, can the forecaster recognize the error and not issue false alarms?

There are also conditions in which regional-scale models may be able to utilize a grid-volume-averaging scheme. It is becoming popular today to use regional-scale and mesoscale models in which convective parameterization schemes are turned off and heating from cumulus convection is treated explicitly (i.e., by the explicitly resolved motions of the model). A regional-scale model so configured could be considered a grid-volume-averaged model. Since regional-scale models typically have horizontal resolution of 50–80 km, the explicitly resolved convection must be due to very-large-scale lifting. The convective systems that may exhibit a significant fraction of the convective fluxes on the mesoscale rather than the cumulus scale are the eyewall region of mature, steady tropical cyclones (Jorgensen, 1984) and, perhaps, large, continental convective complexes (MCCs) during their mature steady phases. Even in these systems a significant fraction of the fluxes and heating appears to occur on the cumulus scale and compete with mesoscale motions for the available moisture supply and moist enthalpy. In the case of developing mesoscale convective systems and less mature cumulus cloud clusters, cumulus-scale heating and fluxes clearly predominate over large-scale, explicitly-resolved heating and fluxes. It is under these conditions that a grid-volume-averaged model using explicit convective heating on the regional scale is clearly inadequate. Thus the need arises for a cumulus parameterization.

26.4.2. Top-Hat-Averaged Models

The top-hat averaging method forms the theoretical basis for current convective parameterization schemes. The method relies on a distinction between the cloudy portion of an averaging domain and the noncloudy region. The method is applied to a single dominant cloud or an ensemble of cumulus elements within a finite averaging volume. The single-dominant cloud application is most appropriate in regional-scale models having "small" horizontal grid scales of 50–80 km, where a complete distribution of cumuli is not likely to be sampled in any given grid volume at any given time. The ensemble of cumulus elements application is most appropriate in coarse-resolution models ($\Delta x = \Delta y \simeq 150$ km) such as synoptic and general circulation models where a complete distribution of clouds from towering cumuli to cumulonimbi can be contained within a sample volume. For a detailed review of cumulus parameterization schemes see Frank (1983) and Cotton and Anthes (1987).

The top-hat approach is most questionable when applied to fine-mesh, regional-scale models of strongly disturbed convective systems such as mature tropical cyclones and MCCs. In regions such as the eyewall of a hurricane, for example, the entire grid volume contains cloudy material. Some of this saturated air ascends in large cumulonimbi; the remainder rises in large-scale ascent. In the inner rainband region of tropical cyclones, air ascends primarily in convective updrafts surrounded by subsaturated cloud-free air or precipitation. At some levels the convective updrafts penetrate stratified cloud layers when the entire region is saturated. Thus the concept of averaging over cloudy regions and cloud-free regions is not appropriate for organized convective systems. Another problem is that the scale of the cumulonimbi and the explicitly resolved predicted scales are not distinctly scale separated. In the case of the eyewall region of mature hurricanes, the scale of cumulonimbi may be 10–15 km, which can overlap the grid scales of a 10–20 km mesh tropical storm model. This is another limitation in the use of the top-hat averaging method.

A common feature of both top-hat and grid-volume-averaged numerical prediction models is that for a given set of unique initial conditions, the data predicted at a given grid point represent a unique area or volume-averaged mean, as well as variances about that mean due to variability at a given time across the grid area or in the grid volume. Thus, the forecaster must recognize that the inherent uncertainties in the model forecast are due primarily to the initial/boundary conditions supplied to the model and to variability across the model grid volume.

26.4.3. Ensemble-Averaging Method

For portions and types of convective systems where neither a grid-volume average nor a top-hat average is appropriate, the generalized ensemble-averaging operator is quite suitable. Owing to its generality, it can also be applied to convective disturbances for which the grid-volume-averaging and top-hat-averaging operators are suitable. However, it must be recognized that use of the generalized ensemble-averaging operator introduces a degree of inherent uncertainty that is not present with the other averaging operators. For example, an ensemble-averaged regional-scale model will predict a set of mean quantities (such as rainfall, winds, temperatures, water contents) that is appropriate for the eyewall region (or some portion) of a tropical cyclone. In addition, the ensemble-averaged model will predict departures or variances from those means that are due to variability at a given time interval over a given averaging volume, but are also due to variability introduced in our selection criteria for look-alike conditions. Thus the variances are also a measure of our inherent uncertainty in defining the existence criteria for a given cloud or cloud distribution. This uncertainty is not limited to cloudy conditions. Large variances in winds, for example, can occur due to clear air turbulence or downslope mountain winds that may develop sporadically under similar mean conditions.

One of the problems with the ensemble-averaging method is that it has been applied only to boundary-layer-confined stratocumulus (Chen and Cotton, 1983; Chen, 1984) and cumuli (Bougeault, 1981). To extend the concept to deep tropospheric convection would be a major step. Either ensemble-averaged data from observed towering cumuli and cumulonimbi or large-eddy-simulation model-predicted ensemble-averaged data are needed to help us formulate ensemble-averaged models for deep convection.

26.5. Implications for Cloud Model Validation or Testing

In a sense, a model of a cloud, regardless of how sophisticated, is nothing more than a complicated hypothesis about the physical processes operating in a cloud system. Thus, if we are to adhere to the scientific method, each model must be tested or verified against observations. The testing or validation process again introduces the question of the appropriate averaging operator. Testing "unique" predictions of a model against "unique" observed case study data is basically an attempt to apply grid-volume-averaging concepts to both the observed and model-predicted data. The problem is that data are rarely sufficient to specify uniquely the initial/boundary conditions needed to initialize a model simulation of a cloud event, especially towering cumuli and ordinary cumulonimbi. Slight variations in mesoscale divergence and wind fields can lead to drastic alterations in model predictions of convective precipitation, etc., so within the range of uncertainty of the observed data, model parameters can often be adjusted to obtain good or poor comparison with observations. Rarely is it possible to obtain representative sampling of a cloud system in a given case study; thus, fragmented samples of a cloud are compared with predicted data from a model driven by many unknown initial/boundary conditions. The procedure can hardly be considered a credible scientific evaluation of a model.

The alternative is to obtain ensemble-averaged data from a set of cloud observations under similar conditions. The key is to define appropriate similar conditions. Warner (1955), for example, computed profiles of liquid water content to adiabatic liquid water content (Q/Q_A) for nonprecipitating cumuli. In this case the similar condition in his ensemble of observations was nonprecipitating cumuli. Other parameters, such as wind shear, static stability, and mesoscale convergence, could be used to identify similar conditions for obtaining a meaningful ensemble-averaged data set.

It is important not only to define appropriate similar conditions for ensemble averaging, but also to identify variables that behave in some consistent way when ensemble averaged. Warner (1970) found that the vertical velocity averaged over the width of the cloud showed no consistent variations with height. The rms vertical velocity averaged over the width of the cloud, however, exhibited a consistent, monotonic increase with height. Thus, the ensemble average of vertical velocity averaged across the width of a cloud would be a poor choice for comparison with model prediction. On the other hand, the rms velocity would be ideally suited for testing model predictability.

Of course, model-predicted data that are also ensemble-averaged can be generated by predicting ensemble-averaged data directly or by performing a number of LES model simulations under similar conditions and ensemble-averaging the results to compare against observations. Often it would be more economical to run model experiments to identify stable ensemble-averaged parameters and the criteria for superficially similar conditions. That would not be an attempt to compare model predictions with observations of a given cloud. Such comparisons are probably futile for towering cumuli and most cumulonimbi.

26.6. Averaging and the Parameterization of Cloud Microphysical and Radiative Transfer Processes

Cloud physicists have formulated models of cloud microphysical processes without any formal definition of the averaging processes. In recent years, focus has shifted from the individual cloud particles to a distribution (ensemble) of particles. This has resulted in the use of the kinetic or stochastic equations for predicting cloud droplet broadening by collision and coalescence and condensation. The kinetic equations represent the average behavior of a distribution of drops and show that true stochastic collection processes can result in departures or fluctuations from a predicted distribution (Gillespie, 1975).

Consideration of fluctuations from a grid-volume or ensemble mean will also result in departures from a predicted average distribution. Thus a grid-volume or ensemble-averaged model should predict both the average distribution or spectral density $\overline{f}(x)$ of drops of mass $x \pm \delta x$ and the fluctuations or variance in the spectral density $\overline{[f''(x)]^2}$. No model has yet attempted to predict both the mean distribution of cloud droplets and ice particles, and the variance from the average distribution. Instead, the task of merely predicting the average distribution of droplets and ice particles is formidable.

Simple parameterized models of cloud microphysics such as Kessler's (1969) and Manton and Cotton's (1977) are parameterizations of the behavior of an ensemble-averaged distribution of droplets/ice particles (Tripoli and Cotton, 1980). For example, in the simple parameterized models for raindrop accretion and settling, both Kessler and Manton and Cotton assumed there exists a stable Marshall and Palmer (1948) type distribution of the form

$$\phi(R) = \frac{N_R}{R_m} \exp\left(\frac{-R}{R_m}\right) , \qquad (26.34)$$

where $\phi(R)$ represents the spectral density of rain drops of $R \pm (\delta R/2)$, R_m represents a characteristic radius of the distribution (note the slope $\lambda =$

$1/R_m$), N_R is the raindrop concentration, and the intercept parameter $N_0 = N_R/R_m$.

Although neither Kessler (1969) nor Manton and Cotton (1977) explicitly specified it, (26.34) can be considered representative only after some form of averaging. Equation (26.34) was used by Kessler under the assumption that N_0 was a constant, whereas Manton and Cotton assumed that the slope, or R_m, of the distribution is constant. Thus, their models can be considered to be ensemble-averaged models of accretion growth under the condition that either N_0 is constant or R_m is constant.

There is another impact of averaging on the predicted collection or accretion growth equations. Consider the equation for the time rate of change of raindrop mixing ratio (r_r) by collecting or accreting cloud droplets having mixing ratio (r_c), where the process is CL_{cr}, Under the assumption that R_m in (26.34) is a constant, Manton and Cotton showed that

$$CL_{cr} = 0.884\overline{E}\left(\frac{g\rho_0}{\rho_1 R_m}\right)^{-0.5} r_c r_r \; , \qquad (26.35)$$

where $\overline{E}$ represents an average collection efficiency, g is the acceleration due to gravity, ρ_1 is the density of water, and ρ_0 is the density of air. After Reynolds averaging, (26.35) becomes

$$CL_{cr} = 0.884\overline{E}\left(\frac{g\rho_0}{\rho_1 R_m}\right)^{-0.5} \left(\overline{r_c}\,\overline{r_r} + \overline{r_c'' r_r''}\right) \; . \qquad (26.36)$$

Note that, after averaging, the accretion equation now contains two contributions, one from the product of the mean mixing ratios of cloud droplets and raindrops and a second from the covariance of the fluctuations from the means of the mixing ratios. In a deep convective cloud having large liquid water contents (i.e., $\overline{r_c} > 1.0$ g kg^{-1}; $\overline{r_r} > 1.0$ g kg^{-1}), the first term on the RHS of (26.36) predominates. However, in stratified cloud systems where the liquid water content is low, such as the stratified portions of mesoscale convective systems and stratocumuli, Chen (1984) found that the covariance term makes an appreciable contribution to (26.36), and under certain situations it predominates. Similar covariance terms appear in the averaged equations for a number of physical processes, including the kinetic collection equations (Berry, 1967).

Averaging affects the modeling of other physical processes such as radiative transfer. Normal parameterizations of short- and long-wave radiative transfer consider radiative transfer through a clear or fully cloudy atmosphere. However, when we average over a given domain, frequently we will diagnose that a given region is partly cloudy. Ensemble-averaged models such as Chen's (1984) and Bougeault's (1981) have diagnostic schemes for predicting cloud fractional coverage and average liquid water contents (including cloudy and cloud-free layers). The results affect the predicted rate of radiative heating at a given level. For example, in Chen's (1984) simulations of marine stratocumulus clouds of significant optical depth, the cloud fractional coverage typically is maintained at 100% through much of the cloud layer but decreases to 5–15% in the topmost 50–100 m where cumulus elements penetrate the capping inversion above the main cloud deck. As a result, instead of having the radiative flux divergence and longwave cooling concentrated within 50 m of the solid cloud deck, the radiative flux divergence is spread through a greater depth.

These are just a few examples of the impact of averaging on the modeling or parameterization of cloud microphysics and radiative transfer. Clearly the averaging operator and space and time scales have a major directing influence on the philosophical approach in modeling these processes.

REFERENCES

Andre, J. C., G. DeMoor, P. Lacarrere, and R. DuVachat, 1976a: Turbulence approximation for inhomogeneous flows. Part I: The clipping approximation. *J. Atmos. Sci.*, **33**, 476–481.

Andre, J. C., G. DeMoor, P. Lacarrere, and R. DuVachat, 1976b: Turbulence approximation for inhomogeneous flows. Part II: The numerical simulation of a penetration convection experiment. *J. Atmos. Sci.*, **33**, 482–491.

Anthes, R. A., 1977: A cumulus parameterization scheme utilizing a one-dimensional cloud model. *Mon. Wea. Rev.*, **105**, 270–286.

Arakawa, A., and W. H. Schubert, 1974: Interaction of a cumulus cloud ensemble with the large-scale environment. Part I. *J. Atmos. Sci.*, **31**, 674–701.

Beniston, M. G., and G. Sommeria, 1981: Use of a detailed planetary boundary layer model for parameterization purposes. *J. Atmos. Sci.*, **38**, 780–797.

Berry, E. X., 1967: Cloud droplet growth by collection. *J. Atmos. Sci.*, **24**, 688–701.

Betts, A. K., 1975: Parametric interpretation of trade-wind cumulus budget studies. *J. Atmos. Sci.*, **32**, 19–34.

Bougeault, Ph., 1981: Modeling the trade-wind cumulus boundary layer. Part II: A high-order one-dimensional model. *J. Atmos. Sci.*, **38**, 2429–2439.

Boussinesq, J., 1877: Essai sur la théorie des eaux courantes. *Mém. Prés. Acad. Sci. Paris*, **23**, 24–46.

Burk, S. D., 1981: An operational turbulence closure model forecast system. Preprints, 5th Conference on Numerical Weather Prediction, Monterey, Calif., American Meteorological Society,

Boston, 309–315.

Chen, C., 1984: The physics of the marine stratocumulus-capped mixed layer. Ph.D. dissertation, Colorado State University, Dept. of Atmospheric Sciences, Fort Collins.

Chen, C., and W. R. Cotton, 1983: A one-dimensional simulation of the stratocumulus-capped mixed layer. *Bound.-Layer Meteor.*, **25**, 289–321.

Cotton, W. R., and R. A. Anthes, 1987: *The Dynamics of Clouds and Precipitating Mesoscale Systems*. Academic Press, New York (to be published).

Cotton, W. R., and G. J. Tripoli, 1978: Cumulus convection in shear flow—Three dimensional numerical experiments. *J. Atmos. Sci.*, **35**, 1503–1521.

Deardorff, J. W., 1970: A three-dimensional numerical investigation of the idealized planetary boundary layer. *Geophys. Fluid Dyn.*, **1**, 377–410.

Deardorff, J. W., 1980: Stratocumulus-capped mixed layers derived from a three-dimensional model. *Bound.-Layer Meteor.*, **18**, 495–527.

Frank, W. M., 1983: The cumulus parameterization problem. *Mon. Wea. Rev.*, **111**, 1859–1871.

Gillespie, D. T., 1975: Three models for the coalescence growth of cloud drops. *J. Atmos. Sci*, **32**, 600–607.

Hill, G. E., 1974: Factors controlling the size and spacing of cumulus clouds as revealed by numerical experiments. *J. Atmos. Sci.*, **31**, 646–673.

Hinze, J. O., 1975: *Turbulence*. Second edition, McGraw-Hill, New York, 790 pp.

Jorgensen, D. P., 1984: Mesoscale and

convective-scale characteristics of mature hurricanes. Ph.D. dissertation, Colorado State University, Fort Collins, 189 pp.

Kessler, E., III, 1969: *On the Distribution and Continuity of Water Substance in Atmospheric Circulations*. Meteor. Monogr. 10, American Meteorological Society, Boston, 84 pp.

Klemp, J. B., and R. B. Wilhelmson, 1978: Simulations of right- and left-moving storms produced through storm splitting. *J. Atmos. Sci.*, **35**, 1097–1110.

Leonard, A., 1974: Energy cascade in large-eddy simulations of turbulent fluid flows. In *Advances in Geophysics*, **18A**, Academic Press, New York, 237–248.

Lilly, D. K., 1967: The representation of small-scale turbulence in numerical simulation experiments. Proceedings, IBM Scientific Computing Symposium on Environmental Sciences, Yorktown Heights, N.Y., 195–210.

Manton, M. J., and W. R. Cotton, 1977: Parameterization of the atmospheric surface layer. *J. Atmos. Sci.*, **34**(2), 331–334.

Marshall, J. S., and W. McK. Palmer, 1948: The distribution of raindrops with size. *J. Meteor.*, **5**, 165–166.

Miller, M. J., and R. P. Pearce, 1974: A three-dimensional primitive equation model of cumulonimbus convection. *Quart. J. Roy. Meteor.*, **100**, 133–154.

Monin, A. S., and A. M. Obukhov, 1954: Basic laws of turbulent mixing in the ground layer of the atmosphere (in Russian). *Tr. Geofiz. Inst., Akad. Nauk SSSR*, **24**, 163–187.

Ogura, Y., and H. Cho, 1973: Diagnostic determination of cumulus cloud populations from observed large-scale variables. *J. Atmos. Sci.*, **30**, 1276–1286.

Panofsky, H. A., and J. A. Dutton, 1984: *Atmospheric Turbulence: Models and Methods for Engineering Applications*. Wiley, New York, 397 pp.

Schlesinger, R. E., 1980: A three-dimensional numerical model of an isolated thunderstorm. Part II: Dynamics of updraft splitting and mesovortex couplet evolution. *J. Atmos. Sci.*, **37**, 395–420.

Smagorinsky, J., 1963: General circulation experiments with the primitive equations. 1: The basic experiment. *Mon. Wea. Rev.*, **91**, 99–164.

Sommeria, G., 1976: Three-dimensional simulation of turbulent processes in an undisturbed trade wind boundary layer. *J. Atmos. Sci.* **33**, 216–241.

Tripoli, G. J., and W. R. Cotton, 1980: A numerical investigation of several factors contributing to the observed variable intensity of deep convection over South Florida. *J. Appl. Meteor.*, **19**, 1037–1063.

Warner, J., 1955: The water content of cumuliform cloud. *Tellus*, **7**, 449–457.

Warner, J., 1970: The microstructure of cumulus cloud. Part III: The nature of the updraft. *J. Atmos. Sci.*, **27**, 611–627.

Yanai, M., S. Esbensen, and J.-H. Chu, 1973: Determination of bulk properties of tropical cloud clusters from large-scale heat and moisture budgets. *J. Atmos. Sci.*, **30**, 611–627.

Zeman, O., and J. L. Lumley, 1976: Modeling buoyancy-driven mixed layers. *J. Atmos. Sci.*, **33**, 1974–1988.

CHAPTER 27

The General Question
of Predictability

Richard A. Anthes

27.1. Introduction and Definition of Predictability

In general usage, "predict" is defined as "to declare in advance; foretell on the basis of observation, experience, or scientific reason" (*Webster's Seventh New Collegiate Dictionary*). This general definition underlies the scientific concept of atmospheric predictability, which has been the subject of a large number of scientific studies in the last 25 years.

One of the first discussions of the limits of atmospheric predictability was that of Schumann (1950), which concerned uncertainties in the subjective forecasts of the time. Schumann noted that earlier scientific opinions considered the atmosphere to be, in principle, perfectly predictable for very long periods of time. For example, he quoted Bjerknes (1919) as stating that "if the initial conditions of the atmosphere were known with sufficient accuracy, and if the equations by which the motions of the atmosphere and the physical changes taking place therein were also known with sufficient accuracy, then the state of the atmosphere could be determined completely by some super-mathematician at any subsequent time."

Serious scientific study of atmospheric predictability did not begin until after routine numerical weather prediction began in 1955. Since that time, it has been realized that, in contrast to Bjerknes' (1919) belief, even with a "perfect" global numerical model (a purely hypothetical model that represents all physical processes exactly as they occur in the real atmosphere), there is an inherent limit in time to the predictability of the atmosphere. This limit exists because the atmosphere can never be observed completely and accurately on all scales of motion (global scale down to the molecular scale). If atmospheric motions and processes were linear, errors at one scale would not affect other scales and (at least theoretically) the observable, large scales of motion would have much greater predictability than they do at present.

However, the nonlinear nature of atmospheric processes allows energy exchanges among all scales of motion, so that uncertainty or error in any one scale will eventually contaminate all scales. Because of the existence of atmospheric instabilities (e.g., convective, baroclinic, barotropic), any error, no matter how small, will eventually grow and contaminate even a perfect model's forecast.

In global atmospheric models, the limit of atmospheric predictability is considered to be the time required for the variance (mean square of the difference of some variable) of a pair of solutions that begin with small differences at the initial time (generally of the magnitude of combined observational and analysis errors) to reach the error variance associated with two randomly chosen atmospheric states (twice the climatological error variance). For example, the curve labeled E_p in Fig. 27.1 shows the ensemble average of the growth of small differences in 500 mb heights between two forecasts from a global model and is assumed to represent the growth of small errors in a perfect model. The time for E_p to reach twice the climatological error variance $(2E_c)$ is considered to be the theoretical limit to the predictability of the 500 mb height field. The value $2E_c$ is the expected error variance between two randomly chosen 500 mb height fields.

In addition to illustrating the concept of atmospheric predictability, Fig. 27.1 depicts the difference between predictability and forecast skill. Predictability can be defined as

$$P(t) = 1 - \frac{E_p(t)}{E_c} \;, \tag{27.1}$$

and forecast skill can be defined as

$$F(t) = 1 - \frac{E(t)}{E_c} \;, \tag{27.2}$$

where $E(t)$ is the error variance associated with a particular forecast method such as a numerical model (Anthes and Baumhefner, 1984). The more rapid growth of a forecast model's error compared with the predictability error growth results in forecast skill that is less than predictability. This occurs because models are not perfect; i.e., the errors in the model's physics and numerics contribute errors in addition to those arising from uncertainties in the initial conditions.

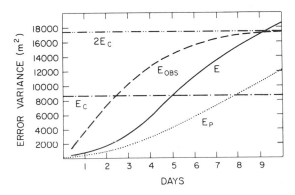

Figure 27.1. Growth of error variance of 500 mb heights. E_c represents the climatological error variance (the error variance associated with a forecast of climatology), E_{obs} a persistence forecast, E a forecast with a model skill approximately equal to that of present operational models, and E_p the predictability error growth, which represents an estimate of the maximum obtainable forecast skill. (From Anthes and Baumhefner, 1984.)

27.2. Review of Predictability Studies

27.2.1. *Predictability of Synoptic and Planetary Scales of Motion*

In studies related to synoptic-scale motions, predictability has referred to the growth of small differences in the structure of two nearly identical atmospheric states. Three methods have been used to estimate the rate of growth of these differences (or errors) and the associated theoretical limits on predictability. The first method measures the divergence of pairs of solutions of a numerical model, pairs with very similar initial conditions (Charney *et al.*, 1966; Smagorinsky, 1969; Lorenz, 1969c; Jastrow and Halem, 1970; Kasahara, 1972; Williamson, 1973). This method suffers from the fact that the growth of errors is model dependent, so that different models yield different estimates of predictability error growth.

A second method, reviewed by Thompson (1984) and Anthes *et al.* (1985), calculates the growth of errors in homogeneous turbulence models. This method suffers from the limitation that the real atmosphere does not always behave like the idealized models, especially in the presence of forcing at the lower boundary and diabatic heating. A third method (Lorenz, 1969b) examines the rate of divergence of pairs of near analogs in the real atmosphere. This third method, which is the most attractive from a conceptual point of view because it makes use of real atmospheric behavior, suffers from the absence of close analogs.

In an early predictability experiment with a general circulation model, Charney *et al.* (1966) found the doubling time of root-mean-square (rms) temperature errors to be about 5 days; this led to an estimate of 3 weeks as the ultimate limit to atmospheric predictability. This 3-week estimate, although since proved overly optimistic, encouraged the planning and execution of the Global Atmospheric Research Program (GARP) and First GARP Global Experiment (FGGE).

Using the European Centre for Medium Range Weather Forecast (ECMWF) operational model, Lorenz (1984) more recently estimated upper and lower bounds to the predictability of 500 mb heights. Lorenz compared pairs of ECMWF operational forecasts over a 100 day period beginning 1 December 1984 to estimate the growth rate of small differences in the initial conditions. The operational forecasts were made 0, 1, 2, ..., 10 days in advance (the 0-day forecast being an analysis). Because the 1 day forecast was quite accurate, the 1 day forecast verifying on Day i (initialized on Day $i-1$) was quite similar to the analysis on Day i.

A subsequent comparison of the forecast initialized at Day $i-1$ with the forecast initialized at Day i yielded an estimate of the error growth rate in a state-of-the-art model. Figure 27.2 shows the rms errors between pairs of forecasts for the same day, averaged over the 100 days of the study. It compares the differences between the 1 day forecast and the analysis, the differences between all forecasts for the same time, and the differences between the forecasts at all time periods and the corresponding analyses (e.g., the 2 day forecast, the 3 day forecast, etc.). The heavy curve in Fig. 27.2

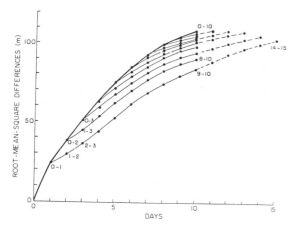

Figure 27.2. Average global root-mean-square height differences, in meters, between j-day and k-day operational forecasts. The ECMWF model for $j < k$ is plotted against k. Values of (j, k) are shown beside some points. The heavy curve connects points where the forecasts are compared against the verifying analysis; this curve represents the performance of the model. Thin curves connect points with equal values of $k - j$. Dashed curves are extrapolations of thin curves. (From Lorenz, 1984.)

thus represents the performance of the model, i.e., the error associated with forecasts at 1, 2, ..., 10 days.

The thin curves of Fig. 27.2 indicate the rate at which the model solution, starting with different initial conditions, diverges. This rate is the error growth rate estimated in classic predictability studies; the smallest error (about 25 m) is quite close to the present analysis error (Baumhefner, 1984b). The doubling time of 3.5 days is considerably less than the doubling time of 5 days found by Charney *et al.* (1966). The difference is probably related to the difference in the characteristics of the initial error and in model differences.

The lowest thin curve in Fig. 27.2 is an estimate of the predictability error growth, which represents the growth of initial small differences in a perfect model. The difference between the upper heavy curve of Fig. 27.2 and the lowest thin curve is an estimate of the model error and is, therefore, an estimate of possible forecast improvements that can still be realized. For example, even if the initial error remains the same, it should be possible for future 10 day forecasts to compare in accuracy with present 6 day forecasts.

27.2.2. Scale-Dependence of Predictability Estimates

The estimates of predictability (Sec. 27.2.1) refer to the growth of the total error in all resolved scales of motion in the models. Careful observers of operational forecasts, however, realize that smaller scales of atmospheric motion are forecast with less skill than are the larger scales. One explanation is that there are inherent differences in the predictability error growth for different scales of motion.

Baumhefner (1984) investigated the growth of errors as a function of horizontal scale. He decomposed the total error growth in a global model into a two-dimensional spherical harmonic wavenumber spectrum; the growth of errors in the various wavenumber groups is illustrated in Fig. 27.3 (the curves for each group are offset by 1 day for clarity). Group 0 represents the zonal vortex, groups 1–4 represent planetary scales of motion (of order 20,000 km), and groups 5–7, 8–11, and 12–18 have horizontal scales of approximately 6000, 4000, and 2500 km, respectively.

The curves in Fig. 27.3 represent the average of two independent predictability experiments; the horizontal error bars on each curve depict the range of error growth over the two experiments. The error bar crossover represents the climatological value of the rms error for each group of scales. The time at which this value is reached can be considered the limit to useful forecasts. The slope of each line in the series of straight lines originating at coordinates (6, 5) reports error growth rate doubling time in days. Earlier in the forecast period, error growth rates correspond to an error doubling in about 2 days, with errors increasing more slowly with increasing time. Using the calculated error growth rates, an assumed initial rms error, and the error bars, Baumhefner (1984) estimated the theoretical limits of predictability for the various scales. His results indicate that predictability decreases with decreasing horizontal scale, a result also obtained by Shukla (1984). For initial errors close to present analysis errors (rms error of 19.0 m), predictability ranges from 8.5 days for the zonal vortex to 3.5 days for the smallest scales of motion. Baumhefner's estimated forecast limit of 7.0 days for the total flow is less than other estimates, including that of Lorenz (1984), the difference probably being due to the characteristics of the initial error and to model differences.

Atmospheric predictability varies not only with horizontal scale, but also with seasons, geographic location, and synoptic pattern. For example, because the natural atmospheric variability (climatological error variance) as

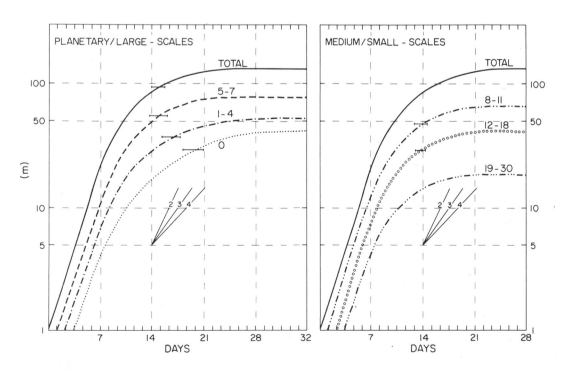

Figure 27.3. Growth of predictability errors: Northern Hemisphere standard deviation of 500 mb errors from predicted values for total field (solid) and various wavenumber groups. (From Baumhefner, 1984b.)

well as the growth rate of small errors varies with geographic regions, the theoretical limit to predictability varies with location. Thus, because the variability is less in low latitudes than in middle latitudes, the limit to deterministic predictability is shorter in the tropics, provided the error growth rates are the same (Shukla, 1984). In predictability experiments with the Goddard Laboratory for Atmospheric Sciences (GLAS) model, Shukla (1984) found that the predictability was greater in the Northern Hemisphere than in the Southern Hemisphere, presumably because of greater surface inhomogeneities in the Northern Hemisphere, which produce stronger forcing at the surface.

Because the growth of initial errors depends upon atmospheric instabilities, predictability varies with the synoptic situation. Thus, initial states that contain regions of strong instability (e.g., baroclinic or convective instability) are less predictable than those that contain fewer or weaker areas of instability.

Finally, the predictability error growth depends on the nature of the initial errors or difference. Daley (1981) showed that errors that were mainly ageostrophic grew at slower rates than initial errors that were in geostrophic balance. Most of the ageostrophic error energy excites gravity wave modes that are damped in the model, whereas most of the geostrophic energy is projected onto the slower, more meteorologically realistic models where more rapid growth can occur.

In summary, the main results of previous predictability studies using global models are listed below:

- The doubling time of small initial errors is 2–5 days.

- Estimates of the limit to the predictability of the total flow range from 8 to 16 days.

- Predictability decreases with decreasing horizontal scale.

- Predictability is less in the tropics than in middle and high latitudes.

- Predictability is higher in the Northern Hemisphere winter than in summer.

- Predictability varies with synoptic type; i.e., some atmospheric states are more predictable than others.

- Ageostrophic, random initial perturbations grow more slowly than geostrophic, systematic perturbations.

27.3. Predictability of Mesoscale Circulations

27.3.1. *Physical Processes Governing Mesoscale Phenomena*

The study of the growth and spread of errors through wavenumber space in homogeneous turbulence models has been an important approach to estimating the limits to predictability of large-scale atmospheric flows (Thompson, 1957; Robinson, 1967, 1971; Lorenz, 1969a; Leith, 1971; Leith and Kraichnan, 1972). As reviewed by Anthes *et al.* (1985), results from these studies have been used to draw pessimistic conclusions concerning mesoscale predictability (Tennekes, 1978).

The principal reason for the limit of predictability for synoptic scales is generally considered to be the nonlinear interactions between different components of the wave spectrum. These interactions depend on the initial distribution of energy in the different wavenumbers and on the number of waves the model can resolve. Uncertainties and errors in the resolvable-scale waves and errors introduced by the neglect of unresolvable scales grow with time and spread throughout the spectrum, eventually contaminating all wavelengths and destroying the forecast.

The more rapid transfer of energy from small to large scales in three-dimensional turbulence, compared with two-dimensional turbulence, indicates less inherent predictability for atmospheric systems that behave as three-dimensional turbulence. Since the mesoscale spans scales of motion ranging from the synoptic scale, which behaves as two-dimensional turbulence, to the microscale, which behaves as three-dimensional turbulence, one would expect that mesoscale atmospheric systems, especially at the smaller scales, would have considerably less inherent predictability than synoptic-scale systems. This is the essence of the argument presented by Tennekes (1978).

The pessimistic conclusion from turbulence studies and the observed atmospheric spectrum is that inevitable errors or initial uncertainties in the small scales of motion will propagate toward larger scales and will reach the mesoscale sooner than the synoptic scale, rendering the former less predictable. Reducing the uncertainty in the smaller scales by increasing the observational density would be very costly and, because of the relatively rapid rate of energy transfer on this scale, would increase the predictability of mesoscale motion only marginally.

Several counterarguments can be made to suggest that prediction of some important mesoscale phenomena is not as hopeless as the above conclusions indicate. First, the observed atmospheric spectra represent a statistical description of atmospheric structure, involving averages over space and time. However, many mesoscale events are highly intermittent, and when they occur, the atmospheric spectrum may not conform to the idealized spectra of three-dimensional turbulence, or to the average atmospheric spectrum. Certain mesoscale circulations, such as fronts and tropical cyclones, may have peculiar dynamic structures that resist the cascade of energy to larger and smaller scales. For example, Lilly (1984) discussed physical reasons why some severe rotating thunderstorms are probably more predictable than ordinary turbulence concepts would suggest.

A second important factor that influences the behavior of many atmospheric mesoscale phenomena, which has not been considered in the predictability studies using turbulence models, is the effect of boundary forcing. Boundary forcing on the synoptic and planetary scales associated with land-sea contrasts and orography appears to be the reason why these scales of motion are more predictable in the Northern Hemisphere than in the Southern Hemisphere (Shukla, 1984). On the mesoscale, surface inhomogeneities including elevation and surface characteristics (albedo, heat capacity, moisture availability) generate many phenomena (such as mountain waves, sea breezes, convection, orographic precipitation, Great Lakes snowstorms, coastal fronts, and orographic damming of shallow, cold air masses) and modulate their behavior. Known surface inhomogeneities, if incorporated properly in numerical models, are likely to increase the predictability of motions they force.

Anthes (1984) classified the development of mesoscale weather systems into two types: (1) those resulting from forcing by surface inhomogeneities, and (2) those resulting from internal modifications of large-scale flow patterns. Land-sea breezes, mountain-valley breezes, mountain waves, heat island circulations, coastal fronts, drylines, and moist convection are often generated by the first mechanisms. Fronts and jet stream phenomena, generated by shearing and deformation associated with large-scale flows, belong to the second class.

A subset of the second class of phenomena includes those mesoscale features that develop in regions of instability produced by large-scale flows; an example is the isolated thunderstorm that develops in a region of large-scale convective instability. Such individual phenomena are likely to have little predictability, even though the development of the large-scale area of instability may have significant predictability. An optimistic hypothesis is that many significant mesoscale atmospheric phenomena evolve from an interaction between large-scale flows and known or predictable surface inhomogeneities, in which case there is hope for skillful forecasts over periods of 1–3 days, based on deterministic methods, provided the synoptic-scale motions are predicted correctly. Because synoptic-scale forecasts at present contain significant errors in this time period, an essential component of a future operational mesoscale modeling system is improved synoptic-scale forecasts by global models.

27.3.2. Preliminary Predictability Studies With a Regional-Scale Numerical Model

The success of numerical models with horizontal resolutions of approximately 40 to 100 km in simulating many mesoscale atmospheric phenomena, such as fronts, drylines, polar lows, orographic waves, mesoscale convective complexes, and mesoscale precipitation features embedded within cyclones, indicates that these models have reached a level of maturity that makes quantitative estimates of predictability possible. However, the use of limited-area models for predictability studies is not a straightforward extension of the methods used in applying global models to the predictability of large-scale motions. In addition to the greater possibility of case dependence and the different role of physics on the smaller scale, the presence of lateral boundary conditions (LBCs) in mesoscale models introduces an added complexity. Not only do errors in the initial data and errors in model physics limit predictability; errors introduced by the LBCs, which are not present in global models, contribute additional errors to the predictions. Conversely, accurate

LBCs supply information to the limited-area models that may improve the predictability by limiting error growth.

Description of the Model

The model used here to investigate the behavior of initial errors is an improved version of the limited-area mesoscale model described by Anthes and Warner (1978). The vertical coordinate is $\sigma = (p - p_t)/(p_s - p_t)$, where p is pressure, p_s is surface pressure, and p_t is the constant pressure at the top of the model (100 mb). The 16 σ levels (0.0, 0.1, 0.2, 0.3, 0.4, 0.5, 0.6, 0.7, 0.78, 0.84, 0.89, 0.93, 0.96, 0.98, 0.998, 1.0) give 15 layers of unequal thickness at which the temperature, moisture, and wind variables are defined. The horizontal grid contains 46 points in the north-south direction and 61 points in the east-west direction.

The parameterization of surface and planetary boundary layer (PBL) processes is described by Zhang and Anthes (1982). Shortwave and longwave radiation are considered in the surface energy budget but not in the free atmosphere. The cumulus parameterization and treatment of nonconvective precipitation follow methods by Kuo (1965, 1974) and Anthes (1977). Additional details, including the initialization procedure and the treatment of lateral boundary conditions, are provided by Anthes *et al.* (1985).

In the following descriptions of predictability experiments, pairs of simulations with slightly different initial conditions are integrated for 72 h and the different solutions are compared. The perturbations are introduced by first adding random perturbations to the relative vorticity field and then calculating consistent perturbations in the model wind and temperature fields. This method, which is described in detail by Anthes *et al.*(1985), attempts to assign most of the perturbation energy to the slow-moving long waves in the model. In all cases, observed (analyzed) LBCs are used.

Case I: 10–12 April 1979

During the spring months of 1979, special observations were collected over the central portion of the United States as part of the Severe Environmental Storm and Mesoscale Experiment (SESAME). This included the 72 h period beginning at 0000 GMT 10 April 1979 during intense organized convection, including tornadoes, that developed over Oklahoma and Texas. The evolution of the large-scale features and their interaction with the mesoscale have been discussed in several papers (Carlson *et al.*, 1980; Moore and Fuelberg, 1981; Anthes *et al.*, 1982; Carlson *et al.*, 1983).

The lower troposphere during the 72 h period was characterized by two large cyclones, one moving slowly eastward from the Rockies and the other moving northeastward along the Atlantic coast. Ahead of the western low, a strong pressure gradient was associated with low-level southerly flow; maximum winds at 850 mb were >25 m s^{-1}. This flow carried extremely moist air (mixing ratios >20 g kg^{-1}) from the Gulf of Mexico northward into the southern plains, supplying the moisture for the intense convection.

At 500 mb, a deep trough was located over the Rockies. Jet streaks propagated around this trough system during the 72 h period. The vertical

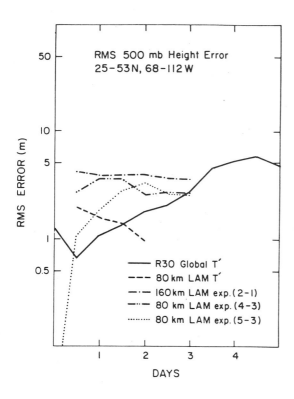

Figure 27.4. Growth of rms differences between two randomly perturbed initial states in a global model of a limited domain over North America (——), and growth of differences in four limited-area model (LAM) experiments: temperature perturbation (- - -); vorticity perturbation at two resolutions (-·· -) (-··· -); small perturbation on boundary only (···). (From Anthes *et al.*, 1985.)

circulations induced by these propagating wind maxima have been linked to the development of the low-level southerly jet stream and possibly the outbreak of the intense convection over Texas and Oklahoma (Moore and Fuelberg, 1981).

Anthes *et al.* (1985) presented results of the nine numerical simulations of this case. Figure 27.4 shows the temporal behavior of the rms height errors from some of the simulations. The two major purposes of these simulations were to investigate the effect of uncertainties in the initial conditions and the effect of the LBCs on the simulations. Simulations were made using a 160 km grid that covered all of North America and a considerable portion of the Atlantic (shown in Fig. 27.5) and using an 80 km grid of the same dimensions (46×61) centered within the larger domain and covering most of the United States. Pairs of simulations that differed only in initial conditions were compared, and pairs that differed only in LBCs were compared.

Experiments 1 and 2 were 160 km simulations on the large domain and differ only in the initial conditions (In Exp. 2, balanced perturbations were added to the initial conditions of Exp. 1.) Experiments 4 and 3, analogous to Exps. 2 and 1, were integrated on the small domain with higher resolution. Experiment 5 was the same as Exp. 3 except that it obtained the time-dependent LBC from Exp. 2 rather than from Exp. 1. The dashed curve represents a pair of 80 km simulations differing only in initial conditions of temperature, random temperature perturbations having been added to one of the simulations. The perturbations in this case are unbalanced

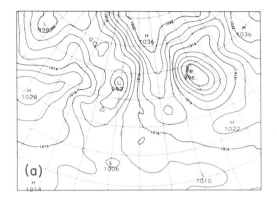

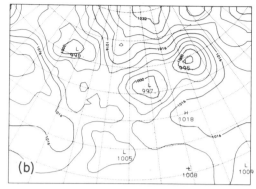

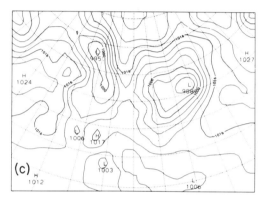

Figure 27.5. Analysis of sea-level pressure (contour interval 4 mb) at (a) 0000 GMT, 22 April 1981; (b) 0000 GMT, 24 April 1981; (c) 0000 GMT, 25 April 1981.

and ageostrophic. The solid curve represents the rms error growth over a region corresponding approximately to the 80 km regional model domain (25°–53°N, 68°–112°W). The global model used in this experiment, the Community Forecast Model (CFM), was developed at the National Center for Atmospheric Research (NCAR). Williamson and Swarztrauber (1984) provided a description of the model.

A surprising result from the SESAME experiments was that, in contrast to previous large-scale studies, the simulations showed little or no growth of differences in initial conditions over the time period 0–72 h. As Fig. 27.4 indicates, in all limited-area simulations with the same LBC, the rms height differences decreased with time. The rms differences in other variables, including temperature, specific humidity, and the horizontal wind components, also showed that little or no growth takes place over the period of integration when only the initial conditions are varied. In contrast, the differences grew in the pair of simulations using a global model as in previous global predictability studies.

Two hypotheses can be advanced to explain these absences of growth of initial errors in the limited-area model. First, the same LBC may be preventing different evolutions of the flow on the interior of the domain by providing identical large-scale information to the periphery of the pairs of simulations. If the large-scale flow, together with the forcing at the Earth's surface through orography and energy fluxes, is controlling the evolution of the mesoscale features as suggested by Anthes (1984), then one would expect little sensitivity of mesoscale forecasts to variations in initial conditions. This hypothesis is supported by experiments discussed by Anthes *et al.* (1985) in which growth of differences did occur in simulations using different LBCs.

A second hypothesis is that the synoptic weather type over the limited area was, by chance, more stable to initial perturbations than typical global circulations, which always contain some regions that are sensitive to initial perturbations. If regions of instabilities to small-scale perturbations existed in the initial large-scale fields, one would expect a much greater sensitivity to variations in the initial conditions.

Because of the practical implications of the results of the SESAME experiments, and in order to understand why no error growth was observed in these simulations, it would be beneficial to repeat the experiments on additional case studies. Some cases would probably show sensitivity to initial conditions. Two questions are important: (1) What percentage of situations is sensitive to variations in initial conditions? (2) What distinguishes the cases that are sensitive from those that are not?

Case II: 22–24 April 1981

During the 72 h period beginning at 0000 GMT 22 April 1981, a cyclone and frontal system moved across the central United States bringing heavy precipitation to the eastern half of the United States. Because the data used were collected by the Environmental Protection Agency in an experiment designated OSCAR (Oxidation and Scavenging Characteristics of April Rains) in the northeastern United States during this period, the case study is designated the OSCAR case. Simulations were performed with control conditions (Exp. 1) and perturbed conditions (Exp. 2).

Figure 27.5 shows analyses of the observed sea-level pressure for the initial time of the simulations, and 48 and 72 h later. At the initial time (Fig. 27.5a), a 997 mb low is located over the Dakotas, a strong ridge of high pressure extends southward from an anticyclone over Hudson Bay and dominates the East Coast, and a 996 mb cyclone is located over the Canadian Maritimes. By 48 h (Fig. 27.5b), the Dakotas low has moved eastward, replacing the ridge of high pressure, and is located over the eastern Great Lakes region; the Maritimes low has remained nearly stationary. At 72 h (Fig. 27.5c), the Dakotas low has nearly merged with the Maritimes low. A pronounced frontal trough of low pressure extends southward along the Atlantic coast. A new cyclone has moved into western North America, and a high pressure ridge covers the central part of the continent.

The observed evolution of the 500 mb flow is shown in Fig. 27.6. During the first 48 h, the northern portion of a trough initially over the western United States (Fig. 27.6a) intensifies as it moves into the Great Lakes region (Fig. 27.6b). The southern portion of this trough becomes a cut-off low over Mexico. A deep cyclone remains nearly stationary over the Canadian Maritimes. At 72 h, the Great Lakes trough is merging with the Maritimes low (Fig. 27.6c). The jet stream enters the northwestern United States, dips southeastward over the upper Midwest, and crosses the Atlantic coast over the Carolinas.

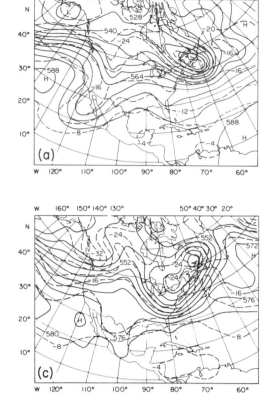

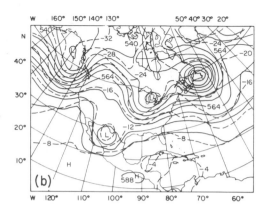

Figure 27.6. Analysis of 500 mb height (solid lines, contour interval 60 m) and temperature (dashed lines, contour interval 4°C) at (a) 0000 GMT, 22 April 1981; (b) 0000 GMT, 24 April 1981; (c) 0000 GMT, 25 April 1981.

Although the main concern of this study is the behavior of small perturbations in the initial conditions of regional-scale models, the 72 h simulation is discussed briefly, first to show that the simulation was reasonably accurate, and second, to point out several interesting mesoscale features that developed in the simulation from the large-scale initial conditions.

The 72 h sea-level pressure forecast (Fig. 27.7) is in general agreement with the observations (Fig. 27.5c), although the model shows a more intense low pressure center over New York State than was observed. The frontal trough of low pressure along the Atlantic coast and the mesoscale ridge of high pressure over the Appalachians are both present in the simulation. To the west, the ridge of the high pressure over the Midwest, the high over Mexico, and the elongated trough of low pressure in the Rockies are all present in the simulation.

At 500 mb (Fig. 27.8), the main circulation features at 72 h are present in the model simulation. This includes a nearly closed low over New York State, a ridge over the upper Midwest, a weak trough over Texas, and a jet stream close to the observed position (compare with Fig. 27.6c). The evolution of the cyclone in the northeastern United States and the associated cold frontal system along the Atlantic coast includes the development of realistic and observed mesoscale features associated with the front. A vertical cross section perpendicular to the front at 72 h (Fig. 27.9) shows a sloping frontal

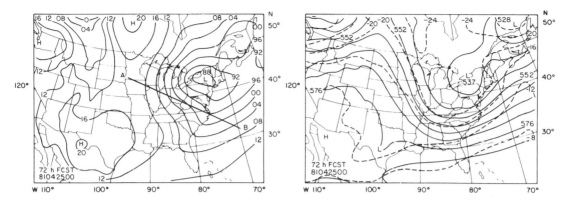

Figure 27.7. Simulated sea-level pressure at 72 h (0000 GMT, 25 April 1981) of Exp. 1 (OSCAR case). The contour interval is 4 mb. See Fig. 27.9 for application of line A-B.

Figure 27.8. Simulated height of 500 mb surface (solid lines in decameters) and 500 mb temperatures (dashed lines in °C) at 72 h (0000 GMT 25, April 1981) of the simulation of Exp. 1.

zone that intersects the surface near the coast, a boundary layer of variable depth, the jet stream, and mesoscale variability in the relative humidity field. In particular, the model simulates moist air ahead of the front and a narrow dry tongue immediately behind the front. This dry tongue spirals into the nearly stationary closed low (Fig. 27.10a). Such a mesoscale wedge of dry air occurs commonly in synoptic situations of this type and is present in the observations of this case (Fig. 27.11), although the model simulation lags the observations by about 9 h. Although the physical mechanisms responsible for the formation of this feature are not completely known, shearing and stretching deformations associated with the large-scale flow are undoubtedly important.

In the second simulation (Exp. 2) of the OSCAR case, balanced perturbations were added to the initial wind and temperature fields. The initial rms differences were 0.5 m s^{-1} for the winds and 0.25°C for the temperature. Because the relative humidity was constant in both sets of initial conditions,

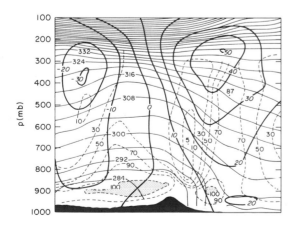

Figure 27.9. Vertical cross section along line A-B in Fig. 27.7 of isentropes (thin solid lines, contour interval 4 K), wind component normal to cross section (thick solid lines, contour interval 10 m s^{-1}), and relative humidity (dashed lines, in %) at 72 h (0000 GMT, 25 April 1981) of the simulation of Exp. 1.

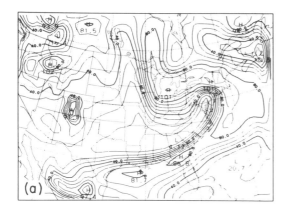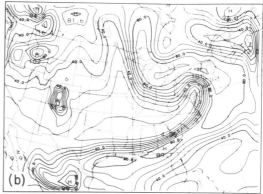

Figure 27.10. Relative humidity at 700 mb and 72 h of simulations for 0000 GMT, 25 April 1981. (a) Control simulation (Exp. 1), and (b) simulation with perturbed initial conditions (Exp. 2).

the variations in temperature also resulted in an initial rms difference of specific humidity of 0.05 g kg^{-1}.

As shown in Fig. 27.12, the initial differences between control and perturbed simulations produce little or no error growth in the OSCAR case, in agreement with the results for the SESAME case. The rms wind errors decrease from 0.5 m s^{-1} to 0.3 m s^{-1} at 72 h; the rms temperature differences decrease from 0.25°C to 0.15°C. The ground temperature and the surface pressure show oscillations but no growth over the period. The only variables that show some growth are those involving moisture; the rms differences in specific humidity grow for the first 20 h and then remain constant. The rms differences in convective and stable precipitation show growth throughout the 72 h period, but the growth rate decreases during the last 24 h.

As in the SESAME case, the details of the OSCAR simulations are quite similar at 72 h. For example, Fig. 27.10b shows the 72 h relative humidity at 700 mb in the perturbed simulation, for comparison with the same field in the control simulation (Fig. 27.10a). The similarity of these, and other fields not shown, together with the absence of growth in the rms differences, indicates that for the OSCAR case, as well as the SESAME case, initial errors do not grow and contaminate the simulations over a 72 h time period.

An estimation of the energy in various horizontal scales of motion of the simulated variables and the energy in the various scales of the differences between two solutions can be obtained by calculating the two-dimensional spectra (Errico, 1985). Consider a variable $A_{I,J}$ defined on the limited-area grid, where $x_J = (J - 1)\Delta x$, $y_I = (I - 1)\Delta y$ for the grid indices I and J in the range $1 \leq I \leq I_M$ and $1 \leq J \leq J_M$, where $I_M = 46$ and $J_M = 61$, in this case. Because the solutions are not periodic over the domain, the linear trend in both the x and y directions must be removed before the Fourier transform is applied, in order that scales of motion larger than that of the domain will not be aliased and appear as fictitious shorter waves. After the linear trend is removed, a Fourier transform on the detrended field A is calculated:

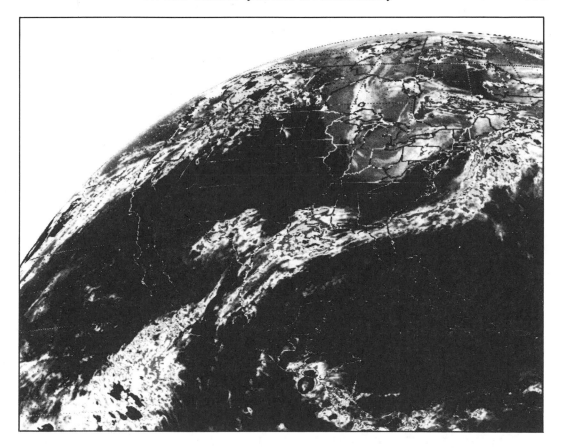

Figure 27.11. Infrared satellite photograph at 1500 GMT, 24 April 1981.

$$S_{K,L} = \frac{1}{(I_M - 1)(J_M - 1)} \sum_{J=1}^{J_M-1} \sum_{I=1}^{I_M-1} A_{I,J} \exp -2\pi i \left[\frac{K(I-1)}{I_M - 1} + \frac{L(J-1)}{J_M - 1} \right] ,$$
$$(27.3)$$

where L is the wavenumber in the x direction and K is the wavenumber in the y direction. The energy $E(k)$ associated with different scales, represented by the two-dimensional wavenumber k, where $k^2 = K^2 + L^2$, is given by

$$E(k) = \sum_K \sum_L S_{K,L}^* S_{K,L} \delta(k^2 - K^2 - L^2) , \qquad (27.4)$$

where δ is the Kronecker delta, S^* is the complex conjugate of S, $K = 0, 1, \ldots, (I_M/2)$, and $L = -(J_M/2), \ldots, -1, 0, 1, \ldots, (J_M/2)$.

Figure 27.13 shows the two-dimensional spectra for the specific humidity q at $\sigma = 0.91$ in the 80 km grid size control simulation of the SESAME case (Exp. 3 in Anthes *et al.*, 1985). Figure 27.13 indicates a rapid decrease of energy associated with decreasing horizontal scale. It also indicates an

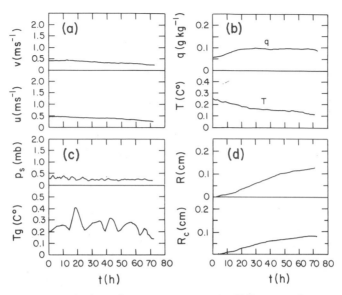

Figure 27.12. Temporal variation of root-mean-square differences between control and perturbed simulations in the OSCAR case: (a) u and v components, (b) mixing ratio and temperature, (c) ground temperature T_g and surface pressure p_s, and (d) total rainfall R and convective rainfall R_c.

increase of energy associated with the smaller scales at 72 h compared with the initial energy. Most of this increase occurs in the first 24 h, since the spectrum at 24 h is very similar to that at 72 h.

The spectrum of the differences of q in a pair of predictability simulations of the SESAME case is shown in Fig. 27.14. Four important observations can be made in Figs. 27.13 and 27.14. First, the spectrum of the differences of q is considerably flatter than the spectrum of q itself, indicating a relatively even contribution of all scales to the differences of q. Second, the error variance in the smaller scales grows by about 2 orders of magnitude during

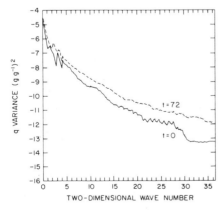

Figure 27.13. Variance of specific humidity q (ordinate in powers of ten) at $\sigma = 0.91$, for time t = 1, 24, and 72 h of Exp. 3 of Anthes *et al.* (1985).

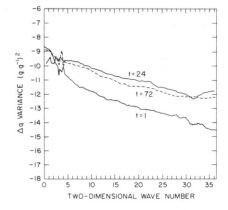

Figure 27.14. Variance of specific humidity (as in Fig. 27.13) for the difference of q in two predictability experiments of the SESAME case.

the first 24 h, but shows no growth during the next 48 h. Third, as indicated by the similar slopes of the spectra at 24 and 72 h, there is no apparent transfer of error energy from one scale to another. Finally, the error variance at all scales remains less than the variance associated with the true (control) solution (note the difference in magnitude of the ordinate).

27.4. Discussion

Predictability studies of large-scale motions indicate that predictability varies with latitude, season, hemisphere, and scale of motion, and from one case study to another. On the average, differences in the initial conditions of global numerical weather prediction models double in about 2–5 days, leading to estimates of the limits to atmospheric predictability of about 2 weeks. This growth of initial differences, or errors, is a consequence of the nonlinear transfer of energy among different scales of motion and the presence of atmospheric instabilities.

In contrast to the behavior of global models, a preliminary study of mesoscale predictability indicates that, in at least some cases, small errors or differences in the initial conditions do not grow over a 72 h period when the LBCs are the same. In two cases, nearly identical 72 h simulations are obtained from slightly different initial conditions. In both cases, similar mesoscale features develop in the simulations. These results indicate that, for these two cases, there is no significant transfer of initial error energy from the smallest scales to the larger scales over the 72 h period. In contrast, it appears that energy is transferred from the larger scale down to the mesoscale, as well as being generated locally by surface inhomogeneities.

An optimistic interpretation of the above results is that improved meso-α scale or regional-scale forecasts are possible in the 72 h time range, provided that large-scale conditions are forecast well and that realistic surface forcing and physical parameterizations are included in high-resolution models. However, there are several important qualifications.

First, it is not yet known how general these results are; it is likely that some cases, perhaps involving weaker large-scale forcing, will be more sensitive to uncertainties in the initial conditions. However, there is some additional evidence for the applicability of these results to a substantial number of synoptic situations. Astling (personal communication) used a high-resolution, primitive-equations model of the lowest 2.2 km of the atmosphere in a predictability study of nocturnal thunderstorms over the central part of the United States. After the wind field of a 6 h forecast (initialized at 1200 GMT 8 May 1979) was perturbed, the model simulation converged to the control (unperturbed) solution in about 8 h. Deaven (personal communication) indicated that the National Meteorological Center (NMC) operational limited-area fine-mesh model also showed little or no error growth over a 48 h period in a case study. Orlanski and Ross (personal communication) indicated that differences introduced into a mesoscale model showed no growth over a 24 h period, and that simulations were not sensitive to small variation in the magnitude of mesoscale gradients of temperature, wind, and moisture.

A second qualification to the result that limited-area models are insensitive to perturbations on the initial conditions concerns the type of initial error. Several studies have shown that numerical simulations are sensitive to systematic errors in representing real atmospheric structure such as jet stream intensity or low-level static stability. Anthes *et al.* (1983) showed that the development of an intense cyclone over the North Atlantic was sensitive to variations in the initial static stability, low-level moisture distribution, and strength of low-level circulation. The maximum differences in that case were larger than those of the predictability experiments summarized here, however (up to 6°C for temperature, 3 g kg^{-1} for moisture, and 8 m s^{-1} for winds). Perhaps more important, the differences that resulted in the more intense storm development were organized in a complementary, rather than random, way (stronger circulation, more low-level moisture, and weaker low-level static stability).

In summary, preliminary predictability studies with a limited-area model indicate that 72 h simulations are not sensitive to random uncertainties in the initial wind, temperature, and moisture fields. Although not emphasized here, other results indicate that simulations on this scale are more sensitive to errors in the large-scale flow and to the model physics, especially the parameterization of the planetary boundary layer and the release of latent heat in cumulus convection.

Considerable further study is needed to verify and extend these results to other synoptic situations and to models having alternative physics, methods of supplying lateral boundary conditions, and variations in the initial perturbations. Finally, these results are not likely to be applicable to meso-β- or meso-γ-scale models when a relatively large fraction of these models' domains may exhibit various instabilities (such as convective instability). Under such conditions, the evolution of the simulation is expected to be considerably more sensitive to the initial conditions.

REFERENCES

Anthes, R. A., 1977: A cumulus parameterization scheme utilizing a one-dimensional cloud model. *Mon. Wea. Rev.*, **105**, 270–286.

Anthes, R. A., 1984: Predictability of mesoscale meteorological phenomena. In *Predictability of Fluid Motions* (*La Jolla Institute–1983*). G. Holloway and B. J. West (Eds.), American Institute of Physics, New York, 247–270.

Anthes, R. A., and D. P. Baumhefner, 1984: A diagram depicting forecast skill and predictability. *Bull. Amer. Meteor. Soc.*, **65**, 701–703.

Anthes, R. A., and T. T. Warner, 1978: Development of hydrodynamic models suit-

able for air pollution and other mesometeorological studies. *Mon. Wea. Rev.*, **106**, 1045–1078.

Anthes, R. A., Y.-H. Kuo, S. G. Benjamin, and Y.-F. Li, 1982: The evolution of the mesoscale environment of severe local storms. Preliminary modeling results. *Mon. Wea. Rev.*, **110**, 1187–1213.

Anthes, R. A., Y.-H. Kuo, and J. R. Gyakum, 1983: Numerical simulations of a case of explosive marine cyclogenesis. *Mon. Wea. Rev.*, **111**, 1174–1188.

Anthes, R. A., Y.-H. Kuo, D. P. Baumhefner, R. M. Errico, and T. W. Bettge, 1985: Predictability of mesoscale atmospheric motions. *Advances in Geo-*

physics, **28**, Part B, 159–202.

Baumhefner, D. P., 1984: The relationship between present large-scale forecast skill and new estimates of predictability error growth. In *Predictability of Fluid Motions (La Jolla Institute–1983)*. G. Holloway and B. J. West (Eds.), American Institute of Physics, New York, 169–180.

Bjerknes, V., 1919: Wettervorhersage. *Meteor. Z.*, **36**, 68.

Carlson, T. N., R. A. Anthes, M. Schwartz, S. G. Benjamin, and D. G. Baldwin, 1980: Analysis and prediction of severe storms environment. . *Bull. Amer. Meteor. Soc.*, **61**, 1018–1032.

Carlson, T. N., S. G. Benjamin, and G. S. Forbes, 1983: Elevated mixed layers in the regional severe storm environment: Conceptual model and case studies. *Mon. Wea. Rev.*, **111**, 1453–1473.

Charney, J. G., R. G. Fleagle, V. E. Lally, H. Riehl, and D. Q. Wark, 1966: The feasiblility of a global observation and analysis experiment. *Bull. Amer. Meteor. Soc.*, **47**, 200–220.

Daley, R., 1981: Predictability experiments with a baroclinic model. *Atmos.-Ocean*, **19**, 77–89.

Errico, R. M., 1985: Spectra computed from a limited-area grid. *Mon. Wea. Rev.*, **113**, 1554–1562.

Jastrow, R., and M. Halem, 1970: Simulation studies related to GARP. *Bull. Amer. Meteor. Soc.*, **51**, 490–513.

Kasahara, A., 1972: Simulation experiments for meteorological observing systems for GARP. *Bull. Amer. Meteor. Soc.*, **53**, 252–264.

Kuo, H. L., 1965: On formation and intensification of tropical cyclones through latent heat release by cumulus convection. *J. Atmos. Sci.*, **22**, 40–63.

Kuo, H. L., 1974: Further studies of the parameterization of the influence of cumulus convection on large-scale flow. *J. Atmos. Sci.*, **31**, 1232–1240.

Leith, C. E., 1971: Atmospheric predictability and two-dimensional turbulence. *J. Atmos. Sci.*, **28**, 145–161.

Leith, C. E., and R. H. Kraichnan, 1972: Predictability of turbulent flows. *J. Atmos. Sci.*, **29**, 1041–1058.

Lilly, D. K., 1984: Some facets of the predictability problem for atmospheric mesoscales. In *Predictability of Fluid Motion (La Jolla Institute–1983)*. G. Holloway and B. J. West (Eds.), American Institute of Physics, New York, 287–294.

Lorenz, E. N., 1969a: The predictability of a flow which possesses many scales of motion. *Tellus*, **21**, 289–294.

Lorenz, E. N., 1969b: Atmospheric predictability as revealed naturally occurring analogues. *J. Atmos. Sci.*, **26**, 636–646.

Lorenz, E. N., 1969c: Three approaches to atmospheric predictability. *Bull. Amer. Meteor. Soc.*, **50**, 345–349.

Lorenz, E. N., 1984: Estimates of atmospheric predictability at medium range. In *Predictability of Fluid Motion (La Jolla Institute–1983)*. G. Holloway and B. J. West (Eds.), American Institute of Physics, New York, 133–139.

Moore, J. T., and H. E. Fuelberg, 1981: A synoptic analysis of the first AVE-SESAME '79 period. *Bull. Amer. Meteor. Soc.*, **62**, 1577–1590.

Robinson, G. D., 1967: Some current projects for global meteorological observation and experiment. *Quart. J. Roy. Meteor Soc.*, **93**, 409–418.

Robinson, G. D., 1971: The predictability of a dissipative flow. *Quart. J. Roy. Meteor. Soc.*, **97**, 300–312.

Schumann, T. E. W., 1950: The fundamentals of weather forecasting. *Weather*, **5**, 220–224.

Shukla, J., 1984: Predictability of a large atmospheric model. In *Predictability of Fluid Motions (La Jolla Institute–1983)*. G. Holloway and B. J. West (Eds.), American Institute of Physics, New York, 449–456.

Smagorinsky, J., 1969: Problems and promises of deterministic extended range forecasting. *Bull. Amer. Meteor. Soc.*, **50**, 286–311.

Tennekes, H., 1978: Turbulent flow in two and three dimensions. *Bull. Amer. Meteor. Soc.*, **59**, 22–28.

Thompson, P. D., 1957: Uncertainty of initial state as a factor in the predictability of large-scale atmospheric flow patterns. *Tellus*, **9**, 275–295.

Thompson, P. D., 1984: A review of the predictability problem. In *Predictability of Fluid Motions (La Jolla Institute–1983)*. G. Holloway and B. J. West

(Eds.), American Institute of Physics, New York, 1–10.

Williamson, D. L., 1973: The effect of forecast error accumulation on four-dimensional data assimilation. *J. Atmos. Sci.*, **30**, 537–543.

Williamson, D. L., and P. N. Swarztrauber, 1984: A numerical weather prediction model—Computational aspects on the CRAY-1. *Proc. IEEE*, **72**, 56–67.

Zhang, D., and R. A. Anthes, 1982: A high-resolution model of the planetary boundary layer sensitivity tests and comparisons with SESAME–79 data. *J. Appl. Meteor.*, **21**, 1594–1609.

CHAPTER 28

Nowcasting Mesoscale Phenomena

John McGinley

28.1. Introduction

Short-range forecasting or nowcasting is based upon the ability of the forecaster to assimilate great quantities of weather data, conceptualize a model of the environment, and extrapolate this forward in time. Frequently this is done just to describe existing conditions, and thus has the term "nowcast" originated. Since the time scales are essentially the same for short-range forecasting and nowcasting (the same processes are used for both), the terms are used here interchangeably.

Mesoscale phenomena are systems of length (L) about 100 km, velocity (U) about 10 m s^{-1} and time scales $(t = L/U)$ of about 10^4 s (3 h). The time is long compared with the normal temporal confines of nowcasting (0–1 h). One might assume then that the nowcaster need not forecast development of mesoscale (or synoptic-scale) systems but need only react to phenomena already present. This must be qualified somewhat. When a mesosystem undergoes a scalar metamorphosis producing responses at smaller scales (e.g., as a jet maximum forces convection), the resulting individual or clustered convective cells or gravitational waves ($L = 10$ km, $U = 10$ m/s, $t = L/U =20$ min) may be liberally encompassed in the lower end of the mesoscale, and come within the nowcast office's area of responsibility. Thus, nowcasting is more than simple "thumb and pencil" extrapolation. It involves close monitoring of all the latest data; predictive skills that recognize and foresee the most subtle changes in the atmosphere; conceptual models that encompass the structure and evolution of the phenomenon; and a communicative system that can reach the user rapidly.

This chapter is concerned with the methodologies of nowcasting rather than the hardware for nowcasting. There is no shortage of articles and papers on system designs, equipment networks, etc., in the nowcasting literature, but less emphasis has been placed upon the techniques that are usable in specific situations, i.e., the "rules of thumb," so desperately needed when quick action is required. Furthermore, the most sophisticated four-dimensional display of high-resolution data is valueless if the forecaster or nowcaster has no intuitive feeling for the small-scale structure of the atmosphere.

However, nowcasting is a continuous learning experience, and skill in nowcasting can be developed in three ways:

- Frequency of analysis—monitoring data hourly.

- Compositing—summarizing features from all sources of data on one presentation.

- Post-nowcast analysis, to assess accuracy of nowcast in terms of occurrence or non-occurrence of predicted weather events.

The nowcaster can develop conceptual models that will improve with time. The availability of on-site computers is eliminating the time-absorbing requirement of hand analysis. However, the quality of nowcasting is based on the nowcaster's recognitive and intuitive skills, and speed.

28.2. Tools for Nowcasting

Fortunately, the major revolution in observational meteorology during the last 25 years has been in the capability to observe events at the mesoscale. In the 1950s and 1960s, only conventional radar and hourly surface observations (and healthy imaginations) were available. The addition of satellite imagery, both visible and infrared, has added enormously to information available at the mesoscale. With this major improvement in data availability one might have expected a major increase in the capability to provide accurate short-term forecasts. Some improvement has been noted, but there are still major problems in forecasting mesoscale events. It might be said that the ability to acquire information has outstripped the ability to assimilate it at the local level. The major problem in observational nowcasting is not in the ability to interpret presentations of each source of data individually (although this certainly is a necessary skill), but is one of assimilation—that is, the combining of all these data sets into one clear coherent picture of a phenomenon, and most important of all, an ability to fit this picture into one of a whole ensemble of conceptual models. Computers are, or soon will be, doing many of the assimilating tasks, but until significant advancement occurs in mesoscale and cloud system models, the burden will fall on the forecaster's conceptual and predictive skills.

28.2.1. Conventional Radar

Weather radar has been generally available since the 1950s. Much has been written on techniques for interpreting radar data. One can determine the intensity of precipitation by its reflectivity in a four-dimensional frame. Recent technology has markedly improved the display of precipitating systems, including time integration for total rainfall determination.

Typical weather radars are fully effective to 200 km, and partially so to 400 km. Thus, a moving, persistent mesosystem may be observed from one location for ~6 h. However, mesosystems rarely propagate in a steady manner without forcing a response at smaller scales. The nowcaster must recognize changes in shape, orientation, and speed of the system, and also quickly identify new echoes developing in potentially dangerous positions relative to the existing mesoscale environment. The radar occasionally gives indications that a mesosystem is in the local area even though precipitation has not begun at the ground. These results from density discontinuities (fine lines) or virga from high clouds.

Particular techniques for using conventional radar to observe specific weather phenomena are discussed in Sec. 28.3.

The PPI Presentation

The plan position indicator (PPI) display respresents a conic slice through the environment. It presents an overall view of the precipitation systems within range and their spatial relationships and horizontal structure. Its disadvantages are beam attenuation by existing echoes, and poor resolution of targets at long distance. The nowcaster should make abundant use of overlays to know precisely echo positions relative to cities, roads, lakes, etc., and note the evolution of echo shape, discontinuities, movement, and intensity. Particular attention should be given to reflectivity gradients, which are central to a number of convective nowcasting methods. Holes in echoes (bounded weak echo regions), echo texture, notches, bowing, hooks, waves, and banding all show that significant dynamic processes are under way. Radar reflectivity patterns are characteristically transient. The key to forecasting is echo persistence; an echo feature that lasts longer than 15 min is usually significant. All PPI information should be summarized on the nowcast composite chart.

The RHI Display

The range height indicator display allows observation of individual elements of the precipitation system, to determine height, vertical shape, weak echo regions, bright bands related to melting, overhangs, and other features. As with the PPI, any unusual echo configuration should be noted. Here, too, persistence is the best test for significance of a feature.

28.2.2. Doppler Radar

Though not now available to all nowcasters, Doppler radar promises to be one of the most useful sources of nowcasting information. Spatial and temporal characteristics match those of conventional radar but with vastly extended capability for determining storm and precipitation structure. Well-documented studies on the use of Doppler radar to delineate storm structure are available (e.g., Lemon *et al.*, 1978). In addition to having all the capabilities of conventional radar, Doppler radar can indicate radial velocities (relative to the radar) of any target and hence give some indication of windflow in the target volume. These may include wind shears, mesocyclones,

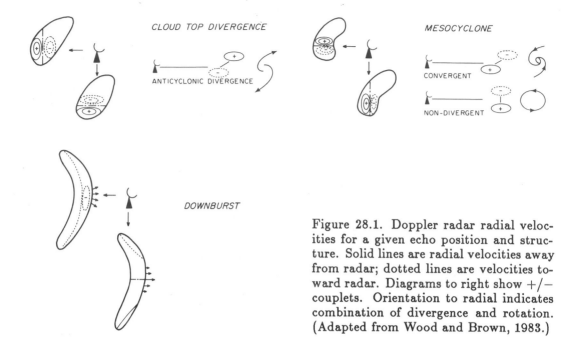

Figure 28.1. Doppler radar radial velocities for a given echo position and structure. Solid lines are radial velocities away
from radar; dotted lines are velocities toward radar. Diagrams to right show $+/-$
couplets. Orientation to radial indicates
combination of divergence and rotation.
(Adapted from Wood and Brown, 1983.)

downbursts, cloud top divergences, and occasionally a tornado itself. Figure 28.1 shows examples of Doppler signatures for a few weather systems
(Wood and Brown, 1983).

Of great interest to the nowcaster is the ability of Doppler radar to sample
the prestorm environment (Wilson and Wilk, 1982). Low-level convergent
boundaries may be located from movement of nonhydrometeoric targets and
the detection of density discontinuities. Equally valuable is Doppler radar's
scanning at elevation to sample mean vertical wind shear profiles and determine the depth of the moist layer in the radar environment. This allows the
nowcaster to monitor these important phenomena, which often determine the
likelihood of convective storm development and severe weather (Fig. 28.2).

28.2.3. Satellite Imagery

Used in conjunction with other sources of information, satellite imagery is
the mainstay of the nowcaster's ensemble of tools. With routine observations
taken every 30 min, with 5 min data available in special circumstances, and
with resolution that ranges from 1 to 6 km, sub-mesoscale elements can be
monitored over wide geographic regions. One difficulty with visible imagery
is that weather systems must contain cloud or dust to make their presence
known. Infrared imagery suffers from poorer resolution and ambiguity in
location of low clouds, but is capable of providing information on ground
surface temperatures, moist layers, cirrus shreds (associated with jet streaks),
and cloud heights. Imagery from water vapor channels can assist in further
definition of jet-streak and vorticity patterns by clearly delineating moist
and dry regions in cloud-free areas (Anthony and Wade, 1983).

In using satellite imagery for nowcasting, some basic guidelines apply:

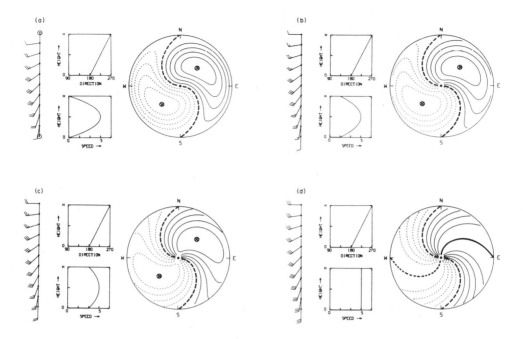

Figure 28.2. Nonuniform wind speed shear with uniform directional shear. Doppler velocity contours (as seen on circular radar display) have solid lines for flow away from radar, short dashes for flow toward radar and heavy long dashes for zero Doppler velocity. Circled x's or heavier solid and short dashed lines indicate location of extreme Doppler velocity values within the display. (From Wood and Brown, 1983; see also Fig. 6.10, this volume.)

- Each cloud band must be monitored and classified according to type and height.

- Cloud patterns should be fitted to all available data such as surface and upper-air patterns.

- A cloud band cannot be assumed to have mesoscale structure without data to support the assumption.

Difficulties can be avoided if satellite data are routinely overlaid on the composite chart (usually an hourly surface chart), and routinely compared with prognoses of wind, vorticity, and temperature fields. Figure 28.3 shows an application of this technique in severe weather forecasting (Miller and McGinley, 1978). The nowcaster marks partial or complete comma structures, cloud lines, low cloud areas, and/or warm, moist regions where fog may form (Gurka, 1978). Sharp upstream boundaries are often indicative of thermal or vertical motion discontinuities. Sharp boundaries may be related to zones of wind deformation. Again, time continuity of features is a critical factor in analysis of satellite imagery. Video loops are a great aid in discerning the origin of cloud and moisture patterns.

Use of satellite imagery in nowcasting thunderstorms, heavy rainfall, and lake snow is discussed in Secs. 28.3, 28.4, and 28.5.

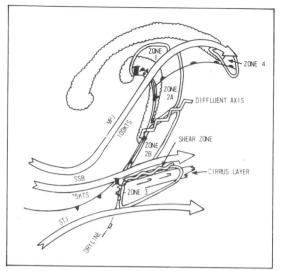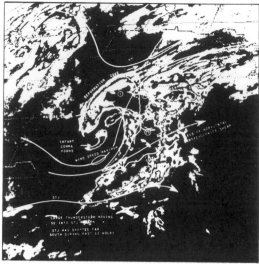

Figure 28.3. (Left) Combination of satellite features with upper-level winds and fronts that delineate severe weather zones (see Miller and McGinley, 1978). (Right) Method applied to actual satellite photo.

28.2.4. *Vertical Soundings*

The vertical sounding is a source of data often neglected by the nowcaster. Owing to the time periods in which the atmosphere is sampled (0400–0700 LST, 1600–1900 LST in the United States), forecasters are reluctant to spend time analyzing vertical profiles that will probably be totally unrepresentative of the atmosphere by the time convection begins. Thus, rather crude indices have developed, to give a summary indication of the potential instability for the day, and often little regard is given to the factors that may greatly alter these indices.

With a little effort (easily amenable to automation) the nowcaster can estimate what a current sounding may look like in the lower 1.5 km at any later time. Frequent monitoring of surface temperatures and moisture, estimates of change in vertical temperatures (from model output), and cloud bases and types can give some idea of what processes are at work. In dynamically weak situations where surface heating can play an important role, it is useful to estimate the amount of heat that will be added to the air column from the ground surface. Energy transfer from the ground to the air is a highly complex process, but there is general agreement from energy budgeting considerations (Haltiner and Martin, 1957; Hess, 1959) that about 14–23% of the incident energy goes to sensible heating of the boundary layer, subject to surface evaporation. As an example, Table 28.1 shows incident solar radiation for Tinker AFB, Okla., for each month. Column 2 shows the portion of this energy that, under ideal conditions, may warm the lowest layer during daylight hours, (assumed to be about 18%, this includes radiative cooling, which continues on a 24 h basis at a rate about $\frac{1}{24}$ of the input

Table 28.1. Ideal clear-sky estimates

Month	Incident energy (J cm^{-2} day^{-1})	Ground heating (J cm^{-2} day^{-1})	"Boxes" (1 box=7 J kg^{-1})
Jan	925	167	23.8
Feb	1238	223	31.8
Mar	1628	293	41.8
Apr	1897	341	48.7
May	2152	387	55.3
Jun	2341	421	60.1
Jul	2286	411	58.7
Aug	2110	380	54.3
Sep	1645	296	42.2
Oct	1381	249	35.6
Nov	1017	183	26.1
Dec	837	150	21.4

energy per hour). The input energy can be converted to area on a thermodynamic diagram. An air column of one square centimeter weighs about a kilogram; thus, J cm^{-2} is approximately J kg^{-1}. The input energy will equal the area increase in the lower part of the atmospheric sounding, assuming an adiabatic lapse rate. For a large Skew-T, Log-P diagram, 1 cm^2 = 28 J kg^{-1}. A short-cut method uses "boxes" formed by the intersection of dry adiabats (at 2°C intervals) and isotherms (at 1°C intervals). In this convention, 1 box (in the lower portion of the chart) equals 7 J kg^{-1}. This box method is applicable to many of the commonly used thermodynamic charts.

The nowcaster's major difficulty is to estimate moderating factors for the ideal, clear-sky estimates in Table 28.1. Factors influencing this heat input are thermal advection, haze and moisture, cloud cover, surface wetness, and snow or ice cover. Estimates of modifying factors are given in Table 28.2. These rough estimates work fairly well for the southern plains. The factors are multiplicative and are useful where strong thermal advection is not a factor. In stagnant situations the energy input can be validated and adjusted by comparing the 1200 GMT sounding with the 0000 GMT sounding. The total number of "boxes" can be found and compared with the expected en-

Table 28.2. Modifying factors for PBL heating

Cloud cover:	
Overcast	0.5
Broken	0.7
Scant	0.9
Haze or moist air	0.8
Surface moisture/water	0.7
Ice/snow cover	0.2
Combinations:	
Ice/snow, overcast	0.1
Surface moisture, haze, scant	0.6

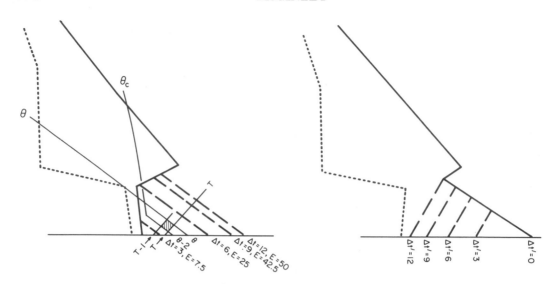

Figure 28.4. (Left) Idealized energy input by surface heating (one energy box). Total heat realized is 50 boxes ($\sim$350 J kg^{-1}); Δt is time from sunrise. E is energy realized for a given Δt (in boxes). (Right) Hypothetical cooling, assuming isothermal lapse rate and energy loss of about 4 boxes per hour. Lapse rate may be between isothermal and moist adiabatic; Δt is time from sunset.

ergy input (Fig. 28.4). For nighttime, the area can be reduced by about $\frac{1}{24}$ of the number of boxes per hour, but the shape of the vertical profile may be difficult to determine. An assumed isothermal profile in the cooled layer often adequately represents the lapse rate. The cumulative number of boxes by hour can be roughly estimated by the formula

$$E(t) = E_T \frac{\left[1 - \cos \pi \left(\frac{t - t_r}{T_s}\right)\right]}{2} \, ,$$

where t is the current time, E is the input energy, E_T is the total input energy, t_r is sunrise time, and T_s is total sun hours.

This method, although crude, offers a better starting position for estimating a maximum temperature than simply assuming a dry adiabatic lapse rate from the 850 mb morning sounding. This technique is useful when dynamical forcing is weak and the possibility of severe convection depends strongly on surface heating. The nowcaster must be able to estimate the current condition of the lowest 1.5 km, including the evolution of the capping layers that are normally present in the prestorm environment. The amount of vertical velocity needed to surmount a 1 box cap or stable layer (negative energy) of 7 J kg^{-1} is nearly 4 m s^{-1}. This is far beyond the range of mesoscale lifting (0.1–0.2 m s^{-1}) and would resist most rising thermals 1–2 m s^{-1}). Thus the role of mesoscale phenomena is not to punch convection through a capping layer, but rather to erode the cap over a comparatively large area through slow, steady lift.

Cap strength can be estimated by computing vertical lift necessary to reduce the cap to a negative energy that can be penetrated by rising ther-

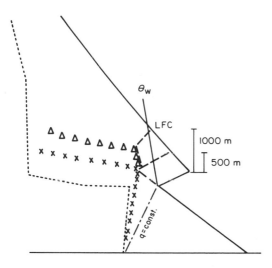

Figure 28.5. Lifting of capping layer, assuming heat input for 9 h. A lift of 500 m allows thermal and convective cloud to reach LFC, but entrainment may destroy cell. A lift of 1000 m eliminates cap totally. Assuming a 750 m lift, mesosystem may accomplish destabilization in 1.5 h.

mals. After estimating a surface energy input for a given forecast time, the nowcaster lifts the stable layer until the cap is less than 7 J kg^{-1} with respect to the convective plumes rising from the surface. This distance can become a nowcasting parameter (Fig. 28.5). If the sounding requires 1 km of lift (Type 1 sounding, Miller, 1972) a typical mesoscale system could destabilize the cap in about 3 h. A synoptic-scale disturbance would do the same thing over a much longer time, 12–24 h. This, of course, does not take into account differential temperature advection, which can further modify the stability and must probably be estimated from model output. It is important to understand that fast-moving systems may not provide enough local lift to weaken a cap.

28.2.5. Hourly Observations

Hourly surface observations, from the network of Federal Aviation Administration, National Weather Service, and other weather stations (with average spacing of 160 km) still are the only multi-parameter quantitative source of information at the mesoscale. Surface observations not only sample the lowest layer, but can provide a wealth of information on the vertical structure of the atmosphere. The current observation and time changes in pressure, temperature, dewpoint, wind direction and speed, cloud base, and cloud type may give hints about the forcing mechanisms in the local environment.

Temperatures and temperature changes should be monitored hourly. Observations from stations that are heating rapidly, cooling, or have stopped heating can hint at the vertical structure of the lower atmosphere. It is usual to assume a dry adiabatic lapse rate from the surface temperature. However, most stations are superadiabatic in the lowest 50 m on days with strong heating, and the temperature profile should reflect about 1–2°C decrease in potential temperature in this layer. In areas of high soil moisture or over water, a lapse rate may be closer to moist adiabatic. When the rate of temperature increase slows, the heat input is probably being mixed through a deep layer. Likewise, rapid temperature rises imply shallow or capped layers. Trends in cooling can help locate fronts, advecting low clouds or moisture, and breezes.

Dewpoint and dewpoint trend data, particularly when used with temperature data, can also give hints about the vertical moisture profile, onset of low cloud or fog layers, and frontal, dryline, or outflow boundary passage. The dewpoint change is often more pronounced than temperature change and therefore more useful in locating weak boundaries. In deducing the structure of the moist layer, the surface mixing ratio is normally assumed constant through the mixed layer. However, with wet ground surfaces or steady moist advection, or in high vegetation areas, a significant moist lapse rate can exist in the lowest 500 m. A good assumption for all cases is about 1 g kg^{-1} lapse per kilometer through the "mixed layer". A rapid drop in an hourly dewpoint observation often indicates that a stable layer has been eliminated by heating. Typically, this will be accompanied by slower rises in temperature. Monitoring the timing of these occurrences in an area can provide clues to the horizontal extent and strength of capping layers.

Visibility and changes in visibility can reveal the existence and destruction of capping layers, onset of fog, and location of moist air. A drop in visibility from 15 to 7 miles could mean moisture is increasing, the cap is intensifying, etc.

The quality of ceiling information varies markedly from site to site. The value of accurate ceiling or cloud base information can reveal whether clouds are being formed from slow lift or from convective processes. If accurate, the cloud base can be compared with a local sounding and used to update it with respect to the convective condensation level (CCL) and temperature and moisture profiles. Figure 28.6 summarizes the use of surface temperature, dewpoint, and cloud observations in deducing vertical temperature profiles.

Sometimes added to surface observations are reports from pilots. This information is particularly useful when bases and tops of clouds, tops of haze layers (often the base of the capping layer), and occasionally temperatures at altitude are given. Separate pilot reports should be monitored for reports of stratus, offshore fog banks, and other visual information. Even a wind observation from a large aircraft can be of some value. These reports should be plotted on a composite chart.

Monitoring pressure and pressure changes is an important part of short-range forecasting. The change in pressure reflects events occurring over a wide range of scales. The equation governing pressure changes gives little insight into causes at mesoscale and below. However, changes not related to the diurnal cycle are indicative of adjustments being made to accommodate dynamic events, i.e., a response to system forcing.

When forecasters use surface observations, the winds are usually of most concern. Analyses of winds and wind trends can lead to diagnosis of convergent/divergent areas and time tendencies, location of outflow boundaries, breezes, low-level jet maxima, and movement of upper-level systems. The wind vector change is the vector sum of the Coriolis acceleration of the ageostrophic wind, the inertial acceleration, and friction or Reynolds stress,

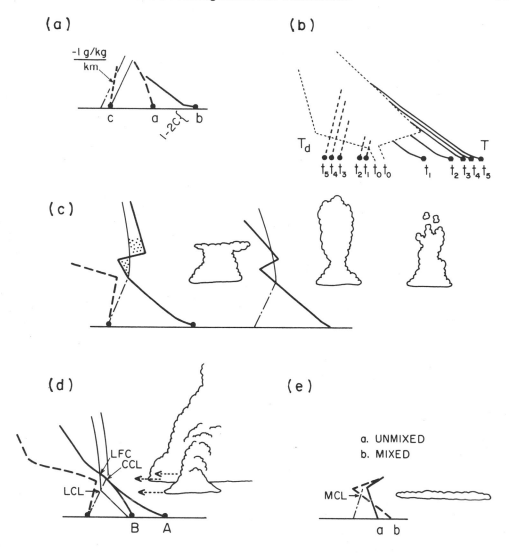

Figure 28.6. (a) Deriving lapse rates from surface observations: a is assumed when surface conditions are moist or if sky conditions are overcast; b illustrates typical super-adiabatic lapse rate; c shows moisture lapse rate. (b) Time evolution of a sounding from surface observations. Moisture decrease at t_3 and slowing of heating rate indicates cap may have been eliminated. (c) Sounding estimation from cloud reports: Base is at CCL for convective cloud, which provides information on moist layer if surface temperature is known. Pinching of cloud indicates penetration of capping layer; evaporating cloud, dry air aloft. (d) Variation in cloud base indicates combined lifting and convective generation. Sounding A outside storm supports cloud base at CCL. Stable sounding B, under storm, supports a lifted cloud base at the LCL, which turns convective at the LFC. (e) Mixing of low layers will generate stratocumulus layers (usually behind fronts).

including acceleration from turbulent transport of momentum. Although often a very noisy field, a plot of wind vector changes can reveal areas where adjustments are taking place. When combined with 1–3-hourly changes of all surface parameters (Sasaki and Tegtmeier, 1974), the wind change chart

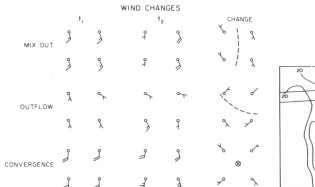

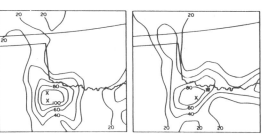

Figure 28.7. Examples of wind changes over 1–3 h. Subtle changes (easily missed) assist in locating a variety of potentially weather-producing features. Observations at t_1 and t_2 give change at hour t_2. Dashed lines indicate hidden boundaries. Circled x indicates convergence center.

Figure 28.8. Horizontal moisture convergence at (a) 2000 GMT and (b) 2200 GMT on 10 April 1979. Units are g kg^{-1} s^{-1} × 10^{-5}. The X in (b) represents a developing tornadic thunderstorm that struck Wichita Falls (W) at 0004 GMT. (From Bothwell and Crawford, 1983.)

becomes a powerful nowcasting tool. Siebers and Schaefer (1983) showed the value of such a product. Convection will frequently develop where there is convergence in the vector change field, implying a negative divergence tendency. Fig. 28.7 gives some examples of wind changes.

Not-easily-diagnosed fields like moisture convergence can be interpreted with objective analysis schemes that run on small, inexpensive computers (Bothwell and Crawford, 1983) (Fig. 28.8). Other products such as the geostrophic wind analysis or Sangster chart, can be compared with actual winds to find regions with intense ageostrophic components (Doswell, 1982).

Surface observations are an important part of nowcasting but must be fully integrated and assimilated with other products to maximize potential. The surface chart should normally serve as the base map for the nowcasting composite chart. It should be kept in mind that the surface observation can give information on the structure of the lowest layers. A 1–3 h change chart of all parameters should be produced, if possible. Significant observations are typically those that go against regional trends, and the chart enhances the visual impact of changes. To lessen the effects of spurious observations, patterns of hourly change should be considered valid only if they persist for 2 or more hours.

28.3. Weather Phenomena and Short-Range Forecasting Techniques

Since most forecasters today trust numerical guidance most of the time for 12–48 h forecasts, nowcasting has become the last frontier of intuitive meteorology. Mesoscale models are being developed for probable future integration into the numerical forecasting system but are not yet available operationally. Even with the vast improvement in numerical forecasting capability, a nowcasting function will be needed for many years to come.

Weather systems and their environments can be diagnosed with routine observations. In the presentation of methodologies and rules of thumb, the emphasis is not on dynamical justification and rigorous proof, but on operational utility and practicality.

28.3.1. *General Convective Weather: Thunderstorms*

For convection to occur, three ingredients are essential: a moisture source, convective instability, and an initiating mechanism. Typically, thunderstorms develop during the afternoon (when convective instability is most likely). A simple conceptual model for a preconvective environment might include

- A moist layer near the ground surface (perhaps at 700 mb in mountain regions).

- Convective instability (with a large equivalent potential temperature at low levels, relative to a minimum aloft usually at middle levels).

- A stable layer or cap acting as a restraining influence (which might be nothing more than a radiation inversion in the surface layer).

- An initiator.

Moist layers are often accompanied by low clouds and high surface dewpoints, and are usually, but not exclusively, generated over a water surface such as the Gulf of Mexico. Moist layers can also originate over wet, rain-soaked areas, occasionally pass over the Rockies from the Pacific, or even wrap around a cyclone and advect in from the north. The nowcaster may determine the boundaries of such moist layers by watching low-cloud areas and warm nighttime IR images related to the longwave-absorbing character of water vapor.

Determining relevant stability and evolution is still one of the most difficult problems in nowcasting. Because of relative motion in adjoining atmospheric layers and system-generated upward and downward motions, stability is constantly changing. Rawinsonde observations and derived stabilities are hours old by the time convection develops. Good nowcasting requires that all possible methods be exploited to determine the present and forecast stability. Darkow and Tansey (1982) discussed the value of using surface observations with upper-level temperature values to monitor stability. Use of the lifted parcel temperature (LPT), the temperature of a parcel lifted from the ground to 500 mb, is another way to resolve the hourly monitoring stability (Hales and Doswell, 1982). As discussed in Sec. 28.2, a great many sources of information are available to the nowcaster in attempting to decipher the structure of the lowest layer, including pilot reports, cloud bases and tops, changes in surface data, and radar.

The concept of capping layers, or lids, is a very useful one, which has been employed in thunderstorm forecasting for many years. Frontal inversions, trade wind inversions, subtropical high-subsidence inversions, warm, dry high-plateau air masses, and even morning inversions can serve as capping layers. The origin of the pre-thunderstorm capping layers of the Great Plains was thought to be subsidence, but Carlson and Farrel (1982) verified that these layers originate with horizontal advection of air from the high-plateau regions of Mexico or the southwestern United States. In the Plains during spring and summer the sounding profile above the low-level moist layer indicates a layer that is nearly adiabatic and quite dry, with nearly constant mixing ratio, similar to the lowest layers of an afternoon sounding from a plateau rawinsonde station. Carlson and Farrel developed the lid strength index (LSI), a stability parameter patterned after the lifted index, which takes into account the strength of the lid. Location of the limits of the lid will allow the nowcaster to narrow the region of concern. Often conventional stability indices will indicate very large instability in regions that are seriously capped. These indices are not necessarily useless, because they do give the convective potential, given sufficient lift. Reducing the convective potential because there is a capping layer is unwise because the total energy will still be realized if the cap is broken. A wiser approach is to treat convective potential and cap strength separately. One good tool for determining the spatial extent of capping layers is to monitor rapid rises in surface temperatures. Frequently the hottest temperatures and the largest increases indicate where the capping layer is strongest. One technique is to look at the distribution of 2–3 h temperature rises and locate sharp horizontal gradients in the change field. The boundaries (particularly the western boundaries) delineate the cap edge and reveal areas that might be ready for explosive thunderstorm development.

The initiating mechanism is often difficult to identify. Clear-cut boundaries such as fronts and convergence lines often produce convection, but more typically storms develop in areas where only subtle features can be found. The initiator actually consists of two components acting at different scales. The first is a destabilizing mechanism operating at mesoscale or even synoptic scale over the appropriate time (3–12 h). The second is the trigger, usually a series of convectively generated thermals originating in the superadiabatic layer near the surface, having vertical motions on the order of 1 m s^{-1}, and possessing enough energy to break through the remnant capping layer which, after lifting, is quite weak. As discussed, even a minimal capping layer is impervious to motion less than 1 m s^{-1}, an order of magnitude larger than that produced by most mesoscale systems. Cap strength can be thought of as the amount of lift necessary to allow penetration of the cap by rising thermals. When a strong mesoscale system interacts with a capping layer in conjunction with intense afternoon heating, development can be explosive. This makes one wonder if it matters whether the mesoscale system is labeled the "destabilizer" or the "trigger." For strong cases it does not, but for weak cases the difference may be important.

The nowcaster must deduce the presence of the destabilizing agent. Some of the tools discussed in Sec. 28.2 can provide hints that serve as precursors of thunderstorm development or nondevelopment. The first sign of a mesoscale destabilizing mechanism is clumped cumulus. Mesoscale lifting should pro-

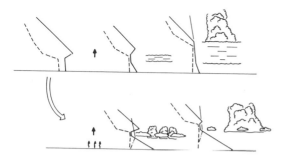 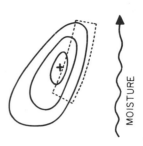

Figure 28.9. (Top) Destabilization of sounding by pure lifting. Layered clouds form in lifted layer and increase in depth until cap is totally erased and convection is initiated. (Bottom) Destabilization by lifting and heating. Layered clouds are broken up by thermals. Cap reduction finally allows thermals to break through and form thunderstorms.

Figure 28.10. Expected area of thunderstorm development (dotted line) 1–3 h after moist convergence pattern (solid line). Area is on moist or downwind side of maximum.

duce layered clouds, and would if strong convective mixing were not also present. Thus the destabilizing region is one where the cumulus are enhanced, preferably for 2 h or more. Doppler radar frequently will indicate a deepening moist layer. Moderate upward motions associated with a typical mesoscale system can destroy an intense Great Plains stable layer (Miller, 1972) in about 2–3 h, if the lift remains over the same area. Lift generated by synoptic-scale warm advection (Maddox and Doswell, 1982) may take 12 h or more (Fig. 28.9).

Automated products such as moisture convergence can reveal the presence of possible zones of destabilization. Experience has shown that the favored area for development is usually on the moist side, or downwind gradient of the moisture convergence maximum (Fig. 28.10). Conventional radar usually offers little help because other indicators will give notice 1–3 h before the first echo. However, radar indications can serve as a confirmation that the other nowcast criteria are valid. The satellite is the best tool for watching for evidence of destabilization. Areas of clumped cumulus, particularly those that encroach into drier upstream air, are likely thunderstorm development regions. Old outflow boundaries, fronts, and drylines offer potential, and clouds along these boundaries should be watched (Fig. 28.11). In general, the stronger the destabilizing agent, the more organized the thunderstorm system. In weaker situations where surface heating has a dominant role, thunderstorm generation will be more random and more difficult to predict. Near-erosion of the cap in very limited areas will frequently lead to development of small narrow convective towers. If additional cells do not generate within 1–2 h, probably no mesoscale forcing is present. For every region of forced mesoscale lift there is an adjoining region of mesoscale subsidence. Here, very strong capping may result. This may account for the waves seen in the cloud patterns on drylines, sharp demarcation of cloud boundaries on fronts, and the often isolated nature of supercell convection.

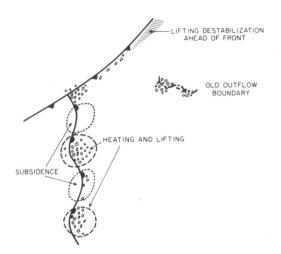

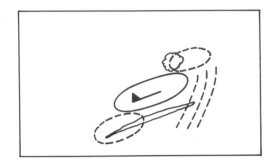

Figure 28.11.   Areas of destabilization
on front, dryline, and outflow boundary.
Clumped cumulus is first indication that
lifting may be occurring.   Areas that
maintain continuity for 2–3 h should be
considered suspect.

Figure 28.12.   Areas of destabilization (dashed) relative to a jet maximum
(solid). On the right-front quadrant, lift
may be due to gravitational modes. Often
cirrus streaks appear south of wind maximum and middle clouds on left-front side,
as illustrated.

Thunderstorms are largely organized by the pattern of destabilization.
One destabilizing mechanism that may be tracked is the wind speed maximum at the jet level. This feature typically produces rapid upward motions on the left-front and right-rear quadrants of the wind speed maximum,
and generates gravity waves in the generally subsiding air on the front-right
quadrant (Golus and Koch, 1983) (Fig. 28.12). In situations where the jet
maximum is rapidly propagating there may be insufficient time for lifting to
break the cap. Gravity waves, thought to be possible triggers of convection,
are difficult to trace in conventional data networks and may give only slight
hints in cloud patterns. The nowcaster may recognize situations that are
prone to gravity wave generation (jet maxima and frontal zones), and act
accordingly. Gravity waves move at 10–60 m s^{-1}, have amplitudes of 0.3
to 2 mb (Atkinson, 1981), and propagate in stably stratified atmospheres,
typically forming in regions where the Richardson number (static stability
divided by shear squared) is less than 0.25. Occasionally, vertical motions
associated with gravity waves can be enough to break weak caps.

Frontal zones undergoing frontogenetic forcing may induce destabilizing
lift in the prefrontal region by direct circulations (Bluestein *et al.*, 1980)
(Fig. 28.13). The most active part of the front is where deformation frontogenesis is most intense, often a region where winds appear to be weak
and cyclone centers are far off. Upward motions of this type may be highly
localized.

The chaotic patterns seen during airmass thunderstorm situations are
indicative of the dominance of surface-generated convection and weak cap-

ping. A large field of uniform convective clouds that forms very early in the afternoon is a sure sign of minimal capping, no mesoscale destabilizing mechanism, and little potential for strong storms. As the environment becomes more stable, development becomes more dependent on destabilization zones and hence, storms appear more organized. Gathering these storms into a limited region often results in merging outflows and organized inflows, and the system may become a single entity (an MCC; Maddox, 1980). The convective system takes on a mesoscale structure and may be self sustaining. Capping, in a sense, acts as a filter on weak convection, allowing low-level energy to build up in the lowest layers through heating, thus creating an increased potential for strong convection. The problem for the nowcaster is to determine the conditions that suppress convection just enough to allow energy to build up, but not enough to prevent thunderstorm development. The 10°C line on the 700 mb chart is often used to indicate the region that has just enough capping to enhance strong storms in the absence of strong forcing. If the 700 mb layer is warmer than this, almost all storms will be suppressed; if it is colder, there will probably be weak, random convection. The 10°C line often delineates the edge of the lid.

After a day of heating and no thunderstorm development, the boundary layer is still at a high level of convective potential. Decoupling of the mean wind flow from the surface layer, combined with differential radiational cooling, results in formation of the low-level jet. Like all jets it has the capacity to generate lifting and gravity waves. When the lowest layers are stabilizng and the layer above the cap is nearly adiabatic, the stage is set to allow gravity waves to propagate horizontally without loss of energy attributable to vertical phase speeds (the waves are said to be ducted). Thus the moist layer that could not be lifted sufficiently during the afternoon, may be destabilized by the vertical motions associated with the jet and/or gravity waves. Typically, development will occur on the downstream side of the low-level wind speed maximum, both in the left-front and right-front sectors. In addition, storms may develop on an isolated basis to the right of the anticyclonic shear zone, along the length of the jet (Fig. 28.14). A likely location is the intersection of the jet and frontal, or jet and outflow boundary. Pressure fall/rise patterns near the jet nose are good indicators. When the jet cannot be located precisely, areas of low stratus clouds advecting northward (typically the jet is on the west edge of the cloud band) may be indicators. The generation of thunderstorms in such a situation usually occurs 3–6 h after sunset. Once the jet establishes itself it is a very good source of moist energy for a developing storm system. The persistence of the jet will often support convective activity until morning.

In summary, to make short-range forecasts of thunderstorms the nowcaster must do the following:

- Accurately analyze the morning soundings in the region of responsibility.
- Determine the stabilities and amount of lift necessary to break the cap (if any).
- Determine the time of the first convective clouds.

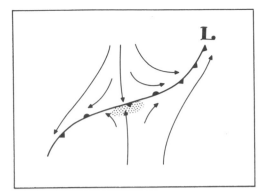

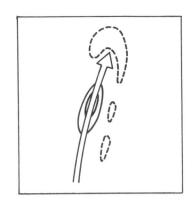

Figure 28.13. Area of potential thunder-storm generation relative to deformation and frontogenesis. This part of the front may have only minimal upper-level forcing and light surface winds.

Figure 28.14.   Area of destabilization (dashed) relative to low-level jet (arrow with solid isotachs). Activation is likely at isolated spots on the anticyclonic shear side of jet, particularly in air that has not realized its potential instability during the afternoon.

- Watch the evolution of seabreezes and urban heating which may locally enhance or suppress the cap (Scofield and Weiss, 1976).

- Look for evidence of moist layer increases (radar, pilot reports, satellite imagery).

- Note the formation and distribution of the first convective clouds, to determine if there is a forcing mechanism at the mesoscale, evident from either satellite or upper-air data.

- Be alert to fronts and jet streaks and the likely pattern of lift.

- On a more local scale, watch old outflow boundaries as sources of destabilization.

- Monitor all active outflows; when active convection exists the outflow can generate enough vertical motion to punch through a capping layer (Purdom, 1976).

- During evening, watch for indications of the establishment and location of the low-level jet.

After a storm has developed, the next question is where it will move. Colquhoun (1980) used a line of best fit to the hodograph plot, and for speed used the mean component along this line through the depth of the storm (Fig. 28.15). The problem is that the nowcaster rarely has a current hodograph that defines winds in the storm's environment. However, this or a similar method is good when conditions are not evolving rapidly. Rotating storms will frequently veer to the right or left, depending on their rotational direction (counterclockwise, and clockwise, respectively). Usually the nowcaster is interested in system movement more than individual cell motion. Cell regeneration on a flank can yield a system movement substantially different from that of the individual cells. However, cell movement is important

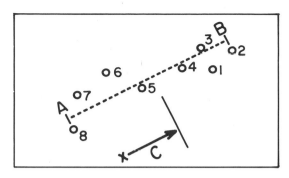

Figure 28.15. A simple method to estimate storm motion, given an environmental hodograph. Circles indicate wind vectors relative to storm (x). Pressure levels are labeled in 100 mb intervals. Direction is line of best fit (AB) through points; speed is mass-weighted component along line AB (vector C).

since the winds relative to the individual convective cells determine the severe storm potential.

28.3.2. Severe Storms

The preconditions for severe storms are similar to those for thunderstorms, with two notable additions: (1) A source of dry air above the moist layer; (2) vertical wind shear. Like the cap, both shear and dry air are filtering mechanisms; that is, they modulate the convective cells so that only the largest and most intense can survive. In a severe threat area, it is often more likely for nothing to happen than for weak storms to develop. Thus, there is a fine line between explosive development and total suppression.

The dry environmental air maximizes the vertical lapse of equivalent potential temperature, thus increasing parcel energies. When this middle-level air is entrained in the storm in an organized way (thus the importance of relative shear), evaporative cooling creates downdrafts that play a significant role in thunderstorm maintenance and severity.

The storm-relative vertical environmental shear is the single most important factor in determining storm motion and potential for severe storms. It is the shear mechanism that can maintain convective instability over long periods in the environment of the severe storm by continually supplying warm moist air at low levels and cool dry air aloft. In a quiescent environment a thunderstorm would quickly stabilize its domain, rapidly ending any convective potential. This is why regions of shear (fronts, jet maxima, outflows, etc.) are associated with severe storms. Certain configurations of shear can also enhance storm splitting, hail development, rotation, and tornado potential. The nowcaster's tool is the relative wind hodograph, which should be updated at all times from any available source of information. This may be difficult, because upper-level wind information is often unavailable for a given locale. However, the Doppler radar operating in clear-air mode (or the wind-profiling system) can give some indications of the vertical wind shear. When storm motion is known, a relative hodograph can be constructed, which will give some advance indication of the potential for severe storms. This would be amenable to automation; relative hodographs could be generated for a number of possible storm motions. Shear configurations are complex, and are discussed in Davies-Jones (1982 and 1983). Certain vertical shears (par-

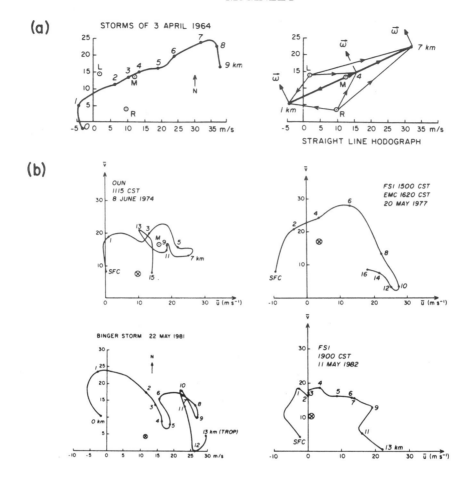

Figure 28.16. Hodographs of winds in storm situations. (a) Left: Hodograph for a day of splitting storms. M is the cloud-layer mean vector; R and L are vector motions of the the right- and left-moving storms. Right: Idealization of hodograph, showing velocity vectors and storm-relative winds at 1, 4, and 7 km for storms moving with velocities R, L, and M. (b) Tornadic hodographs, which differ from severe storm hodographs in that they show large loops (90° of directional shear in the lowest 3–4 km), along with substantial low-level inflow and strong middle-level flow. Storm translation velocities are indicated by circled crosses.

ticularly directional shears) maximize the potential for converting horizontal vorticity into vertical vorticity, and likewise, provide a source of vertical deformation that stretches the air columns between updrafts and downdrafts (Lemon and Doswell, 1979). This results in rapid spin-up, and mesocyclone and tornado formation (Fig. 28.16).

Nowcasting the Generation of Severe Storms

To a great extent many of the general considerations discussed in Sec. 28.3.1 are relevant here. The nowcaster must monitor low-level stability, sounding evolution, surface observations, and clear-air radar as before. In severe-storm situations evidence of mesoscale forcing is usually (but not

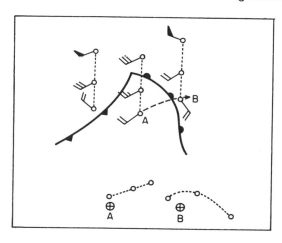

Figure 28.17. Wind hodograph for a storm that crosses a warm frontal boundary. As storm moves from A to B, wind hodograph changes from a straight line to looped pattern, increasing tornado potential. Nowcaster must monitor storm movement across frontal or old outflow boundaries.

always) more clear. Often a maximum in the upper-level winds may be upstream. As discussed before, the ageostrophic adjustments associated with these jet maxima can destabilize regions typically on the left-forward and right-rear flanks. Nowcasting clues concerning jet maxima are usually subtle but include cirrus shreds, isallobaric perturbations, gusty surface winds, and veering surface-wind trends. The classic idea of rapidly moving middle-level clouds as presented by Miller (1972) fits into this general idea, but experience has shown that if middle clouds are too extensive, they inhibit development by slowing surface heating.

Any sources of mesoscale or even synoptic-scale lifting are important possible destabilizing/generating mechanisms (Wilson, 1982). Frontal zones fit this category, particularly when they move into regions of potentially unstable air. The extensive cloud layers associated with frontal zones can inhibit storm potential by reducing surface heating. Warm fronts are especially active as sources of lift and frequently reproduce the optimum vertical shear profile necessary for severe-storm generation. Comparison hodographs for a storm moving across a warm frontal zone are shown in Fig. 28.17 (Maddox *et al.*, 1980). Warm frontal boundaries, and outflow boundaries persistent enough to have developed near-geostrophic wind shear profiles, should be monitored closely.

Also in the category of destabilizing systems are what might be termed thermal/frontal lows, first discussed by Tegtmeier (1974) (Fig. 28.18). When a frontal trough is located in a capped region (seen as a thermal ridge on morning charts), where surface heating is likely to be strong, the possibility exists for a weak low to form on or ahead of the front and progressively strengthen as heating progresses. Pressure falls occur and, in response, winds to the east directionally back, intensifying the convergence on the front north or northeast of the depression. The atmosphere may be locally destabilized in this region and storms are likely. Typically such occurrences are in zones where upper-level dynamics are weak. However, an interesting sidelight is local development of a very favorable hodograph for severe storms. This situation develops after 9–12 h, a long enough time for geostrophic rebalancing to begin (i.e., the upper shear adjusts itself to the thermal pattern). The

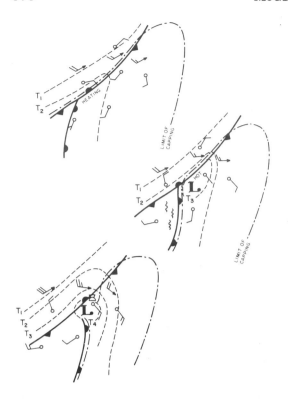

Figure 28.18. Sequence for development of a thermal frontal low. (Top) Weak front is stationary in capped region. Air near the front is heated strongly beneath the cap. (Middle) Mixing erodes the cap in the west, allowing dryline to shift eastward. A low forms in the frontal trough beneath the cap in very hot air. Winds begin to respond to thermal forcing. Winds aloft (barbed arrows) veer slightly in response to the upper return current. (Bottom) Continued pressure falls drive winds into the front northeast of the low. Convergence and forced lift destroy the cap, allowing thunderstorm (CB symbol) to grow. Upper-level winds, responding more to thermal gradients, provide stronger shear and more favorable directional shears.

nowcaster should monitor tendencies of temperature, wind, and pressure to get possible warning of a low and of consequent severe storm initiation.

Closely associated with this type of phenomenon is the inland sea breeze. Here, direct circulations are produced by the juxtaposition of cloud-cover or cooler outflow air to the east and rapidly warming air to the west. The effects are to back the winds, cause enhanced convergence on the western edge of the moist air, and to intensify the cap to the east. Beckman (1976) described outflow boundaries that are initially stationary but begin to move toward hotter air through the afternoon. Often this flow may be upslope, particularly in the Great Plains.

Erosion of the moist layer on its western flank has been described by Schaefer (1973). This process occurs as the shallow moist air, resting on sloping terrain, mixes through the adiabatic layer as surface heating progresses. The nowcaster may see this process as rapidly dropping dewpoints and veering winds. From map to map it appears that the dryline is racing eastward. The eastward progression usually stops where the moist "breeze" can counteract the mixing process. Here the moist layer may deepen, and a steep western edge may result. Storms are likely if the cap over the moist air can be broken or penetrated.

Determination of Severity

When presented with existing storms the nowcaster has ample tools for estimating severity. From conventional radar he knows the height and position of the top, intensity and gradients of reflectivity, water mass, echo

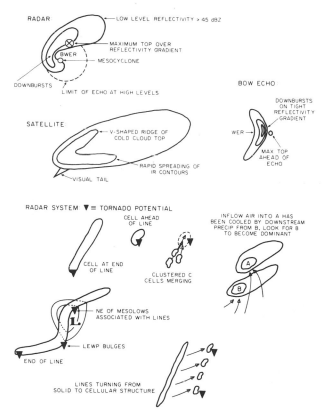

Figure 28.19. Summary of severe storm and system characteristics identified with radar, satellite, and radar systems.

configurations (notches, hooks, weak echo regions), movement, and acceleration. Doppler radar can show mesocyclone or even tornado circulation centers, upper divergence from the cloud top, and estimated straight line wind speeds associated with outflow gusts. The satellite can still be a useful tool; observations of tails (flanking lines) south of large storms, large anvil heights, and rapid anvil spreading are all positive indicators.

Approaches useful for a 0–3 h forecast of storm severity have been compiled from a number of sources (some unpublished) and are listed below (see Fig. 28.19).

- On radar, watch for satisfaction of the criteria for potentially severe hail storms given by Lemon (1977): (a) Reflectivity >45 dBZ in middle levels; (b) at least 6 km overhang outside the limits of a tight reflectivity gradient; (c) maximum top over the region of overhang or bounded weak echo region (BWER). Note that when a cell forms a BWER it has much more severe potential near the eastern side of the pendant or upstream edge of the BWER; tornado touchdown often corresponds to collapse of the BWER (Przybylinski, 1982).

- Significant penetration of the storm top above the equilibrium level (EL) derived from parcel methods (Doswell, 1982) can be applied to an appropriate environmental sounding.

- Watch echoes on the PPI that can maximize relative point inflow. These include echoes at the southern end of solid or broken lines, echoes bulging eastward as in a LEWP, echoes ahead of a line, and echoes developing northeast of a mesoscale or subsynoptic-scale low. The nowcaster should imagine how the environmental flow is entering each storm at low levels and base an estimate of projected strength on the potential instability of the inflow air and rate of inflow. Situations in which upstream storms precipitate into, and hence stabilize the inflow air, will significantly reduce the possibility of severe weather from the cell in question.

- Be alert to clustered small echoes that merge into an organized large cell.

- Lines that initially are solid or broken and then rapidly evolve into a cellular structure indicate enhanced severe potential.

- Cells that veer or move to the right of the flow (or their previous movement direction) can often modify the relative hodograph in a favorable way and begin to produce severe weather. A right-moving cell will result when a storm splits (see Fig. 28.16a).

- When line storms are observed on radar and satellite, watch for certain line echo configurations that may indicate damaging winds. Bow echoes (Fujita, 1978) are often indicative of mesoscale organization, enhanced outflow, and damaging (downburst) winds. Downbursts typically occur in the reflectivity gradient region ahead of the echo. The bow echo may be part of a LEWP. Watch for rapid elongation of echoes or relative penetration of weak reflectivity on the upwind side, along with a maximum top ahead of the reflectivity gradient on the forward edge of the storm. This may be indicated in satellite imagery as warming at anvil level. The downburst activity may be very persistent and located at the nose of the advancing bow echo. The region near the upstream side of the BWER is another favored location for downbursts. Doppler radar, if properly oriented relative to the wind direction, can aid in diagnosing strong outflow winds.

- Winds in single cells can also be significant. Gurka (1976) summarized some important indications in satellite visible imagery and related arc cloud propagation to maximum gust. If the speed of the arc cloud is greater than 15 m s^{-1}, expect gusts of 18 to 25 m s^{-1}.

- The vertically integrated liquid water (VIL) has been shown to be a useful parameter for forecasting (Devore, 1983); values $>45 \text{ kg m}^{-2}$ imply potential for severe storms.

- Adler and Markus (1983) outlined some parameters on satellite imagery: (a) Difference between mean anvil top temperature and minimum anvil temperature, and change in this difference; (b) spreading of isotherms at anvil level; (c) v-shaped ridges at anvil level.

• Moller (1979) identified pressure-fall areas that precede fast- and slow-moving thunderstorm systems. Close monitoring of surface pressure falls could reveal these depressions, which are likely to enhance the flow of moist air into the storm environment.

Nowcasting Hail

Nowcasts of hail can be based on Lemon's (1978) criteria. Potential hail size may be estimated in the prestorm environment by using the parcel energy method and a nomogram (Miller, 1972). This methodology is valuable only as a first guess. Modification for wet bulb zero height typically over-corrects, for often hail is observed in "tropical" air masses. The reason for this may be that when thunderstorms organize at the mesoscale, compensating downdrafts can significantly dry out the middle levels, increasing potential instability and allowing the wet-bulb zero to lower in the surrounding environment. This is very hard to diagnose, but suppressed cloud-free regions that indicate rapid storm development at a later time are potential areas to watch. for. Hail forecasts may improve when simple cloud models can be run operationally using Doppler data and miniprocessors.

Nowcasting Tornadoes

The potential instability criterion is basically the same as for severe storms. The environmental wind profile is critical. Tornadic storms typically are isolated (i.e., can realize a fully three-dimensional structure). Unfortunately a storm environmental sounding is rarely available, but the nowcaster can often deduce what the winds are. The trajectory of existing storms should be monitored for potential changes in the wind. Tornadoes are possible if the storm develops or moves into a region where low-level, storm-relative shear meets these criteria: 90-degree veering in the wind in the lowest 4 km (in other words, a deep layer of directional change) with at least 10 m s^{-1} inflow at 700 mb and moist low-level inflow. These conditions could exist as the storm crosses a warm frontal boundary where shears of this type are common. However, this set of criteria is valid only where low-level air is not so cool as to inhibit the tornado circulation from reaching the ground. As Maddox *et al.* (1980) showed, storms moving nearly parallel to the front will have a longer tornadic potential than storms moving normal to the front. An outflow boundry can serve the same purpose if it has existed long enough for larger scale balancing to produce favorable warm-front-like shears and for the air to have regained sensible heat. Newly generated outflow boundaries usually inhibit tornado formation in the short term by flooding the area with heavy nonbuoyant rain-cooled air, which decouples the tornado cyclone circulation from the surface.

On the individual storm scale, rotation typically begins in upper levels over the low-level reflectivity gradient and spreads downward. The vertical stretching evident in downdraft air on the upwind side may be a contributor. The updraft region has significantly lowered pressures so that air is lifted, condensing below the convective condensation level. This rising, negatively buoyant air forms the wall cloud. As rotation at the lower levels becomes stronger, precipitation from the rain to the north wraps around the west side of the circulation, while dry subsiding air moves in from the southwest. This intense low-level cyclonic shear may be followed by tornado formation. The juxtaposed wraparound rain and dry air form the hook seen on radar but the hook is generally a poor tool for nowcasting since it is relatively rare and the critical event may have already occurred. A better indicator is the inflow notch; the tornado is most likely ahead of the strong reflectivity on the western edge of the notch. Eventually, rain may completely wrap around the tornado, obscuring it from spotters who would otherwise aid the nowcaster. The only nearly foolproof tool is the Doppler radar, which can monitor the existence and evolution of the mesocyclone, and on rare occurrences the tornado itself. Some of the important criteria for mesocyclone identification given in Wood and Brown (1983) are persistent rotation (one-half the period of rotation—about 10–15 min), and vertical continuity.

Even though tornado reports cease and no hook is evident, regeneration is possible if the storm has not become outflow dominated; that is, the out-flow air, heretofore held in check by rapid inflow, suddenly floods the flank of the storm. Sequential development of a new wall cloud and mesocyclone, visible with Doppler radar, occurs southeast (5–10 km) of the old meso-cyclone, usually ahead of the rear flank downdraft in the moist inflow air. The evolution appears much like the occlusion and secondary development process that occurs at synoptic scale. In some cases a third or fourth wall-cloud/mesocyclone/tornado event may occur. This depends on the sustained potential instability of the air south and east of the storm (the right-front quadrant relative to storm motion). Spotters are indispensable in keeping the nowcaster updated on events of this type.

28.4. Flash Floods

Many of the initial conditions for flash floods are similar to those for severe weather. A flood situation results when heavy rain occurs for an extended period in the same drainage basin. Fast-moving systems are not usually candidates for flash flood warnings. However, individual cells that move very fast within systems that are stationary or moving slowly can produce flash floods. The condition necessary for such occurrence is a fixed point for cell regeneration. According to Maddox (1982), this can be a terrain feature, an outflow boundary, or frontal zone. A second ingredient is a source of low-level moisture, usually transported by a low-level jet.

Radar processing systems such as RADAP II can integrate reflectivity over time to produce an estimate of accumulated rainfall in a drainage basin. With automated drainage models, increase of stream or river depth can be estimated quickly and appropriate warnings issued.

The problem for the nowcaster is recognizing transition of convection into a flash-flood-generating system. According to Maddox (1980), the MCC (mesoscale convective complex), one of the major flash-flood-generating systems, begins as a group of convective storms that gradually merge into one

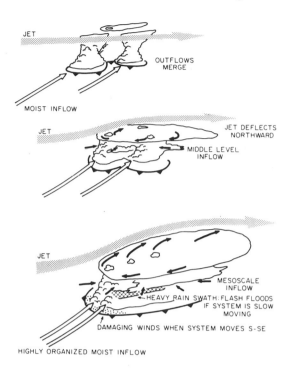

Figure 28.20. Development of flash-flood-producing mesoscale convective system. (Top) Cells merge. (Middle) Cells begin to organize at the mesoscale. (Bottom) Cells regenerate on southwest edge of outflow, allowing heavy rain to occur along cell track. Damaging winds are possible here as well. Note deflection of jet through the 6 h period represented.

large convective area (Fig. 28.20). When the system is organized, general mesoscale uplift begins, which sustains the steady rain and forces low-level flow into the system. Organized, merged outflows and forced steady inflow provide a fixed boundary (or regenerating point) and a steady source of moist air. When such systems form in light wind regimes they must usually move south and east to survive, as new cells form on the southern edge of the outflow and old cells remain nearly stationary. In stronger sheared regimes the system may remain fixed as cells form on the south or southwest edge of the outflow boundary and propagate through the system along an eastbound track. The latter type is more dangerous from a flash flood standpoint; the former may be more likely to produce damaging winds. The convective complex often begins in a strong flow regime. Owing to diabatic heating, the MCC may form a middle- and upper-level thermal gradient on the west or northwest side of the system, and deflect the flow at high levels after a 6–12 h adjustment period. Thus, the MCC may eventually become imbedded in light winds and veer sharply to the right. The nowcaster must pay close attention to weakening severe convective systems experiencing echo broadening, and monitor all data systems for evidence of a low-level jet flowing into the convective area. If cells on radar can be seen developing and moving away from a fixed point, flash floods are likely.

Time integration (manual or digital) of strong radar reflectivities remaining fixed over an area, is usually justification enough to issue a warning (the flash flood criterion, in general, is 3 inches of rain in 3 h or less [Wasserman, 1976]). This may be satisfied if VIP-4 levels persist for 3 h or VIP-6 for 1 h.

The criterion must be modified for the terrain and soil type in the drainage basin.

Oliver and Scofield (1976) used infrared (IR) satellite imagery to estimate rainfall. Their findings were that cold cloud tops (or the tight upwind gradients in IR contouring) on the upwind side can indicate where heavy rain is occurring.

28.5. Heavy Snow

Even in the largest scale cyclone developments that produce heavy snow, the maximum snowbands are usually mesoscale. The mechanisms that cause this are unclear, but with intense wind shears and strong frontal zones, the likelihood of forced mesoscale vertical motions is high. This brings with it the possibility of destabilization and hence, convection into the snow situation. Often thunderstorms are observed in the heavy snow region northwest of the surface cyclone during major blizzards. A strong Gulf moisture source wrapping into the low is likely in these cases.

Snow is a relatively poor radar target. However, heavy snow bands have been related to the tracks of vorticity centers by Beckman (1978). The vorticity center is the focus of many dynamic processes, of which destabilization may be one. The nowcaster should monitor the convective potential for forced lift of air upwind of the warm sector north to northeast of the forecast storm track. Typically these bands are 100–200 km left of the surface low's track.

Another mesoscale problem is the coastal front, well documented in New England, but seen as far south as the Carolinas. Bosart (1973) observed intense snow on the cold side of these boundaries. The effect of the front is to enhance (by 20–30%) the general banded precipitation occurring independently with synoptic-scale systems. This precipitation falls through an elevated moist layer associated with the coastal front, which allows additional growth of the snowflakes west of the frontal region (Marks, 1976).

Lake snows, another convective event, present serious hazards. These snows form under regimes of extreme instability (usually below 3–4 km) due to very cold air traversing the waters (or ice) of a warm lake. Very dry air from the north must linger over the lake long enough to allow vertical mixing of heat and moisture. The windshear and thermal instability of this layer drives the banded pattern of upward motion; additional lift is provided by increased surface friction as the system moves onshore. These are both related to the wind speed. Thus, both the time over the lake and the wind speed are important. One, however, can see that the velocity has conflicting effects. Many studies relate lake snows to the lake fetch only. Lake snows are very localized, and can depend on lakeshore configuration. Weinbeck (1983) studied the phenomenon and showed that it requires a 13°C difference in temperature between the lake surface and 850 mb. The typical lake fetch needed is 90 km. Condella (1983) related the ending of such snows to rapid increases in the stability of the surface-to-850-mb layer.

Nowcasting lake snow is difficult because often little warning can be obtained from radar. Maximum vertical motion and precipitation occur on

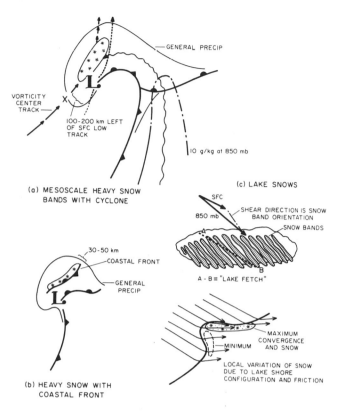

Figure 28.21. Summary of mesoscale snow events.

the lake shore. Satellite IR imagery may show a substantial cloud field and some higher clouds. Typically the clouds are banded along the cloud layer shear direction, and the longest cloud bands are probably related to heaviest snow intensities. Nowcasting lake snow requires close monitoring of satellite imagery and knowledge of lake temperature, the vertical thermal structure of the advancing cold air mass, and local pollutant sources that may enhance the snow. Wind-dependent snow climatologies are useful in showing where maximum snows are likely for various wind directions. Figure 28.21 summarizes lake-snow patterns.

28.6. Future Developments

The future is promising for nowcasting. Many methods discussed here require areal monitoring of vertical stability, vertical wind shear, and motions within storms. Plans exist to test and possibly implement systems that can provide continuous observations to fill these needs. Wind profiler networks (Little, 1982), in conjunction with ground-based and geosynchronous-satellite-based atmospheric radiometer systems, will provide information on wind shear, thermal profiles, and moisture discontinuities (Petersen and Mostek, 1982). The inherent smoothing of the temperature profile is a problem, but structure near stable layers may be recovered by locating discontinuities in the wind profile. Operational availability of visible and IR data, particularly in the 6.7 μm moisture channel and channels that map ozone, will give the nowcaster continuity on jets and jet maxima, vorticity centers, and gradient winds. NEXRAD (next-generation radar) will provide a U.S. network of single Doppler radars with sophisticated processors for rapid assimilation of volume scan data. The network will be employed to define storm structure, to help in rapid location of circulations, downbursts, rainfall intensity, and accumulation, and to monitor intensity trends; all will lead to more timely nowcasts.

A great need exists in development of systems to assimilate the various kinds of data rapidly and to display them in useful formats. The essence of nowcasting is how quickly the forecaster can gather the information needed to develop a three-dimensional picture of the atmosphere in the region of concern. Systems like the PROFS, CSIS, and others employ state-of-the-art computer imagery and software to gather, process, overlay, and display data from a wide variety of sensors in nearly real time. As such networks gradually spread into the field, significant improvement in all types of nowcasts will be evident.

Mesoscale models have performed reasonably well in a limited number of operational tests. By the 1990's mesoscale model output may be available in most weather stations. With model forecasts, the onerous task of identifying mesoscale forcing may be simplified.

One-dimensional cloud models may be useful in interpreting energy availability, wind shear profiles, stability evolution, and hail potential, particularly if the model can be updated with Doppler wind shears, surface temperatures, dewpoints, and radiometer data.

REFERENCES

Adler, R., and M. Markus, 1983: Satellite severe thunderstorm detection: potential and limitations. Preprints, 13th Conference on Severe Local Storms, Tulsa, Okla., American Meteorological Society, Boston, J32–J36 (also in Preprints, 5th Conference on Hydrometeorology, J32–J36).

Anthony, R., and G. Wade, 1983: VAS operational assessment findings for spring 1982–83. Preprints, 13th Conference on Severe Local Storms, Tulsa, Okla., American Meteorological Society, Boston, J23–J28 (also in Preprints, 5th Conference on Hydrometeorology, J23–J28).

Atkinson, B., 1981: *Mesoscale Atmospheric Circulations*. Academic Press, London, 495 pp.

Beckman, S., 1976: Satellite Appl. Note 76/19 KC/SFSS, Kansas City, Mo.

Beckman, S., 1978: Heavy snow along vor-ticity center tracks. Satellite Appl. Note 78/7 KC/SFSS, Kansas City, Mo.

Bluestein, H., E. Berry, J. Weaver, and D. Burgess, 1980: The formation of tornadic storms in NW Oklahoma on 2 May 1979. Preprints, 8th Conference on Weather Forecasting and Analysis, Denver, Colo., American Meteorological Society, Boston, 57–62.

Bosart, L., 1973: Detailed analysis of prescription patterns associated with mesoscale features accompanying U.S. east coast cyclogenesis. *Mon. Wea. Rev.*, 101, 1–12.

Bothwell, P., and Crawford, K., 1983: The operational use of surface objective analysis in evaluating the potential for heavy convection. Preprints, 13th Conference on Severe Local Storms, Tulsa, Okla., American Meteorological Society, Boston, 372–375.

Browning, K.A. (Ed.), 1982: *Nowcasting.* Academic Press, New York, 256 pp.

Carlson, T., and R. Farrell, 1982: The lid strength index as an aid in predicting severe local storms. *Natl. Wea. Dig.*, **8**(2), 27–39.

Colquhoun, J., 1980: A method for estimatng the velocity of a severe thunderstorm using the vertical wind profile in the storm's environment. Preprints, 8th Conference on Weather Forecasting and Analysis, Denver, Colo., American Meteorological Society, Boston, 316–321.

Condella, V., 1983: When snow squalls become no squalls. *Natl. Wea. Dig.*, **8**(2), 22–24.

Darkow, G., and S. Tansey, 1982: Subsynoptic fields of boundary layer heat and moisture from hourly surface data. Preprints, 12th Conference on Severe Local Storms, San Antonio, Tex., American Meteorological Society, Boston, 79–83.

Davies-Jones, R., 1982: A new look at the vorticity equation with respect to tornado genesis. Preprints, 12th Conference on Severe Local Storms, San Antonio, Tex., American Meteorological Society, Boston, 249–252.

Davies-Jones, R., 1983: The onset of rotation in thunderstorms. Preprints, 13th Conference on Severe Local Storms, Tulsa, Okla., American Meteorological Society, Boston, 215–218.

Devore, D. R., 1983: The operational use of digital radar. Preprints, 13th Conference on Severe Local Storms, Tulsa, Okla., American Meteorological Society, Boston, 21–24.

Doswell, C., 1982: The operational meteorology of convective weather, Vol. 1. NOAA Tech. Memo. NWS–NSSFC–5, National Weather Service (NTIS-#PB83-162321).

Fujita, T., 1978: Manual of downburst identification for project NIMROD. Paper No. 156, Department of Geophysical Sciences, University of Chicago, 104 pp.

Golus, R., and S. Koch, 1983: Gravity wave initiation and modulation of strong convection in a CCOPE case study. Preprints, 13th Conference on Severe Local Storms, Tulsa, Okla., American Meteorological Society, Boston, 105–108.

Gurka, J., 1974: Using satellite data for forecasting fog and stratus dissipation.

Preprints, 5th Conference on Weather Forecasting and Analysis, St. Louis, Mo., American Meteorological Society, Boston, 54–57.

Gurka, J., 1976: Satellite and surface observations of strong wind zones accompanying thunderstorms. Preprints, 6th Conference on Weather Forecasting and Analysis, Albany, N.Y., American Meteorological Society, Boston, 252–259.

Hales, J., and C. Doswell, 1982: High resolution of instability using hourly surface lifted parcel temperatures. Preprints, 12th Conference on Severe Local Storms, San Antonio, Tex., American Meteorological Society, Boston, 172–175.

Haltiner, G. R., and D. Martin, 1957: *Dynamical and Physical Meteorology.* McGraw-Hill, New York, 294 pp.

Hess, S., 1959: *Introduction to Theoretical Meteorology.* Holt, Rinehart and Winston, New York, 326 pp.

Lemon, L. R., 1978: On the use of storm structure for hail identification. Preprints, 18th Conference on Radar Meteorology, American Meteorological Society, Boston, 203–206.

Lemon, L. R., and C. Doswell, 1979: Severe thunderstorm evolution and mesocyclone structure as related to tornado genesis. *Mon. Wea. Rev.*, **107**, 1184–1197.

Lemon, L. R., D. Burgess, and R. Brown, 1978: Tornadic storm morphology derived from single-Doppler radar measurements. *Mon. Wea. Rev.*, **106**, 48–61.

Little, G. W., 1982: Ground-based remote sensing for meteorological nowcasting. In *Nowcasting.* K. Browning (Ed.), Academic Press, New York, 65–85.

Maddox, R., 1980: The mesoscale convective complex. *Bull. Amer. Meteor. Soc.*, **61** (11), 1374–1387.

Maddox, R., 1982: Forecasting mesoscale convective complexes. Preprints, 12th Conference on Severe Local Storms, San Antonio, Tex., American Meteorological Society, Boston, 180–183.

Maddox, R., and C. Doswell, 1982: Forecasting severe thunderstorms: A brief evaluation of accepted techniques. Preprints, 12th Conference on Severe Local Storms, San Antonio, Tex., American Meteorological Society, Boston, 105–108.

Maddox, R., L. Hoxit, and C. Chappell, 1980: A study of tornadic thunderstorm interactions with thermal boundaries. *Mon. Wea. Rev.*, **108**, 322–336.

Marks, J., 1976: A study of the mesoscale precipitation patterns associated with the New England coastal front. Preprints, 6th Conference on Weather Forecasting and Analysis, Albany, N.Y., American Meteorological Society, Boston, 285–290.

Miller, R., 1972: Notes on analysis and severe storm forecasting procedures of the Air Force Global Weather Central. AWS TR 200 (Rev.) Air Weather Service, Scott AFB, Ill., 190 pp.

Miller, R., and J. McGinley, 1978: Using satellite imagery to detect and track comma clouds and the application of the zone technique in forecasting severe storms. General Electric, MATSCO, Beltsville, Md. (Available from Air Weather Service, Scott AFB, Ill.).

Moller, A. R., 1979: The climatology and synoptic meteorology of southern plains tornado outbreaks. M.S. thesis, University of Oklahoma, Norman, 70 pp.

Oliver, V., and R. Scofield, 1976: Use of IR imagery to estimate rainfall. Preprints, 6th Conference on Weather Forecasting and Analysis, Albany, N.Y., American Meteorological Society, Boston, 242–245.

Peterson, R. A., and A. Mostek, 1982: The use of VAS moisture channels in delineating regimes of potential convective instability. Preprints, 12th Conference on Severe Local Storms, San Antonio Tex., American Meteorological Society, Boston, 168–171.

Przybylinski, R., 1982: Identifying severe local storms operationally. Preprints, 12th Conference on Severe Local Storms, San Antonio, Tex., American Meteorological Society, Boston, 413–416.

Purdom, J., 1976: Some uses of highresolution GOES imagery in the mesoscale forecasting of convection and its behavior. *Mon. Wea. Rev.*, **104**, 1474–1483.

Sasaki, Y., and S. Tegtmeier, 1974: An experiment of subjective tornadic storm forecasting using hourly surface observations. Preprints, 5th Conference on Weather Forecasting and Analysis, St. Louis, Mo., American Meteorological Society, Boston, 276–279.

Schaefer, J. T., 1973: Motion and morphology of the dryline. NOAA Tech. Memo. ERL-NSSL-66, Environmental Research Laboratories (NTIS#COM-74-10043), 81 pp.

Scofield, R., and C. Weiss, 1976: Application of SMS products and other data for short range forecasting in the Chesapeake Bay region. Preprints, 6th Conference on Weather Forecasting and Analysis, Albany, N.Y., American Meteorological Society, Boston, 67–73.

Siebers, A. L., and J. T. Schaefer, 1983: The temporal shear of the surface wind field—a potential severe thunderstorm forecast tool. Preprints, 13th Conference on Severe Local Storms, Tulsa, Okla., American Meteorological Society, Boston, 257–260.

Tegtmeier, S., 1974: The role of the surface subsynoptic low pressure system in severe weather forecasting. M.S. thesis, University of Oklahoma, Norman, 66 pp.

Wasserman, S., 1976: Use of radar information in determining flash flood potential. Preprints, 6th Conference on Weather Forecasting and Analysis, Albany, N.Y., American Meteorological Society, Boston, 223–227.

Weinbeck, R., 1983: Lake-effect snows of Rochester and Buffalo, N. Y. *Natl. Wea. Dig.*, **8** (3), 42–45.

Wilson, G., 1982: The structure of mesoscale systems influencing severe thunderstorm development during AVE/SESAME 1. Preprints, 12th Conference on Severe Local Storms, San Antonio, Tex., American Meteorological Society, Boston, 192–196.

Wilson, J. W., and K. E. Wilk, 1982: Nowcasting applications of Doppler radar. In *Nowcasting*, K. Browning (Ed.), Academic Press, New York, 87–105.

Wood, V. T., and R. A. Brown, 1983: Single Doppler velocity signatures: An atlas of patterns in clear air/widespread precipitation and convective storms. NOAA Tech. Memo. ERL NSSL-95, Environmental Research Laboratories (NTIS#PB84-155779), 71 pp.

CHAPTER 29

Short-Range Forecasting

C. A. Doswell III

29.1. Introduction

Weather forecasting, as defined in Huschke (1959, p. 623), is a statement of the expected future occurrence of one or more weather elements. Weather elements are taken to include such things as "temperature, humidity, precipitation, cloudiness, brightness, visibility, and wind." Thus, weather forecasting is not just forecasting the variables in the governing equations of meteorology. Inferring the patterns of weather elements from a map depicting those variables is also essential (Tepper, 1959; Sanders, 1971; Weiss and Ferguson, 1982), even with high-resolution data (Tepper, 1959).

The basic dependent variables in the governing equations are pressure p, temperature T, moisture m, and the components of the wind (u, v, w). The independent variables are the three spatial coordinates x, y, z and time t. If X stands for any of the dependent variables, a formal statement of the prediction problem is simply that

$$X(x, y, z, t_1) = X(x, y, z, t_0) + (t_1 - t_0)(\partial X/\partial t)_{t=t_0} + \cdots \qquad (29.1)$$

where t_0 is some initial time and $t_1 > t_0$. The solution can be described as the combination of a diagnosis with a trend. Treating the problem of forecasting in this formal way is the task of numerical models. Although temperature, wind, and humidity can be derived directly with this approach, (29.1) does not produce explicit forecasts of all the weather elements. A major portion of "synoptic meteorology" concerns how to infer those weather elements from the predicted patterns of X at time t. This task probably will remain largely in human hands, especially for short-range forecasting.

The distinction between numerical, objective-model prediction and weather forecasting is illustrated in Fig. 29.1. Both approaches start with the diagnostic step in (29.1), using the data at hand. On the objective side, some of the data are in forms unsuited to the system. For example, radar data do not enter the input stream since reflectivity values are not among the dependent variable set. Also, the surface and rawinsonde data are not ordinarily taken at the exact spatial (and temporal) locations required by the model. The diagnosis remaps the input variables onto a grid (a process called objective analysis) and then subjects them to a special sort of filtering and smoothing ("initialization"), which completes the diagnostic step.

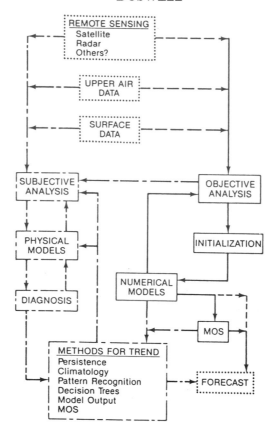

Figure 29.1. The forecasting process (left) for humans and (right) for a numerical prediction model.

The numerical model supplies the trend over some specified time step, which is combined with the diagnosis to advance the model in time. The process is repeated until the specified time limit. The result is a four-dimensional (x,y,z,t) representation of the model variables. The numerical model's dependent variables are then related to the weather elements of concern. A statistical method for determining this relation has been applied to the operational predictions of the National Weather Service's Limited-area Fine-Mesh (LFM) model and is known as MOS (for Model Output Statistics; see Glahn and Lowry, 1972). By following the right-hand side of Fig. 29.1, it is technically possible to produce a weather forecast without human intervention. Although the details are complex, the process follows a more or less linear path, from data to forecast. The only feedback allowed is use of the numerical model's prediction as input for the diagnostic step in the next model run. This system produces one and only one forecast from a given set of data, thus qualifying as an objective system (as defined by Glahn, 1985).

In contrast, the human-based forecast process is more complicated, even if the details are somewhat simplified. That is, a human does not weight equally all the data in four dimensions. Instead, pattern recognition is employed to assimilate the complexity (Allen, 1981). Humans blend experience, theory, concepts, conjectures, and all the available data (regardless of their relevance to the governing equations) into a four-dimensional image of the atmosphere. The resulting image amounts to a hypothesis about how the observed data came to be. There may be more than one image that is consistent with physical theory, but every one must be consistent with the data. Such images are the antithesis of objectivity, since no two humans share the same image and the result is not reproducible.

Pattern recognition is crucial to production of the trend, as well; knowing how the atmosphere will evolve may depend on knowing what processes are responsible for the observed distributions. Production of a trend and the initial diagnosis proceed together during the human-based forecasting process. The diagnostic and forecasting steps feed back on one another through the intermediary of a "physical model," the forecaster's concept of how some atmospheric process (or combination of processes) works. As with the other parts of the subjective process (left side of Fig. 29.1), the physical model is a mixture of experience and theory, which humans use to relate their image of the atmosphere's structure to its behavior (i.e., to the weather).

The subjective forecasting process is not restricted by the limits imposed in using quantitative, objectively treated data. That is, the human can make correct assessments and predictions with limited data—something that no purely objective approach can accomplish. This ability is counterbalanced by the capacity for disastrously incorrect assessments and prediction that no purely objective method would ever make. The ideal would be to have the advantages of the intuitive approach without all the risks. Some way is needed to maximize the human capacity for perception and to minimize incorrect judgments. It is proposed that basic physical understanding makes that possible, although error is not entirely eliminated.

29.2. Extrapolation Bounds of Short-Range Forecasting

29.2.1. Linear Extrapolation

One of the simplest ways to produce a trend is to use linear extrapolation: Assume that the linear trend seen in the most recent diagnosis will continue. Figure 29.2a is a generic diagram of some weather process as it affects some forecast quantity (i.e., a weather element); there is no fixed time scale. The limitations of linear extrapolation are apparent with increasing time, and the bounds on what is considered an acceptable forecast are finally exceeded (Fig. 29.2b). The period of valid linear extrapolation would probably be different in different examples of the same weather process. By considering a large sample of situations involving the same weather process, it is possible to produce an average (or mode, or some other measure of statistical central tendency) time range for which linear extrapolation is valid. Different processes would generally have different periods of valid linear extrapolation. Further, there might be different variability in this quantity for different phenomena.

Zipser (1983) estimated the typical time period of valid linear extrapolation for various weather processes (Table 29.1), and Tepper (1959) developed a comparable breakdown based on spatial scales. Recall that weather

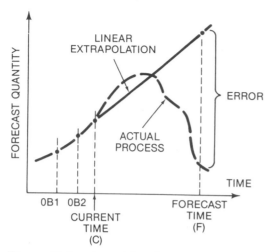

Figure 29.2a. Actual development of a process (heavy dashed line), and linear extrapolation from a set of observations (OB1, OB2, and current). Extrapolation produces a forecast with some error at the forecast valid time (F).

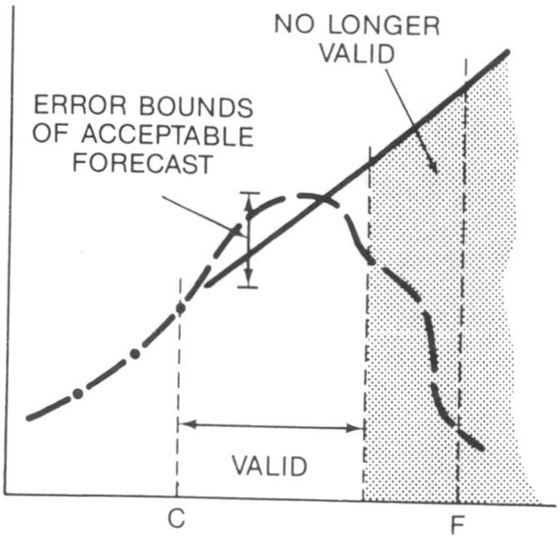

Figure 29.2b. Definition of the period of valid linear extrapolation, given the maximum permissible error for an acceptable forecast. The period depends on the phenomenon, and varies from case to case.

Table 29.1. Examples of typical linear extrapolation time scales for various weather events*

Weather event	Time scale for linear extrapolation validity	Nonlinear predictive capability
Downburst/Microburst	~1 to a few minutes	Very limited
Tornado	~1 to a few minutes	Currently very limited
Thunderstorm, individual	5–20 minutes	Very limited
Severe thunderstorm	10 minutes to 1 hour	Very limited
Thunderstorm organized on mesoscale	~1–2 hours	Some
Flash-flood rainfall	~1 to a few hours	Very limited
High wind, orographic	~1 to a few hours	Some
Lake-effect snowstorms	A few hours	Very limited
Heavy snow/winter storm/blizzard	A few hours	Some
Frost/Freeze	Hours	Some
Low visibility	A few hours	Some
Air-pollution episode	Hours	Some
Wind	Hours	Some
Precipitation	Hours	Some
Hurricane	Many hours	Fair
Frontal passage	Many hours	Fair to good

*Adapted from Zipser (1983).

processes are what produce the weather elements. It may be that the characteristic time scale for variations in a weather element is not necessarily the same as the time scale for the process that produces the element. For example, the life cycle of precipitation during a thunderstorm is not identical to the thunderstorm's life cycle. In fact, the process as a whole must have a longer time scale than any specific weather element it produces.

For most weather-producing processes listed in Table 29.1, it is clear that linear extrapolation beyond a few hours is not very useful; for purposes of this discussion, 1 h is arbitrarily selected as the limit. This limit represents the lower bound for "short-range" prediction. (Linear extrapolation may be useful well beyond this limit for large-scale systems with longer valid extrapolation times. These systems may not be directly responsible for important weather events, but their role in producing smaller scale phenomena can be large.)

29.2.2. Nonlinear Extrapolation

At the upper bound, short-range prediction requires nonlinear forecasting approaches, by definition, but a nonlinear model can produce a worse result than linear techniques (Fig. 29.3). What is sought is shown in Fig. 29.4.

The most obvious choice for nonlinear extrapolation is a numerical model, for example, the LFM. Operational numerical models are subject to constraints that are not imposed on research models, including computational stability under a wide range of initial conditions, short run times, and relative insensitivity to data problems. Thus, operational models tend to lag behind the state of the art in numerical prediction. Present operational models are best suited to synoptic-scale forecasting. As currently configured, they also have a substantial period at the start of the model run when variables are undergoing large oscillations in an effort to achieve internal adjustments to the initial state (Haltiner and Williams, 1980, p. 365ff). For the LFM model, this is about 6 h, a time during which the predictions are of relatively little value. There is considerable experience to suggest that present large-scale

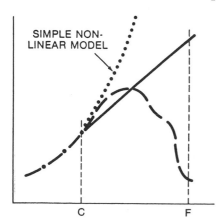

Figure 29.3. An example of a simple nonlinear model that is much worse than a linear one (compare Fig. 29.2a).

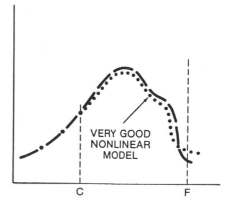

Figure 29.4. An example of a very good nonlinear model. Such models are usually complex and hard to find.

models attain their best performance only after about 12–24 h. Thus, 24 h has been chosen, again somewhat arbitrarily, as the upper bound to short-range forecasting.

29.3. Diagnosis as an Element of Forecasting

29.3.1. Diagnosis of Synoptic-Scale Structure

Given that forecasting begins with diagnosis, there is logic to starting with the synoptic scale. Large-scale systems bring together the ingredients that make possible weather-producing mesoscale phenomena. For example, thunderstorms require moisture, unstable stratification, and a source of lift. Perhaps two of the three ingredients are present at analysis time. Since the addition of the missing ingredient would enhance the thunderstorm potential, the large-scale diagnosis can suggest what to watch for.

It can be argued that automated chart analysis saves time. However, this overlooks the important fact that human analysts vary widely in their capacity to recognize features in computer-drawn patterns. Because of training, experience, knowledge, and other factors (including psychological ones), some individuals can capture the essence of a weather situation with little more than a brief glance; most require more effort.

Somehow, the act of drawing lines on a chart can greatly improve one's comprehension of the data. One way the forecaster can improve perception of a weather situation is to enhance the charts produced by automated means. This might involve any or all of the following: re-contouring in regions of interest at smaller intervals, adding annotations that emphasize the location of features like trough lines, noting past positions of such features, contouring for variables that are not included in the automated analysis.

Maddox (1979) described the process of chart enhancement for the specific problem of heavy convective rainfall. Figures 29.5 and 29.6 show what enhancement can do. The concept can be adapted to any other weather-producing phenomenon.

An essential component of large-scale diagnosis is providing continuity. The time history of features highlighted through chart analysis/enhancement is crucial to developing a four-dimensional conception of the atmosphere. It is not enough to have detailed knowledge of the current structure. One has to know how that structure came about, to determine the trend. As stated earlier, the processes of diagnosis and prognosis, as practiced by humans, are not totally separable. Further, since one is dealing with synoptic-scale structures, linear extrapolation may be valid, but feature continuity is essential to apply that forecasting technique.

Another reason to do detailed large-scale diagnosis is to monitor the performance of numerical prediction models. Although the current models do not provide explicit forecasts of all the weather elements, if the diagnosis suggests how well a numerical model is doing in a given situation, the analyst has acquired important discretionary information for a forecast. To accomplish this qualitative measure of model performance, both chart analysis and comparison with the model(s) must be done routinely.

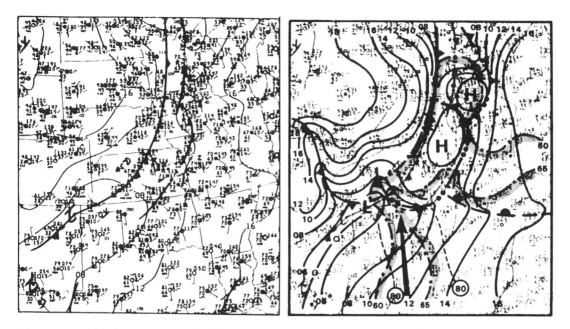

Figure 29.5. Surface analysis (left) as it comes from an analysis center and (right) enhanced to reveal subsynoptic-scale features as given in Maddox (1979).

In a related vein, as Hales (1979) indicated, comparison of the current charts with the initial input to the numerical models can have value. Information about major problems with the input data can use be used to decide how to incorporate the model predictions in the forecast. As Hales showed,

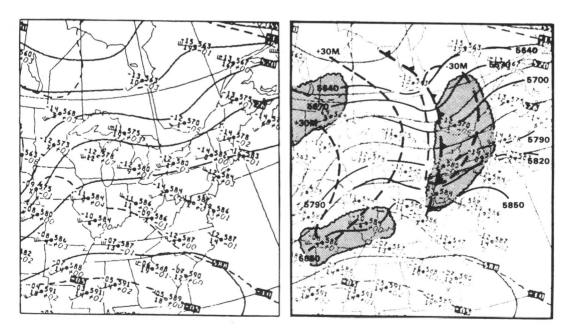

Figure 29.6. (Left) Analysis and (right) enhancement of 500 mb data (compare Fig. 29.5).

human ability to make use of all the data (including radar and satellite data) offers a tremendous advantage at such times. Errors in the initial data generally have diminishing effect on the model predictions after 12–24 h, but the effect may be substantial in short-range forecasting.

Diagnosis on the synoptic scale assumes that the data are reliable. To the extent that the data depart from what one expects to find on this large scale, the synoptic-scale diagnosis is a basis for identifying areas and situations calling for more careful scrutiny. However, the data needed may not always be available.

29.3.2. *Diagnosis of Subsynoptic-Scale Structure*

In this chapter any feature having a characteristic length below that which can be resolved in the upper-air sounding network is termed subsynoptic scale. Processes with subsynoptic spatial and temporal scales account for much of the variation in weather elements, but the data available for resolving them are limited.

To have any hope of doing a proper diagnosis of subsynoptic features, the human analyst must rely on surface observations and remote sensing information (primarily radar and satellite data, at present). Although operational radar and satellite data do not now provide conventional weather measurements (p, T, m, u, v, w), they are used to determine precise location and movement of already identified subsynoptic-scale features.

Knowledge of the local topography is important for blending the data into a coherent diagnosis. Much of this knowledge has to be acquired through experience, but proper diagnosis is impossible without it. For example, if a site is in a valley, under certain circumstances its wind and temperature measurements can be unrepresentative of the local area. If one were to compare these data with those of surrounding stations, one might falsely infer features (e.g., convergence and/or airmass boundaries).

Local topography can also directly affect weather processes. There may be preferred regions where fog forms, or thunderstorms tend to develop, or large-scale winds are channeled and strengthened to damaging proportions. Also, once weather-producing processes begin, the topography can modify them in a predictable fashion. For example, thunderstorm outflows may be impeded or shunted by terrain, or local high spots may experience freezing rain while surrounding regions are receiving only rain.

Another extremely valuable tool in subsynoptic diagnosis, which represents a rational synthesis of all weather-related experience, is the physical model. Indeed, the main difference between synoptic- and subsynoptic-scale diagnosis is in the models used. Having identified areas where large-scale physical models conflict with the data, one must try to use the limited information to form some working hypothesis that can reconcile the conflict. Subsynoptic diagnosis requires a broad knowledge of meteorology and its analytical tools. Some relevant topics are quasi-geostrophic theory, kinematic analysis, boundary layer theory, objective interpolation, information (sampling) theory, precipitation physics, convection dynamics, solenoidal cir-

Figure 29.7. Three blind men's perceptions of an elephant. (Courtesy of S. Barnes, ERL.) Each limited sample is misleading if there is no knowledge of the phenomenon being sampled.

culations (sea/land breezes, mountain/valley flows), the general circulation, and remote sensing.

Subsynoptic models usually use data networks that are unavailable to operational forecasters, and it is possible to argue that those models cannot be applied because the supporting data are not there. The argument has merit, but it does not apply without qualification. To appreciate how one can apply research results without research data, consider the three blind men and the elephant (Fig. 29.7). If the three blind men had a conceptual model of the elephant through prior experience or training, their limited sample would be of vastly more use to them. Although they may not know the elephant's exact weight or height, simple recognition that it is an elephant may be useful in itself.

Many of the subsynoptic models that a forecaster needs to know about are dealt with elsewhere in this volume. More detailed and comprehensive information on topics relevant to short-range forecastings can also be found in other textbooks (e.g., Atkinson, 1981; Browning, 1982; and Pielke, 1984).

29.4. Methods of Short-Range Forecasting

Figure 29.8 suggests the relative importance of various inputs to the forecasting process, as a function of the forecast range. For example, the knowl-

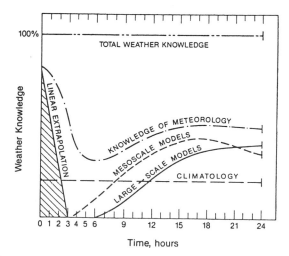

Figure 29.8. Effectiveness of different approaches to short-range (mesoscale) forecasting.

edge obtained from linear extrapolation does not fall completely to zero after 1 h, (the typical time adopted in Sec. 29.2.1). Rather, after an hour, linear extrapolation is contributing about half the total "knowledge" needed to solve the forecast problem, and that fraction is steadily and rapidly decreasing. Recall that linear extrapolation of large-scale features of relevance may still be possible, but direct linear extrapolation has limited value.

The contribution from large-scale models begins only after an adjustment period of about 6 h, becoming the dominant part somewhere between 12 and 24 h. Also shown schematically is the influence of mesoscale models, which have yet to become operational. One intent implicit in the effort to make mesoscale numerical models operational is the closing of the forecasting "gap" suggested in the figure.

"Knowledge of meteorology" represents the potential value of human input, i.e., the physical models. Note the substantial dip in this total knowledge within the same range as the forecasting gap, indicating the lack of detailed scientific understanding of mesoscale weather processes. Nevertheless, this discussion takes as axiomatic that the use of this knowledge offers considerable improvement over pure reliance on the models (referred to by Snellman [1977] as meteorological cancer) and simple linear extrapolation.

29.4.1. Use of Climatology

Figure 29.8 shows climatology as a distinct component of meteorological knowledge. Climatology can make a substantial contribution to filling the forecasting gap. The value of a particular forecasting tool can be measured against climatology. Hence, the advantage of linear extrapolation over climatology virtually vanishes in about 2 h. In effect, the establishment of a climatology for a weather process resets the skill level to some constant value above pure random guessing.

Within the short-range forecasting gap under consideration, improving on the climatological standard may not be easy. For instance, in a very stagnant, summertime large-scale weather pattern, with moisture and instability over large regions, it may be exceedingly hard to find any indication of when and where convection will develop. Under such circumstances, it might be proper to offer a climatologically derived precipitation probability, over the entire forecast area for the period, rather than guessing.

However, this would probably not be a good strategy to follow every day. In the situation just described, there could be old outflow boundaries left over from the previous day's activity, or a weak quasi-stationary frontal zone, or some other feature that careful diagnosis would reveal. Such a feature might provide enough focusing (lift) to offer a chance of improving upon a pure climatology forecast. In weakly forced synoptic situations the scale of the process that produces weather elements at some locations and not at others does not always permit the data at hand to depict that process. If diagnosis fails to reveal any process that would focus subsequent weather events, then and only then should one resort to a purely climatological forecast (or one simply based on objective guidance).

29.4.2. Nonlinear Short-Range Forecasting

In this chapter, simple linear extrapolation is referred to as nowcasting; the term forecasting is reserved for nonlinear methods (as in Zipser, 1983). For purposes of discussion, forecasting can apply to weather events already in progress and weather events yet to commence; nowcasting is possible only for events already in progress.

Applying Meteorological Knowledge to Events in Progress

In essence, the weather in progress is the current manifestation of processes. Scientific knowledge about these processes offers information to apply in making a forecast. That information may concern what weather elements to expect, the amplitude and phase of the variation in weather elements, the typical period over which the processes unfold, typical large-scale conditions under which the processes amplify or decay, how the processes alter environment, and so on.

Even when such knowledge cannot provide detailed, quantitative information, it provides a basis for improving on climatology. As an example of how knowledge is applied, consider Fig. 29.9. The clouds along the Mississippi-Alabama border and over most of Georgia (extending into extreme southeastern Alabama) are Mesoscale Convective Systems (or MCSs—see Rodgers *et al.*, 1985). Forecast personnel in Mississippi, following the progress of the MCS on their eastern border, would recognize that these are not random air mass storms (note the time of day), but an organized process. The forecast problem requires decisions on whether this MCS will continue for several more hours or dissipate, how soon the system will leave the forecaster's area of responsibility, what sorts of weather elements (rain, hail, tornadoes, strong winds, floods) are possible with this system, and how the passage of the MCS will influence subsequent events.

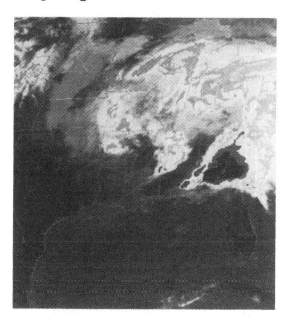

Figure 29.9. Enhanced infrared satellite image for 1600 GMT, 3 May 1984.

Having knowledge of the system's history, the forecaster can employ simple linear extrapolation, since the process is already under way. Knowing whether it has been intensifying or dissipating is clearly helpful. However, there are more complex questions that must be answered by nonlinear methods: Is some large-scale process sustaining the system (and what do the numerical models predict for that process)? Are there any topographic features that might alter the system? How might diurnal cycles within the boundary layer modify the system's evolution? Does the present system fit any physical model whose characteristics can be applied to making a forecast?

The forecaster uses knowledge of meteorology to answer such questions and develop a hypothesis that is consistent with both physical theory and the observations. Knowledge of the characteristics of specific types of weather systems is applied. In the case of convective systems, if all the basic ingredients for convection (moisture, instability, and a source of lift) can be maintained, the system is likely to continue. Since no weather process exists in isolation, the forecaster considers the processes that sustain the MCS (e.g., low-level jets, extratropical cyclones, sea breeze fronts, mesoscale instabilities, solar heating).

In-Depth Diagnosis to Predict Events

When no significant weather-element-producing systems are in progress, the forecaster must decide whether weather-element-producing phenomena will develop at all. However, before events are under way (i.e., in a fair-weather situation) the forecaster has one advantage: the situation offers the maximum amount of time for diagnosis. Furthermore, the really important forecasts are those that deal with changes from fair-weather conditions. Once threatening weather processes commence, in an operational setting the forecaster must cope with the situation, trying to react properly to the events as they happen. Opportunities for in-depth diagnosis can be quite limited. Therefore, it is in the fair-weather situation that a detailed diagnosis is both possible and most important.

This can be illustrated in a continuation (Fig. 29.10) of the case depicted in Fig. 29.9. The westernmost MCS has just about moved out of the state, so a pure extrapolation forecast would call for clearing skies. However, the problem is more complex than simply dealing with the system at hand. The MCS leaving the state has produced a pronounced outflow boundary, east-west across the state. Such a boundary could be a focus for redevelopment of convection. Although no significant weather is actually under way along the boundary, ongoing processes may eventually create weather-element-producing phenomena. The case is more or less typical of operational forecasting. The problem facing the forecaster on the morning of 3 May 1984 can be stated as two questions: (1) Is an identifiable process (or processes) acting (or likely) to redevelop convection on the outflow boundary (or anywhere else)? (2) If convection does redevelop, are there processes that would lead one to anticipate development of hazardous weather, requiring the forecaster to take special action? (Other, routine aspects to the forecast will be influenced substantially by the answers to these questions.)

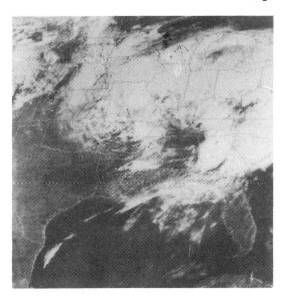

Figure 29.10. Visible satellite image for 1830 GMT, 3 May 1984. Note the outflow boundary across Louisiana, Mississippi, and Alabama in the wake of the mesoscale convective system in east central Alabama.

The answers to the two questions are not completely obvious. However, using operational data the forecaster can anticipate what is possible (and eliminate the unlikely). The morning 850 mb chart (Fig. 29.11), suitably enhanced beyond the standard analysis, contains some clues. A relatively strong cold front is moving southeastward in the north-northwesterly flow behind the extratropical cyclone on the Illinois-Missouri border near St. Louis. There is strong warm thermal advection across Louisiana and Mississippi, but is that associated with a synoptic-scale warm front? If it is, a considerable amount of moisture is north of that thermal boundary. A hypothesis more consistent with the data is that the morning convection has created this boundary, with the large-scale warm front well north, into the Ohio Valley. Note that the 850 mb flow is bringing warm air (from the thermal ridge in western Louisiana and Arkansas) into the moisture (ahead of the dryline). This is a favorable configuration for convection (see Doswell *et al.*, 1985). The dryline at 850 mb has remained well ahead of the advancing cold front.

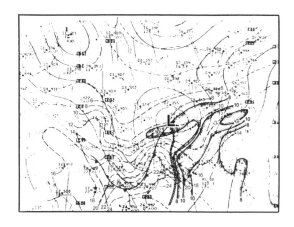

Figure 29.11. Enhanced 850 mb analysis for 1200 GMT, 3 May 1984. Dashed lines are isotherms at 2°C intervals; solid shaded lines are isodrosotherms at 2°C intervals for values ≥ 8°C.

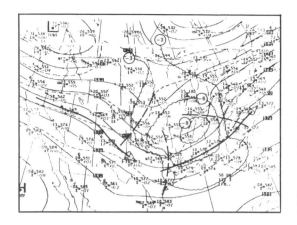

Figure 29.12. Enhanced 500 mb analysis for 1200 GMT, 3 May 1984. Short-wave trough axes are denoted by heavy dashed lines; the dashed line with barbs represents a thermal trough axis; 12 h height changes (dam) are shown by scalloped lines; jet stream axes are indicated by heavy solid lines with arrowheads.

At 500 mb (Fig. 29.12), one short-wave trough is rotating northeastward across the Ohio Valley and another lies over eastern Oklahoma and Texas. The data do not resolve separate height fall centers with these troughs, but the thermal pattern indicates some 500 mb cold advection in Louisiana and Mississippi. Combining these data with those at 850 mb gives a Total Totals Index (see Miller, 1972) of 42 for Jackson, Miss., and 46 for Bootheville, La. The Total Totals Index is a rough measure of convective instability, and these values indicate a threat of convection, but do not suggest much chance of severe weather.

At the surface (Fig. 29.13), the pattern is rather confusing, owing to the morning's convective activity. The standard analysis has mislocated the front, focusing on the windshifts behind the dryline. Further, the surface warm front has been analyzed in the approximate position of the convective outflow boundary. There is a double, surface low structure associated with convective systems south of the Great Lakes.

The approaching short-wave trough, the considerable moisture south of the boundary, and the moderate instability all make new convective development rather likely. However, the intensity is not yet obvious. Clearing skies in the wake of the MCS may allow solar heating to decrease the stability beyond that indicated by the soundings at 1200 GMT. The advancing

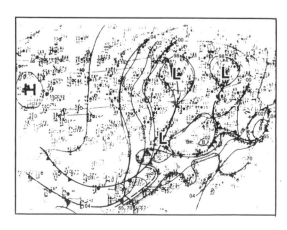

Figure 29.13. Enhanced surface analysis for 1500 GMT, 3 May 1984. Outflow boundaries are indicated by dash-double dotted lines with barbs: heavy dashed lines show troughs, the one in southeastern Texas being the surface reflection of a dryline aloft; solid, shaded lines are the 65° and 70°F isodrosotherms.

short-wave trough may back the winds at low levels ahead of the dryline, bringing more moisture into the threat area. The thermal advection at low levels may add to the instability by warming the low layers, or act as a lifting mechanism.

However, negative possibilities can be seen. The passage of the thermal trough at 500 mb might act to decrease the instability somewhat. As this feature passes, the cold air at low levels could sweep through Mississippi (before heating has a chance to destabilize the situation), thus removing the outflow boundary as a focusing mechanism. The front might replace the outflow boundary as a lifting mechanism, but be out of the state before convection develops. Also, it is not very unstable, and a lot of things would have to happen in just the right way for severe weather to materialize.

What about flash flooding? For convection to produce heavy rain does not require that storms be severe in an official sense (i.e., 50 kt straight winds, 3/4-inch-diameter hail, and/or tornadoes). The morning thunderstorms have produced substantial rain, so the ground may be locally saturated, enhancing the flood threat. However, the synoptic pattern is quite energetic, with progressive features, so a slow-moving system is unlikely. Assuming that the cold front will pass through the state within 12–24 h, repeated passage of storms over the same general area is also unlikely. Thus, there is not a very substantial threat of widespread flooding, although localized problems are a possibility.

Actual events bore out this assessment (Fig. 29.14). New thunderstorms developed along the outflow boundary in Louisiana and Mississippi. Isolated large hailstones were reported in Mississippi during the afternoon, but flooding was not a serious problem.

Because the afternoon thunderstorms did not exist at the time of the morning forecast, direct linear extrapolation could be of no value. Had one

Figure 29.14. Visible satellite image for 2030 GMT, 3 May 1984.

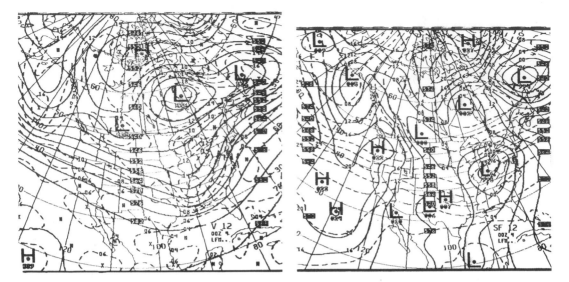

Figure 29.15. LFM 12 h predictions valid at 0000 GMT, 4 May 1984, for (left) 500 mb and (right) surface. At 500 mb, solid lines are height contours (dam) and dashed lines are absolute vorticity $(\times 10^{-5} s^{-1})$. At the surface, the solid lines are sea-level isobars (mb, last two digits only) and dashed lines are 1000–500 mb thickness contours (dam).

not anticipated new convection, one might have overreacted to the development of thunderstorms in the afternoon. The morning diagnosis made it clear than new convection would develop, but no process could be seen that suggested severe convection. A judicious approach would be to monitor the unfolding events, using the data at hand, to detect departures from the anticipated scenario.

Objective Guidance

Numerical model guidance can influence the forecaster's expectations. The LFM 12 h forecast valid for that evening (Fig. 29.15) verified rather well, in synoptic-scale terms. Since the previous LFM 12 h forecast (not shown) had verified rather well, one could have some confidence in the way the LFM model was handling the situation. Of course, the model does not explicitly predict convection, but it does provide information about large-scale processes relevant to convection (e.g., short wave troughs, fronts). Details of the surface analysis, like the precise location of the cold front or the complicated double low structure in the Ohio Valley, are not likely to be perfectly predicted, in large part because they depend on processes that are not dealt with in large-scale models. In this case, the model output indicates that the synoptic-scale evolution is going to proceed as expected.

More generally, the value of numerical model predictions in a given situation depends on (1) how accurately the model has predicted the large-scale pattern, and (2) how important the large-scale pattern is in determining the significant weather of the day. When the performance of the model is suspect (perhaps because of recent poor predictions), or initialization is obviously bad, the forecaster must decide whether the model solution can be "patched up," perhaps with some simple adjustments. Another of the several operational models may offer a more reasonable prediction. Indeed, one important factor in evaluating model performance is the intercomparison of solutions from different models, but time is often too short for detailed model intercomparison in operational forecasting. Outright rejection of all numerical model solutions is rare, and subjective adjustments are typical operating procedure.

For the case under consideration, large-scale aspects of model performance are apparently well handled. The problem is how to infer the anticipated weather elements from a prediction of the large-scale pattern. The surface predictions of the models are most amenable to modification, because the greatest detail is available at the surface. Detailed hourly diagnosis permits quality checks of the model forecasts and provides details that the model cannot resolve (e.g., convectively driven features such as outflow boundaries).

It is also of interest to consider how objective guidance is applied to making a forecast. The MOS technique is the primary objective tool for translating predictions of operational model variables into forecasts of weather elements. Because MOS is a statistical approach that relates model-predicted variables to observed events, the resulting prediction equations have some characteristics of which forecasters need to be aware (see the discussion by Lowry, 1980). For instance, since the numerical models may perform better for certain variables than for others, it is likely that the variables that are well-treated in the models will be selected in preference to those with poor treatment. This may create MOS prediction equations that do not include variables known to be relevant to predicting a certain phenomenon.

Table 29.2 lists terms of a typical MOS probability of precipitation (PoP) equation (adapted from Schwartz, 1984). Note that humidity variables dominate the equation. If there is reason to believe that the numerical model prediction of humidity is substantially in error for the time and location in question, then the value of the MOS PoP is almost certainly questionable.

Other potential problems are discussed in Lowry (1980), Maddox and Heckman (1982), and Schwartz (1984). Most of the potential for improvement on MOS guidance by forecasters lies in the statistical nature of MOS. That is, in certain situations MOS guidance is likely to be faulty. For example, winter MOS temperature guidance has a tendency to predict too high a minimum temperature in areas with snow cover (snow can be a very efficient heat radiator, depending on factors like its density and how recently it fell).

Forecasters aware of these characteristics can modify the objective guidance to account for them. In effect, MOS has a great deal of skill with weather forecasts for situations with a typical structure. However, statistical methods have difficulty dealing with situations that are on the margins of the statistical distribution. Often, hazardous and unusual weather presents the greatest opportunity to improve on guidance. These situations may be relatively rare at any given location at a given time, but arise somewhere within the country virtually every day (Fig. 29.16).

Table 29.2. MOS-derived LFM predictors in the probability-of-precipitation prediction equation valid 24 h after synoptic time for Indianapolis, Ind.*

Predictor	Coefficient
Mean relative humidity (18)	0.00056
Mean relative humidity (24)	0.00955
Mean relative humidity (18) b=85%	−0.09252
Precipitation amount (24) b=0.01 in	−0.01089
700 mb height (18)	−0.00074
Mean relative humidity (18) b=65%	−0.10220
500 mb dewpoint (12)	0.00638
500 mb v-wind (24) b=10 m s^{-1}	−0.06799
Mean relative humidity (12) b=70%	0.04592
Mean relative humidity (18) b=80%	0.64001
1000–850 mb thickness (18)	0.00262
700 mb dewpoint (24)	−0.01519

*Adapted from Schwartz (1984).

Note: Predictors are in order of selection by the screening regression algorithm. Numbers in parentheses following the predictor denote the time in hours into the LFM forecast, and b=binary cutoff threshold (binary predictors only.)

A careful diagnosis of the weather patterns must be combined with a knowledge of the types of situations for which the objective technique is likely to fail (see, for example, Charba, 1979). Forecasters need to know how objective guidance is derived, upon what data the guidance depends, and the strengths and weaknesses of the guidance. In an operational setting, this knowledge may not be acquired easily.

Objective guidance is often used as a standard by which to judge the quality of subjective forecasts. The nature of subjective forecasting precludes making the correct modifications to guidance on every occasion. However, if

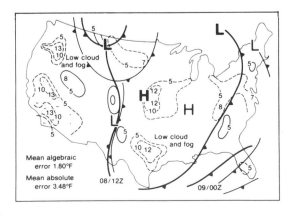

Figure 29.16. MOS maximum temperature forecast errors for 8 February 1981, including surface features at 1200 GMT on the 8th (heavy) and 0000 GMT on the 9th (light). Solid contours are MOS forecasts that were too warm; dashed contours are MOS forecasts too cold. Intervals are 5°F; local extrema are indicated.

to avoid mistakes, forecasters make no modifications at all, many opportunities to provide better service to the forecast consumer are lost.

29.4.3. Structured Forecasting

Order is imposed on forecasting activities to ensure that they are productive and that important factors of the diagnosis/prognosis are not overlooked. The order may take the form of a Standard Operating Procedure (SOP), a parameter checklist (e.g., Miller, 1972), a decision tree (e.g., Cahir *et al.*, 1981), or some other guide. Allen (1981) refers to these as decision aids. The use of such aids has many positive aspects: It allows inexperienced forecasters to participate productively; it offers a method for documenting and preserving the lessons learned by experienced personnel, it provides a basis for guiding research. Glahn (1985) summarized many of the possible approaches to structured forecasting by objective methods. Structures are discussed here as examples of what can be done to incorporate subjective judgment into structured methods.

Regardless of type, a structured forecasting system must be continually monitored. Situations in which the technique fails should be noted. Flaws may surface only in rare circumstances. No structured approach should be finalized in the sense of not allowing for change. Structure has value, but it should never become inflexible. Forecasters require enough freedom to exercise their meteorological knowledge. Probably the best path is between the extremes of no structure and too much rigidity.

Checklists

The parameter checklist draws upon the experience of forecasters to create a list of variables ordered according to their importance to the forecast problem. The forecaster determines a value (current and/or projected) for each parameter by assessing its contribution toward occurrence (or nonoccurrence) of the event. Often, the parameters are arranged in terms of their importance in past events of the same type. The well-known Miller (1972) checklist for severe thunderstorm forecasting is reproduced in Table 29.3.

When the checklist is complete, an important question remains unanswered. Unless indications about the likelihood of the events are nearly unanimous, it is not always clear how to interpret the results. That is, some parameters may favor the occurrence of the event, some may indicate that the event is unlikely, and some may be neither favorable nor unfavorable. In fact, this is a typical outcome. Since it is possible to make more than a single forecast from the data, the checklist is not totally objective. How does one make a decision in this situation? The order of the checklist guides the decision; interpretation is tilted in the direction implied by the most significant parameters.

Another problem that arises with checklists is the ordering. It is often done subjectively, and no quantitative weights are assigned to each parameter. Further, it is not obvious that the importance (relative or absolute) of each parameter is constant from case to case. Maddox and Doswell (1982)

Table 29.3. Parameter checklist for severe convective weather

Rank	Parameter		Weak	Moderate	Strong
1	500 mb vorticity		Neutral or negative vort. advection	Contours cross vort. pattern $\leq 30°$	Contours cross $> 30°$
2	Stability	Lifted Index	-2	-3 to -5	-6
		Totals	< 50	50–55	> 55
3	Middle level	Jet	35 kt	35–50 kt	50 kt
		Shear	15/90 n mi	15–30/90 n mi	30/90 n mi
4	Upper level	Jet	55 kt	55–85 kt	85 kt
		Shear	15/90 n mi	15–30/90 n mi	30/90 n mi
5	Low-level jet		20 kt	25–34 kt	35 kt
6	Low-level dewpoint		8°C	8°–12°C	12°C
7	850 mb max-temp field		E of moist ridge	Over moist ridge	W of moist ridge
8	700 mb no-change line		Winds cross line $\leq 20°$	Winds cross line 20–40°	Winds cross line $> 40°$
9	700 mb dry-air intrusion		Not available, or available but weak wind field	Winds from dry to moist intrude at an angle of 10°–40° and are at least 15 kt	Winds intrude at an angle of 40° and are at least 25 kt
10	12 h surface pressure falls		< 1 mb	1–5 mb	> 5 mb
11	500 mb height change		30 m	30–60 m	> 60 m
12	Height of wet-bulb zero above surface		Above 11000 ft Below 5000 ft	9000–11000 ft 5000– 7000 ft	7000–9000 ft
13	Surface pressure over threat area		1010 mb	1010–1005 mb	1005 mb
14	Surface dewpoint		55°F	55°–64°F	65°F

showed this with respect to the Miller checklist. By assigning an implicit weight to each parameter, one effectively has defined what a "typical" event looks like. No matter how carefully this is done (up to and including statistical methods), one still has to deal with a substantial number of important but atypical cases.

Decision Trees

Decision trees offer many of the same advantages and disadvantages associated with parameter checklists. Decision trees simulate the mental pro-

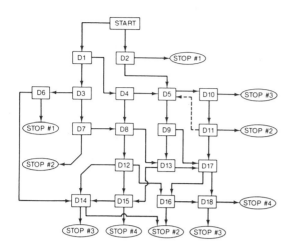

Figure 29.17. Sample schematic of a decision tree.

cesses of a human forecaster. But there is a limited number of outcomes, so the decision tree combines subjectively with more objective methods. In operation, the decision tree presents a series of questions about the current situation, each requiring a decision, which determines the next question. Questions might deal with parameter values (as in a checklist) but may also require decisions involving pattern recognition. The answers may be binary (i.e., yes or no), or they may involve more than two branches. This system allows one to account for different variables (or categories of variables) depending on the situation, an advantage over simple checklists.

In a hypothetical, more flexible decision tree based on Miller's (1972) severe weather forecasting guide, for example, it would be possible to focus on different parameters (and/or different values of the parameters) for five basic severe weather synoptic patterns. The ordering of parameters could also be different for each basic pattern. In such a decision tree, the first decision would concern which pattern type most closely fits the current situation. Subsequent questions would depend on which branch was entered.

The ordering of parameters is far more obvious in a decision tree than in a checklist. Certain questions arise only if the answers to previous questions permit. It is possible to construct alternate paths to a particular intermediate decision box, but all possible paths must terminate in a final decision; infinite loops are not permitted.

Figure 29.17 illustrates several aspects of decision trees. In this example, there are four possible final decisions, e.g., (1) no event, (2) marginal event, (3) moderate event, and (4) severe event. There are 19 decision boxes (including "start"), some of which can be accessed by only one path. The path indicated by the dashed line from Decision 11 (D11) back to Decision 5 (D5) offers the possibility of an endless loop, unless the questions are arranged to prevent it. That is, if one takes the D11–D5 path, the results of the second trip through D5 must lead to D9 or Stop #3.

Expert Systems

Recently, "expert systems" (or "artificial intelligence") have been developed (see Gaffney and Racer, 1983) that produce a forecast roughly comparable with a screening regression, but based on input from an expert in the problem. The expert weighs input variables, perhaps from a checklist, to provide a quantitative estimate of the event's likelihood. This approach is objective in the sense that once the weights have been assigned, the algorithm produces only one interpretation of the data; however, the weighting is subjective. It is possible for such a system to "learn" from past success or failure. As in screening regression, the expert system's end result usually is a probability of event occurrence.

29.4.4. Empirical Forecasting

An undeniably important forecasting tool is experience. An excellent way to document and share the benefits of experience is the "local study." It may be as simple as a climatology for specific weather elements (severe thunderstorms, fog, flash floods, windstorms, etc.) or as complex as a detailed analysis of the physics of some local phenomenon. For a small commitment of resources the potential value in forecasting can be substantial. Perhaps just as important is the fact that such studies can increase understanding of how local weather systems operate.

29.4.5. The Role of Verification

The process of improving the short-range forecast must begin with an appreciation of the present forecast quality. This appreciation may not be easily gained. Every verification scheme contains, implicitly or explicitly, a statement of what constitutes an ideal forecast. For example, statistics of forecast temperature errors carry the implicit assumption that the ideal temperature forecast is without error. Probability forecasts have different implications (e.g., it should precipitate 20% of the time, on the average, in all situations where the forecast precipitation probability is 20%).

The problems in designing forecast verification schemes are discussed by Colls *et al.* (1981), Mason (1982), and Murphy and Daan (1985). One should try to develop methods that contain implicit ideals that can be achieved. For example, as pointed out by Weiss (1977), the National Weather Service severe weather outlook categories originally had unrealistically high event densities. When this was corrected, another problem was uncovered: one could maximize one's verification score by recognizing the characteristics of the verification technique (see Winkler and Murphy, 1968, for a discussion on "proper" scoring rules). Specifically, the scheme verified equally when all the events occurred within a small cluster and when the events were more or less evenly distributed throughout the outlook area. Thus, it was possible to increase the likelihood of a good score by never issuing a high density category outlook. This problem was recognized and corrected (Weiss *et al.*, 1980), but it illustrates the difficulty of devising a verification technique without loopholes.

Verification schemes serve a variety of purposes—to determine how accurate forecasts are, how useful they are to the users, what situations typically are poorly forecast, what forecasting strategies tend to work, which individuals are in need of training, and so on. Comparative, or relative, verification is a tool of great scientific value, since it allows choices to be made among competing models. Every model, be it numerical, mathematical, statistical, empirical, conceptual, or whatever, makes predictions. If competing models are tested under equal conditions, the results can offer guidance about which model performs better. Such comparisons are an essential part of the scientific method (Popper, 1962).

It is easy to see why comparative forecast verification is employed. On one level, forecasts are simple predictions and it is natural (and scientific) to submit them to verification. However, when verification statistics are compared (for individuals, stations, states, etc.), caution must be exercised. Verifications must be for the same places, the same time periods, the same scoring rules, etc., if the comparison is to be valid. One might argue that over a long enough sample, the differences tend to even out, but this is not always the case. For instance, comparison of the forecasting skill of a group of forecasters in San Diego, Calif., with another at Amarillo, Tex., would have to account for the differences in weather variability between the two locations. Comparisons are possible only if the effect of unequal conditions is minimized. Because of the difficulty in making a single approach serve all the purposes of verification, various coexisting schemes are needed.

29.5. An Example of Short-Range Forecasting

A local severe thunderstorm situation does not exhaust the range of issues involved in short-range forecasting or in forecasting convection, but it is fairly typical of day-to-day convective problems. Although severe convection developed in the case discussed, it was not particularly intense or widespread. For additional discussion see Doswell (1984).

29.5.1. The Convective Outlook

Anticipation of a day's convective activity begins no later than the arrival of the upper-air data late in the previous day (0000 GMT for the United States). These data constitute a 24 h forecast. If there is ongoing convective weather when the evening data arrive, the forecaster may be unable to make a proper diagnosis. If the convection declines, there may be opportunities for attention to the outlook. However, during the warm season intense convective mesosystems often persist throughout the night (Maddox, 1980), so a detailed analysis may never be accomplished.

The current state of meteorological science requires that a 24 h forecast be a large-scale forecast. It is possible to speculate about the influence of mesoscale systems, but there is no way to provide mesoscale detail in an operational 24 h forecast. Instead, the forecaster makes the best possible large-scale forecast, recognizing that large-scale systems provide the ingredients for a weather element, while mesoscale systems determine whether

Figure 29.18. Convective outlook product from the National Severe Storms Forecast Center at 0800 GMT, valid from 1200 GMT, 6 May 1983, to 1200 GMT, 7 May 1983.

the potential of those ingredients is realized. Since mesoscale detail is not included, the convective outlook may not be a "weather" forecast, as defined in Sec. 29.1.

On 6 May 1983, the National Weather Service's Severe Local Storm (SELS) forecast unit issued a convective outlook at 0800 GMT (Fig. 29.18). This product would have been available to the previous shift of forecasters and might have been a topic of discussion at the shift change briefing. A forecaster in Kansas could have been thinking about the problem of possible convection during the day shift of the previous day, so the new SELS outlook could confirm or contradict an earlier assessment.

Input data for that SELS outlook include the previous evening's LFM run, which the day shift forecaster may not have seen. At the start of the

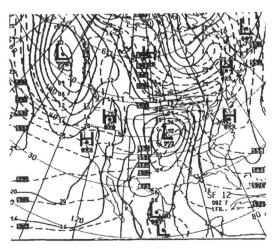

Figure 29.19. LFM 12 h surface prediction for 0000 GMT, 7 May 1983 (details as in Fig. 29.15).

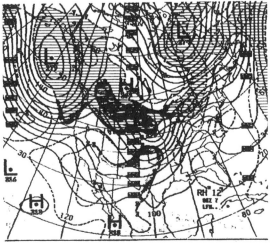

Figure 29.20. LFM 12 h 700 mb prediction valid 0000 GMT, 7 May 1983. Solid lines are geopotential heights (dam); dashed lines show surface to 500 mb mean relative humidity (% × 0.1). Hatching indicates relative humidity ≥ 70%.

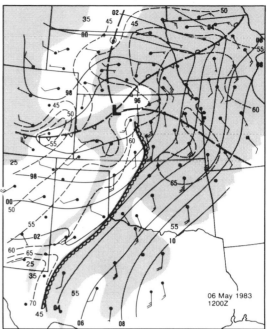

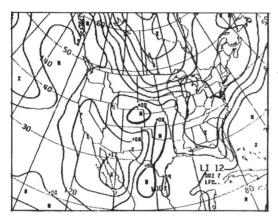

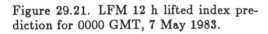

Figure 29.21. LFM 12 h lifted index prediction for 0000 GMT, 7 May 1983.

Figure 29.22. Surface analysis for 1200 GMT, 6 May 1983. Thin solid lines are sea-level isobars (mb, last two digits only). Thin dashed lines are isotherms (°F). Isodrosotherms are along the boundaries of stippling (°F).

day shift, there may or may not be time to consider that earlier model forecast run, the data upon which it was based, and the apparent events of the previous evening and night. ("Apparent" is emphasized because the convective events of a given day may not be well-documented until months later.) Within a few hours, the morning LFM prognoses begin to arrive and analysis of the 1200 GMT upper-air data starts.

For this case, the LFM 12 h forecast valid at 0000 GMT shows cyclogenesis in the central plains (Fig. 29.19), bringing a dry intrusion into Kansas as the cyclone develops (Fig. 29.20), but having a rather modest instability ahead of it (Fig. 29.21). The morning surface data (Fig. 29.22) show why the instability is limited: there is relatively little low-level moisture. Furthermore, Topeka's 1200 GMT sounding reveals a strong inversion between 850 and 700 mb (Fig. 29.23).

The synoptic pattern is generally favorable for severe convection, which is consistent with the SELS outlook. The potential for severe convection is present, but moisture is marginal and the strength of the restraining inversion is not known. According to the narrative (Fig. 29.24) that accompanies the SELS outlook, the increasing southerly flow accompanying cyclogenesis may bring in more moisture at low levels and the dryline may provide enough lift

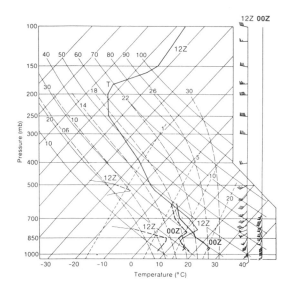

Figure 29.23. Soundings at 1200 GMT, 6 May 1983 (thin) and at 0000 GMT, 7 May 1983, for Topeka, Kans., on a skew-T, log-p diagram.

to allow the release of the convective instability. These will be the subjects of the forecaster's continuing attention.

Although not shown here, the last several LFM forecast runs had been verifying rather well, so the most recent run is likely to be correct, at least in its basic features. The details in the morning surface analysis have to be monitored during the day shift, especially those related to features discussed in the SELS outlook narrative.

29.5.2. Severe Weather Watch

From the morning diagnosis, a forecaster in Kansas would have a basis for writing forecasts that include some potential for severe weather within the Kansas area of responsibility. Because the threat of severe weather is not unambiguous, the forecasts probably do not contain the strongest possible wording. Nevertheless, the forecaster needs to be prepared to react to a developing situation.

Severe convection has broken out during the afternoon in Nebraska and Iowa, in association with the developing cyclone (as described in the outlook). However, things are relatively quiet along the dryline in Kansas. The 2100 GMT surface data (Fig. 29.25), late in the forecaster's day shift, show that the morning LFM large-scale forecast is in good agreement with the unfolding events. Surface dewpoints in the southerly flow ahead of the dryline

```
UPR AIR PATTERN IS FCST TO AMPLIFY OVR CNTRL PLAINS BY TNGT WITH ASSOCD SFC LO
MOVG INTO NRN MO BY 07/12Z.  CURRENT TSTM ACTVTY MOVG ACRS NEB AND NERN KS
EXPCD TO REINFORCE DVLPG WRMFNTL BNDRY ALG NRN KS-EXTRM NRN MO LN WHICH WILL
LIKELY PROVIDE FOCUS FOR RDVLPMNT LATER THIS AFTN.   LO LVL MSTR REMAINS
MARGINAL ATTM...HOWEVER AN INCRS IN SFC DEW POINTS SHOULD OCCUR AS THEY SLY LO
LVL FLO STRENGTHENS THRU THE EASTERN PLAINS.   ISOLD SVR TSTMS ALSO PSBL SWD
ALG DRYLINE LATE THIS AFTN AND ERY TNGT TO THE N OF THE PROGGD UPR LVL JET
ACRS OK.
```

Figure 29.24. Narrative accompanying convective outlook shown in Fig. 29.18.

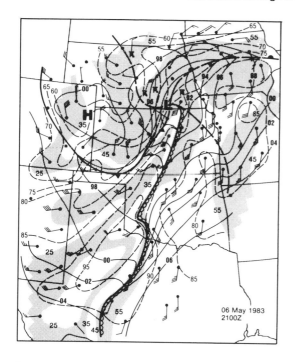

Figure 29.25. Surface analysis for 2100 GMT, 6 May 1983.

have crept above 55°F, but only towering cumulus have developed along the dryline proper (Fig. 29.26). Thus, until 2100 GMT the restraining inversion has been too strong to be overcome by the relatively weak dryline forcing.

But what about the cold front sweeping out of northwest Kansas? The 1200 GMT surface analysis showed no sign of the intense gradient or the strong northwesterly winds. At 1956 GMT, SELS issued Tornado Watch

Figure 29.26. Visible satellite image for 2130 GMT, 6 May 1983. Note the small towering cumulus in northeastern Kansas.

```
BULLETIN - IMMEDIATE BROADCAST REQUESTED
TORNADO WATCH NUMBER 136
NATIONAL WEATHER SERVICE KANSAS CITY MO
256 PM CDT FRI MAY 06 1983

A..THE NATIONAL SEVERE STORMS FORECAST CENTER HAS ISSUED A TORNADO WATCH FOR

        EXTREME SOUTHEASTERN NEBRASKA
        PARTS OF EXTREME WESTERN MISSOURI
        MOST OF EASTERN KANSAS

FROM 400 PM CDT UNTIL 1000 CDT THIS FRIDAY AFTERNOON AND EVENING.

TORNADOES...LARGE HAIL AND DAMAGING THUNDERSTORM WINDS ARE POSSIBLE IN THESE
AREAS.

THE TORNADO WATCH AREA IS ALONG AND 80 STATUTE MILES EITHER SIDE OF A LINE
FROM BEATRICE NEBRASKA TO 30 MILES SOUTH SOUTHEAST OF CHANUTE KANSAS.

REMEMBER...A TORNADO WATCH MEANS CONDITIONS ARE FAVORABLE FOR TORNADOES AND
SEVERE THUNDERSTORMS IN AND CLOSE TO THE WATCH AREA.  PERSONS IN THESE AREAS
SHOULD BE ON THE LOOKOUT FOR THREATENING WEATHER CONDITIONS AND LISTEN FOR
LATER STATEMENTS AND POSSIBLE WARNINGS.

C...TORNADOES AND A FEW SVR TSTM WITH HAIL SFC AND ALF TO 3 IN. EXTRM TURBC
AND SFC WND GUSTS TO 65 KT. A FEW CBS WITH MAX TOPS TO 500. MEAN WIND VECTOR
24030.

D...STG DRY PUNCH MOVG NEWD OUT OF SWRN KS EXPCD TO INITIATE TSTMS ALG DRY LN
SWD FROM SFC LOW PSN NR CNK. REST LOW LVL MOIST AXIS FROM CNU TO GVW. SFC LOW
TRACK EXPCD EWD INTO NERN KS.
```

Figure 29.27. Tornado Watch No. 136. Note especially section D.

No. 136 (which became valid at 2100 GMT for most of eastern Kansas), on the basis of a surge of dry air (Fig. 29.27), but no mention was made of this cold air surge. As discussed in Doswell (1984), this cold front apparently had complex origins, but the first clue to its presence appeared in the pressure change analysis (Fig. 29.28a) at 1500 GMT, and it evolved as shown in Fig. 29.28b.

At 2100 GMT the forecaster faced an interesting situation. A tornado watch sets in motion a series of standard operating procedures to deal with the evolving severe weather potential, but nothing had developed yet.

Shortly after 2100 GMT, the cold air surge collided with the dryline west of Topeka, Kansas. This clearly was the forcing needed to break the capping inversion, because the towering cumulus along the dryline rapidly erupted into a line of severe thunderstorms (Fig. 29.29). As the intersection point traveled south along the dryline, the most intense severe weather followed it into northern Oklahoma. Farther south, in central and southern Oklahoma, the weakening "frontal" surge was no longer able to overcome the inversion and severe thunderstorm activity ceased.

This situation exemplifies the important role of mesoscale processes in the creation of weather elements. It is difficult to anticipate events unfolding on short time scales (say, 1–6 h). However, monitoring of the data is most likely to be fruitful when one can recognize the significance of the unfolding

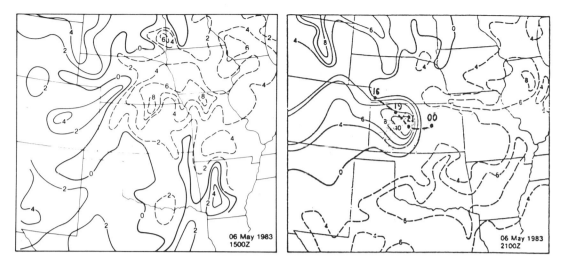

Figure 29.28. Analysis of changes in altimeter setting (left) at 1500 GMT and (right) at 2100 GMT. Solid lines show rises and dashed lines show falls (in hundredths of an inch of mercury). The track of the rise center is superimposed on the 2100 GMT analysis; dots show its inferred location at the indicated GMT times.

events. In this case, the anticipated activity along the dryline developed more or less as expected, but for unexpected reasons. It is conceivable that without the surge of cold air to enhance the lift along the dryline, the activity might never have developed, and the potential for severe convection would have gone untapped.

Considering the complexity of mesoscale processes and the ways they can interact with processes on other scales, it may be concluded that many

Figure 29.29. Visible satellite image for 0000 GMT, 7 May 1983.

forecasts end up verifying well for reasons that were not foreseeable 24 h in advance. Conversely, many essentially correct forecasts may go awry because of mesoscale details.

Although the treatment of this case study necessarily has been brief, short-range forecasting calls upon a wide range of knowledge and skills. The most essential component on the human side of the forecasting process is the diagnostic step, which allows one to apply basic physical understanding to the data at hand. Far from being an onerous burden, weather map analysis is critical if humans are to contribute value over and above that of computerized forecasts.

REFERENCES

Allen, G., 1981: Aiding the weather forecasting: Comments and suggestions from a decision analytic perspective. *Aust. Meteor. Mag.*, **29**, 25–29.

Atkinson, B.W., 1981: *Mesoscale Atmospheric Circulations.* Academic Press, London, 496 pp.

Browning, K.A. (Ed.), 1982: *Nowcasting.* Academic Press, New York, 256 pp.

Cahir, J.J., J.M. Norman, and D.A. Lowry, 1981: Use of a real-time computer graphics system in analysis and forecasting. *Mon. Wea. Rev.*, **109**, 485–500.

Charba, J.P., 1979: Recent performance of operational two-six hour objective forecasts of severe local storms on outbreak days. Preprints, 11th Conference on Severe Local Storms, Kansas City, Mo., American Meteorological Society, Boston, 600–607.

Colls, K.E., I.B. Mason, and F.A. Daw, 1981: A forecast verification procedure for public weather forecasts. *Aust. Meteor. Mag.*, **29**, 9–23.

Doswell, C.A., III, 1984: Mesoscale aspects of a marginal severe weather event. Preprints, 10th Conference on Weather Forecasting and Analysis, Clearwater, Fla., American Meteorological Society, Boston, 131–137.

Doswell, C.A., III, L.R. Lemon, and R.A. Maddox, 1981: Forecaster training—a review and analysis. *Bull. Amer. Meteor. Soc.*, **62**, 983–988.

Doswell, C.A., III, F. Caracena, and M. Magnano, 1985: Temporal evolution of 700–500 mb lapse rate as a forecasting tool: A case study. Preprints, 14th Con-

ference on Severe Local Storms, Indianapolis, Ind., American Meteorological Society, Boston, 398–401.

Gaffney, J.E., Jr., and I.R. Racer, 1983: A learning interpretive decision algorithm for severe storm forecasting support. Preprints, 13th Conference on Severe Local Storms, Tulsa, Okla., American Meteorological Society, Boston, 274–276.

Glahn, H.R., 1985: Statistical weather forecasting. In *Probability, Statistics, and Decision Making in the Atmospheric Sciences*, A.H. Murphy and R.W. Katz (Eds.), Westview Press, Boulder, Colo., 289–335.

Glahn, H.R., and D.A. Lowry, 1972: The use of model output statistics (MOS) in objective weather forecasting. *J. Appl. Meteor.*, **11**, 1203–1211.

Hales, J.E., 1979: A subjective assessment of model initial conditions using satellite imagery. *Bull. Amer. Meteor. Soc.*, **60**, 206–211.

Haltiner, G.J., and R.T. Williams, 1980: *Numerical Prediction and Dynamic Meteorology.* John Wiley and Sons, New York, 477 pp.

Huschke, R.E., 1959: *Glossary of Meteorology.* American Meteorological Society, Boston, 639 pp.

Lowry, D.A., 1980: How to use and not to use MOS guidance. Preprints, 8th Conference on Weather Analysis and Forecasting, Denver, Colo., American Meteorological Society, Boston, 11–12.

Maddox, R.A., 1979: A methodology for forecasting heavy convective precipitation and flash flooding. *Natl. Wea. Dig.*, **4**, 30–42.

Maddox, R.A., 1980: Mesoscale convective complexes. *Bull. Amer. Meteor. Soc.*, **61**, 1374–1387.

Maddox, R.A., and C.A. Doswell III, 1982: An examination of jet stream configurations, 500 mb vorticity advection, and low-level thermal advection patterns during extended periods of intense convection. *Mon. Wea. Rev.*, **110**, 184–197.

Maddox, R.A., and B.E. Heckman, 1982: The impact of mesoscale convective weather systems upon MOS temperature guidance. Preprints, 9th Conference on Weather Forecasting and Analysis, Seattle, Wash., American Meteorological Society, Boston, 214–218.

Mason, I., 1982: A model for assessment of weather forecasts. *Aust. Meteor. Mag.*, **30**, 291–303.

Miller, R.C., 1972: Notes on Analysis and Severe Storm Forecasting Procedure of the Air Force Global Weather Central. Air Weather Service Tech. Rep. 200 (rev.), Air Wea. Serv. Hq., Scott AFB, Ill., 190 pp.

Murphy, A.H., and H. Daan, 1985: Forecast evaluation. In *Probability, Statistics, and Decision Making in the Atmospheric Sciences*, A.H. Murphy and R.W. Katz (Eds.), Westview Press, Boulder, Colo., 379–437.

Pielke, R.A., 1984: *Mesoscale Meteorological Modeling*. Academic Press, London, 612 pp.

Popper, K.R., 1962: *Conjectures and Refutations: The Growth of Scientific Knowledge*. Basic Books, New York, 417 pp.

Rodgers, D.M., M.J. Magnano, and J.H. Arns, 1985: Mesoscale convective complexes over the United States during 1983. *Mon. Wea. Rev.*, **113**, 888–901.

Sanders, F., 1971: Analytic solutions of the nonlinear omega and vorticity equations for a structurally simple model of disturbances in the baroclinic westerlies. *Mon. Wea. Rev.*, **99**, 393–407.

Sanders, F., and L.F. Bosart, 1985: Mesoscale structure in the megalopolitan snowstorm of 11–12 February 1983. Part I: Frontogenetical forcing and symmetric instability. *J. Atmos. Sci.*, **42**, 1050–1061.

Schwartz, B.E., 1984: Typical warm season MOS guidance errors. Preprints, 10th Conference on Weather Forecasting and Analysis, Clearwater Beach, Fla., American Meteorological Society, Boston, 50–56.

Snellman, L.W., 1977: Operational forecasting using automated guidance. *Bull. Amer. Meteor. Soc.*, **58**, 1036–1044.

Tepper, M., 1959: Mesometeorology—the link between macroscale atmospheric motions and local weather. *Bull. Amer. Meteor. Soc.*, **40**, 56–72.

Weiss, S.J., 1977: Objective verification of the severe weather outlook at the National Severe Storms Forecast Center. Preprints, 10th Conference on Severe Local Storms, Omaha, Nebr., American Meteorological Society, Boston, 395–402.

Weiss, S.J., D.L. Kelly, and J.T. Schaefer, 1980: New objective verification techniques at the National Severe Storms Forecast Center. Preprints, 8th Conference on Weather Forecasting and Analysis, Denver, Colo., American Meteorological Society, Boston, 412–418.

Weiss, S.J., and E.W. Ferguson, 1982: An experiment in medium range quantitative severe local storm forecasting: Preliminary results. Preprints, 12th Conference on Severe Local Storms, San Antonio, Tex., American Meteorological Society, Boston, 116–119.

Winkler, R.L., and A.H. Murphy, 1968: "Good" probability assessors. *J. Appl. Meteor.*, **7**, 751–758.

Zipser, E.J., 1983: Nowcasting and very-short-range forecasting. In *The National STORM Program: Scientific and Technological Bases and Major Objectives*, Univ. Corp. for Atmos. Res., Boulder, Colo., 6-1 to 6-30.

CHAPTER 30

An Overview of
Numerical Weather Prediction

Bruce B. Ross

30.1. A Brief History of Operational Numerical Weather
Prediction in the United States

Starting with the first barotropic model experiments of Charney and
Von Neumann in 1950 (Charney *et al.*, 1950), numerical models of the atmosphere have developed into the primary means by which forecasters are
able to predict synoptic-scale weather beyond 6 h. The success of the early
numerical experiments of Charney led, in 1954, to the formation of the Joint
Numerical Weather Forecasting Unit in Washington, D.C., with the purpose
of developing operational versions of these research models. The first operational numerical model, a geostrophic barotropic model of the Northern
Hemisphere, was introduced as an objective forecasting tool for National
Weather Service (NWS) forecasters in 1958. This model (Cressman, 1958)
contained much of the same physics as had been used by Charney and his
associates in 1950. However, because of improvements in computer speed
and capacity,* enhanced communication capabilities, and refinements in the
original barotropic model, two-dimensional barotropic forecasts out to several days could be provided in a timely fashion to field forecasters. Significant
improvements in 36 h, 500 mb predictions over subjective forecasts resulted
from the introduction of this first objective forecast.

Figure 30.1 summarizes the increase in 36 h forecasting skill between
1955 and 1981 as operational forecasting models were improved. The initial
improvements that the barotropic model made in the 500 mb forecasts were
followed by improvements in the surface forecasts as well when a three-level
baroclinic filtered-equation (geostrophic) model (an outgrowth of Charney's
[1954] *n* level research model) became operational in 1962. During this period, an increasing number of rawinsonde stations, located in North America, on some 20 ships, and in foreign countries, were providing upper-air

---

* In the 1950 experiments, the ENIAC computer had required 36 h to complete a 24 h
forecast (Platzman, 1979).

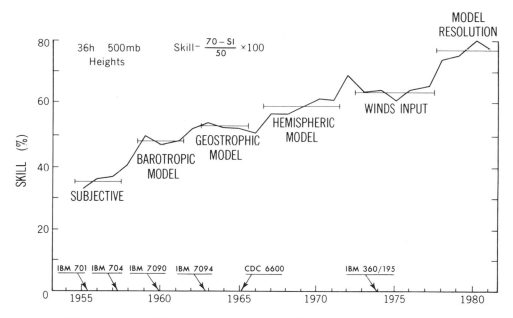

Figure 30.1. The accuracy of short-range weather prediction from the time prior to numerical weather prediction to 1981. The curve indicates the skill, averaged annually, of subjective and objective forecasts in predicting 500 mb heights at 36 h. The skill parameter is related, by the formula shown, to the so-called S_1 score (Teweles and Wobus, 1954) which is a measure of the normalized error in horizontal pressure gradients. Important changes in operational models and installation dates for different computer systems at NMC are indicated. (Data from the NMC; figure courtesy of William Bonner and Kiku Miyakoda.)

data twice daily for use as initial conditions for the hemispheric model forecasts. The six-level, hemispheric, primitive equation (PE) model (Shuman and Hovermale, 1968), which became operational in 1966, likewise benefited from the earlier research of others as well as the extensive development work at the National Meteorological Center (NMC), which was formed in 1961.

The Limited-area Fine-Mesh (LFM) model, introduced in 1971, was a limited-domain version of this hemispheric PE model. The LFM model provided increased horizontal resolution over North America, the region of primary importance for short-range forecasts of synoptic weather systems. This model represented the first operational version of a regional forecasting system and contained some features of what we would now call mesoscale numerical models. As an outgrowth of the limited-area model, the higher resolution Movable Fine-Mesh (MFM) model was introduced in 1975 for use in an "on-call" basis to forecast hurricane movement as well as intense precipitation patterns in severe weather conditions.

About 1972, observed winds were incorporated directly into models' initial conditions; prior to this, initial wind fields were derived from observed geopotential heights by means of a balance equation. A data assimilation system was made operational in 1974 to improve the ongoing analysis of observational data and their incorporation into the global PE model. In 1978, the horizontal grid size of the hemispheric model was halved to match that of the LFM model. Finally, in 1980, the grid-point hemispheric PE model was replaced by a 12 level global spectral model that predicts large-scale features for periods of 5–10 days.

30.2. Comparison of Mesoscale and Global-Scale Numerical Models

The evolution of operational numerical weather prediction (NWP) from larger to smaller grid scales partially reflects the increased computer power that has allowed global models to resolve more details of atmospheric flow fields. However, more specialized limited-area models, such as the LFM and MFM operational models, have also permitted the simulation and forecasting of subsynoptic and mesoscale weather phenomena.

In fact, just as early operational numerical models were preceded by experimental research models, so also the increased emphasis of operational NWP on smaller scales has been preceded by an increased research interest in mesoscale modeling. The last ten years have seen considerable advancement in the simulation of many mesoscale phenomena in a research environment. The diversity of mesoscale modeling efforts has reflected the great diversity of mesoscale phenomena. Hence, the different types of mesoscale models have traditionally been categorized as follows. (Relevant review papers are listed after each; also see Haltiner and Williams, 1980, for a review of numerical weather prediction, and Pielke, 1984, for a general review of mesoscale modeling.)

- Regional-scale phenomena (Kreitzberg, 1979; Anthes, 1983; UCAR, 1983).
- Hurricanes (Anthes, 1982; Ooyama, 1982).
- Orographically forced flows (Pielke, 1981; UCAR, 1983).
- Convective clouds (Cotton, 1975; Schlesinger, 1982; Farley and Orville, 1982; Clark, 1982).

The classifications are somewhat arbitrary. For example, regional-scale simulations will certainly contain terrain effects. Likewise, hurricane models, which use several different grid sizes within the model domain, could conceivably resolve certain aspects of convective cloud phenomena if high-resolution regions of the model had sufficiently small grid size. This section addresses the primary features that distinguish mesoscale models from their more classical global-scale counterparts.

30.2.1. Phenomena To Be Modeled

The phenomena that are to be simulated determine the attributes required of the numerical model. Hence, if one wishes to forecast the evolution of baroclinic waves (with wavelength >2000 km) for a period of several days to a week, a global-scale model such as the NMC spectral PE model would be required, capable of resolving the smallest important baroclinic waves and with representation of physical processes such as radiation, cumulus convection, and boundary layer effects in a manner appropriate to the large space and time scales of the waves.

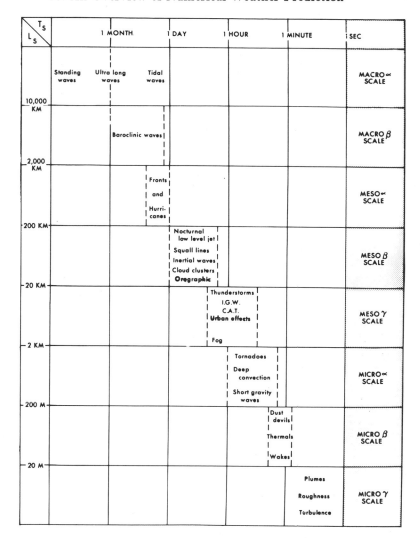

Figure 30.2. Scale definition proposed by Orlanski, with examples of atmospheric processes and their corresponding time and horizontal space scales. Representative physical time scales are given at the top. (After Orlanski, 1975; see also Fig. 2.12, this volume.)

When mesoscale phenomena are to be modeled, however, the spatial and temporal scales to be resolved may be smaller by an order of magnitude or more. Figure 30.2 shows the differences in time and space scales between macroscale and mesoscale weather systems and processes. A numerical model that is to represent these phenomena must be correct not only in its spatial resolution and domain design, but also in the types of physical processes that are included and the way in which they are represented (or parameterized).

30.2.2. Grid Resolution and Domain Size

As in models representing baroclinic waves, the grid size (i.e., the distance between adjacent points in the gridded domain) of a mesoscale model should ideally be determined by the smallest scale associated with the phenomenon of interest. It is desirable that the smallest wave-like features be 4–6 times the grid size in order to reduce numerical (truncation) errors in wave propagation and advection to acceptable levels (see Pielke, 1981). (Such criteria may be difficult to follow when convection is present, since the most unstable scales for convection decrease to zero in a hydrostatic model [Orlanski, 1981].) For example, if one wishes to model the meso-α-scale structure of a hurricane (without trying to resolve the structure of the hurricane eye) with a scale of several hundred kilometers, then a horizontal grid size of roughly 50 km is needed. The integration of a global-scale model with 50 km horizontal resolution and a vertical resolution of 6–10 levels would be a major undertaking even for present-day supercomputers. The obvious solution is to use a limited-area domain, and communicate information about conditions outside the domain through lateral boundary conditions. (Nesting techniques are discussed in Sec. 30.3.)

30.2.3. Parameterization of Physical Processes

Many of the physical processes that are treated in synoptic-scale numerical models should also be included in their meso-α (regional)-scale and meso-β-scale counterparts. However, because of the reduced spatial and temporal scales of the mesoscale model, the processes will have increased importance in the simulation and may require a more detailed treatment than in corresponding large-scale versions. (See Anthes [1983] for a review of different treatments of physical processes in regional-scale models.)

Figure 30.2 helps to explain the increased importance of subgrid-scale processes in mesoscale models. Resolvable scales in synoptic and global systems are confined to the macro-α and macro-β length scales; the predominant subgrid effects, cumulus convection and turbulence, occur on scales of meso-γ and smaller. Hence, these parameterized effects are energetically quite separate in spatial scales from the important processes treated by the synoptic-scale model and can probably be treated effectively in a statistical sense. In a mesoscale model, on the other hand, resolved scales, such as the meso-α and meso-β, border on those categories that must still be parameterized. Hence these subgrid phenomena must be represented in a more detailed manner by the mesoscale model, because they are more energetically significant to the model representation.

Cumulus parameterization provides a good example of the need for different formulations between synoptic-scale and mesoscale models. Cumulus convection may be treated in synoptic-scale models as large cumulus cloud ensembles within a grid box (Arakawa and Schubert, 1974), since the grid spacing is hundreds of kilometers. In a mesoscale model with a grid size of 50 km, however, this statistical representation of convection is not so useful (since cumulus cloud elements each will have a horizontal scale of a kilometer or more and a spatial separation of the same order). A satisfactory parameterization of cumulus convection has yet to be demonstrated for mesoscale models, although a number of different methods have been proposed. In fact, ultimately the best approach to cumulus convection in a mesoscale model may not be to parameterize it, but rather to let the model attempt to resolve it if sufficiently high resolution is feasible (Rosenthal, 1978; Ross and Orlanski, 1978).

As a second example, in a synoptic-scale model used only for short-range forecasts, details of the vertical dynamic and thermodynamic structure of the planetary boundary layer may be of only secondary importance. As shown by Carlson *et al.* (1983), however, such details may be quite important for the mesoscale simulation of convective outbreaks. Therefore, although the representation of boundary layer and surface processes can be fairly simple for a short-range synoptic forecasting model (surface effects become important in synoptic models as the temporal forecast range is extended), a fairly elaborate representation of surface and boundary layer physics may be necessary for mesoscale forecasting of the environment of convective events.

30.2.4. Data Sources

For the simulation of synoptic-scale baroclinic waves, a satisfactory representation of the initial atmospheric state can be derived from the operational rawinsonde network, supplemented by aircraft and satellite data over the data-sparse ocean regions. However, the horizontal resolution of such a data set is sufficient only to resolve coarse mesoscale features. Hence, if it is necessary to initialize a mesoscale forecast with detailed mesoscale features (in analogy with the global model's resolution of baroclinic waves), high resolution observations are needed. For example, detailed observations from aircraft are potentially important for defining the initial storm structure in operational hurricane track forecasts.

Although such mesoscale observations are highly desirable and can be obtained from ground- and satellite-based remote sensing systems, their extensive use in operational forecasting systems is unlikely in the near future. Until such mesoscale data sources become available, idealized initial conditions must be used, as they are in many research models and in the hurricane track forecasting models, or else the model must be allowed to generate its own mesoscale features from larger-scale initial conditions, as has been demonstrated in a number of mesoscale simulations (e.g., Anthes *et al.*, 1982; Orlanski and Polinsky, 1984).

30.2.5. Initialization Procedures

The initialization of mesoscale models is discussed in Ch. 8 and Ch. 25 of this volume. Several points are made here, however, to contrast the initialization of mesoscale and of large-scale models: (1) Whereas initialization procedures that assume geostrophic balance conditions may be quite suitable for modeling synoptic weather conditions, such balance assumptions become progressively less appropriate as the horizontal scale L of the phenomenon decreases. The Rossby number, $Ro \equiv V/(fL)$, which is the ratio of the neglected advection and tendency terms to the Coriolis terms in the momentum equation, is an indicator of the validity of this geostrophic approximation (f is the Coriolis parameter and V is a characteristic wind speed). When

Ro $<<$ 1, advection terms may be neglected and the geostrophic approximation is useful; for Ro $\sim$ 1, the approximation is not valid. For a wind speed
$V \sim 20 \text{ m s}^{-1}$ and $f \sim 10^{-4}\text{s}^{-1}$, the Rossby number will be 0.2, which is
of order one, when the length scale $L \sim 100$ km. Figure 30.2 shows this
to be in the range of meso-β phenomena. (2) When moist convection is an
important feature of the initial conditions, it may be necessary to include
latent heating effects in the initialization.

In summary, methods such as normal model initialization and procedures
using geostrophic balance may have value for large meso-α circulations. However, satisfactory techniques have yet to be developed that permit initialization of meso-β and small meso-α phenomena, particularly those in which
latent heating and gravity wave effects are important.

30.3. The Concept of a Nested Grid

The nested-grid concept is fundamental to all the mesoscale models discussed here. Because of the finite size and speed of computers, numerical
models are capable of resolving phenomena over only a limited range of horizontal scales. For example, regional-scale numerical models, which also are
referred to as meso-α-scale or subsynoptic-scale models, use a horizontal grid
spacing on the order of 50–200 km, so as to be capable of resolving phenomena with scales of several hundred kilometers in the small-scale end of the
meso-α range (Fig. 30.2). Since the domain of such models would typically
extend over at least several thousand kilometers, it is reasonable to expect
that such models would be capable of simulating a variety of phenomena,
ranging in size from the domain size (> 2000 km) down to 3–4 times the grid
spacing (~ 200 km).

In fact, if the computer resources were available, one could create a
mesoscale model spanning the Northern Hemisphere with a 50 km resolution.
Such a model would be capable of resolving all scales from $L > 10,000$ km
(planetary waves) down to $L \sim 200$ km (small meso-α phenomena such as
mesoscale convective systems). However, it is not cost effective to use a
50 km resolution over Europe and Asia if one is interested only in forecasting or simulating mesoscale features over North America. The planetary and
synoptic-scale waves over the Eastern Hemisphere would certainly be important to a medium-range forecast (2–5 days) of North America, but these
waves ($L > 2000$ km) could be very well resolved by a grid size of 200 km or
more.

Hence, a reasonable strategy would be to use a grid that varies in the
horizontal from a fine mesh over the mesoscale area to a coarser mesh over
the rest of the globe. Ideally, one could use a mesh that varied smoothly
from 50 km in the primary region of interest, North America in this case, to
200 km in the regions far from this zone. Another procedure, which is more
convenient, is to define a zone of constant grid resolution that is located or
"nested" within a coarser mesh outer domain.

The ability of a finite-difference model to represent a propagating disturbance depends on the local grid size (particularly when the scale of the
disturbance is 4–6 times the grid size). Therefore, the effect of the interface

between fine and coarse grids upon a disturbance propagating through it will be larger or smaller as the ratio between the inner and outer grid sizes is chosen to be larger or smaller. To reduce this interface effect, one could use a telescoping nested grid, in which each successively coarser grid has a mesh size that is double the size of the next interior mesh.

Phillips (1979) used this basic approach in the Nested-Grid Model (NGM) that is the regional forecasting model in NMC's new Regional Analysis and Forecasting System. The NGM consists of three grids, denoted A, B, and C (Fig. 30.3), as defined on a polar stereographic projection of the Northern Hemisphere.

The grid-nesting approach used in the NGM involves a so-called two-way interaction. This terminology is used because information is communicated not only down-scale from the larger grids A and B to smallest grid C, for example, but also up-scale from grid C to grids B and A. That is, not only will changes in synoptic-scale features in region A be sensed by grid points in region C, but also mesoscale disturbances within region C will influence the larger scale solution in region A.

Most mesoscale models in use today, other than those simulating hurricanes, use only one-way interaction. In this simpler nesting procedure, the nested model receives information from the larger domain model in which it is nested; however, it is unable to influence the outer solution. In fact, the large-domain solution need not be integrated with the one-way nested solution but rather may be run by itself at an earlier time. The previous NMC operational forecasting system with the LFM and global PE models was an example of one-way interaction (Fig. 30.4). After the global PE model was run, its predicted fields were used to provide lateral boundary conditions for the nested LFM model forecast during the next forecast cycle.

Although one-way interaction produces a physical approximation that causes some errors in the numerical solution (Phillips and Shukla, 1973), it also provides certain conveniences to the modeler. The nested mesoscale model can be run repeatedly using different numerical and physical parameters but with the larger-scale host model only run once. The forecast from the host model (e.g., the LFM or spectral PE forecast) is needed so that time-varying lateral boundary conditions can be constructed that are used by the nested simulation. In fact, researchers frequently use observations alone, taken over the time period of the simulation, to prescribe these lateral boundary fields. Note that, in the latter case, the model is no longer a predictive system, since it then requires knowledge about the observed fields along the model boundaries after the initial time.

30.4. Examples of Mesoscale Simulations and Forecasts

Several examples of mesoscale simulations/forecasts will serve to indicate the philosophy behind the design and use of two types of models: (1) regional-scale (meso-α) models treating a springtime severe storm outbreak and (2) nested hurricane models. In the latter case, two different types of simulations/forecasts are presented: (a) simulated changes in the structure of an idealized storm during landfall and (b) examples of storm-track

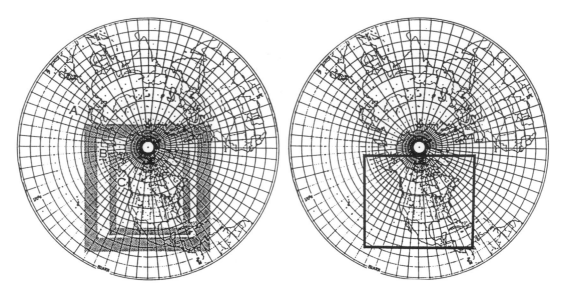

Figure 30.3. The three-grid configuration of the Nested Grid Model (NGM), an example of two-way interaction. The outer grid A is hemispheric. Grids B and C have two and four times, respectively, the resolution of Grid A. The shaded areas indicate zones where adjacent grids overlap. (After Hoke, 1984.)

Figure 30.4. Forecast domain (indicated by heavy line) of the Limited-area Fine-Mesh (LFM) model, an example of a one-way interacting model. Boundary conditions for the LFM model are derived from the global spectral model forecast made during the previous prediction cycle. (After Hoke, 1984.)

forecasts. (Simulations of two other classes of mesoscale phenomena, namely terrain-forced circulations and isolated convective storms, are described in Ch. 22 and Ch. 15 respectively.)

All the models treated here use the hydrostatic approximation (i.e., the vertical momentum equation reduces to a balance between buoyancy and vertical pressure gradient forces; advection and tendency terms are neglected). Most have a primary meso-α-scale outer domain (with resolved horizontal scales ~200–2000 km), although several solutions resolve meso-β scales (i.e., 20–200 km).

30.4.1. Comparison of Regional-Scale Models

Regional-scale numerical models, that is, models that resolve subsynoptic or meso-α-scale phenomena, should be capable of treating a variety of phenomena, ranging from synoptically forced phenomena, such as fronts and frontal circulations, to terrain-induced processes, such as mountain waves and land/sea-breeze effects. Ideally, such models should be able to represent warm-season mesoscale convective systems as well as wintertime snowstorms. For a regional-scale model to simulate all these diverse phenomena properly would seem to require complex and general portrayals of the different physical processes, such as subgrid-scale convection and turbulence, surface energy balance, and radiation, suited to each flow regime and phenomenon to be modeled. However, Anthes *et al.* (1982) demonstrated that a model with rather simple physics can produce realistic simulations of a variety of different mesoscale processes in the same solution. Similarly, Orlanski and Polinsky (1984) produced good simulations of both winter- and springtime precipitation events, using a model with simplified physical processes. However, although such successes are evident in the research literature, one should expect that different models, using different initialization procedures and data and different lateral boundary information, will produce significantly different results. This point is emphasized in the examples that follow.

The LFM model was one of the earliest regional-scale numerical models; its resolution of roughly 160 km in middle latitudes permits it to resolve meso-α phenomena with horizontal scales ~600 km. Operational use of the LFM model has been shown to produce a significant improvement in forecasting skill over North America (Shuman, 1978). However, although predictions of surface temperature and 500 mb winds have improved, accuracy of precipitation forecasts, particularly those indicating locations where rainfall should exceed 1 inch in a 24 h period, has remained relatively unchanged.

Analysis of seasonal variations in the success of LFM rainfall predictions shows that forecast skill is best for winter and worst for summer (Fawcett, 1977). This trend is explainable by the fact that wintertime precipitation is stable and is forced on scales that are meso-α or larger; on the other hand, rainfall during the warm season is often convective, the most unstable scales occurring in the range of meso-γ or small meso-β. Thus, while the winter precipitation scales are largely resolvable by a regional-scale model, the most unstable summertime scales are not and may need to be parameterized. In the latter case, one would hope that the convective system as a whole, which is made up of individual convective elements, is organized by resolvable dynamical processes, such as frontal lifting. If it is not, as in the case of summertime isolated thunderstorms, then the resulting convection will not be predictable, in a deterministic sense, by regional-scale models.

Although a numerical forecast may predict the unstable environment properly, the forecast model may still not be able to produce observed severe weather or precipitation in a given case. In such a case, the failure of the model to produce convective activity at a given location may be due to its inability to resolve the triggering mechanism that initiated the convection. As discussed in Sec. 30.2., any numerical prediction uses incomplete mesoscale initial conditions and approximate subgrid-scale parameterizations to forecast mesoscale phenomena of interest. Hence, once cannot expect a model to provide a deterministic forecast of mesoscale precipitation events when the triggering event, such as a propagating gravity wave or some other meso-β or meso-γ disturbance, is not resolvable by the model or, if it is resolvable, is not represented properly by the initial and/or boundary conditions. This point needs to be emphasized in any discussion of numerical prediction of mesoscale precipitation.

It is instructive to compare precipitation forecasts produced by several different mesoscale models for a single case, the first observing day of SESAME-AVE (Severe Environmental Storms and Mesoscale Experiment–Atmospheric Variability Experiment). For the period of this case (1200

Table 30.1. Characteristics of simulations and forecasts of the first observing day of SESAME-AVE

	Model	Integration period (GMTh/day)	Convective parameterization	Initial condition data source
I	LFM-II [NMC Operational]	(a) 12/10–12/11 (b) 12/9–12/11	Convective Adjustment	NMC observed
II	Anthes et al. (1982) Benjamin & Carlson (1986) [NCAR-PSU]	12/10–12/11	Kuo-type (Anthes)	NMC observed
III	Kaplan et al. (1982) [MASS]	12/10–12/11	Resolved	NMC observed
IV	Mills & Hayden (1983) [ANMRC]	(a) 12/10–03/11 (b) 21/10–03/11	Kuo-type	(a) NMC observed (b) Satellite obs.
V	Orlanski et al. (1983) [GFDL/HIBU]	12/9–12/11	Convective Adjustment	(a) NMC observed (b) FGGE observed
VI	Orlanski & Polinsky (1984) [GFDL/MAC]	12/9–12/11	Resolved	NMC observed
VII	Kalb (1984) [Drexel/LAMPS]	21/10–06/11	Plume-cloud type (Kreitzberg-Perkey)	SESAME observed
VIII	Kuo & Anthes (1984) [NCAR-PSU]	12/10–12/11	Kuo-type (Anthes)	SESAME observed
IX	Chang et al. (1984) [Drexel/LAMPS]	12/10–12/11	Plume-cloud type (Kreitzberg-Perkey)	NMC observed
X	Ross (1983) [GFDL/BES]	12/10-12/11	Resolved	SESAME observed

GMT, 10 April–1200 GMT, 11 April 1979), at least eight different models have been run to produce ten different simulations/forecasts. Table 30.1 summarizes the salient characteristics of each simulation/forecast.

Grid and Domain Sizes

The models may be roughly divided into two groups according to grid size: (1) those in the range of 67–160 km, models I, II, IV, V, VI, and IX, and (2) those in the range of 20–50 km, models III, VII, VIII, and X. Using the criterion that a model resolves phenomena with horizontal scales greater than or equal to roughly four times the horizontal grid size, one could categorize the former group as meso-α models. On the other hand, models in the latter group are resolving some meso-β-scale phenomena (i.e., scales less

Table 30.1. Continued

Model	Lateral boundary data source	Horizontal grid size [middle-latitude] (km)*	Domain array size	No. of vertical levels
I	NMC hemispheric PE forecast	110	79×67	7
II	NMC observed	111	37×37	11
III	NMC observed	41	157×117	14
IV	NMC observed	67	101×71	10
V	GFDL global spectral PE forecast	156	49×27	9
VI	NMC observed	160	21×21	17
		60	51×51	
VII	SESAME observed (21, 00, and 12GMT only)	35	44×50	15
VIII	SESAME observed	50	45×34	11
IX	NMC observed	140	41×35	15
X	SESAME observed	40	21×21	17
		20	41×41	

* Cases II, IV, V, VI, VIII, and X use models with staggered grid meshes, i.e., grid configurations in which not all variables are at the same spatial location. For these models, the grid size shown here, which is the distance between the nearest like-variable points, is an overestimation of the actual effective grid size when compared with the other, unstaggered-grid models in this table.

than 200 km). Regarding domain size, the models in the former group all cover at least two-thirds of the United States. In the latter group, the large domain size of case III allows it to cover most of North America, whereas cases VII and VIII include only the south-central United States and case X covers only part of Oklahoma and Texas. All models except the LFM-II (model I) contain at least nine vertical levels.

Integration Periods

The integration periods of six of the ten cases extend over the 24 h SESAME-I period, 1200 GMT, 10 April, to 1200 GMT, 11 April. The two meso-α GFDL simulations, V and VI, have 48 h periods from 1200 GMT, 9 April, to 1200 GMT, 11 April, as does one of the LFM solutions. Cases IV and VII are integrated for only a portion of the 24 h period.

Parameterization Schemes

Several different parameterization schemes are employed (Table 30.1). Models I and V use a convective-adjustment-type scheme (see, e.g., Miyakoda, 1973). Solutions II and VIII use a Kuo-type parameterization (Anthes, 1977). Case IV also uses a Kuo scheme for one solution but typically involves solutions without convective parameterization. The model used in cases VII and IX employs a parameterization based on a plume-type cloud model (Kreitzberg and Perkey, 1976). Finally, three of the solutions, III, VI, and X, do not specifically parameterize convection but rather include only resolved moist convection.

Data Sources

Four different data sources are used for initial (and lateral boundary) conditions. (The detailed procedures by which the observed data are converted to the model grids may differ considerably among the cases and may have a significant effect on the resulting simulations. See general discussion of the procedures in Chs. 8, 24, 26, this volume. However, such details are beyond the limits of the present discussion.) The designation NMC indicates the standard operational data set used by the National Meteorological Center to initialize its forecast models. Because these are operational data, the data set is restricted to include only those upper-air reports that met the cut-off time for reporting. Solutions I, II, III, IV, Va, VI, and IX use these data. In several of these cases, significant as well as mandatory levels are used from the soundings. Solution II also uses subjective enhancement to emphasize mesoscale features. The second data set, designated FGGE, is used only in solution Vb. These data were collected as part of the 1979 First GARP Global Experiment (FGGE) and were prepared at the Geophysical Fluid Dynamics Laboratory (GFDL) through a dynamic assimilation method (Ploshay *et al.*, 1983). The data encompass the upper-air reports included in the NMC operational analysis as well as late reports not meeting the operational cut-off time and special observations from normally data-sparse regions over the ocean. The third data set, designated SESAME, was obtained from the SESAME field experiment and was employed in solutions VII, VIII, and X. The upper-air data used in these experiments were obtained from a rawinsonde network over the central United States with average spacing of roughly 250 km and with soundings made every 3 h from 1200 GMT 10 April, to 1200 GMT, 11 April (Alberty *et al.*, 1979). Note that solution VII uses reports only at 2100, 0000, and 1200 GMT, whereas solutions VIII and X use all sounding times. Finally, in case IVb, high-resolution satellite sounding

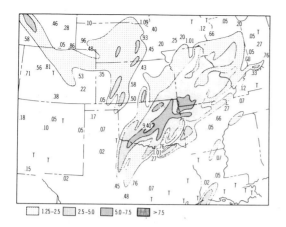

Figure 30.5. Analysis of observed 24 h accumulated precipitation for the period 1200 GMT, 10 April, to 1200 GMT, 11 April 1979. Maximum (9.40 cm) occurred over central Oklahoma. Contours are in centimeters. (Analysis courtesy of the Heavy Precipitation Forecasting Group of NMC.)

data from two orbital passes (2140 and 2200 GMT) over the United States are used to enhance temperature and moisture fields in the model solution at 2100 GMT, 10 April.

Synoptic Conditions

The synoptic conditions associated with the 10–11 April case have been described by several authors (e.g., Moore and Fuelberg, 1981) and are discussed only briefly here as they relate to the discussion that follows. At 1200 GMT, 10 April, a deepening surface low was located over Colorado. A surface cold front extended south into eastern New Mexico, and a warm front ran east-west along the gulf coast. Warm moist air was being advected northward from the Gulf of Mexico beneath a capping inversion over eastern Texas. The warm front moved northward during the period, reaching the Texas-Oklahoma border about 0000 GMT, 11 April, while the cold front slowly moved into western Texas.

Two separate convective systems can be identified to have occurred during the period. The first developed about 1800 GMT, 10 April, to the west and south of the Red River Valley, which forms the Texas-Oklahoma border. The southernmost part of this system then moved east, producing numerous tornadoes and hail storms (see Alberty *et al.*, 1979). A second system began over west-central Texas about 0130 GMT, 11 April, and developed into a slow-moving squall line that extended to the northeast and persisted for the next 9 hours. The combined effect of these two systems was to produce the 24 h accumulation of precipitation shown in Fig. 30.5, the heaviest rainfall having occurred over central Oklahoma about 0300 GMT, 11 April. Of particular note in Fig. 30.5 is the orientation of rainfall fields in a streaked pattern from southwest to northeast over Texas, Oklahoma, and Missouri. This alignment reflects the orientation of the 11 April squall line, which was determined by the strong middle- and upper-level winds over the region. This pattern will be shown to differ quite dramatically from most of the model rainfall patterns which show more of an east-west alignment.

30.4.2. Comparison of Results in Modeled Precipitation

The model precipitation results have been taken from the work of several different groups and therefore tend to differ with regard to the period of accumulation (as well as the contour interval used for each display). The reader should note that the maps of "observed" rainfall differ considerably from each other. This reflects the well-known fact that precipitation patterns, particularly in convective situations like this, have spatial and temporal scales that make them difficult, if not impossible, to analyze in a coherent and objective way from relatively sparse rain-gauge network data (see e.g., Krietzberg, 1979; Wilson and Brandes, 1979). (The Heavy Precipitation Forecasting Group at NMC produced Fig. 30.5 through the use of satellite cloud photography and radar summaries in addition to rain-gauge reports.) Accordingly, the comparison made here can only be qualitative with regard to differences between the simulations and "reality."

Also note that the results of case IV are not discussed here, because Mills and Hayden (1983) did not present explicit precipitation results, except to indicate regions of vertical motion over Oklahoma that might produce precipitation in their model.

Case I

Two LFM precipitation forecasts (case I of Table 30.1) are shown in Fig. 30.6 for the same 24 h period as that in Fig. 30.5. No results from a global PE model are given here; however, relative to the discussion of Sec. 30.2, the LFM results provide some indication of what such a PE prediction might be like. Figure 30.6a shows the accumulated rainfall from the LFM-II forecast, initialized at 1200 GMT, 10 April, most of the rainfall having occurred during the last 12 hours. Figure 30.6b shows the 24–48 h accumulation from the LFM forecast initialized at 1200 GMT, 9 April. The 0–24 h forecast (Fig. 30.6a) provides a reasonable forecast of rainfall amounts

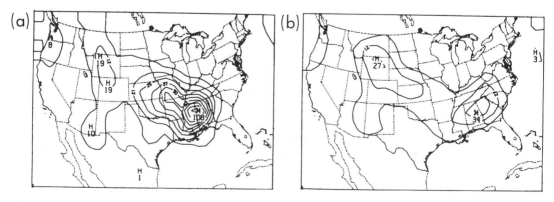

Figure 30.6. Precipitation amounts for the period 1200 GMT, 10 April, to 1200 GMT, 11 April, as predicted by two LFM–II forecasts: (a) 0–24 h forecast, initialized 1200 GMT, 10 April, and (b) 24–48 h forecast, initialized 1200 GMT, 9 April. Labeled contours are in millimeters; the outermost contour is 2.5 mm. H indicates local maximum; X indicates figure maximum. (Data courtesy of NMC.)

over Oklahoma, Texas, and Missouri, but predicts a maximum of 10.8 cm over Alabama and Mississippi where no rainfall occurred. A maximum in the southeast also was forecast in the 24-48 h result, which showed precipitation less than 1 cm in Oklahoma and Missouri. In addition, both model results suggest a southeast-to-northwest alignment of the rainfall distribution, in contrast to the southwest-to-northeast pattern of the observations.

Case II

Figure 30.7b shows the 24 h accumulated precipitation from solution II by Anthes *et al.* (1982), which uses the NCAR-PSU mesoscale model (Anthes and Warner, 1978) with enhanced NMC observed data for initial and boundary conditions (started at 1200 GMT, 10 April). The analyzed precipitation map, Fig. 30.7a (with logarithmic contouring scale), shows agreement with Fig. 30.5 with regard to pattern over the eastern portion of the heavy rainfall area, although more significant differences are evident to the west. Model results indicate precipitation >1.48 cm (level 5 of their scale) in a region on the east side of the domain oriented approximately west to east. Comparable rainfall was observed in this region over Oklahoma and Missouri but not to the east and occurred in a swath running southwest to northeast. (There is a suggestion of the squall line in the level-3 and -4 contours of the model results over Texas as there was in Fig. 30.7a.) The large rainfall

(a) OBSERVATIONS

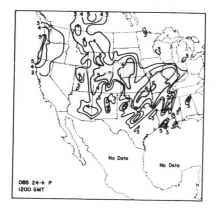

(b) SIMULATION (NMC Data)

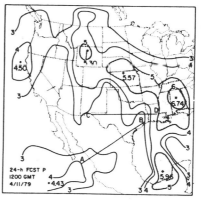

(c) SIMULATION (SESAME Data)

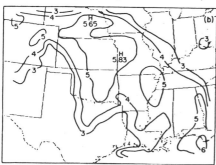

Figure 30.7. (a) Observed 24 h accumulated precipitation, and corresponding accumulated precipitation from (b) case II, which used NMC data set, and from (c) case VIII, which used SESAME data set. Both simulations used the NCAR-PSU model. Contour interval designations 3, 4, 5, and 6 correspond to precipitation amounts of 0.19, 0.54, 1.48, and 4.05 cm, respectively. (After Anthes *et al.*, 1982; Kuo and Anthes, 1984.)

maximum over Mississippi was hypothesized by Anthes *et al.* (1982) and later demonstrated by Benjamin (1983) to be caused by erroneous low-level convergence in the initial conditions along the gulf coast, which was believed to be due to differing data densities from land to sea. (Figure 30.7c shows results from Case VIII, discussed below.)

Case III

Solutions for case III are shown in Fig. 30.8. Kaplan *et al.* (1982) used their MASS mesoscale model to produce this 24 h simulation using NMC observed data for initial conditions and apparently also for lateral boundary conditions. The analysis of observed rainfall amounts for each 12 h period used only rain gauge data from first-order NWS stations and therefore contains considerably less detail than the analyses in Figs. 30.5 and 30.7. The model results, which were produced by resolvable moist convection only (more recent versions of the MASS model use a convective parameterization scheme), show small precipitation amounts during the first 12 h of the solution. This result reflects the tendency of numerical models to under-predict precipitation during the early stages of an integration as well as the bias of resolvable convection in a mesoscale model to respond more slowly than parameterized convection (because of the model's reliance on resolved vertical motion to advect moisture aloft and the requirement for the entire grid box to reach saturation before convection can occur). The model precipitation during the next 12 h, when the major precipitation occurred, shows heaviest amounts over Missouri, but relatively weaker amounts over Okla-

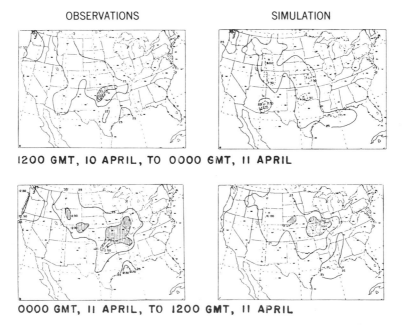

Figure 30.8. Observed accumulated precipitation (first-order NWS rainfall measurements only) and results from MASS model (Solution III) for successive 12 h periods. Contour intervals: 0.25,12.5,25.0,37.5 mm. (After Kaplan *et al.*, 1982.)

homa where the heaviest precipitation was observed to occur. No significant rainfall occurs in the southeast, in agreement with observations.

Case V

Figure 30.9 shows the 24 h amounts for 1200 GMT, 10 April, to 1200 GMT, 11 April, from forecast solutions Va and Vb reported by Orlanski *et al.* (1983). As with the LFM model, these are true forecasts rather than simulations; they were obtained from the limited-area GFDL/HIBU (Hydrometeorological Institute and Belgrade University [Yugoslavia]) model (see, e.g., Mesinger and Strickler, 1982) nested within the GFDL global spectral model (Gordon and Stern, 1982). Both the spectral and limited-area models were initialized at 1200 GMT, 9 April. In the first run, solution Va, the NMC operational analysis was used for initialization of the HIBU and spectral models; in the second, solution Vb, the special FGGE data set, analyzed by GFDL, was employed. Both solutions were run to 48 h. The precipitation forecasts for the last 24 h of both runs show great similarity to each other and to the LFM result over the southeast United States. Again, this result does not appear in the observed rainfall, obtained by a hand analysis of the primary NWS station data.

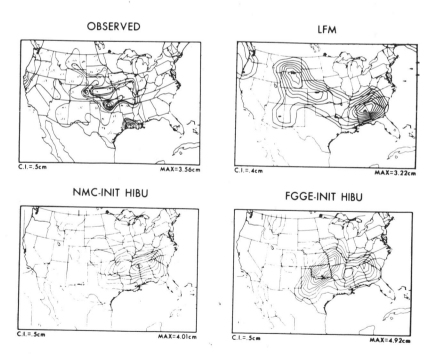

Figure 30.9. Accumulated precipitation for period from 1200 GMT, 10 April, to 1200 GMT, 11 April: as observed (from first-order NWS stations); according to LFM-II prediction (Cases 1), started 1200 GMT, 9 April; as forecast by GFDL/HIBU model using NMC initial conditions (Solution Va); and as forecast by GFDL/HIBU model using GFDL–FGGE initial conditions (Solution Vb). Contour interval and maximum are shown below each frame. (After Orlanski *et al.*, 1983.)

Figure 30.10. Accumulated precipitation (1200 GMT, 10 April, to 1200 GMT, 11 April) from GFDL/MAC model (Solution VI). Contours are in centimeters. (Figure courtesy of Orlanski and Shaginaw.)

Orlanski *et al.* (1983) attributed this erroneous precipitation in the southeast to the model's over-intensification of the warm front that moved north from the Gulf of Mexico. The modeled rainfall occurred along and ahead of this front. Orlanski's suggestion may also be the physical explanation of why Benjamin (1983) was able to eliminate this precipitation by changing his treatment of data along the Gulf Coast. In fact, when he started his run—1200 GMT, 10 April—the warm front was positioned along the coast line. Therefore, by reducing the data contrast, presumably of both wind and temperature, across the coast line, he was effectively reducing the initial intensity of the warm front.

The new and unique feature of these results, compared with earlier model results, is the large precipitation band in solution Vb running from southwest to northeast over Texas, Oklahoma, and Missouri (Fig. 30.9c). This is the first strong indication of the squall line character apparent in Fig. 30.5. The unique feature of this solution is the use of the FGGE data set, which includes improved observations over the data-sparse Pacific Ocean. The effect of this seems to have been to increase the upper-level winds over this region, thereby intensifying the surface cold front and its associated cross-stream circulation. This apparently provided the lifting for the convection and probably also increased the low-level transport of moist air from the Gulf that fueled the convective system. (In fact, whereas previous solutions tended to underestimate the effect of the cold front, this solution tends to overestimate it.)

Case VI

The 24 h precipitation pattern from solution VI was produced by Orlanski and Shaginaw (see Orlanski and Polinsky, 1984) using the GFDL/MAC model (Ross and Orlanski, 1982) with initial and boundary conditions derived from the NMC analysis. The western portion of the major precipitation zone (Fig. 30.10) has the same location as the observed pattern (as does the local rainfall maximum over Kansas), but the modeled pattern extends more to the east. The anomalous precipitation to the east has been reduced considerably in this solution, compared, for example, with the LFM solution (Fig. 30.6a), suggesting that the influence of the warm front has been reduced. However, the major region of rainfall still occurs too far to the east; also, the absence of a southwest-to-northeast pattern indicates that the cold frontal intensity over Oklahoma and Texas remains too weak.

Case VII

Solution VII, performed by Kalb (1984, 1985) using the Drexel/LAMPS (Limited Area Mesoscale Prediction System) model (Perkey, 1976), is different from the previous solutions in several ways. First, the model was initialized at the asynoptic time, 2100 GMT, 10 April, using only SESAME data for the initial conditions and for the boundary conditions at 0000 and 1200 GMT of 11 April. In addition, only mandatory-level, observed geopotential height data were used in a nonlinear balance equation to determine the wind field above 1250 m (the nondivergent part of the observed winds was used below 1250 m). This was done in order to simulate a procedure of using satellite temperature soundings to produce the initial mass and momentum fields for the model.

A major effect of the choice of the initialization procedure and starting time was to produce several short waves in the simulation. One of these waves persisted and moved northward over Oklahoma and Kansas during the 9 h simulation. Figure 30.11 shows a composite of 700 mb geopotential heights and precipitation rates from the solution at 0300 GMT, 11 April, and the observed radar summary at 0235 GMT. Although no accumulated precipitation is shown, the two precipitation rates, both resolved and parameterized (convective), indicate the effect of the short wave and are supported to some extent by the radar summary. The squall line shown in the radar summary is also suggested by a tongue of increased relative humidity (see Kalb, 1985) but is not evident in the rainfall rates.

Cases VIII, IX, and X

Case VIII involves another simulation that uses the NCAR-PSU model. However, this solution uses a smaller domain (primarily encompassing the coverage of the regional-scale SESAME network) with a grid resolution of 50 km, rather than the 111 km used in case II. In addition, both the initial and boundary data used in the simulation were obtained from the 3 h SESAME rawinsonde observations. Figure 30.7c shows the accumulated precipitation from this solution. The broad zone of weaker rainfall (levels 3 and 4) still shows the same orientation as in case II. However, the regions of heavier rainfall (level 5, >1.48 cm) now have a north-south orientation over Oklahoma and Missouri. This result is also found in case X, which also used the SESAME data set.

Case X (Fig. 30.12) was performed by Ross (1983) using the GFDL/BES model (similar to the GFDL/MAC model) on an 800 × 800 km domain over

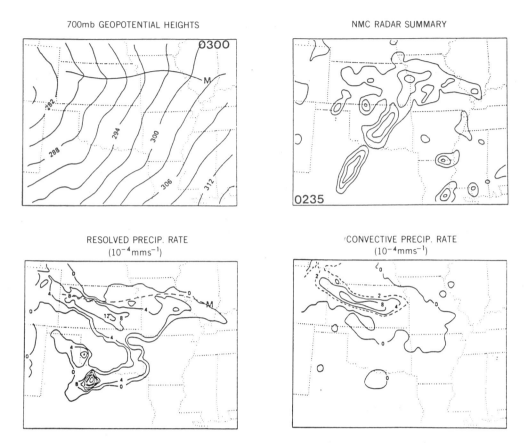

Figure 30.11. Geopotential heights at 700 mb for 0300 GMT, 11 April, of Solution VII (contours are in tens of meters); NMC radar summary at 0235 GMT; precipitation rate resolved by model at 0300 GMT; convected precipitation rate parameterized by model at same time. (After Kalb, 1985.)

southern Oklahoma and northeastern Texas. The precipitation accumulated at 3 h intervals is compared with observations at several different times during the 24 h integration. Simulations with 40 km and 20 km grid size are compared with observations. The observations were objectively analyzed from hourly rain-gauge data and reflect the great spatial variability of the rainfall data. The squall line structure is clearly evident for the period 0300–0600 GMT, 11 April, although the simulation provides a more continuous rainfall areal coverage than the rather sparse observations do. Ross found that the use of SESAME winds and the associated temperature field, and proximity of the boundary conditions to the location of the squall line were both necessary to represent the cold front system properly and thereby to produce a realistic squall line structure in the simulation. This supports the conclusions made earlier with regard to solutions Va and Vb: the enhanced FGGE data set produced the proper orientation of the heavy precipitation zone over Oklahoma whereas the operational (NMC) data set did not. Also this agrees with the improved precipitation results from case VIII (Fig. 30.7c), which used the SESAME data, compared with results from case II (Fig. 30.7b),

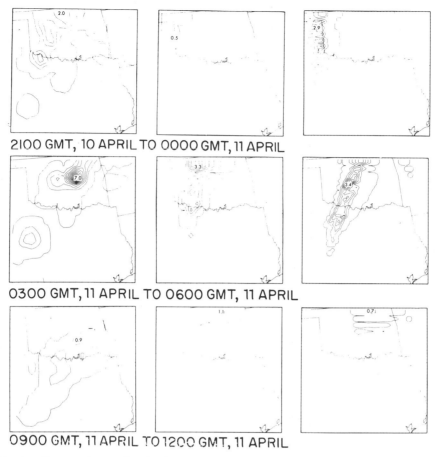

2100 GMT, 10 APRIL TO 0000 GMT, 11 APRIL

0300 GMT, 11 APRIL TO 0600 GMT, 11 APRIL

0900 GMT, 11 APRIL TO 1200 GMT, 11 APRIL

Figure 30.12. Comparison of 3 h accumulated precipitation at three different times: as observed (left column), as modeled for Case X with 40 km grid size (center column); and from the same model with 20 km grid size (right column). Contour intervals are 0.5 cm. (After Ross, 1983.)

which used NMC data. (The greater proximity of the upstream boundary in the former case probably also played a role in this improved simulation.) Although the correct mass and momentum fields were shown to be important in defining the forcing for the squall line in case X, the relative humidity field was found to control the structure and intensity of the resulting squall line.

The apparent sensitivity of rainfall results to the initial (and boundary) momentum fields agrees with the findings of the meso-α simulations in case IX as described by Chang *et al.* (1984). Using the Drexel/LAMPS model employed in case VII, but with a 140 km grid size, Chang *et al.* performed two simulations of the 24 h SESAME period that were identical except for the wind fields used to initialize the model. By altering the first-guess wind field in their initialization for 1200 GMT, 10 April, primarily by excluding a single wind observation over northwestern Mexico, they produced a considerable intensification in the jet streak over California and Mexico in the model's initial conditions. Compared with the solution based on a complete initial wind data set, the solution with the overly intense jet produced heavier precipitation over Oklahoma and enhanced the unverified rainfall over the southeastern states (the latter result being similar to what was found in many of the other solutions reviewed above). The authors concluded that the prediction of the severe storm environment can be very sensitive to uncertainties in the initial wind field used for the forecast.

Conclusions From Precipitation Results

There is some difficulty in comparing modeled precipitation patterns with observations, partly because of the difficulty in producing a representative picture of the actual precipitation amounts that occurred. One would expect some differences between modeled and observed rainfall results, because of the considerable model variation in grid size, model physics, etc. Certainly, actual accumulated rainfall magnitudes differ widely as do details of rainfall patterns. However, although there are significant differences between model formulations, several features are common to many of them, namely, the erroneous rainfall in the southeast and the absence of an intense squall line over the south-central United States. These similarities suggest the possibility of similar biases in many of the different mesoscale models as well as a possible strong dependence of model solutions upon potentially inadequate initial and lateral boundary conditions. Work is still needed to identify these biases and the sensitivity of model precipitation and associated forcing mechanisms to different initial, boundary, and surface data.

30.4.3. Hurricane Modeling

Tropical cyclones are ideal candidates for nested-mesh models because they are compact and most of the smaller scale convective features and large gradients are near the vortex center. Because of this characteristic structure of hurricanes, several research groups have developed three-dimensional mesoscale models that use telescoping nested-grid systems, in which the finest mesh is near the center and meshes grow progressively coarser outward from the center (Mathur, 1974; Ley and Elsberry, 1976; Jones, 1977; Ookochi, 1978; Kurihara et al., 1979). Most of these models also have the capability of shifting the interior grids relative to coarser outer grids so as to maintain the high-resolution mesh over the hurricane center as the storm moves. As an example, the telescoping nested-grid model, used by the Hurricane Group at GFDL (Kurihara et al., 1979; Kurihara and Bender, 1980) is described here in detail.

GFDL's Nested-Grid Hurricane Model

In this triply nested model, the grid is defined on constant latitude and longitude lines. The three meshes, A, B, and C, are progressively finer— 1, 1/3, and 1/6 degree, respectively—and have corresponding domain sizes (Fig. 30.13). The outermost lateral boundaries (for mesh A) are either open (i.e., use conditions from an externally defined state through one-way interaction) or cyclic (i.e., disturbances passing out through the boundary on one side will enter the domain on the opposite side). Each internal boundary

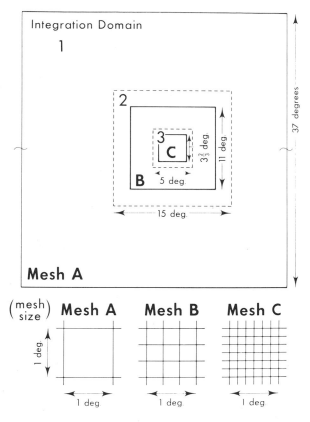

Figure 30.13. Horizontal structure of the triply-nested-mesh system, showing (above) domain and (below) grid sizes (not same scale as domain sizes). (After Kurihara and Bender, 1980.)

between mesh domains has an interface zone (between the dashed and solid lines in the figure), several coarse grid elements wide, in which the two-way interaction takes place during the time-marching of the model. The fluxes of physical quantities such as momentum and moisture through the interface are conserved by this procedure.

The movable meshes B and C are positioned with respect to the center of gravity of the surface pressure depression. For example, when this center moves in mesh C by more than the grid size Δ_B of the next coarser mesh (B), the fine mesh C is moved by the distance Δ_B in the direction of movement so as to recenter the storm. This movement is accomplished by converting coarse to fine grid points at the leading edge and replacing fine with coarse points at the trailing edge. Mesh B is moved relative to A in a similar manner, as recentering becomes necessary. As one might expect, computational noise is generated during this procedure. This noise is suppressed in the GFDL model through the use of a time-integration method that selectively damps high-frequency noise, and by nonlinear horizontal viscosity and occasional spatial smoothing.

Comparison of propagating vortex solutions with and without the nesting procedure demonstrates several advantages of nesting. First, as shown in the landfall simulation (next section), the detailed structure of the hurricane, including the wind field, surface pressure, and precipitation patterns, is represented much more satisfactorily in the nested model. In addition, the motion of the storm as a whole is altered significantly when the movable-nested-mesh system is used (Kurihara and Bender, 1980). This improvement seems to be due to the improved numerical as well as physical representation of the storm. Although such advantages are clearly desirable in a research mode, they must be weighted against the considerable increase in computational expense that would be incurred if a high-resolution nested model were used operationally.

Simulation of Landfall

The landfall of a hurricane is discussed in Ch. 14. The simulation of an idealized landfall event by the GFDL hurricane model is described here as an example of hurricane modeling in a research mode. Issues concerning initialization of the model for a real-data case and complications due to contrasting air masses, topography, etc., are not addressed in this study but are obviously important for an operational forecast. Other research simulations of landfall have been done by Moss and Jones (1978), Tuleya and Kurihara (1978), and Chang (1982).

GFDL's simulation (Tuleya *et al.*, 1984) deals with a simplified landfall situation. However, the capability of the simulation to be rerun with different conditions, such as reduced land surface temperature or varying land surface roughness, permits evaluation of the importance of each effect. In the solution presented here, the land surface is characterized as having the same moisture availability as the ocean surface, but a reduced surface temperature of 298 K (ocean value = 302 K) and an increased surface roughness length z_o of 25 cm. (The ocean value is typically less than 1 cm.) The large-scale wind field in the simulation consists of an approximately constant easterly wind of 10 m s^{-1}.

The hurricane is initialized in the model by adding an appropriate nondivergent vortex wind field to the basic background conditions. At the beginning of the simulation, ocean surface conditions are used over the entire domain; also, cyclic east-west boundaries are applied during an initial adjustment period. After the model has been integrated for 55 h of model time, the cyclic conditions on eastern and western boundaries are replaced by open boundaries; also, land surface conditions are imposed at all latitudes west of a longitude that is 8° west of the storm center. Landfall then occurs at 76.3 h.

As the cyclone encounters the change in surface conditions, the sudden increase in surface roughness, with its associated increase in drag, produces a drop in the low-level wind speed and a larger inflow angle of parcel trajectories into the hurricane. (These features were observed in Hurricane Frederic by Powell, 1982.) Figure 30.14 shows the evolution of the wind field in the lowest level above the surface as the storm encounters land. At 73 h, which

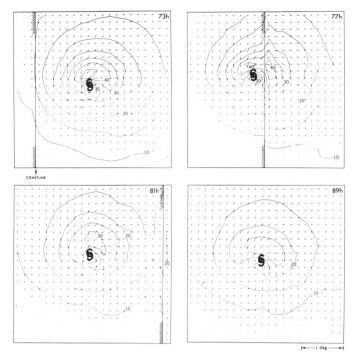

Figure 30.14. Analysis of winds at the first level above the surface within the mesh C domain of the landfall experiment at 73, 77, 81, and 89 h. The storm center defined by the surface pressure is indicated by a heavy dot. Contours indicate wind speed (m s^{-1}). Vectors show wind direction and magnitude. Land is to the west (left) of the indicated coastline. (After Tuleya *et al.*, 1984.)

is prior to landfall, the wind maximum is to the right and somewhat forward of the storm center (surface pressure minimum). Just after landfall (77 h), several wind peaks are evident. In fact, the location of the peak wind speed oscillates after this time but tends to occur, on average, at a right-rear position relative to the storm center. Winds decrease below hurricane force (which is defined as >33 m s^{-1}) by 81 h (hurricane force winds penetrate 165 km inland) and continue to drop thereafter. However, comparison of the hurricane motion with a control run that includes only ocean surface conditions shows only minor deflection (~50 km at 90 h) of the storm track due to landfall.

Figure 30.15 shows an analysis of rainfall rates and surface isobars for the times represented in Fig. 30.14. Because of the increased convergence within the storm, caused by the sudden change in surface drag, average precipitation rates increase slightly at landfall. Subsequently the area of heavy precipitation (>2.9 cm h^{-1}) begins to decrease 2 h later. The rainfall rates and areal coverage are similar to those estimated over land by Parrish *et al.* (1982) for Hurricane Frederic. However, their observation of a 50% increase in the area of convection at landfall is larger than that shown in the present simulations, probably because of differences in land surface conditions.

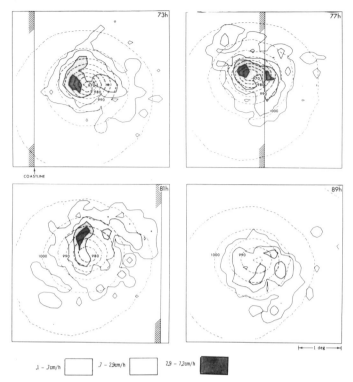

Figure 30.15. Analysis of rainfall rates for the landfall experiment, for the domain and times represented in Fig. 30.14. Dashed lines show isobars of surface pressure in 5 mb intervals. (After Tuleya *et al.*, 1984.)

The realistic simulation of the landfall event can be attributed to the 4° decrease in the land surface temperature as well as the increased surface roughness. The decay, indicated by the rise in surface pressure in Fig. 30.15, is due to the associated reduction in surface evaporation. Although the avail-ability of surface moisture was the same over land as over ocean, the cooler land temperature decreased the surface evaporation rate, thereby reducing the water vapor available to the storm. In addition, the cool land conditions caused a decrease in the conditional instability of the storm environment and an alteration of the planetary boundary layer. Hence, the energy source that maintains the storm, the latent heating due to this convective instability, is reduced and the storm decays as it moves inland.

Recent experiments (Bender *et al.*, 1985) have shown that topographical influences may enhance the decay rate through their disorganizing effect on the storm system. Topography also intensifies the precipitation rate at landfall and affects the storm track.

An Operational Hurricane Forecast Model

The above description of a hurricane simulation indicates what may be possible operationally in the future. However, current operational modeling does not attempt to predict intensity or to resolve detailed features of the hurricane, such as winds and precipitation patterns. The primary and vitally important objective of present-day operational models is the prediction of storm movement. Storm track forecast models are operational in three different organizations: the National Meteorological Center (Hovermale and Livezey, 1977); the U.S. Navy's Joint Typhoon Warning Center (Harrison and Fiorino, 1982); and the Japan Meteorological Agency (Ookochi, 1978). Following is a description of how NMC produces a storm track forecast, using its Movable Fine-Mesh (MFM) model.

The MFM model is a limited-area model designed for on-call use, to resolve severe mesoscale weather phenomena such as hurricanes and other heavy-precipitation weather systems. A typical configuration for the model consists of 10 vertical levels, a horizontal domain of 3000 × 3000 km, and 60 km grid spacing. The model uses fields from the NMC global spectral (or hemispheric grid-point) forecast to provide lateral boundary conditions in a one-way interaction mode. Also, in order to keep the hurricane away from the lateral boundaries, the mesh moves (relative to the global model) in a manner similar to that described for the GFDL model.

Given the limited computer resources available at NMC prior to 1983 and the typical paucity of detailed mesoscale observations of hurricanes in an operational time frame, NMC initialized the MFM model with a hurricane vortex derived from an idealized conceptual model rather than from observations. (The 60 km resolution of the model is unable to resolve the detailed inner core of the hurricane, anyway.)

The procedure used in 1975–76 to initialize the model was as follows:

- The standard NMC global analysis is interpolated to the MFM model; any small-scale features due to the presence of the hurricane are filtered out. This establishes the basic steering flow in the initial conditions.
- A two-dimensional (2-D), axisymmetric hurricane model, similar in its physics to the 3-D MFM model, is run using a representative tropical sounding and warm ocean until a quasi-steady hurricane is produced (1–2 minutes of IBM 360/195 time). This storm is in balance with respect to the 2-D model and the MFM model physics. However, it does not contain asymmetric features such as large-scale vertical wind shear and beta effects, i.e., effects due to changes in the Coriolis parameter across the storm. The latter may be included to some extent by modifying the 2-D wind field through a gradient wind relation.
- The resulting mass and momentum fields are then added, as perturbations, to the smoothed steering flow provided in step 1, and the 3-D MFM model is then run for the 48 h period of the forecast (using boundary conditions provided by the NMC global forecast).

As one might expect, the merging of the modeled vortex from step 2 with the background steering flow of step 1 produces an adjustment period during the early stages of the MFM forecast. Certain characteristics of the initial storm, such as its intensity and asymmetric features, affect the subsequent storm-track prediction. Figure 30.16 shows an example of the influence of these initial-condition features on track forecasts of Hurricane Eloise. The first forecast, made operationally in 1975, included an overly strong and symmetric initial vortex. The resulting model vortex (solid line) moved too rapidly northward, compared with observations (indicated by the hurricane symbols in Fig. 30.16) and was not pushed eastward by a propagating middle-latitude trough passing through the region until it was only several hundred kilometers from landfall. In a later experimental forecast, the intensity of the initial vortex was reduced, and the wind field was corrected somewhat for β-effect asymmetries. These changes caused the model hurricane to move northward more slowly (dashed line). As a result, the trough encountered the storm more to the south, thereby pushing it farther to the east and producing a very accurate prediction of landfall location. (However, the predicted landfall time seems to be too late.)

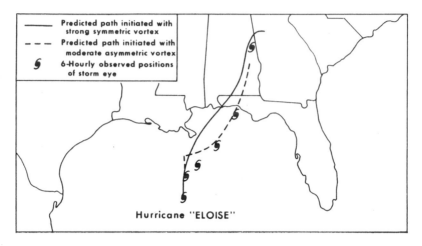

Figure 30.16. Observed track of Hurricane Eloise compared with two predicted tracks based on two different initial vortex conditions. (From a figure provided by John Hovermale.)

This example indicates how some of the many factors involved in a hurricane forecast can alter the storm-track prediction. Some of these factors can be controlled through improvements in the initialization procedure and enhancements of the dynamical model, such as increasing grid resolution. For example, a more recent version of the operational MFM model uses a three-dimensionally derived vortex to reduce initial adjustment effects in the forecast and thereby to improve short-range (0–24 h) predictions. Finally, a good forecast of the synoptic-scale conditions, used to prescribe lateral boundary conditions, is important in this case as in regional model predictions for forecast periods beyond a day or so.

REFERENCES

Alberty, R. L., D. W. Burgess, C. E. Hane, and J. F. Weaver, 1979: SESAME 1979 operations summary. NOAA Environmental Research Laboratories, Boulder,

Colo., 253 pp.

Anthes, R. A., 1977: A cumulus parameterization scheme utilizing a one-dimensional cloud model. *Mon. Wea.*

Rev., **105**, 270–286.

Anthes, R. A., 1982: *Tropical Cyclones— Their Evolution, Structure and Effects.* Meteor. Monogr. **41**, American Meteorological Society, Boston, Mass., 208 pp.

Anthes, R. A., 1983: Regional models of the atmosphere in middle latitudes. *Mon. Wea. Rev.*, **111**, 1306–1335.

Anthes, R. A., and T. T. Warner, 1978: Development of hydrodynamic models suitable for air pollution and other mesometeorological studies. *Mon. Wea. Rev.*, **106**, 1045–1078.

Anthes, R. A., Y.-H. Kuo, S. G. Benjamin, and Y.-F. Li, 1982: The evolution of the mesoscale environment of severe local storms: Preliminary modeling results. *Mon. Wea. Rev.*, **110**, 1187–1213.

Arakawa, A., and W. H. Shubert, 1974: Interaction of a cumulus cloud ensemble with the large-scale environment, Part I. *J. Atmos. Sci.*, **31**, 674–701.

Bender, M. A., R. E. Tuleya, and Y. Kurihara, 1985: A numerical study of the effect of a mountain range on a landfalling tropical cyclone. *Mon. Wea. Rev.*, **113**, 567–582.

Benjamin, S. G., 1983: Some effects of surface heating and topography on the regional severe storm environment. Ph.D. dissertation, The Pennsylvania State University, College Park, Pa., 265 pp.

Benjamin, S. G., and T. N. Carlson, 1986: Some effects of surface heating and topography on the regional severe storm environment. Part I: Three-dimensional simulations. *Mon. Wea. Rev.*, **114**, 307–329.

Carlson, T. N., S. G. Benjamin, G. S. Forbes, and Y.-F. Li, 1983: Elevated mixed layers in the regional severe storm environment: Conceptual model and case studies. *Mon. Wea. Rev.*, **111**, 1453–1473.

Chang, C. B., D. J. Perkey, and C. W. Kreitzberg, 1984: Impact of initial wind field on the forecast of a severe storm environment. Proceedings, 10th Conference on Weather Forecasting and Analysis, 1984, Clearwater Beach, Fla., American Meteorological Society, Boston, Mass., 513–520.

Chang, S. W., 1982: The orographic effects induced by an island mountain range on propagating tropical cyclones. *Mon.*

Wea. Rev., **110**, 1255–1270.

Charney, J. G., 1954: Numerical prediction of cyclogenesis, *Proc. Natl. Acad. Sci.*, **40**, 99–110.

Charney, J. G., R. Fjortoft, and J. von Neumann, 1950: Numerical integration of the barotropic vorticity equation. *Tellus*, **2**, 237–254.

Clark, T. L., 1982: Cloud modeling in three spatial dimensions. *Hailstorms of the Central High Plains*, Vol. 1, National Hail Research Experiment, C.A. Knight and P. Squires (Eds.), Colorado Associated University Press, Boulder, 225–247.

Cotton, W. R., 1975: Theoretical cumulus dynamics. *Rev. Geophys. Space Phys.*, **13**, 419–448.

Cressman, G. P., 1958: Barotropic divergence and very long atmospheric waves. *Mon. Wea. Rev.*, **86**, 293–297.

Farley, R. D., and H. D. Orville, 1982: Cloud modeling in two spatial dimensions. *Hailstorms of the Central High Plains*, Vol. 1, National Hail Research Experiment, C.A. Knight and P. Squires (Eds.), Colorado Associated University Press, Boulder, 207–223.

Fawcett, E. B., 1977: Current capabilities in prediction at the National Weather Service's National Meteorological Center, *Bull. Amer. Meteor. Soc.*, **58**, 143–149.

Gordon, C. T., and W. F. Stern, 1982: A description of the GFDL global spectral model, *Mon. Wea. Rev.*, **110**, 625–644.

Haltiner, G. J., and R. T. Williams, 1980: *Numerical Prediction and Dynamic Meteorology.* 2nd ed., Wiley, New York, 477 pp.

Harrison, E. J., Jr., and M. Fiorino, 1982: A comprehensive test of the Navy Nested Tropical Cyclone Model. *Mon. Wea. Rev.*, **110**, 645–650.

Hoke, J. E., 1984: Forecast results for NMC's new regional analysis and forecast system. Proceedings, 10th Conference on Weather Forecasting and Analysis, Clearwater Beach, Fla., American Meteorological Society, Boston, Mass., 418–423.

Hovermale, J. B., and R. E. Livezey, 1977: Three-year performance characteristics of the NMC hurricane model. Proceedings, 11th Technical Conference on Hurricanes and Tropical Meteorology, Miami, Fla., American Meteorological Society, Boston, Mass., 122–124.

Jones, R. W., 1977: A nested grid for a three-dimensional model of a tropical cyclone. *J. Atmos. Sci.*, **34**, 1528–1553.

Kalb, M. W., 1984: Initialization of a mesoscale model for April 10, 1979, using alternate data sources. NASA Contractor Report 3826, Marshall Space Flight Center, Ala., 230 pp.

Kalb, M. W., 1985: Results from a limited-area mesoscale numerical simulation for 10 April 1979. *Mon. Wea. Rev.*, **113**, 1644–1662.

Kaplan, M. L., J. W. Zack, V. C. Wong, and J. J. Tuccillo, 1982: Initial results from a mesoscale atmospheric simulation system and comparison with the AVE-SESAME I data set. *Mon. Wea. Rev.*, **110**, 1564–1590.

Kreitzberg, C. W., 1979: Observing, analyzing, and modeling mesoscale weather phenomena. *Rev. Geophys. Space Phys.*, **17**, 1852–1871.

Kreitzberg, C. W., and D. J. Perkey, 1976: Release of potential instability. Part I: A sequential plume model within a hydrostatic primitive equation model. *J. Atmos. Sci.*, **33**, 456–475.

Kuo, Y.-H., and R. A. Anthes, 1984: Accuracy of diagnostic heat and moisture budgets using SESAME-79 field data as revealed by observing system simulation experiments. *Mon. Wea. Rev.*, **112**, 1465–1481.

Kurihara, Y., and M. A. Bender, 1980: Use of a movable nested-mesh model for tracking a small vortex. *Mon. Wea. Rev.*, **108**, 1792–1809.

Kurihara, Y., G. J., Tripoli, and M. A. Bender, 1979: Design of a movable nested-mesh primitive equation model. *Mon. Wea. Rev.*, **107**, 239–249.

Ley, G. W., and R. L. Elsberry, 1976: Forecast of Typhoon Irma using a nested-grid model. *Mon. Wea. Rev.*, **104**, 1154–1161.

Mathur, M. B., 1974: A multiple-grid primitive equation model to simulate the development of an asymmetric hurricane (Isbell, 1964). *J. Atmos. Sci.*, **31**, 371–393.

Mesinger, F., and R. F. Strickler, 1982: Effect of mountains on Genoa cyclogenesis. *J. Meteor. Soc. Japan*, **60**, 326–338.

Mills, G. A., and C. M. Hayden, 1983: The use of high horizontal resolution satellite temperature and moisture profiles to initialize a mesoscale numerical weather prediction model—A severe weather event case study. *J. Climate Appl. Meteor.*, **22**, 649–663.

Miyakoda, K., 1973: Cumulative results of testing a meteorological-mathematical model: The description of the model. *Proc. Royal Irish Acad.*, **73(A)**, 99–130.

Moore, J. T., and H. E. Fuelberg, 1981: A synoptic analysis of the first AVE-SESAME '79 period. *Bull. Amer. Meteor. Soc.*, **62**, 1577–1590.

Moss, M. S., and R. W. Jones, 1978: A numerical simulation of hurricane landfall. NOAA Tech. Memo. ERL NHEML-3, Environmental Research Laboratories, Boulder, Colo. (NTIS#PB-290 39817GA), 15 pp.

Ookochi, Y., 1978: Preliminary test of typhoon forecast with a moving multi-nested grid (MNG). *J. Meteor. Soc. Japan*, **56**, 571–583.

Ooyama, K. V., 1982: Conceptual evolution of the theory and modeling of the tropical cyclone. *J. Meteor. Soc. Japan*, **60**, 369–380.

Orlanski, I., 1975: A rational subdivision of scales for atmospheric processes. *Bull. Amer. Meteor. Soc.*, **56**, 527–530.

Orlanski, I., and L. Polinsky, 1984: Predictability of mesoscale phenomena. In Nowcasting II, Mesoscale Observations and Very Short-Range Forecasting (Proceedings, Second International Symposium on Nowcasting, 3–7 September 1984, Norrköping, Sweden), ESA SP-208, European Space Agency, Noordwijk, Netherlands, 271–280.

Orlanski, I., D. Miller, and K. Miyakoda, 1983: The impact of initialization analyses in the forecasting of precipitation patterns. WMO Workshop on Very Short-Range Forecasting Systems Research Aspects, Boulder, CO, 15–17 August 1983, Short- and Medium-Range Weather Prediction, Research Publication Series, No. 5, World Meteorological Organization, 59–62.

Parrish, J. R., R. W. Burpee, F. D. Marks, Jr., and R. Grebe, 1982: Rainfall patterns observed by digitized radar during the landfall of Hurricane Frederic (1979).

Mon. Wea. Rev., **110**, 1933–1944.

Perkey, D. J., 1976: A description and preliminary results from a fine-mesh model for forecasting quantitative precipitation. *Mon. Wea. Rev.*, **104**, 1513–1526.

Phillips, N. A., 1979: The nested grid model. NOAA Technical Report NWS 22, National Weather Service (NTIS#PB–299046/3GA), 80 pp.

Phillips, N. A., and J. Shukla, 1973: On the strategy of combining coarse and fine grid meshes in numerical weather prediction. *J. Appl. Meteor.*, **12**, 763–770.

Pielke, R. A., 1981: Mesoscale numerical modeling. *Adv. Geophys.*, **23**, 185–344.

Pielke, R. A., 1984: *Mesoscale Meteorological Modeling*. Academic Press, New York, 632 pp.

Platzman, G. W., 1979: The ENIAC computations of 1950—Gateway to numerical weather prediction. *Bull. Amer. Meteor. Soc.*, **60**, 302–312.

Ploshay, J. J., R. K. White, and K. Miyakoda, 1983: FGGE level III-B daily global analyses, Part I (Dec. 1978–Feb. 1979). NOAA Data Report GFDL–1, Environmental Research Laboratories (NTIS#PB73–221051), 278 pp.

Powell, M. D., 1982: The transition of the Hurricane Frederic boundary wind field from the open Gulf of Mexico to landfall. *Mon. Wea. Rev.*, **110**, 1912–1932.

Rosenthal, S. L., 1978: Numerical simulation of tropical cyclone development with latent heat released by the resolvable scales. I: Model description and preliminary results. *J. Atmos. Sci.*, **35**, 258–271.

Ross, B. B., 1983: Sensitivity of meso-β scale precipitation forecasts to specification of initial and boundary data. Abstracts, 1st Conference on Mesoscale Meteorology, Norman, Okla., CIMMS, Univ. Oklahoma, Norman.

Ross, B. B., and I. Orlanski, 1978: The circulation associated with a cold front. Part II: moist case. *J. Atmos. Sci.*, **35**, 445–465.

Ross, B. B., and I. Orlanski, 1982: The evolution of an observed cold front. Part I: Numerical simulation. *J. Atmos. Sci.*, **39**, 296–327.

Schlesinger, R. E., 1982: Three-dimensional numerical modeling of convective storms: A review of milestones and challenges. Proceedings, 12th Conference on Severe Local Storms, San Antonio, Tex., American Meteorological Society, Boston, 506–515.

Shuman, F. G., 1978: Numerical weather prediction. *Bull. Amer. Meteor. Soc.*, **59**, 5–17.

Shuman, F. G., and J. B. Hovermale, 1968: An operational six-layer primitive equation model. *J. Appl. Meteor.*, **7**, 525–547.

Teweles, S., and H. Wobus, 1954: Verification of prognostic charts. *Bull. Amer. Meteor. Soc.*, **35**, 455–463.

Tuleya, R. E., and Y. Kurihara, 1978: A numerical simulation of the landfall of tropical cyclones. *J. Atmos. Sci.*, **35**, 242–257.

Tuleya, R. E., M. A. Bender, and Y. Kurihara, 1984: A simulation study of the landfall of tropical cyclones using a movable nested-mesh model. *Mon. Wea. Rev.*, **112**, 124–136.

UCAR, 1983: The National STORM Program—Scientific and technological bases and major objectives. R. A. Anthes (Ed.), University Corporation for Atmospheric Research, Boulder, Colo., to NOAA, Contract NA81RAC00123, 520 pp.

Wilson, J. W., and E. A. Brandes, 1979: Radar measurements of rainfall summary. *Bull. Amer. Meteor. Soc.*, **60**, 1048–1058.

CHAPTER 31

The Use of Computers for the Display of Meteorological Information

Thomas W. Schlatter

31.1. Introduction

Computer displays are especially useful for nowcasting and very-short-range forecasting of regional and local weather events. Nowcasting is the extrapolation of current weather to some future time, based on the behavior of existing phenomena as described by intensive observations. Extrapolation is considered to involve no physics, dynamics, or the application of numerical or conceptual models. How long it is successful depends heavily upon the phenomenon itself, the location, and, more than likely, the season. Very-short-range forecasting is the anticipation of events beyond the period during which extrapolation usually works but not beyond 12 h. These definitions, proposed by Zipser (1983), are adopted here.

A focus on regional and local weather events, that is, mesoscale phenomena, demands an observing and display system that delivers data frequently and at high spatial resolution. Such a system, in turn, requires specialized observation, wide-band communications, substantial computing power, and versatile display hardware. Remarkable advances in communication (microwave relays, commercial satellites, and fiber optics) and computing power (integrated circuits) have made available a wealth of meteorological data and means of presentation to the forecaster.

31.2. New Observing Systems

Remote sensing techniques have changed forever the way we view the atmosphere. Information from conventional radars and Geostationary Operational Environmental Satellites (GOES) is already well integrated into National Weather Service (NWS) operations and plays an important role in many televised weathercasts. Less familiar information is also being collected (or thought about) which may eventually find its way into operations.

31.2.1. Remote Sensing from Space

Both geosynchronous and polar-orbiting satellites measure radiation in different wavelength intervals (channels); from such measurements, vertical temperature profiles and the total precipitable water may be estimated. Radiometers measuring only at infrared wavelengths can sound the atmosphere only in clear areas or above clouds.

Of equal importance are tropospheric wind profiles. Atlas and Korb (1981) discussed a possible way of measuring winds from satellite by means of Doppler lidar. An on-board laser generates pulses of energy at visible or infrared wavelengths. A small fraction of the emitted energy is scattered back toward the satellite by atmospheric aerosols. By measuring the travel time of a pulse from the satellite to the scattering volume and back again and the shift in wavelength caused by the motion of the aerosols, one can measure the component of wind in the direction of the pulses at a given range. By looking forward and later backward at the same volume of atmosphere, an orbiting satellite could measure the horizontal wind. This sounder would have the same limitation as infrared temperature sounders: it could not obtain measurements in or below clouds.

31.2.2. Remote Sensing from the Ground

Scientists from NOAA's Aeronomy Laboratory and Wave Propagation Laboratory have developed a system for obtaining vertical profiles of temperature, moisture, and wind (Hogg *et al.*, 1983). The system has two major components: a radiometer measuring passively and a UHF or VHF Doppler radar actively probing the troposphere with energy pulses and measuring the backscattered energy. Both components obtain useful measurements except in moderate or heavy precipitation (see Ch. 4, this volume).

Sodar, the use of sound pulses to study atmospheric structure near the ground, is a measuring technique dating from the 1940s (Little, 1969). Since about 1970, it has contributed toward greater understanding of surface-based and elevated inversions, thermal plumes, gravity waves, the development of the daytime planetary boundary layer, and the vertical variation of wind close to the ground. Forecasters who would profit from more detailed information on the character and depth of the boundary layer when trying to forecast thunderstorms and their interactions may one day be helped by sodar data.

The U.S. National Weather Service plans to install a network of 10-cm Doppler radars (NEXRAD) by the end of the decade. In addition to reflectivity, these radars can measure the radial motion of hydrometeors and are good at detecting horizontal shear, convergence, and rotation, at least indirectly. Consequently, they will be the primary tool for detecting severe thunderstorms. On occasion, they can also measure radial motions in clear air, presumably because insects, dust, or large fluctuations in refractive index provide detectable targets.

Doppler lidar, already mentioned in connection with remote sensing by satellite, is being used in field experiments by NOAA's Wave Propagation Laboratory to measure winds in clear air (Hall *et al.*, 1984; Hardesty *et al.*,

1983). Operating at an average power of 1 W and a wavelength of 10.6 μm, its range is somewhat less than 20 km, compared with a range of about 300 km for a 10 cm weather radar. It "sees" aerosols rather than precipitation and so can measure the radial component of wind in the clear air far more effectively than a radar, but it cannot penetrate clouds. Whereas ground clutter makes observations of low-level outflows from storms difficult with a Doppler radar, a Doppler lidar can aim its pulses only a few meters above the ground without suffering from ground clutter. The capabilities of radar and lidar are thus complementary.

31.3. Current Capabilities in a National Weather Service Forecast Office

Walk into any Weather Service Forecast Office (WSFO) in the contiguous United States, and you will see the first results of technological change. AFOS, the Automation of Field Operations and Services (Klein, 1976), was installed beginning in the late 1970s as the replacement for the facsimile and teletype machines, to streamline message composition and transmission, to increase the speed and capacity for communicating weather information nationwide, and to bring at least limited computing power on site.

AFOS delivers a very broad selection of mostly large-scale weather information to forecasters, both diagnostic information and the output of the National Meteorological Center's (NMC) numerical models. It also provides the facility for composing local, zone, and aviation forecasts, special weather statements, and warnings. It has several display screens, each provided with up to three overlays, independent control of the intensity and line-drawing style (solid, dashed, or dotted) for each overlay, and the ability to magnify selected features. AFOS can display alphanumeric information (numbers and letters) and contours in black and white, but it cannot display images such as a visible satellite picture. Operators can create procedures (sequences of frequently used commands) and run applications programs such as regional data plots, sounding analysis, and graphic animation.

AFOS delivers comprehensive surface reports, national radar summaries and weather depiction charts every hour, and surface plots and analyses every 3 h, but forecasters get most of their local information from the following additional sources. The United Press International's Unifax provides detailed visible and infrared hard-copy images of clouds every 30 min. A water vapor image arrives every 6 h. Data from NWS or FAA (Federal Aviation Administration) radars are fed to WSFOs for display on a commercial color monitor. These typically provide PPI (Plan Position Indicator) images at up to four ranges and as frequently as every 5 min.

During severe weather, many WSFOs use a hotline to the nearest radar site. The radar operator tells the forecaster about features that cannot be seen on the local radar display, such as high reflectivity in the upper troposphere or weak echo vaults. In addition, trained spotters, safety and law enforcement officers, and the general public relay reports of severe weather directly to the WSFO by radio or telephone.

Among the shortcomings of the current system for data collection and display at WSFOs are the lack of color displays (except for radar), lack of high-resolution radar data, inability to animate satellite images, and the need for the forecaster to move about the office to consult different information sources. NWS recognizes the need to integrate all important information sources on a versatile forecaster workstation. It has detailed plans for an Advanced Weather Interactive Processing System for the 1990s (AWIPS-90). Builders of this system will undoubtedly benefit from the experience of groups now testing experimental workstations in the United States— the subject of Sec. 31.4.

31.4. Six Systems for Processing and Displaying Meteorological Data

Six experimental systems, most used for research and development, are described here. These representative interactive graphic systems in meteorology are a good indication of what may be available for operations in 5–10 years.

31.4.1. McIDAS

McIDAS, the Man-computer Interactive Data Access System, was first developed in the early 1970s at the Space Science and Engineering Center (SSEC) at the University of Wisconsin (Madison) primarily for the display of geosynchronous satellite and conventional meteorological data. As the first color graphics system for animating satellite images and overlaying plotted or contoured data, McIDAS has served as a model for the development of other systems. McIDAS alone, remote terminals connected to McIDAS, and even progeny of McIDAS such as CSIS (see Sec. 31.4.2) are now at more than a dozen locations in the United States. McIDAS has been described in detail by Suomi *et al.*(1983).

31.4.2. CSIS

CSIS is the Centralized Storm Information System, a spinoff of McIDAS, installed in Kansas City at the National Severe Storms Forecast Center (NSSFC), the origin of all severe weather watches, and at the Satellite Field Services Station. The result of a joint effort by NWS, SSEC, the National Aeronautics and Space Administration (NASA) and the National Environmental Satellite, Data, and Information Service (NESDIS), CSIS was installed in 1982 and has been tested in operational settings ever since (Anthony *et al.*, 1982; Mosher and Schaefer, 1983).

During outbreaks of severe weather, forecasters need rapid and selective access to high-resolution data. Delays are intolerable. CSIS has clearly demonstrated that the difference between peak and average computer load in an operational environment greatly exceeds that in a research environment, where demand is more uniform.

31.4.3. AOIPS, VASP, and VAP

AOIPS (Atmospheric and Oceanographic Information Processing System), VASP (VISSR Atmospheric Sounder Processing System), and VAP (VAS Assessment Processor) are three interconnected computer systems developed at the NASA Goddard Space Flight Center. VISSR stands for the Visible and Infrared Spin-Scan Radiometer aboard the GOES satellites. VAS is a second-order acronym for VISSR Atmospheric Sounder.

The oldest of the three systems, AOIPS, has been described by Billingsley (1976). It was originally developed to extract cloud information from VISSR data and, as a by-product, wind vectors from cloud motions. Later it was expanded to handle ingest of VISSR data in real time, to generate stereographic images, and to process and remap radar images. The VASP is dedicated to processing VAS multi-spectral images and retrieving temperature soundings from radiation measurements in 12 different channels. The VAP assimilates data collected or generated by AOIPS and VASP and supports a variety of graphic devices.

31.4.4. NEDS

NEDS is the Naval Environmental Display Station developed by the Fleet Numerical Weather Central, U.S. Navy, in Monterey, Calif. (Thormeyer, 1978). NEDS obtains its information through the global high-speed Automated Weather Network operated by the U.S. Air Force and from the Navy's own Environmental Data Network; it provides oceanographic and meteorological analyses and forecasts at a number of naval installations worldwide.

The Navy is now developing and testing a successor to NEDS called SPADS, a Satellite (data) Processing And Display System (Schramm *et al.*, 1982). This system will support data acquisition from GOES or polar orbiters, renavigation and remapping of images, superposition of political and geographic boundaries and hand- or machine-drawn contours on the satellite image, and objective analysis.

31.4.5. ADVISAR

ADVISAR is the All Digital Video Imaging System for Atmospheric Research developed by the Department of Atmospheric Science at Colorado State University (CSU) in Fort Collins, primarily for research in mesoscale meteorology (Green and Kruidenier, 1982). CSU operates a ground readout station with antennas pointing toward GOES. Researchers specify the geographic window they wish to view, and a sectorizer limits data collection to that area. VISSR data may be viewed in real time or, as is usually the case, they may be archived on nine-track tape for later study. Other meteorological data, including radar data, are available through external computer links.

31.4.6. POWS

POWS refers to the PROFS Operational Workstation, a facility designed as a model for future local or regional weather services. PROFS, the Program for Regional Observing and Forecasting Services, is a NOAA program created in the late 1970s. The workstation was originally developed for the detection and very-short-range prediction of hazardous convective weather (Reynolds, 1983). It was to be compatible (not redundant) with AFOS. Now the application of the workstation has been broadened to include nonconvective weather, and the restriction to non-AFOS functions has been removed. The workstation displays a wide variety of data—conventional surface and upper air reports, soundings from WPL's ground-based profiling system, surface mesonet data, images from geosynchronous satellites, radar reflectivity and radial velocity data, and precipitation and lightning data—on several different map projections covering areas as large as the United States or as small as a few counties.

31.5. Desirable System Attributes

Systems for displaying meteorological information differ in capabilities and computing power, but their designers are all trying to incorporate a number of desirable attributes: color, animation, combinations of diverse data on a single display, user-friendliness, speed, resolution, and flexibility. Sometimes it is necessary to choose between attributes or to trade one advantage (or disadvantage) for another.

31.5.1. Color

Humans are still much better and faster than computers in recognizing patterns and extrapolating their movement. The use of color, especially in pictorial data such as radar or satellite images, can be crucial in highlighting patterns. "False color" images are commonly used in the interpretation of radar and satellite data. A "color table" gives the correspondence between the measured value (of radar reflectivity or cloud brightness, for example) and the color that represents the value. Building an appropriate color table is still an art, but some guidelines have been discovered through study of the physiological response of the human eye to color (Murch, 1984):

- Avoid the simultaneous display of highly saturated, spectrally extreme colors. The eye cannot focus simultaneously on colors far removed from each other in the visible spectrum—red and blue, for example. In order to prevent frequent refocussing, either use colors spectrally close together or reduce the brightness.

- Pure blue should be avoided for text, thin lines, and small shapes because, for one reason, the eye will have difficulty focusing on them. On the other hand, blue is an excellent background color.

- Avoid adjacent colors that differ only in the amount of blue. Edges that differ only in the amount of blue appear indistinct.

- Older operators need higher brightness levels to distinguish colors. With age, the eye loses overall sensitivity.

- It is difficult to focus on edges created by color alone. Brightness differences at an edge aid in clear focusing. Thus, multicolored images should be differentiated on the basis of brightness as well as color.

- Avoid red and green in the periphery of full-screen displays. These colors do not show up well in peripheral vision. Yellows and blues are better peripheral colors.

- Opponent colors (red and green or yellow and blue) go well together in simple color displays such as sets of overlaid contours.

- For color-blind observers, avoid single-color distinctions.

31.5.2. *Animation*

Animation enhances the ability to detect changes, quantify developments and rates of change, and see relationships between different elements in the field of view. The evolution of surface convergence lines, the differential motion of clouds, or the development of thunderstorms is plainly evident, given the right kinds of data. Because animation is so important, the forecaster should be able to manipulate data displays in these ways:

- Loop forward or backward or in both directions in sequence.
- Control the frame rate.
- Pause at the beginning or end of a loop or not at all.
- Control the length of the pause.
- Manually step through each frame of a loop, forward or backward.

Most display devices store internally all the images required for a loop. For example, if the storage space for the screen is 1024 pixels on a side and if 12 bits are available at each pixel location for specifying color, one can load four frames, each 512 pixels on a side and 12 bits "deep" into the display buffer. Looping occurs through pixel replication; that is, each of the four images in the display buffer is blown up to full-screen size. The color at each pixel in the original 512 × 512 image simply fills a four-pixel square when that image is expanded to fill the 1024 × 1024 screen.

For more money, one can buy more storage space in the display device. Ultimately, however, the display space is limited, and trade-offs must be made between spatial resolution on the screen, the number of frames that can be loaded into the device, and the dynamic range of the image, that is, the number of colors used to specify different values of the observed field. For example, GOES infrared data contain 8 bits for the specification of radiative temperature, enough for 256 different values. To retain the dynamic range in the original data, one must use 256 different colors in the image or, equivalently, 8 bits for the specification of color.

To fix ideas, suppose that a 1024 × 1024 screen with 12-bit planes is available. To load an 8-frame loop of infrared images, yet allow for two overlays such as 500 mb height contours, political boundaries, or data plots, the storage in a display device might be allocated as shown in Fig. 31.1. Each image is 512 pixels on a side and 4 bits deep, allowing for color representation of 16 different values. Two single-color overlays take up two more bit planes. That accounts for four images with overlays in the first 6-bit planes. The

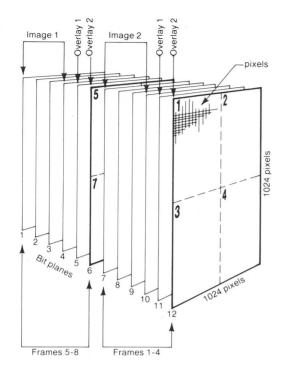

Figure 31.1. The use of 12-bit planes for storing a loop of 8 frames, each with a 512 × 512 4-bit image and two overlays.

remaining 6-bit planes are allocated similarly to accomodate the last four images of the loop and their overlays.

A laser video disk is different from the hardware just described. It can store many images and display them rapidly in any programmable sequence. It has the disadvantage that, once written, an image cannot be altered; thus, overlays cannot be added or removed.

31.5.3. Combination of Diverse Data on a Single Display

The ability to combine many different types of data in a single display is a powerful diagnostic aid. There are several ways to do this on a workstation, the most obvious being a simple overlay of plotted data and contours on an image. Examples abound:

- 500-mb height field on an infrared image of the United States (for matching cloud motion with direction of geostrophic wind).

- Surface mesonet data on a radar reflectivity image or a Doppler velocity image (for identification of thunderstorm inflows and outflows).

- Lightning strike data on a satellite image of clouds (for early detection of electrical activity in a line of towering cumulus clouds).

- Streamflow and precipitation data on a gray-scale map of topography (for evidence of excessive rains in hilly terrain).

- Watch and warning areas plotted on an image of radar reflectivity (for movement of storms through the boxes and for monitoring new development).

It should be easy to turn overlays on and off. For example, when a severe weather warning is issued, it may be critical to use the overlay with the names of cities and counties. After the warning has been issued, the overlay may be turned off so as not to interfere with the view of radar echoes.

It should also be easy to combine images. Color-coded reflectivity data on a gray-scale map of topography or on a gray-scale visible satellite image are useful combinations when flash floods threaten. A "tint" image, in which the gray-scale visible satellite image is tinted by a color corresponding to the cloud-top temperature (from infrared data), is also useful. In general, one would like to combine any two scale-compatible images and mix them in any ratio by intensifying one as the other fades out.

The need to combine different kinds of data raises a problem. Most low-volume data such as surface observations and upper air reports come in latitude-longitude coordinates. Radar data come in range-azimuth-elevation coordinates. Satellite data come in line-element coordinates, which are related to the scan strategy of the radiometer rather than the location of the area viewed. In order to combine different data on the same display, one must decide on a common coordinate system.

Choosing a map projection is a major decision, usually dictated by the computing resources. For example, it is customary to remap the low-volume data into the coordinate system of the high-volume data (plot 500 mb data in the satellite projection) for the reason that using fewer coordinate transformations will save computing time. A dilemma arises, however, if one wants to display a large number of vector-graphic overlays (contours or streamlines) on GOES satellite images. If the GOES satellite were in a perfect orbit, it would remain in precisely the same spot in the sky. Then, any vector-graphics product could be remapped once and for all. However, its position in the sky wanders enough that, for mesoscale work, the overlay of county and state boundaries must be recomputed each time an image arrives in order to avoid apparent motion of landmarks. Thus, at least for detailed work, one must map all the vector graphics anew with the arrival of each new image because the projection changes measurably.

The Appendix describes a computationally efficient method for remapping satellite images into a standard projection. The argument for doing so is that the spatial distortions inherent in all satellite images well away from the subpoint are corrected for. The argument against doing so is that cloud patterns near the limb of the Earth are smeared over a large area in the remapped image, giving the false impression that the clouds were viewed from above rather than from the side.

In designing a composite image of radar and satellite data, the programmer has three choices: (1) Map the radar data into the satellite projection. This choice has merit if only the precipitation echoes, which cover a small fraction of the PPI image, are to be remapped. (2) Map the satellite data into the radar projection. This may be desirable if the frequency of radar data greatly exceeds that of the satellite data. (3) Map both kinds of data into a standard map projection.

Option 3 seems to require two remappings for each image, but under normal circumstances it is no more expensive than option 2. As Saucier (1955) pointed out, conformal map projections are favored by meteorologists because the shapes of small objects are preserved on the map. Of the conformal projections commonly used (Mercator, polar stereographic, and Lambert), the Lambert is best for middle latitudes because the map factor is most nearly constant across the map. For small areas such as those covered by a radar PPI with 150 km radius, it varies by less than 1%.

This fact prompted a study to determine whether a properly centered and properly scaled PPI image could be displayed on a Lambert projection without remapping. The average map factor was computed for a Lambert projection big enough to contain the 240-km-range circle for the Limon, Colo., radar. The radar at known latitude and longitude (39°N and 104°W) could be located precisely on the map. Then the average map factor was used to find the approximate positions on the map for various combinations of range-azimuth coordinates. These positions were compared with the corresponding "exact" positions obtained by means of spherical trigonometry. At an elevation angle of zero, the position errors, contoured in Fig. 31.2, are very small, less than 100 m. At elevation angles greater than a few degrees, the errors are much greater (several kilometers at a range of 240 km) unless the slant range is first projected onto the surface of the Earth. The conclusion was that it is unnecessary to remap radar data in middle latitudes, provided that the display area on a Lambert projection is reasonably small. The position errors for a polar stereographic projection are about 10 times greater than those shown in Fig. 31.2.

Once satellite and radar data are displayed on a common map projection, it is disconcerting to notice a mismatch between reflectivity echoes and visible cloud locations. Sometimes as large as 20–30 km, the mismatch has two causes, both unrelated to remapping procedures or choice of map projection. The first cause stems from errors in the orbit and attitude information transmitted with the satellite images. This problem would be solved, given more accurate navigation information from the central ground station.

The second cause, parallax, becomes serious only when tall clouds are viewed obliquely by the satellite. Parallax is the apparent location error of a cloud top relative to the ground, resulting from the satellite's nonvertical viewing angle. The latitude and longitude of any feature viewed by the satellite is computed under the assumption that the feature is at sea level. If a tall cloud is viewed obliquely, the location of its top will be displaced in a direction away from the satellite subpoint. The amount of displacement grows with the zenith angle of the satellite and the height of the cloud. For example, if the satellite is over the Equator at 70°W, its zenith angle as measured at 39°N, 94°W (near Kansas City) is 47.4°. A cumulonimbus anvil 10 km high will thus be displaced 10.9 km toward the northwest from its correct position, and the positions of the cloud and its radar echo will disagree by that amount from the effect of parallax alone. The most common manifestation of parallax is the appearance of shadows on the "wrong" side of small cumulus clouds in high-resolution visible images.

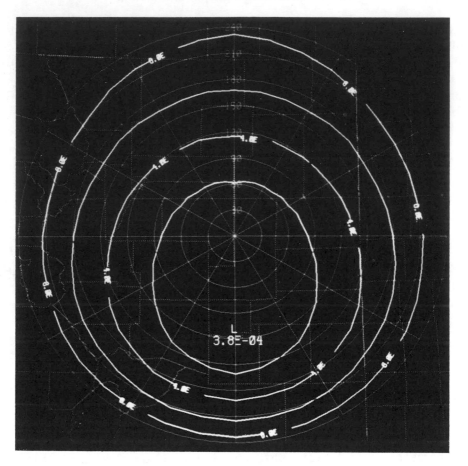

Figure 31.2. The position errors that result from placement of a properly scaled and centered radar PPI image for Limon, Colo., in range-azimuth coordinates, on a Lambert conformal projection of a portion of eastern Colorado. The contour interval is 0.02 km.

It is important to remember that the problems of misnavigation and parallax show up in both the original and remapped images. Remapping neither cures nor exacerbates them.

There is a clear value in the ability to combine different kinds of data on a single display for mesoscale forecasting; however, few meteorologists want to be restricted to a single screen. Side-by-side comparisons have been an important part of forecasting ever since facsimile maps were hung on the wall. Designers of new display systems should keep this in mind.

31.5.4. *User-Friendliness*

In a landmark article Foley *et al.* (1984) systematically described those characteristics that make workstations "user-friendly." They recommended that the workstation design be judged in terms of the following:

- The time required by the user to accomplish any task for which the system is designed.

- The accuracy with which the user can accomplish the task.

- The pleasure the user derives from the process.

Secondary concerns include learning and recall time, demands on the user's memory, protection from errors, susceptibility to fatigue, naturalness (how well the procedures follow familiar behavior) and boundedness (how big a space the user has to work in, physically and mentally).

Foley *et al.* (1984) differentiated between interaction tasks, which reference the objects displayed on the screen in some way, and controlling tasks, which move or modify those objects. They listed these principal interaction tasks:

- Select from a group of alternatives, e.g., display commands.

- Indicate a position on the interactive display and place a marker there.

- Orient a symbol on the screen by rotation or, in a more complicated case, show an object as it would appear when viewed from a particular point in space.

- Generate a path by specifying sequential positions.

- Specify a numerical value.

- Enter a text string; e.g., annotate a diagram.

Friendliness in an interactive task may be illustrated by a brief discussion of menus and selection. A menu is a list of the available workstation products and functions. It may employ words, abbreviations, or symbols. There are many ways to choose a menu item, all easy, each with advantages and disadvantages.

Products can be requested by typing letter commands on a keyboard. This method of communicating with the workstation usually becomes tiresome for inexperienced users, especially when the range of available products is large, but it is actually preferred by some workstation veterans. Some facilities make use of a touch screen, on which the names of all available features are displayed in colored boxes. The pressure of a finger over the appropriate box is sufficient to bring the desired product to the screen, but the number of "misses" (finger touching area outside the box) can be large. Other facilities employ a data tablet on which choices are listed. The forecaster selects a product by touching a stylus, a pen-like device, to the appropriate spot on the tablet. This is less direct than using a finger. Yet another possibility is to move a cursor to the desired menu choice and press a button. A disadvantage of this method is the necessity to find the cursor before it can be positioned. Function keys can be used for choosing, but this method is cumbersome when the number of choices is large.

Besides offering ease of use, a user-friendly workstation is communicative. The forecaster needs detailed information on system status at all times. Here are some examples of status information in a menu-driven workstation:

- A visual cue indicating that a requested product is still loading and that, short of clearing the screen, the operator cannot get immediate response from the workstation on another request.

- A brighter color for boxes that are activated. For example, if surface mesonet data are currently displayed on the screen, the corresponding box on the menu will be colored light blue; all other boxes (for products not currently displayed) will be dark blue.

- A reminder that an application program, started earlier, is still executing or has finished, and that the results are ready for display.

- Times of the most recently generated product of a given type. Knowing this, the forecaster does not have to waste time calling up products already seen.

A portion of the screen should be reserved for diagnostic or instructional messages. The forecaster needs to know when the workstation is operating normally, when failure occurs, and when it is unable to respond to user input. The forecaster also needs prompting when an application program requires certain actions (such as making the cursor visible on the screen so that a storm can be tracked) or certain information (such as the temperature limits defining the portion of an infrared image that is to be colored bright red). Forecasters also appreciate information allowing them to correct mistakes, for example, notification that the scale or map projection for a requested overlay is incompatible with an image already on the screen.

A sample menu layout is diagrammed in Fig. 31.3. Different parts of the screen are reserved:

- Time and date (actual or playback time).

- Choice of menus by scale (local, national, etc.).

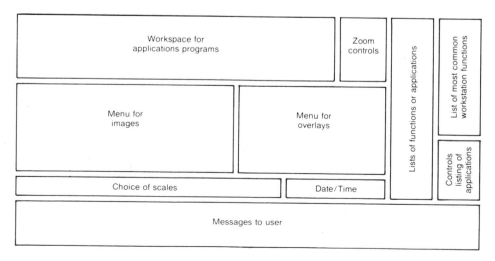

Figure 31.3. The allocation of space for a menu used at PROFS during a forecasting exercise in the spring of 1984.

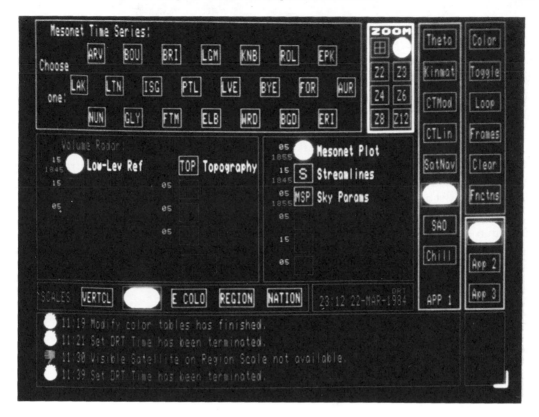

Figure 31.4. An actual menu corresponding to the layout in Fig. 31.3.

- Menu of products for a given scale (images and overlays).

- List of available functions (loop, zoom, roam, etc.) and applications programs (e.g., the operator can ask for a plot of temperature vs. time at a specific station).

- Workspace for applications programs (algorithm asks a question; user responds).

An actual menu for winter forecasting is reproduced in black and white in Fig. 31.4. Whenever a choice is made, the appropriate box lights up so that the user always knows which options are activated. In this example, local-scale products are shown. The images and graphical overlays are listed in separate places. The user has just decided to plot a mesonet time series (MesoTS) of data from one of the mesonet stations listed. Later, the user will be given a choice of parameters to plot—temperature, dewpoint, pressure, and others. The most recent diagnostic messages, none currently applicable, are shown at the bottom.

31.5.5. Speed

The need to detect severe convective storms and warn people in their path places heavy demands on a workstation computer. Downbursts, a hazard primarily to landing or departing aircraft, last for only a few minutes. The lifetime of the average tornado is hardly longer. Vigorous convection can result in a severe thunderstorm where only a towering cumulus cloud existed 15 min earlier. It is thus important to display radar data at frequent intervals within a minute or so of the scan's completion. Additionally, the delay between the request of a product and its display must be very short. A delay of 1–2 s is good and 4–5 s is tolerable; longer delays handicap forecasters.

The system design should distinguish between products generated by a clock-driven scheduler and placed automatically on the disk awaiting display, and products generated only upon request of the forecaster. The former are generally the most often viewed and are the most critical when severe weather threatens. The latter, called applications, are specialized products, such as a sounding analysis for a distant raob site, a time sequence of surface observations, or an analysis of moisture convergence. Forecasters are willing to wait considerably longer than 5 s for an application, especially if they can perform other tasks while they wait.

31.5.6. Resolution

Forecasters have an insatiable appetite for data. They want more data more quickly and with as much original detail as possible. There are practical limits to resolution, however, including the following, listed sequentially according to the data flow:

- Resolution of the sensor
- Efficiency of any on-site data preprocessor
- Communication efficiency from sensor to central processor
- Volume of raw data to be stored on disk
- Computation speed of central processor
- Resolution of the display device

The GOES satellite and a Doppler radar provide two examples of these limitations. Suppose a visible image is to be displayed at full resolution on a screen using 512 pixels in each direction. The first problem is to point an antenna at the satellite and read the stream of information, which is broadcast at the rate of 28 megabits per second. In real time, there is no alternative to this step. After the image is centered, a "window" of 512 lines is extracted from the full disk image and, within each line, the appropriate 512 elements of brightness data. Since each visible element is about 2 km on a side at 40°N, the image (in satellite coordinates) would cover an area very roughly 1000 km on a side. For rapid interval data— a new image, say, every 5 min—the windowing, image composition, and generation of a map background would have to be performed in less than 5 min. If processing power were not adequate, options would include making the window smaller, reducing spatial detail by dropping every other datum in the image, or processing only every other image.

Doppler radars, such as those envisioned for use by the late 1980s through the Next Generation Weather Radar (NEXRAD) Program of NWS (Ray and Colbert, 1982), generate a huge volume of data. The CP–2, a close relative of the proposed NEXRAD, is now operated experimentally by the National Center for Atmospheric Research. It can scan a full circle every 22 s, producing four 8-bit values for each of 360 azimuths and 1000 range gates per azimuth. This data rate far exceeds the capacity of, say, a 56 kilobit-per-second telephone line between the radar site and a remote computer. For this communication constraint, a preprocessor at the radar site could reduce the data rate to one value each of reflectivity and radial velocity for each 2 deg in azimuth and every fourth range gate. The phone line could then accomodate the resulting data rate—about 33 kilobits per second. Moreover, the remaining information would be appropriate for a display having 512 × 512 resolution because values from the 500 range bins spanning the PPI would nearly reach across the image.

The subject of temporal resolution is controversial, not so much on the frequency of images as on the number required to portray the life cycle of a particular phenomenon. One point of agreement is that the frequency of images should depend strongly upon the lifetime of the phenomenon. For monitoring a mesoscale convective complex, which typically lasts 5–10 h, 30 min frequency of satellite images is reasonable. For studying individual cells within the complex, 5 min frequency is desirable. For watching overshooting tops, 30 s frequency may be necessary.

There is little agreement on how many frames should constitute a loop. Only qualitative statements can be made: (1) the greater the nonlinearity of the phenomenon, the greater the required number of frames (interacting gust fronts that generate new convection are more nonlinear than the movement of a cold front); (2) fewer frames are required for extrapolation of a storm track than for a study of the storm's evolution; and (3) the longer the loop, the greater the storage requirement and the longer it takes to load into the display buffers.

31.5.7. Flexibility

A basic decision must be made early in the design of any workstation. Will the workstation be used to forecast local and regional weather in the same geographic area (as at a WSFO) or will it be used to survey the potential for hazardous weather anywhere in the country (as at the NSSFC in Kansas City)? The answer determines whether the map projections and backgrounds and satellite "windows" can be defined ahead of time. In any case, the workstation should allow examination of weather data on several different scales.

The choice of display area is often dictated by the size and longevity of the weather system of interest, but even if the problem of the day is a forecast of local afternoon convection, most meteorologists prefer to look at the large-scale information first in order to put the problem in context (see Ch. 29, this volume) and to look at possible mesoscale forcing mechanisms later. A practical advantage of viewing large-scale information is the superposition of

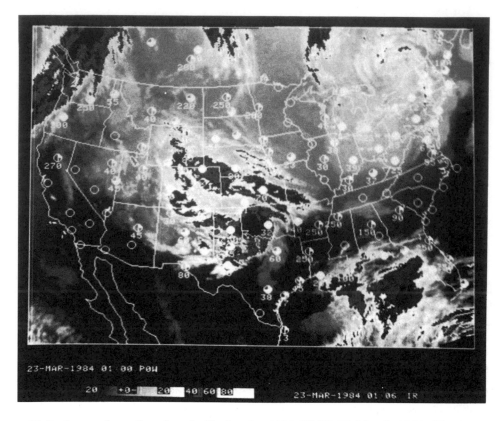

Figure 31.5. A sample national-scale display for 0100 GMT, 23 March 1984. The gray scale depicts cloudtop temperatures from a GOES infrared image. A weather depiction plot from AFOS is overlaid.

national guidance from NMC on satellite images from GOES. As an example, see Fig. 31.5.

The choice of area for the regional scale (approximately 1000 × 1000 km) is dictated primarily by the need to show the forecaster enough upstream data for a good 12 h forecast. Plots and analyses of surface aviation observations show well on this scale. Half-resolution (2 km) GOES visible data in a 512 × 512 pixel image are appropriate for the regional scale. It is easy to check the correspondence between landmarks and the overlaid political boundaries—a measure of the accuracy of the orbit and attitude parameters transmitted from the satellite along with the VAS images, but when smaller areas are viewed, the odds decrease that one will find a good landmark. Fig. 31.6 shows a regional-scale GOES image with overlays.

Forecasters need local-scale products for monitoring convection or its precursors, tracking storms, and issuing warnings. The viewing area is about 250 km on a side, which allows plotting of meso-β surface data without crowding and display of radar and visible satellite data at or near maximum resolution. A typical local-scale product is shown in Fig. 31.7.

There are exceptions to the above choices of display area. For example, mosaics of radar data are most informative when displayed on a scale in

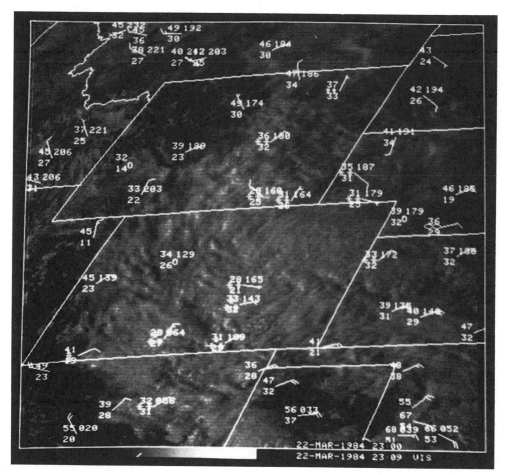

Figure 31.6. A sample regional-scale display for 2300 GMT, 22 March 1984. Clouds cover most of the central Rocky Mountain states in this GOES visible image. The overlay is a plot of surface data.

between local and regional. Data from dense networks of rainfall and stream gauges in flash-flood-prone areas are best displayed on drainage maps, which cover smaller areas than the local-scale map.

Flexibility in display areas should be matched by broad functionality of the workstation. Although the selection of routinely generated products may be quite large, inevitably the forecaster wants specialized information as well. The system must be able to supply this information through application programs that act upon the stored data; it must also be able to accommodate new programs that are suggested by the accumulated experience of the users.

A hard-copy device is a valuable adjunct to the workstation. Many forecasters prefer to analyze plotted data by hand even though an objective analysis could do the job satisfactorily. Hand analysis forces examination of individual reports that might otherwise be ignored; it also can lead to a conceptual model of the flow, which emphasizes the mesoscale far more than centrally prepared guidance can.

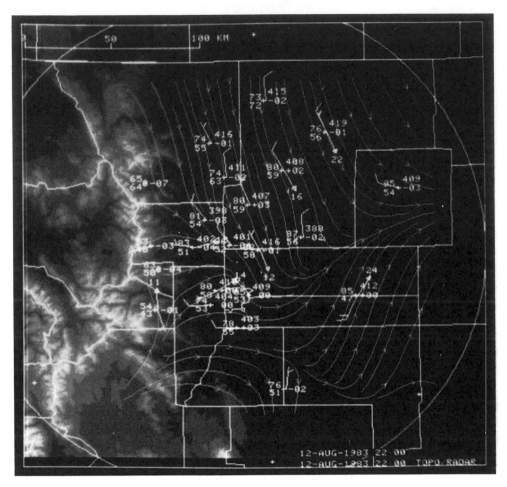

Figure 31.7. A local-scale product for 2200 GMT, 12 August 1983, showing the gray-scale topography of the Colorado Rocky Mountains and adjacent plains. There are three overlays: county boundaries, data plotted from the PROFS surface mesonetwork, and streamlines derived through objective analysis of wind observations.

31.6. The Need for Education and Research

With rapid technological advances in observing methods, communications, display hardware, and computing, ever more sophisticated systems for viewing and manipulating meteorological data are being built which will revolutionize weather service capabilities. Any good system will provide to the forecaster color, animation, combinations of different but complementary data, ease of use, speed, adequate resolution, and flexibility. The difficulty and expense of providing all these elements are justified on the presumption that the improvements in detection and prediction of mesoscale weather events will follow.

Unfortunately, a state-of-the-art display system is no guarantee of improvement. Meteorologists need to become familiar with new data sources, methods of measurement, probable sources of error, and relevance of the data to the meteorological situation (i.e., radar reflectivity data may not be particularly informative on a clear day). Exposure to the data through a workstation may well be the quickest way to become skilled in the interpretation of the new information.

Even with a state-of-the-art workstation and knowledge of the data and their limitations, the forecaster is still handicapped by incomplete understanding of mesoscale phenomena. Practical experience and a solid grounding in atmospheric physics and dynamics are not always sufficient to explain puzzling events. On the other hand, high-resolution data and a sophisticated tool for viewing them may bring to light previously unnoticed phenomena; the resulting curiosity is often the first step toward understanding. Mesoscale convective complexes were always an important feature of summertime weather east of the Rocky Mountains, but they were not discovered until the 1970s when researchers noticed them in GOES images and began to wonder about their origin and effect on the surrounding troposphere.

Meteorological workstations are a valuable tool in the hands of knowledgeable and experienced forecasters. When they stimulate thought or suggest new avenues of investigation, they serve their purpose well.

Appendix: Efficient Remapping of Satellite Images

The display of many types of data in a common projection often demands coordinate transformations. Nowhere is it more important to devise efficient transformation than in remapping the satellite projection. One useful algorithm is sketched here. The idea behind it is applicable to remapping in general.

Suppose a visible or infrared image is to be displayed on the monitor in a standard map projection, for example, a Lambert conformal projection with two standard parallels. Locations on the screen can be referenced by counting pixels, starting at the lower left. Suppose there are MM pixels in the horizontal direction and NN pixels in the vertical.

Subdivide the screen into squares M pixels on a side. The corners of the squares will be referenced by (II, JJ), where II ranges from 1 to MM in steps of M, and JJ ranges from 1 to NN in steps of M. Unless $MM = mM + 1$ and $NN = nM + 1$ (m and n are integers), there will be rectangles instead of squares along the top and right borders of the image, but this causes no problems. The proposed subdivision of the image is illustrated in Fig. 31.A1.

Step 1, which can be done once and for all provided the map projection doesn't change, involves finding the longitudes and latitudes corresponding to the points (II, JJ) on the screen. Most map projection packages provide the user with values of rectangular coordinates (u, v) corresponding to the four borders of the image. With these numbers, it is a simple matter to transform from the integer coordinates (II, JJ) to the continuous coordinates (u, v).

It is also straightforward to convert from (u, v) to longitude/latitude coordinates (λ, ϕ). For the Lambert projection, the conversion equations are

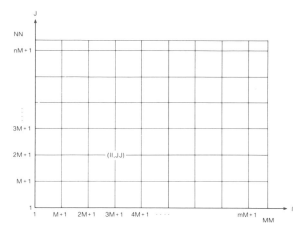

Figure 31.A1. The labeling convention for the grid of the screen on which satellite data are displayed.

$$R = (u^2 + v^2)^{1/2}$$

$$\phi = \frac{\pi}{2} - 2\arctan(R^{1/n})$$

$$\lambda = \lambda_0 + \frac{1}{n}\arcsin\left(\frac{u}{R}\right) \qquad (31.\text{A}1)$$

$$n = \ln\left(\frac{\cos\phi_1}{\cos\phi_2}\right) \Big/ \ln\left[\frac{\tan(\frac{\pi}{4} - \frac{\phi_1}{2})}{\tan(\frac{\pi}{4} - \frac{\phi_2}{2})}\right] .$$

The value n is called the cone constant; it depends only upon the standard parallels ϕ_1, and ϕ_2 for the Lambert projection. (The scale is true at these latitudes.) λ_0 is the longitude that points vertically on the map. The longitudes and latitudes (λ_i, ϕ_j) corresponding to the (II, JJ) on the display screen may be stored permanently or at least until the standard map projection is changed.

Step 2 is the transformation of the longitudes and latitudes (λ_i, ϕ_j) to pixel coordinates (KK, LL) on the VISSR image. The VISSR camera scans in lines from north to south and within lines from west to east. Consequently, LL increases from top to bottom and KK from left to right on the VISSR image.

The transformation from (λ_i, ϕ_j) to (KK, LL) is performed each time a new image is received. Computer routines written by Hambrick and Phillips (1980) based upon equations presented in Mottershead and Phillips (1976) are used at PROFS. By means of vector algebra, the equations are derived in an inertial reference frame, fixed with respect to the stars, and for an Earth assumed to have the shape of an oblate spheroid. The mapping equations involve parameters specified in the 129 "documentation" bytes that are part of the stream of information coming from GOES with each line of infrared brightness temperatures. Thus Step 2 must wait until image transmission begins. The equations are complicated enough that it is impractical to map every pixel directly from one image to another; only the points (λ_i, ϕ_j) are mapped exactly.

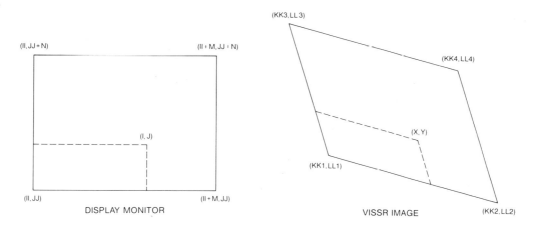

Figure 31.A2. A rectangle in the target projection on the display monitor (left), and the area remapped to the GOES satellite projection (right). Labeled points are in corresponding geographic locations.

The transformation from (λ_i, ϕ_j) to (KK, LL) essentially maps a square grid on the Lambert projection into a distorted grid on the visible or infrared image. If M is sufficiently small, a grid square maps into a parallelogram, the shape a square (or rectangle) assumes when viewed obliquely (Fig. 31.A2). If, on the other hand, M is too large, the Earth's curvature and the effect of perspective may cause the shape of the distorted grid elements to depart from a parallelogram. By comparing the lengths of opposite sides of the remapped figure, one can determine when M is small enough.

Step 3 is the interpolation of locations within each parallelogram in the VISSR image, based upon corresponding locations on the display monitor. To be specific, suppose that the rectangle at the left of Fig. 31.A2 maps into the figure at the right. The line and element numbers (X, Y) are obtained through bilinear interpolation. The image brightness at (X, Y) can now be used at location (I, J) on the display monitor.

If the resolution of the VISSR image is significantly greater than the resolution of the display monitor, then a simple average of pixels in the VISSR image is appropriate for transfer to the monitor. If the resolution of the VISSR image is significantly less than the resolution of the display monitor (as may occur with infrared images), then pixels of the VISSR image may map into parallelograms on the screen, giving the transformed image an unsightly appearance. Filtering can improve the aesthetics of the image but can also lead to artifacts not in the original data. If the resolution of the VISSR image is comparable with that of the screen, neither loss of resolution nor blockiness is a problem, but a few pixels on the VISSR image may not be mapped at all, or the same pixel on the VISSR image may occasionally map to adjacent pixels on the screen. In a 512×512 image, the eye does not discern these quirks.

This method of remapping makes full use of the resolution of the display monitor. The exact transformation is done for a very small fraction of the pixels, $(m + 1) \times (n + 1)/(MM \times NN)$. The great majority of pixels are mapped by linear interpolation, which is computationally efficient.

REFERENCES

Anthony, R. W., W. E. Carle, J. T. Schaefer, R. L. Livingston, A. L. Siebers, F. L. Mosher, J. T. Young, and T. M. Whittaker, 1982: The centralized storm information system at the NOAA Kansas City complex. Proceedings, 9th Conference on Weather Forecasting and Analysis, Seattle, Wash., American Meteorological Society, Boston, 40–43.

Atlas, D., and C. L. Korb, 1981: Weather and climate needs for lidar observations from space and concepts for their realization. *Bull. Amer. Meteor. Soc.*, **62**, 1270–1285.

Billingsley, J. B., 1976: Interactive image processing for meteorological applications at NASA/Goddard Space Flight Center. Preprints, 7th Conference on Aerospace and Aeronautical Meteorology and Symposium on Remote Sensing from Satellites, Melbourne, Florida, American Meteorological Society, Boston, 268–275.

Foley, J. D., V. L. Wallace, and P. Chan, 1984: The human factors of computer graphics interaction techniques. *IEEE Computer Graphics and Applications*, **4**, 13–48.

Green, R. N., and M. A. Kruidenier, 1982: Interactive data processing for mesoscale forecasting applications. Preprints, Conference on Weather Forecasting and Analysis, Seattle, Wash., American Meteorological Society, Boston, 60–64.

Hall, F. F., R. M. Huffaker, R. M. Hardesty, M. E. Jackson, T. R. Lawrence, M. J. Post, R. A. Richter, and B. F. Weber, 1984: Wind measurement accuracy of the NOAA pulsed infrared Doppler lidar. *Applied Optics*, **23**, 2503–2506.

Hambrick, L. N., and D. R. Phillips, 1980: Earth location image data of spin-stabilized geosynchronous satellites. NOAA Technical Memorandum NESS 111, Washington, D.C. (NTIS#PB81–120321), 49 pp., Appendices A–F.

Hardesty, R. M., K. Elmore, and M. E. Jackson, 1983: Comparisons of lidar and radar wind measurements made during the JAWS experiment. Preprints, 21st Conference on Radar Meteorology. Edmonton, Alberta, Canada, American Meteorological Society, Boston, 584–589.

Hogg, D. C., M. T. Decker, F. O. Guiraud, K.B. Earnshaw, D. A. Merritt, K. P. Moran, W. B. Sweezy, R. G. Strauch, E. R. Westwater, and C. G. Little, 1983: An automatic profiler of temperature, wind and humidity in the troposphere. *J. Climate Appl. Meteor.*, **22**, 807–831.

Klein, W. H., 1976: The AFOS program and future forecast applications. *Mon. Wea. Rev.*, **104**, 1494–1504.

Little, C. G., 1969: Acoustic methods for remote probing of the lower atmosphere. *Proc. IEEE*, **57**, 571–578.

Mosher, F. R., and J. T. Schaefer, 1983: Lessons learned from the CSIS. Preprints, 9th Conference on Aerospace and Aeronautical Meteorology, Omaha, Nebr., American Meteorological Society, Boston, 73–78.

Mottershead, C. T., and D. R. Phillips, 1976: Image navigation for geosynchronous meteorological satellites. Preprints, 7th Conference on Aerospace and Aeronautical Meteorology and Symposium on Remote Sensing from Satellites, Melbourne, Fla., American Meteorological Society, Boston, 260–264.

Murch, G. M., 1984: Physiological principles for the effective use of color. *IEEE Computer Graphics and Applications*, **4**, 49–53.

Ray, P. S., and K. Colbert, 1982: Proceedings of the NEXRAD Doppler Radar Symposium/Workshop, 22–24 September, Norman, Oklahoma. Cooperative Institute for Mesoscale Meteorological Studies (University of Oklahoma-NOAA), Norman, Okla., 235 pp.

Reynolds, D. W., 1983: Prototype workstation for mesoscale forecasting. *Bull. Amer. Meteor. Soc.*, **64**, 264–273.

Saucier, W. J., 1955: *Principles of Meteorological Analysis.* University of Chicago

Press, Chicago, 29–35.

Schramm, W. G., P. Zeleny, R. E. Nagle, and A. I. Weinstein, 1982: The Navy SPADS, a second generation environmental display system. Preprints, 9th Conference on Weather Forecasting and Analysis, Seattle, Wash., American Meteorological Society, Boston, 72–75.

Suomi, V. E., R. Fox, S. S. Limaye, and W. L. Smith, 1983: McIDAS III: a modern interactive data access and analysis system. *J. Climate Appl. Meteor.*, **22**, 766–778.

Thormeyer, C. D., 1978: Use of the Naval Environmental Display Station (NEDS) in the solution of Navy environmental forecast problems. Preprints, Conference on Weather Forecasting and Analysis and Aviation Meteorology, Silver Spring, Md., American Meteorological Society, Boston, 211–214.

Zipser, E. J., 1983: Nowcasting and very-short-range forecasting. The National STORM Program: Scientific and Technological Bases and Major Objectives, Ch. 6, University Corporation for Atmospheric Research, Boulder, Colo.

ACKNOWLEDGMENTS

The work described in Chapter 2 was sponsored by NSF under Grant ATM-8109828, by NOAA/NESDIS, under Grant NABOAAD00001, and by NASA under Grant NGR 14–001–008. A portion of Chapter 9 was supported by NSF Grant ATM–83043734. Preparation of Chapter 10 was funded by the NASA Mesoscale Atmospheric Processes Research Program and the Air Force Office of Scientific Research through Contract AFOSR–ISSA–84–12. Chapter 11 received support from NSF Grant ATM–8304330. Chapter 22 received support from NSF Grants ATM–5304042 and ATM–8414181, from the National Park Service Grant NA85RAH05045, Amendment 1, Item 9, and from NASA Contract NAG 5–359. Computations reported in Chapter 22 were done at NCAR, which is supported by NSF, and on the CSU CYBER 205 with partial support from the Control Data Corporation. Partial support for Chapter 24 was provided by NASA Grant NAS8–34833 and by NSF Grant ATM–8311593. The original research reported in Chapter 26 was funded by NSF under Grant ATM–N00014–83–K–0321. The work described in Chapter 27 was partially supported by the Environmental Protection Agency under Interagency Agreement DW 930144–01–0 but has not been subject to EPA review procedures. Partial support for the manuscript was provided by NOAA under contract 40RAN601580 and the Florida State Department of Meteorology and Supercomputer Computations Research Institute, which is partially funded by the U. S. Department of Energy through Contract No. DE–FC05–85ER250000.

ACRONYMS AND INITIALISMS

ADVISAR	All Digital Video Imaging System for Atmospheric Research
AFOS	Automation of Field Operations and Services
AOIPS	Atmospheric and Oceanographic Information Processing System
ASCOT	Atmospheric Studies in Complex Terrain
ASDAR	Aircraft-to-Satellite Data Relay
ASOS	Automated Surface Observing System
ATS	Advanced Technology Satellite
AVE	Atmospheric Variability Experiment
AVHRR	Advanced Very High Resolution Radiometer
AWIPS	Advanced Weather Interactive Processing System
AWOS	Automated Weather Observation System
BWER	Bounded Weak Echo Region
CAT	Clear Air Turbulence
CCL	Convective Condensation Level
CFL	Courant-Friedrichs-Lewy
CFM	Community Forecast Model
CISK	Conditional Instability of the Second Kind
CSI	Critical Success Index
CSIS	Centralized Storm Information System
CSU	Colorado State University
CUE	Coastal Upwelling Experiment
DCS	Data Collection System
DIAL	Differential Absorption Lidar
DMSP	Defense Meteorological Satellite Program
DOD	Department of Defense
ECMWF	European Centre for Medium-Range Weather Forecasting
EL	Equilibrium Level
ERL	Environmental Research Laboratories
FAA	Federal Aviation Administration
FAR	False Alarm Ratio
FGGE	First GARP Global Experiment
FM-CW	Frequency Modulated, Continuous Wave
FSSP	Forward Scattering Spectrometer Probe
GARP	Global Atmospheric Research Program
GATE	GARP Atlantic Tropical Experiment
GFDL	Geophysical Fluid Dynamics Laboratory

GLAS	Goddard Laboratory for Atmospheric Sciences
GMS	Geostationary Meteorological Satellite
GOES	Geostationary Operational Environmental Satellite
GSM	Global Spectral Model
HIRS	High-resolution Infrared Radiation Sounder
IFFA	Interactive Flash Flood Analyzer
INSAT	Indian National Satellite
IR	Infrared
JAWOS	Joint Automated Weather Observing System
JDOP	Joint Doppler Operational Project
JNWPU	Joint Numerical Weather Prediction Unit
K-H	Kelvin-Helmholtz
LAM	Limited-Area Model
LBC	Lateral Boundary Condition
LCL	Lifting Condensation Level
LES	Large-Eddy Simulation
LEWP	Line Echo Wave Pattern
LFC	Level of Free Convection
LFM	Limited-area Fine-Mesh
LHS	Left-Hand Side
LLJ	Low-Level Jet
LMW	Level of Maximum Wind
LORAN	Long-Range Aid to Navigation
LPT	Lifted Parcel Temperature
LSI	Lid Strength Index
LST	Local Standard Time
LT	Lead Time
LTE	Local Thermodynamic Equilibrium
MCC	Mesoscale Convective Complex
McIDAS	Man-computer Interactive Data Access System
MCS	Mesoscale Convective System
METROMEX	Metropolitan Meteorological Experiment
MFM	Movable Fine-Mesh
MOS	Model Output Statistics
m.s.l.	mean sea level
MSU	Microwave Sounding Unit
NASA	National Aeronautics and Space Administration
NCAR	National Center for Atmospheric Research

NEDS	Naval Environmental Display Station
NESDIS	National Environmental Satellite, Data, and Information Service
NEXRAD	Next-Generation Weather Radar
NGM	Nested–Grid Model
NIMROD	Northern Illinois Meteorological Research On Downbursts
NMC	National Meteorological Center
NOAA	National Oceanic and Atmospheric Administration
NSSFC	National Severe Storms Forecast Center
NSSL	National Severe Storms Laboratory
NSSP	National Severe Storms Project
NWP	Numerical Weather Prediction
NWS	National Weather Service
OI	Optimum Interpolation
OSCAR	Oxidation and Scavenging Characteristics of April Rains
PAM	Portable Automated Mesonet
PBE	Potential Buoyant Energy
PBL	Planetary Boundary Layer
PE	Primitive Equation
PMS	Particle Measuring System
POD	Probability of Detection
POP	Probability of Precipitation
POWS	PROFS Operational Work Station
PPI	Plan Position Indicator
PRF	Pulse Repetition Frequency
PROFS	Program for Regional Observing and Forecasting Services
RADAP	Radar Data Processing
RAFS	Regional Analysis and Forecasting System
RAMOS	Remote Automated Meteorological Observing System
RHI	Range-Height Indicator
RHS	Right-Hand Side
rms	root mean square
SB	Sea Breeze
SELS	Severe Local Storm [forecast unit]
SEM	Space Environment Monitor
SESAME	Severe Environmental Storms And Mesoscale Experiment
SMS	Synchronous Meteorological Satellite
SOP	Standard Operating Procedure
SPADS	Satellite (Data) Processing And Display System

SSEC	Space Science and Engineering Center
SSU	Stratospheric Sounding Unit
TIROS	Television and Infrared Observing Satellite
TOVS	TIROS Operational Vertical Sounder
TVS	Tornado Vortex Signature
UA	Upper Air
UHF	Ultra-High Frequency
VAD	Velocity-Azimuth Display
VAP	VAS Assessment Processor
VAS	VISSR Atmospheric Sounder
VASP	VISSR Atmospheric Sounder Processing
VHF	Very High Frequency
VIL	Vertically Integrated Liquid
VIP	Vertically Integrated Precipitation
VIS	Visible
VISSR	Visible and Infrared Spin-Scan Radiometer
WER	Weak Echo Region
WKB	Wentzel-Kramers-Brillouin
WPL	Wave Propagation Laboratory
WSFO	Weather Service Forecast Office

AUTHORS

Richard A. Anthes
National Center for Atmospheric Research
P. O. Box 3000
Boulder, CO 80307

Diana L. Bartels
NOAA/ERL Environmental Sciences Group
Weather Research Program
325 Broadway
Boulder, CO 80303

Howard B. Bluestein
School of Meteorology
University of Oklahoma
Norman, OK 73019

Joe F. Boatman
NOAA/ERL Air Resources Laboratory
325 Broadway
Boulder, CO 80303

Donald Burgess
NOAA/ERL National Severe Storms
    Laboratory
1313 Halley Circle
Norman, OK 73069

Robert W. Burpee
NOAA/ERL Atlantic Oceanographic and
    Meteorological Laboratory
Hurricane Research Division
4301 Rickenbacker Causeway
Miami, FL 33149

Charles F. Chappell
NOAA/ERL Environmental Sciences Group
Weather Research Program
325 Broadway
Boulder, CO 80303

William R. Cotton
Dept. of Atmospheric Science
Colorado State University
Fort Collins, CO 80523

Charles A. Doswell
NOAA/ERL Environmental Sciences Group
Weather Research Program
325 Broadway
Boulder, CO 80303

Dale R. Durran
Department of Meteorology
University of Utah
Salt Lake City, UT 84112

Kerry A. Emanuel
Center for Meteorology and Physical
    Oceanography
Dept. of Earth, Atmospheric, and Planetary
    Sciences
Massachusetts Institute of Technology
Cambridge, MA 02139

Joseph Facundo
NOAA National Weather Service
8060 13th Street
Silver Spring, MD 20910

T. T. Fujita
Department of Geophysical Sciences
University of Chicago
Chicago, IL 60637

Joseph H. Golden
NOAA National Weather Service
8060 13th Street
Silver Spring, MD 20910

Carl Hane
NOAA/ERL National Severe Storms
    Laboratory
1313 Halley Circle
Norman, OK 73069

William H. Hooke
NOAA/ERL Environmental Sciences Group
325 Broadway
Boulder, CO 80303

John B. Hovermale
National Meteorological Center
Washington, DC 20233

    Current address:
    Naval Environmental Prediction Research
        Facility
    Monterey, CA 93940

Kenneth W. Howard
NOAA/ERL Environmental Sciences Group
Weather Research Program
Boulder, CO 80303

Daniel Keyser
Laboratory for Atmospheres
NASA/Goddard Space Flight Center
Greenbelt, MD 20771

Joseph B. Klemp
National Center for Atmospheric Research
P. O. Box 3000
Boulder, CO 80307

Vincent Lally
National Center for Atmospheric Research
P. O. Box 3000
Boulder, CO 80307

Douglas Lilly
School of Meteorology
University of Oklahoma
Norman, OK 73019

Robert Maddox
NOAA/ERL Environmental Sciences Group
Weather Research Program
Boulder, CO 80303

   Current address:
   NOAA/ERL National Severe Storms
     Laboratory
   1313 Halley Circle
   Norman, OK 73069

John McGinley
Cooperative Institute for Mesoscale
   Meteorological Studies
University of Oklahoma
Norman, OK 73019

   Current address:
   NOAA/ERL Environmental Sciences
     Group
   Program for Regional Observing and
     Forecasting Services
   325 Broadway
   Boulder, CO 80303

Ronald D. McPherson
National Meteorological Center
Washington, DC 20233

Donald J. Perkey
Dept. of Physics and Atmospheric Science
Drexel University
Philadelphia, PA 19104

Roger A. Pielke
Dept. of Atmospheric Science
Colorado State University
Fort Collins, CO 80523

Raymond T. Pierrehumbert
NOAA/ERL Geophysical Fluid Dynamics
   Laboratory
Princeton University
Princeton, NJ 08542

James F. W. Purdom
NOAA/NESDIS Satellite Applications
   Laboratory
Fort Collins, CO 80523

Glen Rasch
NOAA/NWS Western Region
Box 11188, Federal Building
125 S. State Street
Salt Lake City, UT 84147

Peter S. Ray
NOAA/ERL National Severe Storms
   Laboratory
Norman, OK 73069

   Current address:
   Dept. of Meteorology and Supercomputer
     Computations Research Institute
   Florida State University
   Tallahassee, FL 32306

Roger F. Reinking
NOAA/ERL Environmental Sciences Group
Weather Modification Program
325 Broadway
Boulder, CO 80303

Dennis M. Rodgers
NOAA/ERL Environmental Sciences Group
Weather Research Program
325 Broadway
Boulder, CO 80303

Bruce B. Ross
NOAA/ERL Geophysical Fluid Dynamics
   Laboratory
Princeton University
Princeton, NJ 08542

Richard Rotunno
National Center for Atmospheric Research
P. O. Box 3000
Boulder, CO 80307

Joseph T. Schaefer
NOAA/NWS Central Region
Scientific Services Division
Kansas City, MO 64106

Thomas W. Schlatter
NOAA/ERL Environmental Sciences Group
Program for Regional Observing and
   Forecasting Services
325 Broadway
Boulder, CO 80303

Roderick A. Scofield
NOAA/NESDIS Satellite Applications
  Laboratory
Washington, DC 20233

Moti Segal
Dept. of Atmospheric Science
Colorado State University
Fort Collins, CO 80523

Robert Serafin
National Center for Atmospheric Research
P. O. Box 3000
Boulder, CO 80307

Daniel Smith
NOAA/NWS Southern Region
819 Taylor Street
Fort Worth, TX 76102

Dennis W. Thomson
Department of Meteorology
503 Walker Building
Pennsylvania State University
University Park, PA 16802

Morris L. Weisman
National Center for Atmospheric Research
P. O. Box 3000
Boulder, CO 80307

Fred L. Zuckerberg
NOAA/NWS Eastern Region
585 Stewart Avenue
Garden City, NY 11530